Handbook of Perovskite Solar Cells, Volume 1

Organic–inorganic hybrid metal halide perovskite materials have attracted significant attention due to their advantages of low cost, tunable band gap, solution processing, high molar extinction coefficient, low exciton binding energy, and high carrier mobility. Perovskite absorber layers play a decisive role in the realization of high-power conversion efficiency in perovskite solar cells (PSCs). This book systematically and comprehensively discusses device structures, working principles, and optimization strategies of perovskite absorber layers for PSCs to help foster commercialization of these environmentally friendly power sources. It describes strategies to optimize the quality of perovskite films, including composition engineering, dimensional engineering, solvent engineering, strain engineering, additive engineering, and interface engineering.

This volume:

- Introduces crystal structures of perovskites, configurations of PSCs, and their working principles.
- Discusses the modulation of perovskite compositions and dimensionality towards highly stable and efficient perovskite photovoltaics.
- Details the advancements of low-dimensional PSCs including phase stability of perovskite films and strategies for modulating phases.
- Summarizes progress in solvent engineering, additive engineering, and strain engineering in efficient and scalable perovskite photovoltaics.
- Describes the complex crystallization dynamics of perovskites, interface engineering, and synergistic modulation of grain boundaries and interfaces in PSCs.
- Highlights advances in ion migration and mitigation in halide perovskite solar cells and origins and elimination of hysteresis.

This book is aimed at researchers, advanced students, and industry professionals in materials, energy, and related areas of engineering who are interested in development and commercialization of photovoltaic technologies.

Advances in Materials Science and Engineering

Series Editor: Sam Zhang

Nanostructured and Advanced Materials for Fuel Cells
San Ping Jiang and Pei Kang Shen

Hydroxyapatite Coatings for Biomedical Applications
Sam Zhang

Carbon Nanomaterials: Modeling, Design, and Applications
Kun Zhou

Materials for Energy
Sam Zhang

Protective Thin Coatings and Functional Thin Films Technology, Two-Volume Set
Sam Zhang, Jyh-Ming Ting, Wan-Yu Wu

Fundamentals of Crystallography, Powder X-ray Diffraction, and Transmission Electron Microscopy for Materials Scientists
ZhiLi Dong

Materials for Devices
Sam Zhang

Handbook of Perovskite Solar Cells, Volume 1: Fundamentals and Absorber Layer Optimization for Efficiency and Stability
Edited by Jiangzhao Chen and Sam Zhang

Handbook of Perovskite Solar Cells, Volume 2: Functional Layer Optimization and Diverse Device Types for Efficiency and Stability
Edited by Jiangzhao Chen and Sam Zhang

Handbook of Perovskite Solar Cells, Volume 3: Critical Challenges and Opportunities in Commercialization
Edited by Jiangzhao Chen and Sam Zhang

Materials in Advanced Manufacturing
Yinquan Yu and Sam Zhang

For more information about this series, please visit: www.routledge.com/Advances-in-Materials-Science-and-Engineering/book-series/CRCADVMATSCIENG

Handbook of Perovskite Solar Cells, Volume 1

Fundamentals and Absorber Layer Optimization for Efficiency and Stability

Edited by
Jiangzhao Chen and Sam Zhang

CRC Press
Taylor & Francis Group
Boca Raton London New York

CRC Press is an imprint of the
Taylor & Francis Group, an **informa** business

Chapter 9 Synergistic Modulation of Grain Boundaries and Interfaces in Perovskite Solar Cells 272

Dong Wei, Ru Li, Guilin Chen, Dongmei He, Xuxia Shai, and Jiangzhao Chen

Chapter 10 Strain Engineering in Perovskite Solar Cells 310

Yuanyuan Zhao, Lei Gao, Yinping Teng, Yusheng Cao, Liqiang Bian, and Qunwei Tang

Chapter 11 Ion Migration and Mitigation in Halide Perovskite Solar Cells 360

Huan Li and Longbin Qiu

Chapter 12 Origins and Elimination of Hysteresis in Perovskite Solar Cells 414

Rui Lin, Yibing Wu, and Xinhua Ouyang

Index 439

About the Editors

Jiangzhao Chen (陳江照) is a professor at Faculty of Materials Science and Engineering in Kunming University of Science and Technology. He received his BS and PhD degrees in Applied Chemistry from Northeast Forestry University in July 2011 and in Optical Engineering from Huazhong University of Science and Technology in June 2016, respectively. From October 2016 to February 2019, he worked as a postdoctoral researcher at Sungkyunkwan University. From March 2019 to November 2019, he worked as a postdoctoral researcher at University of Hong Kong. From December 2019 to December 2023, he worked as a professor at School of Optoelectronic Engineering, Chongqing University. His current research interests focus on perovskite solar cells.

Sam Zhang Shanyong (張善勇), FRSC, FTFS, FIoMMM, academically known as Sam Zhang, was born and brought up in the famous "City of Mountains" Chongqing, China. He received his Bachelor of Engineering in Materials in 1982 from Northeastern University (Shenyang, China), Masters of Engineering in Materials in 1984 from Iron and Steel Research Institute (Beijing, China), and PhD degree in Ceramics in 1991 from The University of Wisconsin-Madison, USA. He was a tenured full professor (beginning 2006) at the School of Mechanical and Aerospace Engineering, Nanyang Technological University Singapore. In January 2018, he joined the School of Materials and Energy, and served as the founding director of the Centre for Advanced Thin Films and Devices, Southwest University, China. In 2023, he joined the School of Aeronautics, Harbin Institute of Technology as Chair Professor and also served in the Harbin Institute of Technology's Zhengzhou Research Institute.

Contributors

Ibrahim Albrahee
Key Laboratory of Advanced Energy
Storage Materials and Technology
Shihezi University
Xinjiang, China

Liqiang Bian
College of Mechanical and Electronic
Engineering
Shandong University of Science and
Technology
Qingdao, PR China

Bing Cai
School of Materials and Energy
Yunnan University, Kunming
Yunnan, China

Qingli Cao
Green Catalysis Center and College
of Chemistry
Zhengzhou University
Zhengzhou, China

Yusheng Cao
College of Mechanical and Electronic
Engineering
Shandong University of Science and
Technology
Qingdao, PR China

Guilin Chen
College of Physics and Energy
Fujian Normal University
Fuzhou, PR China

Jiangzhao Chen
Faculty of Materials Science and
Engineering
Kunming University of Science and
Technology
Kunming, People's Republic of China

Guoyu Ding
Key Laboratory of Materials for
High-Power Laser
Shanghai Institute of Optics and Fine
Mechanics
Chinese Academy of Sciences
Shanghai, China

Yachao Du
ShaanXi Normal University
Xi'an, People's Republic of China

Jiandong Fan
Institute of New Energy Technology
College of Information Science and
Technology
Jinan University
Guangzhou, China

Lei Gao
College of Mechanical and Electronic
Engineering
Shandong University of Science
and Technology
Qingdao, People's Republic of China

Dongmei He
Key Laboratory of Optoelectronic
Technology and Systems
(Ministry of Education)
College of Optoelectronic Engineering
Chongqing University
Chongqing, People's Republic
of China

Juan Hou
Key Laboratory of Advanced
Energy Storage Materials and
Technology
Shihezi University
Xinjiang, China

Huan Li
Shenzhen Key Laboratory of Intelligent
Robotics and Flexible Manufacturing
Systems
Department of Mechanical and Energy
Engineering
SUSTech Energy Institute for Carbon
Neutrality
Southern University of Science and
Technology
Shenzhen, China

Pengwei Li
Green Catalysis Center and College of
Chemistry
Zhengzhou University
Zhengzhou, China

Ru Li
Key Laboratory of Optoelectronic
Technology and Systems (Ministry
of Education)
College of Optoelectronic Engineering
Chongqing University
Chongqing, People's Republic of China

Wenzhe Li
Institute of New Energy Technology
College of Information Science and
Technology
Jinan University
Guangzhou, China

Yang Li
Bingtuan Energy Development Institute
and Key Laboratory of Advanced
Energy Storage Materials and
Technology
Shihezi University
Shihezi City, Xinjiang, China

Rui Lin
Fujian Agriculture and Forestry
University
Fuzhou, People's Republic of China

Ming Luo
State Key Laboratory of
Photovoltaic Science and
Technology
Shanghai Frontiers Science
Research Base of Intelligent
Optoelectronics and
Perception
Department of Material Sciences,
Institute of Optoelectronics
Fudan University
Shanghai, People's Republic
of China

Zhipeng Miao
Green Catalysis Center and College
of Chemistry
Zhengzhou University
Zhengzhou, China

Xinhua Ouyang
Fujian Agriculture and Forestry
University
Fuzhou, People's Republic
of China

Jieshan Qiu
College of Chemical Engineering
State Key Laboratory of Chemical
Resource Engineering
Beijing University of Chemical
Technology
Beijing, China

Longbin Qiu
Shenzhen Key Laboratory of
Intelligent Robotics and Flexible
Manufacturing Systems
Department of Mechanical and Energy
Engineering
SUSTech Energy Institute for Carbon
Neutrality
Southern University of Science and
Technology
Shenzhen, China

Xuxia Shai
Institute of Physical and Engineering
    Science/Faculty of Science
Kunming University of Science and
    Technology
Kunming, China

Yuchuan Shao
Key Laboratory of Materials for
    High-Power Laser
Shanghai Institute of Optics and
    Fine Mechanics, Chinese
    Academy of Sciences
Shanghai, China

Pengju Shi
School of Engineering
Westlake University and Institute of
    Advanced Technology
Westlake Institute for Advanced Study
Hangzhou, China

Qunwei Tang
Institute of Carbon Neutrality
College of Chemical and Biological
    Engineering
Shandong University of Science
    and Technology
Qingdao, People's Republic
    of China

Yinping Teng
College of Mechanical and Electronic
    Engineering
Shandong University of Science and
    Technology
Qingdao, People's Republic of China

Haonan Wang
Key Laboratory of Materials for
    High-Power Laser
Shanghai Institute of Optics and Fine
    Mechanics
Chinese Academy of Sciences
Shanghai, China

Rui Wang
School of Engineering
Westlake University and Institute
    of Advanced Technology
Westlake Institute for Advanced
    Study
Hangzhou, China

Dong Wei
College of Physics and Energy
Fujian Normal University
Fuzhou, People's Republic
    of China

Yibing Wu
Fujian Agriculture and Forestry
    University
Fuzhou, People's Republic of
    China

Baomin Xu
Department of Materials Science and
    Engineering
Southern University of Science
    and Technology
Shenzhen, China

Jingjing Xue
State Key Laboratory of Silicon
    Materials and School of Materials
    Science and Engineering
Zhejiang University
Hangzhou, China

Hong Zhang
State Key Laboratory of Photovoltaic
    Science and Technology
Shanghai Frontiers Science Research
    Base of Intelligent Optoelectronics
    and Perception
Department of Material Sciences,
    Institute of Optoelectronics
Fudan University
Shanghai, People's Republic
    of China

Lixin Zhang
Bingtuan Energy Development Institute
and Key Laboratory of Advanced
Energy Storage Materials and
Technology
Shihezi University
Shihezi City, Xinjiang, China

Yiqiang Zhang
Green Catalysis Center and College of
Chemistry
Zhengzhou University
Zhengzhou, China

Kui Zhao
ShaanXi Normal University
Xi'an, People's Republic of China

Yuanyuan Zhao
College of Mechanical and Electronic
Engineering
Shandong University of Science and
Technology
Qingdao, People's Republic
of China

Yifan Zheng
Key Laboratory of Materials for
High-Power Laser
Shanghai Institute of Optics and Fine
Mechanics
Chinese Academy of Sciences
Shanghai, China

1 A Brief Introduction to Perovskite Materials and Perovskite Solar Cells

Bing Cai and Jieshan Qiu

1.1 HISTORICAL BACKGROUND OF PEROVSKITES

1.1.1 SOLAR ENERGY

In human history, the use of fire marked the beginning of human civilization. Since then, the development and utilization of various energy sources have been promoting the development and progress of human society. Especially since the Industrial Revolution, the large-scale exploitation and utilization of fossil fuels such as coal, oil, and natural gas has greatly contributed to the modernization of human society and facilitated the rapid development of science and productivity. However, with the dramatic increase in fossil energy consumption in the second half of the 20th century, two major problems, the energy crisis and environmental pollution, gradually came to light. To change the situation, the development of green and renewable clean energies has become an inevitable trend in the development of energy strategies in various countries. New energy technologies, including hydro, wind, biomass, geothermal, tidal, and solar, have been developed. Among them, solar energy is promising to dominate the future energy supply with a wide range of applications.[1]

Solar energy comes from nuclear fusion reactions inside the Sun, which continues to release large amounts of solar energy over long periods of time, reaching Earth in the form of radiation. From the perspective of energy conversion, various renewable energy sources (wind, water, biomass, etc.) and traditional fossil energy are almost indirectly related to solar energy. The total amount of solar energy obtained by the Earth is as high as 173,000 TW, which is only $1/(2.2 \times 10^9)$ of the total radiation energy of the sun (about 3.75×10^{26} W), but it is still high enough to meet all the needs on the earth including human activities.[2] It is estimated that only 0.1% of the Earth's surface area is needed to be covered by solar cells with an average conversion efficiency of 10%, which can meet all the energy needs of current human society.[3]

In addition to the huge reserves, solar energy has the following obvious advantages: (i) fewer geographical restrictions: except for a few cloudy areas and the north and south poles, solar energy is distributed everywhere on the earth and can be used locally without transportation. (ii) Pollution-free: the solar energy utilization process does not produce greenhouse gases, chemical pollutants, or radioactive waste. (iii) Convenience: solar energy can be directly converted into electricity without the use

DOI: 10.1201/9781003400486-1

of large-scale power stations. Therefore, it can be applied to a variety of electronic devices such as watches, calculators, and even solar-powered aircraft, satellites, and space stations. However, there are still several disadvantages in the development and utilization of solar energy: (i) low-energy density: although the total amount of solar radiation is large, the energy density of the earth's surface is relatively low. Thus, solar devices need a large radiation area to obtain enough energy. (ii) Large power fluctuations: limited by natural conditions such as day and night, seasons, and weather, the solar radiation arriving at a fixed location fluctuates greatly. (iii) Low efficiency and high cost: Due to the limitations on thermodynamic and other aspects, the power conversion efficiency (PCE) of solar devices is lower than traditional heat engines. In addition, although the operation and maintenance costs are very low, the cost of solar device manufacturing and power plant construction are still very high. As a result, the current economics of large-scale solar power are not yet competitive with conventional energy technologies.

1.1.2 SOLAR CELLS

There are three main conversion modes for the utilization of solar energy: photo-thermal, photochemical, and photoelectric conversion.[4] Photothermal conversion is one of the oldest methods, which means using solar radiation to heat other objects. Photochemical conversion refers to the use of solar energy to prepare energetic materials such as hydrogen for further storage and utilization. Photoelectric conversion, or photovoltaic, refers to the use of solar cells to directly convert solar energy into electricity for further use. The advent of solar cells marks a new stage in the development and utilization of solar energy by humans.

The story of the solar cell began in 1839, when the French experimental physicist E. Becquerel put two platinum sheets into a halide solution in an experiment and found that when the solution was illuminated with light, an electric current could be detected between the two platinum sheets and the voltage changed with time.[5] It was the first time that a photovoltaic phenomenon was observed by humans. Until the year of 1954, D. M. Chapin, C. S. Fuller and G. L. Pearson of Bell Laboratories invented a monocrystalline silicon solar cell with a PCE of 6%.[6] It is at this point that the first practical solar cell finally steps onto the stage of history. Since then, PCEs of silicon solar cell has gradually increased to more than 26% with technological advances.[7] Today, the silicon solar cell industry is quite mature, accounting for more than 90% of the photovoltaic market share (Figure 1.1), but the development of the silicon solar cell industry is limited by complex manufacturing processes. For ultra-high purity (> 99.9999%) photovoltaic grade silicon wafers, high temperature purification is a key process with high energy consumption, which seriously hinders the further reduction of manufacturing costs. As a result, researchers have been exploring green and low-cost photovoltaic materials to replace conventional crystalline silicon. In the past few decades, several new types of solar cells have been developed, such as copper indium gallium selenium solar cells (CIGS),[8] cadmium telluride solar cells (CdTe),[9] dye-sensitized solar cells (DSSC),[10] and organic photovoltaic cells (OPV).[11] However, due to limitations in material cost or device performance, these solar cells are difficult to meet the requirements of competing with traditional silicon solar cells. Until

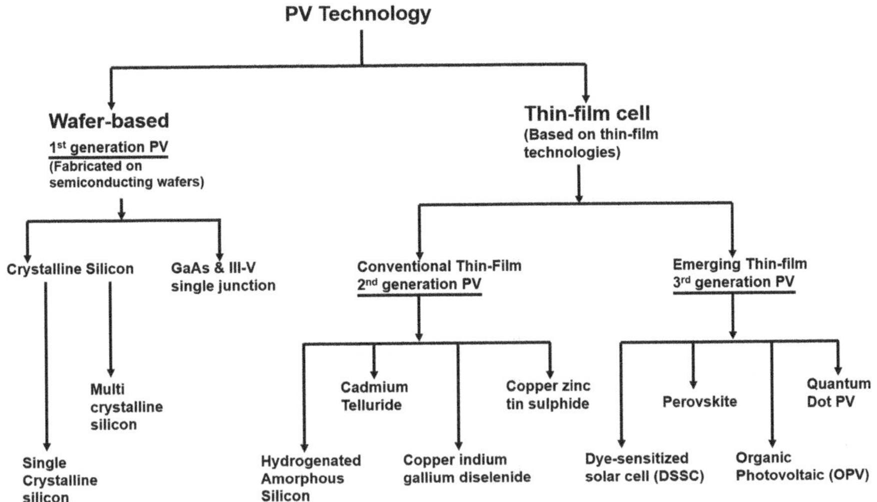

FIGURE 1.1 Classification of different generations of solar cells and photovoltaic technologies. Reproduced with permission.[12] Copyright 2018, Elsevier.

recent years, a new type of solar cells based on organometal halide perovskites have emerged and developed rapidly, showing a high application potential and bringing new development opportunities for the innovation in the photovoltaic industry.

1.1.3 Perovskite Materials

Perovskite material refers to an ABX_3-type compound, in which cations in A and B positions and anions X of different sizes coexist stably. The first perovskite material calcium titanate ($CaTiO_3$) was discovered in 1839 by mineralogist Gustav Rose in the Ural Mountains of Russia and named after the famous Russian mineralogist Count Lev A. Perovskiy (1792–1856).[13] Since then, many ABX_3-type compounds similar to $CaTiO_3$ have been reported, such as $BaTiO_3$, $PbTiO_3$, $SrTiO_3$, $BiFeO_3$, etc., which are referred to as perovskite materials. Due to the wide variety of elements that can be accommodated at A, B, and X sites, the number of perovskite materials that can be composed is also very large. Meanwhile, there are various lattice structure distortion ways to form different crystal phases that exhibit unique physical and chemical properties.[14] Table 1.1 lists the timeline in perovskite material research since its discovery.

1.1.4 Perovskite Solar Cells (PSCs)

Although perovskites have been widely used in many different research fields, their journey in the field of photovoltaics started relatively late. The photocurrent in the perovskite $BaTiO_3$ was discovered in 1956, and the photovoltaic effect was later discovered in perovskite materials such as $LiNO_3$ and is related to the ferroelectric properties of these materials.[16] However, the efficiencies generated in these early studies

TABLE 1.1

Major Discoveries and Breakthroughs in the Research of Perovskite Materials and Solar Cells.[13,15]

Year	Short description	Lead researcher(s)
1839	Discovery and naming of perovskite ($CaTiO_3$) in a skarn ("Schisto chloritico") sample from the Ural Mts., Russia	Rose
1893	Synthesis of lead halide perovskite $CsPbX_3$ (X = Cl, Br, I)	Wells et al.
1925	First description of the perovskite crystal structure (done on metamorphic samples from Zermatt, Switzerland)	Barth
1940s	Discovery of ferroelectric ceramics with a high dielectric constant ($BaTiO_3$)	Vul and Gol'dman; Hippel at al.
1957	Confirmation of the perovskite structure of $CsPbCl_3$ and $CsPbBr_3$	Møller
1978	Synthesis of organic-inorganic hybrid perovskite $MAMX_3$ (MA = CH_3NH_3; M = Sn, Pb)	Weber et al.
1994	Development of hybrid organic–inorganic halide perovskites for thin-film transistor applications	Mitzi et al.
2005	Development of a perovskite-sensitized solar cell (based on the hybrid halide perovskite $CH_3NH_3PbBr_3$ and $CH_3NH_3PbI_3$)	Miyasaka et al.
2009	First report on perovskite solar cells (PCE = 3.8%)	Miyasaka et al.
2012	Development of a solid-state perovskite solar cell with spiro-OMeTAD as hole transport material (PCE = 9.7%)	Park et al.
2012	Development of a solid-state perovskite solar cell based on $CH_3NH_3PbI_{3-x}Cl_x$ loaded on mesoporous Al_2O_3 (PCE = 10.9%)	Snaith et al.

were low (< 1%). Fortunately, halide perovskite materials with desired semiconducting properties for photovoltaic applications attract the attention of researchers, which is expected to become a potential photovoltaic material. The discovery of these halide perovskites dates back to 1893. Wells et al. synthesized and systematically studied lead halide perovskite $CsPbX_3$ (X = Cl, Br, I).[17] In 1957, Møller confirmed that $CsPbCl_3$ and $CsPbBr_3$ have a perovskite structure, which exists in a tetragonal distorted structure and transforms into a pure cubic phase at high temperatures.[18,19] These cesium-lead halide ionic crystals can be synthesized by a simple solution method, which may motivate the replacement of Cs^+ with organic cations. In 1978, Weber introduced methylamine ions ($CH_3NH_3^+$, MA^+) into the crystal structure of the perovskite to form the organic-inorganic hybrid perovskite $MAMX_3$ (M = Sn, Pb).[20] Unlike conventional hybrid materials, these organic-inorganic hybrid perovskites are homogeneous at the molecular scale and can therefore well integrate the advantages of organic and inorganic materials. At the end of the 20th century, Mitzi et al. synthesized a variety of hybrid halide perovskites with different organic cations and applied them to devices including light-emitting diodes (LEDs), initially indicating the potential applications of these perovskite materials in solar cells.[21]

In 2005, Miyasaka et al. attempted to introduce $MAPbX_3$ (X = Br, I) into dye-sensitized solar cells to replace organic dyes as light-absorbing materials, successfully obtaining a PCE of more than 3% and reported in 2009.[22] However, due to the use of liquid electrolytes, perovskite materials dissolve rapidly in them, resulting in a very short device lifetime. In 2011, Park et al. increased the load of $MAPbI_3$ perovskite on porous titanium oxide by using a higher concentration of precursor solution, further improving the device performance to 6.5%, but the instability problem remained unresolved.[23] In 2012, Park et al. replaced the liquid electrolyte with solid organic hole transport materials in order to avoid the dissolution of perovskite, and the PCE was increased to 9.7% with greatly improved stability.[24] In the same year, Snaith et al. used porous alumina instead of titanium oxide to support $MAPbI_{3-x}Cl_x$ perovskite, delivering a PCE of 10.9%.[25] At this point, the photovoltaic field has finally birthed a new star. In 2013, the journal *Science* named PSCs one of the top ten scientific and technological advances of the year.[26] PSCs have become a famous research hotspot and have received wide attention. Since then, with the continuous efforts of researchers around the world, the certified PCE of PSCs has been rapidly improved to 26.1% in a decade, closing to the record of crystalline silicon cells (26.8%) which have developed for more than half a century (see Figure 1.2).[7] After solving the problems of stability and large-scale preparation in the future, PSCs are very promising to achieve practical application and compete with the existing commercial photovoltaic technologies. Interestingly, with the success in photovoltaic, the fever of perovskite materials has even expanded to other applications such as LEDs, photodetectors, X-ray detectors, storage devices, etc.[27–30]

1.2 CRYSTAL STRUCTURES OF PEROVSKITES

The ideal perovskite materials have the general formula of ABX_3, which has an equiaxed crystal system structure (*Pm3m* space group).[31] In an ideal cubic structure, A cations are at the hexahedral vertices, and B cations are in the center and coordinated with X anions as BX_6 octahedra (Figure 1.3).

To preliminarily deduce the crystallographic stability and probable structure of a perovskite material, a tolerance factor t and an octahedral factor μ are usually considered,[32] which are respectively defined as:

$$t = \left(r_A + r_X \right) / \left[\sqrt{2} \left(r_B + r_X \right) \right] \tag{2.1}$$

$$\mu = r_B / r_X \tag{2.2}$$

where r_A, r_B, and r_X are the ionic radii of those atoms located at the A, B, and X sites, respectively. For a stable perovskite structure, $0.81 < t < 1.11$ and $0.44 < \mu < 0.90$ must be required. For example, the most studied perovskites in PSCs $MAPbI_3$ ($r_A = 1.80$ Å, $r_B = 1.19$ Å, $r_X = 2.20$ Å), whose t and μ values are 0.834 and 0.541, respectively, completely meets the requirement.[14] To design a new and stable perovskite crystal structure, we can adjust t and μ values simply by entirely or partly introducing ions with different sizes (Figure 1.4).

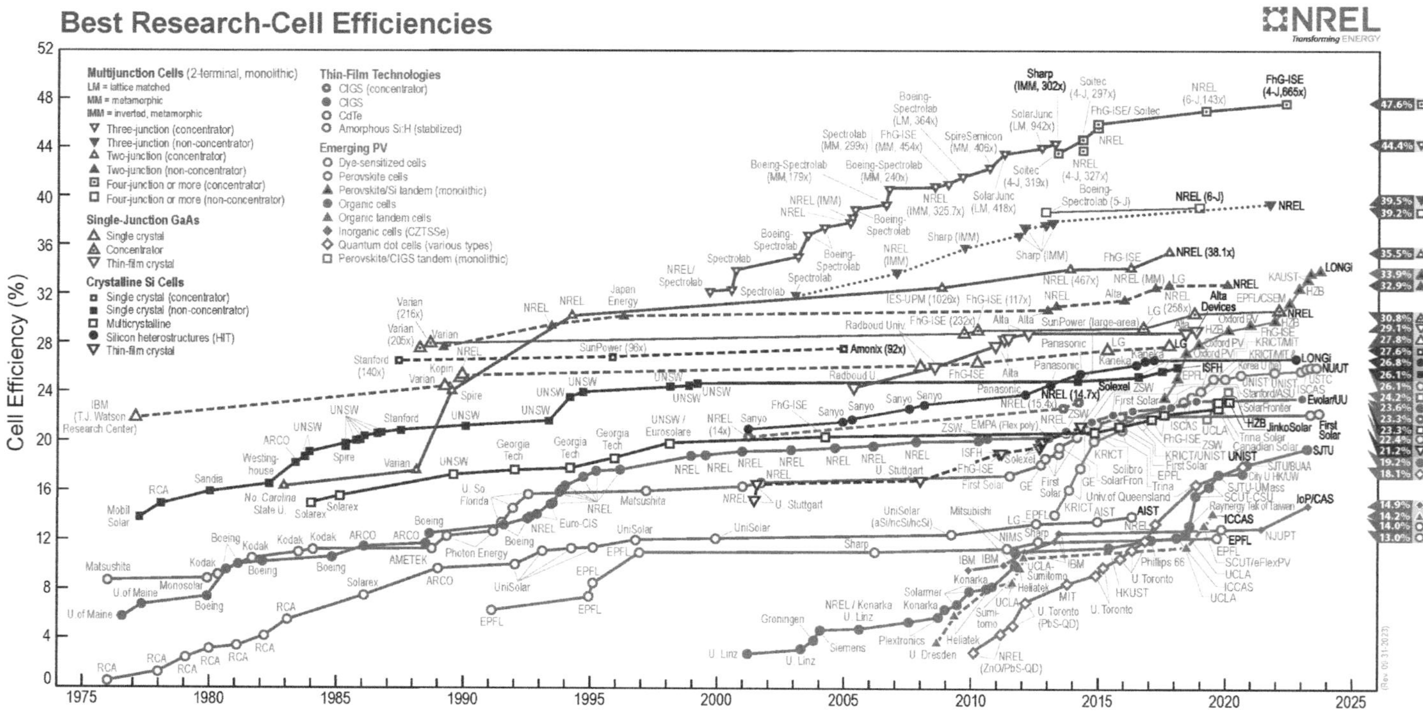

FIGURE 1.2　The best certified research-cell PCEs. This plot is courtesy of the National Renewable Energy Laboratory (NREL), Golden, CO, USA.[7]

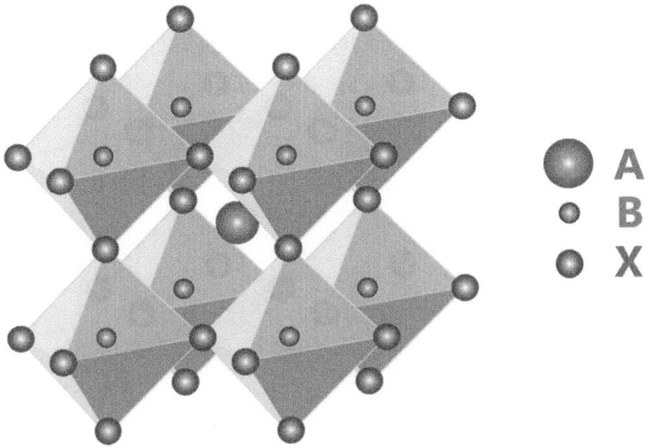

FIGURE 1.3 Structural features of the ABX_3 perovskites.

Among various photovoltaic active perovskites, $MAPbI_3$ is the earliest and most extensively studied one since its first report by Kojima et al. in 2009.[24] Nowadays, it is considered as a standard model for the photovoltaic–promising perovskite materials. There are three different phases of $MAPbI_3$ in different temperature ranges: orthorhombic (below -113°C), tetragonal (-113 to ~57°C), and pseudocubic (above 57°C).[20,33,34] Considering the actual operating condition of solar cells around room temperatures, we don't need to worry about the low-temperature orthorhombic phase. The reversible tetragonal to cubic phase transition around 57°C, which can be easily realized under the radiation of the sun, may introduce a large volume change due to a big change of the lattice parameter. As a result, mechanical stress would build up which limits the performance and lifetime of the PSCs.[35] Moreover, caused by its low formation energy (0.11 ~ 0.14 eV) and low thermal conductivity (~0.5 W $K^{-1}m^{-1}$ at room temperature), $MAPbI_3$ will decompose significantly when it is stored at 85°C even in an inert atmosphere.[36,37] It is also worth noting that almost all the deposited perovskite layers used in solar cells are polycrystalline films. Although single crystals with large sizes for perovskite $MAPbX_3$ (X = Cl, Br or I) had been synthesized successfully,[38] the fabrication of PSCs with these single crystals is currently complicated and full of challenges for large-scale production.

To obtain perovskites with better thermal stability than $MAPbI_3$, it is necessary to exchange MA^+ ($CH_3NH_3^+$) with other stable cations, such as FA^+ (formamidinium, $NH_2CH = NH_2^+$).[39] The t and μ factors of $FAPbI_3$ are 0.987 and 0.541, respectively, indicating that the $FAPbI_3$ crystal would be more stable than $MAPbI_3$. However, the larger radius of FA^+ also raises the energy barrier for the FA^+ intercalation between the PbI_6 octahedral networks. As a result, during the formation of perovskite thin films, higher annealing temperature is required for $FAPbI_3$ (~150°C) than $MAPbI_3$ (90~110°C).[40] The main problem for $FAPbI_3$ is the instability of its cubic α–phase (black) at room temperature, which tends to transform into a hexagonal polymorph (δ–phase) with an undesirable yellow color.[41,42] To improve the performance and

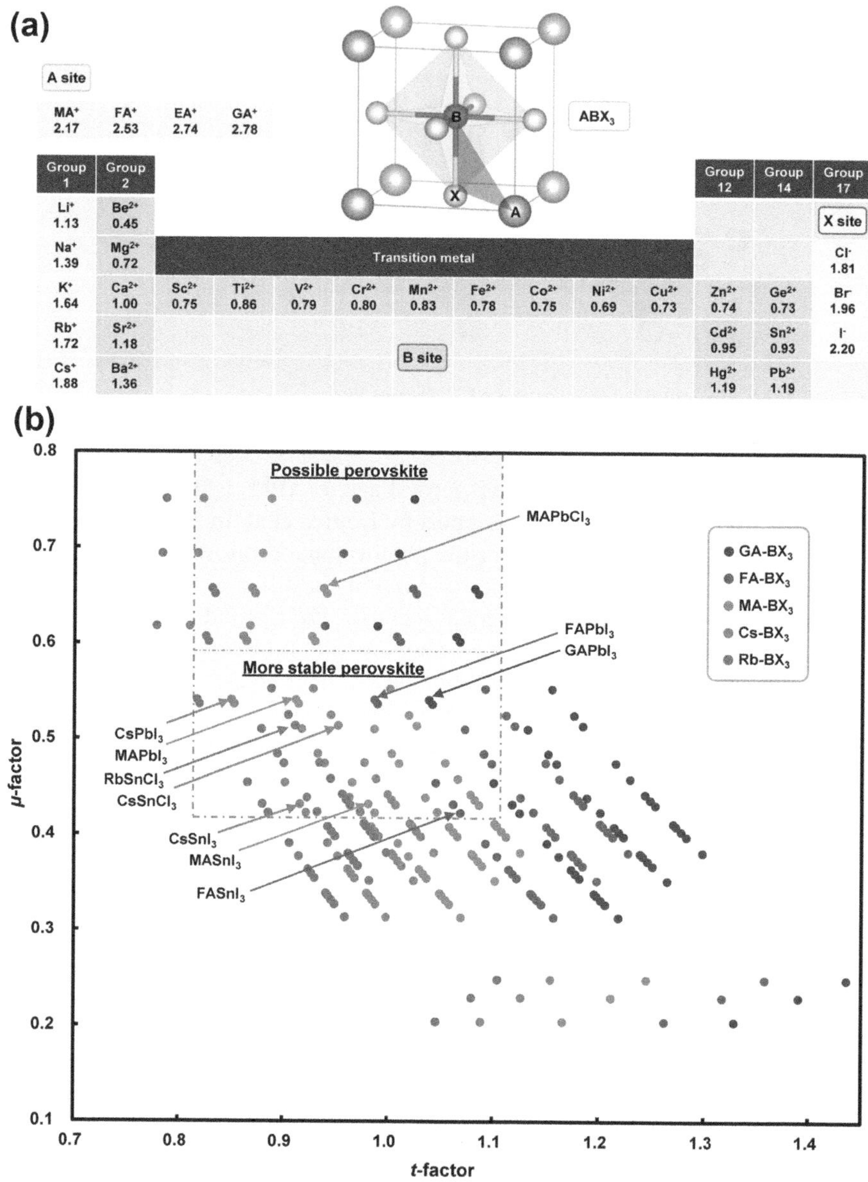

FIGURE 1.4   (a) Ionic radii (Å) of elements and molecules in ABX_3. (b) Calculated octahedral (μ) and tolerance factors (t) for various organic-inorganic hybrid perovskites. Reproduced with permission.[31] Copyright 2020, T. Oku.

phase stability of $FAPbI_3$, smaller MA^+ is re-added to form a stable mixed-cation perovskite, because of a stronger interaction with the PbI_6 octahedra.[43,44] Besides MA^+, the introduction of Cs^+, Rb^+, or Br^- can also play a similar function of stabilizing black-phased FA perovskites, thus these formulations are widely used in high-performance PSCs.[44–46]

Moreover, by completely replacing the organic part of the A-site with Cs^+, a more thermally stable all-inorganic halide perovskite, $CsPbX_3$ (X = Cl, Br, I), can be obtained (stable even at 400°C).[47,48] However, similar to pure $FAPbI_3$, $CsPbI_3$ also suffers from poor phase stability: moisture in the air causes phase transition from the black α–phase to the undesired yellow δ–phase rapidly even at room temperature.[49–52] Many efforts have been made to overcome this obstacle, such as composition engineering, crystal size control, introducing intermediate phase, fabrication processes, and surface passivating via organic polymers.[52–57]

When the A cations with larger size are introduced in perovskites, the crystal lattice will be expanded and the 3D network will be stretched to form a two-dimensional (2D) layered structure and even become one-dimensional (1D) and zero-dimensional (0D) structures as shown in Figure 1.5.[58–61] 2D perovskites based on large organic cations, including PEA^+ (phenylethylammonium cation), BA^+ (n–butylammonium cation), EDA^{2+} (ethylenediamine cations), PEI^+ (branched polyethylenimine cations), etc., have attracted substantial attention for their excellent stability.[62–65] However, comparing with the 3D perovskites, the exciton binding energy for the 2D perovskites is much larger and the out plane charge-transporting property is weaker, resulting in poor device performance and limiting their further application.[66,67] A compromise solution would be to combine 2D and 3D perovskites, which would maintain a balance between stability and optoelectronic performance. Large organic cations are added directly to the 3D perovskite precursor solution to deposit hybrid films with 2D/3D bulk heterojunctions.[68–70] 2D components such as $EDAPbI_4$ and PEA_2PbI_4 doped in the precursor solution will stabilize the as-prepared 3D perovskite or passivate grain boundaries while providing moisture protection and facilitating the separation and collection of photogenerated charges. Besides bulk heterostructures,

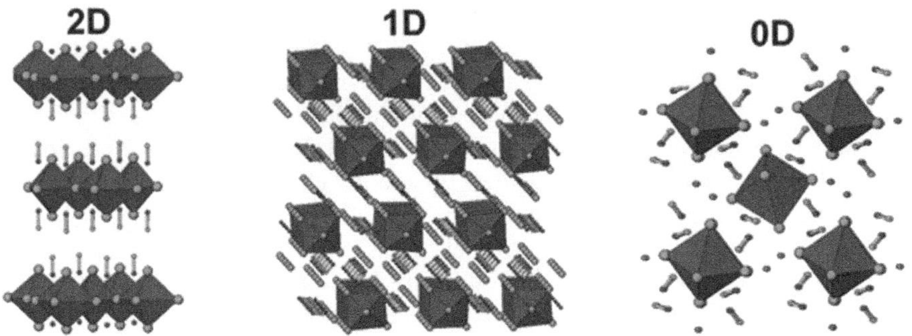

FIGURE 1.5 Schematic diagram of the crystal structures of 2D, 1D, and 0D perovskites. Reproduced with permission.[61] Copyright 2014, Royal Society of Chemistry.

another strategy is 2D/3D bilayer structures, which involve growing 3D perovskites on top of 2D perovskite layers or post-processing to form 2D perovskites on top of 3D perovskite films.[71–75] It has been reported that the 2D layer not only protects the 3D perovskite film from moisture but also blunts defects on the surface of the 3D perovskite layer.[76]

In terms of commercial applications, one of the main concerns about PSCs is the toxicity of Pb compounds released from decomposed perovskite films, which can dissolve in water (k_{sp} = 9.8 × 10–9).[77,78] Although researchers have long been aware of the toxicity of perovskites, replacement of Pb by safe elements have been less successful. Many bivalent cations, including Group 14 elements (Sn and Ge, closest to Pb), alkaline earth metals, and various transition metals have been tried as alternatives to Pb.[79–84] Some "perovskite-like" lead-free halide compounds with slightly altered stoichiometric compositions, such as $Cs_2InAgCl_6$, Cs_2SnI_6, and $MA_3Bi_2I_9$, have also been developed but with much lower PCEs (less than 4%).[85–87] Sn^{2+}-based $ASnI_3$ (A= MA, FA, Cs, etc.) has an ideal optical gap (1.2~1.4 eV) and is the earliest and most promising Pb-free perovskite. However, when exposed to ambient conditions, Sn^{2+} readily oxidizes to Sn^{4+}, resulting in "self-doping," which limits carrier diffusion length and device performance.[83,88] As a compromise, the PSCs with Pb^{2+} partially replaced by Sn^{2+} ions can obtain relatively high performance while decreasing the Pb^{2+} content.[89–92]

1.3　BASIC PROPERTIES OF PEROVSKITES

Almost all classical photovoltaic semiconductors, including Si, CdTe, and GaAs, are covalent compounds. However, metal halide perovskite is a special ionic crystal with both ionic and semiconductor properties. Although they were discovered very early, the solid-state physics of these materials has only been extensively studied in recent years following the rise of PSCs. The semiconductor properties, carrier transport mechanisms, and photovoltaic applications of metal halide perovskites have been systematically studied. The important properties that support the excellent performance of PSCs are as follows:

1. Tunable band gap:

The band gap of semiconductor materials is of great significance in photovoltaic cells and other optoelectronic devices. While in perovskite materials, the band gap can be easily adjusted by changing the composition in the ABX_3 structure (Figure 1.6).[93] For example, by changing the halogen anion (X = Cl, Br, or I) in $MAPbX_3$, the band gap can be continuously tuned between 1.6 and 3.1 eV, and the luminous color can cover almost the entire visible light region, indicating that mixed halide perovskites can form good solid solutions as ionic crystals.[94,95] Interestingly, the band gaps of $MAPbX_3$ (X = Cl, Br, or I) single crystals are found to be 2.97, 2.24, and 1.53 eV, respectively, which is a little smaller than the corresponding data for the polycrystalline films, possibly because of a weak indirect band gap absorption around their band edge.[96] It was demonstrated by Chen et al. with single-crystal $MAPbI_3$-based PSCs, which showed obviously broader spectral response (EQE cutoff at ~820 nm) than the

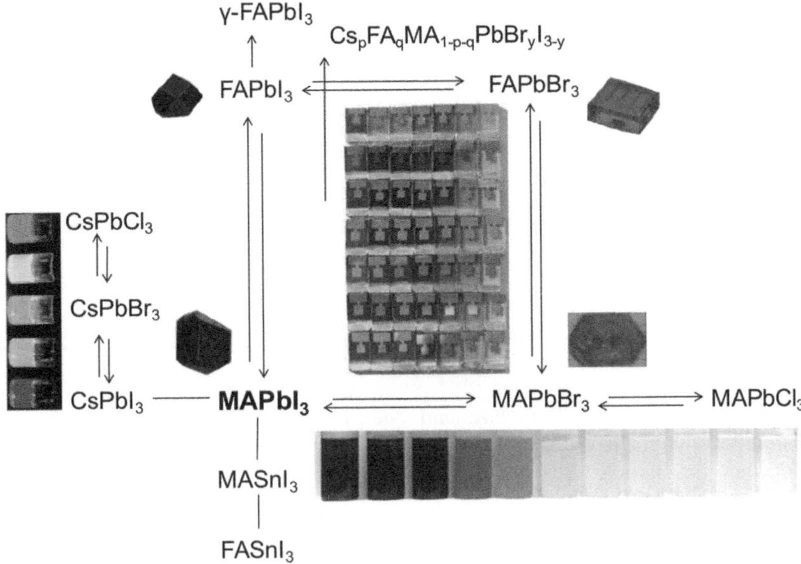

FIGURE 1.6 The absorption tunability of hybrid perovskite materials. Reproduced with permission.[93] Copyright 2017, Royal Society of Chemistry.

polycrystalline thin film (~780 nm).[97] In addition, the band gap modulation could also be achieved by changing the cation ratio in A site (e. g. MA^+/FA^+) or the metal cation ratio in B site (Pb^{2+}/Sn^{2+}) as mentioned in the last section.

2. High optical absorption coefficient:

Due to the high symmetry of the ABX_3 crystal structure, most of the halide perovskites exhibit the optical features of a semiconductor with direct band gap. Figure 1.7 shows the band structure of $MAPbI_3$. Based on the study of Kondo et al. and Brivio et al., the valence band (VB) of $MAPbI_3$ consists of approximately 70% I 5p orbitals and 25% Pb 6s orbitals (lone pair) with strong coupling between them,[98–100] while the conduction bands (CB) consist of a mixture of Pb 6p and other orbitals. The direct band gap and the p–p electron transition from VB to CB, enabled by the Pb 6s orbital lone pair, contribute to the high optical absorption coefficient of $MAPbI_3$ ($> 10^5$ cm^{-1}).[101]

3. High defect-tolerance:

In other ionic semiconductors (such as CdTe and GaAs), localized nonbonding orbitals surrounding the ion vacancies create trap states deep within the band gap. However, in halide perovskites, the trap states formed by defects in lead halide perovskites are either in-band (VB or CB) or exist as shallow traps near CB and VB

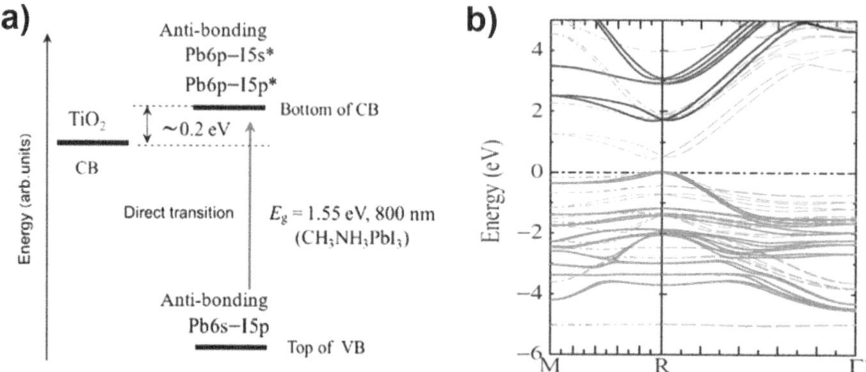

FIGURE 1.7   (a) Band gap structure and energy levels of MAPbI$_3$. Reproduced with permission.[15] Copyright 2019, American Chemical Society. (b) Electronic structure of MAPbI$_3$ based on the quasiparticle self-consistent GW approximation. Solid line above zero, gray dotted line, and solid line below zero depict bands of Pb 6p, Pb 6s, and I 5p respectively. Zero denotes the valence band maximum. Points denoted as M and R are zone-boundary points. Reproduced with permission.[100] Copyright 2014, American Physical Society.

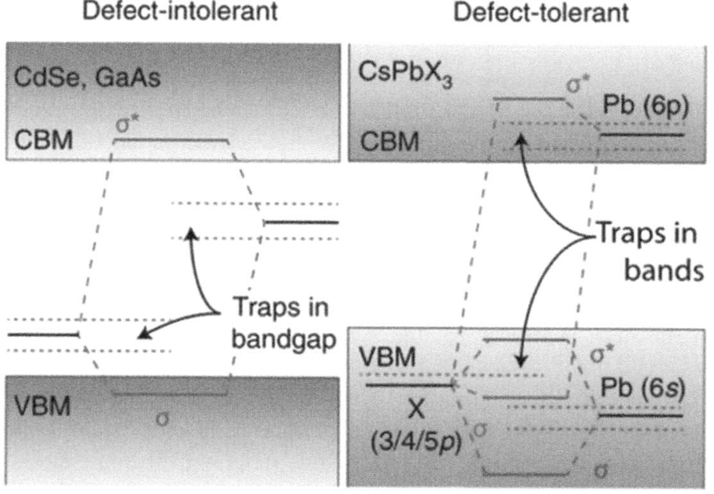

FIGURE 1.8   Schematic diagram of the band structures for conventional defect-intolerant semiconductors (left) and defect-tolerant perovskites (right). Reproduced with permission.[102] Copyright 2018, Springer Nature.

(Figure 1.8).[102,103] Carriers trapped in shallow defects can be easily detrapped, without reducing the photocurrent. As a result, perovskite materials have a good tolerance to defects, resulting in long carrier diffusion distances, which range from 1 μm (polycrystalline thin film) to over 100 μm (single crystalline) measured by the PL lifetime.[104,105]

4. Well-balanced charge transfer:

As an intrinsic semiconductor, $MAPbI_3$ has bipolar carrier mobility and exhibits similar effective mass values for electrons (0.23) and holes (0.29).[106] This rare property promotes a well-balanced transfer of electrons and holes, which facilitates the reduction of recombination and energy loss due to charge accumulation on one side.

In addition to the basic properties described earlier, various perovskite materials have a variety of unique physical and chemical properties, which require a large amount of energy to be explored by researchers. Recently, machine learning (ML), as a branch of artificial intelligence (AI), has received much attention and been used as a powerful tool to advance the research on materials science, including perovskites.[108] Machine learning can be used not only to explore potential stable new materials for PSCs through high-throughput screening but also to analyze and summarize data on perovskite device performance in order to suggest directions for improvement.[109–111] As shown in Figure 1.9, the systematic workflow of ML includes steps like samples data processing, feature engineering, model building and model evaluation, etc.[107] For example, Wu et al. used ML in a multistep screening process and created a large database of 230,808 metal halide perovskites, as shown in Figure 1.10.[109] Despite some impressive advances in machine learning-assisted perovskite research, there are still significant limitations. In addition to further improving algorithms based on the properties of perovskite materials and devices, the main problem at present is that the databases used to train machine learning models are not large enough, leading to overfitting. In addition, it is necessary to attach importance to the combination of machine learning and experiments, using experimental results for cross-validation and iteration, to facilitate the rapid development of perovskite materials and devices.

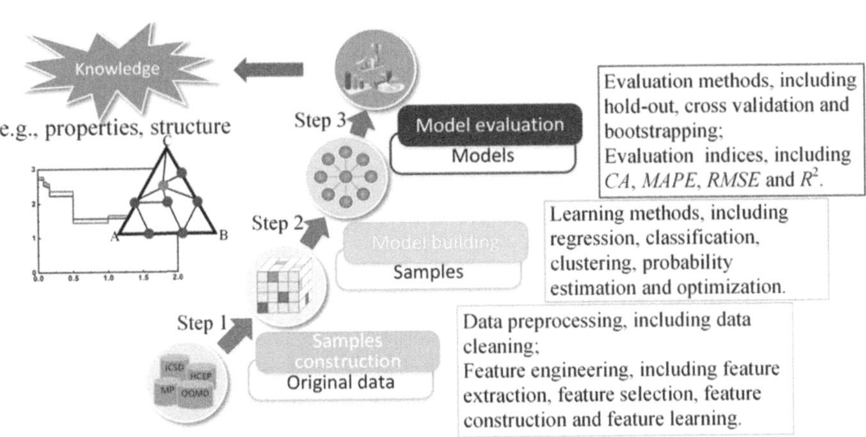

FIGURE 1.9 Schematic of the general process for machine learning in materials science. Reproduced with permission.[107] Copyright 2017, Elsevier.

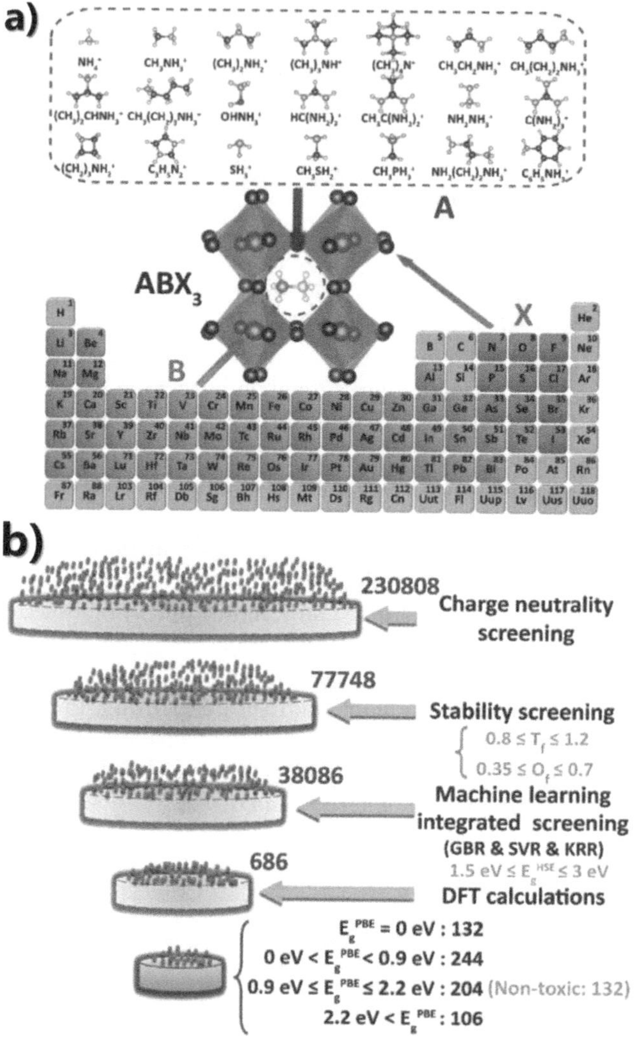

FIGURE 1.10   (a) Alternatives of A, B and X-site ions for the database of ML. (b) Schematic design framework for predicting novel perovskite materials based on the combination of ML and DFT calculations for PSCs. Reproduced with permission.[109] Copyright 2019, Elsevier.

1.4   PREPARATION METHODS OF PEROVSKITES

The quality of perovskite films is one of the decisive factors in determining the performance of PSC. To date, researchers have developed a variety of methods to improve the morphology, uniformity, crystallinity, and phase purity of perovskite films, which largely rely on film deposition techniques.[112] In this section, we will discuss various deposition techniques of perovskite thin films.

1.4.1 Evaporation Deposition

The evaporation processes are widely used to fabricate commercial semiconductor devices including thin-film solar cells, which involve sublimating the precursors in a vacuum chamber and depositing them onto a rotating substrate. In a typical dual-source co-evaporation process (Figure 1.11a), metal halides and organic ammonium salts deposited on the substrate undergo an initial reaction followed by thermal annealing process to promote crystal growth. The final film quality and morphology are highly dependent on the evaporation rate and proportion of the precursor, as these factors are closely related to the crystal growth and nucleation mechanisms.[113] In 2013, Snaith's research group successfully introduced this method into the preparation of PSCs by co-evaporating $PbCl_2$ and MAI to the substrates, resulting in homogeneous and compact $MAPbI_{3-x}Cl_x$ perovskite films, demonstrating that evaporation is a promising method that can ensure the compactness of as-prepared films.[114] However, it is difficult to accurately control the deposition rates of multiple evaporation sources at the same time. This can be overcome by employing sequential deposition methods as shown in Figure 1.11b. For example, Feng et al. deposited high-quality FA-Cs perovskite films on glass and flexible substrates by sequentially evaporating PbI_2, FAI and CsI.[115] Yi et al. evaporated inorganic metal salts (CsI, PbI_2 and $PbCl_2$) and organic ammonium salt (FAI) separately in different vacuum chambers to avoid the

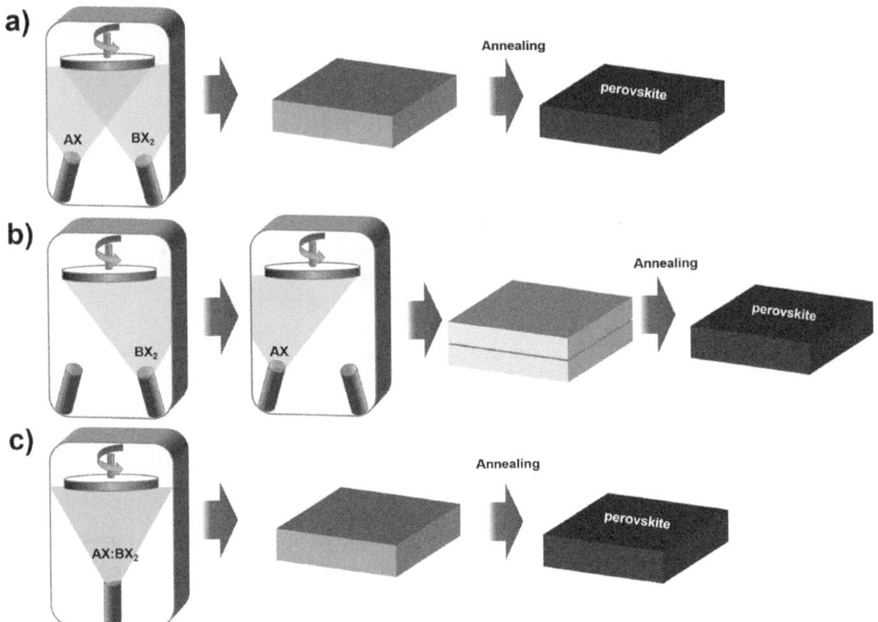

FIGURE 1.11 Schematic diagram of evaporation process for depositing perovskite films. a) Dual-source co-evaporation, b) dual-source sequential evaporation, c) single-source evaporation.

risk of cross-contamination.[116] This modified vacuum deposition method can produce perovskite films with high crystallinity and homogeneity, resulting in PSCs with a record PCE of 24.42%. For all-inorganic perovskites (such as $CsPbX_3$) without highly volatile organic components, it is also possible to evaporate the pre-mixed powder of perovskite precursors with a single source (Figure 1.11c). Compact and uniform $CsPbX_3$ films with high crystallinity can be obtained by tuning the precursor ratio and post-annealing conditions, resulting in PCEs exceeding 8%.[117,118] Due to the high cost of vacuum equipment, complicated process, and long deposition time, vacuum deposition has not been widely promoted in PSC research.

1.4.2 SINGLE-STEP SOLUTION-PROCESSED DEPOSITION

One-step solution deposition of halide perovskite films can be traced back to the 1990s, when Mitzi et al. dissolved organic components and tin halide into a hydro-iodic acid solution and obtained perovskite structural crystals by heating and filtering.[119] Due to its simple process and low cost, single-step solution deposition is still the most widely used technique.[112] The precursor solution is formed by mixing organic and inorganic salts in aprotic polar solvents such as DMF, DMSO, NMP, 2ME, etc.[120] As shown in Figure 1.12a, the precursor solution is coated on the substrate by spin-coating (or blading, printing, etc.), and then with the volatilization of the solvent, the solute is precipitated from the solution and crystallized, and finally the perovskite film is obtained. However, most of the solvents used in precursor solutions have high boiling points and low vapor pressures at room temperature, thus the natural volatilizing process is slow and uneven, resulting in non-uniform films with

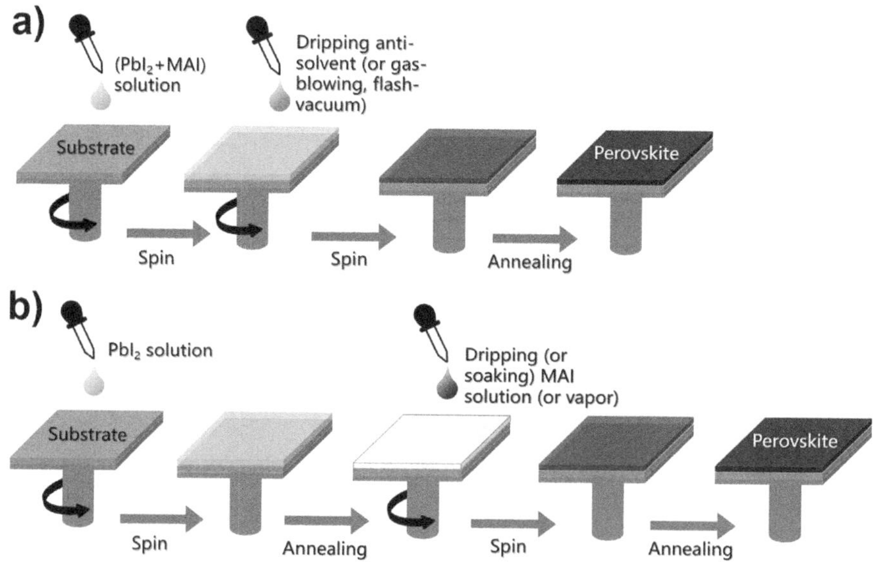

FIGURE 1.12 Typical solution-processed deposition methods of $MAPbI_3$: a) one-step spin-coating, b) two-step spin-coating.

pinholes. Cheng et al. and Seok et al. reported antisolvent-induced rapid crystallization methods that involve spin-coating a solution of perovskite precursors and then immediately dripping antisolvents (such as chlorobenzene, toluene, and diethyl ether) to induce rapid crystallization.[121–123] In this way, the crystallinity and uniformity of the prepared perovskite films are greatly improved. This method has been widely used in the laboratory research of small-area devices, but it is difficult to directly extend to large-area thin films. The researchers have developed several alternative methods such as blowing air[124] and flash vacuum[125] to assist crystallization, which still need to be further optimized.

1.4.3 Two-Step Solution-Processed Deposition

At early times, the surface coverage of perovskite films produced by the one-step deposition method is poor. To avoid this, Mitzi et al. reported a two-step deposition method of first spin-coating (or evaporating) PbI_2 onto the substrate and then reacting with MAI solution to convert into $MAPbI_3$ perovskite films.[126] In 2013, Grätzel et al. introduced this method to the fabrication of PSCs.[127] As shown in Figure 1.12b, spin-coated PbI_2 film was dipped into MAI/IPA solution, resulting in $MAPbI_3$ perovskite film with better coverage than the previous one-step method, with PCEs exceeding 15%. Im et al. studied the effect of MAI concentration on surface morphology and obtained a decrease in grain size with increasing MAI concentration.[128] Later, some improved methods were reported, which changed the process of soaking in organic salt solution to solution spin coating or even directly contact with MAI vapor or powder.[129–131] It is worth noting that the MAI concentration of the soaking solution is very low, and most of the PbI_2 crystals reacting with MAI follows the "in-situ transformation," while in the spin-coating process, the high concentration of MAI destabilizes the PbI_6 octahedral structure and leads to the formation of an I-rich complex via dissolution, followed by further conversion to $MAPbI_3$ crystals (Figure 1.13).[132]

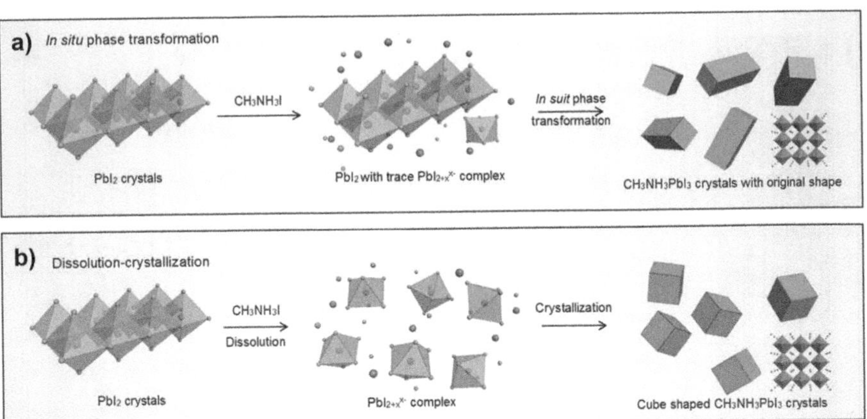

FIGURE 1.13 Schematic diagram of formation mechanisms for the two-step deposited perovskite films: a) in-situ transformation, b) dissolution-crystallization. Reproduced with permission.[132] Copyright 2014, American Chemical Society.

Therefore, it is easier to achieve uniform crystallization and less PbI$_2$ residue by the spin-coating method. Compared to the one-step method, the two-step method allows a wider processing window, and the conditions of each deposition step can be optimized individually, so the nucleation and crystallization of perovskites can be better controlled, resulting in higher repeatability.[131] However, there are still some major shortcomings: the organic components in the second step of the reaction are typically excessive, making it difficult to precisely control the chemical composition of the final film. In addition, the inorganic film deposited in the first step inevitably forms crystals, hindering the subsequent molecular exchange reactions with organic components, resulting in PbI$_2$ residues that cannot be fully converted. Although it has been reported that an appropriate amount of PbI$_2$ residue can play a passivation role and improve the performance of PSCs, it is not conducive to the long-term stability of perovskites.[133]

1.5 WORKING PRINCIPLES AND DEVICE STRUCTURES OF PSCs

In 2009, Miyasaka et al. reported the first PSCs with a device structure of dye sensitized solar cells.[22] Perovskite absorbers are loaded on the surface of mesoporous TiO$_2$ (mp-TiO$_2$) (Figure 1.14) and photoexcited by absorbing light that matches its bandgap wavelength, and then electrons are injected from the perovskites into the CB of TiO$_2$ and transported to the FTO substrate. From there, the electrons flow to the counter electrode (cathode) in an external circuit and combine with oxidized species in the electrolyte. The reduced species then diffuses into the vicinity of the oxidized perovskite to reduce it. A PCE of 3.8% was obtained on MAPbI$_3$-sensitized solar cells. However, more than 80% of the MAPbI$_3$ particles were resolved in the liquid electrolyte within 10 min, which severely hindered further investigation.[23] In 2012, Park et al. used spiro-OMeTAD as a hole transport layer instead of a liquid electrolyte to assemble a solid-state PSC as shown in Figure 1.15.[24] The stability of

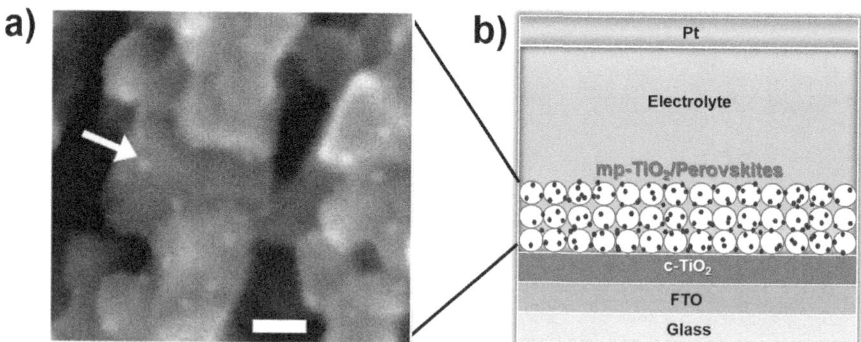

FIGURE 1.14   a) SEM image of perovskite nanocrystalline deposited on the TiO$_2$ surface. The arrow indicates a particle, and the scale bar shows 10 nm. Reproduced with permission.[22] Copyright 2009, American Chemical Society. b) Schematic diagram of a perovskite sensitized solar cell with liquid electrolyte.

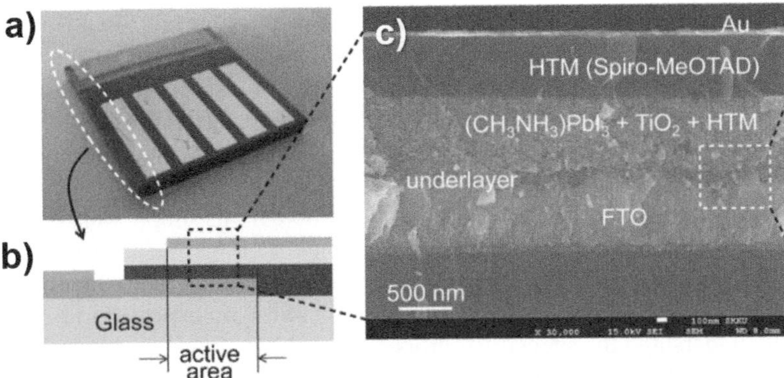

FIGURE 1.15 Solid-state MAPbI$_3$-sensitized solar devices. a) Digital photo image, b) cross-sectional structure of the device, c) cross sectional SEM image of the device. Reproduced with permission.[24] Copyright, Springer Nature.

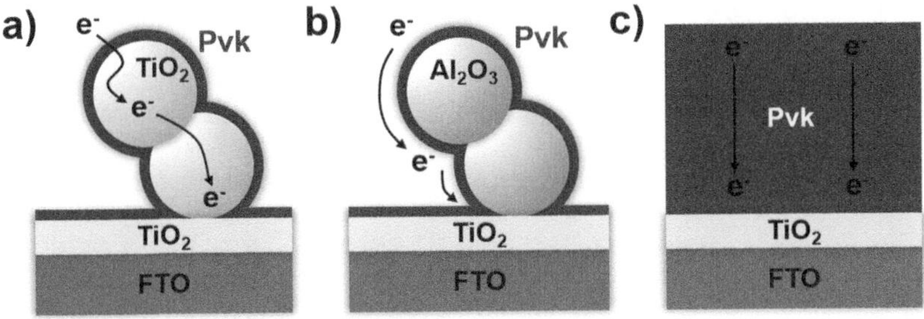

FIGURE 1.16 Schematic diagram of charge transfer and transport for a) mp-TiO$_2$-based PSC, b) mp-Al$_2$O$_3$-based PSC, and c) planar structured PSC without mesoporous layer.

spiro-OMeTAD based PSCs has been greatly improved (more than 500 h), and the PCEs has been increased to 9.7%.

In sensitized solar cells, the contact area between the light-absorbing materials (dye, perovskites, quantum dots, etc.) and the electron transport materials (metal oxide mesoporous scaffold, such as TiO$_2$, SnO$_2$, ZnO, etc.) is very large, which is conducive to the rapid extraction of photogenerated electrons.[134]] However, defects in the contact surface also lead to severe recombination, which limits further improvements in device performance. In 2012, Snaith et al. replaced mp-TiO$_2$ with porous aluminum oxide (mp-Al$_2$O$_3$) to support MAPbCl$_x$I$_{3-x}$, yielding a PCE of 10.9%.[25] Since the photogenerated electrons cannot be injected into the insulated Al$_2$O$_3$, they transport through the perovskite film itself (Figure 1.16). Later in 2013, they further reported a planar structured PSCs without any mesoporous layer,[114] indicating that in these devices, mesoporous scaffold is not indispensable, and the working mechanism is different from the conventional sensitized solar cells.

Nowadays, it is widely accepted that PSCs work similarly to n-i-p and p-i-n heterojunction solar cells, in which perovskites acts as an intrinsic absorber sandwiched between two charge transport layers (p and n). Perovskites absorb light to produce electrons and holes, which are selectively collected by the n-type and p-type charge transport layer, respectively. Electrons flow through the external circuit to the p-type layer and bind to the holes. Figure 1.17 shows the device structure and energy level diagram of a typical n-i-p PSCs. Although this simple working principle is now widely accepted, a clear and concrete understanding of the processes of carrier production, separation, transport, and even accumulation is still lacking. Further in-depth studies of in-situ carrier and lattice dynamics are expected to help us more precisely understand the special PV properties of PSCs.

At present, although a variety of device structures have been developed, as shown in Figure 1.18, to take full advantage of the excellent properties of perovskites, the

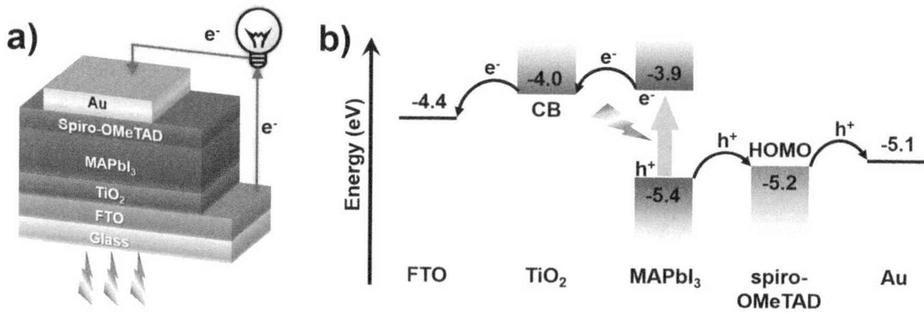

FIGURE 1.17 a) The architecture and b) energy diagram of a typical n-i-p structured PSCs.

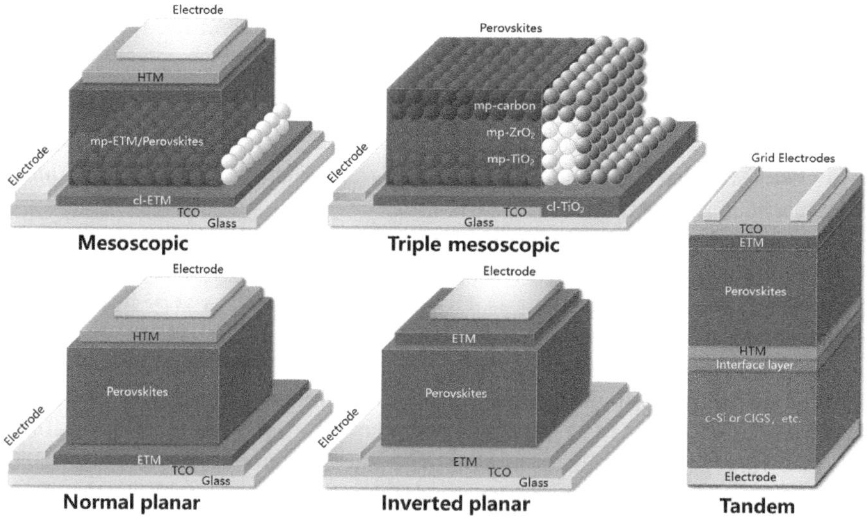

FIGURE 1.18 PSCs based on five different device configurations: mesoscopic structure, triple mesoscopic structure, normal and inverted planar structures, and tandem structure. Reproduced with permission.[14] Copyright 2019, John Wiley and Sons.

mesoscopic structure inherited from DSSCs remains one of most important options for high-performance PSCs.[135–138] In a typical mesoporous PSCs (n–i–p, or formal type), a thin layer of electron transport material (usually TiO_2 nanoparticles) with mesoporous pores is used as a scaffold for perovskite absorber penetration. Typically, a thick layer of perovskite covers the scaffold, completely separating the hole transport material (usually spiro-OMeTAD or PTAA) from the electron transport layer. Mesoporous layers not only contribute to the formation of homogeneous high-quality perovskite films but also facilitate the separation and transport of electric charges.

With the advancement of perovskite film deposition and passivation technology, the planar structure PSCs without mesoporous scaffold layer also achieved high PCEs (> 25%). In a typical planar PSCs, the compact perovskite film is sandwiched-between a dense electron transport layer (e.g., TiO_2, SnO_2, C_{60}, PCBM, etc.) and a hole transport layer (e.g., spiro-OMeTAD, PTAA, PEDOT:PSS, etc.).[139–142] Due to the simpler device structure, planar PSCs are very promising for the mass production. Interestingly, based on different materials and deposition techniques, the device structure can be n-i-p (normal) or p-i-n (inverted), which further expands the field of application.[114,139,142,143]

Compared with the precious metal electrode (Au or Ag), the carbon electrode has the advantages of high stability, water resistance, inertia to ion migration, and low material cost. Mei et al. invented a device structure of PSCs[144]: three mesoporous layers (TiO_2, ZrO_2, and carbon, respectively as electron transport layer, insulation layer, and counter electrode) were deposited sequentially by screen printing method, and then the perovskite precursor solution was infiltrated into the pores of the three layers to obtain a complete device. With the optimization of electrode structure, composition engineering, fabrication techniques, and interface engineering, the performance and stability of these PSCs were much improved.[145–149] Due to their excellent stability and low material cost, which does not require any precious metal back contacts or expensive organic hole transport materials, triple-mesoscopic PSCs have better scale-up feasibility. In order to further promote the application of triple-mesoscopic PSCs, more efforts should be put into improving the PCEs and catching up with devices based on precious metal electrodes.

According to the calculation by Shockley and Queisser in 1961, the theoretical limit of PCEs for a single solar cell is 33.7% (optimized band gap 1.34 eV, AM 1.5G light illumination).[150] To break this S-Q limit, tandem architecture is a promising photovoltaic technology, owing to the efficient utilization of incident sunlight via constructing multiple light-absorbing materials with desired band gaps together. In this configuration, each subcell absorbs different portion of the sunlight spectrum to circumvent the energy loss from the thermalization process in a single-junction solar device that when a photon with energy higher than the bandgap is absorbed, the excited electrons will relax to the conduction band with the excess energy lost.[151] One of the advantages of metal halide perovskite semiconductors is that they have tunable band gaps between 1.1 and 3.1 eV, which holds great promise in tandem devices when combined with other photovoltaic materials (such as CIGS, CdTe, GaAs, CZTS, etc.) or perovskite itself.[152–154] In particular, it can be compatible with silicon solar cells (1.1eV), the market-leading and most developed photovoltaic technology, to form an ideal 1.1/1.75eV tandem solar cell. Progress on PSC tandem devices has been impressively rapid. In March 2015, Mailoa et al. reported the first demonstration of a double-ended perovskite/silicon tandem solar cell with a PCE of 13.7%.[155] Recently (June 2023), Saudi Arabia's King Abdullah University of Science and Technology (KAUST) reported a certified PCE of 33.7%, which is a new world record for tandem solar cells consisting of a silicon bottom cell and a perovskite top cell[7]. Tandem devices present an attractive opportunity for PSCs to expand the booming silicon cell market rather than compete directly with it. With the continuous advancement of research and improvement of performance, commercial PSC tandem devices with PCEs of more than 30% are very promising.

REFERENCES

1. DOE releases solar futures study providing the blueprint for a zero-carbon grid, www.energy.gov/articles/doe-releases-solar-futures-study-providing-blueprint-zero-carbon-grid?utm_source=kawo&utm_medium=weibo&utm_campaign=none&utm_content=app.kawo.com%2F614c2866c6aa711476db4486
2. Vast amounts of solar energy radiate to the earth, but tapping it cost-effectively remains a challenge, https://phys.org/news/2011-10-vast-amounts-solar-energy-earth.html
3. Solar energy potential and utilization, www.e-education.psu.edu/earth104/node/950
4. W. Wang, M. O. Tadé, Z. Shao, Research progress of perovskite materials in photocatalysis- and photovoltaics-related energy conversion and environmental treatment, *Chemical Society Reviews* **2015**, *44*, 5371.
5. A. E. Becquerel, Recherches sur les effets de la radiation chimique de la lumiere solaire, au moyen descourants electriqurs, *Comptes Rendus de l'Académie des Sciences Paris* **1839**, *9*, 561.
6. D. M. Chapin, C. S. Fuller, G. L. Pearson, A new silicon *p-n* junction photocell for converting solar radiation into electrical power, *Journal of Applied Physics* **1954**, *25*, 676.
7. Best research-cell efficiencies, www.nrel.gov/pv/assets/pdfs/best-research-cell-efficiencies.pdf
8. H. W. Schock, R. Noufi, CIGS-based solar cells for the next millennium, *Progress in Photovoltaics* **2000**, *8*, 151.
9. A. Romeo, *et al.*, Development of thin-film Cu(In, Ga)Se$_2$ and CdTe solar cells, *Progress in Photovoltaics* **2004**, *12*, 93.
10. S. Mathew *et al.*, Dye-sensitized solar cells with 13% efficiency achieved through the molecular engineering of porphyrin sensitizers, *Nature Chemistry* **2014**, *6*, 242.
11. L. Hong, *et al.*, 18.5% Efficiency organic solar cells with a hybrid planar/bulk heterojunction, *Advanced Materials* **2021**, *33*, 2103091.
12. M. I. H. Ansari, A. Qurashi, M. K. Nazeeruddin, Frontiers, opportunities, and challenges in perovskite solar cells: A critical review, *Journal of Photochemistry and Photobiology C: Photochemistry Reviews* **2018**, *35*, 1.
13. A. R. Chakhmouradian, P. M. Woodward, Celebrating 175 years of perovskite research: A tribute to Roger H. Mitchell, *Physics and Chemistry of Minerals* **2014**, *41*, 387.
14. P. Wang, Y. Wu, B. Cai, Q. Ma, X. Zheng, W. H. Zhang, Solution-processable perovskite solar cells toward commercialization: Progress and challenges, *Advanced Functional Materials* **2019**, *29*, 1807661.
15. A. K. Jena, A. Kulkarni, T. Miyasaka, Halide perovskite photovoltaics: Background, status, and future prospects, *Chemical Reviews* **2019**, *119*, 3036.
16. Y. Yuan, Z. Xiao, B. Yang, J. Huang, Arising applications of ferroelectric materials in photovoltaic devices, *Journal of Materials Chemistry A* **2014**, *2*, 6027.
17. H. L. Wells, Über die Cäsium- und Kalium-Bleihalogenide, *Zeitschrift für Anorganische Chemie* **1893**, *3*, 195.
18. C. K. Møller, A phase transition in caesium plumbochloride, *Nature* **1957**, *180*, 981.

19. C. K. Møller, Crystal structure and photoconductivity of cæsium plumbohalides, *Nature* **1958**, *182*, 1436.
20. D. Weber, $CH_3NH_3PbX_3$, ein Pb(II)-system mit kubischer Perowskitstruktur/$CH_3NH_3PbX_3$, a Pb(II)-system with cubic perovskite structure, *Zeitschrift für Naturforschung B* **1978**, *33*, 862, 1443.
21. D. B. Mitzi, K. Chondroudis, C. R. Kagan, Organic-inorganic electronics, *IBM Journal of Research and Development* **2001**, *45*, 29.
22. T. Miyasaka, A. Kojima, K. Teshima, Y. Shirai, Organometal halide perovskites as visible-light sensitizers for photovoltaic cells, *Journal of the American Chemical Society* **2009**, *131*, 6050.
23. J. H. Im, C. R. Lee, J. W. Lee, S. W. Park, N. G. Park, 6.5% efficient perovskite quantum-dot-sensitized solar cell, *Nanoscale* **2011**, *3*, 4088.
24. H.-S. Kim, *et al.*, Lead iodide perovskite sensitized all-solid-state submicron thin film mesoscopic solar cell with efficiency exceeding 9%, *Scientific Reports* **2012**, *2*, 591.
25. M. M. Lee, J. Teuscher, T. Miyasaka, T. N. Murakami, H. J. Snaith, Efficient hybrid solar cells based on meso-superstructured organometal halide perovskites, *Science* **2012**, *338*, 643.
26. Science's top 10 breakthroughs of 2013, www.science.org/content/article/sciences-top-10-breakthroughs-2013
27. S. A. Veldhuis, P. P. Boix, N. Yantara, M. Li, T. C. Sum, N. Mathews, S. G. Mhaisalkar, Perovskite materials for light-emitting diodes and lasers, *Advanced Materials* **2016**, *28*, 6804.
28. L. Dou, Y. Yang, J. You, Z. Hong, W.-H. Chang, G. Li, Y. Yang, Solution-processed hybrid perovskite photodetectors with high detectivity, *Nature Communications* **2014**, *5*, 5404.
29. W. Heiss, C. Brabec, Perovskites target X-ray detection, *Nature Photonics* **2016**, *10*, 288.
30. S.-J. Chang, S.-Y. Chen, P.-W. Chen, S.-J. Huang, Y.-C. Tseng, Pulse-driven nonvolatile perovskite memory with photovoltaic read-out characteristics, *ACS Applied Materials & Interfaces* **2019**, *11*, 33803.
31. T. Oku, Crystal structures of perovskite halide compounds used for solar cells, *Reviews on Advanced Materials Science* **2020**, *59*, 264.
32. V. M. Goldschmidt, Die Gesetze der Krystallochemie, *Die Naturwissenschaften* **1926**, *14*, 477.
33. A. Poglitsch, D. Weber, Dynamic disorder in methylammoniumtrihalogenoplumbates (II) observed by millimeter-wave spectroscopy, *The Journal of Chemical Physics* **1987**, *87*, 6373.
34. N. Onoda-Yamamuro, T. Matsuo, H. Suga, Calorimetric and IR spectroscopic studies of phase transitions in methylammonium trihalogenoplumbates (II)†, *Journal of Physics and Chemistry of Solids* **1990**, *51*, 1383.
35. A. Dualeh, P. Gao, S. I. Seok, M. K. Nazeeruddin, M. Grätzel, Thermal behavior of methylammonium lead-trihalide perovskite photovoltaic light harvesters, *Chemistry of Materials* **2014**, *26*, 6160.
36. B. Conings, *et al.*, Intrinsic thermal instability of methylammonium lead trihalide perovskite, *Advanced Energy Materials* **2015**, *5*, 1500477.
37. A. Pisoni, J. Jaćimović, O. S. Barišić, M. Spina, R. Gaál, L. Forró, E. Horváth, Ultra-low thermal conductivity in organic–inorganic hybrid perovskite $CH_3NH_3PbI_3$, *The Journal of Physical Chemistry Letters* **2014**, *5*, 2488.
38. Y. Liu, *et al.*, Two-inch-sized perovskite $CH_3NH_3PbX_3$(X = Cl, Br, I) crystals: Growth and characterization, *Advanced Materials* **2015**, *27*, 5176.
39. J.-W. Lee, D.-J. Seol, A.-N. Cho, N.-G. Park, High-efficiency perovskite solar cells based on the black polymorph of $HC(NH_2)_2PbI_3$, *Advanced Materials* **2014**, *26*, 4991.
40. A. Binek, F. C. Hanusch, P. Docampo, T. Bein, Stabilization of the trigonal high-temperature phase of formamidinium lead iodide, *The Journal of Physical Chemistry Letters* **2015**, *6*, 1249.

41. C. C. Stoumpos, C. D. Malliakas, M. G. Kanatzidis, Semiconducting tin and lead iodide perovskites with organic cations: Phase transitions, high mobilities, and near-infrared photoluminescent properties, *Inorganic Chemistry* **2013**, *52*, 9019.
42. T. M. Koh, *et al.*, Formamidinium-containing metal-halide: An alternative material for near-IR absorption perovskite solar cells, *The Journal of Physical Chemistry C* **2014**, *118*, 16458.
43. N. J. Jeon, J. H. Noh, W. S. Yang, Y. C. Kim, S. Ryu, J. Seo, S. I. Seok, Compositional engineering of perovskite materials for high-performance solar cells, *Nature* **2015**, *517*, 476.
44. T. J. Jacobsson, J. P. Correa-Baen, M. Pazoki, M. Saliba, K. Schenk, M. Grätzel, A. Hagfeldt, Exploration of the compositional space for mixed lead halogen perovskites for high efficiency solar cells, *Energy & Environmental Science* **2016**, *9*, 1706.
45. Q. Jiang, *et al.*, Enhanced electron extraction using SnO_2 for high-efficiency planar-structure $HC(NH_2)_2PbI_3$-based perovskite solar cells, *Nature Energy* **2017**, *2*, 16177.
46. M. Saliba, *et al.*, Incorporation of rubidium cations into perovskite solar cells improves photovoltaic performance, *Science* **2016**, *354*, 206.
47. M. Kulbak, S. Gupta, N. Kedem, I. Levine, T. Bendikov, G. Hodes, D. Cahen, Cesium enhances long-term stability of lead bromide perovskite-based solar cells, *The Journal of Physical Chemistry Letters* **2016**, *7*, 167.
48. M. Kulbak, D. Cahen, G. Hodes, How important is the organic part of lead halide perovskite photovoltaic cells? Efficient $CsPbBr_3$ cells, *The Journal of Physical Chemistry Letters* **2015**, *6*, 2452.
49. G. E. Eperon, G. M. Paterno, R. J. Sutton, A. Zampetti, A. A. Haghighirad, F. Cacialli, H. J. Snaith, Inorganic caesium lead iodide perovskite solar cells, *Journal of Materials Chemistry A* **2015**, *3*, 19688.
50. S. Dastidar, *et al.*, High chloride doping levels stabilize the perovskite phase of cesium lead iodide, *Nano Letters* **2016**, *16*, 3563.
51. J. Lin, *et al.*, Thermochromic halide perovskite solar cells, *Nature Materials* **2018**, *17*, 261.
52. C. F. J. Lau, *et al.*, $CsPbIBr_2$ perovskite solar cell by spray-assisted deposition, *ACS Energy Letters* **2016**, *1*, 573.
53. W. K. Zhou, *et al.*, Light-independent ionic transport in inorganic perovskite and ultrastable Cs-based perovskite solar cells, *The Journal of Physical Chemistry Letters* **2017**, *8*, 4122.
54. A. Swarnkar, *et al.*, Quantum dot-induced phase stabilization of alpha-$CsPbI_3$ perovskite for high-efficiency photovoltaics, *Science* **2016**, *354*, 92.
55. P. F. Luo, W. Xia, S. W. Zhou, L. Sun, J. G. Cheng, C. X. Xu, Y. W. Lu, Solvent engineering for ambient-air-processed, phase-stable $CsPbI_3$ in perovskite solar cells, *The Journal of Physical Chemistry Letters* **2016**, *7*, 3603.
56. B. Li, Y. A. Zhang, L. Fu, T. Yu, S. J. Zhou, L. Y. Zhang, L. W. Yin, Surface passivation engineering strategy to fully-inorganic cubic $CsPbI_3$ perovskites for high-performance solar cells, *Nature Communications* **2018**, *9*, 1076.
57. Y. Wang, T. Zhang, M. Kan, Y. Zhao, Bifunctional stabilization of all-inorganic α-$CsPbI_3$ perovskite for 17% efficiency photovoltaics, *Journal of the American Chemical Society* **2018**, *140*, 12345.
58. S. M. Wang, D. B. Mitzi, C. A. Feild, A. Guloy, Synthesis and characterization of $[NH_2C(I)=NH_2]_3MI_5$ (M = Sn, Pb) – Stereochemical activity in divalent tin and lead halides containing single 110 perovskite sheets, *Journal of the American Chemical Society* **1995**, *117*, 5297.
59. D. B. Mitzi, S. Wang, C. A. Feild, C. A. Chess, A. M. Guloy, Conducting layered organic-inorganic halides containing (110)-oriented perovskite sheets, *Science* **1995**, *267*, 1473.
60. J. Calabrese, N. L. Jones, R. L. Harlow, N. Herron, D. L. Thorn, Y. Wang, Preparation and characterization of layered lead halide compounds, *Journal of the American Chemical Society* **1991**, *113*, 2328.

61. T. C. Sum, N. Mathews, Advancements in perovskite solar cells: Photophysics behind the photovoltaics, *Energy & Environmental Science* **2014**, *7*, 2518.
62. I. C. Smith, E. T. Hoke, D. Solis-Ibarra, M. D. McGehee, H. I. Karunadasa, A layered hybrid perovskite solar-cell absorber with enhanced moisture stability, *Angewandte Chemie International Edition* **2014**, *53*, 11232.
63. D. H. Cao, C. C. Stoumpos, O. K. Farha, J. T. Hupp, M. G. Kanatzidis, 2D homologous perovskites as light-absorbing materials for solar cell applications, *Journal of the American Chemical Society* **2015**, *137*, 7843.
64. T. Y. Zhang, M. I. Dar, G. Li, F. Xu, N. J. Guo, M. Grätzel, Y. X. Zhao, Bication lead iodide 2D perovskite component to stabilize inorganic alpha-$CsPbI_3$ perovskite phase for high-efficiency solar cells, *Science Advances* **2017**, *3*, e1700841.
65. K. Yao, X. F. Wang, Y. X. Xu, F. Li, L. Zhou, Multilayered perovskite materials based on polymeric-ammonium cations for stable large-area solar cell, *Chemistry of Materials* **2016**, *28*, 3131.
66. H. H. Tsai, *et al.*, High-efficiency two-dimensional Ruddlesden-Popper perovskite solar cells, *Nature* **2016**, *536*, 312.
67. X. Zhang, *et al.*, Stable high efficiency two-dimensional perovskite solar cells via cesium doping, *Energy & Environmental Science* **2017**, *10*, 2095.
68. J. W. Lee, *et al.*, 2D perovskite stabilized phase-pure formamidinium perovskite solar cells, *Nature Communications* **2018**, *9*, 3021.
69. M. H. Li, *et al.*, Highly efficient 2D/3D hybrid perovskite solar cells via low-pressure vapor-assisted solution process, *Advanced Materials* **2018**, *30*, 1801401.
70. Z. P. Wang, Q. Q. Lin, F. P. Chmiel, N. Sakai, L. M. Herz, H. J. Snaith, Efficient ambient-air-stable solar cells with 2D-3D heterostructured butylammonium-caesium-formamidinium lead halide perovskites, *Nature Energy* **2017**, *2*, 17135.
71. Y. H. Hu, J. Schlipf, M. Wussler, M. L. Petrus, W. Jaegermann, T. Bein, P. Muller-Buschbaum, P. Docampo, Hybrid perovskite/perovskite heterojunction solar cells, *ACS Nano* **2016**, *10*, 5999.
72. J. Z. Chen, J. Y. Seo, N. G. Park, Simultaneous improvement of photovoltaic performance and stability by in situ formation of 2D perovskite at $(FAPbI_3)_{0.88}(CsPbBr_3)_{0.12}$/CuSCN interface, *Advanced Energy Materials* **2018**, *8*, 1702714.
73. P. Chen, Y. Bai, S. C. Wang, M. Q. Lyu, J. H. Yun, L. Z. Wang, In situ growth of 2D perovskite capping layer for stable and efficient perovskite solar cells, *Advanced Functional Materials* **2018**, *28*, 1706923.
74. G. Grancini, *et al.*, One-year stable perovskite solar cells by 2D/3D interface engineering, *Nature Communications* **2017**, *8*, 15684.
75. Y. Qian, *et al.*, Passivating perovskites in air via an alternating cation interlayer phase formed by benzylamine vapor fumigation, *Advanced Functional Materials* **2023**, *33*, 2214731
76. I. Metcalf, *et al.*, Synergy of 3D and 2D perovskites for durable, efficient solar cells and beyond, *Chemical Reviews* **2023**, *123*, 9565.
77. J. M. Kadro, A. Hagfeldt, The end-of-life of perovskite PV, *Joule* **2017**, *1*, 29.
78. B. Hailegnaw, S. Kirmayer, E. Edri, G. Hodes, D. Cahen, Rain on methylammonium lead iodide based perovskites: Possible environmental effects of perovskite solar cells, *The Journal of Physical Chemistry Letters* **2015**, *6*, 1543.
79. W. Ke, M. G. Kanatzidis, Prospects for low-toxicity lead-free perovskite solar cells, *Nature Communications* **2019**, *10*, 965.
80. T. M. Koh, *et al.*, Formamidinium tin-based perovskite with low E-g for photovoltaic applications, *Journal of Materials Chemistry A* **2015**, *3*, 14996.
81. M. R. Filip, F. Giustino, Computational screening of homovalent lead substitution in organic-inorganic halide perovskites, *The Journal of Physical Chemistry C* **2016**, *120*, 166.

82. X. P. Cui, *et al.*, Cupric bromide hybrid perovskite heterojunction solar cells, *Synthetic Metals* **2015**, *209*, 247.

83. F. Hao, C. C. Stoumpos, D. H. Cao, R. P. H. Chang, M. G. Kanatzidis, Lead-free solid-state organic-inorganic halide perovskite solar cells, *Nature Photonics* **2014**, *8*, 489.

84. T. Krishnamoorthy, *et al.*, Lead-free germanium iodide perovskite materials for photovoltaic applications, *Journal of Materials Chemistry A* **2015**, *3*, 23829.

85. G. Volonakis, *et al.*, $Cs_2InAgCl_6$: A new lead-free halide double perovskite with direct band gap, *The Journal of Physical Chemistry Letters* **2017**, *8*, 772.

86. Z. W. Xiao, H. C. Lei, X. Zhang, Y. Y. Zhou, H. Hosono, T. Kamiya, Ligand-hole in sni_6 unit and origin of band gap in photovoltaic perovskite variant Cs_2SnI_6, *Bulletin of the Chemical Society of Japan* **2015**, *88*, 1250.

87. B. W. Park, B. Philippe, X. L. Zhang, H. Rensmo, G. Boschloo, E. M. J. Johansson, Bismuth based hybrid perovskites $A_3Bi_2I_9$ (A: Methylammonium or Cesium) for solar cell application, *Advanced Materials* **2015**, *27*, 6806.

88. N. K. Noel, *et al.*, Lead-free organic-inorganic tin halide perovskites for photovoltaic applications, *Energy & Environmental Science* **2014**, *7*, 3061.

89. F. Hao, C. C. Stoumpos, R. P. H. Chang, M. G. Kanatzidis, Anomalous band gap behavior in mixed Sn and Pb perovskites enables broadening of absorption spectrum in solar cells, *Journal of the American Chemical Society* **2014**, *136*, 8094.

90. Z. B. Yang, A. Rajagopal, C. C. Chueh, S. B. Jo, B. Liu, T. Zhao, A. K. Y. Jen, Stable low-bandgap Pb-Sn binary perovskites for tandem solar cells, *Advanced Materials* **2016**, *28*, 8990.

91. Y. L. Li, *et al.*, 50% Sn-based planar perovskite solar cell with power conversion efficiency up to 13.6%, *Advanced Energy Materials* **2016**, *6*, 1601353.

92. W. Q. Liao, *et al.*, Fabrication of efficient low-bandgap perovskite solar cells by combining formamidinium tin iodide with methylammonium lead iodide, *Journal of the American Chemical Society* **2016**, *138*, 12360.

93. J. P. Correa-Baena, A. Abate, M. Saliba, W. Tress, T. J. Jacobsson, M. Grätzel, A. Hagfeldt, The rapid evolution of highly efficient perovskite solar cells, *Energy & Environmental Science* **2017**, *10*, 710.

94. J. H. Noh, S. H. Im, J. H. Heo, T. N. Mandal, S. I. Seok, Chemical management for colorful, efficient, and stable inorganic-organic hybrid nanostructured solar cells, *Nano Letters* **2013**, *13*, 1764.

95. A. Sadhanala, *et al.*, Blue-green color tunable solution processable organolead chloride-bromide mixed halide perovskites for optoelectronic applications, *Nano Letters* **2015**, *15*, 6095.

96. Y. C. Liu, *et al.*, Two-inch-sized perovskite $CH_3NH_3PbX_3$ (X = Cl, Br, I) crystals: Growth and characterization, *Advanced Materials* **2015**, *27*, 5176.

97. Z. L. Chen, *et al.*, Thin single crystal perovskite solar cells to harvest below-bandgap light absorption, *Nature Communications* **2017**, *8*, 1890.

98. T. Kondo, *et al.*, Resonant third-order optical nonlinearity in the layered perovskite-type material $(C_6H_{13}NH_3)_2PbI_4$, *Solid State Communications* **1998**, *105*, 503.

99. K. Tanaka, T. Takahashi, T. Ban, T. Kondo, K. Uchida, N. Miura, Comparative study on the excitons in lead-halide-based perovskite-type crystals $CH_3NH_3PbBr_3$ $CH_3NH_3PbI_3$, *Solid State Communications* **2003**, *127*, 619.

100. F. Brivio, K. T. Butler, A. Walsh, M. van Schilfgaarde, Relativistic quasiparticle self-consistent electronic structure of hybrid halide perovskite photovoltaic absorbers, *Physical Review B* **2014**, *89*, 155204.

101. M. A. Green, A. Ho-Baillie, H. J. Snaith, The emergence of perovskite solar cells, *Nature Photonics* **2014**, *8*, 506.

102. Q. A. Akkerman, G. Rainò, M. V. Kovalenko, L. Manna, Genesis, challenges and opportunities for colloidal lead halide perovskite nanocrystals, *Nature Materials* **2018**, *17*, 394.

103. W.-J. Yin, T. Shi, Y. Yan, Superior photovoltaic properties of lead halide perovskites: Insights from first-principles theory, *The Journal of Physical Chemistry C* **2015**, *119*, 5253.

104. S. D. Stranks, *et al.*, Electron-hole diffusion lengths exceeding 1 micrometer in an organometal trihalide perovskite absorber, *Science* **2013**, *342*, 341.

105. Q. Dong, Y. Fang, Y. Shao, P. Mulligan, J. Qiu, L. Cao, J. Huang, Electron-hole diffusion lengths > 175 μm in solution-grown $CH_3NH_3PbI_3$ single crystals, *Science* **2015**, *347*, 967.

106. G. Giorgi, J.-I. Fujisawa, H. Segawa, K. Yamashita, Small photocarrier effective masses featuring ambipolar transport in methylammonium lead iodide perovskite: A density functional analysis, *The Journal of Physical Chemistry Letters* **2013**, *4*, 4213.

107. Y. Liu, T. Zhao, W. Ju, S. Shi, Materials discovery and design using machine learning, *Journal of Materiomics* **2017**, *3*, 159.

108. N. Parikh, *et al.*, Is machine learning redefining the perovskite solar cells? *Journal of Energy Chemistry* **2022**, *66*, 74.

109. T. Wu, J. Wang, Global discovery of stable and non-toxic hybrid organic-inorganic perovskites for photovoltaic systems by combining machine learning method with first principle calculations, *Nano Energy* **2019**, *66*, 104070.

110. Ç. Odabaşı, R. Yıldırım, Performance analysis of perovskite solar cells in 2013–2018 using machine-learning tools, *Nano Energy* **2019**, *56*, 770.

111. Z. Liu, N. Rolston, A. C. Flick, T. W. Colburn, Z. Ren, R. H. Dauskardt, T. Buonassisi, Machine learning with knowledge constraints for process optimization of open-air perovskite solar cell manufacturing, *Joule* **2022**, *6*, 834.

112. H. Sun, P. Dai, X. Li, J. Ning, S. Wang, Y. Qi, Strategies and methods for fabricating high quality metal halide perovskite thin films for solar cells, *Journal of Energy Chemistry* **2021**, *60*, 300.

113. S. Ullah, *et al.*, Evaporation deposition strategies for all-inorganic $CsPb(I_{1-x}Br_x)_3$ perovskite solar cells: Recent advances and perspectives, *Solar RRL* **2021**, *5*, 2100172.

114. M. Z. Liu, M. B. Johnston, H. J. Snaith, Efficient planar heterojunction perovskite solar cells by vapour deposition, *Nature* **2013**, *501*, 395.

115. J. Feng, *et al.*, High-throughput large-area vacuum deposition for high-performance formamidine-based perovskite solar cells, *Energy & Environmental Science* **2021**, *14*, 3035.

116. H. Li, *et al.*, Sequential vacuum-evaporated perovskite solar cells with more than 24% efficiency, *Science Advances* **2022**, *8*, eabo7422.

117. J. Li, *et al.*, Fabrication of efficient $CsPbBr_3$ perovskite solar cells by single-source thermal evaporation, *Journal of Alloys and Compounds* **2020**, *818*, 152903.

118. M. Tai, *et al.*, Efficient inorganic cesium lead mixed-halide perovskite solar cells prepared by flash-evaporation printing, *Energy Technology* **2019**, *7*, 1800986.

119. D. B. Mitzi, C. A. Feild, W. T. A. Harrison, A. M. Guloy, Conducting tin halides with a layered organic-based perovskite structure, *Nature* **1994**, *369*, 467.

120. L. Chao, T. Niu, W. Gao, C. Ran, L. Song, Y. Chen, W. Huang, Solvent engineering of the precursor solution toward large-area production of perovskite solar cells, *Advanced Materials* **2021**, *33*, 2005410.

121. M. D. Xiao, *et al.*, A fast deposition-crystallization procedure for highly efficient lead iodide perovskite thin-film solar cells, *Angewandte Chemie International Edition* **2014**, *53*, 9898.

122. N. J. Jeon, J. H. Noh, Y. C. Kim, W. S. Yang, S. Ryu, S. Il Seol, Solvent engineering for high-performance inorganic-organic hybrid perovskite solar cells, *Nature Materials* **2014**, *13*, 897.

123. N. Ahn, D.-Y. Son, I.-H. Jang, S. M. Kang, M. Choi, N.-G. Park, Highly reproducible perovskite solar cells with average efficiency of 18.3% and best efficiency of 19.7% fabricated via lewis base adduct of lead(II) iodide, *Journal of the American Chemical Society* **2015**, *137*, 8696.

124. J. Ding, *et al.*, Fully air-bladed high-efficiency perovskite photovoltaics, *Joule* **2019**, *3*, 402.

125. L. Xiong, *et al.*, A vacuum flash-assisted solution process for high-efficiency large-area perovskite solar cells, *Science* **2016**, *353*, 58.

126. K. N. Liang, D. B. Mitzi, M. T. Prikas, Synthesis and characterization of organic-inorganic perovskite thin films prepared using a versatile two-step dipping technique, *Chemistry of Materials* **1998**, *10*, 403.

127. J. Burschka, N. Pellet, S.-J. Moon, R. Humphry-Baker, P. Gao, M. K. Nazeeruddin, M. Grätzel, Sequential deposition as a route to high-performance perovskite-sensitized solar cells, *Nature* **2013**, *499*, 316.

128. J. H. Im, I. H. Jang, N. Pellet, M. Grätzel, N. G. Park, Growth of $CH_3NH_3PbI_3$ cuboids with controlled size for high-efficiency perovskite solar cells, *Nature Nanotechnology* **2014**, *9*, 927.

129. Z. G. Xiao, *et al.*, Efficient, high yield perovskite photovoltaic devices grown by interdiffusion of solution-processed precursor stacking layers, *Energy & Environmental Science* **2014**, *7*, 2619.

130. Q. Chen, *et al.*, Planar heterojunction perovskite solar cells via vapor-assisted solution process, *Journal of the American Chemical Society* **2014**, *136*, 622.

131. Z. Yang, B. Cai, B. Zhou, T. T. Yao, W. Yu, S. Z. Liu, W. H. Zhang, C. Li, An up-scalable approach to $CH_3NH_3PbI_3$ compact films for high-performance perovskite solar cells, *Nano Energy* **2015**, *15*, 670.

132. S. Yang, *et al.*, Formation mechanism of freestanding $CH_3NH_3PbI_3$ functional crystals: In situ transformation vs dissolution–crystallization, *Chemistry of Materials* **2014**, *26*, 6705.

133. Q. Chen, *et al.*, Controllable self-induced passivation of hybrid lead iodide perovskites toward high performance solar cells, *Nano Letters* **2014**, *14*, 4158.

134. A. Hagfeldt, G. Boschloo, L. C. Sun, L. Kloo, H. Pettersson, Dye-sensitized solar cells, *Chemical Reviews* **2010**, *110*, 6595.

135. N. J. Jeon, *et al.*, A fluorene-terminated hole-transporting material for highly efficient and stable perovskite solar cells, *Nature Energy* **2018**, *3*, 682.

136. Y. H. Lv, *et al.*, High performance perovskite solar cells using TiO_2 nanospindles as ultrathin mesoporous layer, *Journal of Energy Chemistry* **2018**, *27*, 951.

137. Y. H. Lv, *et al.*, High-efficiency perovskite solar cells enabled by anatase TiO_2 nanopyramid arrays with an oriented electric field, *Angewandte Chemie International Edition* **2020**, *59*, 11969.

138. Y. Ding, *et al.*, Single-crystalline TiO_2 nanoparticles for stable and efficient perovskite modules, *Nature Nanotechnology* **2022**, *17*, 598.

139. Q. Jiang, *et al.*, Planar-structure perovskite solar cells with efficiency beyond 21%, *Advanced Materials* **2017**, *29*, 1703852.

140. J. Y. Jeng, Y. F. Chiang, M. H. Lee, S. R. Peng, T. F. Guo, P. Chen, T. C. Wen, $CH_3NH_3PbI_3$ Perovskite/fullerene planar-heterojunction hybrid solar cells, *Advanced Materials* **2013**, *25*, 3727.

141. H. R. Tan, *et al.*, Efficient and stable solution-processed planar perovskite solar cells via contact passivation, *Science* **2017**, *355*, 722.

142. X. P. Zheng, *et al.*, Defect passivation in hybrid perovskite solar cells using quaternary ammonium halide anions and cations, *Nature Energy* **2017**, *2*, 17102.

143. H. P. Zhou, *et al.*, Interface engineering of highly efficient perovskite solar cells, *Science* **2014**, *345*, 542.

144. A. Y. Mei, *et al.*, A hole-conductor-free, fully printable mesoscopic perovskite solar cell with high stability, *Science* **2014**, *345*, 295.

145. Y. G. Rong, *et al.*, Hole-conductor-free mesoscopic $TiO_2/CH_3NH_3PbI_3$ heterojunction solar cells based on anatase nanosheets and carbon counter electrodes, *The Journal of Physical Chemistry Letters* **2014**, *5*, 2160.

146. X. B. Xu, *et al.*, Hole selective NiO contact for efficient perovskite solar cells with carbon electrode, *Nano Letters* **2015**, *15*, 2402.

147. M. Xu, *et al.*, Highly ordered mesoporous carbon for mesoscopic $CH_3NH_3PbI_3/TiO_2$ heterojunction solar cell, *Journal of Materials Chemistry A* **2014**, *2*, 8607.

148. Y. S. Sheng, *et al.*, Enhanced electronic properties in $CH_3NH_3PbI_3$ via LiCl mixing for hole-conductor-free printable perovskite solar cells, *Journal of Materials Chemistry A* **2016**, *4*, 16731.

149. Y. G. Rong, *et al.*, Challenges for commercializing perovskite solar cells, *Science* **2018**, *361*, eaat8235.

150. W. Shockley, H. J. Queisser, Detailed balance limit of efficiency of p-n junction solar cells, *Journal of Applied Physics* **1961**, *32*, 510.

151. G. E. Eperon, M. T. Horantner, H. J. Snaith, Metal halide perovskite tandem and multiple-junction photovoltaics, *Nature Reviews Chemistry* **2017**, *1*, 0095.

152. Q. F. Han, *et al.*, High-performance perovskite/Cu(In, Ga)Se$_2$ monolithic tandem solar cells, *Science* **2018**, *361*, 904.

153. S. Ašmontas, M. Mujahid, Recent progress in perovskite tandem solar cells, *Nanomaterials* **2023**, *13*, 1886.

154. K. Xiao, *et al.*, H. Tan, All-perovskite tandem solar cells with 24.2% certified efficiency and area over 1 cm^2 using surface-anchoring zwitterionic antioxidant, *Nature Energy* **2020**, *5*, 870.

155. J. P. Mailoa, *et al.*, A 2-terminal perovskite/silicon multijunction solar cell enabled by a silicon tunnel junction, *Applied Physics Letters* **2015**, *106*, 121105.

2 Component and Dimension Engineering towards Highly Stable and Efficient Perovskite Photovoltaics

Jiandong Fan and Wenzhe Li

2.1 INTRODUCTION

2.1.1 PEROVSKITE AND ITS CRYSTALLOGRAPHIC STRUCTURES

Diverse perovskite materials have attracted much interest from the scientific community towards various applications, including solar cells,[1] laser,[2] light emitting diodes (LED),[1a, 3] water splitting,[4] photodetectors,[5] field-effect transistors,[6] nonvolatile memory,[7] capacitors,[8] battery,[9] optical amplifiers,[10] lasing,[11] and laser cooling.[12] They are now considered to be the most exceptional materials due to their high efficiency and easy fabrication and the materials used to form perovskite are extensively available and inexpensive.[13] Metal halide perovskites exhibit high absorption coefficients (in $CH_3NH_3PbI_3$ of order 10^4 cm^{-1}), high charge-carrier mobilities (exceeding 10 cm^2 V^{-1} s^{-1}), long minority carrier diffusion lengths ($\geq$ 1 µm), and low trap densities (lower than 10^{16} cm^{-3}). These unique properties enable halide perovskites to be excellent semiconductors for photoelectronic devices.

Hybrid perovskites are an emerging class of semiconducting materials exhibiting outstanding optoelectronic properties. The term "perovskite" implies the general class of materials derived from an elemental composition of ABX_3 and demonstrates the crystal structure of perovskite crystal $CaTiO_3$, where A and B are cations and X are anions that octahedrally coordinate to B. The large A-site cations and the smaller B-site cations allow $[BX_6]^{4-}$ octahedra to corner-share in a 3D framework, with the A-site cations located in the framework cavities.

2.1.2 COMPONENTS AND STRUCTURE DIVERSITY OF PEROVSKITES

Three lattice positions in perovskites are referred to as the A, B, and X sites. In 3D metal halide perovskites, the B-site metal is typically a divalent cation (e.g., Pb^{2+}, Sn^{2+}, Ge^{2+}, Mg^{2+}, Ca^{2+}, Sr^{2+}, Cu^{2+}, Ni^{2+}), the X-site is a halide or pseudohalide

   DOI: 10.1201/9781003400486-2

anion (e.g., I^-, Br^-, Cl^-, etc.), and the A-site is a monovalent cation (e.g., Cs^+, $CH_3NH_3^+$ (MA^+), $HC(NH_2)_2^+$ (FA^+), etc). Aside from the traditional 3D perovskite structure, accessing the low-dimensional structural family induced by various larger organic group A-sit cations, such as two-dimensional (2D) plane and one-dimensional (1D) chained architectures, as well as zero-dimensional (0D) isolated octahedra, adds further tuning through chemical and structural engineering (as shown in Figure 2.1a)[14]. Modification of composition and structure in low-dimensional perovskites enables tuning of electronic bandgaps, exciton binding energies, and electronic transport properties.

2.1.3 PEROVSKITE STRUCTURAL STABILITY

The highest symmetry phase perovskite owns the Pm3m space group in which the ions are perfectly packed and their ionic radii follow $r_A + r_X = \sqrt{2}\left(r_B + r_X\right)$, where rA, rB, and rX are the ionic radii of A, B, and X, respectively. To study the stability of perovskites, Goldschmidt proposed the tolerance factor t and the octahedral factor μ. The tolerance factor indicates the state of distortion by $t = \dfrac{r_A + r_X}{\sqrt{2}\left(r_B + r_X\right)}$, and the octahedral factor is determined by the ratio of, rA/rB the BX_6 octahedron. In general, hybrid perovskites present $0.44 < \mu < 0.90$ and $0.81 < t < 1.11$. For the perovskite cubic structure, t falls between 0.9 and 1.0, whereas the 0.71–0.90 range exhibits tetragonal and orthorhombic structures. Consequently, partial or full cation and/or anion substitution can be employed to regulate the physical properties of perovskite.

The structural stability of perovskites can be primarily judged by the Goldschmidt tolerance factor (t), which is an empirical index widely used to predict the formation of different crystal structures of ABX_3. The value of t varies with the size of the ions in ABX_3. Since ionic radii of organic cations (A) cannot be determined accurately, a certain amount of uncertainty lies in the calculated tolerance factors. Nevertheless, the values can be used to compare cations of reasonably different ionic radii. In an ideal case, the t factor of perovskite should be close to 1, so that it can maintain a high symmetry cubic structure with high stability.

If not, its crystallographic structure with high symmetry will be distorted, and thereafter, the symmetry is lowered.[15] For the values of $t < 0.8$ and $t > 1$, it reduces the possibility of formation of perovskite structures (as shown in Figure 2.1c). Therefore, it can be expected that if t is close to the middle of the range from 0.8 and 1, other values give non-perovskite zones.[16]

In particular, $FAPbI_3$, t value is closed to 1, which is the near upper boundary for perovskite structure, and therefore, $FAPbI_3$ is prone to stabilize in its perovskite phase with superior photo-active ability. $CsPbI_3$ with $t \approx 0.8$, is at the edge of the lower boundary for perovskite structures, and it normally crystallizes into a δ-phase. $MAPbI_3$ with $t \approx 0.9$ is close to the middle of the perovskite zone and forms a black photoactive perovskite phase. By compositional engineering of perovskite, mixing different cations and anions improves the structural stability of perovskites, essentially by adjusting the value of t to stabilize the perovskite phase. The addition of Cs^+ or MA^+ to $FAPbI_3$ moves the t value down from 1 to stabilize the cubic phase of $FAPbI_3$.

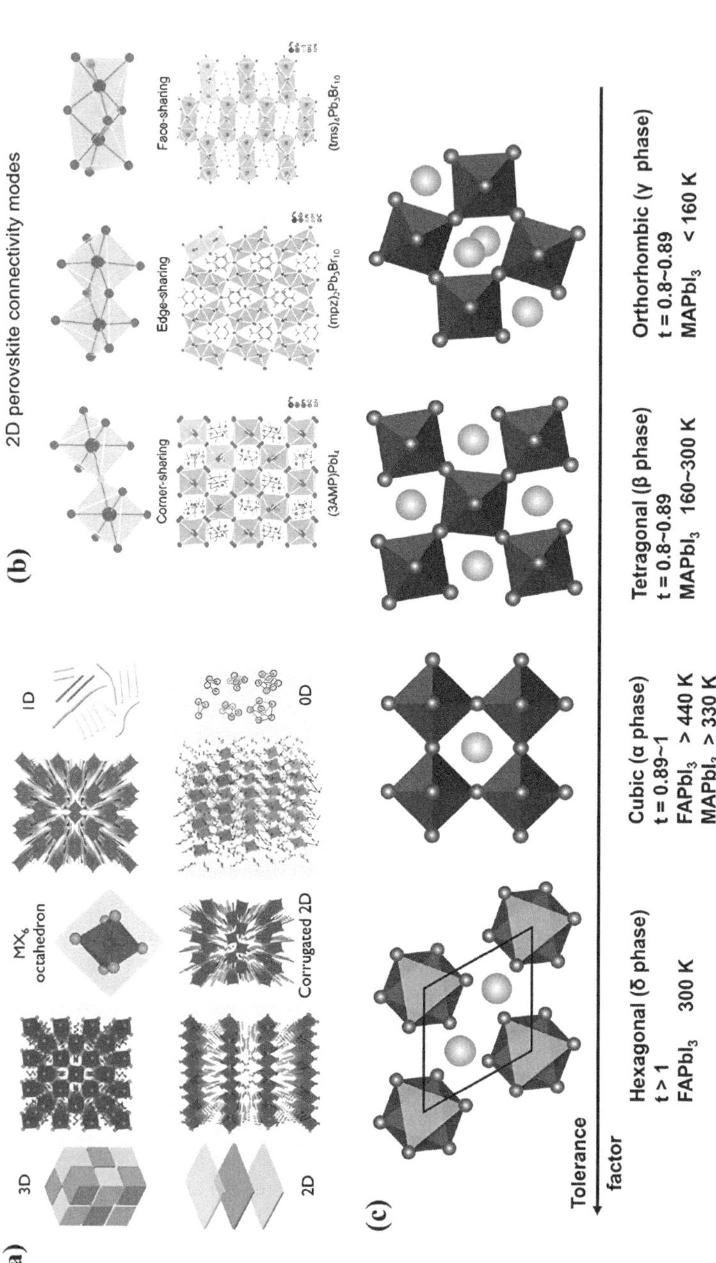

FIGURE 2.1 (a) Typical structures of 3D, 2D, corrugated 2D, 1D, and 0D perovskites (spheres: metal centers; spheres: halide atoms; spheres: nitrogen atoms; gray spheres: carbon atoms; spheres: oxygen atoms; polyhedrons: metal halide octahedra; hydrogen atoms are hidden for clarity), as well as their corresponding conventional materials with different dimensionalities. 2D, 1D, and 0D perovskites can therefore be considered as bulk assemblies of 2D quantum wells, 1D quantum wires, and 0D molecules/clusters. (b) Connectivity modes in 2D perovskite and their representative examples: corner-sharing, (3AMP)PbI₄ (3AMP = 3-(aminomethyl)piperidinium); edge-sharing, (mpz)₂Pb₃Br₁₀ (mpz = 1-methylpiperazine); and face-sharing, (tms)₄Pb₃Br₁₀ (tms = trimethylsulfonium; (CH₃)₃S⁺). (c) Possible crystal symmetries/phases of organic-inorganic halide perovskites (OIHPs), depending on tolerance factor. Copyright 2023 Wiley-VCH GmbH.[14]

Unfortunately, it is not enough to predict that the stability of hybrid halide perovskites depends only on the tolerance factor due to some issues: (i) For the organic cations, defining the A-site real ionic radius is an obvious trouble; (ii) the relatively lower decomposition temperature of an organic group that located at the A site probably leads to the thermodynamical stability problems; and (iii) especially for the iodide anion, the lower electronegativity of the heavyweight halides suggests that the unpolarizable hard spheres of the ions are less effective. Hence, it is needed to know how well the tolerance factor is relevant to the structure stability in the case of perovskite materials.[17]

Some researchers utilized the octahedral factor ($\mu = R_B/R_X$) for the purpose of structure stability, where stable structure seems to have $\mu > 0.41$.[18] It is realized that the larger A-site and/or smaller B-site cations have a high octahedral factor as shown in the case of radioactive elements, Cs^+ ion, which is the largest metallic cation with a radius of 1.88 Å in 12 coordination. Although the perovskite composed of Cs^+ ion and a supposed metallic cation at the B site can allow the radius to meet the stability limit of the octahedral factor, the formed perovskite material only has a tolerance factor of 0.93. In this scenario, in the case of inorganic perovskites with $\mu > 0.41$, the achievable maximum t value would be 0.93. A complex cation should be incorporated inside the perovskite lattice structure if we aim to obtain a higher t value. There are limited candidates available for the metallic cation that occupies the B site in the case of a stable perovskite structure with an octahedral factor $\mu > 0.41$. It should be noted that the reported μ value may demonstrate a slightly discrepant range with respect to different calculation approaches, e.g., $0.44 \leq \mu \leq 0.90$.[15] In any case, it is necessary to consider both the tolerance factor and octahedral factor to evaluate the structure stability of perovskites.

2.2 COMPONENT ENGINEERING

Component engineering is a major strategy to regulate the tolerance factor and octahedral factor and thereafter the stability of perovskite. This compositional play with the perovskite has not only exerted a strong influence on device efficiency but also raised the stability substantially. Although such compositional mixing does not always succeed in forming a homogeneous solid solution, certain combinations of cations in the A-site and halides in the B-site of the perovskites have demonstrated superiority over single cation/halide perovskites in terms of both efficiency and stability. Based on the ionic size and geometrical tolerance factor $\left(t = \dfrac{r_A + r_X}{\sqrt{2}\left(r_B + r_X\right)} \right)$, which is an empirical index widely used for predicting perovskite crystal structure, different cations such as methylammonium (MA^+), formamidine (FA^+), Cs^+, and Rb^+ and anions like I^-, Br^-, and Cl^- and their combinations have been explored in the past few years.

2.2.1 A-SITE CATIONS

It is widely accepted that the cations in the A-site are considered not to contribute directly towards the band structure, but they play a significant role in providing structural stability by charge compensation within the PbI_6 octahedra, largely based on their electrostatic (van der Waals) interactions with the inorganic cage.[19] Nevertheless, any change in the size of cations in the A-site can either contract or expand the crystal lattice, thereby altering the optical properties of the perovskite. Smaller cations like Cs^+ and Rb^+ are expected to contract the lattice and thus increase the band gap, while larger cations, like formamidine (FA^+), are supposed to expand the lattice and decrease the E_g. FA^+ was the first cation that was used instead of methylammonium (MA^+). FA^+, having a larger ionic radius ($r = 0.253$ nm) than MA^+ ($r = 0.217$ nm), expands the crystal a bit, resulting in decreased Pb–I bond distance, which eventually lowers the band gap. It has been proposed that, when an MA^+, which has almost ten times greater dipole moment than FA^+, is incorporated, it exhibits stronger interactions with the $[PbI_6]^{4-}$ octahedra and thus stabilizes the 3D arrangement of α-$FAPbI_3$ with little lattice shrinkage or changes in the optical properties.[20] Therefore, partial replacement of MA^+ with FA^+ in MA-based perovskites also improves the thermal stability of the perovskites ($FA_xMA_{1-x}PbI_3$) remarkably.

In addition, the incorporation of the inorganic cation Cs became a choice. Indeed, it has been found that a certain amount of substitution of MA^+ with Cs^+ simultaneously improves both the device efficiency and the thermal stability.[21] There are several reports on the comparisons of the moisture resistance of $FAPbI_3$ with that of $(Cs/FA)PbI_3$ hybrid perovskite, showing results that strongly endorse the fact that moisture stability is improved by Cs^+ inclusion in the perovskite, where small and large tolerance factors respectively for $CsPbI_3$ ($t = 0.85$) and $FAPbI_3$ ($t = 0.98$) stabilize the perovskites in orthorhombic (yellow) and hexagonal structures, respectively.

Clearly, tuning the tolerance factor by mixing ions of different sizes has been a successful strategy to stabilize the perovskites in cubic structures. For example, the effective tolerance factor can be tuned to form a cubic structure ($0.9 \leq t \leq 1$) by alloying $CsPbI_3$ with $FAPbI_3$ ($FA_xCs_{1-x}PbI_3$) over a range of proportions. It has been also observed that $FAPbI_3$ perovskite thin films when exposed to a humid environment for 18 d, degrade fast, whereas $FA_{0.85}Cs_{0.15}PbI_3$ thin films remain stable correspondingly.

In addition, Cs^+, Rb^+, and K^+ have also attracted attention, as $FAPbI_3$ perovskites including either Rb or K have demonstrated improvements in the PV performance of the cells. Although its location in the crystals is still a matter of controversy, it is believed that a small amount (x ≤ 0.05) of Rb^+ can be included in $FAPbI_3$, and higher concentrations lead to phase segregation.[22] Devices based on Rb-mixed $FAPbI_3$ (i.e., $Rb_{0.05}FA_{0.95}PbI_3$) outperform those based on $FAPbI_3$, and more importantly, the stability of this $Rb_{0.05}FA_{0.95}PbI_3$ film, against humid conditions, is superior to that of $Cs_{0.05}FA_{0.95}PbI_3$. Besides, K^+, being even smaller than Rb^+, has also been used in mixed cation compositions.

Aside from the smaller-molecular-based component engineering, the large-organic-molecular-groups-induced multi-dimensional perovskites will be carefully discussed in the next section, which will be elucidated combined with the crystallographic structure of low-dimensional perovskites.

2.2.2 B-Site Cations

Despite Pb-based perovskites displaying potentially commercialized possibility, there are still some limitations towards its final application, such as its intrinsic toxicity of Pb species. It has been established that the remarkable photoelectronic properties of Pb-based perovskites are not only associated with the high symmetry of structure but also with the strong Pb 6s-I 5p anti-bonding coupling. In this scenario, the metal cations of Sn^{2+}, Ge^{2+}, Bi^{3+}, and Sb^{3+}, etc., which have ns^2 lone pairs and can form octahedral structures with halogen anions, are employed to be the candidates for lead substitution.

In particular, substituting Pb with Sn in the halide perovskites has been demonstrated to narrow the bandgap to 1.2–1.4 eV. Mixed Sn/Pb perovskites with the composition $CsPb_{1-x}Sn_xIBr_2$ have been fabricated via one-step antisolvent spin-coating methods.[23] Unfortunately, the stability and device efficiency of Sn-based devices are still largely lower than their counterparts of the Pb-based perovskites, which suffer from the severe oxidation of Sn^{2+}. To overcome this oxidation issue, SnF_2 is widely used to perform as the reductant. Aside from the SnF_2, $SnCl_2$, $SnBr_2$, and SnI_2 are also scientifically explored, and $SnCl_2$ was verified to be the most outstanding beneficial additive for addressing the oxidation issue.

Likewise, Sn^{2+}, Ge^{2+}, Mg^{2+}, Ca^{2+}, Sr^{2+}, Cu^{2+}, and Ni^{2+} substitution alters the driving force for phase segregation and increases the barrier for ionic diffusion, a factor that may contribute to enhanced phase stability. Again, both the efficiency and stability of lead-free perovskites are currently inferior to their counterparts of the Pb-based perovskites.

2.2.3 X-Site Cations

Like different cations in the A-site, different halides, like Cl, Br, and even non-halide/pseudohalide ions like SCN^-, have been incorporated into pure $MAPbI_3$, $FAPbI_3$, or mixed-cation (FA^+/MA^+, $FA^+/MA^+/Cs^+$, FA^+/Cs^+, $FA^+/MA^+/Cs^+/Rb^+$, etc.) lead iodide perovskites. But, unlike cations, different halides mixed in the X-site impart a dramatic effect on optical and electronic properties, absorption and emission spectra (band gap), and carrier lifetime and diffusion length. Although the ambiguity of inclusion of Cl^- in pure iodide perovskites like $MAPbI_3$ remains unresolved, a majority of reports claim that Cl^- easily sublimes out of the perovskite film during preparation and, therefore, does not exist in the final film,[24] despite the starting solution containing Cl-based precursors ($PbCl_2$ or MACl). On the contrary, a number of studies also show evidence of the existence of a trace amount of Cl in $MAPbI_3$,[25] even though the amount measured was always substantially lower than that used in the starting materials. Nevertheless, distinct effects of Cl^- present in the precursor solution on the quality, morphology, and crystallinity of the perovskite films have been observed in all the studies. Besides, the longer diffusion length of electrons in $MAPbI_{3-x}Cl_x$ than in $MAPbI_3$ films[26] has been credited with improving the performance of the cells. However, as the carrier lifetime or diffusion length is dependent on the morphology (grain size and grain boundaries) of polycrystalline films, it is not easy to separate the effect of Cl^- from the effects of morphology on the electronic properties. Hence, it seems that Cl^- in the precursor solution has an indirect but positive effect on the overall performance of the device.

Unlike Cl^-, the presence and effect of Br^- in the X-site in mixed Br^-/I^- perovskites have been observed clearly and directly. The influence of Br on the optical and electronic properties has been remarkable and recorded well. The incorporation of smaller Br^- ions in $MAPbI_{3-x}Br_x$ increases the band gap of the mixed halide perovskite, and this increment in E_g follows a quadratic relation with the concentration of Br^-.[27] In a similar manner, FA-based mixed I^-/Br^- ($FAPbI_{3-x}Br_x$) perovskites with a range of Br concentrations ($x = 0-1$) were studied, and a similar effect of increasing band gap with Br^- incorporation was observed.[28] But the interesting and surprising fact about $FAPbI_{3-x}Br_x$ is that it does not form any crystalline phase (amorphous) for $2.3 \leq x \leq 2.5$.[28] The origin of this amorphous regime is not yet understood. However, like in the case of $MAPbI_{3-x}Br_x$, photoinduced phase segregation was also observed in $FAPbI_{3-x}Br_x$.[29]

Out of all possible combinations, the mixed perovskite has become the most popular in recent years. It is expected that deeper understanding related to the roles of different cations or anions in the mixed perovskites will come through more studies in the future, but one thing that has been commonly noticed in most of these mixed perovskite studies is the enhanced structural/intrinsic stability, which is doubtlessly an important development.

2.3 MOLECULAR-LEVEL MULTI-DIMENSIONAL PEROVSKITES

In the structure of perovskites, the basic building blocks, i.e., metal halide BX_6 octahedra, are arranged in different ways to exhibit zero-dimensional (0D), one-dimensional (1D), two-dimensional (2D), and three-dimensional (3D) structures, which is induced by means of A-site incorporation of various large-size organic-groups (Table 2.1). It should be noted that, here, the 2D, 1D, and 0D perovskites on molecular level definitions are different from the morphological 2D nanosheets, 1D nanowires, and 0D nanoparticles based on 3D ABX_3. Reducing perovskite materials' dimensionality also allows them to modulate their optical properties[30] (Figure 2.2).[31]

Low-dimensional perovskites are formed by slicing the 3D compounds in different crystallographic directions, where the organic cation and reaction stoichiometry play an important role in determining the crystal orientation. Tuning of the dimensionality affects physical properties: the bandgap of the compounds increases as the dimensionality of the structure is reduced, as in thin quantum well semiconductors. The structural freedom in reduced dimensional perovskites creates the possibility tuning the photophysical and electronic properties. The ABX_3 structure has rigid structural constraints due to the corner-shared $[BX_6]^{4-}$ octahedral crystal structure, whereas the reduced dimensional perovskites allow for increased structural flexibility, influenced by the length of A-site interlayer organic cations. As the ABX_3 perovskite structure is cut into layers, the size limitation and the tolerance factor seen in the bulk perovskites can be widely engineered. For example, in 2D derivatives of the perovskites, the interlayer A cation lengths are more easily tuned, and in the 0D derivatives, size restrictions are not applicable, as the isolated $[BX_6]^{4-}$ octahedral.

2.3.1 3D METAL HALIDE HYBRIDS

A class of perovskite bulk materials, which consists of a framework of corner-sharing metal halide octahedra that extend in all three dimensions with small cations fitting into the void spaces between the octahedra, is called 3D halide perovskites. The

TABLE 2.1
A-Site Cations for LD Perovskites

2D

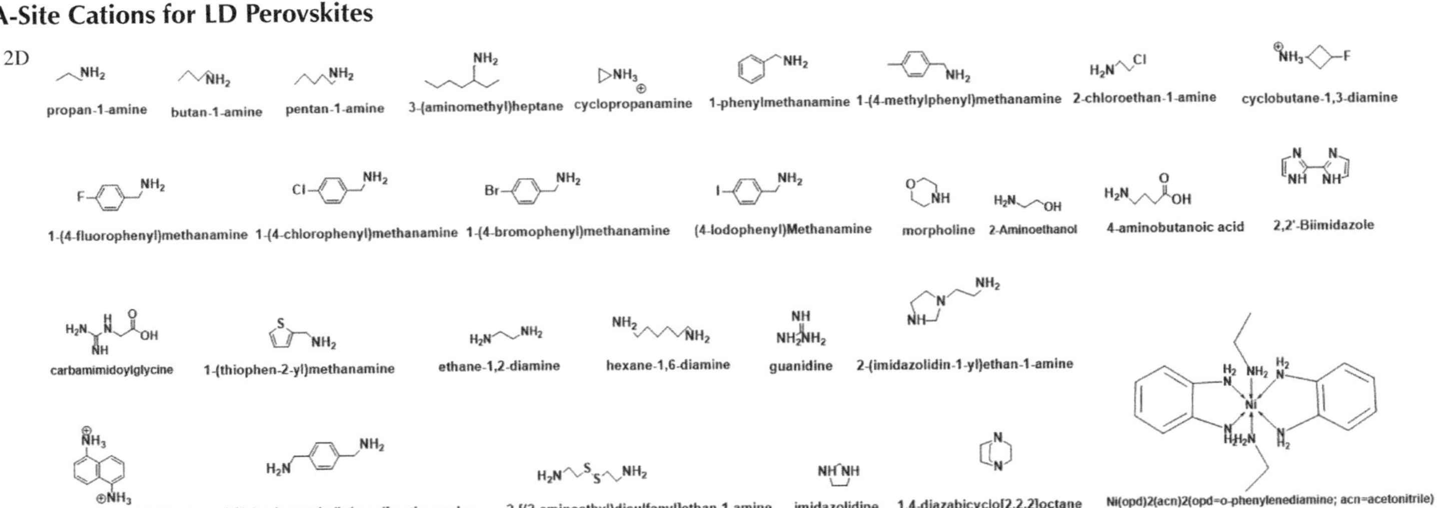

propan-1-amine butan-1-amine pentan-1-amine 3-(aminomethyl)heptane cyclopropanamine 1-phenylmethanamine 1-(4-methylphenyl)methanamine 2-chloroethan-1-amine cyclobutane-1,3-diamine

1-(4-fluorophenyl)methanamine 1-(4-chlorophenyl)methanamine 1-(4-bromophenyl)methanamine (4-Iodophenyl)Methanamine morpholine 2-Aminoethanol 4-aminobutanoic acid 2,2'-Biimidazole

carbamimidoylglycine 1-(thiophen-2-yl)methanamine ethane-1,2-diamine hexane-1,6-diamine guanidine 2-(imidazolidin-1-yl)ethan-1-amine

naphthalene-1,5-diamine 1-[4-(aminomethyl)phenyl]methanamine 2-[(2-aminoethyl)disulfanyl]ethan-1-amine imidazolidine 1,4-diazabicyclo[2.2.2]octane Ni(opd)2(acn)2(opd=o-phenylenediamine; acn=acetonitrile)

(Continued)

TABLE 2.1 (*Continued*)

A-Site Cations for LD Perovskites

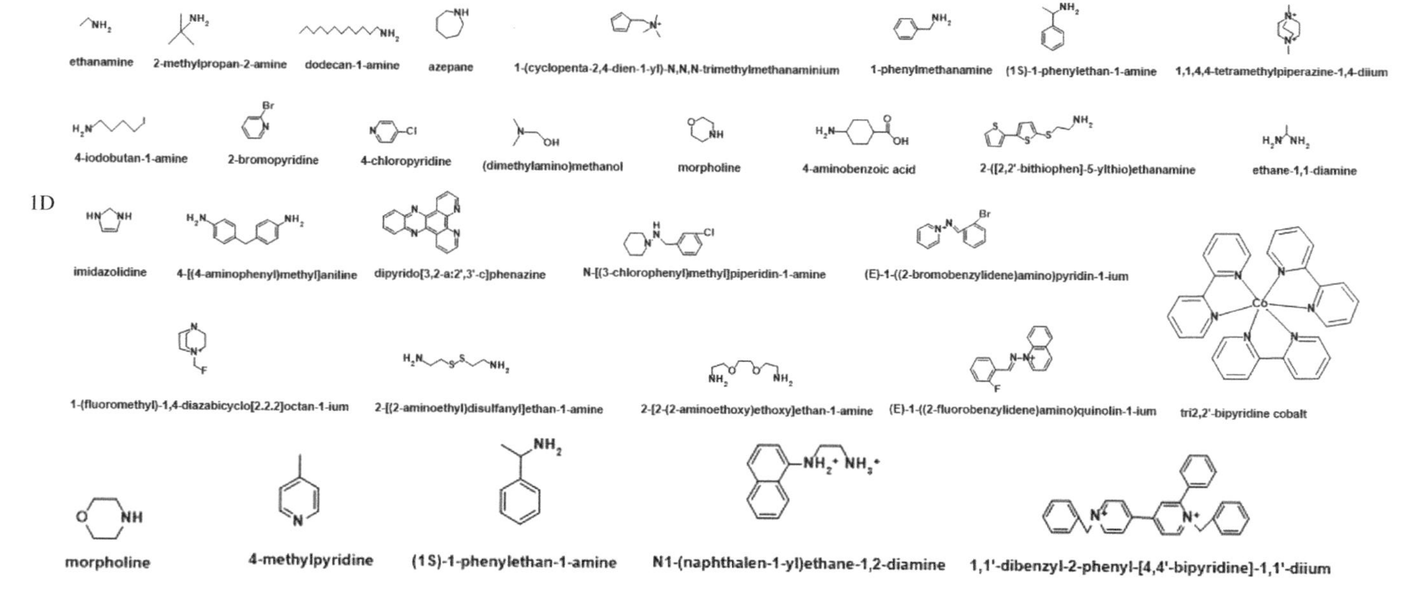

1D

0D

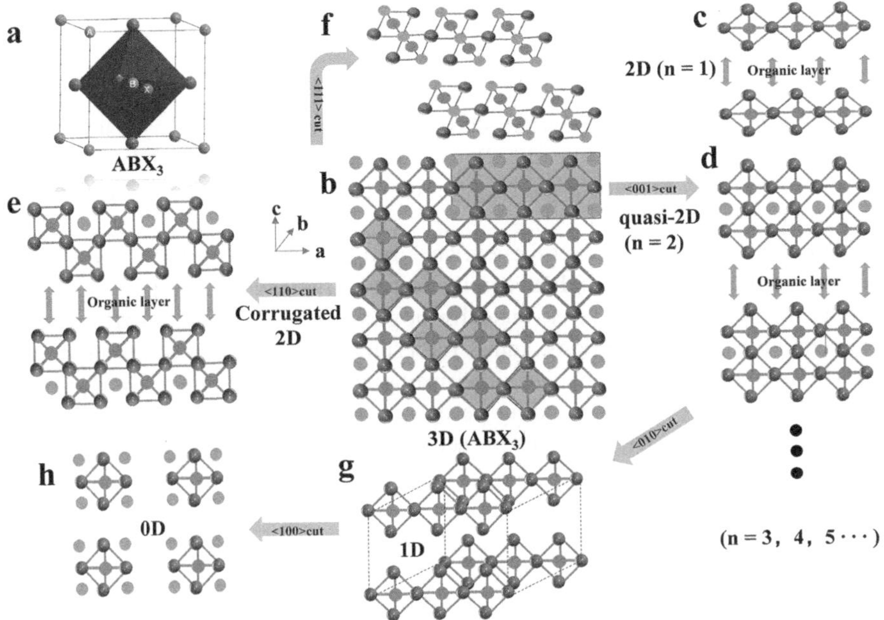

FIGURE 2.2 Halide perovskite structures with different dimensionalities at the molecular level. (a) The unit cell of 3D perovskite. (b) The <010> orientation projection of 3D halide perovskites. (c) The crystal structure of <001>-oriented 2D perovskite ($n = 1$). (d) The crystal structure of <001>-oriented quasi-2D perovskite ($n = 2$). (e) The crystal structure of <110>-oriented 2D perovskite ($n = 2$). (f) The crystal structure of <111>-oriented 2D perovskite ($n = 2$). (g) The crystal structure of 1D perovskite with octahedra connecting in a chain form. (h) The crystal structure of 0D perovskites with the isolated octahedra. Copyright 2021, the authors.[31]

general chemical formula of 3D metal halide perovskite is ABX_3, in which A stands for small cations such as methylammonium cations ($CH_3NH_3^+$), B stands for bivalent metal cations (e.g., Pb^{2+}, Sn^{2+}), and X stands for halide anions (i.e., Cl^-, Br^-, or I^-).

Early examples of perovskite solar cells were demonstrated by using 3D $MAPbI_3$ compounds in 2009. The 3D perovskites exhibit a sharp optical absorption onset, with an absorption coefficient (α) exceeding 10^4 cm^{-1} near the band edge.[32] Replacing MA^+ with other organic or inorganic cations tunes the bandgap of 3D lead halide perovskites. A new choice of A-site cation changes the bandgap because of the modification of the lattice constant. FA^+ is larger than MA^+, which in turn is larger than Cs^+, resulting in an increase of bandgap going from FA^+ to Cs^+ through MA^+. The A-site cation also influences the extent of metal-halide orbital overlap. This change in metal-halide bonding has a direct impact on valence and conduction band positions.[33] The metal halide units result in the formation of electronic bands with potentially large bandwidths. Owing to its low exciton binding energy and long-range carrier transport nature, 3D halide perovskites have emerged as one of the most promising materials for next-generation solar cells.

2.3.2 2D Metal Halide Hybrids

The octahedra connected in layers or corrugated sheets, which are sandwiched between large-sized organic A-site cations are called 2D and quasi-2D perovskites. Single or multiple corner-sharing layers separated by large-size organic cations are known as Ruddlesden-Popper (RP) perovskite, Dion-Jacobson (DJ) perovskite, and Alternating Cations in the Interlayer Space (ACI) perovskite (shown in Figure 2.3a).[14]

2D perovskite has the general chemical formula as $A_{n-1}A'_2B_nX_{3n+1}$, where A represents small cations fitted in between the space of two layers, A' represents organic cations between different layers (usually large ligands with long alkyl chains), B represent bivalent metal cations, and X represents different halides. Different metal halide monolayer sheets in between the insulating A' organic layers are represented by n. For $n = \infty$ the corresponding perovskite refers to the conventional 3D perovskites. The crystal structure is expressed with reference to the orientation and configuration of the organic cation.[34]

2.3.3 1D Metal Halide Hybrids

The octahedra of metal halide perovskites connected in a chain (sharing corners, edges, or faces) surrounded by organic cations are called 1D perovskites. The corresponding configuration of the 1D perovskite could be either linear or zigzag and their chemical formulas are variable, which depends upon the connectivity of the corners and the organic cations. These anionic metal halide chains and surrounding organic cations can be assembled to form single crystalline bulk structures. Unlike morphological 1D metal halide perovskites being nanomaterials, 1D organic metal halide hybrids are large crystals indeed with the 1D structure at the molecular level,[35] which can be considered as bulk assemblies of metal halide quantum wires.

The first examples of 1D metal halide hybrids were reported in the 1990s when Mitzi and coworkers prepared a family of metal halide hybrid structures $FA_2MA_mSn_mI_{3m+2}$.[36] Whilst $m = 1$, the corrugated 2D structures break down into 1D chains with corner-sharing of $[SnI_6]^{4-}$ octahedra. As perovskite-related materials, 1D organic metal halide hybrids have attracted great attention in recent years. A very recent work by Kanatzidis and coworkers has shown that a single crystal can consist of two types of 1D metal halide chains.[37] The recent interest in photoactive 1D metal halide hybrids was greatly promoted as a 1D lead halide hybrid $C_4N_2H_{14}PbBr_4$, in which the edge-sharing octahedral chains $[PbBr_4]^{2-}$ are entirely surrounded by the organic cations $[C_4N_2H_{14}]^{2+}$.[38] This structure can be considered as the bulk assembly of core-shell quantum wires.

2.3.4 0D Metal Halide Hybrids

The perovskite materials in which the halide octahedral anions or metal halide clusters are completely surrounded and isolated by the organic large-size cations are called zero-dimensional perovskites. The octahedra of molecular perovskite are surrounded periodically in the crystal lattice together with the assistance of organic cations to form bulk materials. The general chemical formula of 0D perovskites is

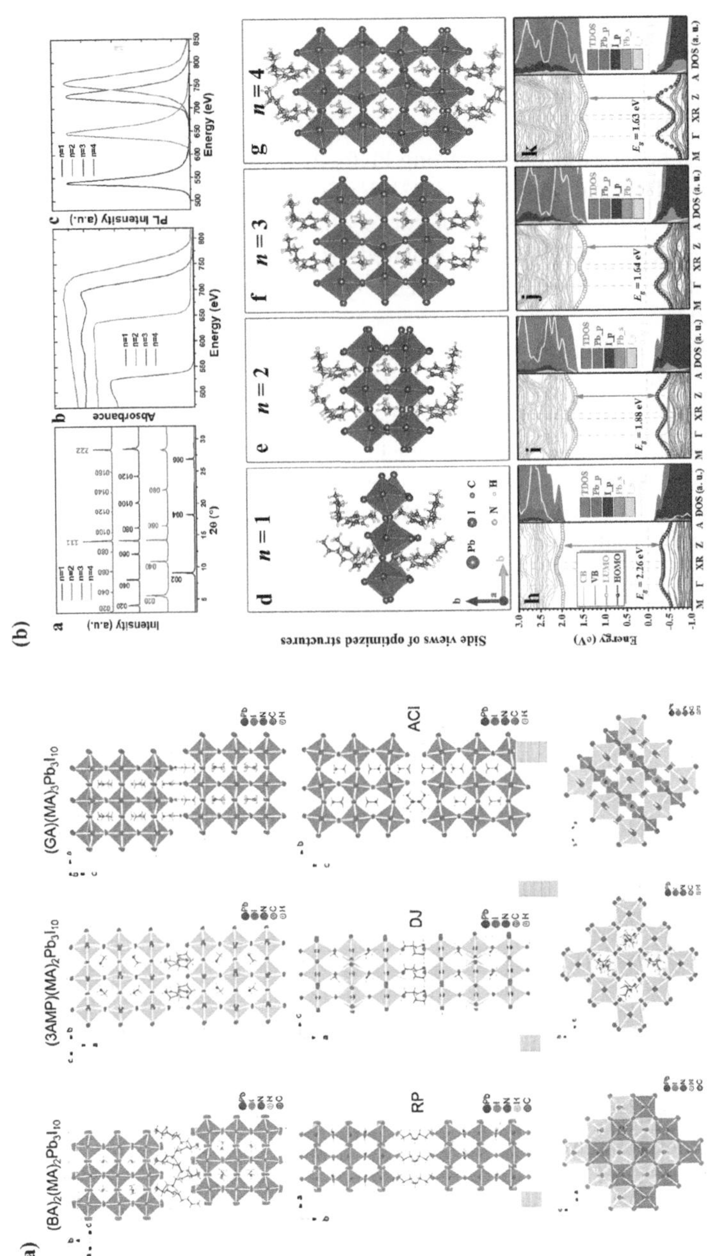

FIGURE 2.3 (a) Structural comparison among the $n = 3$ crystal structures of RP phases, DJ phases, and ACI phases: $(PEA)_2(MA)_2Pb_3I_{10}$ PEA = Phenylethylamine; $(BA)_2(MA)_2Pb_3I_{10}$ BA = Butylamine; $(3AMP)(MA)_2Pb_3I_{10}$; and $(GA)(MA)_3Pb_3I_{10}$ GA = Guanidinium. (b) Chemical structures of the new RP type 2D perovskites; XRD patterns of single crystal $BM_2MA_{n-1}Pb_nI_{3n+1}$ ($n = 1, 2, 3, 4$) BM = Butyl-3-methylimidazolium; absorption spectra and PL spectra of single crystal $BM_2MA_{n-1}Pb_nI_{3n+1}$ ($n = 1, 2, 3, 4$), n represents the number of inorganic layers; the band structures and density of states of $(BM)_2(MA)_{n-1}PbnI_{3n-1}$ ($n = 1-4$), respectively; Side views of optimized structures of $(BM)_2(MA)_{n-1}Pb_nI_{3n-1}$ ($n = 1-4$), respectively; he Fermi level is set to zero; the E_g value for $(BM)_2(MA)_{n-1}Pb_nI_{3n-1}$ ($n = 1-4$), respectively; he Fermi level is set to zero; the CB, VB, LUMO, and HOMO are conduction band, valence band, lowest unoccupied molecular orbital, and highest occupied molecular orbital. Copyright 2023 Wiley-VCH GmbH.[14]

A_4BX_6, where A represents monovalent organic cations and $[BX_6]^{4-}$ represents metal halide octahedral. Structurally distorted perovskites can be considered 0D analogues.

2.3.5 Electronic Structures of Multi-Dimensional Perovskites

Theoretical studies have shown that the superior photovoltaic properties of lead halide perovskite absorbers are partially attributed to their unique electronic features, including the Pb $6s^2$ lone pair that causes strong Pb 6s–I 5p antibonding coupling and the high symmetry of the perovskite structure.[39] Herein, as shown in Figure 2.1b perovskites refer to the structures consisting of corner-sharing and non-corner-sharing, e.g., edge-sharing, face-sharing, or even isolated $[BX_6]^{4-}$ octahedra, respectively. Again, these perovskites can also be categorized by their structural dimensionality, i.e., 3D, 2D, 1D, and 0D.

The electronic dimensionality is herein defined as the connectivity of the orbital that comprises the conduction band minimum (CBM) and balance band maximum (VBM), which refers to the extension of the connected orbitals along the three directions that were defined by high-symmetry points in Brillouin-zone (Figure 2.4)[31]. The electronic dimensionality is favorable to an in-depth understanding of the electronic structure, e.g., bandgaps, defect levels, and mobility. The perovskites with higher structure dimensionality but lower electronic dimensionality usually demonstrate inferior potential as an absorber, owning to the enhanced electron/hole effective mass and barriers to isotropic current flow. In this scenario, the electronic dimensionality could elucidate the perovskite photophysical properties much better in comparison to the structural dimensionality.

In the case of ideal 3D $CsPbI_3$ perovskite, the rotation of the $[PbI_6]^{4-}$ octahedra would not change the structural dimensionality, but it indeed results in a reduced electronic dimensionality. In the case of $Cs_2AgBiBr_6$, double perovskite, which usually obtains a 3D dimensional structure and 0D orbital arrangement considering that the CBM-associated charge distribution is located at the region within the $[BiBr_6]^{3-}$ octahedra. The result displays that the $BiBr_6$ orbital states are isolated by the $[AgBr_6]^{5-}$ octahedra, which gives rise to the formation of the 0D electronic dimensionality for $Cs_2AgBiBr_6$ double perovskite.

The conductive ability of the perovskite materials depends on not only the physical crystal connectivity but also the electronic dimensionality. Therefore, it is essential to understand in depth the electronic dimensionality of multi-dimensional perovskites towards further enhancing the device performance.

2.4 LD-3D PEROVSKITES AND SOLAR CELLS

It is well known that the effective method to balance the efficiency and stability of solar cells is based on a hybrid LD-3D structure. The water and oxygen shielding capacity of LD perovskite plays an important role, e.g., internal packaging, enhancing the stability of the LD-3D perovskite solar cells. In addition, the specific skeleton dimension orientation can provide effective charge transport. Particularly, the molecular conjugation involved in the A-site of LD perovskite could obtain auxiliary

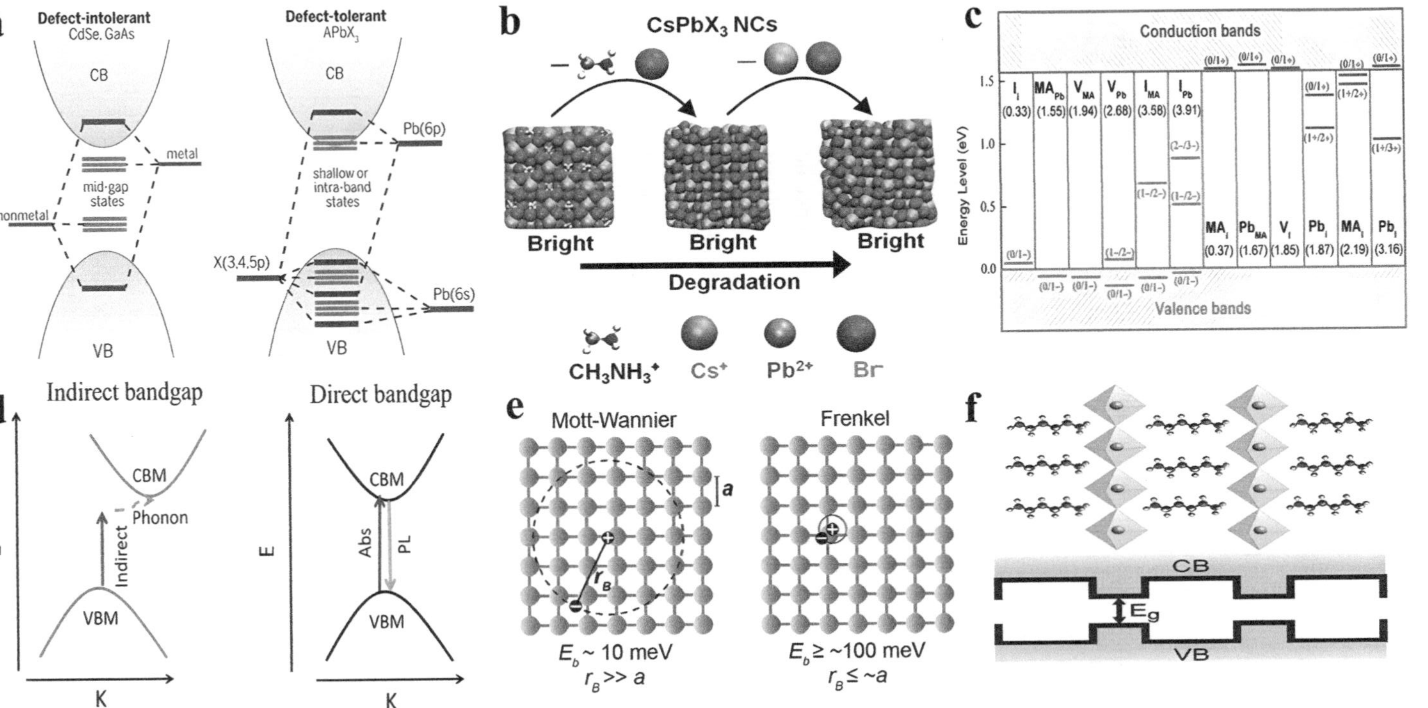

FIGURE 2.4 (a) Electronic band structure of conventional semiconductor (for example, CdSe, GaAs) and lead halide perovskites. (b) A schematic representation that CsPbBr$_3$ still keeps bright emission even with the removal of the surface atoms and ligands. (c) Calculated transition energy of point defects in MAPbI$_3$. The values in the parentheses represent the formation energies of neutral defects. MA$^+$ is indicated. (d) Absorption and recombination processes between indirect bandgap and direct bandgap of semiconductor. (e) Mott-Wannier excitons and Frenkel excitons in an arbitrary atomic lattice. The lattice constant a and Bohr radius r$_B$ are indicated. Reproduced with permission. (f) Schematic representation of the natural quantum-well structures, wherein the inorganic layers act as "wells" and the organic molecules as "barriers." Copyright 2021, the authors.[31]

transmission channels. In this section, the solar cells based on the hybrid LD-3D perovskites will be discussed in detail.

2.4.1 2D/3D PEROVSKITE SOLAR CELLS

Larger organic cations divide the metal-halide octahedra to create 2D perovskites. Due to the hydrogen bonding between organic cations and octahedra and the organic layer's water-repellent characteristic, the environmental stability of 2D perovskites is much higher than that of 3D perovskites. The bulky organic spacer present in the 2D perovskite structure makes it more hydrophobic than 3D perovskite, which has been found to improve the ambient stability of 2D PSCs. It is thought that the bulky organic cation plays a significant role in photocurrent hysteresis and cell deterioration under bias or heat stress. The presence of the bulky organic cation can physically hinder ion mobility in 3D perovskite. These features make using pure-phase 2D or low n-value quasi-2D perovskites as the photo-active layer in perovskite solar cells (PSCs) particularly tempting, although some of these materials' optoelectronic properties are subpar to 3D perovskites for reaching high efficiencies. In particular, the thin insulating layers of the organic spacers provide a negative impact on the conductivity of 2D perovskite films and hinder charge extraction.

Many examples of mixed 2D-3D PSCs outperforming reference devices based on 3D perovskite films can be found in the literature, and mixed-dimensional perovskite compositions have produced some of the greatest recorded PSCs efficiencies to date. There are a variety of reasons and potential causes for this performance improvement, which may also be influenced by the spatial distribution of the 2D perovskite within the film and at the surface. The passivation of trap-states at grain boundaries or within the bulk of 3D perovskite and/or the interfacial trap-states at the interfaces between 3D perovskite and neighboring charge transport layers were the main causes of the power conversion efficiency (PCE) improvement. In addition to this chemical passivation method, certain studies have also revealed 2D perovskite-induced electronic passivation, which, through advantageous energy band alignment, enables increased charge extraction to charge transport layers. As shown in Figure 2.5, according to the principle of Lewis acid-base coordination, Li et al. used the electron-rich structure of 1,8-naphthyridine (1,8-ND) to form a complex with Pb^{2+} and at the same time prepared a surface modification with effective defect passivation.[40]

2.4.1.1 Optimization of Film Growth

The device efficacy of PSCs is greatly influenced by a variety of perovskite film quality factors, including crystallinity, crystal orientation, exciton binding energy, surface morphology, roughness, and grain size. It has been stated that improving various aspects of 3D perovskite film quality for better device performance involves bulk incorporation or surface treatment with 2D perovskite precursors. It has been extensively reported that bulk 2D incorporation enhances the crystallinity of 3D perovskite films, which reduces defect-related recombination. BA^+ is one of the commonly used A-site ions in 2D-3D perovskite films. The addition of BA^+ cation to 3D perovskite precursor is in favor of crystal orientation with the (100) direction parallel to the substrate, optimizing the morphologies of the thin films.[41] The high

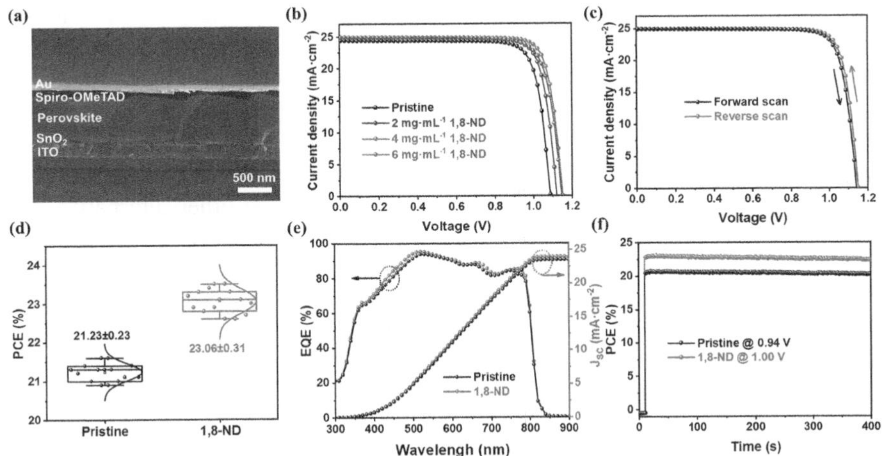

FIGURE 2.5 (a) Cross-view scanning electron microscope (SEM) image of PSCs. J-V curves of (b) PSCs modified with different concentrations of 1,8-ND and (c) PSCs modified by 4 mg mL^{-1} 1,8-ND under forward and reverse scan. (d) Statistical distributions of PCE, (e) external quantum efficiency (EQE) curves, and (f) stabilized power output (SPO) curves of pristine and 1,8-ND modified PSCs. Copyright 2022, Elsevier B.V. All rights reserved.[40]

orientation of the thin films is beneficial to the enhancement of stabilities. In addition, 2-(methylthio)ethylammonium (MTEA) chloride and n-propylammonium (PA) iodide are even better at controlling crystal orientation and favor to a uniform and macrocrystalline thin film. Generally, the 2D perovskite in precursor solution has a relatively low solubility and the initial nuclei orientation of 2D perovskite occurs at the liquid-air interface of the precursor solution, which tends to orient growth to cover the substrates. And then, the 3D perovskite growth can be directionally induced by highly oriented 2D perovskite. Therefore, the 2D introduced in 3D precursor tends to obtain highly oriented hybrid perovskite thin films. The resulting (100) oriented film was found to have a lower defect density compared to randomly oriented perovskite film. It should be noted that the I$^-$ replaced by SCN$^-$ anion can obtain a 2D perovskite with the SCN$^-$ anion contributes to more uniform and few dead areas. And the introduction of SCN$^-$-mixed 2D perovskite is an effective way to enhance the orientation of the perovskite thin films.

A mixed growth of 2D and 3D perovskite with high orderly properties can form quasi-2D structures. The Chun-Chao Chen group studied the growth process of quasi-2D perovskite.[42] The distribution of the quasi-2D film prepared by the antisolvent method is according to the polarity of the antisolvent. Due to the strong polarity of isopropyl alcohol (IPA), large-sized crystal nuclei are dissolved instead of small-sized crystal nuclei. The heterogeneous nucleation at the bottom of the n-butanol (nBA) treatment is more obvious than that of IPA. Therefore, the nBA treatment trends to obtain thin films with large grain size. In addition, IPA has a higher nucleus density than nBA, which promotes more (2-thiophenemethylammonium) TMA to take part in the development of the 2D perovskite at the top side of thin films.

Contrarily, only a small number of TMA cations are consumed on top of nBA will leads to a significant amount of TMA cations at the interface between the substrate and the perovskite layer (heterogeneous nucleation). Consequently, in the case of nBA, the quantum well distribution is sandwich-like, but in IPA, it is reverse-graded. The photoelectric properties of the 3D-like phase in iPA are extremely similar to those of the 3D perovskite, which is conducive to efficient PSCs devices.

In lead-free-based PSCs, Sn^{2+}, Ge^{2+}, and Bi^{3+} are some other alternative metal cations that have been investigated but sadly without much success because of their poor stability and/or worse power conversion characteristics. Tin is considerably one of the most promising candidates of Pb-based perovskite in solar cells, however, the Sn^{2+} valence state is very unstable in contrast to the tetravalent state. Thus, in tin-based perovskite, a high concentration of 2D components is involved in the 3D thin films, aiming to improve the thin film stability. In contrast, an excitonic band is consistently detected in the Pb^{2+} counterpart regardless of the "n" value. Tin is added to the structure and thus appears to be a chemical way to reduce energy losses by encouraging exciton dissociation. It is noted that the solvent properties affect the film orientation. In contrast to N, N-dimethylformamide (DMF), which causes to become orientated perpendicular to the substrate, the perovskite layers are aligned parallel to the substrate when utilizing dimethylsulfoxide (DMSO). Importantly, the low-dimension perovskite can improve the quality of the thin films. For example, the 4-(aminomethyl) pyridine (4AP) cation introduced in $FASnI_3$ can obtain 2D-3D hybrid structure perovskite thin films, in which the 2D component belongs Dion–Jacobson (DJ) perovskite. Interestingly, as the content of low-dimensional components increases, the grain boundaries gradually decrease. Another example in Figure 2.6, the 3,5-diflubenzylamine (DFBA) as an A-site insertion group inside the perovskite lattice can obtain a quasi-2D structure ($n = 1$ or 2).[43] The thin film displays the plane family as highly orienting/crystallizing[44]. The surface of 3,5-diflubenzylamine (DFBA) perovskite thin film is homogeneous and hole-free. Moreover, the degree of dispersion of DFBA grain size decreases to 200~800 nm, whereas the control thin films reflect a high range from 200 to 1,400 nm. The addition of DFBA molecules may trigger secondary nucleation on the thin films. Additionally, the surface distribution of the DFBA perovskite samples is more uniform, which proves that the DFBA perovskites effectively passivate the thin film defects.

2.4.1.2 Interface Passivation

The defect states in that position can adversely affect the performance of device. Suppressing interfacial states in perovskite film is thus an important step towards enhancement of efficiency. In the case of 2D/3D perovskite films, the incorporation of 2D at the surface of 3D perovskite provides a passivation effect. The long chains in the A-site of 2D components have been observed to concentrate at interfaces, reducing the overall roughness of the 3D perovskite film, which may suppress current leakage in the form of shunting. It is beneficial to suppress short circuits caused by contact of functional layers. It should be noted that 2D perovskite usually exhibits higher exciton binding energy and poorer charge transport properties than 3D perovskite. However, it is plausible that these results may be related to a recent finding that 2D perovskite is significantly more conductive at its layer edges compared to its

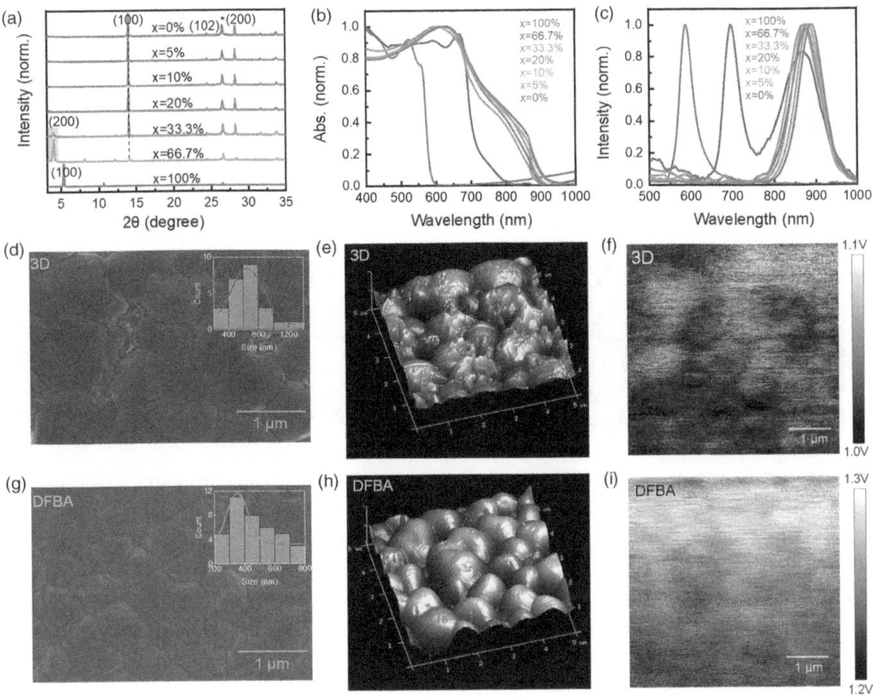

FIGURE 2.6 (a) X-Ray diffraction (XRD) pattern, (b) UV-vis spectroscopy, and (c) photoluminescence (PL) for DFBA perovskite films with different x(DFBAI) ratios. SEM of (d) 3D and (g) 10% DFBA perovskite films, inset of (d) and (g) are the corresponding distribution diagram of perovskite grain size. Atomic force microscope (AFM) images of (e) 3D and (h) DFBA perovskite films. Kelvin probe force microscope (KPFM) images of (f) 3D and (i) DFBA perovskite films.[43]

bulk terrace regions. Therefore, the increased charge separation and collection of photo-generated electrons close to the grain-boundary regions have also been linked to 2D perovskite phases at the grain boundaries of 3D perovskite.

On 3D perovskite surfaces, it has been observed that a thin coating of the organic halide salt phenethylammonium iodide (PEAI) reduces iodine vacancies by raising the lead to iodine ratio (Pb:I), which causes the passivation of the corresponding surface point defects. Additionally, it has been observed that compared to its corresponding 2D perovskite phase, an unreacted PEAI surface layer offers more effective surface passivation (PEA_2PbI_4), as shown in Figure 2.7.[45] A similar passivation effect has also been reported with 1-naphthylmethylammonium iodide (NMAI). In the case of short circuit current density (J_{sc}) enhancement, *tert*-butylammonium iodide (tBAI) is investigated to understand, which has a higher optical bandgap than the underlying 3D perovskite. However, the semi-insulating nature of this layer is expected to reduce J_{sc}. Going beyond the widely used mono-ammonium cation species, surface treatment of 3D films with the bi-ammonium cation *p*-phenyl dimethylammonium iodide (PDMAI) has also been reported to prove an effective passivation, which is thought

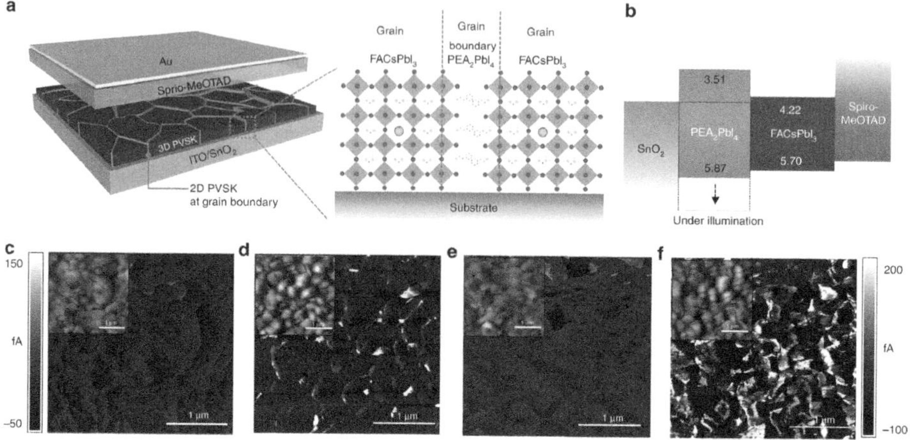

FIGURE 2.7 (a) Schematics of the device incorporating polycrystalline 3D perovskite film with 2D perovskite at grain boundaries and b band structure of each layer in device analyzed by ultraviolet photoelectron spectroscopy (UPS) and Tauc plots. Conductive atomic force microscopic (c-AFM) images of (c, e) bare $FAPbI_3$ and (b, d) with 2D perovskite films on SnO2 coated ITO glass. The measurement was carried out with bias voltage of 100 mV under (c, d) room light or (d, f) low intensity light illumination provided by the AFM setup. Inset of each image shows corresponding topology of the films. Scale bar at left side is for (c) and (d) while at right side is for (e) and (f).[45]

to be better adhesion to the 3D perovskite film via ionic or hydrogen bonding. The aromatic rings in PDMAI capture the electronic defects on the 3D perovskite surface and enhance the interface passivation.

Considering the cationic salt passivation on the surface of the perovskite thin films to form an ultra-thin and inhomogeneous layer, IPA solvent can dissolve FA^+ or MA^+ cations on the 3D perovskite surface and promote 2D perovskite generation. Improved charge transport in the mixed 2D-3D PSCs has a relation to this surface mixing of 2D and 3D perovskite phases. The insufficient surface coverage of 2D perovskite enables effective localized charge extraction through pinholes or other similar passivated regions of the 3D perovskite. As recently shown by altering the ratio of two spacer cations in a mixed-cation 2D perovskite, the result reveals that surface passivation is sensitive to the spacer cation. In contrast to 2D films containing only one cation species, surface intermixing of $n\text{-}BA^+$ and $i\text{-}BA^+$ (iso-butylammonium) cations increases the distribution of the 3D perovskite phase at the interface between perovskite and HTL. Compared to the films passivated with only one cation species, better interfacial charge transport (greater FF) can be found in PSCs passivated with the mixed cation 2D perovskite, which is attributed to the increased 3D phase ratio in the interface layer.

Figures 2.7c and 2.7d show higher currents in the grain boundaries/near grain boundaries in 2D PVSK thin films, while relatively uniform current films are observed in pure $FAPbI_3$. Figures 2.7e and 2.7f indicate that the current is further enhanced at the grain boundaries/near grain boundaries for 2D PVSK, while the current in pure $FAPbI_3$ thin films increases uniformly.

2.4.2 1D/3D Perovskite Solar Cells

Compared with 2D perovskite, 1D perovskite with linearly arranged $[BX_6]^{4-}$ can improve the skeleton strength of perovskite lattice and the inactive organic cations can wrap and protect the metal halides, leading to higher performance than that of 2D/3D based perovskite solar cells. However, due to the presence of large organic cations, the 1D perovskite also suffers from certain undesirable properties to 1D/3D multidimensional perovskite-based devices, such as decreased carrier mobility and a weaker carrier separation capacity. As a result, 1D/3D heterojunction-based PSCs are still mainly unexplored.

2.4.2.1 Mixed 1D/3D Perovskite Solar Cells

Fan et al. first proposed the 1D/3D hybrid perovskite in solar cells. The 2-(1H-pyrazol-1-yl)pyridine (PZPY) is introduced into the $Cs_{0.04}MA_{0.16}FA_{0.8}PbI_{0.85}Br_{0.15}$ precursor and then spin-coated it on substrates to obtain 1D/3D heterojunction films.[46] The grain size decreases as the content of 1D perovskite increases because the molecule may distort and/or distribute the crystal lattice in the interjunction domain in the self-assembly process. The PCE is reversible under temperature cycling (25–85°C) at an RH of 55% for 30 h, while the control device can only maintain five cycles, which reflects a self-healing property of the 1D-/3D-based perovskite solar cells. Additionally, hydrazinium cations (HA^+) mixed into FA-based perovskites can obtain a 1D/3D multidimensional structure. Impressively, the trap state density of 1D/3D perovskite thin films decreases from 1.06×10^{16} to 3.10×10^{15} cm^{-3}, compared to the pure $FAPbI_3$ perovskite, and the carrier lifetime is enhanced from 434.6 to 1216.57 ns. More importantly, the corresponding device realized a PCE of 21.20% and maintained 90% of its initial efficiency for 2520 h under ambient conditions (~25°C temperature and ~30% humidity), which is far better than the pure FA-based perovskite-based device with a PCE of 19.36% and a stability of ~500 h. Liu et al. creatively self-built an in-situ characterization system to investigate the impact of external conditions (water, illumination, and external electric fields) on 1D/3D perovskite films in order to clarify the dynamic process and mechanism of multidimensional perovskite decomposition with bipyridine (BPy) introducing, as shown in Figure 2.8.[47] Because of the molecular interaction of hydrogen bonds and its function as an electron donor, the BPy in 1D perovskite can reduce ion migration and quench oxygen radicals by improving the migration activation energy. Finally, the champion device displays an improvement of PCE to 21.18% and an enhancement of the sensitive environment of oxygen, light, and water.

Moreover, guanidinium (GA) was made into an FA-based perovskite the precursor and prepared PSCs by one-step coating. The introduction of GA can generate the $GA_xMA_{1-x}PbI_3$ perovskite phase with the 3D $MAPbI_3$ perovskite. It should be noted that the $GA_xMA_{1-x}PbI_3$ 3D perovskite can be formed by adding a suitable concentration of GA in the precursor solution. An overabundant GA will cause the creation of the 1D phase of yellowish δ-$FAPbI_3$, which can passivate the trap states of thin

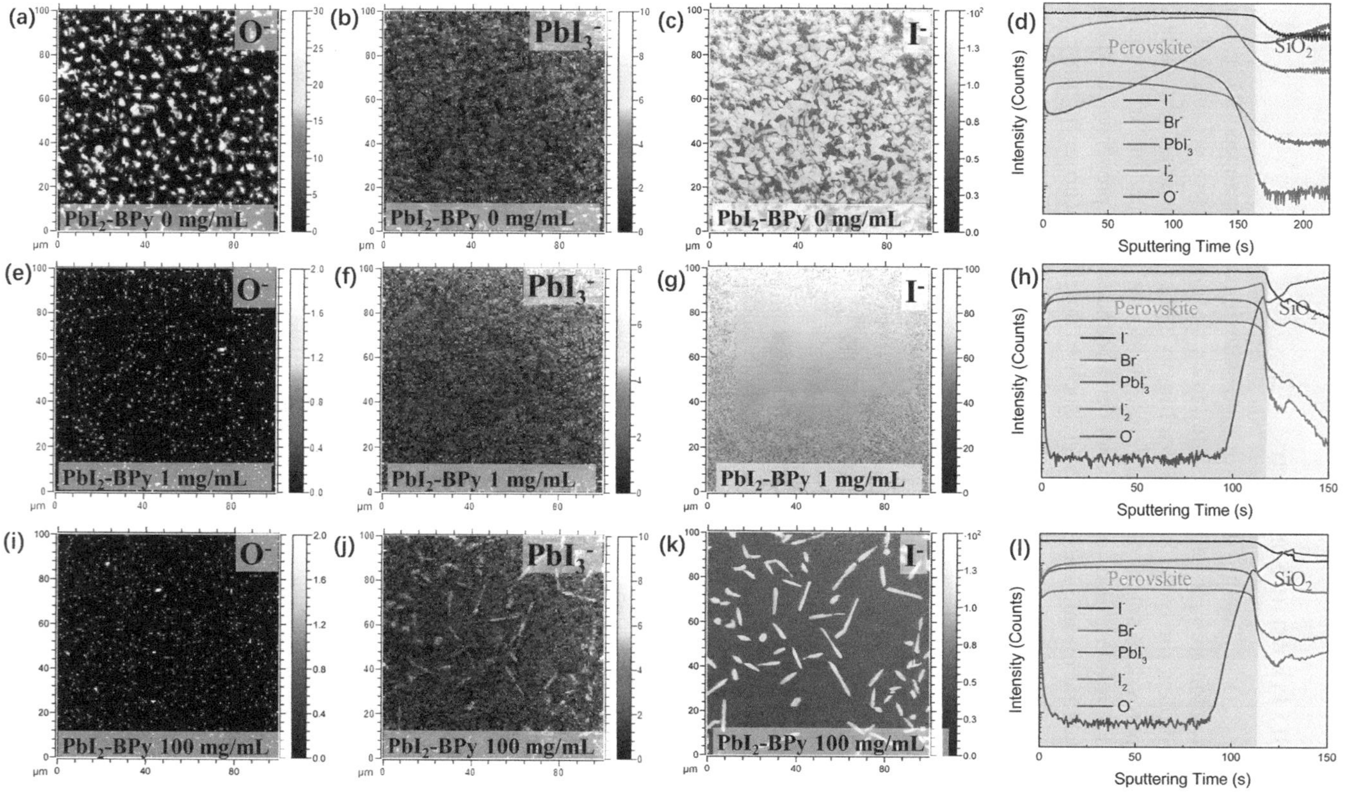

FIGURE 2.8 Time of flight secondary ion mass spectrometry (ToF-SIMS) measurements of perovskite thin films doped with different PbI$_2$-BPy concentration (0, 1, and 100 mg mL^{-1}). (a, e, i) The corresponding images of O$^-$ distribution. (b, f, j) The corresponding images of PbI$_3^-$ distribution. c, j, k) The corresponding images of I$^-$ distribution. (d, h, l) The evolution of ion content as the function of sputtering time with the corresponding concentration.[47]

films. By optimizing the GA concentration, a longer carrier lifetime can be obtained, compared to the perovskite film without GA involved. Consequently, the passivation effect and the stronger interaction of hydrogen bonds provided by GA boost the PCE of 20.29% and enhances stability in ambient, thermal, and light atmospheres.

2.4.2.2 1D/3D Bilayer Perovskite Solar Cells

The 1D/3D bilayer perovskite can be obtained commonly by post-coating organic halide solution. The strategy is beneficial to larger decay time and higher PL intensity, suggesting a reduction of the non-radiative recombination losses. For example, with the surface treatment of an optimal amount of ethylammonium iodide (EAI), imidazolium iodide (IAI), and guanidinium iodide GAI, the hysteresis of the devices was greatly reduced and the corresponding PCE was obviously enhanced from 20.5% to 22.3%, 22.1%, and 21.0%, respectively. The 1D $TAPbI_3$ (TA = Thiazole ammonium) capping layer serves to passivate three-dimensional (3D) perovskite films, which promotes charge transport, improves carrier lifetime, and prevents iodide ion migration of the 3D $(MA, FA)PbI_3$ perovskite thin film. The optimized device realizes a PCE of 18.97% and retains 92% of the PCE in air (20% humidity and 20°C temperature) for 2 months. The 1D perovskite grows on the 3D perovskite as epilayers or capping layers, forming 1D/3D heterojunction structure. Additionally, the 4-tert-butyl-pyridinium iodide (TBPI) can used as a modifier on the 3D perovskite surface and forming the $TBPPbI_3$ capping layer.

The 1D/3D bilayer perovskite can also be fabricated by one-step method. The organic halide can dissolve in antisolvent. The antisolvent quenches the 3D perovskite precursor on the liquid film surface, with the organic halide reacting with the lead source on the surface. For instance, spin-coating antisolvent with pyrrolidine iodide (PyI) involved can obtain 1D $PyPbI_3$/3D $MAPbI_3$ bilayer perovskite heterojunction structure. Remarkably, the A-site organic components also can be in-situ cross-linked under thermal treatment. Propargyl ammonium (PA^+) is a thermal active polymeric monomer and also can be reacted as an A-site component.[48] This passivation strategy significantly improves the interfacial carrier transport. The cross-linkable alkynyl terminal is surrounded by the 1D perovskite skeleton, allowing the carrier to effectively transport through the 1D chains of face-sharing octahedra $[BX_6]^{4-}$. The hole mobility value is 1.89×10^{-3} cm^{-2} V^{-1} s^{-1} for the cross-linked 1D/3D perovskite, which is much higher than that of the 1D/3D perovskite (8.42×10^{-4} cm^{-2} V^{-1} s^{-1}) and the control sample (6.70×10^{-4} $cm^{-2}V^{-1}s^{-1}$). Moreover, the 1D/3D cross-linked strategy releases residual tensile strain in perovskite films, which is a source of instability. Consequently, the corresponding devices achieve a champion PCE of 21.19% and remarkable stability, maintaining 93% of their initial efficiency after 3,055 h of continuous illumination under maximum power point (MPP) operating conditions. All of these 1D perovskite top layers act as barrier layers to inhibit ion migration and charge carrier recombination under sensitive atmospheres, i.e., moisture, oxygen, and UV light.

2.4.3 0D/3D Perovskite Solar Cells

The 0D perovskites are an important component of low-dimensional structured perovskites. The formulas of 0D perovskites are A_4BX_6 or $A_3B_2X_9$. The $[BX_6]^{4-}$ octahedra in the 0D perovskite are totally segregated from each other and their photo-exciton can be trapped in their own location, which is beneficial to a high photoluminescence quantum yield (PLQY), meanwhile suffering from low carrier mobility. Thus, the property of 0D perovskite with low carrier mobility is inappropriate for solar cells, but in favor of light emission. Based on this, the numerous researches on 0D/3D $Cs_4PbX_6/CsPbX_3$ hybrid perovskite are mainly produced as luminous application and obtained positive results. The 0D-structured Cs_4PbX_6 can passivate the surface states of 3D-structured $CsPbI_3$ or $CsPbBr_3$ due to their good lattice matching. Additionally, the inorganic dopant is favorable to the thermal stability of the hybrid perovskite.

The tuning ratio of CsI in the precursor can be in-situ prepared 0D/3D multidimensional perovskite films. It is reported that when the ratios of CsI and PbI are, respectively, kept as 4:1 and 1:1, the derived films will be composed of colorless 0D Cs_4PbI_6 and yellowish non-perovskite phase (δ-$CsPbI_3$) due to instability of γ-$CsPbI_3$ at room temperature. However, when the precursor ratio of 1.2:1, the stable γ-$CsPbI_3$ phase can be obtained accompanied with the coexistence of 0D Cs_4PbI_6 in the film. The 0D Cs_4PbI_6 locates at the surface of the γ-$CsPbI_3$ grain, which can passivate the defects and stabilize the γ-$CsPbI_3$ phase stability. Further, the phase of 0D $Cs_4Pb(IBr)_6$ and 3D $CsPbI_{3-x}Br_x$ in the films is studied. The facilitated crystal orientation of the $CsPbI_{3-x}Br_x$ 3D perovskite is (100) planes, which further favored vertical carrier transport in the PSCs. The (040) 0D//(002) 3D interface observed in the HRTEM images clearly demonstrates the smooth transition region from the 0D $Cs_4Pb(IBr)_6$ to the 3D $CsPbI_{3-x}Br_x$. The lattice misfit value ($d = (d(040)_{0D}-d(002)_{3D})/d(002)_{3D}$) of the (040)//(002) interface for $Cs_4Pb(IBr)_6$ and $CsPbI_{3-x}Br_x$ is calculated to be only 1.3%, which indicates a small interfacial energy. As a result, the derived PSCs can yield an impressive PCE of 16.39% with improved stability.[49]

Expect for one-pot preparation, the post treatment of organic halide can also promote the generation of 0D/3D structure. For instance, the benzyldodecyldimethylammonium bromide (BDABr) post treatment leads to a surface reconstruction via in-situ 3D-to-0D phase transformation.[50] The optimal BDA-$CsPbI_3$ films with 0D-perovskite passivation exhibits reduced non-radiative recombination and promotes charge carrier transfer. The trap density calculated to be 5.70×10^{15} cm^{-3} for $CsPbI_3$ sample and 3.88×10^{15} cm^{-3} for BDA-$CsPbI_3$, which can confirm the fewer defect sites. The passivation of the inorganic perovskite is beneficial to high efficiency. As a result, the champion BDA-$CsPbI_3$ device shows a much-improved PCE of 20.63%, compared to the PCE of 18.89% of the $CsPbI_3$ control device. Additionally, the unencapsulated devices are tested at MPP under 1 sun in nitrogen atmosphere. The BDA-$CsPbI_3$ device retained over 90% of the original efficiency after 500 h, while the $CsPbI_3$ device dropped to almost 60%. This demonstration of utilizing organic cations for phase transition, surface reconstruction, and the consequent defect passivation successfully tackled the challenge of the hardship on the secondary growth of inorganic perovskite film, especially for the materials of soft crystalline nature.

2.4.4 INTERFACIAL CARRIER TRANSPORT

The carrier transport cross through the interface is sensitive to the conductivity and trap state, thus the charge extraction across the perovskite/charge transport layer interface may be compromised by the low conductivity of the organic spacers utilized to generate 2D perovskites. Also, an unfavorable energy band alignment between 3D bulk films and 2D surface layers can still negatively impact charge transfer in mixed dimensional PSCs. Bulk incorporated 2D perovskite, i.e., PEA_2PbI_4, localized at the grain boundaries is beneficial to band bending at the grain boundary region under illumination, which could facilitate efficient transfer of photo-generated electrons from the grain core to the grain boundary, thus reducing charge recombination. The PEA_2PbI_4 at the interface is shown to be in charge of rising the conduction band of the perovskite film, improving electron collection from the perovskite to the electron transport layer (ETL) and preventing the charge recombination from the ETL flowing back to the perovskite. Moreover, the 2D perovskite works as a dipole layer that can reduce surface recombination.

In general, 2D perovskites exhibit a higher exciton binding energy than 3D perovskites, which means a harder generation of excitons in 2D than 3D perovskites. However, in 2D/3D hybrid perovskite thin films, the exciton binding energy can be lowered by using a combination of multiple A-site organic component, i.e., a mixed composition of 2D perovskite based on BA^+ and 2,2,2-trifluoroethylamine (F_3EA^+) cations has been reported to lower the exciton binding energy from 147 to 128 meV. The lowered exciton binding energy is favor to free carrier generation and transport. It should be noted that the π-systems in A-site can further lower the exciton binding energy, e.g., replacement of the BA^+ cation with PEA^+ lowered the exciton binding energy without any modification to the assembled 2D perovskite thin films.

Polycrystalline perovskite film inevitably exhibits grain-boundaries and heterojunction interfaces because of its low-temperature, solution-processed method of film formation. Especially, the solution-processed 3D perovskite films typically contain structural defects on the surface, which contribute to non-radiative carrier recombination at the perovskite-charge transport layer interfaces. Interface recombination is believed to be a main contribution to open circuit voltage (V_{OC}) loss in PSCs, limiting performance to well below their Shockley-Queisser efficiency potential. The interfaces, i.e., top surface and buried interface, are sensitive to the trap states. For the top surface traps, Fan's group regulated the interface between inorganic perovskite $CsPbIBr_2$ and carbon electrode by difluorobenzylamine (DFBA). The strongly electronegative F in 26-DFBA makes the lone pair electrons on the amine group more concentrated and coordinated with the unsaturated Pb^{2+} and Pb clusters. The reduction of the amount of surface trap states resulted in the V_{OC} of the target devices improvement from 1.05 to 1.14 V. Seo's group [2-(9H-carbazol-9-yl)ethyl]phosphonic acid (CEPA) was introduced as an interface modifier at the interface of perovskite and the hole transporting material. The average V_{OC} of the device was increased considerably by 0.05–0.06 V as compared to the pristine device. More interestingly, a new molecule 2-thiopheneethylamine thiocyanate (TEASCN) could be constructed to bilayer quasi-2D structure perovskite on the top surface of Sn-Pb based perovskite thin films, which boosted the *VOC* from 0.779 to 0.845 V. Different from the researches on the top surface in PSCs, the investigating of the buried interface is more difficult. A lift-off strategy is developed to expose and uncover the buried interfaces of polycrystalline perovskite films. A large amount of flaky lead-halide grains was observed from SEM images, which accelerated the annihilation of the photo-generated charge carriers and non-radiative recombination. In order to reduce non-radiative recombination at the buried perovskite interface, Sargent et al. devised a large organic cation GA that may be build a 2D buffer layer GA_2PbI_4 at the buried interface.[49] Particularly at higher GA concentrations (GA0.1), excess GA molecules are randomly distributed throughout the film when the exposed preferred oxygen binding sites are saturated. They could be incorporated into the perovskite lattice to create mixed-cation structures or connect to the 3D phase's grain boundaries to build 2D/3D structures.

2.4.5 STABILITY OF LD/3D PEROVSKITES AND SOLAR CELLS

The LD perovskites are highly feasible materials for solar cells, which are majorly crystalline and possess wide tunable properties (Figure 2.9).[51] However, the stability problems are still not effectively resolved. Majorly, there are three serious instability issues in PSCs i.e., moisture, thermal treatment, and degradation of the photocurrent with respect to time during the light soaking. The most of the LD perovskites with large organic molecules are water separate component. The incorporation of low LD perovskite is favor to water stability. More important, the encapsulation schemes are commercial method to avoid the perovskite reactions with water and oxygen. In the case of thermal stability, the thermal induced ion migration and degradation in 3D perovskite material or devices is directly correlated with MA cations and iodide ions. By some means, the interfacial modification and direct replacement of MA cations can greatly enhance the thermal stability of the perovskite. For the third, the light degradation meanly occurs at TiO_2/perovskite interface under the UV-light irradiation, which can be regulated by interface modifiers or replaced TiO_2 by SnO_2. Additionally, the J_{SC} degradation is presumed by metastable devices which can be improved by using mixed cations perovskites or using additives such as ionic liquids.

2.5 CONCLUSIONS AND OUTLOOK

Dimensional engineering of perovskite films has attracted a lot of attention from researchers recently because it has the potential to address some of the problems with traditional 3D perovskite, i.e., stability, film-forming regulation, and band structure alignment. Particularly the device stability issues that have prevented PSCs from becoming commercial despite a decade of intensive research. Mixed LD-3D perovskite films can provide a significant improvement in device performance. Therefore, dimensional engineering of PSCs will continue to be a popular research area in the near future. We highlight some of the most interesting research topics in this section. First, novel organic spacers have good thermal stability, UV-light stability, and moisture resistance, i.e., alkyl ammonium or aromatic amine salts with stereo-cyclic structure and polyamine structures. Second, the organic components with large spatial hindrance would inhibit the charge transport crossing the Van der Waals gap. Amine salts with conjugated structure extending to amino groups and π-π stacking between organic molecules are favor to the charge transport. Third, the organic molecules including elements of S, I, P, are profitable for strong interaction to the halide vacancies and low band gap, which is beneficial to compensating for the light absorption of traditional 3D perovskite.

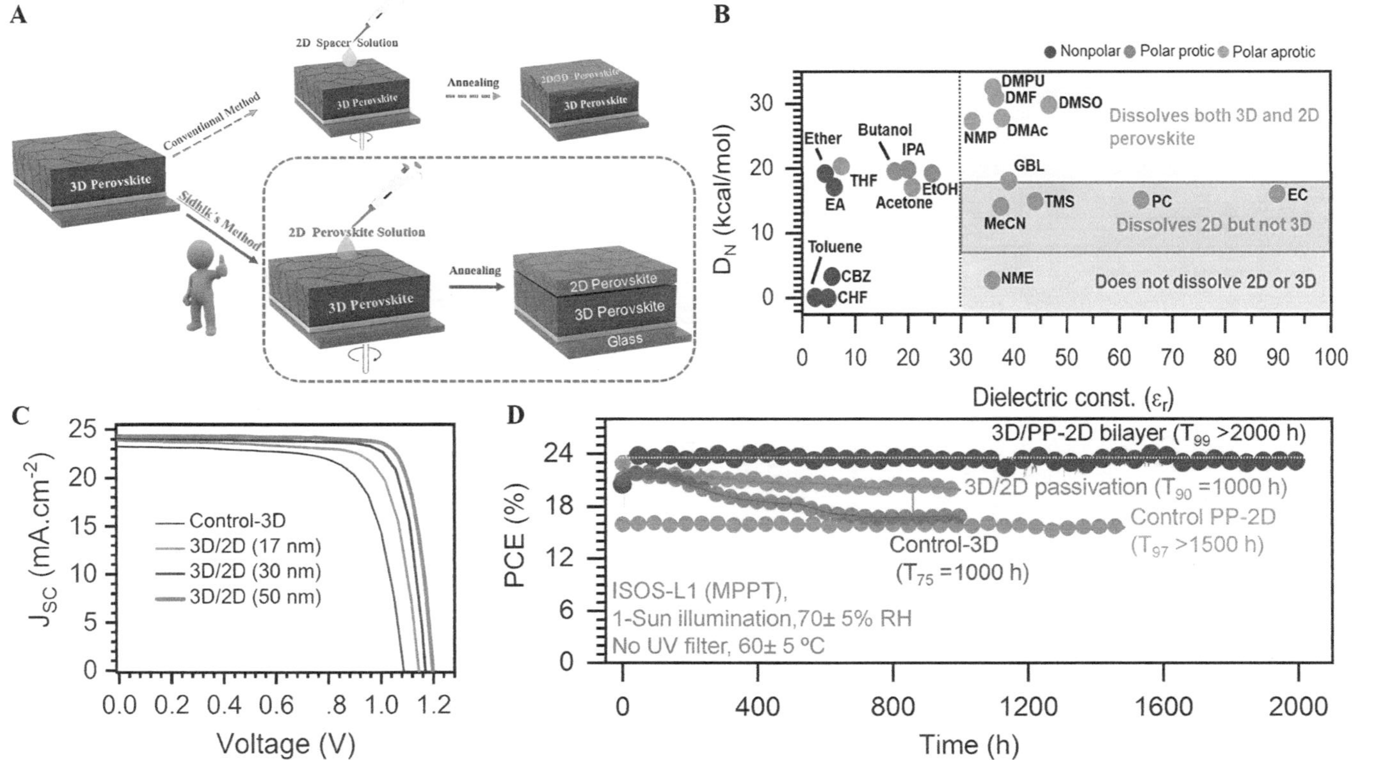

FIGURE 2.9 Preparation of 2D/3D perovskite bilayer heterojunctions and performance comparison of different methods and solvents. (a) Schematic illustration of preparing 3D/2D perovskite bilayer heterojunctions with different methods. (b) Comparison of different solvents based on dielectric constant ($\mathcal{E}r$) and Gutmann number (DN). (c) Current–voltage curves of the champion 3D/2D perovskite bilayer heterojunction PSCs. (d) Stability assessment of the 3D-/2D-based modules with different perovskite films at maximum power point tracking in an ambient atmosphere.[51]

From the perspective of the LD-3D structural materials, the growth kinetics for the high-quality thin films require more studies, e.g., formation sequence of LD and 3D perovskites and distribution of LD perovskite in thin films. Likewise, reports on the relationship between the structure of organic spacers and the dimensionality of their formed LD perovskite are still rare, and more in-depth research is needed in this field. Therefore, in order to improve PSCs device efficiency and stability with LD-3D, a comprehensive strategy of structure design for combination stability and charge transport needs to be further explored.

REFERENCES

1. a) J. H. Heo, *et al.*, Efficient inorganic–organic hybrid heterojunction solar cells containing perovskite compound and polymeric hole conductors, *Nature Photonics* **2013**, *7*, 486; b) J. Burschka, *et al.*, Sequential deposition as a route to high-performance perovskite-sensitized solar cells, *Nature* **2013**, *499*, 316; c) J. H. Park, *et al.*, Efficient $CH_3NH_3PbI_3$ perovskite solar cells employing nanostructured p-type NiO electrode formed by a pulsed laser deposition, *Advanced Materials* **2015**, *27*, 4013.
2. a) T. Jeon, *et al.*, Laser crystallization of organic-inorganic hybrid perovskite solar cells, *ACS Nano* **2016**, *10*, 7907; b) S. Chen, *et al.*, A photonic crystal laser from solution based organo-lead iodide perovskite thin films, *ACS Nano* **2016**, *10*, 3959; c) H. Yu, *et al.*, Organic-inorganic perovskite plasmonic nanowire lasers with a low threshold and a good thermal stability, *Nanoscale* **2016**, *8*, 19536.
3. H. P. Kim, *et al.*, High-efficiency, blue, green, and near-infrared light-emitting diodes based on triple cation perovskite, *Advanced Optical Materials* **2017**, *5*, 1600920.
4. A. R. Bin, M. Yusoff, J. Jang, Highly efficient photoelectrochemical water splitting by a hybrid tandem perovskite solar cell, *ChemComm* **2016**, *52*, 5824.
5. H. Zhou, *et al.*, Self-powered all-inorganic perovskite microcrystal photodetectors with high detectivity, *The Journal of Physical Chemistry Letters* **2018**, *9*, 2043.
6. Y. Mei, C. Zhang, Z. Vardeny, O. Jurchescu, Electrostatic gating of hybrid halide perovskite field-effect transistors: Balanced ambipolar transport at room-temperature, *MRS Communications* **2015**, *5*, 297.
7. C. Gu, J.-S. Lee, Flexible hybrid organic-inorganic perovskite memory, *ACS Nano* **2016**, *10*, 5413.
8. S. Zhou, L. Li, H. Yu, J. Chen, C. P. Wong, N. Zhao, Thin film electrochemical capacitors based on organolead triiodide perovskite, *Advanced Electronic Materials* **2016**, *2*, 1600114.
9. N. Vicente, G. Garcia-Belmonte, Methylammonium lead bromide perovskite battery anodes reversibly host high li-ion concentrations, *The Journal of Physical Chemistry Letters* **2017**, *8*, 1371.
10. Y. Zhou, *et al.*, Ultra-broadband optical amplification at telecommunication wavelengths achieved by bismuth-activated lead iodide perovskites, *Journal of Materials Chemistry C* **2017**, *5*, 2591.
11. Y. Jia, R. A. Kerner, A. J. Grede, A. N. Brigeman, B. P. Rand, N. C. Giebink, Diode-pumped organo-lead halide perovskite lasing in a metal-clad distributed feedback resonator, *Nano Letters* **2016**, *16*, 4624.
12. S.-T. Ha, C. Shen, J. Zhang, Q. Xiong, Laser cooling of organic–inorganic lead halide perovskites, *Nature Photonics* **2016**, *10*, 115.
13. E. Yablonovitch, Lead halides join the top optoelectronic league, *Science* **2016**, *351*, 1401.
14. Y. Liu, *et al.*, Structurally dimensional engineering in perovskite photovoltaics, *Advanced Energy Materials* **2023**, *13*, 2300188.

15. H. Yin, Y. Xian, Y. Zhang, W. Li, J. Fan, Structurally stabilizing and environment friendly triggers: Double-metallic lead-free perovskites, *Solar RRL* **2019**, *3*, 1900148.

16. A. K. Jena, A. Kulkarni, T. Miyasaka, Halide perovskite photovoltaics: Background, status, and future prospects, *Chemical Reviews* **2019**, *119*, 3036.

17. W. Travis, E. Glover, H. Bronstein, D. Scanlon, R. Palgrave, On the application of the tolerance factor to inorganic and hybrid halide perovskites: A revised system, *Chemical Science* **2016**, *7*, 4548.

18. G. Volonakis, *et al.*, $Cs_2InAgCl_6$: A new lead-free halide double perovskite with direct band gap, *The Journal of Physical Chemistry Letters* **2017**, *8*, 772.

19. Q. Chen, *et al.*, $Cs_2InAgCl_6$: A new lead-free halide double perovskite with direct band gap, *Nano Today* **2015**, *10*, 355.

20. A. Binek, F. C. Hanusch, P. Docampo, T. Bein, Stabilization of the trigonal high-temperature phase of formamidinium lead iodide, *The Journal of Physical Chemistry Letters* **2015**, *6*, 1249.

21. J. Gong, P. Guo, S. E. Benjamin, P. G. Van Patten, R. D. Schaller, T. Xu, Cation engineering on lead iodide perovskites for stable and high-performance photovoltaic applications, *Journal of Energy Chemistry* **2018**, *27*, 1017.

22. D. J. Kubicki, D. Prochowicz, A. Hofstetter, S. M. Zakeeruddin, M. Grätzel, L. Emsley, Phase segregation in Cs-, Rb-and K-doped mixed-cation $(MA)_x (FA)_{1-x} PbI_3$ hybrid perovskites from solid-state NMR, *Journal of the American Chemical Society* **2017**, *139*, 14173.

23. N. Li, Z. Zhu, J. Li, A. K. Y. Jen, L. Wang, Inorganic $CsPb_{1-x}Sn_xIBr_2$ for efficient wide-bandgap perovskite solar cells, *Advanced Energy Materials* **2018**, *8*, 1800525.

24. F. X. Xie, D. Zhang, H. Su, X. Ren, K. S. Wong, M. Grätzel, W. C. Choy, Vacuum-assisted thermal annealing of $CH_3NH_3PbI_3$ for highly stable and efficient perovskite solar cells, *ACS Nano* **2015**, *9*, 639.

25. E. L. Unger, *et al.*, Chloride in lead chloride-derived organo-metal halides for perovskite-absorber solar cells, *Chemistry of Materials* **2014**, *26*, 7158.

26. F. X. Xie, H. Su, J. Mao, K. S. Wong, W. C. Choy, Evolution of diffusion length and trap state induced by chloride in perovskite solar cell, *The Journal of Physical Chemistry C* **2016**, *120*, 21248.

27. J. H. Noh, S. H. Im, J. H. Heo, T. N. Mandal, S. I. Seok, Chemical management for colorful, efficient, and stable inorganic–organic hybrid nanostructured solar cells, *Nano Letters* **2013**, *13*, 1764.

28. G. E. Eperon, S. D. Stranks, C. Menelaou, M. B. Johnston, L. M. Herz, H. J. Snaith, Formamidinium lead trihalide: A broadly tunable perovskite for efficient planar heterojunction solar cells, *Energy & Environmental Science* **2014**, *7*, 982.

29. E. T. Hoke, D. J. Slotcavage, E. R. Dohner, A. R. Bowring, H. I. Karunadasa, M. D. McGehee, Reversible photo-induced trap formation in mixed-halide hybrid perovskites for photovoltaics, *Chemical Science* **2015**, *6*, 613.

30. L. N. Quan, *et al.*, Ligand-stabilized reduced-dimensionality perovskites, *Journal of the American Chemical Society* **2016**, *138*, 2649.

31. X. Li, *et al.*, Lead-free halide perovskites for light emission: Recent advances and perspectives, *Advanced Science* **2021**, *8*, 2003334.

32. S. De Wolf, *et al.*, Organometallic halide perovskites: Sharp optical absorption edge and its relation to photovoltaic performance, *The Journal of Physical Chemistry Letters* **2014**, *5*, 1035.

33. B. Philippe, *et al.*, Valence level character in a mixed perovskite material and determination of the valence band maximum from photoelectron spectroscopy: Variation with photon energy, *The Journal of Physical Chemistry C* **2017**, *121*, 26655.

34. D. B. Mitzi, Synthesis, structure, and properties of organic-inorganic perovskites and related materials, *Progress in Inorganic Chemistry* **1999**, *48*, 1–121.

35. P. Chen, Y. Bai, M. Lyu, J. H. Yun, M. Hao, L. Wang, Progress and perspective in low-dimensional metal halide perovskites for optoelectronic applications, *Solar RRL* **2018**, *2*, 1700186.

36. a) D. Mitzi, S. Wang, C. Feild, C. Chess, A. Guloy, Conducting layered organic-inorganic halides containing <110>-oriented perovskite sheets, *Science* **1995**, *267*, 1473; b) D. B. Mitzi, Templating and structural engineering in organic-inorganic perovskites, *Journal of the Chemical Society, Dalton Transactions* **2001**, *1*, 1–12; c) S. Wang, D. B. Mitzi, C. A. Feild, A. Guloy, Synthesis and characterization of $[NH_2C(I):NH_2]_3MI_5$ (M= Sn, Pb): Stereochemical activity in divalent tin and lead halides containing single. ltbbrac. 110. rtbbrac. Perovskite sheets, *Journal of the Chemical Society* **1995**, *117*, 5297.

37. L. Mao, *et al.*, Structural diversity in white-light-emitting hybrid lead bromide perovskites, *Journal of the Chemical Society* **2018**, *140*, 13078.

38. Z. Yuan, *et al.*, One-dimensional organic lead halide perovskites with efficient bluish white-light emission, *Nature Communications* **2017**, *8*, 14051.

39. W.-J. Yin, T. Shi, Y. Yan, Superior photovoltaic properties of lead halide perovskites: Insights from first-principles theory, *The Journal of Physical Chemistry C* **2015**, *119*, 5253.

40. G. Li, *et al.*, Surface defect passivation by 1,8-Naphthyridine for efficient and stable formamidinium-based 2D/3D perovskite solar cells, *Chemical Engineering Journal* **2022**, *449*, 137806.

41. Y. Lin, *et al.*, Enhanced thermal stability in perovskite solar cells by assembling 2D/3D stacking structures, *The Journal of Physical Chemistry Letters* **2018**, *9*, 654.

42. J. Liang, *et al.*, A finely regulated quantum well structure in quasi-2D Ruddlesden-Popper perovskite solar cells with efficiency exceeding 20%, *Energy & Environmental Science* **2022**, *15*, 296.

43. M. Sun, *et al.*, Difluorine-substituted molecule-based low-dimensional structure for highly stable tin perovskite solar cells, *Solar RRL* **2022**, *6*, 2200672.

44. Z. Xu, *et al.*, 3D-to-2D dimensional reduction for exploiting a multilayered perovskite ferroelectric toward polarized-light detection in the solar-blind ultraviolet region, *Angewandte Chemie International Edition* **2020**, *59*, 21693.

45. J.-W. Lee, *et al.*, 2D perovskite stabilized phase-pure formamidinium perovskite solar cells, *Nature Communications* **2018**, *9*, 3021.

46. J. Fan, *et al.*, Thermodynamically self-healing 1D–3D hybrid perovskite solar cells, *Advanced Energy Materials* **2018**, *8*, 1703421.

47. P. Liu, *et al.*, Lattice-matching structurally-stable 1D@ 3D perovskites toward highly efficient and stable solar cells, *Advanced Energy Materials* **2020**, *10*, 1903654.

48. N. Yang, *et al.*, An in situ cross-linked 1D/3D perovskite heterostructure improves the stability of hybrid perovskite solar cells for over 3000 h operation, *Energy & Environmental Science* **2020**, *13*, 4344.

49. B. Chen, *et al.*, Passivation of the buried interface via preferential crystallization of 2D perovskite on metal oxide transport layers, *Advanced Materials* **2021**, *33*, 2103394.

50. Y. Chen, X. Wang, Y. Wang, X. Liu, Y. Miao, Y. Zhao, Functional organic cation induced 3D-to-0D phase transformation and surface reconstruction of $CsPbI_3$ inorganic perovskite, *Science Bulletin* **2023**, *68*, 706.

51. X. Lian, H. Zhang, X. Mo, J. Chu, Fully solution-processed phase-pure 3D/2D perovskite bilayer heterojunctions, *Communications Chemistry* **2023**, *6*, 11.

3 Low-Dimensional Perovskite Solar Cells

Pengwei Li, Yiqiang Zhang, Zhipeng Miao, and Qingli Cao

Low-dimensional perovskite has attracted widespread attention due to its inherent stability compared to its three-dimensional (3D) counterparts. These materials have extensive customizability in composition, structure, and bandgap, providing an interesting platform for revealing structure-property relationships in solid-state chemistry and physics. In the field of photovoltaics, the low-dimensional perovskite solar cells (PSC) prepared have achieved high stability and sustainable breakthroughs in power conversion efficiency (PCE). This chapter focuses on studying low-dimensional halide perovskite from a structural perspective, including Ruddlesden-Popper (RP) phase, Dion-Jacobson (DJ) phase, interlayer alternating cation (ACI) phase, and mixed organic ligand phase, due to the diversity and versatility of spacers. Subsequently, the effects of the types, chemical composition, and physical properties of spacers on low-dimensional perovskite are discussed. Then, the characteristics of low-dimensional perovskite thin films closely related to carrier transport in solar cells are evaluated (phase separation, grain orientation, crystallization kinetics, etc.) Finally, the current challenges and future research prospects of efficient and stable LD PSC were discussed.

3.1 INTRODUCTION

Three-dimensional (3D) perovskite solar cells (PSCs) have made incredible advances in photoelectric conversion efficiency (PCE) over the past decade[1–8] and are now recognized as the most promising photovoltaic materials. At present, the PCE of authoritative 3D PSCs has exceeded 26.1%[9], which is comparable to that of mono-crystalline silicon solar cells. These unique achievements of PSCs are mainly attributed to the efficient carrier transport performance, high absorption coefficient, low trap density, and simple solution processing technology of 3D perovskites[10–15]. However, because 3D perovskites are easy to decompose and decay in the natural environment, it is difficult to directly face a variety of harsh and harsh environmental conditions in the real world, which greatly limits the practical application and material industrialization of 3D perovskites[16–18].

Researchers have adopted a series of strategies to overcome its stability issue, including interface engineering[2,19–23], doping process[24–27], hybrid A cationic scheme[28–30], and packaging technology[31–33]. However, these measures have not fundamentally changed

DOI: 10.1201/9781003400486-3

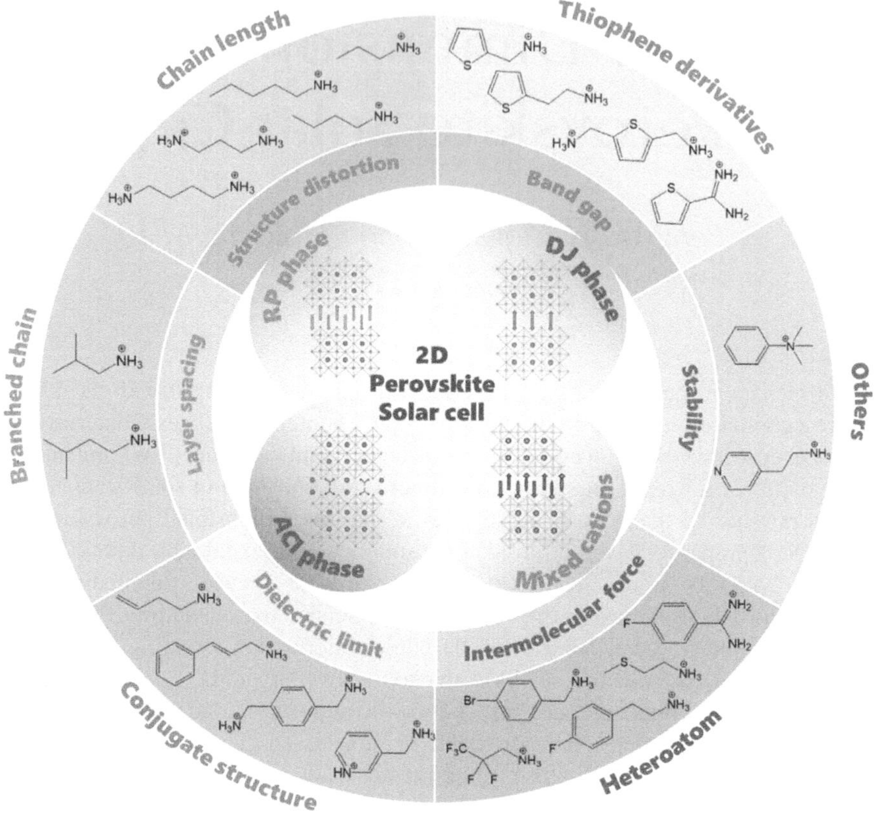

the structure of 3D perovskite, and once the 3D perovskite is eroded by water and oxygen, its single and weak structure will decompose rapidly like a broken embankment, so the problem of 3D perovskite instability has not been fundamentally solved. In the past few years, researchers have reduced the dimension of perovskites from 3D to two-dimensional (2D) to obtain efficient and stable PSCs[34–36]. 2D perovskites can be understood as nanosheet layers of perovskites with atomic or/and molecular sizes formed by "cutting" 3D perovskites structure in specific directions with organic ligands.

Such novel nanostructures are formed by combining them through interactions between organic ligands and 3D frameworks to form 2D perovskites (Figure 3.1a). 2D perovskites have stronger water stability than 3D perovskites due to the hydrophobic nature of organic ligands and the large formation energy of nanostructures. In addition, unlike traditional 3D perovskites, 2D perovskites have high versatility and flexible and adjustable photoelectric properties. With the continuous innovation of material optimization strategies, the PCE of 2D PSCs has also increased to more than 20%[37]. This shows that 2D perovskites as absorbing layers have great potential in the preparation of efficient and long-term stable PSCs.

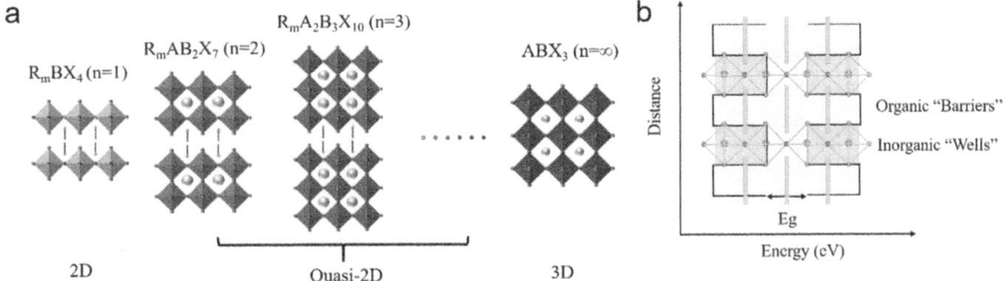

FIGURE 3.1 (a) Schematic diagram of 2D, quasi-2D, and 3D perovskites structure; (b) QW schematic of 2D perovskites, where the organic ligand acts as a quantum "barrier" and the inorganic layer acts as the "wells".

Although the PCE of 2D PSCs has been greatly improved, there is still a large lag compared with 3D PSCs. This is mainly due to the strong quantum confinement effect and dielectric confinement effect of the layered structure, as well as the quantum well (QWs) structure with wide distribution of multiple random wells in perovskite films, resulting in poor carrier mobility and diffusion length of 2D perovskites. 2D perovskites are generally composed of the QWs structure, where the "well" is the 2D perovskite inorganic layer and the "barrier" is the organic layer of the 2D perovskite, as shown in Figure 3.1b[38–41]. Therefore, the quantum confinement effect greatly increases the exciton binding energy (E_b) of 2D perovskite. In addition, photogenerated 2D perovskite carriers are confined in a "well," which limits the external conduction of carriers. In addition, the dielectric constant of conventional 2D perovskite organic ligands is much smaller than that of the inorganic layer of perovskite, which leads to strong dielectric confinement and increases the E_b. Large exciton binding can lead to the formation of photogenerated carriers in 2D perovskites in the form of excitons rather than free electrons and holes. Therefore, the key bottleneck to improving the PCE of 2D PSCs is to adjust the physicochemical properties between organic ligands and inorganic ligands. At present, various organic ligands and their unique properties (including content, functional group type, interaction, etc.) have been widely explored in the preparation of 2D perovskite structure and film-forming characteristics of high-efficiency photovoltaic devices. Therefore, this chapter summarizes the types, composition, and properties of 2D perovskite organic spacer cations, which can provide guidance for the preparation of high-performance 2D PSCs.

In this chapter, the structure of 2D perovskites are first classified (based on the splicing types between organic spacer cations and inorganic plates, namely Ruddlesden-Popper (RP) type, Dion-Jacobson (DJ) type, interlaminar space alternating cation (ACI) type, and mixed organic ligand type and their unique properties of the structural aspects are also discussed. Subsequently, this chapter continues to emphasize the effects of different properties of organic ligands, such as volume, chain length, conjugation, solubility, dielectric constant, etc. on the specific photophysical properties of 2D perovskites and photovoltaic performance of solar cells. In particular, the diversity of organic ligand properties has a crucial impact on the properties of the corresponding perovskite crystals and also gives perovskite films different photoelectric properties and environmental stability, which will be discussed in depth here. The last part of this chapter summarizes the existing research results and looks forward to structural design, customized methods of organic ligand molecules, and intrinsic physics properties optimization of 2D perovskites.

3.2 PROPERTIES OF 2D PEROVSKITES

Studies have shown that different types of 2D perovskites can be formed using macromolecular organic amines cut along the <100>, <110>, and <111> directions of the 3D perovskite structure, namely <100>-, <110>-, or <111>-oriented 2D perovskites[42]. Among them, due to the different types of organic amines, the <100>-orientation 2D perovskites can be subdivided into three structural types: RP type[43], DJ type[44], and ACI type[45], as shown in Figure 3.2. The common chemical expressions are $R_2A_{n-1}B_nX_{3n+1}$, $R'A_{n-1}B_nX_{3n+1}$, and $(A'A)_{n-1}B_nX_{3n+1}$, where R is a large volume of organic monoamine cations (such as BA^+ = butylamine), R' is a large volume of organic diamine cations (such as $3\text{-}AMP^{2+}$ = 3-(aminomethyl)piperidine), A' is a special organic monoamine (usually GA^+ = guanidine), A is a small volume of monoamines (such as MA^+ = methylamine), B is a divalent metal cation (such as Pb^{2+}, Sn^{2+}), X is a halogen anion (such as Cl^-, Br^-, I^-), and n is the number of inorganic layers between adjacent organic layers, which can be adjusted by controlled stoichiometry[46,47].

The 2D perovskite layers are connected by a combination of van der Waals and/or hydrogen bonds. The hydrophobicity of spaced cations prevents water molecules from invading 2D perovskites structure, which explains excellent humidity stability of 2D perovskites. The dielectric confinement effect of 2D perovskites leads to large E_b and a complex band gap, which can be adjusted by changing the number of accumulation layers. It is generally believed that $n = 1$ represents pure 2D perovskites, which has a high E_b and is not suitable as the active layer of PSCs. With the increase of the number of inorganic layers (n-value), the E_b and band gap of 2D perovskites

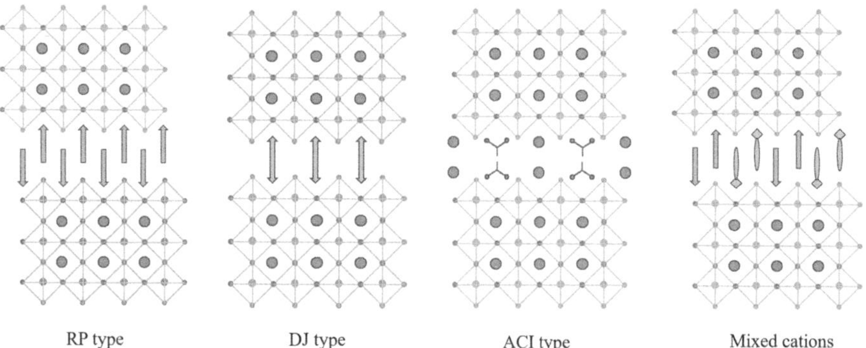

FIGURE 3.2 Schematic diagram of 2D perovskites structure. RP type with van der Waals gap with monoammonium cations, DJ type with diammonium cations, ACI phase featuring alternating cation interlayer, and mixed cations phase with more than one cation, respectively.

decrease. The phase of $1 < n \leq 5$ is often referred to as quasi-2D perovskites, which are widely used as the active layer of 2D PSCs. Quasi-2D perovskites converge to a 3D structure at $n = \infty$. It is worth noting that the thermodynamic stability of high-n structure is slightly different, which is not conducive to obtaining pure quasi-2D perovskites. On the one hand, the composition of the perovskite precursor solution dominates the n-value of the 2D perovskite films, and the internal charge transport of the 2D perovskite films with multiple n-value is anisotropic[48], that is, the charge mobility along the inorganic layer direction is much higher than the charge mobility perpendicular to the inorganic plate direction. Therefore, 2D perovskite film orientation regulation is particularly important for obtaining high-performance PSCs. On the other hand, the selection and design of idealized organic ligands can regulate the structure and properties of 2D perovskites. For example, the selection of organic ligands with a high dielectric constant is expected to improve charge transfer efficiency perpendicular to the direction of the inorganic layer, which will be discussed in the next section of thischapter. The adjustability and versatility of this structure make 2D perovskites have a wider range of photoelectric properties than 3D perovskites, so they have a wider range of applications in the field of photovoltaics[49]. The influence of organic ligands with different intrinsic physical and chemical properties on 2D perovskites mainly has three aspects: crystal structure (inorganic octahedral distortion, layer spacing, etc.), optical properties (band gap, etc.), and film properties (crystallinity, orientation, stability, etc.). In the next section, this chapter will summarize the role of organic ligands from these three aspects and clarify the intrinsic properties of organic ligands and their influence on the structure of 2D perovskites, laying the foundation for the design of novel organic amines.

3.3 2D RP TYPE PEROVSKITES

RP type perovskites are the most common of the above three structure and the most widely studied. The RP type perovskite layer has two layers of monovalent organic cations between the inorganic layers, and the van der Waals forces between the organic layers bind them together. This weak van der Waals interaction makes the accumulation of RP type perovskite inorganic layers more flexible and free. Structurally, it can be observed that the RP type perovskite inorganic layer has a relatively staggered configuration and generally presents (1/2, 1/2) in-plane displacement along the plane of the inorganic layer, which makes the inorganic layer have large distortion. Due to the development and design of organic ligands, RP type perovskites are becoming more and more diverse, and a lot of effort has been made in synthesizing RP type perovskites to prepare high-performance 2D perovskite materials. At present, RP type PSCs have the highest PCE in the 2D perovskite family. As shown in Figure 3.3, the unique properties of different types of organic ligands are classified and discussed here.

3.3.1 LINEAR MONOAMINE LIGANDS

Perovskites with linear monoamine organic ligands have typical 2D perovskite characteristics and are widely used in the field of solar cells. For example, in 1994,

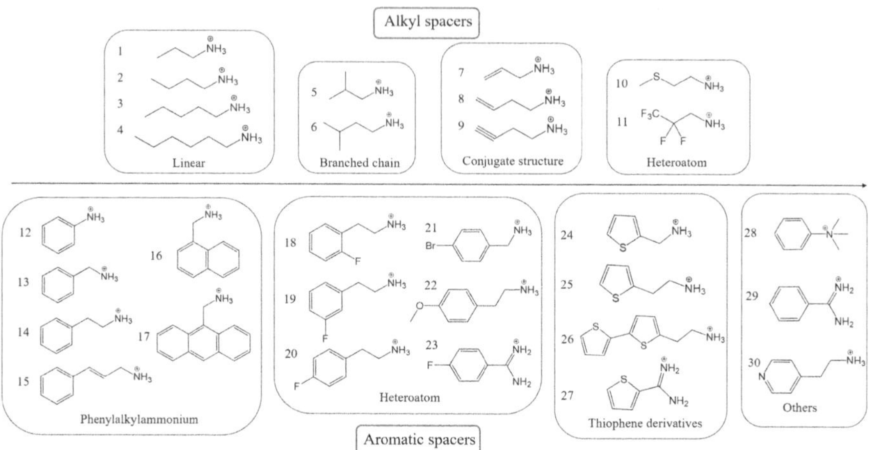

FIGURE 3.3 Organic ligands for 2DRP PSCs. 1: Propylamine[50]; 2: Butylamine (BA)[43,51]; 3: Amylamine (AA)[52,53]; 4: Hexylamine (HA)[52,53]; 5: Isobutylamine (IBA)[54,55]; 6: Isoprenolamine[54]; 7: Allylamine[56]; 8: 3-Buten-1-amine[57]; 9: 3-Butyn-1-amine[57]; 10: 2-Methylthioethylamine[58]; 11: Pentafluoropropylamine[59]; 12: Aniline[60]; 13: Benzylamine[61–63]; 14: Phenethylamine[64–66]; 15: 3-Phenyl-2-propenylamine[67,68]; 16: 1-Naphthylmethylamine[69]; 17: 9-Anthracenemethylamine[69]; 18: 2-Fluorophenethylamine[70]; 19: 3-Fluorophenethylamine[70]; 20: 4-Fluorophenethylamine[70–74]; 21: 3-Bromobenzylamine[75]; 22: 4-Methoxyphenethylamine[71]; 23: p-Fluorobenzamidine[76]; 24: 2-Thiophenemethylamine[77–79]; 25: 2-Thiopheneethylamine[79,80]; 26: Dithiophenethylamine[81]; 27: 2-Thiophenecarboxamide[82]; 28: Phenyltrimethylamine[83]; 29: Benzamidine[76]; 30: 4-(Ethylamine) pyridine(4-AEP)[84,85].

Mitzi et al. first reported the synthesis and analysis of a series of 2D perovskites with different n values based on BA[86]. Since then, BA-based 2DRP perovskites have undergone decades of development[87–89]. In 2015, BA-based 2D perovskites were first applied as absorbers to the stability of PSCs, with a PCE of 4.02%[51]. With the continuous research and development of solvent engineering[47,90–93], hot spinning process[94], additive process[95], and other technologies, the PCE of PSCs has been greatly improved, and the PCE of 2D PSCs based on BA has reached 18%[96]. Inspired by the BA-based catalog of 2DRP perovskites, many researchers have designed different organic ligands based on BA to explore the effects of different organic amine structure on 2D perovskite properties. In this section, we use BA as a benchmark to explore the different characteristics of these novel organic amines applied to 2D PSCs. Many factors affect the properties of linear monoamine molecules and their films. Specifically, this section clarifies and summarizes the relationship between these structures and properties from the aspects of chain length, short branched chain, conjugate structure, and heteroatoms, respectively, so as to provide a reference for the design and expected effects of novel organic ligands.

3.3.1.1 Length of the Carbon Chains

Alkyl amines of different chain lengths have been extensively studied[50,52, 97,98]. Among them, propylamine (PA) is a linear alkyl amine with the shortest chain length and can form a 2D structure[99]. PA-based $(PA)_2MA_2Pb_3I_{10}$ ($n = 3$) and $(PA)_2MA_3Pb_4I_{13}$($n = 4$) signal crystals have been successfully synthesized[50]. However, when trying to lower the value of n ($n = 1$ and 2), a series of new stepped structures are formed instead of the normal low-n structure. The main reason is that there is no A-position cation to support the octahedral cage, and the PA cation is too small to effectively template the 2D structure. Organic amines of different chain lengths affect the distortion of the inorganic $[PbI_6]^{4-}$ structure (usually expressed as Pb-I-Pb bond angle) through interaction with inorganic layers, and the band gap is largely determined by the Pb-I-Pb bond angle[42,44]. At the same time, the different chain lengths of organic amines have a certain influence on the interlayer spacing, which in turn affects the band gap and optical properties of perovskite. The previous characteristics are particularly evident in $n = 1$ perovskites. With the increase of the n value, the influence of organic amine insulation is almost negligible. When the n-values are the same, membranes with organogenic amines with different chain lengths have similar absorption spectra and photoluminescence spectra (Figure 3.4a). This means that the increase in the chain length of organic amines has little effect on its band gap and optical properties (Figure 3.4b) but affects its stability and film-forming capacity. On the other hand, as the chain length increases, the proportion of the softer organic part relative to the harder inorganic layer part increases, so Young's modulus and hardness of the corresponding crystal decrease, thereby increasing the flexibility and softness of the material structure. This has a positive impact on the fabrication of flexible devices. It is worth noting that the stability of perovskite films does not increase with the increase of organic amine chain length. The results show that the film with amino (AA) as the substrate has higher stability than the film with BA and hexylamine (HA) as the substrate. However, the source of the unique stability of AA organic ligands remains to be further explored. In summary, due to the low correlation between chain length and optical properties, the study of alkyl organic amines with different chain lengths mainly focuses on the effect of different chain lengths on film-forming performance and stability.

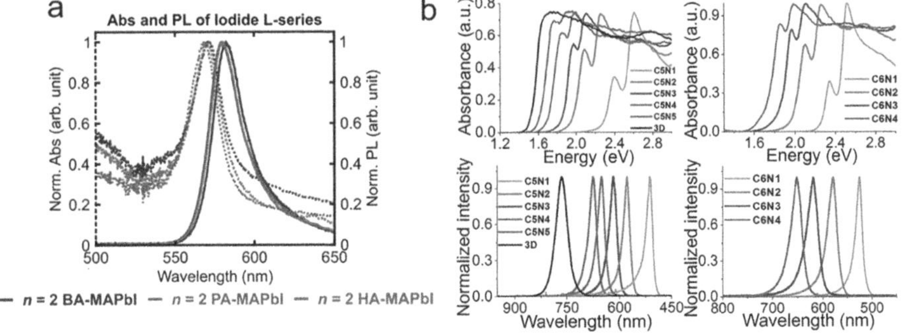

FIGURE 3.4 (a) Optical absorption (punct) and PL (solid) spectra of 2D lead halide perovskite sheets peeled off by $BA_2MAPb_2I_7$, $PA_2MAPb_2I_7$, and $HA_2MAPb_2I_7$ at room temperature[97]; (b) Systematic evolution of light absorption and emission spectra for compounds C5N1−5, C6N1−4 and $MAPbI_3$ (C5 = AA, C6 = HA, N is the number of inorganic layers)[52].

For the preparation of perovskite films and solar cells, Wu et al. reported the effects of a series of aliphatic alkylamine ligands (ethylamine (EA), PA, BA, AA, and HA) with different chain lengths on perovskite films and devices from the perspective of molecular interactions (Figure 3.5a)[53]. Different organic amine-based films were prepared by stoichiometry of the target components. The results showed that except for the EA ligand, the rest could form a uniform film layer. The films based on AA and HA are striped, which is beneficial to increase the contact area between the perovskite and transport layer interface and promote charge extraction. The van der Waals forces between molecules increase with increasing chain length, and proper van der Waals interactions are essential for crystallization.

At the same time, the appropriate intermolecular van der Waals force promoted the aggregation of perovskite precursor solutions, promoted the growth of perovskite membrane vertical 2D and 3D phase components, and thus changed the vertical phase distribution of perovskite film. However, if the interaction between ligands is too strong, such as HA, strong aggregation in the precursor may trigger too rapid growth of the 3D layer, resulting in poor orientation. Therefore, AA-based films have the highest crystallinity and orientation (Figure 3.5b, e). At the same time, Figure 3.5c and d show the absorption spectra of films based on different n values and different amines, which indicates that AA-based films contain fewer low n phases, and low n has adverse effects on carrier transport in optoelectronic devices. The E_b of AA-based perovskite ligands decreases, the dielectric constant increases, and the photoelectric performance is improved, which promotes the generation and separation of charges. Finally, AA-based perovskites have efficient charge transfer and weak well well-assisted recombination. After a slow back-annealing treatment[100], the PCE of AA-based PSCs ($n = 4$) reached the highest level of 18.42%, the open-circuit voltage (V_{OC}) was 1.25 V, the short-circuit current density (J_{SC}) was 18.57 mA cm^{-2}, and the fill factor (FF) was 80%. These results suggest that medium-chain length organic amines have a critical influence on the film-forming process. In addition, as mentioned earlier, the chain length of the organic ligands failed to play a dominant role in stability. It is universally acknowledged that the preparation of high-quality films from organic ligands of suitable chain length is considered to have excellent stability.

3.3.1.2 Branch Chains of Carbon Chains

BA sparked a boom in 2D perovskite research, and crystals of short branched alkyl amines were synthesized and decomposed[54]. Branched isobutylamine (IBA) and iso-amylamine (IAA) are essentially the same length as straight-chain alkylamines PA and BA. But due to the presence of branched chains, IBA and IAA are larger in size. Surprisingly, bulky branched-chain organic amines do not cause larger lattice distortion but result in larger Pb-I-Pb bond angles, as shown in Figure 3.6a–c. This means that organic amines branched-chain-based 2D perovskites have a smaller band gap than linear alkyl amines at the same length (Figure 3.6d–g).

To study the effect of branched-chain alkyl amines on photovoltaic performance, Liang et al. studied perovskite films and devices based on IBA and BA spacers[101,55]. It can be seen that IBA-based perovskites have a lower band gap and a higher optical absorption range, which is mainly due to the large Pb-I-Pb bond angle and the

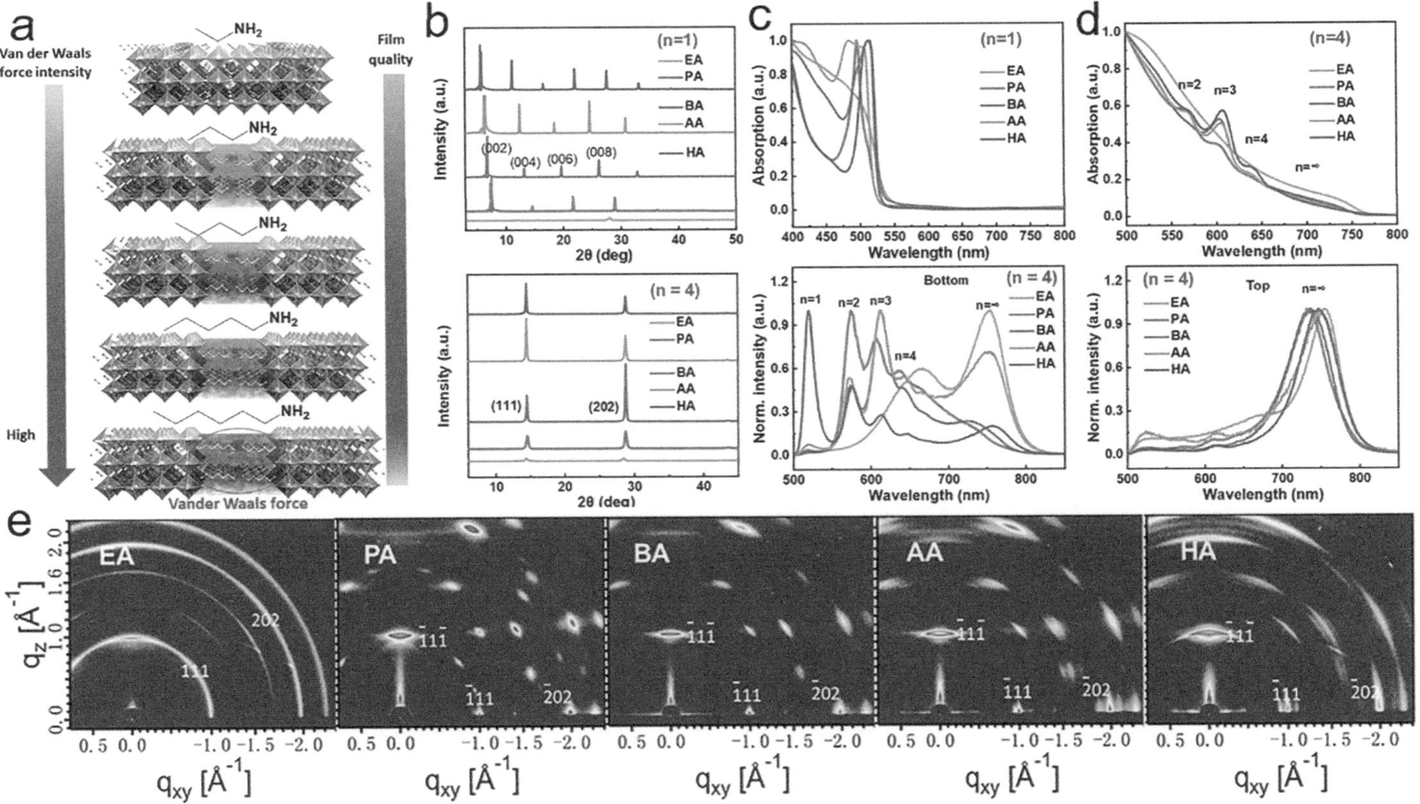

FIGURE 3.5 (a) Schematic structure of 2DRP perovskites integrated by different organic ligands; (b) X-ray diffractometer measured on various 2DRP perovskite films of $n = 1$ and $n = 4$; (c) UV-vis absorption spectra of $n = 1$ and $n = 4$ perovskite films obtained using transmission mode, respectively; (d) PL spectra of 2DRP perovskite films ($n = 4$) with different excitation configurations; (e) grazing incidence wide-angle X-ray scattering (GIWAXS) of different 2DRP perovskite films[53].

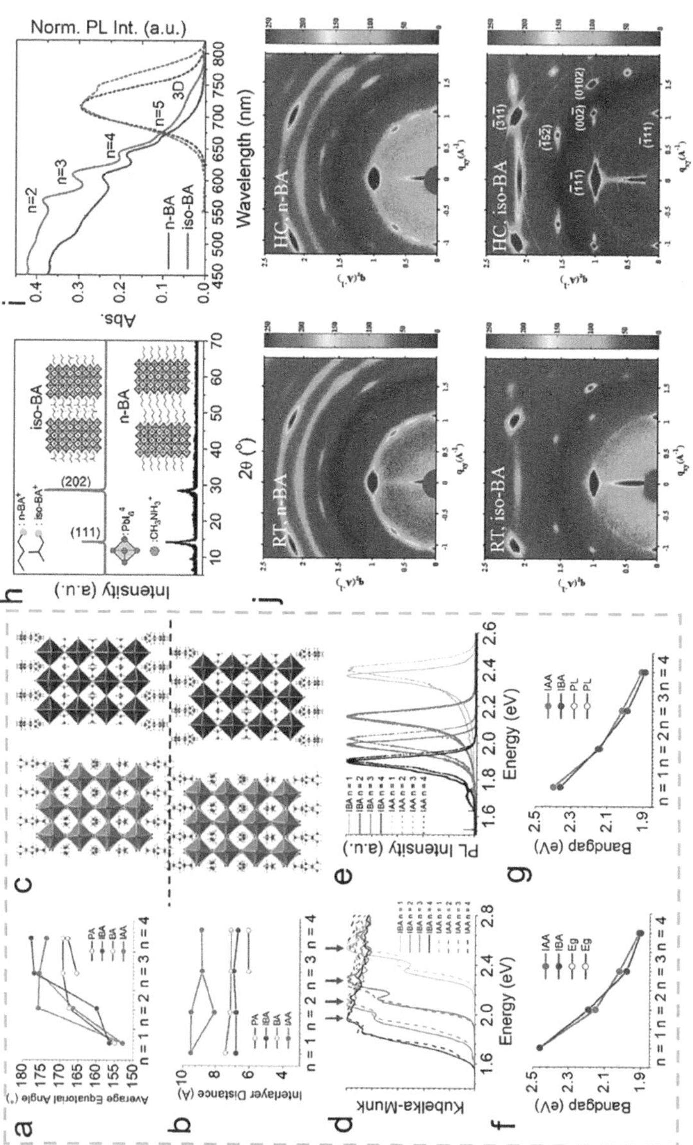

FIGURE 3.6 (a) The average equatorial angle of branched and straight-chain amine-based perovskites with different n; (b) interlayer spacing of branched and straight-chain amine-based perovskites with different n; (c) comparison of interlayer spacing between BA (left) and IAA (right), the thick line in the middle highlights the inability of branched-chain amines to cross; (d) the reflection spectrum of each system transformed by the Kubelka-Munk equation; (e) PL spectra for each system of IBA and IAA[54]; (f, g) comparison of band gap and PL of IBA and IAA systems, respectively; (h) XRD profiles of n-BA and iso-BA (IBA) perovskite films with corresponding chemical and crystal structure; (i) UV-visible absorption (solid) and PL (dashed) spectra of n-BA and iso-BA perovskite films obtained, which show characteristic exciton bands of 2D QW of different thicknesses[101]; (j) GIWAXS images of RT and HC(n-BA)$_2$(MA)$_3$Pb$_4$I$_{13}$, RT and HC(iso-BA)$_2$(MA)$_3$Pb$_4$I$_{13}$ perovskites[55].

reduction of the inorganic octahedral structural spacing due to short branching chains (Figure 3.6h, i). It is worth mentioning that IBA-based perovskite membranes have better crystallinity and orientation compared to BA class, even with room temperature (RT) deposition methods (Figure 3.6j). The results show that under the RT deposition method, the PCE of the 2D perovskite solar cell based on IBA is 8.82%, which is three times that of the BA-based perovskite solar cell (2.71%). The PCE of the 2D PSCs of IBA prepared by the hot casting method was further increased to 10.63%. Internal charge transfer or energy transfer between the QWs mediated by long-chain spacing cations has also been demonstrated in 2D perovskites[102–104]. Later studies of photoconductivity dynamics showed that IBA-based perovskites have a charge/energy transfer rate of about twice that of BA perovskites. In addition, IBA-based perovskite membranes have a more balanced charge transfer, which inhibits charge accumulation. Comparative studies of IAA and IBA films and devices show that IBA films of suitable length and volume have better performance and higher PCE than IAA. This suggests that the presence of branched chains shortens the inorganic layer spacing of 2D perovskites and potentially enhances light absorption and charge transfer.

3.3.1.3 Carbon Heterochain Series Monoamine Ligands

The incorporation of heteroatoms diversifies the properties of alkyl amines. The application of alkylamine heteroatoms in 2D perovskites was reviewed to optimize carrier transport performance[105,106,56,57]. On the one hand, the introduction of heteroatoms can change the interaction between molecules (between organic amines and inorganic layers, organic amines and solvents, respectively) and then affect the structure of perovskite and film quality (crystallinity, orientation, phase distribution, etc.). On the other hand, heteroatom alkyl amines can increase the dielectric constant of organic amines, thereby alleviating dielectric and quantum confinements. Recently, Huang et al. introduced 2-(methylsulfio)ethanamine hydrochloride (MTEACl) (sulfur-containing atoms) as an organic ligand to improve film quality and cross-layer charge transport[105]. They combined theory with experiments, aiming to enhance the interaction between the inorganic layers of 2D perovskites, resulting in high-quality perovskite films and high-performance PSCs. First, the S-S interaction was discovered by X-ray photoelectron spectroscopy (XPS) and X-ray absorption near-edge structure (XANES) spectroscopy of $(MTEA)_2(MA)_4Pb_5I_{16}$ perovskite films (Figure 3.7a, c), and then the S-S interaction was confirmed by energy and electronic structure calculations based on first-principles density functional theory (DFT; Figure 3.7d). S-S bond interactions promote charge transfer in 2D perovskites in many ways. On the one hand, it can stabilize and tighten the frame of the 2D perovskite phase, improve the vertical orientation, and promote vertical charge transport. On the other hand, this interaction also enhances the interlayer charge transport and further improves the charge mobility. High-quality $(MTEA)_2(MA)_4Pb_5I_{16}$ perovskite films were obtained, and the inorganic perovskite components in the crystal plane had strong preferential out-of-plane alignment. The PCE of this device is 18.06%, and the J_{SC} is 21.77 mA cm^{-2}, which is significantly higher than that of $(BA)_2(MA)_4Pb_5I_{16}$ device (PCE is 15.94%, J_{SC} is 20.25 mA cm^{-2}).

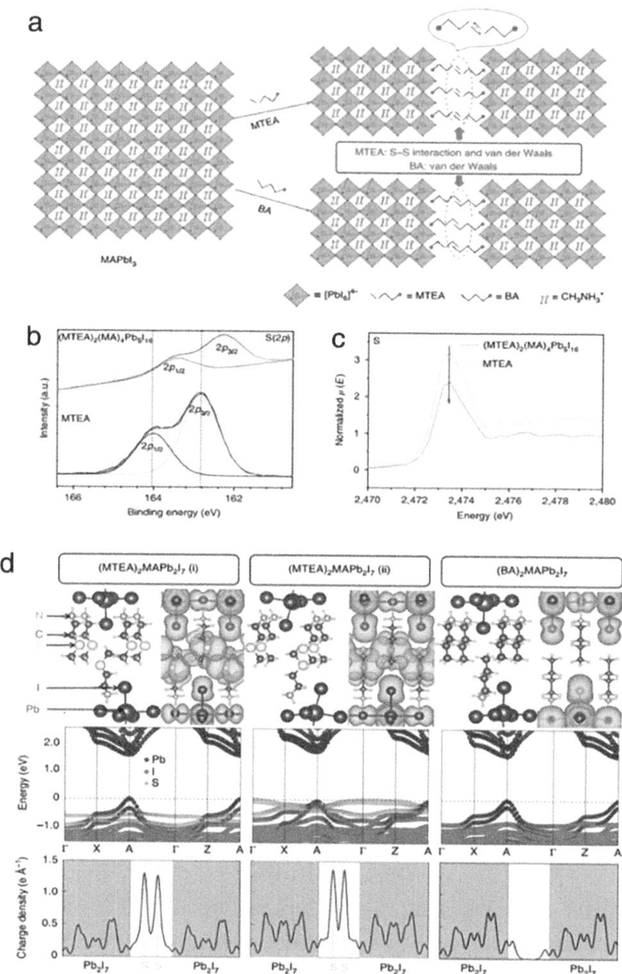

FIGURE 3.7 Schematic diagram of the crystal structure of (a) 2DRP perovskite $(MTEA)_2(MA)_{n-1}Pb_nI_{3n+1}$ and $(BA)_2(MA)_{n-1}Pb_nI_{3n+1}$; (b) XPS spectral comparison of MTEACl film and $(MTEA)_2(MA)_4Pb_5I_{16}$ perovskite film S (2p); (c) XANES spectral comparison of S elements in MTEACl films and $(MTEA)_2(MA)_4Pb_5I_{16}$ perovskite films; (d) first-principles DFT calculation of 2DRP perovskite $(MTEA)_2MAPb_2I_7$ and its counterpart $(BA)_2MAPb_2I_7$[105].

In addition, Xing et al. introduced hydrophobic polyfluorinated cations $(CF_3CF_2CH_2NH_3^+, 5F-PA^+$, Figure 3.8a) into 2D perovskites to clarify how heteroatoms (F) affect 2D structure and device performance[106]. Although the introduction of 5F-PA$^+$ can significantly improve the environmental stability of PSCs without sacrificing efficiency, 5F-PA$^+$, as an organic ligand, cannot form a 2D perovskite phase due to the large enthalpy of 2D perovskite formation of polyfluorine 5F-PAI. DFT calculations further confirm this. However, it is worth noting that 5F-PA$^+$ cations can form a 2D protective layer at the grain boundary of 3D perovskites (Figure 3.8b, c),

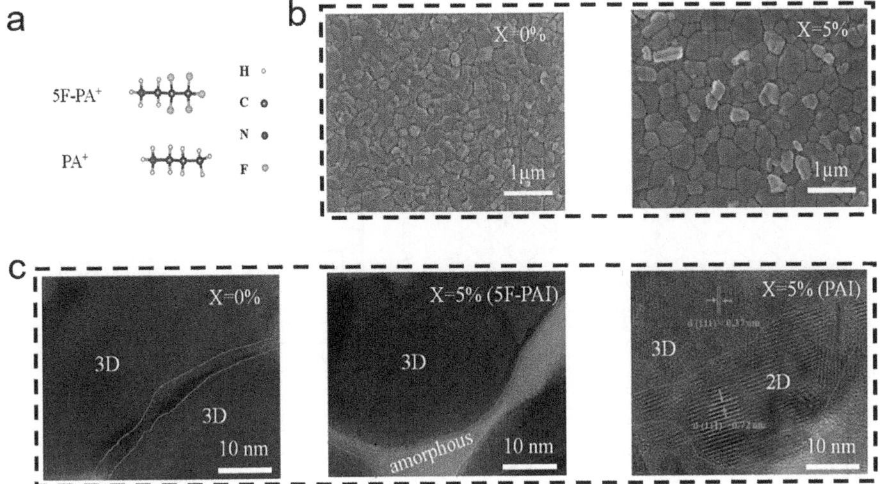

FIGURE 3.8 (a) Molecular structure of 5F-PA⁺ and PA⁺; (b) top view scanning electron microscopy (SEM) images of raw perovskites and 5% 5F-PAI incorporated into perovskites; (c) transmission electron microscopy (TEM) images of virgin perovskite, 5% 5F-PAI incorporated into perovskite and 5% PAI incorporated into perovskite[106].

which can largely prevent moisture corrosion of the perovskite layer and effectively inhibit non-radiative recombination. The heteroatoms in linear alkyl amines have multiple functions for 2D perovskites structure, and future research needs to explore more organic ligands with heteroatoms to achieve different functions of 2D perovskites.

3.3.1.4 Forces between Linear Monoamine Ligands

2DRP perovskites actually involve two charge transport channels: (i) along inorganic components; (ii) transport between inorganic $[PbI_6]^{4-}$ by bulky alkylammonium[107,108]. The introduction of insulating long-range large-volume organic amine ligands such as BA hinders the transfer of charge between layers. Therefore, many attempts have focused on the latter charge transport method, hoping to further improve the efficiency of charge transfer through cross-layer transfer. Therefore, alkylamine cations with conjugated structure have been introduced by many researchers because of their conspicuous charge transport in adjacent the QWs. It can be seen that in quasi-2D perovskite films, there are the QWs composed of different n-values (Figure 3.9b). The polydispersion of QW and the resulting non-flat energy landscape are expected to adversely affect the performance of photovoltaic materials. To this end, Proppe et al. prepared $n = 10$ perovskite membranes based on different organic amines by comparing allylamine (ALA) cations containing double bonds with two of the most commonly used organic amines[58] BA and phenethylamine (PEA; Figure 3.9a). The results show that compared with BA and PEA membranes, the number of low-n of the QWs of ALA-based perovskite membranes is greatly reduced, and the QWs distribution is narrower (Figure 3.9c, d). Furthermore, GIWAXS shows that ALA-based films with low n-valued films ($n = 3$) have a higher degree of orientation compared to

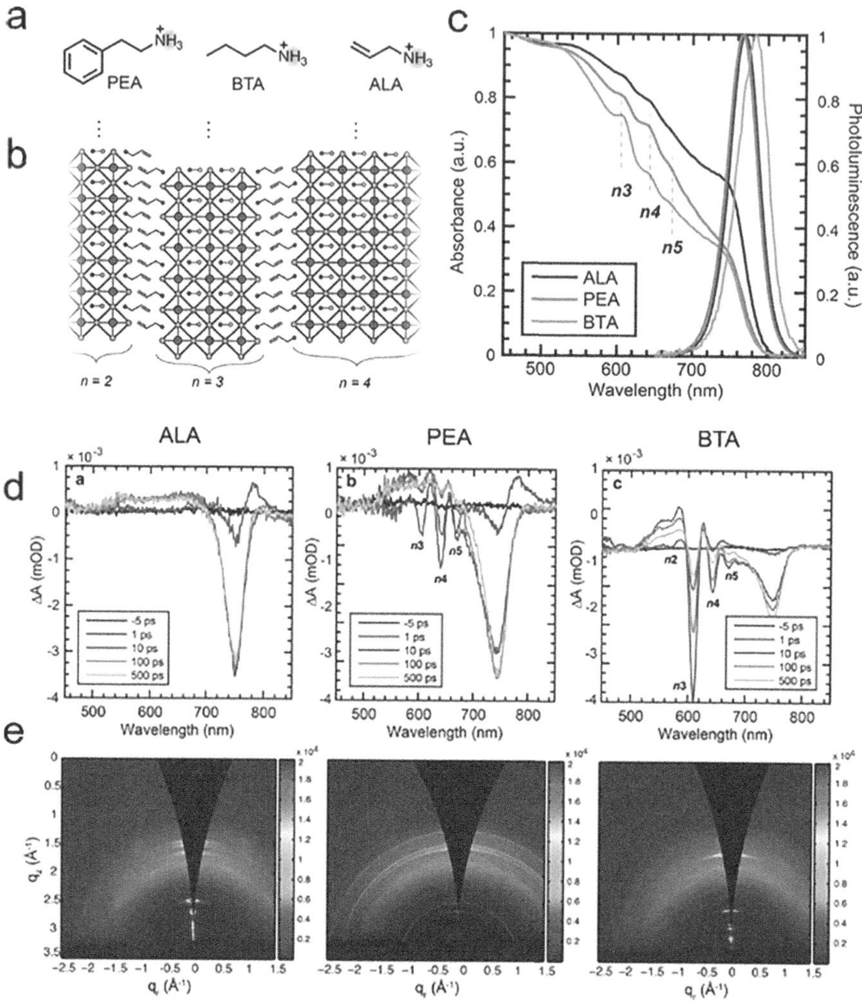

FIGURE 3.9 (a) Structure of ALA, PEA, and BTA organic ligands; (b) schematic diagram of a 2D perovskite QW where $n = 2$, $n = 3$, and $n = 4$ wells; (c) normalized absorption and emission spectra of $n = 10$ films using three different ligands to determine the exciton peaks of QWs $n = 3$, $n = 4$, and $n = 5$; (d) TA spectra for $n = 10$ ALA, PEA and BTA films at different delay times, respectively; (e) $n = 3$ GIWAXS images for ALA, PEA, and BTA, respectively[58].

BA and PEA (Figure 3.9e). It can be qualitatively assumed that the flat energy pattern based on ALA greatly improves carrier mobility. It is speculated that the higher n distribution in the ALA membrane is due to differences in intermolecular bonds at the QW interface. Since molecular interactions between ALA-based perovskite the QWs can only be enhanced by carbon-carbon double bonds, this may reduce the formation energy of low-n in the QWs, making the distribution of the QWs narrower and closer to the preset ratio. However, compared with the effect of conjugated organic amines on cross-layer charge transport, the ALA benchmark 2D perovskite membrane has a more prominent effect on the distribution of the QWs.

In addition, Chao et al. constructed 2DRP perovskites (Figure 3.10a) with 1-amino-3-butene hydrochloride (BEACl) and 3-butan-1-amine hydrochloride (BYACl) as short-course organic amine ligands instead of common butylamine hydrochloride (BACl)[59]. BEA and BYA with unsaturated bonds have different conjugation degrees and higher permittivity, which can theoretically promote the formation and transport of free carriers. At the molecular scale, BEA and BYA cations have shorter lengths (6.53 and 6.61 Å) compared to BA (6.75 Å) ligands (Figure 3.10b), enhancing the interaction between adjacent QWs, indicating a reduced charge transfer threshold. As shown in Figure 3.10c, the $(BEA)_2(MA)_3Pb_4I_{13}$ film exhibits dense, highly crystalline, pinhole-free characteristics. This may be mainly due to the fact that the BEA chain is shorter than the BA chain, which reduces steric hindrance during

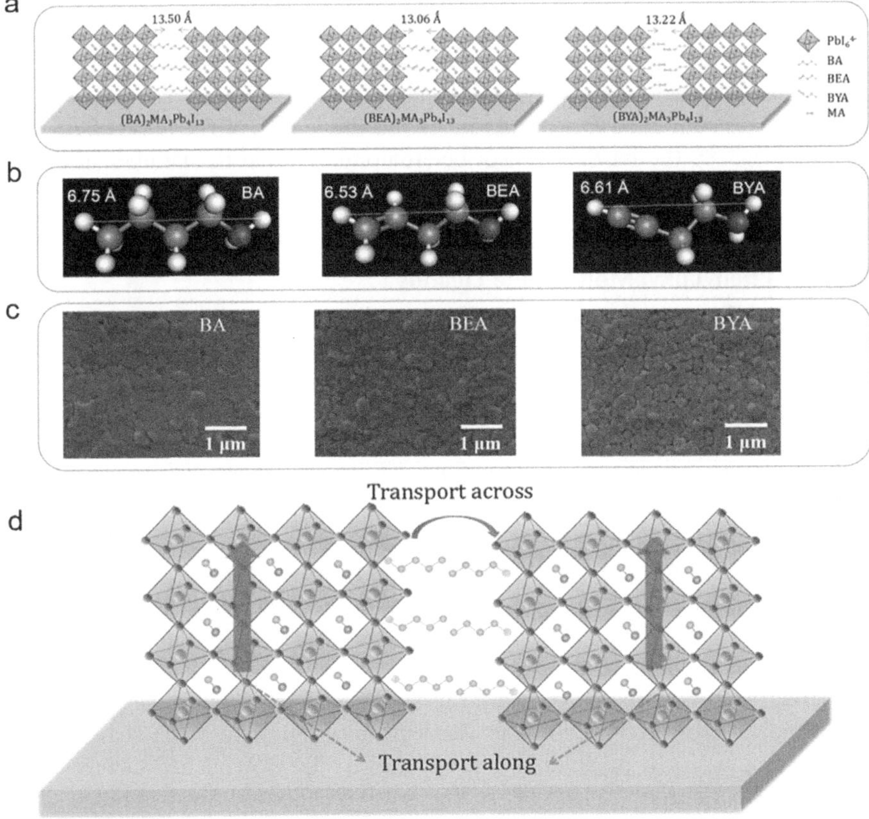

FIGURE 3.10 (a) Schematic diagram of 2DRP perovskites with BA, BEA, and BYA; (b) schematic diagram of the molecular structure of organic amines with BA, BEA, and BYA; (c) top view SEM image of perovskite films with BA, BEA, and BYA; (d) illustration of charge transport in 2DRP perovskite[59].

perovskite self-assembly and favors high-quality crystallization of 2DRP perovskite. The results show that compared with $(BA)_2MA_3Pb_4I_{13}$ (15.1%) and $(BYA)_2MA_3Pb_4I_{13}$ devices, the PCE of $(BEA)_2(MA)_3Pb_4I_{13}$ devices is 16.1%, and it has better long-term stability. Combined with the previous theory, it can be concluded that the use of conjugated amines can effectively improve the charge transport between inorganic layers (Figure 3.10d).

3.3.2 BENZENE RING SERIES MONOAMINE LIGANDS

In 2014, multilayer perovskites based on aromatic cationic PEA were first applied to PSCs[64], triggering a research boom in 2D PSCs. Unlike linear alkyl amines, aromatic compounds have a larger volume and a higher dielectric constant, which is thought to reduce dielectric and quantum confinement effects. PEA[+] and BA[+] are the two most commonly used spaced cations in 2D perovskites, but there are significant differences in structure. BA has a soft alkyl long chain that provides greater flexibility for the stacking of basic structural units[65]. The 2D perovskite based on the PEA spacer layer inhibits the slippage of the inorganic plate along the van der Waals gap, which is due to the rigid benzene ring in the PEA that limits the structural freedom, making the 2D structure more stable. We note that organic ligands containing benzene rings are converted from PEA, including the length of the alkyl chain of the benzene ring[60–62], the introduction of heteroatoms[70–75], and the number of benzene rings[62,69,109]. These spacer layers represent an extended direction to study properties such as 2D perovskite structure and carrier behavior.

3.3.2.1 Phenylalkyl Monoamine Ligands

Kamminga et al. intend to synthesize 2D perovskite single crystals with phenyl alkyl amines of different alkyl chain lengths[61]. The PL spectra of these single crystals clearly show that increasing the alkyl chain length of the ligand, i.e., from benzylamine to 4-phenylbutylamine (referred to as PMA, PEA, PPA, and PBA), results in a significant blueshift in the PL spectrum (Figure 3.11a). However, unlike alkyl amines, not all phenylalkylamine-based crystals structure are composed of octahedral inorganic layers that share angles. When the alkyl amine chain length is one or two carbons, a normal pure 2D perovskite structure can be formed. For aniline organic ligands, $(PMA)_2PbI_4$ exhibits significant interfacial octahedral distortion with a Pb-I-Pb bond angle of approximately 158°. In $(PMA)_2PbI_4$, the organic ligand increases the length of one carbon atom, reducing the Pb-I-Pb bond angle in the plane to about 153°, which means that octahedral distortion is greater due to the increase of alkyl chains compared to $(PMA)_2PbI_4$. Note that the alkyl amines on the benzene ring contain three or four carbons, giving up a non-shared angular structure (Figure 3.11b, c). Studies have shown that edge-sharing and surface-sharing adversely affect charge transport[50]. In this sense, 2D perovskites with edge-sharing and surface-sharing characteristics are not suitable for the manufacture of PSCs. So far, <100>-oriented single crystals based on long alkyl chain phenyl alkylamine (more than two carbons) have not been synthesized[110]. However, whether long alkyl phenylalkyl amine can be used to prepare low-dimensional polyimide remains to be explored. In addition, Du et al. also studied the effect of alkyl amine chains on

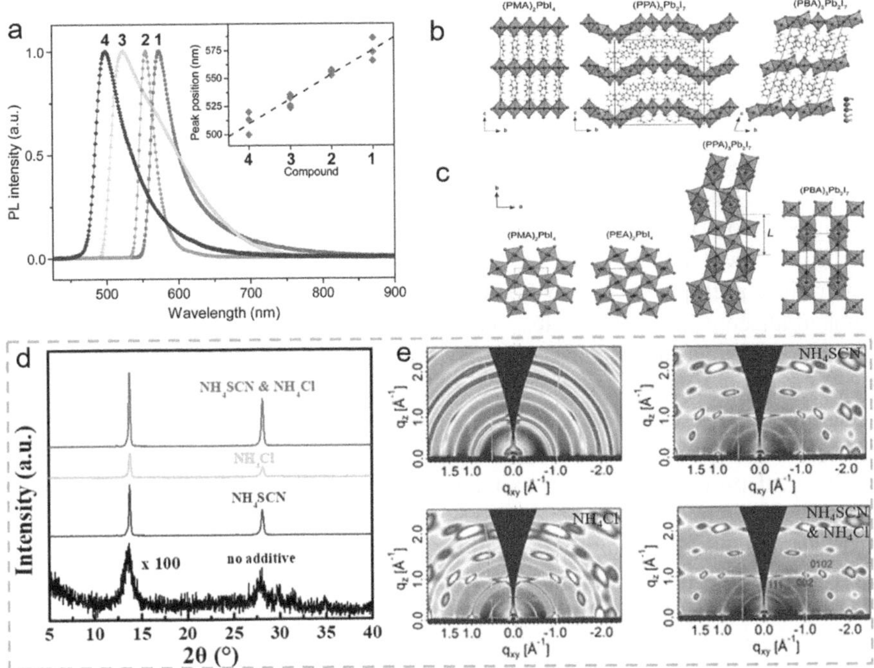

FIGURE 3.11 (a) Selected normalized PL spectra of layered lead iodide hybrids with ligands of increased length (1–4), compounds 1–4 refer to $(PMA)_2PbI_4$, $(PEA)_2PbI_4$, $(PPA)_3Pb_2I_7$, and $(PBA)_3Pb_2I_7$, the inset shows the PL peak position of multiple crystals, and the dotted trend line shows a blue shift with the increase of alkyl chain length; (b) 100 K polyhedral models of $(PMA)_2PbI_4$, $(PPA)_2PbI_4$, and $(PBA)_3Pb_2I_7$ projected in the [100] direction; (c) a polyhedral model of $(PMA)_2PbI_4$, $(PEA)_2PbI_4$, $(PPA)_3Pb_2I_7$ and $(PBA)_3Pb_2I_7$ projected in the direction of [001]; $(PMA =$ benzylamine, $PPA =$ 3-amphetamine, $PBA =$ 4-phenylbutylamine)[61]; (d, e) XRD and GIWAXS plots of PEA-based perovskite films made with various additives: no additives, NH_4SCN as additives, NH_4Cl as additives, NH_4SCN and NH_4Cl as additives[66].

benzene rings[62]. They introduced the concept of cation permeation, which is defined as the average distance between the nitrogen atom of an organic cation and the plane of the axial halogen atom of a perovskite sheet. Different carbon chain lengths on the benzene ring affect the penetration depth of organic amines, resulting in different degrees of $[PbI_6]^{4-}$ distortion, which in turn affects the optical properties of perovskite.

The preparation of perovskite films and devices by phenylalkylamines with different chain lengths is not as superior as that of alkylamine systems. Rodríguez-Romero et al. reported the synthesis of 2D perovskites using aniline as an organic ligand and its application in optoelectronic devices[60]. The use of conjugated aniline cations as organic ligands can significantly improve the charge transport performance of 2D perovskite, and the PCE of solar cells is higher than that of the BA group, indicating that conjugated cations are of great significance in improving the performance of

2DRP PSCs. In addition, aniline-based 2DRP polyvinyl chloride exhibits relatively low PCE due to its absorption layer being low n-value perovskite[63]. In addition to the saturated alkane branch on the benzene ring, phenylalkyl amines containing the unsaturated alkyl branch have also been developed and used in 2D perovskites and devices[67,68]. The crystal structure shows that there are multiple hydrogen bond interactions between PPeA and the inorganic layer due to the presence of double bonds in the alkyl chain, stabilizing the structure of the 2D perovskite. At the same time, the introduction of FA inhibits the distortion of the inorganic layer and reduces the band gap of the 2D perovskite, thereby improving the optical properties. During film preparation, multiple hydrogen bonds between PPeA and inorganic layers help to connect adjacent inorganic plates, thereby inhibiting crystal growth parallel to the substrate and tending to be vertical. By adjusting the perovskite composition, the $(PPeA)_2(Cs_{0.05}(FA_{0.88}MA_{0.12})_{0.95})_3Pb_4(I_{0.88}Br_{0.12})_{13}$ perovskite solar cell device obtained a champion PCE of 14.76% under RT spin coating conditions. This study further highlights the important influence of conjugated structure on the structure and physicochemical properties of perovskite. In addition to capturing phenylalkylamines of different chain lengths, 2D perovskites using phenylalkylamines containing different benzene ring numbers as ligands have been reported. Xu et al. developed polycyclic aromatic ammonia, 1-naphthalenemethylethylamine (NpMA), and 9-anthracenemethylethylamine (AnMA) as ligands to prepare 2D PSCs ($n = 4$), and the PCE increased by 17.25% and 14.47%, respectively[69]. It was confirmed that there was a strong hydrogen bond between polycyclic aromatic ammonia and the inorganic layer, which enhanced the stability of the 2D polystyrene composite. High-resolution transmission electron microscopy (HRTEM) showed that 2D lamellar and 3D phases coexisted and were randomly distributed in the film. Ultrafast transient absorption (TA) spectroscopy confirms the ultrafast exciton migration process (within 7 ps) between the two phases. This provides a reasonable explanation for the high efficiency of 2D PSCs. In addition, studies have also shown that quaternary amine ions can be used as passivation components of 3D perovskites to improve the performance of PSCs[111–113]. Inspired by this script, Dong et al. developed a novel anilinequaternary amine ion phenyltrimethylamine (PTA) as an organic ligand for the preparation of 2D perovskites[83]. By adding MACl to adjust the crystallization and orientation of the film, the PCE of $(PTA)_2(MA)_3Pb_4I_{13}$ PSCs was increased by 11.53%, mainly due to the better charge transfer ability and low-defect-induced recombination in the device. At present, the research of p-phenylalkylamine mainly focuses on PEA organic ligands, and by optimizing the in-plane growth of perovskite particles, the PCE of PEA-based 2DRP PSCs is increased to 14.1% (Figure 3.11d, e)[66], and the highest efficiency of PEA-based 2D PSCs reaches 18% by introducing vacuum polarization treatment and strengthening nucleation during crystallization[114]. The application of phenylalkyl amines with different alkyl chain lengths and different benzene ring numbers in perovskite films and devices requires further systematic research.

3.3.2.2   Benzene Derivatives Alkyl Monoamine Ligands

Heteroatom-containing aromatic cations (such as O^{71}, $F^{70–74,115–118}$, $Cl^{72,115,116,119}$, $Br^{115,116,119}$, $I^{115,119}$, etc.) have important effects on the structure or physicochemical properties of 2D perovskites and have been directly or indirectly characterized in various reports. For example, in the crystal structure of $n = 1$ phase 2D perovskites, electron-absorbing halogens in organic cations were found to reduce the distortion of inorganic sheets, thereby greatly improving UV-visible absorption (Figure 3.12a–c)[72]. Note that spaced cations with greater polarity are thought to reduce quantum and dielectric confinement in 2D perovskites[120]. Halogen atoms with high electronegativity have high dipole moments due to their strong ability to absorb electrons, which reflects their molecular polarity and promotes the transfer of charge within the molecule. Therefore, the introduction of halogens into aromatic ligands reduces the E_b of 2D perovskites, which can effectively improve the dissociation and transport capacity of carriers. In addition, the halogens in aromatic ligands can also improve the crystallinity and orientation of 2D perovskite grains, thereby further enhancing charge transfer.

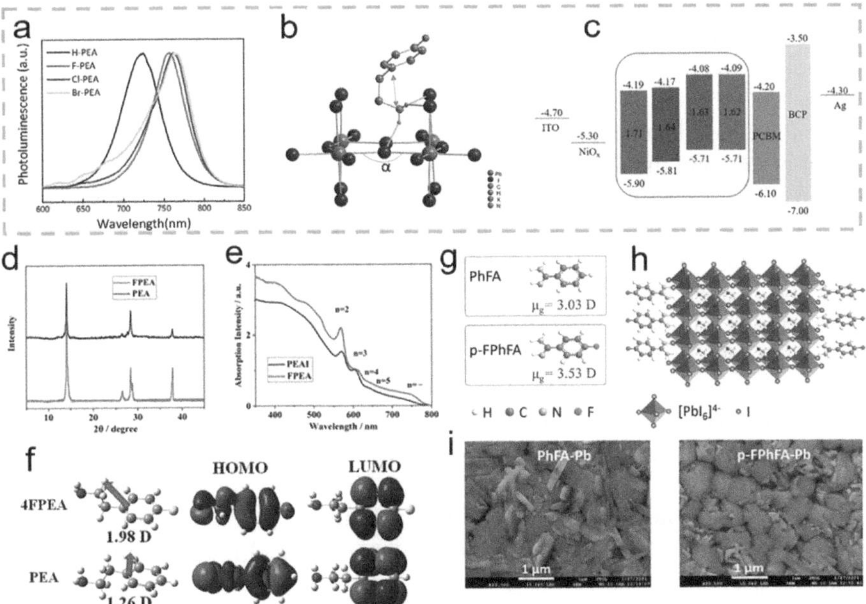

FIGURE 3.12 (a) Steady-state PL spectra of H-PEA(PEA) and X-PEA (X = F, Cl, Br) RP perovskite films with $n = 3$ on a glass substrate; (b) schematic diagram of the structure of X-PEA$_2$PbI$_4$ (X = F, Cl, Br) with Pb-I-Pb angles and intramolecular hydrogen bonds; (c) band alignment of H-PEA$_2$PbI$_4$- and X-PEA$_2$PbI$_4$ (X = F, Cl, Br)-based PSCs, from left to right: H-PEA$_2$PbI$_4$, F-PEA$_2$PbI$_4$, Cl-PEA$_2$PbI$_4$, and Br-PEA$_2$PbI$_4$[72]; (d, e) XRD patterns and UV-vis absorption spectra of perovskite films based on PEA$_2$MA$_4$Pb$_5$I$_{16}$ (black line) and (F-PEA)$_2$MA$_4$Pb$_5$I$_{16}$ (red line)[74]; (f) Calculate the molecular dipole moment of the 4FPEA cation and its electron density distribution in the lowest unoccupied molecular orbital (LUMO) and highest occupied molecular orbital (HOMO) compared to PEA cations, with white, blue, gray, and black spheres responding to fluorine, nitrogen, carbon, and hydrogen atoms, respectively[73]; (g) chemical structure of PhFA and p-FPhFA molecules and calculated molecular dipole moments; (h) schematic diagram of the structure of p-FPhFA-Pb 2DRP perovskite ($n = 5$); (i) top view SEM image of 2DRP perovskite PhFA-Pb and p-FPhFA-Pb ($n = 5$) perovskite films[76].

In addition, Huang's team introduced 3-bromobenzyl iodide amine (3BBAI, Br is heteroatomic) as an organic ligand to prepare 2D PSCs, and the PCE of the obtained solar cells reached an impressive 18.2% and had long-term stability[75]. The researchers believe that the introduction of 3BBAI first induces the directional growth of low-n-phase perovskites at the bottom perpendicular to the substrate and then orderly crystallizes large n components on top of the film. The distribution characteristics of large N perovskites in the upper layer and small N components in the vertical orientation of the bottom layer are conducive to the capture of solar photons and the transport of photogenerated carriers[51,94]. This achievement can be attributed to the high electronic order and excellent crystallinity of heteroatoms in 3BBAI organic ligands. Halogens are used as heteroatom organic ligands of aromatic hydrocarbons, and F atoms are the most mentioned. Fluorination of aromatic organic ligands has proven to be an effective molecular design strategy, which can not only change the energy level of 2D perovskites[121,122] but also promote charge dissociation and transfer in 2DRP PSCs[123,124], thereby improving the PCE of devices[125]. For example, Zhang et al. used 4-fluorophenethylamine (4F-PEA) to prepare additive-free 2D PSCs[74], with a PCE of 13.64%. Shao and colleagues then used NH4SCN as an additive to further increase the PCE of 4F-PEA-based 2DRP PSCs to 17.3%[73]. Both studies showed that 4F-PEA-based membranes have higher crystallinity, larger grain size, and better vertical orientation than PEA-based 2D perovskite control membranes, resulting in low trap density and efficient charge transport (Figure 3.12d, e). Combining DFT calculations (Figure 3.12f), dielectric constant (ε) measurements, and Kelvin probe force microscopy (KPFM), the authors confirmed that the substitution of F in PEA cations essentially causes a strong molecular dipole moment for charge separation, which fundamentally reduces the dielectric confinement effect, thereby reducing the E_b and improving more efficient charge transfer between adjacent the QWs. Recently, Li et al. reported that the aromatic formamidine ligand benzamidine (PhFA) and its fluoropyridine p-fluorobenzamidine (p-FPhFA) were used to construct 2D perovskites for solar cells (Figure 3.12i)[76]. The effects of ligand fluorination on the quality, photophysical, and photoelectric properties of 2D perovskite films were systematically studied. This indicates that the fluorinated p-FPhFA ligand has a higher molecular dipole moment, which DFT calculations confirm (Figure 3.13a), resulting in a lower E_b for 2D perovskites, facilitating efficient charge separation compared to PhFA-based perovskites. In addition, the 2D perovskite membrane of the p-FPhFA spacer exhibits higher quality characteristics vertically than the PhFA spacer (Figure 3.13b). The authors believe that this favorable factor is due to the high dipole moment of p-FPhFA, which enhances the interaction between the inorganic plate and the spacer cation. Finally, high-quality perovskite films inhibit non-radiative recombination, greatly improving carrier mobility and lifetime. The PCE of p-FPhFA-based solar cells ($n = 5$) is 17.37%, which is much higher than that of PhFA-based devices (12.92%).

In addition to the different types of heteroatoms in the aromatic ligand, the position of substituting H also has a great influence on the structure and properties of the prepared films and devices. Hu et al. reported that selective monofluorination of aromatic groups (2-fluorophenethylamine (oF1PEA), 3-fluorophenethylamine (mF1PEA), and 4-fluorophenethylamine (pF1PEA)) in the phenylethylamine

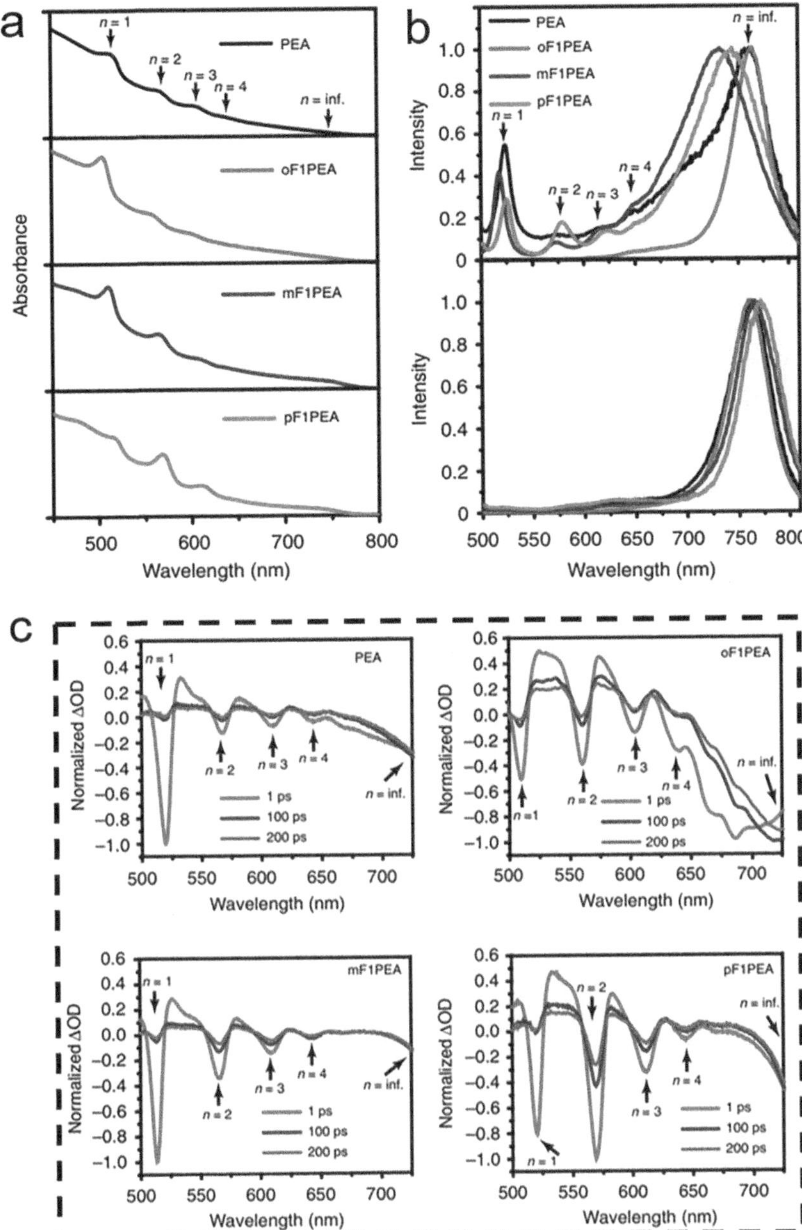

FIGURE 3.13 (a) UV-visible absorption of 2D perovskite films of PEA, oF1PEA, mF1PEA, and pF1PEA; (b) normalized PL spectra of the reverse (glass) side and front (air) side of the 2D perovskite family of PEA; (c) TA spectra of 2D perovskite films of PEA, oF1PEA, mF1PEA and pF1PEA[70].

molecule can significantly affect the efficiency of photovoltaic devices[70]. In particular, the PL and TA spectra show that fluorine atom substitution at different locations results in different phase distributions in the corresponding perovskite membranes, as shown in Figure 3.13c. Perovskite membranes based on PEA, mF1PEA, or pF1PEA have a continuous phase distribution from the top high n phase to the bottom low n phase, which is thought to favor energy transfer (and/or charge transfer)[46,81,126,127]. However, films based on oF1PEA enable random phase distribution and disordered energy landscape, resulting in poor PV device performance. In addition, the three organic ligands changed different surface morphology and crystal orientation. mF1PEA- and pF1PEA-based perovskite films exhibit homogeneity, are pinhole-free, and have high grain orientation. The perovskite film based on oF1PEA is of poor quality. The results show that pF1PEA-based 2D PSCs have the highest efficiency, more than 10%, followed by mF1PEA (more than 10%), PEA (more than 7%), and oF1PEA (less than 1%). The position of fluorine atoms on PEA (ortho, meta-, and para-elevation) causes changes in the accumulation of large organic cations, accompanied by varying degrees of structural disorders. Through the single crystal structure analysis and DFT calculation of the correlated $n = 1$ 2D perovskite, the superposition law of these substituted organic cations was established. It was found that the differences in the formation energy of different perovskite compounds were well correlated with the changes in the accumulation and disorder of spaced cations, which were caused by organic cation interactions (such as dipole-dipole interactions), which was the root cause of the differences observed on photovoltaic devices in these different PEA cation-based 2D perovskite films. In addition to halogen atoms, substitutions of other groups have also been studied. For example, Fu et al. used 4F-PEA with hydrophobic fluorine atoms and 4-methoxyphenethylamine (MeO-PEA) containing larger substituents as organic ligands to construct 2D perovskites[71]. The results showed that both larger cations and more hydrophobic cations could improve the moisture resistance of perovskites, especially for MeO-PEA-based perovskites. However, larger cations destroy the accumulation between organic cations, resulting in poor crystallinity and weak perovskite grain orientation, which adversely affect the performance of the device. Thus, 2D PSCs ($n = 5$) based on 4F-PEA achieved a 14.5% improvement in PCE, while MeO-PEA-based devices showed only 9.9% efficiency. In addition to the effects on film morphology and grain crystallization, aromatic cations with conjugated structure are more widely used in the interlayer transfer of charge of 2D perovskites, the weakening of the dielectric confinement effect, and the reduction of E_b. Therefore, the optimization of 2DRP perovskites with aromatic cations of different heteroatoms is an effective way to realize high-performance PSCs, so the further large-scale expansion of this material and device is an expanding topic.

3.3.3 HETEROCYCLIC MONOAMINE LIGANDS

The successful application of phenyl organic amines has led to the study of other aromatic organic amines. However, research on aromatic heterocycles in 2D perovskites has focused mainly on DJ type perovskites (such as piperidine derivatives), which will be discussed in more detail in the next section.

Research on heterocyclic organic amines in RP type PSCs is relatively limited, mainly focusing on the use of thiophene-containing organic amines. Low-thiothiophene derivatives are thought to alter the band arrangement, thereby affecting the optical properties of 2DRP perovskites, which have been confirmed by numerous studies[128–130]. Due to the special interaction between the electronic structure of the thiophene ring and the electronic structure of the inorganic perovskite layer, researchers have made several attempts to introduce thiophene derivatives into 2D PSCs.

In 2018, Chen et al. first introduced a thiophene derivative (2-thiophenemethylamine (ThMA$^+$)) as an organic ligand to develop a novel 2D perovskite[77]. MACl-assisted crystallinity and crystalline orientation growth technology were used to obtain ThMA$^+$-based 2D perovskite films with nanorod-like morphology. The crystal size of these films is significantly increased, the out-of-plane orientation is enhanced, and the carrier mobility is large and balanced. The results show that the PCE of (ThMA)$_2$(MA)$_2$Pb$_3$I$_{10}$ PSCs is 15.42%, and it has long-term stability. It is particularly well known that the partial or complete replacement of MA with formamidine (FA) reduces the band gap of 3- and 2D perovskites and improves light absorption, thereby achieving high photocurrent density and high efficiency[131–136]. Inspired by this, Chen and colleagues continued to study ThMA-based 2DRP perovskites, replacing MA with larger FAs to improve light absorption[78]. The PCE of ThMA-based 2D PSCs increased from 15.42% to 16.18%. In this work, the use of 4-(trifluoromethyl)benzyl iodide amine (tFM-PMAI) organic salt-assisted crystal growth technology effectively improved the quality of the film (Figure 3.14 a–c). This optimization is mainly due to the co-regulation of the crystallization process of IPA and tFM-PMAI aligned with 2DRP perovskite. It is worth noting that with the increase of carrier mobility, the trap density and non-radiative composite loss of quasi-2DRP perovskites are effectively suppressed. Therefore, the PCE of the optimized device of (ThMA)$_2$(FA)$_4$Pb$_5$I$_{16}$ is further increased to 19.06%.

Qin et al. prepared 2DRP perovskite materials and solar cells using another thiophene ligand, 2-thiopheneethanamine (TEA)[79]. Coordination engineering using single-crystal derived precursors to control the phase purity and arrangement of 2D perovskite films is different from previous studies using additives. In this study, dimethylacetamide (DMAc) with different co-solvents was used as a solvent to dissolve 2D perovskite single crystals, thereby regulating the growth of high-quality films. The coordination capacity of different co-solvents significantly affects the interaction between the 2DRP perovskite unit and DMAc in the precursor, thereby determining the properties of the colloidal precursor and the quality of the final 2D perovskite membrane. The results show that the 4:1 mixed solvent based on dimethylformamide (DMF): benzene and 8vol% HI additive has high crystallinity and grain orientation, and the PCE reaches 14.68%. Similarly, Yan et al. also researched the preparation of 2DRP perovskites based on TEA organic ligands[80]. It is proposed that the TEA-based 2D perovskite matrix is embedded in the 3D phase, and a 2D/3D heterojunction structure is spontaneously formed. As shown in Figure 3.14 d–f, the spectra of TEA-based 2D perovskite films provide convincing evidence for the existence of 3D phases, and TEA-based 2D perovskite films exhibit more pronounced 3D-phase perovskite characteristics compared to PEA. The spatial distribution of

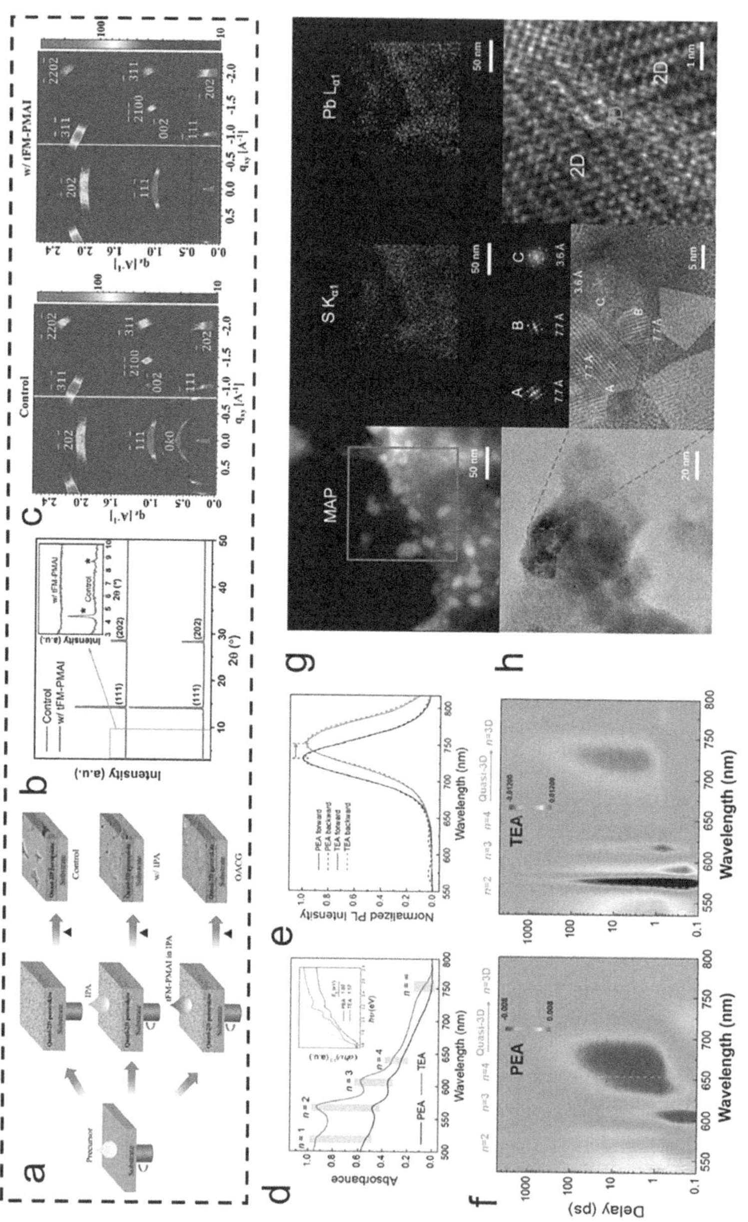

FIGURE 3.14 (a) Procedure for preparing quasi-2DRP perovskite films using different processing methods; XRD patterns of (b) $(ThMA)_2(FA)_4Pb_5I_{16}$ control and OACG-treated quasi-2DRP perovskite films[78]; (c) $(ThMA)_2(FA)_4Pb_5I_{16}$ control and GIWAXS images of OACG-treated quasi-2DRP perovskite films[78]; optical absorption (d) and PL spectra (e) of pure PEA (black) and TEA (red) films ($n = 4$) and TA spectra (f) of their PEA and TEA samples; (g) STEM image and corresponding EDS element mapping showing the distribution of sulfur (S) and lead (Pb) in $(TEA)_2(MA)_3Pb_4I_{13}$–based perovskite ($n = 4$); transmission electron microscopy and magnified HR-TEM image of (h) $(TEA)_2(MA)_3Pb_4I_{13}$–based perovskite[80].

2D and 3D phases in TEA-based 2D perovskites was studied by scanning transmission electron microscopy (STEM) and corresponding energy-dispersive X-ray spectroscopy (EDS). High-resolution transmission electron microscopy (HR-TEM) imaging further clarifies the relative spatial organization of the prepared membrane (Figure 3.14 g, h). The formation of this heterojunction structure is thought to be due to the strong interaction between Pb and S that causes the 3D phase to compress epitaxial growth at the grain boundary of the 2D phase, resulting in a longer exciton diffusion length and carrier lifetime than PEA analogues, and the PCE of the 2D perovskite solar cell of $(TEA)_2(MA)_3Pb_4I_{13}$ is 11%. Studying the interaction between heteroatoms and inorganic perovskite networks in organic amines has important guiding significance for the design of new organic amines. In addition to the study of thiophene with different alkyl chain lengths, Dong et al. designed a novel thiophene-containing organic amine-2-thiophenecaramidine (ThFA) and prepared 2D perovskite signal crystals and films to explore their physicochemical properties[82]. The precursor organic salt-assisted crystal growth (PACG) technology was used to deposit the prepared 2D perovskite film with IPA solution (ThFAI, MAI, and MACl mixed in the same proportion) as an antisolvent, which increased the PCE of PSCs from 7.23% to 16.72%. The fundamental reason for this is that the PACG-treated film exhibits significantly improved film quality, excellent carrier mobility, and low well-state density. Organic amines on dithiophene ethylamine (BTEA) polythiophene rings have also been studied[81]. This chapter mainly introduces a new back annealing technique for the preparation of 2D perovskite films with vertical phase separation to extend carrier diffusion length, the structural characteristics of which are not fully summarized.

In addition to thiophene derivatives, some other heterocyclic derivatives are also used in 2DRP perovskites. For example, Li et al. used 4-(aminoethyl)pyridine (4-AEP) as an organic ligand to prepare 2DRP perovskite films and solar cells to explore their specific characteristics[84]. The results showed that only amino groups were successfully protonated, while nitrogen in pyridine was not successfully protonated. Therefore, 4-AEP is considered to eventually form RP-phase 2D perovskite, however, further crystallographic characterization is still required for the final structure. Perovskite films of $(4\text{-}AEP)_2MA_{n-1}Pb_nI_{3n+1}$ (n = 1, 3, 4, and 5) and $(PEA)_2MA_4Pb_5I_{16}$ (control group) were prepared, and the effect of 4-AEP was discussed. Before and after annealing, there was no significant change in the color of PEA-based 2D perovskite film, indicating that $(PEA)_2MA_4Pb_5I_{16}$ crystallization rate was faster. However, $(4\text{-}AEP)_2MA_4Pb_5I_{16}$ presents a uniform and transparent film before annealing and gradually turns brown during the annealing process, indicating that 4-AEP may play the role of slow crystallization of 2D perovskite. It is speculated that nitrogen atoms on the pyridine group cooperate with Pb^{2+} to form a colorless amorphous intermediate, delaying the crystallization of 2D perovskite, and this hypothesis is verified by Fourier transform infrared spectroscopy (FTIR) and XPS. This delayed crystallization eventually led to the production of high-quality perovskite films. The results show that the PCE of the solar cells prepared with $(4\text{-}AEP)_2MA_4Pb_5I_{16}$ is 11.68%, and it has long-term stability. Li et al. prepared a high n-value 2DRP perovskite (n = 9) film using 4AEP as an organic ligand, and the ligand optimized perovskite crystallization[85]. The results show that the optimized device efficiency is 9.05% and the stability is improved, which is mainly attributed to the intermolecular interaction between 4-AEP molecules through hydrogen bonding bridging charge transport grooves, which enhances the structural stability.

In addition to the aforementioned studies, the study of aromatic heterocyclic organic amines in 2DRP perovskites can pay more attention to the intermolecular interactions caused by heteroatoms in the aromatic ring, such as the interaction of adjacent bilayer organic ligand layers, the interaction of heteroatoms with inorganic perovskite networks, and the interaction of heteroatoms with solvents. The phase distribution is adjusted on the basis of controlled crystallization to stabilize the structure.

3.4   2DDJ TYPE PEROVSKITES

In recent years, 2DDJ phase perovskites have developed into promising materials for solar cell applications, and in just a few years, the PCE has risen sharply from 7.32% to more than 17%[36], comparable to 2DRP phase perovskites. Compared to 2DRP perovskites with two monovalent spaced cations, the 2DDJ structure contains one divalent organic amine ion, and each organic cation can connect two adjacent inorganic layers, which eliminates the van der Waals gap between the two monovalent cations and enhances the interaction between adjacent inorganic layers. From a crystallographic point of view, the feathery organic plates of 2DDJ phase perovskites are displaced relative (1/2, 0 shifted) or completely unshifted (0, 0 shift) along the ab surface, which is different from the RP phase (1/2, 1/2 shift). 2DDJ perovskites are expected to surpass 2DRP type perovskites and become candidates for efficient and stable PSCs[36]. However, 2DDJ type perovskites also encounter quantum and dielectric confinement effects and multidimensional the QWs films, which make exciton dissociation and transport difficult. Fortunately, due to its unique structure, 2DDJ type perovskites have special physical and chemical properties, which open the way for the application of organic diamine ligands in highly efficient and stable PSCs. There are few organic amines for the preparation of 2DDJ type perovskite, mainly alkyldiamine, piperidine derivatives and pyridine derivatives, phenylenediamines, thiophene derivatives, etc., as shown in Figure 3.15, which will be detailed in this section.

3.4.1   LINEAR DIAMINE LIGANDS

There are quantum and dielectric confinement effects in typical layered 2D perovskites, which are due to the high conductivity of inorganic layers isolated by low dielectric constant organic cations with large insulation volumes, resulting in a significant decrease in the performance of photovoltaic devices. As mentioned in the RP type alkylamine section earlier, improving interlaminar charge transport can effectively improve this situation. Unlike the 2DRP type perovskites claimed to increase the dielectric constant of organic amines to reduce dielectric confinement, DJ PSCs with alkyldiamine cations also achieve charge transport by shortening the interlayer spacing through molecular design. Ma et al. use smaller propane 1,3-diamine (PDA) cations to replace larger insulating organic cations (e.g., BA) to shorten the distance

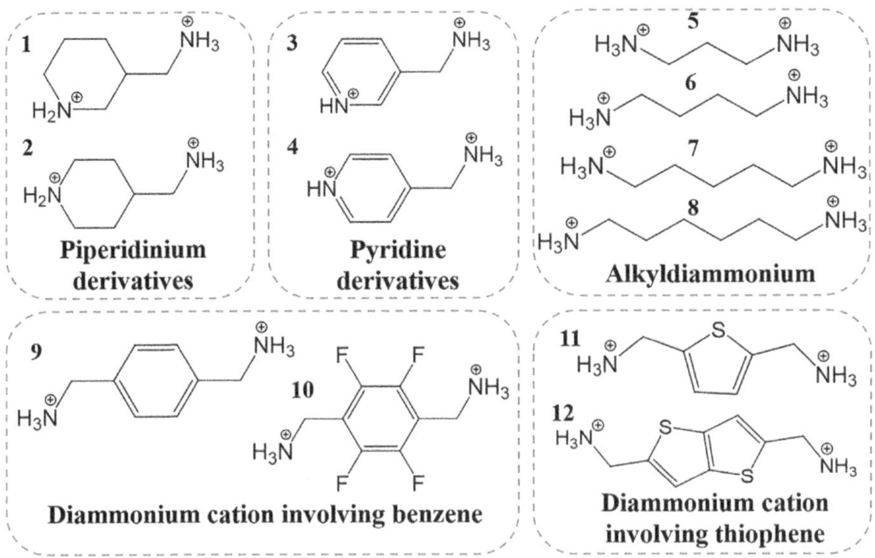

FIGURE 3.15 Organic ligands for 2DDJ PSCs. 1: 3-(Aminomethyl)piperidinium[44,137–139]; 2: 4-(Aminomethyl)piperidinium[44,138,140]; 3: 3-(Aminomethyl)pyridine[141]; 4: 4-(Aminomethyl)pyridine[141]; 5: propylenediamine[142,143]; 6: Butylenediamine[144–146]; 7: pentanediamine[144,145]; 8: Hexamethylenediamine[144,145]; 9: 1,4-Phenylenedimethylamine[147–150]; 10: 2,3,5,6-Tetrafluoro-1,4-phenylenedimethylamine[151,140]; 11: 2,5-Thiophenedimethylamine[152]; 12: Thiophene[3,2-b]thiophene-2,5-dimethylamine (TTDMA)[153].

between inorganic layers, thereby weakening the quantum confinement effect[142]. The (0k0) reflection peak in the X-ray diffraction (XRD) spectrum can reflect the characteristics of 2D perovskites and can distinguish different n values of 2D perovskites. Based on the different n values of the perovskites and the size of the inorganic octahedral (6.2 Å), it can be determined that the distance between adjacent inorganic layers of PDMA-based 2D perovskites is 2 Å (Figure 3.16a), which is significantly shorter than the van der Waals distance (4 Å), resulting in strong interactions between adjacent layers of perovskite. Electrochemical impedance spectroscopy and carrier behavior measurements further confirm a significant enhancement in charge transport. In the end, $PDAMA_3Pb_4I_{13}$ obtained the champion PCE of more than 13% and greatly improved stability. Similarly, Ahmad et al. used PDAs as organic ligands to deploy the crystallinity, the QWs-width distribution, and charge mobility of 2DDJ perovskites[143]. Due to the stable structure and hydrophobicity of $(PDA)MA_{n-1}Pb_nI_{3n+1}$, solar cells based on PDA organic ligands can maintain long-term stability under 65% humidity, continuous illumination, and high-temperature conditions. The alkylamine chain length of the 2D perovskite formation has a great influence on the structural stability, morphology, and device characteristics of the prepared membrane. Currently, PDA is considered to be the smallest alkyldiamine that can form DJ type perovskites structure. However, data on the exact crystal structure (PDA) $MA_{n-1}Pb_nI_{3n+1}$ have not been reported.

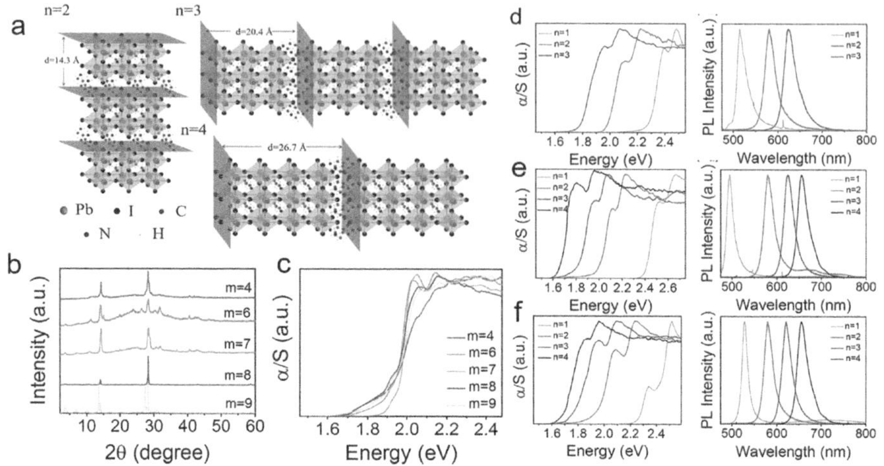

FIGURE 3.16 (a) Diffraction plane diagram of $PDAMA_{n-1}Pb_nI_{3n+1}$ perovskite film with $n = 2$–4[142]; (b, c) XRD and optical absorption spectra of $(NH_3C_mH_{2m}NH_3)(CH_3NH_3)_2Pb_3I_{10}$ ($m = 4, 6, 7, 8,$ and 9) films on glass substrates; (d) light absorption spectra and PL spectra of $(NH_3C_7H_{14}NH_3)(CH_3NH_3)_{n-1}Pb_nI_{3n+1}$ ($m = 7, n = 1$–3); (e) light absorption spectra and PL spectra of $(NH_3C_8H_{16}NH_3)(CH_3NH_3)_{n-1}Pb_nI_{3n+1}$ ($m = 8, n = 1$–4); (f) light absorption spectra and PL spectra of $(NH_3C_9H_{18}NH_3)(CH_3NH_3)_{n-1}Pb_nI_{3n+1}$ ($m = 9, n = 1$–4)[144].

Recently, Kanatzidis' team reported $(NH_3C_mH_{2m}NH_3)(CH_3NH_3)_{n-1}Pb_nI_{3n+1}$ ($m = 4$–9; $n = 1$–4), a series of 2DDJ type perovskites, which were successfully synthesized using complementary solution methods and solid milling methods, depending on chain length and solubility[144]. Organic diamine ligands of different chain lengths exist in the crystal structure in a folded conformation, making their interlayer distance shorter than that of linear chains. Figure 3.16d–f shows the UV-vis absorption spectra and PL spectra of DJ type perovskites with alkyldiamine as organic ligands, showing different optical properties identified by different chain lengths. In addition, perovskite films based on different organic amines were prepared and their optical properties were characterized (Figure 3.16b, c). Compounds with $m = 8$ ($n = 3$) exhibit excellent photovoltaic properties.

Subsequently, Zheng et al. studied alkyldiamines with different chain lengths[145], namely 1,3-propanediamine (PDA), 1,4-butanediamine (BDA), 1,5-pentanediamine (PeDA), and 1,6-hexanediamine (HDA). The effect of large organic amine ligands with different chain lengths on quantum dot width and distribution was further confirmed. The UV/vis absorption spectra and PL spectra of perovskite films based on these organic amines are shown in Figure 3.17a. The general blueshift of UV-vis absorption indicates that the band gap of DJ type perovskites is increased compared with that of 3D phase perovskites. In the case of PDA and BDA ligands, the QWs in the film are mainly distributed in the high-n-value perovskite phase, while the amplitude of the low-n-value of the QWs is small, indicating that the crystals have a low degree of disorder, special crystallinity, and vertical growth of grains (Figure 3.17b–e). This phenomenon is in stark contrast to films based on PeDA and HDA, which produce peaks at n as low as 2 and 3, respectively. To further clarify the bottom-up the QWs distribution of perovskite films based on these four organic ligands, we used ultrafast TA spectroscopy to probe the front (upper) and posterior (bottom) sides of the films with an excitation wavelength of 480 nm. In PDA- and BDA-based perovskite films, only dominant bleaching peaks at about 760 nm were observed, which is responsible for the large n-valued DJ type. In the PeDA and HDA perovskite films, there are four distinct bleaching peaks in the TA spectra, consistent with the peaks in the steady-state absorption spectrum, indicating that the distribution of QW in the PeDA and HDA perovskite films is disordered. In summary, the QWs distribution of PDA- and BDA-based perovskite films is more uniform than that of PeDA- and HDA-based perovskite films. The authors suggest that a possible plausible explanation is that the low solubility of PeDA and HDA tends to favorably form low n-valued perovskite phases during film preparation. Finally, the optimal PCE of PSCs based on PDA and BDA is 14.16% and 16.38%, respectively, which is much higher than that of PSCs based on PeDA (12.95%) and HDA (10.55%). This is mainly due to the preferential orientation of the crystal and the narrow phase distribution that increase the carrier diffusion length mobility. Notably, they offer championship efficiency and improved stability. In our previous work, BDA was introduced into 2D Sn-based perovskites as organic ligands to mitigate structural distortions and inhibit the oxidation of Sn^{2+}[154]. The PCE of PSCs prepared with $(BDA)FA_2Sn_3I_{10}$ in the N_2 environment is 6.43%, which has significant stability. Notably, another piece of work from our group shows that BDA can also establish ACI-type perovskites (Pb as divalent metal cations), which will be discussed in the next section.

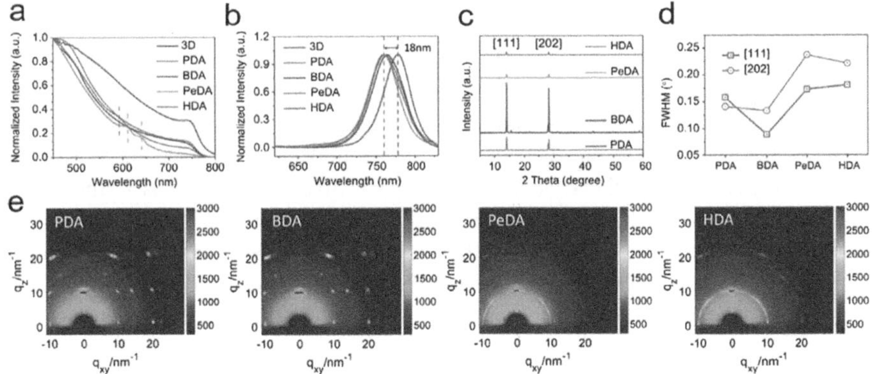

FIGURE 3.17 (a, b) Absorbance and PL spectra of PDA, BDA, PEDA, and HDA-based perovskite films and compared with their 3D counterparts; (c) XRD diffraction patterns of PDA, BDA, PEDA, and HDA-based perovskite films with distinct [111] and [202] peaks; FWHM for (d) [111] and [202] peaks in (c); (e) GIWAXS diagram of perovskite films based on PDA, BDA, PeDA, and HDA[145].

Current research on alkyldiamines is limited to straight-chain adiamines with different carbon chain lengths. Further development of alkyldiamines should be widely improved. In terms of improving device performance, the interaction of alkyl amines with precursor solutions will be emphasized first to slow down the premature precipitation of the organic amine moiety and achieve a phase pure film. Then, the introduction of heteroatoms or conjugated structure, combined with the characteristics of short layer spacing of DJ type perovskite, can be considered to jointly promote charge transfer between inorganic layers. Finally, the single-crystal structure analysis of 2D perovskites helps to understand the form of organic amines present in the film. Therefore, the crystal structure research based on partial alkyldiamine needs to be further improved.

3.4.2 Benzene Ring Series Diamine Ligands

The 2DDJ structure and benzene ring with conjugated structure have excellent performance in theory and have attracted widespread attention in low-dimensional perovskites, and organically combining the two is a promising method. However, research in this area is relatively lacking. So far, diamine ligand DJ type perovskite materials involving benzene, such as 1,4-phenyldimethylamine (PDMA), have been poorly reported. Li et al. were the first to use PDMA as an organic ligand to prepare 2DDJ perovskite ($n = 3$)[147]. The authors demonstrate the effect of ligands on the structure and photoelectric properties of perovskite films (PDMA). The PCE of solar cells in the FA2Pb3I10 formula exceeds 7%, and it has outstanding moisture stability and photostability. Subsequently, Cohen et al. prepared 2DDJ perovskites ($n = 10$) mixed with A-cations and halogens using PDMA and studied their structure and membrane properties[148]. Surprisingly, the 2D perovskite membrane using PDMA showed a high degree of orientation without repeated treatment, which indicates the influence of PDMA molecules on the crystallization process of perovskite. The efficiency of 2D PSCs reaches 15.6%, which is one of the high efficiencies of 2DDJ perovskite. The advantages of using PDMA are summarized as follows: (i) as a diamine ligand, a more ordered DJ type structure can be formed; (ii) PDMA molecules shorten the distance between inorganic layers; (iii) conjugated benzene rings enhance charge transport at adjacent quantum dots. Recently, Yu et al. also studied 2DDJ type perovskite $(PDMA)(MA)_{n-1}Pb_nI_{3n+1}$[149]. Structural analysis found that the width of the QWs affects the separation distance. As shown in Figures 4.4c and d, as the molar ratio of PDMA decreases (the increase of n value), perovskites compress from constrained to unconstrained structure, as confirmed by the estimation of the free charge to exciton equilibrium ratio by HRTEM (Figures 3.18a, b) and Saha-Langmuir theory. This unconstrained structure has a short inorganic layer spacing and small structural distortion, resulting in lower E_b. At present, PDMA-based DJ perovskites have the highest PCE, reaching 15.81%[150]. However, a comprehensive understanding of the inherent rigidity of benzene rings and 2DDJ type perovskites structure greatly limits the freedom of inorganic layer arrangement, which can be a difficult problem to solve.

Recently, Di Wang et al. developed a phenylfluoridediamine, i.e., 2,3,5,6-tetrafluo ro-1,4-benzenedimethylamine (4 FPhDMA) as an organic ligand[151]. Compared with

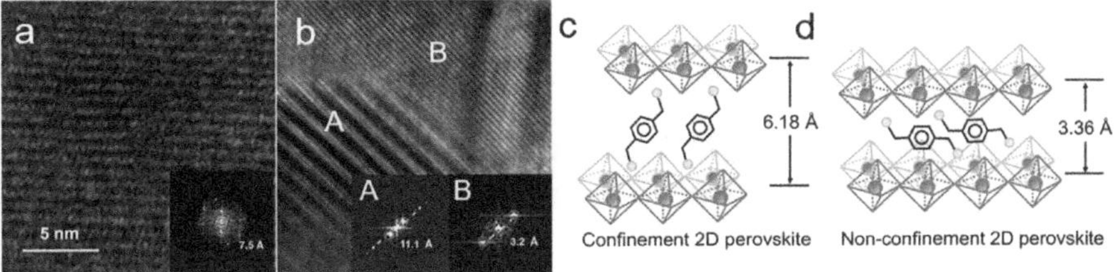

FIGURE 3.18 (a–d) HR-TEM image, crystal structure schematic diagram of (PDMA) MAPb₂I₇ (a, c) and (PDMA)MA₅Pb₆I₁₉ (b, d) samples, illustrations showing FFTs of (A) 2D phase and (B) 3D phase, respectively[149].

unfluorinated analogues, fluorinated organic ligands obtain higher quality and more stable membranes, and the final photoconversion efficiency reaches 15.24% ($n = 10$). Similarly, Liu's team used 4FPhDMA for research. They found that 4-F-PhD-based films had higher stability due to unconventional, non-covalent interactions between organic ligands and inorganic layers, for example, CH⋯F and F⋯I electrostatic interaction, exhibiting high dissociation energy, ultimately reaching an efficiency of 16.62% ($n = 4$).

3.4.3 HETEROCYCLIC DIAMINE LIGANDS

In addition to the benzene cyclic diamine ligands described above, researchers have also developed heterocyclic diamine ligands. These heterocyclic diamine ligands are further introduced into 2D perovskites to form 2DDJ perovskite materials with different properties.

3.4.3.1 Piperidine Derivatives Organic Ligands

In 2018, Kanatzidis et al. first reported the synthesis of 2DDJ type perovskites based on 3-(aminomethyl)piperidine (3AMP) and 4-(aminomethyl)piperidine (4AMP) and their application as light collectors in PSCs[44]. The influence of functional group position on perovskite structure is discussed. Structurally, the main difference between 3AMP and 4AMP is the position of the $-CH_2NH_3^+$ group (relative to the 3- and 4-positions of piperidine nitrogen), which mainly affects the crystal structure by forming different hydrogen bond patterns between amino groups and octahedron $[PbI_6]^{4-}$. This can be reflected by the deformation of the inorganic layer, which has a large impact on the optical and electronic properties. As shown in Figures 4.5a and b, the color of the 3AMP series crystals gradually darkens from $n = 1$ (red) to $n = 4$ (black). Similarly, 4AMP has the same trend, but the crystal starts with lighter colors than the $n = 1$ (orange) and $n = 2$ (red) members of 3AMP. As shown in Figure 3.19c and d, the (3AMP)PbI₄ crystal differs from the complete overlap between the (4AMP) PbI₄ layers and from the (3AMP)PbI₄ layer due to a slight out-of-plane tilt. With the increase of n value, the crystal structure of the two amines is mainly reflected in the difference of the Pb-I-Pb bond angle (Pb-I-Pb bond angle is divided into two types, along the direction of the longest crystal axis and along the equatorial direction of

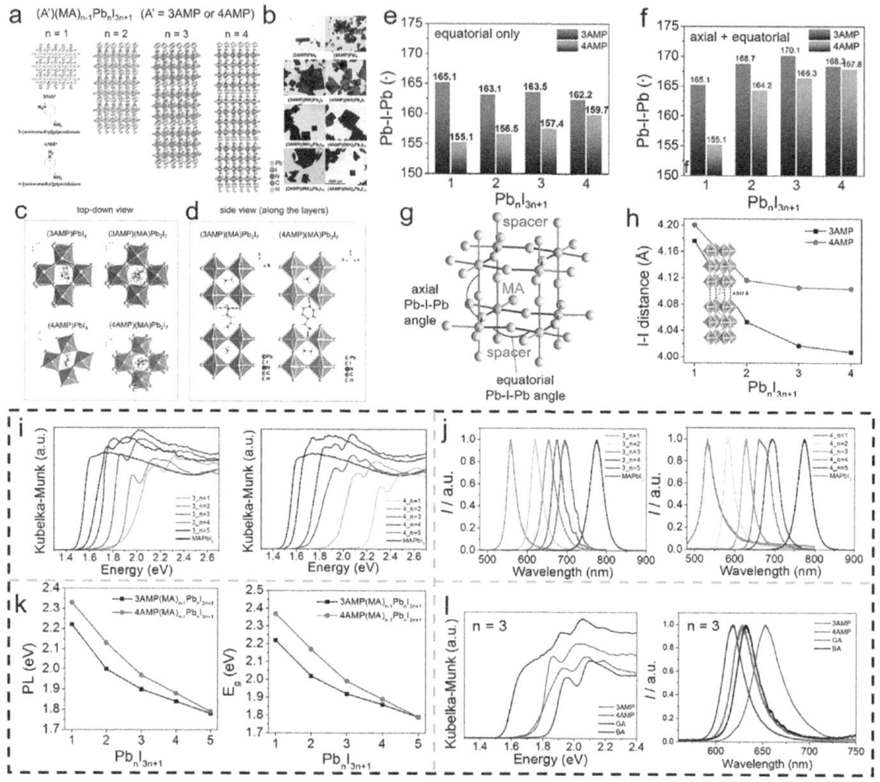

FIGURE 3.19　(a) $(A')(MA)_{n-1}Pb_nI_{3n+1}$ (A' = 3AMP or 4AMP) general crystal structure; (b) optical images of 3AMP and 4AMP crystals. The scale bar in the lower right corner is suitable; (c) top view of $(3AMP)PbI_4$, $(4AMP)PbI_4$, $(3AMP)(MA)Pb_2I_7$, and $(4AMP)(MA)Pb_2I_7$; (d) side view of $(3AMP)(MA)Pb_2I_7$ and $(4AMP)(MA)Pb_2I_7$, hydrogen bonds marked in red; (e) the mean equatorial Pb-I-Pb angle of the 3AMP and 4AMP series from $n = 1$ to 4; (f) average axial and equatorial angles of the 3AMP and 4AMP series, from $n = 1$ to 4; (g) definition of axial and equatorial Pb-I-Pb angles; (h) I$\cdots$I distance trends in 3AMP and 4AMP series range from $n = 1$ to 4, where 3AMP series are closer together; (i) light absorption spectra for the 3AMP and 4AMP series from $n = 1$ to 5; (j) steady-state PL spectra of the 3AMP and 4AMP series, from $n = 1$ to 5; (k) absorption and PL spectra of 3AMP and 4AMP series energies from $n = 1$ to 5; (l) comparison of light absorption spectra and PL spectra between $(3AMP)(MA)_2Pb_3I_{10}$, $(4AMP)(MA)_2Pb_3I_{10}$, $(GA)(MA)_3Pb_3I_{10}$ and $(BA)_2(MA)_2Pb_3I_{10}$[44].

the inorganic plane, reflecting out-of-plane distortion and in-plane distortion, respectively), which is related to the difference in their hydrogen bond networks. For crystals with $n = 2$ (Figure 3.19c, d), it is proposed that 3AMP cations and 4AMP cations form hydrogen bonds with terminal I⁻ and bridge I⁻, respectively, resulting in different degrees of distortion in the crystal structure plane. In addition, I-bonding to bridging has a greater effect on the Pb-I-Pb angle, indicating that the (4AMP)(MA) Pb_2I_7-based crystal structure has greater in-plane distortion. As the layer thickness increases from $n = 1$ to $n = 4$, the mean (axial and equator) Pb-I-Pb angles of the two systems gradually increase (Figure 3.19e–h). With the increase of n, the difference in mean (axial and equatorial) Pb-I-Pb angles decreased, indicating that the influence of organic cations on high n-value perovskite inorganic plates gradually weakened. This version also applies to bandgap and PL emission energy, as shown in Figure 3.19j. On the other hand, the layer spacing of the 3AMP series is also shorter than that of the 4AMP, which has an important impact on the electronic band structure. Similar to other 2D perovskites, the light absorption spectra and PL spectra of Figure 3.19i–k confirm that the band gap of both perovskites decreases with the increase of the QWs thickness. DFT calculations also confirmed the drift of low bandgap and two-series perovskite bandgap variation based on 3AMP perovskite.

The optical properties of perovskites with different organic ligand layers are closely related to their structural properties. UV-vis absorption spectra and PL spectral pairs of typical organic amines based on different structure of $n = 3$ (BA, GA, 3AMP, and 4AMP are organic ligands) are shown in Figure 3.19i–l. Compared with RP type perovskite, DJ type perovskite has red shift luminescence energy and a wide light absorption range. The properties of ACI perovskites are between RP and DJ perovskites. Due to the attraction characteristics of 3AMP- and 4AMP-based DJ type perovskites, the PCE of $(3AMP)(MA)_3Pb_4I_{13}$-based devices is the highest, which is 7.32%, which is higher than that of $(4AMP)(MA)_3Pb_4I_{13}$ devices (less than 5%). This work sets a precedent for the study of DJ type PSCs.

Subsequently, Kanatzidis et al. continued to study 3AMP-based DJ type perovskites[137]. They demonstrated the compositional engineering of cations at position A, and by introducing the large cation FA^+, the Pb-I-Pb angle can be further expanded, thereby reducing the distortion of the inorganic skeleton. The subject of the study was the effect of different proportions of mixed A-position cations on perovskite structure and photovoltaic properties (Figure 3.20a, b). Due to the small structural distortion, the introduced FA^+ cation has a narrow band gap of 1.48 eV $((3AMP)(MA_{0.75}FA_{0.25})_3Pb_4I_{13})$. The results show that photelectric conversion deficiency of $(3AMP)(MA_{0.75}FA_{0.25})_3Pb_4I_{13}$ is 12.04%, which is higher than that of $(3AMP)(MA)_3Pb_4I_{13}$. In addition, Kanatzidis et al. mixed $3AMP^{2+}$ and $4AMP^{2+}$ as organic ligands to explore the photophysical properties of $(3AMP)_a(4AMP)_{1-a}(FA)_b(MA)_{1-b}Pb_2Br_7$ by varying the ratio of $3AMP^{2+}$ and $4AMP^{2+}$ and the ratio of cations at position A, as shown in Figure 3.20c-h[138]. This shows that increasing the proportion of $4AMP^{2+}$ components can reduce the Pb-Br-Pb angle (Figure 3.20h), deform the inorganic skeleton, and increase the band gap. Conversely, a larger proportion of FA^+ components results in a larger Pb-Br-Pb angle, resulting in a narrower band gap (Figure 3.20f). This conclusion was confirmed by PL spectroscopy and DFT calculations (Figure 3.20g). Therefore, piperidine derivatives as ligands to constitute 2DDJ perovskites can demonstrate the importance of adjusting photoelectric properties through fine design of structural features, providing guidance for efficient 2D perovskite photovoltaic devices.

3.4.3.2 Pyridine Derivatives Organic Ligands

Encouraged by the 3AMP and 4AMP organic ligands, Kanatzidis' team further adopted two new aromatic ligands, 3AMPY = 3-(aminomethyl)pyridine and

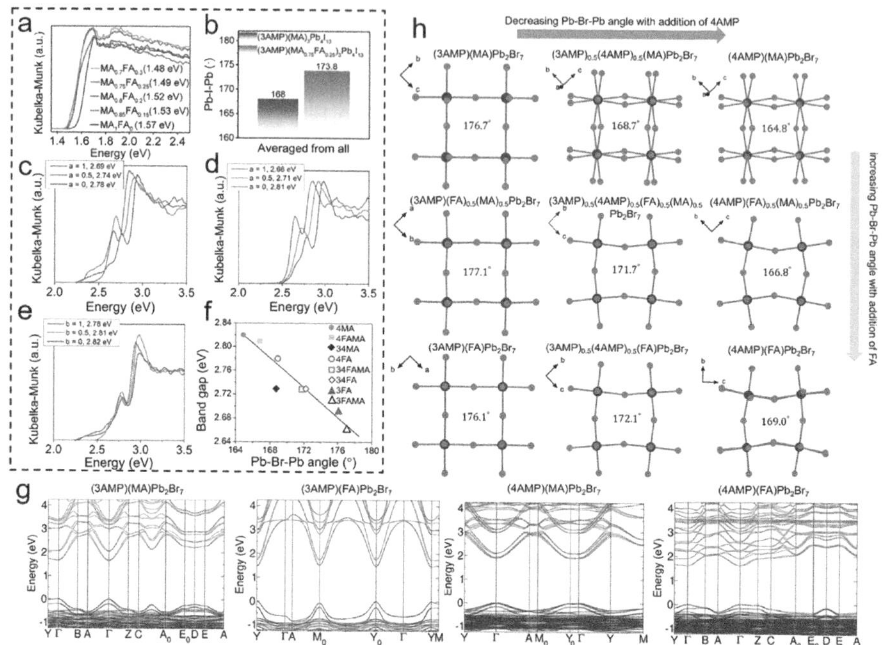

FIGURE 3.20 Absorption spectra of (a) $(3AMP)(MA_{1-x}FA_x)_3Pb_4I_{13}$ ($x = 0$–0.3) crystals; (b) $(3AMP)(MA)_3Pb_4I_{13}$, and $(3AMP)(MA_{0.75}FA_{0.25})_3Pb_4I_{13}$ average Pb-I-Pb angle[137]; light absorption spectra of (c) $(3AMP)_a(4AMP)_{1-a}(FA)Pb_2Br_7(a = 0, 0.5, \text{ or } 1)$, (d) $(3AMP)_a(4AMP)_{1-a}(FA)0.5(MA)_{0.5}Pb_2Br_7$ ($a = 0, 0.5, \text{ or } 1$), and (e) $(4AMP)(FA)_b(MA)_{1-b}Pb_2Br_7$ ($b = 0, 0.5, \text{ or } 1$); (f) the correlation between the Pb-Br-Pb angle and the band gap in $(3AMP)_a(4AMP)_{1-a}(FA)_b(MA)_{1-b}Pb_2Br_7$ (a and $b = 1, 0.5, \text{ or } 0$); (g) The calculated band structure and band gap are $(3AMP)(MA)Pb_2Br_7$, 1.66 eV (Γ), $(3AMP)(FA)Pb_2Br_7$, 1.46 eV (Y), $(4AMP)(MA)Pb_2Br_7$, 1.89 eV (Γ), and $(4AMP)(FA)Pb_2Br_7$, 1.62 eV (Γ); (h) top view of the crystal structure of the nine members of the extended solid solution $(3AMP)_a(4AMP)_{1-a}(FA)_b(MA)_{1-b}Pb_2Br_7$ (a and $b = 1, 0.5, \text{ or } 0$). The mean Pb-Br-Pb angle in the inorganic frame of each composition is marked. Atomic symbols: big spheres (Br), small spheres (Pb), and organic cations[138].

4AMPY = 4-(aminomethyl)pyridine, to explore the effect of pyridine derivatives containing conjugated structure on the crystal structure and optical properties of DJ type perovskites[141]. At the same time, the influence of the location of the functional group was also studied. For the first time, DJ perovskite single crystals based on two aromatic diamines ($n = 1$–4) were synthesized, revealing crystal structure characteristics. The location of functional groups has a significant effect on the crystal structure, which is very similar to aliphatic analogues, as shown in Figure 3.21a. The conjugation effect attenuated the freedom of cation atoms to move, so only the more symmetric 4AMPY cation maintained the typical DJ structure (eclipse conformation) where inorganic layers stack each other (0, 0 displacement). Due to the planarity of the rigid aromatic rings, the 3AMPY-based inorganic layers are slightly staggered (0.238, 0.238 shifted) to accommodate asymmetric 3AMPY cations. Compared with 4AMPY, the average Pb-I-Pb angle of 3AMPY-based

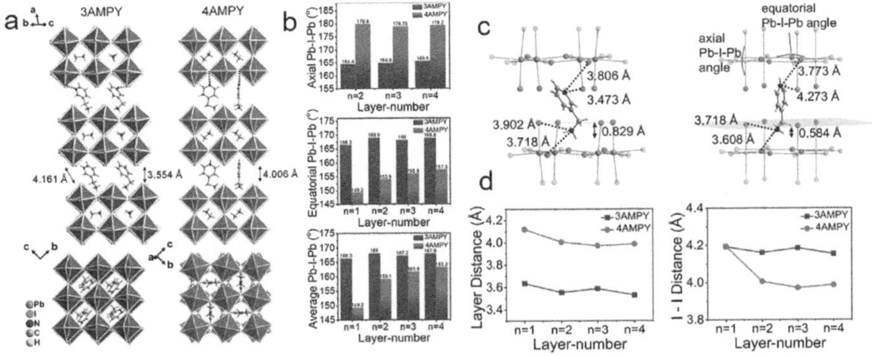

FIGURE 3.21 (a) Side and top views of (3AMPY)(MA)Pb$_2$I$_7$ and (4AMPY)(MA)Pb$_2$I$_7$; (b) (xAMPY)(MA)$_{n-1}$Pb$_n$I$_{3n+1}$ (x = 3 or 4, n = 1–4) axial, equatorial, and mean Pb-I-Pb angles; (c) comparison of the structure of (3AMPY)(MA)Pb$_2$I$_7$ and (4AMPY)(MA)Pb$_2$I$_7$, with the dotted line indicating the closest distance of NH⋯I; (d) the plane defined by the terminal iodide is the same as the nearest I⋯ Interlayer distance I distance (xAMPY)(MA)$_{n-1}$Pb$_n$I$_{3n+1}$ (x = 3 or 4, n = 14).

perovskites is larger, and the deformation of the inorganic layer is smaller, as shown in Figure 3.21b, c. In addition, 3AMPY-based 2DDJ type perovskites have a deeper amino penetration depth compared to 4AMPY, resulting in shorter interlaminar spacing (Figure 3. 21d) and narrower band gaps. With the increase of the n value, the Pb-I-Pb angle in the 2D perovskite structure of the two series increases, indicating that the structural distortion gradually weakens. It was worth noting that the aromaticity of the pyridine ring in 3AMPY and 4AMPY made the C-C bond shorter compared with its aliphatic counterpart, causing the proximity of the adjacent inorganic slabs.

The particularity of the 3AMPY and 4AMPY structure makes the prepared 2DDJ perovskite have unique photophysical and photochemical properties. The UV-vis absorption spectrum and PL spectrum of 2DDJ perovskites in Figure 3.22a–c confirm that 3AMPY has broad application prospects in solar cells. The authors further compared the optical properties of 2D perovskites (n = 4) based on xAMPY (x = 3, 4) and its aliphatic counterparts, as well as RP type BA cations. (3AMP)(MA)$_3$Pb$_4$I$_{13}$ and (3AMP)(MA)$_3$Pb$_4$I$_{13}$ show minimal band gap due to small distortion and short interlaminar spacing (Figure 3.22d, f). For (4AMP)(MA)$_3$Pb$_4$I$_{13}$ and (4AMP)(MA)$_3$Pb$_4$I$_{13}$, although the reduction of layer spacing results in a decrease in band gap, this can be offset by a more twisted structure that gives their PL emission peaks a high energy close to (BA)$_2$(MA)$_3$Pb$_4$I$_{13}$. In addition to the different optical properties, aromatic and adipose lines also exhibit diversity on E$_b$. As shown in Figure 3.22e and g, the E$_b$ can be roughly estimated by subtracting the energy of the exciton emission peak relative to the band gap[43,144]. The results show that the aromatic system obtains a smaller E$_b$ than the aliphatic system, which is mainly due to the increase of the dielectric constant of the aromatic cation, which reduces the dielectric mismatch in the quantum hydrazine structure and weakens the dielectric

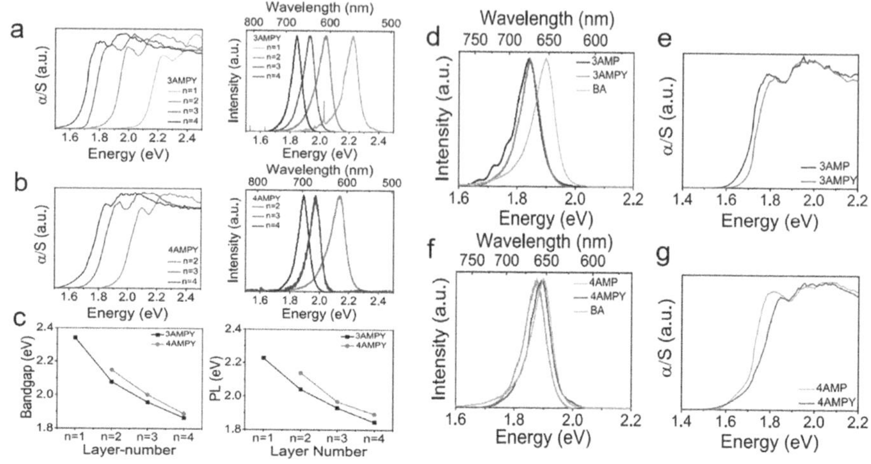

FIGURE 3.22 (a) $(3AMPY)(MA)_{n-1}Pb_nI_{3n+1}$ (n = 1–4) optical absorption spectra and steady-state PL spectra; (b) $(4AMPY)(MA)_{n-1}Pb_nI_{3n+1}$($n$ = 2–4) light absorption spectra and steady-state PL spectra; (c) comparison of bandgap and PL maximums for the 3AMPY and 4AMPY series; (d, e) PL spectra comparison of $(xAMPY)(MA)_3Pb_4I_{13}$, $(xAMP)(MA)_3Pb_4I_{13}$ (x = 3, 4) and $(BA)_2(MA)_3Pb_4I_{13}$ crystals; (f, g) absorption comparison of $(xAMPY)(MA)_3Pb_4I_{13}$, $(xAMP)(MA)_3Pb_4I_{13}$ (x = 3, 4)[141].

confinement effect. Finally, $(3AMPY)(MA)_3Pb_4I_{13}$ PSCs obtained the best PCE and long-term stability of 9.20%.

Piperidine derivatives and pyridine derivatives are protoorganic ligands for the preparation of 2DDJ type perovskites and are used in the photovoltaic field. Obtaining accurate single crystal structure is one of the most effective ways to study material properties. Currently, 2DDJ perovskite single crystals with a maximum of n = 7 based on 4AMP[140] have been synthesized. The synthesis of high n-value single crystals provides structural support for high-performance devices. Therefore, to improve the PCE of 2DDJ PSCs, it is necessary to create excellent process conditions.

3.4.3.3 Thienylalkyldiamine Organic Ligands

Thiophene derivatives have been certified as organic ligands, and the prepared 2DRP perovskites exhibit unique properties and are very suitable for the preparation of photovoltaic devices. Therefore, divalent thiophene derivatives have also been successfully used as organic ligands in 2DDJ perovskites. For example, thiophene-containing 2,5-thiophenedimethylamine (ThDMA) (Figure 3.23a, b) was first introduced into 2DDJ perovskites by Liu's team[152]. Ultraviolet photoelectron spectroscopy (UPS) determined that the VBM of the 2DDJ perovskite $(ThDMA)(MA)_4Pb_5I_{16}$ was 5.54 eV (CBM 3.96 eV), which matched well with the carrier transport layer energy level. The authors then systematically investigated the effects of precursor solutions DMF and dimethyl sulfoxide (DMSO) on film quality and verified by HRTEM and fast Fourier transform (FFT) analysis, as shown in Figure 3.24a–e.

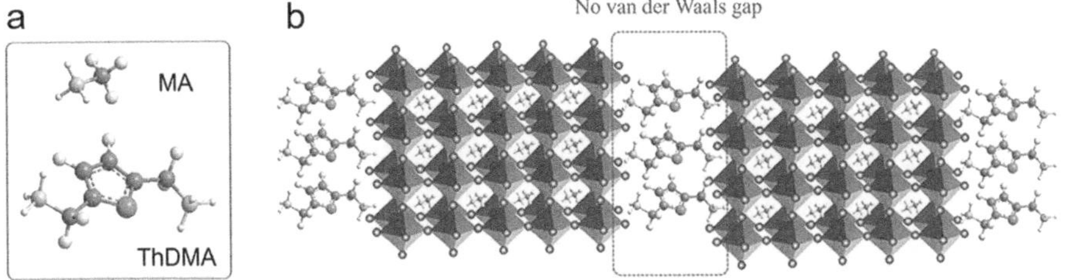

FIGURE 3.23 (a) 3D structure of MA and ThDMA; (b) Schematic diagram of the crystal structure of $(ThDMA)(MA)4Pb_5I_{16}$.

Based on the properties of these films, a charge transport model in optoelectronic devices is proposed (Figure 3.24). The results show that the PCE of 2DDJ PSCs prepared by the one-step method of DMF/DMSO (v/v, 9: 1) mixed solvent reaches an impressive 15.75%. Importantly, the photoimmersion stability and thermal stability of ThDMA-based DJ PSCs were significantly improved. Taking it a step further, Liu's team successfully developed a thiophene-based molten ligand, thiophene[3,2-b] thiophene-2,5-dimethyliodide (TTDMAI), as a bulky organic ligand, ultimately achieving a photoconversion efficiency of up to 18.82%. Due to the lengthening of the π conjugate length of the TTDMA ligand, vertically oriented crystal growth is obtained for films based on the TTDMA ligand compared to ThDMA ligands. This improvement comes from the template effect of strong molecular interactions. Their work explains the great potential of thiophene-containing organic ligands in templated 2D perovskites. However, relatively few studies have been done on thienyl-diamine ligands and their effects on the properties of 2D perovskites, and further development and research are needed.

3.5 2DACI TYPE PEROVSKITES

2DACI perovskites have two different alternating cations in interlaminar space. Structurally, ACI-type perovskites exhibit the stacking characteristics of (1/2.0) off-set shift, mixing the chemical formula of RP type perovskites and the structural characteristics of DJ type perovskites. In 2DACI-type perovskites, A-position cations (MA^+, FA^+, Cs^+) are not only present in cubic octahedral cages but also alternate with another large block of cations between inorganic layers, resulting in different central positions of unit cells, as shown in Figure 3.2. In particular, 2DACI perovskites make inorganic layer spacing short, giving it a more suitable band gap close to 3D. However, compared to RP type and DJ type perovskites, ACI-spaced perovskites have been less studied. To our knowledge, GA is the only organic ligand widely believed to be able to construct 2DACI perovskites because its structural formula is reminiscent of iodoformamidine, which stabilizes the structure of 2DACI perovskite. The unique and excellent photoelectric properties of 2DACI perovskites make it one of the most promising candidates for 2D PSCs.

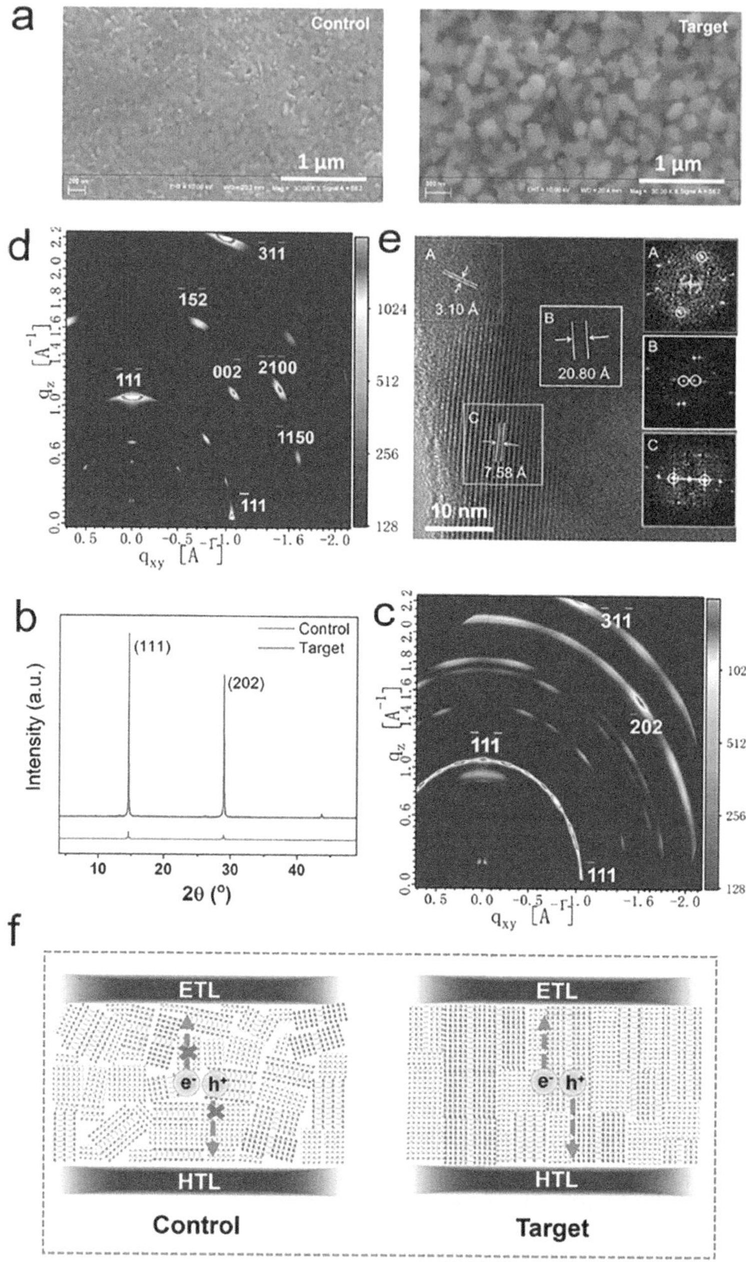

FIGURE 3.24 (a) SEM top view image of control ((ThDMA)(MA)$_4$Pb$_5$I$_{16}$ perovskite, DMF as solvent) and target (ThDMA)(MA)$_4$Pb$_5$I$_{16}$ perovskite optimized DMF/DMSO ratio (9:1, v/v as solvent) films; (b) XRD patterns of control (DMF as solvent) and target (DMF/DMSO ratio (9: 1, v/v) as solvent) films; (c, d) GIWAXS data for control and target films; (e) HRTEM image of the target film, the illustration is an FFT image of the corresponding area; (f) schematic diagram of the morphological and charge transport model of the control and target devices[152].

In 2017, Kanatzidis' team synthesized a 2DACI type perovskite based on GA cations for the first time and used it to prepare 2D PSCs[45]. The crystals structure at different GA n values ($n = 1$–3) is shown in Figure 3.25a. All compounds crystallize in orthogonal space groups. The odd n-value (1, 3) perovskites use the centrally symmetric *Imma* space group, and the even-n-value perovskites use the polar non-central symmetric *Bmm2* space group. The inorganic layer spacing of GA-based ACI type perovskites is about 3 Å, which is shorter than that of typical 2DRP (BA-based, 7 Å) and 2DDJ (3AMP-based, 4 Å) type inorganic layer spacing. In addition, ACI type perovskites have less distortion and more regular cation stacking, which makes its optical performance more suitable for solar cells than the 2DRP class. As shown in Figure 3.25b, the absorption edge and exciton peak position of ACI perovskites have undergone significant redshift compared to RP perovskite. A possible explanation is that the symmetry (degree of distortion) of the two series perovskites is different, that is, ACI type perovskites are distorted in only one direction in the ab plane, while RP type perovskites are distorted in both directions in the ab plane. This symmetry characteristic suggests that a less distorted crystal structure narrows the band gap suitable for 3D halide perovskites and may also be suitable for 2D perovskites, with ACI and RP perovskites being prime examples. The luminescence spectra of ACI and RP series perovskites have the same trend as the absorption spectra (Figure 3.25d), and the luminescence wavelength increases with the increase of the width of the QWs. The PL redshift emission and absorption spectra of ACI perovskites show that the conjugated properties of GA cations can enhance the photovoltaic performance of 2D perovskites. DFT calculations further refined the band structure of ACI series perovskites (Figure 3.25c), confirming the experimental results. Finally, PSCs based on $GAMA_3Pb_3I_{10}$ compounds were prepared and characterized with a PCE of 7.26% and FF of 80%.

To further improve the PCE of 2DACI PSCs, Ma et al. used DMSO as a co-solvent to inhibit the formation of low n phases in ACI type perovskites, thereby improving carrier lifetime and charge transport[155]. The optimized PCE of 2DACI PSCs based on $GAMA_4Pb_4I_{13}$ perovskite is 12.8%. Similarly, ACI perovskites face quantum and dielectric constraints, as well as multidimensional the QWs distributed in fabricated films, where carriers are always confined to inorganic plates.

To study charge transport patterns in 2DACI perovskite films, Gu et al. reported a nanoscale hybrid multidimensional $(GA)(MA)_3Pb_3I_{10}$ perovskite film with vertically stacked crystals and preferred crystal orientation[156]. High-resolution scanning transmission electron microscopy (HR-STEM) studied the distribution of the QWs in the nanoscale domain in 3D perovskite nanonetworks (Figure 3.26a–f), which confirmed the formation of nanoscale heterojunctions. Such heterojunctions realized ultrafast (~0.3ps) and effective carrier localization from low-dimensional ACI perovskites to 3D perovskite and the extraction of 3D perovskite to the transport layers as proved by TA measurement. As shown in Figure 3.26g–i, there were four distinct photobleaching peaks of 580 nm (PB1, $n = 1$), 617 nm (PB2, $n = 2$), 650 nm (PB3, $n = 3$), and 740 nm (PB4, $n \geq 4$), corresponding to the multiphase distribution of the film, where $n = 1$, 2, and 3 phase excited carriers are localized to $n \geq 4$ phases in a short time (~0.3 ps).

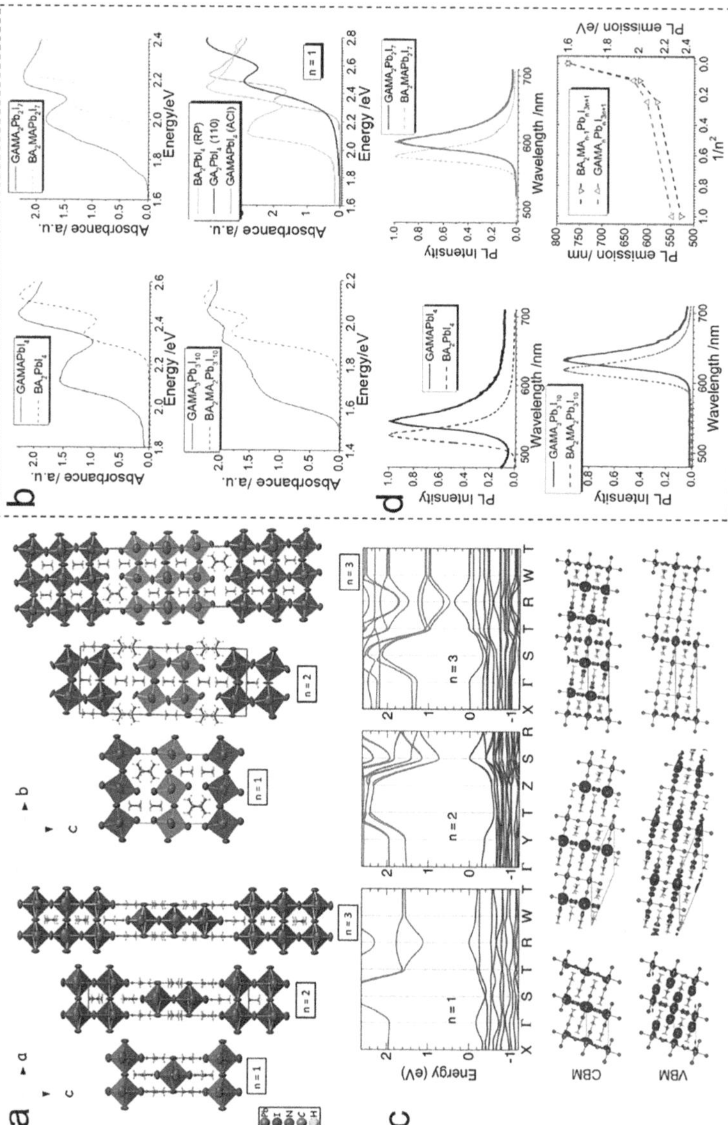

FIGURE 3.25 (a) Unit cell view of $(GA)(MA)_n Pb_n I_{3n+1}$ ($n = 1-3$) perovskites along the crystallographic b-axis and crystallographic a-axis; (b) $(GA)(MA)_n Pb_n I_{3n+1}$ ($n = 1-3$) optical absorption spectra of perovskites and selected absorption spectra of members $n = 1$ of ACI perovskite GAMAPbI$_4$, RP perovskite BA$_2$PbI$_4$ and (110) cut perovskite GA$_2$PbI$_4$; (c) $n = 1-3$ members of the electronic band structure and the local state density around the conduction band minimum (CBM) and valence band maximum (VBM); (d) steady-state PL spectra and graphical representation of $(GA)(MA)_n Pb_n I_{3n+1}$ ($n = 1-3$) ACI perovskite (solid) and their comparison with the corresponding n members of $(BA)_2(MA)_{n-1} Pb_n I_{3n+1}$ series RP perovskites[45].

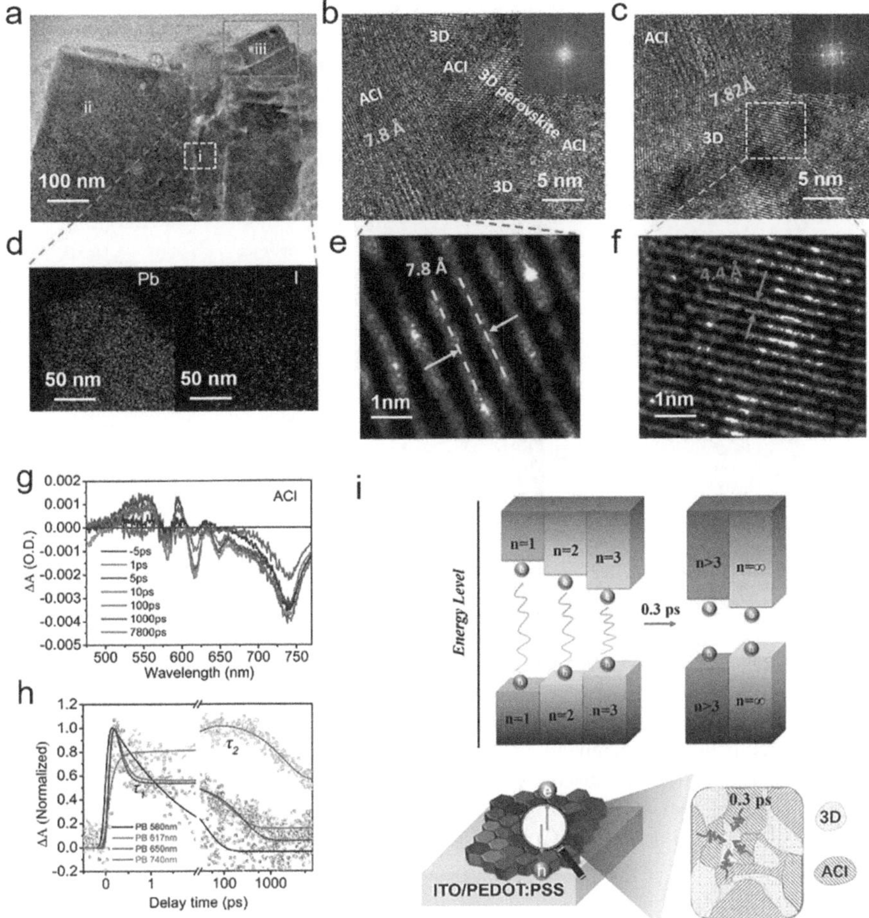

FIGURE 3.26 (a) The low magnification image shows the high crystallinity of grains; Areas (i) and (ii) highlighted were investigated in (b) and (c), respectively, and FFT analysis for the selected area was shown in the insertion; (d) EDX mapping images showing Pb and I, FFT-filtered HRTEM images are shown in (e) and (f), respectively, with (b) pink box and (c) area within yellow box; (g) femtosecond TA spectra for chirp-corrected ACI perovskite films at selected probe delay times; (h) normalized bleaching kinetics corresponding to PB1 (580 nm), PB2 (617 nm), PB3 (650 nm) and PB4 (740 nm) after ACI perovskite excitation at 425 nm (1 KHz, 100 fs, ~ 1 μJ/cm²/pulse); (i) Schematic diagram of charge carrier localization from small n-phase to large n-phase and extraction of charge carriers from ACI perovskite[156].

Recently, Liu et al. have studied the crystallization process[157], using time-resolved grazing incidence wide-angle X-ray scattering (GIWAXS) to study the evolution of in-situ structure during the crystallization of films deposited by different methods (Figure 3.27a, b), proving that the formation of ACI type is a complex process, including crystal solvates and intermediate phases such as 2D GA_2PbI_4 perovskite. Due to the complex interactions between solvents, lead halide and organic components,

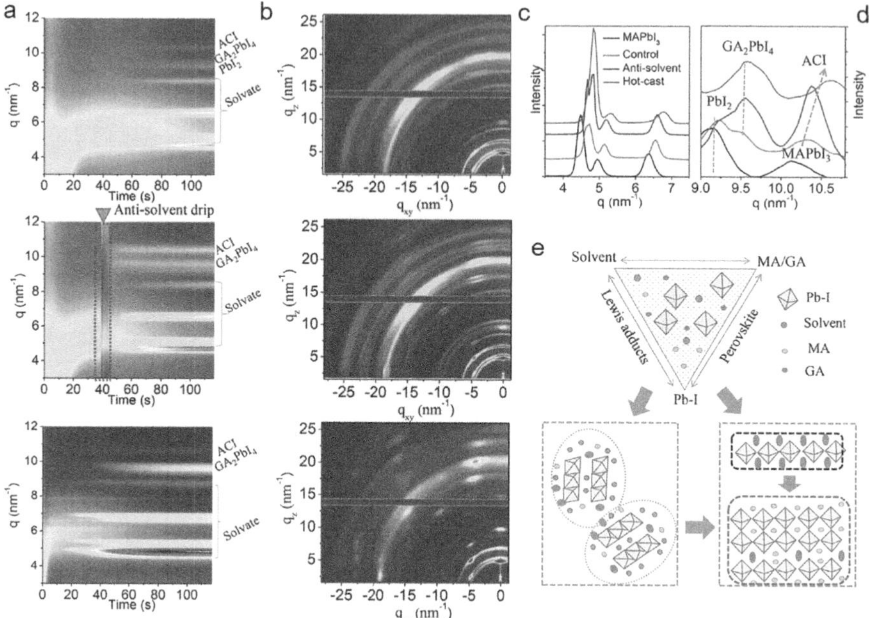

FIGURE 3.27 (a) In-situ GIWAXS analysis, showing the dynamic transition from disordered precursors to perovskites in control, antisolvent treated, and hot-cast samples, respectively; (b) 2D GIWAXS image taken at the end of spin coating, showing different textures of the control film, the antisolvent (chlorobenzene) treated film and the hot spin coating film, respectively. The red horizontal line is caused by the detector; (c, d) corresponding strength and q of diffraction characteristics of MAPbI$_3$, control film, antisolvent treated film, and heat cast film; (e) schematic model illustrating ACI perovskites self-assembly[157].

different deposition processes had an impact on the crystallization process. These interactions lead to the creation of different interphases, which are partially broken down during annealing, providing a framework for the growth of ACI perovskites (Figure 3.27c–e). At present, the secondary crystallization of perovskite membranes is regulated by guanidine bromide (GABr) treatment, and the highest efficiency (~ 19.3%) of ACI perovskite-based 2D PSCs is obtained, which shows the superiority and great potential of ACI perovskite[37]. In our previous work, a novel ACI perovskite was prepared using BDA as an organic ligand[158].

While previous reports have suggested that BDA uses 2DDJ type perovskites as ligands, our work found that organic diamine (BDA^{2+}) and monoamine (MA$^+$) cations alternately arranged in space with the formula (BDA)$_{0.5}$MA$_n$Pb$_n$I$_{3n+1}$, as shown in Figure 3.28a. Unlike GA$^+$, the amino groups at both ends of the BDA^{2+} ligand chain are connected to the inorganic layer by hydrogen bonding, forming a new structure, named *B*-ACI. This new perovskite combines the wider light absorption range of conventional ACI perovskites with lower E$_b$. Compared to typical RP perovskites, typical DJ perovskites have a more stable layered structure. DFT calculations show that (BDA)$_{0.5}$MA$_3$Pb$_3$I$_{10}$ is a direct bandgap semiconductor with

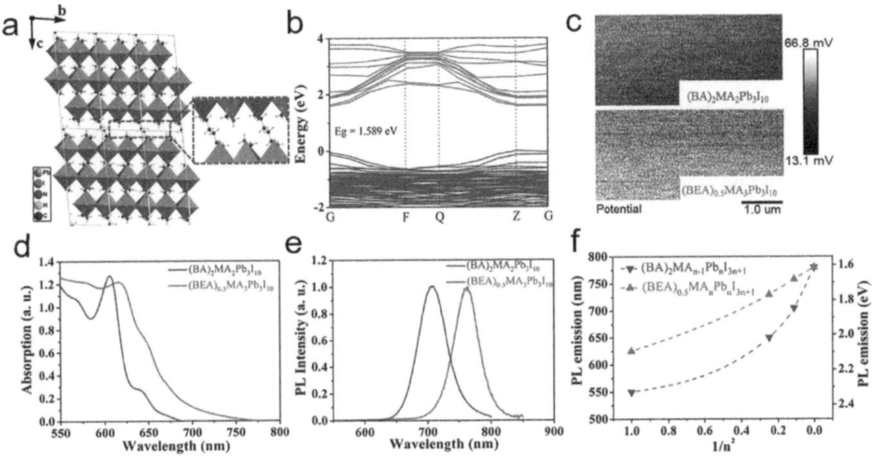

FIGURE 3.28 (a) Crystal structure of *B*-ACI $(BDA)_{0.5}MA_3Pb_3I_{10}$ perovskite, BDA^{2+} and MA^+ cations alternately arranged in interlaminar space; (b) electronic band structure of *B*-ACI $(BDA)_{0.5}MA_3Pb_3I_{10}$ perovskite, indicating a direct bandgap semiconductor of *B*-ACI perovskite; (c) surface potential mapping of *B*-ACI and 2DRP perovskite films by scanning Kelvin probe microscopy, with the average SP of *B*-ACI films increasing to approximately 56 mV compared to 2DRP films (approximately 36 mV); (d, e) UV-vis absorption and PL spectra of $(BDA)_{0.5}MA_3Pb_3I_{10}$ and $(BA)_2MA_2Pb_3I_{10}$ films; (f) graphical representation of $(BDA)_{0.5}MA_nPb_nI_{3n+1}$ (n = 1–3) *B*-ACI perovskites and their comparison with the corresponding n values of $(BA)_2MA_{n-1}Pb_nI_{3n+1}$ series 2DRP perovskites.

a bandgap of 1.589 eV (Figure 3.28b). To highlight the advantages of the novel ACI perovskite, the optical properties of $(BDA)_{0.5}MA_3Pb_3I_{10}$ with RP perovskite $(BA)_2MA_2Pb_3I_{10}$ were compared (Figure 3.28d-f). Absorption spectroscopy and PL spectra showed that the band edge of *B*-ACI had a significant redshift, which contributed to the influence of dielectric properties on carriers. And compared with $(BA)_2MA_2Pb_3I_{10}$, the E_b of $(BDA)_{0.5}MA_3Pb_3I_{10}$ is smaller and the average surface potential is larger (Figure 3.28c), which is favored in the dissociation of photogenerated carriers.

In addition, they used a femtosecond thermal analyzer to study the QWs distribution and carrier transport mechanism of *B*-ACI perovskites. Figure 3.29a shows the diversity of quantum wave distributions in RP perovskite control films, but $(BDA)_{0.5}MA_3Pb_3I_{10}$ films have narrow quantum wave distributions and flat energy landscapes, resulting in reduced energy losses. TA kinetic analysis (Figure 3.29b) also verified that the carrier transport pathway in *B*-ACI perovskite membranes is more efficient (Figure 3.29c). As a result, the PSCs with *B*-ACI phase perovskite $(BDA)_{0.5}MA_3Pb_3I_{10}$ as the active layer yield a champion PCE of 14.86%. This work shows that *B*-ACI perovskites are expected to combine the advantages of 2DDJs and 2DACI perovskites to overcome the compromise of performance stability in low-dimensional PSCs.

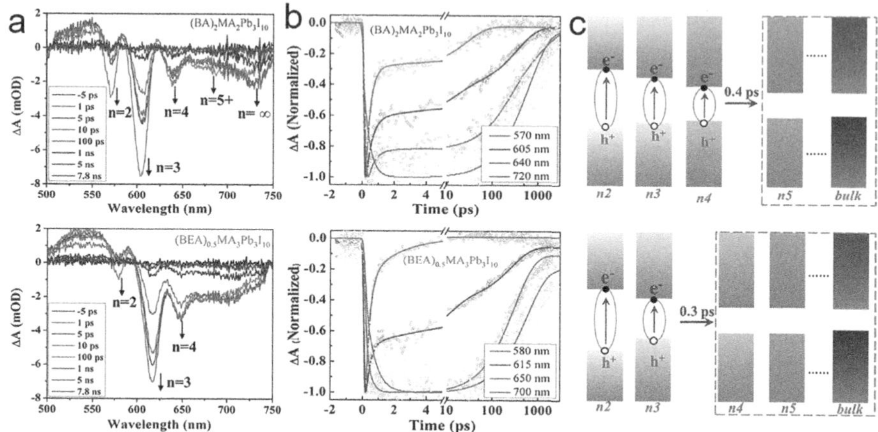

FIGURE 3.29 (a) $(BDA)_{0.5}MA_3Pb_3I_{10}$ and $(BA)_2MA_2Pb_3I_{10}$ films TA spectra at different delay times, the films are photoexcited at 425 nm (1 kHz, 100 fs and 1 μJ cm^{-2}) to excite all possible QWs present; (b) TA kinetics probed in the $n = 2$, 3, and 4 and $n \approx \infty$ bands, with the solid line being the fit of the kinetics by the exponential function; (c) schematic diagram of the ribbon structure and carrier transport pathway of mixed perovskite QW, $(BA)_2MA_2Pb_3I_{10}$ film can achieve stepped transport, while $(BDA)_{0.5}MA_3Pb_3I_{10}$ film has lateral transmission[158].

3.6 MIXED LIGAND TYPE 2D PEROVSKITES

The three perovskites structure mentioned above all have only one large block of organic amine in the spacer layer. In 3D perovskites, the researchers applied component engineering techniques to fine-tune A-position cations by mixing A-position organic amine components to create highly efficient PSCs. Thus, this knowledge gives researchers the opportunity to recognize that hybrid organic-isolated cations (more than one type of ligand) occupy spatial locations to further improve the performance of 2D PSCs.

BA organic ligands have an unshakable dominance in 2D perovskites, and BA-based hybrid systems are also the most studied among mixed organic ligand cations[159–166]. Lian et al. introduced a second-interval cationic fraction of PEA to replace BA to form a 2DRP perovskite[159]. The results show that the addition of a second interval cation can induce precursor solution aggregation by inducing nucleation and reducing nucleation sites, resulting in high-quality films with large grain sizes and favorable charge transport with vertical crystal orientation (Figure 3.30a–d). The PCE of 2D PSCs after the optimization of PEA addition reached 14.09%. Similarly, Zhou et al. also used a mixture of PEA and BA as spaced organic cations to achieve uniform high-speed transport of carriers in 2D perovskites[165]. They found that alkylamine cationic BA can promote directed growth of perovskite crystals by facilitating the assembly of precursors, while unsaturated alkylamine cation PEA greatly reduces E_b by inhibiting dielectric confinement effects, thereby improving carrier transport in 2D PSCs (Figure 3.31a). Finally, the PCE of the hybrid cation

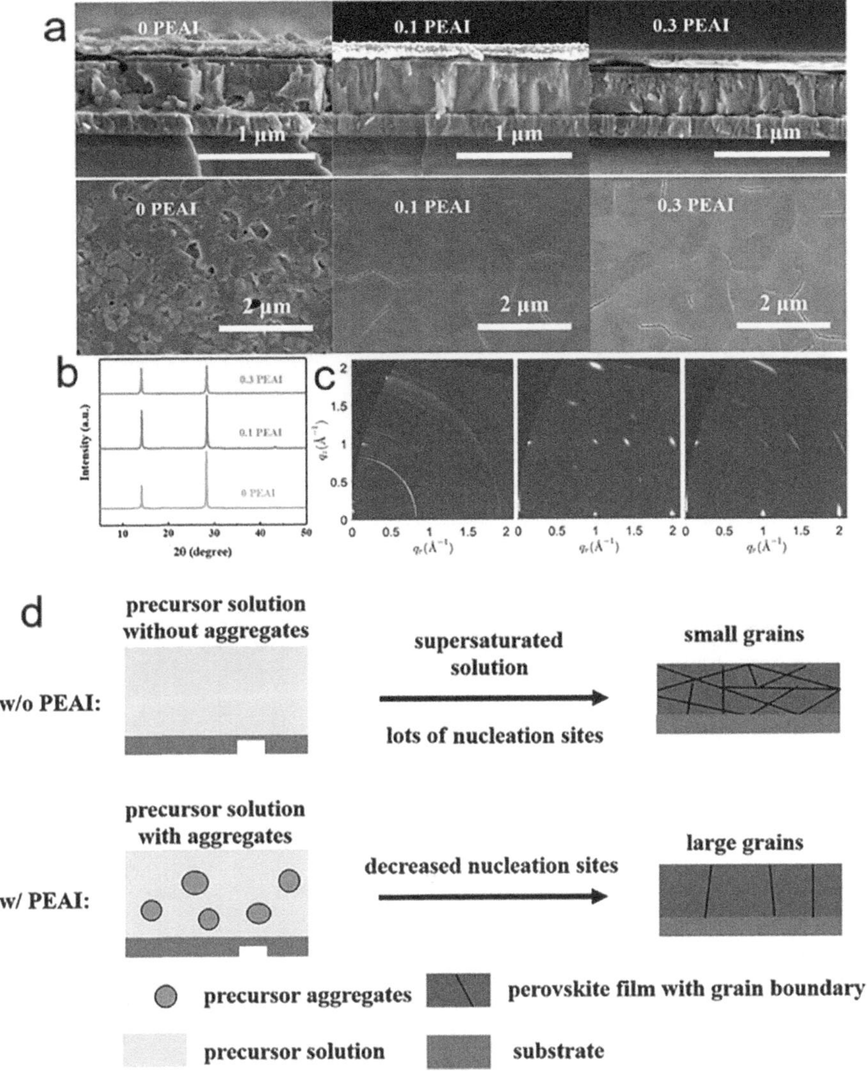

FIGURE 3.30   (a) Cross-sectional SEM images of $BA_2MA_4Pb_5I_{16}$ perovskite films with different amounts of PEAI, top-view SEM images and (b) their XRD maps; (c) 2D GIWAXS pattern of $BA_2MA_4Pb_5I_{16}$ perovskite films containing 0 PEAI, 0.1 PEAI, and 0.3 PEAI; (d) illustration of the formation of 2D perovskite films without or with PEAI[159].

device reaches 15.46%, which also proves that the design can be generalized to the mixing of other organic amine ions. Notably, the previous two studies show that 2D hybrid organic spacer perovskites that replace BA with 10% PEA have excellent film properties, thereby improving the efficiency of PSCs. In addition, Tan et al. added the polyfluorinated organic ligand 2,2,2-trifluoroethanamine (3FEA) to BA-based 2D perovskites[160], effectively reducing dielectric confinement, in which highly

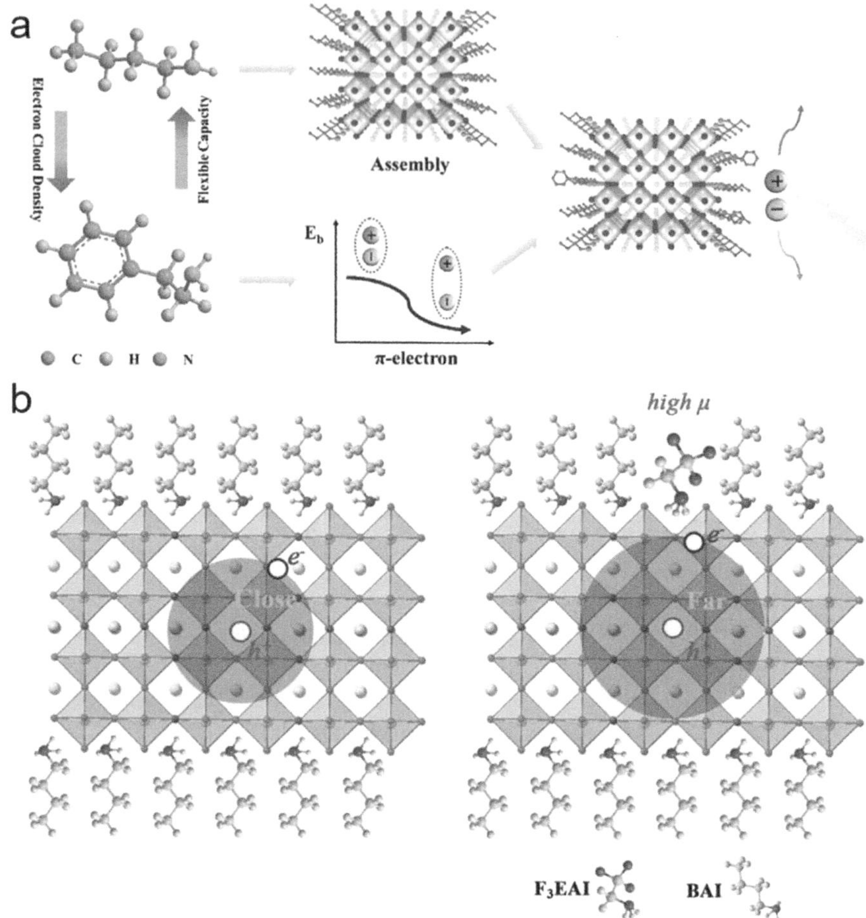

FIGURE 3.31 (a) Schematic diagram of the mechanism of introducing PEA⁺ into BA substrate perovskite[165]; (b) schematic diagram of the influence of F₃EAI on bounded excitons in layered 2D perovskite [(BA)$_{1-x}$(F₃EA)$_x$]$_2$(MA)$_3$Pb$_4$I$_{13}$ material[160].

electronegative fluorine promotes the separation and transfer of photogenerated carriers (Figure 3.31b). This version proves the positive effects of heteroatom-containing organic amines. In another study, Chen et al. optimized physical properties using 4-phenylbutane-1-amine (PBA), which has a longer alkyl chain length than PEA, mixed with BA as an organic ligand[166]. The nucleation effect of PBA on films is similar to that of PEA, which reduces the nucleation site of the precursor, so as to have the advantages of large-grain, optimized-grain orientation, reduced trap density, and reduced charge recombination. (PBA$_{0.5}$BA$_{0.5}$)$_2$MA$_3$Pb$_4$I$_{13}$ PSCs achieved 16% high efficiency. In addition to the organic ligands mentioned earlier, medium-sized GA has also been studied as a second ligand cation[162]. The results show that the introduction of GA cations into BA-based 2D perovskites can improve the crystallinity and out-of-plane orientation of the film due to the increased interaction between inorganic layers. In addition, due to the medium size of GA cations, hybrid cationic systems can effectively reduce lattice distortion, prevent phase segregation, and give high stability to 2D structure.

In addition to the introduction of a second isolated cation in the BA system, it is foreseeable that the mixing of different types of organic ligands plays an indispensable role in promoting the evolution of 2D PSCs[167]. The effect of extra interval cations can be summarized as; (i) induction of precursor aggregation, reduction of nucleation sites, improvement of crystallinity and vertical orientation of perovskite grains, thereby improving PCE; (ii) strengthening the interaction with organic plates, reducing the deformation of inorganic layers, stabilizing the 2D perovskite structure, and improving the stability of solar cells; (iii) the introduction of organic amines containing large electronegative atoms can promote the dissociation of photogenerated carriers; (iv) the introduction of organic cations with a large dielectric constant can weaken the dielectric confinement, resulting in a decrease in the E_b. The mixed-interval cationic system combines the advantages of all bulk organic amine cations. Therefore, the selection and design of multiple organic spacer components based on structure and dielectric properties can accelerate exciton dissociation and carrier transport, so as to prepare efficient and stable 2D PSCs.

3.7   CONCLUSION

In the past few years, researchers have made comprehensive and in-depth improvements to 2D perovskites, aiming to develop the superior properties and practical applications of this material. However, the PCE of 2D PSCs still lags behind the most advanced multi-cation 3D PSCs. The reasons for this are complex. First, 2D perovskites are strongly constrained systems containing quantum and dielectric constraints, which leads to the formation of strongly bound excitons. In addition, the introduction of multiple organic ligands leads to changes in the stacking of inorganic layers (structural distortion), resulting in poor optical absorption and increased E_b. In addition, the anisotropy of 2D perovskite crystals and the films prepared by the solution methods typically withstand a mixture of multiple the QWs with random well-width (n-value) distributions. These disorders often adversely affect charge transfer. These shortcomings are mainly caused by the iteration of organic spacer cations, which is the root cause of the poor performance of 2D perovskite photovoltaics. In this chapter, various organic ligands affecting different physicochemical properties of 2D perovskites are reviewed. The effects of these ligands on film structure distortion, quantum and dielectric confinement effects, E_b, the QWs distribution, and solar cell performance are systematically discussed. The organic ligand layer can be summarized as follows: (i) chain length. It mainly affects the solubility of organic amines in precursor solutions, affects the phase distribution of films and the width of the QWs, and the characteristics of the QWs determine charge transfer. (ii) Branching. The branched chains increase the volume of organic spacer that raises the uncertainty of hosting the organic cation in 2D perovskites structure, stretching or compressing the inorganic framework, resulting in changes in the optical properties of the perovskite.

(iii) Conjugate structure. Organic ligands with conjugated structure typically have a high permittivity, which reduces quantum and dielectric confinement effects. Thus, it can improve the carrier cross-layer transport. (iv) Heteroatoms and heterocycles. Heteroatoms and heterocycles typically increase intermolecular interactions to stabilize 2D perovskites structure, providing polarity of organic amine molecules to reduce the dielectric confinement and facilitate carrier separation.

According to the relationship between structure and performance, reasonable design and selection of new organic amines is the key to the preparation of high-performance polystyrene materials. In this context, we propose design principles for novel organic amines that benefit photovoltaic applications: (i) appropriate charge. The number of amino groups that can be protonated determines the type of structure of the final 2D perovskite. Therefore, a new organic amine should first determine the amount of amino group that can be protonated. Considering previous reports, organic ligands containing more than three amino groups tend to form zero- or one-dimensional perovskites, which further limits the application of photovoltaics. (ii) Compatibility. Proper organic amine volume is essential to stabilize the structure of 2D perovskites. An organic amine that is too small may not be able to insert and separate the inorganic layer, but too large a volume (maximum cross-sectional area of about 40 Å^2) will cause serious structural distortion of the $[PbI_6]^{4-}$, resulting in lattice mismatch of adjacent inorganic layers, which seriously affects the efficiency of photovoltaic devices. (iii) High dielectric constant. As mentioned earlier, the high dielectric constant of organic ligands can reduce the quantum and dielectric confinement effect caused by the mismatch of dielectric constant between the no-machine well and the surrounding organic ligand and reduce the E_b. iv) The solubility is appropriate. In 2D perovskites, the solubility of organic amines is closely related to film-forming properties. Too high or too low solubility will cause organic amines to fail to precipitate at the same time as the inorganic layer due to the deviation of the eutectic point, resulting in multiple the QWs in the membrane. (v) Appropriate intermolecular interactions. The intermolecular interactions, including the intermolecular forces of organic amines and the hydrogen bonds between organic amines and inorganic layers, have a great influence on the stability of perovskite structure and the film-forming mechanism. Excessive intermolecular interactions not only cause distortion of inorganic layers and octahedrons but also affect the free stacking of components in the film-forming process. Conversely, weak interactions cause inorganic layers to slip, making the 2D structure unstable. Based on the aforementioned design principles, we believe that the PCE of 2D PSCs can be comparable to 3D PSCs, which provides great potential for the development and industrialization of PSCs in the future.

REFERENCES

1. A. Kojima, K. Teshima, Y. Shirai, T. Miyasaka, *Journal of the American Chemical Society* **2009**, *131*, 6050.
2. H. Zhou, *et al.*, *Science* **2014**, *345*, 542.
3. M. Gratzel, *Accounts of Chemical Research* **2017**, *50*, 487.
4. J.-P. Correa-Baena, M. Saliba, T. Buonassisi, M. Graetzel, A. Abate, W. Tress, A. Hagfeldt, *Science* **2017**, *358*, 739.

5. H. J. Snaith, *The Journal of Physical Chemistry Letters* **2013**, *4*, 3623.
6. J. Burschka, N. Pellet, S. J. Moon, R. Humphry-Baker, P. Gao, M. K. Nazeeruddin, M. Gratzel, *Nature* **2013**, *499*, 316.
7. Q. Jiang, *et al.*, *Nature Photonics* **2019**, *13*, 460.
8. S. Bai, *et al.*, *Nature* **2019**, *571*, 245.
9. NREL's 'best research-cell efficiencies' chart, https://www.nrel.gov/pv/cell-efficiency.html
10. G. Xing, *et al.*, *Science* **2013**, *342*, 344.
11. D. Shi, *et al.*, *Science* **2015**, *347*, 519.
12. Q. Dong, Y. Fang, Y. Shao, P. Mulligan, J. Qiu, L. Cao, J. Huang, *Science* **2015**, *347*, 967.
13. H. Oga, A. Saeki, Y. Ogomi, S. Hayase, S. Seki, *Journal of the American Chemical Society* **2014**, *136*, 13818.
14. W. J. Yin, T. Shi, Y. Yan, *Advanced Materials* **2014**, *26*, 4653.
15. L. Zheng, Y. Ma, S. Chu, S. Wang, B. Qu, L. Xiao, Z. Chen, Q. Gong, Z. Wu, X. Hou, *Nanoscale* **2014**, *6*, 8171.
16. T. Leijtens, *et al.*, *Advanced Energy Materials* **2015**, *5*, 1500963.
17. F. Arabpour Roghabadi, M. Alidaei, S. M. Mousavi, T. Ashjari, A. S. Tehrani, V. Ahmadi, S. M. Sadrameli, *Journal of Materials Chemistry A* **2019**, *7*, 5898.
18. Y. Zhou, Y. Zhao, *Energy & Environmental Science* **2019**, *12*, 1495.
19. G. Grancini, *et al.*, *Nature Communications* **2017**, *8*, 15684.
20. P. Huang, Q. Y. Chen, K. C. Zhang, L. G. Yuan, Y. Zhou, B. Song, Y. F. Li, *Journal of Materials Chemistry A* **2019**, *7*, 6213.
21. A. Agresti, *et al.*, *ACS Energy Letters* **2017**, *2*, 279.
22. W. Chen, L. M. Xu, X. Y. Feng, J. S. Jie, Z. B. He, *Advanced Materials* **2017**, *29*, 1603923.
23. Y. Zhang, X. Liu, P. Li, Y. Duan, X. Hu, F. Li, Y. Song, *Nano Energy* **2019**, *56*, 733.
24. K. Wang, *et al.*, *Nano Energy* **2019**, *58*, 175.
25. L. Wang, *et al.*, *Science* **2019**, *363*, 265.
26. W. Xiang, *et al.*, *Joule* **2019**, *3*, 205.
27. J. Duan, Y. Zhao, X. Yang, Y. Wang, B. He, Q. Tang, *Advanced Energy Materials* **2018**, *8*, 1802346.
28. W. Yan, H. Rao, C. Wei, Z. Liu, Z. Bian, H. Xin, W. Huang, *Nano Energy* **2017**, *35*, 62.
29. N. J. Jeon, J. H. Noh, W. S. Yang, Y. C. Kim, S. Ryu, J. Seo, S. I. Seok, *Nature* **2015**, *517*, 476.
30. A. D. Jodlowski, *et al.*, *Nature Energy* **2017**, *2*, 972.
31. J. Idigoras, *et al.*, *ACS Applied Materials & Interfaces* **2018**, *10*, 11587.
32. M. Wong-Stringer, *et al.*, *Advanced Energy Materials* **2018**, *8*, 1801234.
33. E. Bi, *et al.*, *Nature Communications* **2017**, *8*, 15330.
34. G. Grancini, M. K. Nazeeruddin, *Nature Reviews Materials* **2018**, *4*, 4.
35. X. Li, J. M. Hoffman, M. G. Kanatzidis, *Chemical Reviews* **2021**, *121*, 2230.
36. P. Huang, S. Kazim, M. Wang, S. Ahmad, *ACS Energy Letters* **2019**, *4*, 2960.
37. Y. Huang, *et al.*, *Journal of the American Chemical Society* **2021**, *143*, 3911.
38. S. Ahmad, C. Hanmandlu, P. K. Kanaujia, G. V. Prakash, *Optical Materials Express* **2014**, *4*, 1313.
39. T. Ishihara, J. Takahashi, T. Goto, *Solid State Communications* **1989**, *69*, 933.
40. X. Hong, T. Ishihara, A. V. Nurmikko, *Physical Review B* **1992**, *45*, 6961.
41. E. A. Muljarov, S. G. Tikhodeev, N. A. Gippius, T. Ishihara, *Physical Review B* **1995**, *51*, 14370.
42. B. Saparov, D. B. Mitzi, *Chemical Reviews* **2016**, *116*, 4558.
43. C. C. Stoumpos, *et al.*, *Chemistry of Materials* **2016**, *28*, 2852.
44. L. L. Mao, *et al.*, *Journal of the American Chemical Society* **2018**, *140*, 3775.
45. C. M. M. Soe, *et al.*, *Journal of the American Chemical Society* **2017**, *139*, 16297.

46. M. Yuan, *et al.*, *Nature Nanotechnology* **2016**, *11*, 872.
47. C. M. M. Soe, *et al.*, *Advanced Energy Materials* **2018**, *8*, 1700979.
48. R. L. Milot, *et al.*, *Nano Letters* **2016**, *16*, 7001.
49. Y. Lin, *et al.*, *The Journal of Physical Chemistry Letters* **2018**, *9*, 654.
50. J. M. Hoffman, *et al.*, *Journal of the American Chemical Society* **2019**, *141*, 10661.
51. D. H. Cao, C. C. Stoumpos, O. K. Farha, J. T. Hupp, M. G. Kanatzidis, *Journal of the American Chemical Society* **2015**, *137*, 7843.
52. I. Spanopoulos, *et al.*, *Journal of the American Chemical Society* **2019**, *141*, 5518.
53. G. Wu, *et al.*, *Matter* **2021**, *4*, 582.
54. J. M. Hoffman, C. D. Malliakas, S. Sidhik, I. Hadar, R. McClain, A. D. Mohite, M. G. Kanatzidis, *Chemical Science* **2020**, *11*, 12139.
55. Y. Chen, *et al.*, *Advanced Energy Materials* **2017**, *7*, 1700162.
56. N. Mercier, S. Poiroux, A. Riou, P. Batail, *Inorganic Chemistry* **2004**, *43*, 8361.
57. S. Sourisseau, N. Louvain, W. Bi, N. Mercier, D. Rondeau, F. Boucher, J.-Y. Buzare, C. Legein, *Chemistry of Materials* **2007**, *19*, 600.
58. A. H. Proppe, R. Quintero-Bermudez, H. Tan, O. Voznyy, S. O. Kelley, E. H. Sargent, *Journal of the American Chemical Society* **2018**, *140*, 2890.
59. L. Chao, T. Niu, Y. Xia, X. Ran, Y. Chen, W. Huang, *The Journal of Physical Chemistry Letters* **2019**, *10*, 1173.
60. J. Rodríguez-Romero, B. C. Hames, I. Mora-Seró, E. M. Barea, *ACS Energy Letters* **2017**, *2*, 1969.
61. M. E. Kamminga, *et al.*, *Chemistry of Materials* **2016**, *28*, 4554.
62. K. Z. Du, Q. Tu, X. Zhang, Q. Han, J. Liu, S. Zauscher, D. B. Mitzi, *Inorganic Chemistry* **2017**, *56*, 9291.
63. L. Mao, *et al.*, *Chemistry of Materials* **2016**, *28*, 7781.
64. I. C. Smith, E. T. Hoke, D. Solis-Ibarra, M. D. McGehee, H. I. Karunadasa, *Angewandte Chemie International Edition* **2014**, *53*, 11232.
65. X. Zhang, *et al.*, *Advanced Energy Materials* **2018**, *8*, 1702498.
66. W. Fu, J. Wang, L. Zuo, K. Gao, F. Liu, D. S. Ginger, A. K. Y. Jen, *ACS Energy Letters* **2018**, *3*, 2086.
67. J. Xi, *et al.*, *Journal of the American Chemical Society* **2020**, *142*, 19705.
68. C. Ran, *et al.*, *Joule* **2019**, *3*, 3072.
69. Z. Xu, *et al.*, *ACS Nano* **2020**, *14*, 4871.
70. J. Hu, *et al.*, *Nature Communications* **2019**, *10*, 1276.
71. W. Fu, H. Liu, X. Shi, L. Zuo, X. Li, A. K. Y. Jen, *Advanced Functional Materials* **2019**, *29*, 1900221.
72. H. Pan, X. Zhao, X. Gong, Y. Shen, M. Wang, *The Journal of Physical Chemistry Letters* **2019**, *10*, 1813.
73. J. Shi, Y. Gao, X. Gao, Y. Zhang, J. Zhang, X. Jing, M. Shao, *Advanced Materials* **2019**, *31*, 1901673.
74. F. Zhang, *et al.*, *Journal of the American Chemical Society* **2019**, *141*, 5972.
75. R. Yang, *et al.*, *Advanced Materials* **2018**, *30*, 1804771.
76. Q. Li, Y. Dong, *et al.*, *ACS Energy Letters* **2021**, *6*, 2072.
77. H. Lai, *et al.*, *Journal of the American Chemical Society* **2018**, *140*, 11639.
78. H. Lai, D. Lu, Z. Xu, N. Zheng, Z. Xie, Y. Liu, *Advanced Materials* **2020**, *32*, 2001470.
79. Y. Qin, *et al.*, *Advanced Energy Materials* **2020**, *10*, 1904050.
80. Y. Yan, S. Yu, A. Honarfar, T. Pullerits, K. Zheng, Z. Liang, *Advanced Science* **2019**, *6*, 1900548.
81. L. Yan, J. Hu, Z. Guo, H. Chen, M. F. Toney, A. M. Moran, W. You, *ACS Applied Materials & Interfaces* **2018**, *10*, 33187.
82. Y. Dong, D. Lu, Z. Xu, H. Lai, Y. Liu, *Advanced Energy Materials* **2020**, *10*, 2000694.
83. Z. Li, *et al.*, *Nano Letters* **2019**, *19*, 5237.

84. Y. Li, H. Cheng, K. Zhao, Z. S. Wang, *ACS Applied Materials & Interfaces* **2019**, *11*, 37804.

85. J. Li, *et al.*, *Organic Electronics* **2019**, *67*, 122.

86. D. B. Mitzi, C. A. Feild, W. T. A. Harrison, A. M. Guloy, *Nature* **1994**, *369*, 467.

87. J. Liu, J. Leng, K. Wu, J. Zhang, S. Jin, *Journal of the American Chemical Society* **2017**, *139*, 1432.

88. Y. Lin, *et al.*, *Nature Communications* **2019**, *10*, 1008.

89. T. He, *et al.*, *Nature Communications* **2020**, *11*, 1672.

90. Y. Liao, *et al.*, *Journal of the American Chemical Society* **2017**, *139*, 6693.

91. Y. Xu, M. Wang, Y. Lei, Z. Ci, Z. Jin, *Advanced Energy Materials* **2020**, *10*, 2002558.

92. J. M. Hoffman, *et al.*, *Advanced Materials* **2020**, *32*, 2002812.

93. C. J. Dahlman, *et al.*, *Chemistry of Materials* **2019**, *31*, 5832.

94. H. Tsai, *et al.*, *Nature* **2016**, *536*, 312.

95. X. Zhang, G. Wu, S. Yang, W. Fu, Z. Zhang, C. Chen, W. Liu, J. Yan, W. Yang, H. Chen, *Small* **2017**, *13*, 121.

96. X. Li, *et al.*, *Advanced Energy Materials* **2020**, *10*, 2001832.

97. W. Paritmongkol, N. S. Dahod, A. Stollmann, N. Mao, C. Settens, S.-L. Zheng, W. A. Tisdale, *Chemistry of Materials* **2019**, *31*, 5592.

98. J.-T. Lin, *et al.*, *Angewandte Chemie International Edition* **2021**, *60*, 7866.

99. P. Cheng, *et al.*, *ACS Energy Letters* **2018**, *3*, 1975.

100. G. B. Wu, *et al.*, *Advanced Materials* **2019**, *31*, 1903889.

101. K. Zheng, *et al.*, *Journal of Materials Chemistry A* **2018**, *6*, 6244.

102. J. X. Liu, J. Leng, K. F. Wu, J. Zhang, S. Y. Jin, *Journal of the American Chemical Society* **2017**, *139*, 1432.

103. J. C. Blancon, *et al.*, *Science* **2017**, *355*, 1288.

104. G. Xing, *et al.*, *Nature Communications* **2017**, *8*, 14558.

105. B. Luo, *et al.*, *The Journal of Physical Chemistry Letters* **2019**, *10*, 5271.

106. A. Lemmerer, D. G. Billing, *Crystengcomm* **2010**, *12*, 1290.

105. H. Ren, *et al.*, *Nature Photonics* **2020**, *14*, 154.

106. C. Liang, *et al.*, *Journal of Materials Chemistry A* **2020**, *8*, 5874.

107. C. Ma, D. Shen, T.-W. Ng, M.-F. Lo, C.-S. Lee, *Advanced Materials* **2018**, *30*, 1800710.

108. R. K. Misra, C. Bat-El, L. Iagher, L. Etgar, *Chemsuschem* **2017**, *10*, 3712.

109. J. V. Passarelli, D. J. Fairfield, N. A. Sather, M. P. Hendricks, H. Sai, C. L. Stern, S. I. Stupp, *Journal of the American Chemical Society* **2018**, *140*, 7313.

110. E. I. Marchenko, *et al.*, *Chemistry of Materials* **2020**, *32*, 7383.

111. E. H. Jung, *et al.*, *Nature* **2019**, *567*, 511.

112. X. Zheng, *et al.*, *Nature Energy* **2017**, *2*, 17102.

113. Y. Wang, T. Zhang, M. Kan, Y. Zhao, *Journal of the American Chemical Society* **2018**, *140*, 12345.

114. J. Zhang, *et al.*, *Joule* **2019**, *3*, 3061.

115. M.-H. Tremblay, J. Bacsa, B. Zhao, F. Pulvirenti, S. Barlow, S. R. Marder, *Chemistry of Materials* **2019**, *31*, 6145.

116. Z. T. Xu, D. B. Mitzi, C. D. Dimitrakopoulos, K. R. Maxcy, *Inorganic Chemistry* **2003**, *42*, 2031.

117. D. B. Mitzi, D. R. Medeiros, P. R. L. Malenfant, *Inorganic Chemistry* **2002**, *41*, 2134.

118. J. Hu, *et al.*, *ACS Materials Letters* **2019**, *1*, 171.

119. T. Schmitt, *et al.*, *Journal of the American Chemical Society* **2020**, *142*, 5060.

120. B. Cheng, *et al.*, *Communications Physics* **2018**, *1*, 80.

121. S. Dai, *et al.*, *Journal of the American Chemical Society* **2017**, *139*, 1336.

122. X. Li, *et al.*, *Advanced Energy Materials* **2018**, *8*, 1800815.

123. W. Zhao, S. Li, H. Yao, S. Zhang, Y. Zhang, B. Yang, J. Hou, *Journal of the American Chemical Society* **2017**, *139*, 7148.

124. M. L. Tang, Z. Bao, *Chemistry of Materials* **2011**, *23*, 446.

125. D. Bi, *et al.*, *Advanced Materials* **2016**, *28*, 2910.
126. O. F. Williams, Z. Guo, J. Hu, L. Yan, W. You, A. M. Moran, *The Journal of Chemical Physics* **2018**, *148*, 134706.
127. J. Wang, J. Leng, J. Liu, S. He, Y. Wang, K. Wu, S. Jin, *The Journal of Physical Chemistry C* **2017**, *121*, 21281.
128. D. B. Mitzi, K. Chondroudis, C. R. Kagan, *Inorganic Chemistry* **1999**, *38*, 6246.
129. C. Liu, W. Huhn, K. Z. Du, A. Vazquez-Mayagoitia, D. Dirkes, W. You, Y. Kanai, D. B. Mitzi, V. Blum, *Physical Review Letters* **2018**, *121*, 146401.
130. Y. Gao, *et al.*, *Nature Chemistry* **2019**, *11*, 1151.
131. H. Min, *et al.*, *Science* **2019**, *366*, 749.
132. N. Pellet, P. Gao, G. Gregori, T. Y. Yang, M. K. Nazeeruddin, J. Maier, M. Gratzel, *Angewandte Chemie International Edition* **2014**, *53*, 3151.
133. W. S. Yang, J. H. Noh, N. J. Jeon, Y. C. Kim, S. Ryu, J. Seo, S. I. Seok, *Science* **2015**, *348*, 1234.
134. N. Zhou, *et al.*, *Journal of the American Chemical Society* **2018**, *140*, 459.
135. Y. Y. Jiang, *et al.*, *ACS Energy Letters* **2019**, *4*, 1216.
136. L. Gao, *et al.*, *Angewandte Chemie International Edition* **2019**, *58*, 11737.
137. L. Mao, *et al.*, *Journal of the American Chemical Society* **2020**, *142*, 8342.
138. X. Li, *et al.*, *Journal of the American Chemical Society* **2019**, *141*, 12880.
139. W. Ke, *et al.*, *Journal of the American Chemical Society* **2020**, *142*, 15049.
140. G. Lv, *et al.*, *Nano Letters* **2021**, *21*, 5788.
141. X. Li, *et al.*, *Journal of the American Chemical Society* **2018**, *140*, 12226.
142. S. Ahmad, *et al.*, *Joule* **2019**, *3*, 794.
143. Y. Zheng, *et al.*, *Solar RRL* **2019**, *3*, 1900090.
144. L. Mao, *et al.*, *Chem* **2019**, *5*, 2593.
145. P. Li, *et al.*, *Angewandte Chemie International Edition* **2020**, *59*, 6909.
146. T. Niu, *et al.*, *The Journal of Physical Chemistry Letters* **2019**, *10*, 2349.
147. B.-E. Cohen, Y. Li, Q. Meng, L. Etgar, *Nano Letters* **2019**, *19*, 2588.
148. S. Yu, Y. Yan, M. Abdellah, T. Pullerits, K. Zheng, Z. Liang, *Small* **2019**, *15*, 1905081.
149. X. Zhang, T. Yang, X. Ren, L. Zhang, K. Zhao, S. Liu, *Advanced Energy Materials* **2021**, *11*, 2002733.
150. D. Wang, S.-C. Chen, Q. Zheng, *Journal of Materials Chemistry A* **2021**, *9*, 11778.
151. W. Ke, L. Mao, C. C. Stoumpos, J. Hoffman, I. Spanopoulos, A. D. Mohite, M. G. Kanatzidis, *Advanced Energy Materials* **2019**, *9*, 1803384.
152. D. Lu, G. Lv, Z. Xu, Y. Dong, X. Ji, Y. Liu, *Journal of the American Chemical Society* **2020**, *142*, 11114.
153. Z. Xu, D. Lu, X. Dong, M. Chen, Q. Fu, Y. Liu, *Advanced Materials* **2021**, *33*, 2105083.
154. Y. Li, *et al.*, *Nano Letters* **2019**, *19*, 150.
155. C. Ma, M.-F. Lo, C.-S. Lee, *Journal of Materials Chemistry A* **2018**, *6*, 18871.
156. H. Gu, *et al.*, *Nano Energy* **2019**, *65*, 104050.
157. Y. Zhang, *et al.*, *Journal of the American Chemical Society* **2019**, *141*, 2684.
158. P. Li, *et al.*, *Advanced Materials* **2019**, *31*, 1901966.
159. X. Lian, *et al.*, *Angewandte Chemie International Edition* **2019**, *58*, 9409.
160. S. Tan, *et al.*, *Advanced Energy Materials* **2018**, *9*, 1803024.
161. S. Chen, N. Shen, L. Zhang, W. Kong, L. Zhang, C. Cheng, B. Xu, *Journal of Materials Chemistry A* **2019**, *7*, 9542.
162. M. Long, *et al.*, *ACS Energy Letters* **2019**, *4*, 1025.
163. J. Meng, *et al.*, *Physical Chemistry Chemical Physics* **2019**, *22*, 54.
164. J. Qiu, *et al.*, *ACS Energy Letters* **2019**, *4*, 1513.
165. N. Zhou, *et al.*, *Advanced Energy Materials* **2019**, *10*, 1901566.
166. S. Chen, *et al.*, *Advanced Functional Materials* **2020**, *30*, 1907759.
167. G. Liu, *et al.*, *Journal of Materials Chemistry A* **2020**, *8*, 5900.

4 Regulating Phase Transition to Enhance the Performance of Perovskite Solar Cells

Yang Li, Yifan Zheng, Ibrahim Albrahee, Guoyu Ding, Juan Hou, Haonan Wang, Lixin Zhang, and Yuchuan Shao

4.1 PHASE TRANSITION OF HALIDE PEROVSKITES

4.1.1 PROCESS OF PHASE TRANSITION OF PEROVSKITES

The crystal structure, which is generally determined by the composition, ionic radius, and valence states of the constituent elements, plays a significant role in the fundamental properties of materials, such as optical, electronic, and mechanical properties. The last decade has witnessed significant developments in perovskite solar cells (PSCs), due to their suitable bandgaps and outstanding electronic and optical properties. Halide perovskites typically have an ABX_3 structure, where the A-site is the organic cation, the B-site is metal, and the X-site is a halide. Methylammonium (MA) was initially used as the A-site cation, and it is now partially or fully replaced by the large formamidinium (FA) components due to its relative higher thermal stability and broadened light absorption[1]. With efforts from all over the world, FA-dominant perovskites significantly prompted the efficiency of PSCs, now reaching 26.1%[2]. Perovskites have different structures, such as cubic, tetragonal, and orthorhombic. The cubic phase, which is essential for fabricating highly efficient FA-based PSCs, is easily converted to other useless phases. Overcoming the limitations of phase transitions is a crucial issue in obtaining high-quality perovskite films and high-performance PSCs. Here we briefly summarize the phase structures of perovskite. Subsequently, the origin of the phase transition in perovskites was intensively analyzed. Such factors, including temperature, pressure, composition, humidity, strain, and light, can have a large impact on the phase transition process and should be noted for the fabrication of highly efficient PSCs.

4.1.2 ORIGINS OF PHASE TRANSITION IN PEROVSKITES

Perovskites with the chemical formula of ABX_3 undergo several temperature-dependent phase transitions, corresponding mostly to rotation and distortion of

DOI: 10.1201/9781003400486-4

"

the $[BX_6]^{4-}$ octahedra. Typically, at a high temperature, the material maintains an extremely symmetric cubic structure (α-phase); as the temperature decreases, the cubic structure is distorted and the crystal symmetry is lowered, resulting in the phase transition to tetragonal, orthorhombic, or even hexagonal structure[3]. Experimentally, there is also a cursory phase category for perovskites: the black phase (e.g., cubic structure) and the yellow phase (e.g., hexagonal structure), which can be visibly distinguished by their color due to the distinct energy bands difference. The black phase is required for PSCs because the yellow phase is not photo-active due to its limited absorption range. In addition to the temperature, a shift in the ionic radius can modify the tolerance factor, leading to a shift in the perovskite phase structure[4]. For example, the pure $MAPbI_3$ commonly exhibits a tetragonal structure at RT as the A-site cations are not sufficiently large to fill up the empty voids between the $[PbI_6]^{4-}$ octahedral structures. While $FAPbI_3$ still turn into a hexagonal phase (δ-phase) at RT where the framework is broken since the FA^+ ions are too large. Tolerance factor is used for describing the influence of ion radius on the crystal structure of perovskite[5]. Cubic and tetragonal structures of the black phase can be obtained at RT and are widely used in highly efficient PSCs. As shown in Figure 4.1, regulating the

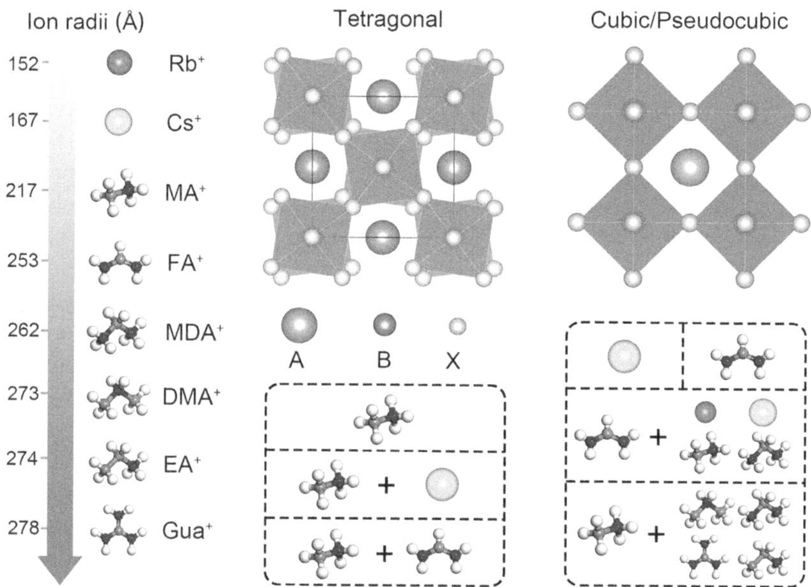

FIGURE 4.1 Illustrative diagram displaying the formation of tetragonal and cubic/pseudocubic phase perovskites, contrasted by different A-site cations. On the left is a gradient bar indicating ion radii in Angstroms, with various cations (Rb^+, Cs^+, MA^+, FA^+, MDA^+, DMA^+, EA^+, Gu^+) depicted alongside their molecular structures. To the right are two panels representing the tetragonal and cubic/pseudocubic perovskite structures. The tetragonal structure shows a three-dimensional network with spheres at the center surrounded by octahedra linked at corners with spheres at their vertices. The cubic/pseudocubic panel shows a similar arrangement but in a more symmetrical cubic layout. Beneath both panels are combinations of different cations that can form each structure, denoted by their molecular representations and an "A" for the A-site cation, "B" for the B-site cation, and "X" for the halide ion.

composition of perovskite not only increases the configuration entropy but also regulates the tolerance factor[6].

Polymorphism is observed in numerous classes of crystalline materials[7,8]. This phenomenon is based on the ability of a material to exist in several crystalline phases, with different structural arrangements of atoms or molecules[9,10]. Both organic and inorganic materials may exhibit polymorphism, which has been shown to greatly impact their optoelectronic properties[11]. It is thus intriguing to explore the possibility of combining two polymorphs of the same material to form a phase heterojunction (PHJ). A promising candidate for the demonstration of this concept is the material class of metal halide perovskites, exhibiting favorable optoelectronic properties, which have led to their successful integration into high-performance photovoltaic devices[12,13].

Perovskites exhibit multiple polymorphs depending on their composition. For example, the inorganic $CsPbI_3$ perovskite exhibits four different polymorphs[14], three of which are photoactive and have been exploited for the fabrication of photovoltaic devices[15]. α-$CsPbI_3$ can be formed by spin-coating a 1:1 CsI: PbI_2 solution in N, N-dimethylformamide (DMF) and heating to 335°C. With cooling to RT, the perovskite would not directly turn to the yellow phase (δ-$CsPbI_3$). Instead, $CsPbI_3$ tends to form metastable polymorphs, where the elevated-symmetry cubic phase (α-$CsPbI_3$) initially distorts to a tetragonal phase (β-$CsPbI_3$), followed by a further shift to the orthorhombic phase (γ-$CsPbI_3$). Their choice of the β-$CsPbI_3$ and γ-$CsPbI_3$ phases was motivated by their relatively low-temperature processing, unlike the α-$CsPbI_3$ that requires annealing at $T > 300$°C. Considering that solution processing of all polymorphs of $CsPbI_3$ requires the use of the same type of polar solvent[16], a hybrid deposition approach was utilized that combines both solution processing and thermal evaporation. Specifically, the narrower energy band of β-$CsPbI_3$ was deposited by a single-step solution deposition method, followed by vapor deposition of the γ-$CsPbI_3$ (Figure 4.2a). Experiments of energy-dispersive X-ray spectroscopy elemental mapping revealed that the distribution of Pb, Cs, and I is completely homogeneous across the layers, suggesting that these minute clusters are $CsPbI_3$ domains formed during evaporation. Cross-sectional SEM imaging was performed to examine the horizontal microstructure of the PHJ100 film (Figure 4.2b), which shows that the β-$CsPbI_3$ layer exhibits large vertical grains with few boundaries in the direction of transport. To further confirm the structure of the γ-$CsPbI_3$/β-$CsPbI_3$ PHJ, grazing-incidence XRD (GIXRD) experiments were performed on the PHJ100 sample by varying the incidence angle from 0.2° to 5° (Figure 4.2c). The lower incidence angles are more surface sensitive, and the XRD patterns (measured around 28°) show a stronger contribution of γ-$CsPbI_3$ (004) diffraction peak, with a smaller contribution of the β-$CsPbI_3$ (220) peak.

Recently, interfacial engineering through surface passivation has emerged as an effective method to tune the perovskite phase and enhance the performance of PSCs. For example, CsI has been used to passivate the perovskite surface, constraining the perovskite phase and dramatically enhancing its crystallinity. To analyze the influence of CsI passivation on the structure of the perovskite absorber layer, X-ray diffraction pattern (XRD) measurements were conducted (Figure 4.3a). CsI passivation was found to entirely remove the signal of PbI_2 at 12.6° in the XRD pattern. This suggests a full conversion of the reactant into the perovskite phase.

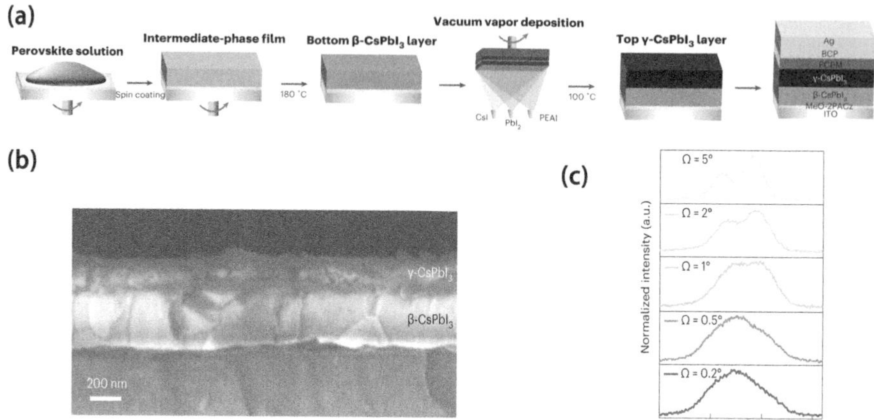

FIGURE 4.2 The image presents the fabrication process and crystallinity characterization of CsPbI$_3$ based PSCs. (a) A schematic of the fabrication process illustrates the perovskite solution being coated and annealed to form an intermediate phase film, followed by vacuum vapor deposition, creating top γ-CsPbI$_3$ and bottom β-CsPbI$_3$ layers. (b) A cross-sectional SEM image shows a clear distinction between the dense γ-CsPbI$_3$ and the porous β-CsPbI$_3$ layers within the solar cell structure. (c) GIXRD patterns for the PHJ100 sample are displayed, showing peaks corresponding to different orientations, with the peak intensities normalized to the largest peak in each set.

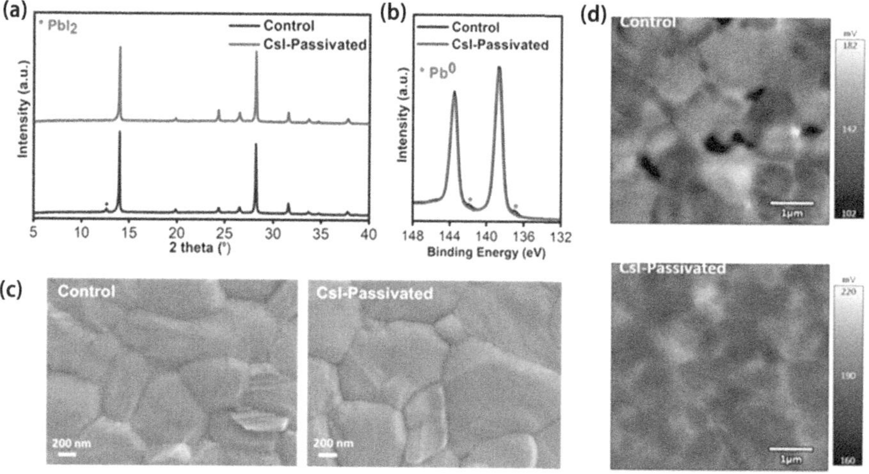

FIGURE 4.3 This composite image illustrates the comparative analysis of control and CsI-passivated perovskite films. (a) Two XRD patterns are displayed, one for the control and the other for the CsI-passivated film, indicating a prominent PbI$_2$ peak in the control sample which is suppressed in the passivated sample. (b) XPS spectra show the binding energy peaks for lead (Pb) 4f in both control and CsI-passivated samples, with a noticeable shift in the passivated sample suggesting a change in the chemical environment of Pb. (c) Top-view SEM images reveal the surface morphology of the films; the CsI-passivated film shows much larger grain sizes as compared to the control. (d) KPFM measurements illustrate the surface potential, where the CsI-passivated film exhibits a more uniform potential across the surface, contrasting with the heterogeneous potential seen in the control film.

Also, to probe whether undesired perovskite phases were formed upon the passivation with $CsPbI_3$, XRD measurements of the perovskite films at different temperatures (RT, 70°C, 120°C, and 160°C) were performed after the CsI post-treatment. Furthermore, X-ray photoelectron spectroscopy (XPS) measurements (Figure 4.3b) demonstrated that upon CsI passivation, any residual metallic Pb is seen in the control films and is entirely transformed into Pb^{2+}[17]. This is a key finding, as metallic lead aggregates form deep-level traps, a known source of non-radiative recombination and degradation[18]. Figure 4.3c shows the surface morphology of both the control and CsI-passivated perovskite films. CsI-passivated samples are uniform and highly crystalline with substantially larger perovskite grains as compared to the non-treated perovskite (control). In Figure 4.3d, Kelvin probe force microscopy (KPFM) measurements of control and CsI-passivated films demonstrate the striking effect of CsI in mitigating surface potential inhomogeneities of the perovskite films.

Based on the previous discussion of polymorphism via doping or surface passivation, it has been found that the phase of perovskites determines the performance of PSCs. Various factors can affect the phase state of perovskites, including temperature, pressure, composition, humidity, strain, light, etc. These factors can greatly influence the perovskite phase transition process, which controls the quality of the perovskite crystal and thus impressively affects the performance of PSCs.

4.1.2.1 Temperature Change

The widely used $MAPbI_3$ has two phase transitions: one is the phase transition between the octahedral structure and the tetragonal structure at around 161 K and the other is the phase transition between the tetragonal structure and the cubic structure at around 328 K[19]. Based on the analysis of thermally stimulated current (TSC), the phase transition at around 161 K has been shown to induce the formation of carrier traps[20–22]. Since this phase-transition temperature is much lower than the operating temperature of PSCs in naturally occurring terrestrial environments, the influence of the low-temperature phase transition on device performance should be negligible. However, the high-temperature phase transition is at a temperature just slightly higher than RT[23]. Therefore, how the high-temperature phase transition affects the device performance and stability must be understood to develop PSCs that can pass strict lifetime tests under high temperatures up to 85°C for future practical applications.

Figure 4.4a shows the differential scanning calorimetry (DSC) properties of the perovskite powders[24]. Similarly to the reported DSC properties of $MAPbI_3$ single crystals, a reversible phase transition between the tetragonal structure and the cubic structure was observed at 54.6 and 56.2°C for the exothermic and endothermic processes, respectively[25]. To understand the thermal properties of the perovskite alloy compounds, thermos gravimetric analysis (TGA) and DSC were performed on $MA_{0.6}FA_{0.4}PbI_3$, $MAPbI_{2.6}Br_{0.4}$, $MA_{0.6}FA_{0.4}PbI_{2.8}Br_{0.2}$, and $MAPbI_3$ powders prepared by drying the precursor solutions used for the device fabrication at 100°C for 30 min in a nitrogen-filled glove box[26]. The TGA results in Figure 4.4b show that $MA_{0.6}FA_{0.4}PbI_3$ and $MA_{0.6}FA_{0.4}PbI_{2.8}Br_{0.2}$ have similar thermal decomposition temperatures while the decomposition temperature of $MAPbI_3$ is as high as 300°C. On the other hand, $MAPbI_{2.6}Br_{0.4}$ has a relatively low decomposition temperature of around 200°C that is overlapped with pure $MAPbBr_3$, which may be caused by

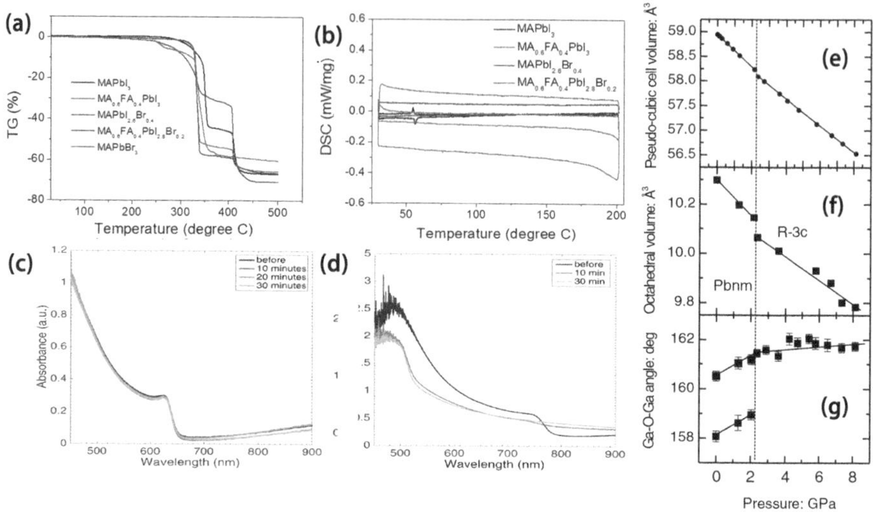

FIGURE 4.4 This multi-part figure showcases various properties of perovskite materials under different conditions. (a) Displays the thermogravimetric analysis (TGA) curves for various perovskite films, showing weight loss over a temperature range up to 500°C. Different compositions, including $MAPbI_3$, $MAPbBr_3$, $FAPbI_3$, and a mixed perovskite, are compared. (b) Shows DSC traces for the same perovskite powders, highlighting their thermal transitions. Absorption spectra are presented in (c) for $CsPbI_2Br$ and (d) for $MAPbI_3$, each showing the light absorption efficiency over a range of wavelengths. Finally, panels (e), (f), and (g) display in-situ single-crystal X-ray structure determinations for $LaGaO_3$ under pressure. These graphs depict the unit-cell volume, octahedral volume, and Ga-O-Ga angles variation with increasing pressure, respectively, showing how the structure becomes more compact as pressure is applied.

decomposition of $MAPbI_{2.6}Br_{0.4}$ to $MAPbI_3$ and $MAPbBr_3$. At elevated temperatures, chemical decomposition and crystalline phase transitions can easily occur. The remaining three perovskite alloys have no detectable phase transition between RT and 200°C, suggesting that possible phase-transition-related degradation may not exist in PSCs during device operation at temperatures around 60°C[27].

Several temperature-dependent studies have been performed on $MAPbI_3$, providing significant information on the phase changes of this material[28,29]. Above 330 K, the structure is cubic whereas at RT a phase transition towards a tetragonal structure is involved where the PbI_6-octahedra rotates around the c-axis[30]. Continuing decreasing the temperature, another phase transition is observed at approximately 160 K[31–33] (tetragonal-to-orthorhombic transition). Various experiments have demonstrated drastic changes in the performance of PSCs under thermal stress[34,35]. For commercial certification, PSCs must be able to operate at temperatures as low as -40°C and as high as 85°C[36]. These instabilities do not appear for large single crystals. For instance, a powder of $MAPbX_3$ is thermally stable up to 200°C, but achieving thermal stability at temperatures considerably lower than 200°C in $MAPbX_3$ thin films is challenging because of the steep surface-to-volume ratio[37]. At RT, $MAPbI_3$ films are stable, but a temperature increase of 20° shifts the structure of $MAPbI_3$ from its photo-active phase to its tetragonal photo-inactive phase. Also, chemical degradation occurs when $MAPbI_3$ is exposed to moisture, rapidly decaying into PbI_2, H_2O, and $MAPbI_3$[38]. Formamidinium-based perovskites are considerably more stable than methyl ammonium-based perovskites at elevated temperatures, but the photo-inactive phase is favored at RT. When exposed to moisture and polar solvents formamidinium-based perovskite undergoes rapid degradation[39].

Recent work has demonstrated that perovskites based on cesium are more stable than hybrid inorganic perovskites. Figure 4.4c, d show the absorption spectra of $CsPbI_2Br$ and $MAPbI_3$ after heating to 180°C for 10, 20, and 30 mins. After ten mins of heating, the absorption onset at 780 nm vanished, indicating chemical degradation, while $CsPbI_2Br$ experienced negligible changes[40].

4.1.2.2 Pressure Change

Applying pressure to a material can induce structural changes to alter and tune its electronic and optical properties[41]. The response of lead halide perovskites under compression is therefore a subject of considerable interest. Piezochromic transitions under pressure have been observed in several hybrid perovskites, which signal changes in the fundamental gap and electronic structure. It is most likely caused by the phase transition initiated by the rotation of the octahedra and the bond contraction under pressure. Jaffe et al. reported the orthorhombic structure of $MAPbI_3$ transformed to a cubic phase (Im3) at a pressure of 0.3 GPa and amorphization ensued above 2.7 GPa[42, 43]. Perovskites in which the A and B cations have the same formal charge, including $NdNiO_3$, $LaAlO_3$, $LaGaO_3$, $YAlO_3$, $GdFeO_3$, and $GdAlO_3$, in which the BX_6 octahedra are therefore more compressible than the AX_{12} sites. As a consequence, the tilts of the octahedral decrease with increasing pressure and thus evolve back towards the cubic aristo type[44]. Experimental difficulties in $LaAlO_3$ have thus far prevented the determination of the structure up to the phase transition at 14 GPa. The single rotation angle of the octahedral decreases continuously with lower pressure and extrapolates to zero in the neighborhood of the known transition pressure[45]. The reduction in unit cell volume with increasing pressure is provided by the compression of the octahedra, which is softer than the structure as a whole, since the reduction in tilt angle contributes, in the absence of additional distortions, to the increase in volume. These principles can be explored in additional detail with $LaGaO_3$ that has Pbnm symmetry at ambient conditions and exhibits the same phase transition to R-3c symmetry at both elevated temperature[46] and extreme pressures (Figure 4.4, e–g).

4.1.2.3 Composition Change

The orthorhombic-to-tetragonal phase transition in perovskites can significantly alter their optical, electrical properties and affect the corresponding applications. Here they report a systematic investigation of the size-dependent orthorhombic-to-tetragonal phase transition using a combination of temperature-dependent optical, electrical transport, and transmission electron microscopy studies. Previous studies have shown that the physical size of a material can be an essential variable in determining the phase transition points in addition to the compositions[47]. Numerous compositions have been reported for hybrid halide perovskites, with composition engineering objectives of tailoring absorption spectra, charge transport properties, and morphology in addition to crystal structure stabilization.

Comprehensive investigations of optical and electronic properties as well as photovoltaic performance of PSCs based on $FA_xMA_{1-x}PbI_3$[48] and $FA_xMA_{1-x}PbI_{3-y}Br_y$[49] have been reported. For a commonly used triple-cation (Cs, FA, MA) combination, a comprehensive study of the composition dependences for bandgap, carrier lifetime, and photovoltaic performance parameters was conducted, and the composition regions expected to yield superior performance were identified[50]. In addition, triple-cation perovskites such as (Rb, FA, MA) perovskites have also been reported[51]. In general, mixed-composition perovskite devices have demonstrated obvious improvements in stability, not only upon storage but also under continuous illumination and/or thermal stress[52,53]. In contrast to $MAPbI_{3-x}Cl_x$, $FA_{0.83}Cs_{0.17}Pb(I_{0.6}Br_{0.4})_3$ based devices exhibited stable performance under continuous illumination for more than 600 h without encapsulation[54]. It should also be noted that there have been many reports concerning the outdoor stability of the mixed FA-based PSCs[55-57]. For instance, devices with a $(FAPbI_3)_{0.85}(MAPbBr_3)_{0.15}$ active layer exhibited stable outdoor performance for more than 1,000 h (winter, Barcelona, Spain)[58].

To establish that partially replacing methylene diammonium dichloride ($MDACl_2$) with hexamethylenetetramine (HMTA) could enhance the phase stability of $FAPbI_3$-M, polarization transfer (DEPT-135, 1H–13C) NMR spectroscopy (Figure 4.5a, b, respectively) was conducted, inferring that HMTA-H^+ and NH^{4+} are kinetic products of $MDACl_2$ degradation, while THTZ-H^+ is a thermal dynamically favored product[59]. The mechanics of the formation of THTZ-H^+ from MDA^{2+} are shown in Figure 4.5c (highlighted red-purple). 1H magic angle spinning (MAS) NMR spectra of $FAPbI_3$, $FAPbI_3$-M, and $FAPbI_3$-H crystals were shown in Figure 4.5d. While the spectra are dominated by intense FA^+ signals (6 – 9 ppm), several additional, well-resolved signals (highlighted in blue) are present in both $FAPbI_3$-M and $FAPbI_3$-H spectra that are not present in neat $FAPbI_3$ crystals. These signals correspond to the 1H environment in organic species other than FA^+. To confirm that the additive detected in the solid state is actually THTZ-H^+, 13C MAS NMR (Figure 4.5e) was performed. As in the case of 1H, fresh signals (154.5, 59.4 ppm) in $FAPbI_3$-M and $FAPbI_3$-H were observed, which are identical in both materials but are absent in reference to $FAPbI_3$. This result corroborates that the same species are present in $FAPbI_3$-M and $FAPbI_3$-H despite their differing growth environments. Figure 4.5f shows the 1H–1H correlation spectroscopy of $FAPbI_3$-H. A strong crossover between FA^+ and THTZ-H^+ was observed, as well as among GBL and both cations. As a result, a mixed phase containing THTZ-H^+ and FA^+ is presented in $FAPbI_3$-H and $FAPbI_3$-M, rather than an isolated THTZ-H^+ secondary phase.[127] I nuclear quadruple resonance (NQR) spectroscopy was conducted on the three materials (Figure 4.5g) to better isolate the independent effects of THTZ-H^+ and chloride incorporation.

4.1.2.4 Humidity Change

In order to obtain high-quality perovskite thin films with enhanced optoelectronic properties, it is crucial to understand the moisture-induced phase transitions.

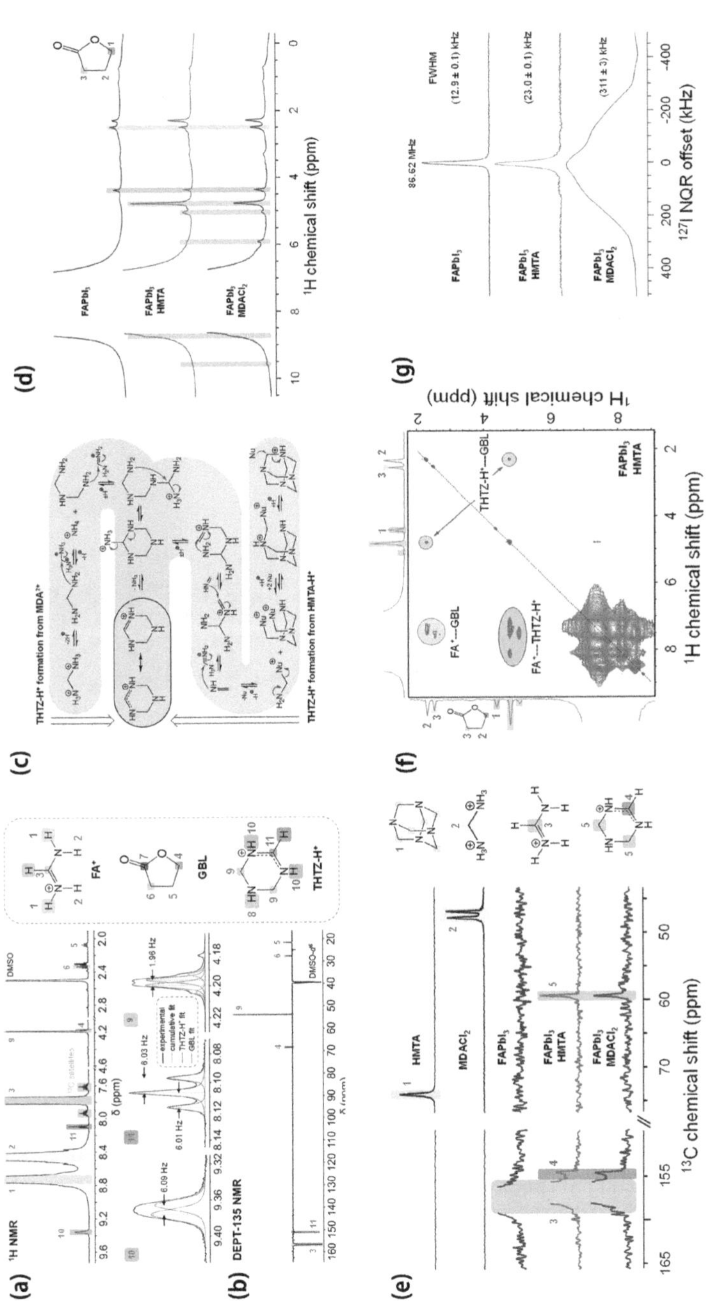

FIGURE 4.5 The image is a compilation of NMR and NQR spectral analyses of FAPbI$_3$-M single crystals. (a) Shows a ^{1}H NMR spectrum with inset details highlighting the expansion of signals for THTZ-H$^+$. (b) Presents a DEPT-135 NMR spectrum identifying various carbon types within the crystal structure. (c) Illustrates the proposed mechanisms for the formation of THTZ-H$^+$ from MDA^{2+} or HMTA-H$^+$. (d) Features a ^{1}H MAS NMR spectrum comparing neat FAPbI$_3$, FAPbI$_3$-H, and FAPbI$_3$-M crystals, with some signals in blue being unassigned. (e) Exhibits a ^{13}C echo-detected NMR spectrum contrasting FAPbI$_3$ crystals with different additives. (f) Displays a ^{1}H-$_1$H spin-diffusion NMR spectrum for FAPbI$_3$-H, and (g) shows ^{127}I NQR spectra of FAPbI$_3$ in its various forms. Each spectrum provides insights into the composition and molecular interactions within the crystals.

A mechanistic understanding of perovskite film formation with reduced trap density by exposing the deposited film to air before thermal annealing was reported. Unlike previous reports, exposure to ambient-air, rather than annealing, plays a key role in dissociation and the conversion of the intermediate phase to the perovskite phase[60]. The composition of the intermediate phase was clarified and the transition of this phase was attributed to the moisture-induced decomposition of the organic product, i.e., methylammonium acetate (MAAc). Upon thermal annealing, the ambient exposed film is transformed into a compact perovskite film with large crystalline domains and a deep trap density that is one order of magnitude lower than that of a film directly annealed in an inert atmosphere[61] (shown in Figure 4.6). Moderate humidity levels or short exposure to excessive humidity led to electron transfer from water molecules to perovskite lattices. However, exposure to elevated humidity levels for longer durations leads to degradation of perovskites via different routes[62]. In general, there are two main pathways on the degradation of $MAPbI_3$ in the presence of

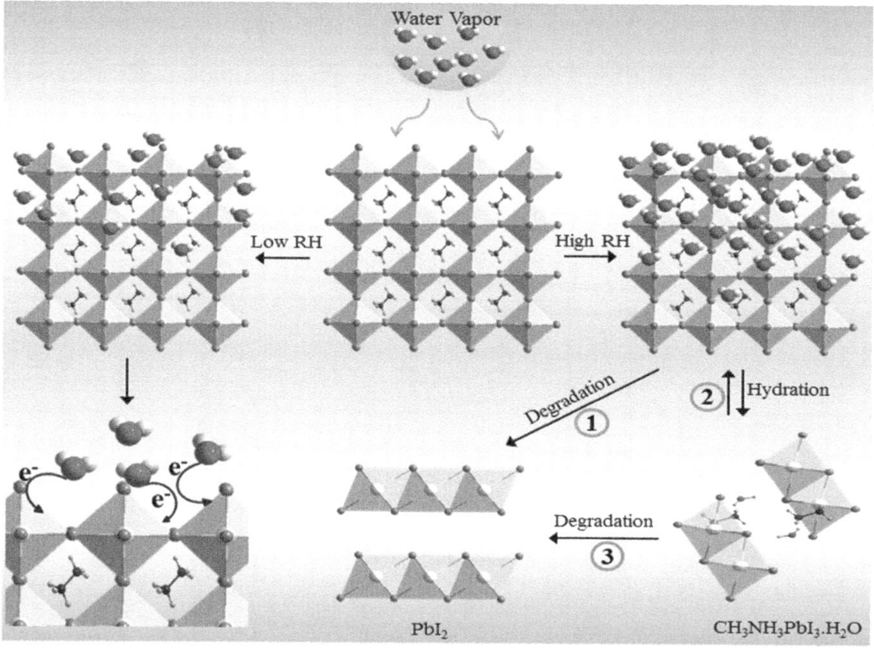

FIGURE 4.6 The diagram illustrates the interaction of perovskite structures with moisture and the subsequent effects on the material. At low RH, the perovskite structure is intact, depicted by blue octahedra and red spheres. As water vapor is introduced, shown at high RH, the structure begins to absorb water, leading to hydration and structural alteration. The process is visualized in two steps: (1) desorption of an electron from the perovskite, represented by small black spheres leaving the structure and (2) hydration where water molecules, shown as red spheres with two lighter atoms, integrate into the structure. The final stage (3) shows degradation, where the perovskite framework breaks down into PbI_2 and $CH_3NH_3PbI_3 \cdot H_2O$, highlighting a significant change in the structural integrity as a result of interaction with water vapor.

humidity. The first pathway is governed by catalytic action, where water triggers the expulsion of MA^+, resulting in degradation of the perovskite into PbI_2 and NH_3 gas. In the second pathway, $MAbI_3 \cdot H_2O$ and $(MA)_4PbI_6 \cdot 2H_2O$ are formed in the presence of moisture, eventually resulting in the degradation of perovskite[63]. According to previous reports, hydration of perovskite ($MAPbI_3 \cdot H_2O$ or $(MA)_4PbI_6 \cdot 2H_2O$) is reversible when the humidity level is moderate or the sample is exposed afterward to a dry gas such as N_2[64]. In particular, $MAPbI_3 \cdot H_2O$ is known to be metastable and capable of losing the water molecules spontaneously, transforming back to $MAPbI_3$[65].

4.1.2.5 Strain Change

Strain is defined as a deformation of the crystal structure caused by an applied stress. In turn, if the lattice length decreases under stress, the strain is compressive. Strain can be generated internally and externally, depending on the source of stress: (i) internal strain is intrinsic in perovskite crystals and is caused by the non-periodicity of the crystal lattice in the absence of any external stress. The crystal symmetry disruption in this case typically stems from (a) $[BX_6]^{4-}$ octahedra tilting and shift in B–X–B bond angle, thus deviating from the ideal cubic symmetry; (b) heterogeneous crystallization of polycrystalline perovskite films. (ii) External strain is extrinsic in perovskite crystals and is characterized by the distortion of crystal lattice periodicity in the presence of external effects: (a) lattice and thermal expansion mismatch between perovskite film and the adjacent substrate during the thermal annealing process; (b) external stress conditions (e.g., light, temperature, external pressure, applied bias)[66,67].

Due to the ion migration nature of perovskite materials, strain effects on perovskite films have recently been recognized as one of the key factors affecting their optoelectronic properties and device stability. Using piezo response force microscopy (PFM), Huang et al. demonstrated how strain is created in halide perovskite films upon cubic to tetragonal phase transition after perovskite annealing, leading to the formation of stripe-shaped domains[68]. In addition, Rothmann et al. have utilized transmission electron microscopy (TEM) under low dosage and rapid acquisition conditions to investigate the creation of strain during phase transition of $MAPbI_3$ films[69]. It was reported that tensile strain in perovskite films can reach a value high enough to deform copper (excess of 50 MPa in magnitude), which can provide a dramatic driving force to the formation of vacancies[70], facilitate phase segregation, accelerate ion migration and subsequent phase transition, and eventually decompose the perovskite films while exposed to humidity, heat, and illumination[71,72].

Deviations of the $[BX_6]^{4-}$ octahedral orientation in ideal cubic symmetry cause changes in B–X bond length and interaction between their electronic orbitals, resulting in the alternation of the electronic band structure of perovskite[73]. As shown in Figure 4.7a, among the three cations suitable for incorporation into the cubic lead-halide perovskite structure, cesium (Cs^+) has the smallest ionic radius, followed by methylammonium (MA^+) and formamidinium (FA^+). Although these steric effects play the dominant role in the octahedral tilt of inorganic halide perovskites (e.g., $CsPbX_3$), the hydrogen bonding between organic A-cations and the halides is also considered to be responsible for the octahedral tilt in hybrid halide perovskites such as $FAPbX_3$ or $MAPbX_3$ (Figure 4.7b)[74]. Such strain induced by the $[PbX_6]^{4-}$ octahedral tilt could

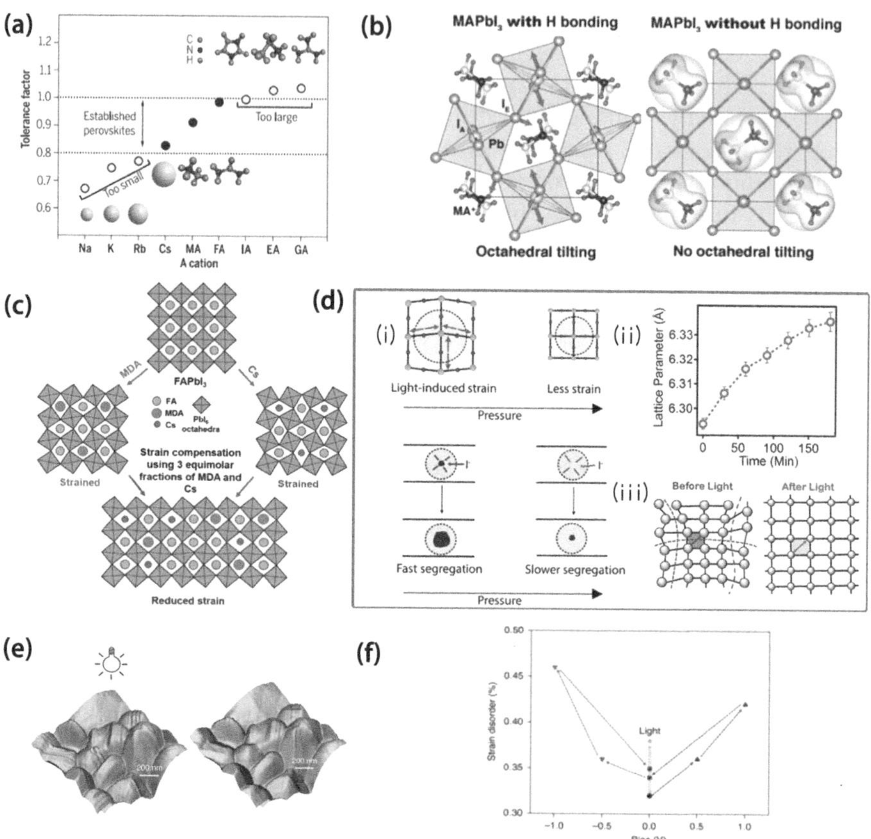

FIGURE 4.7 The figure presents a multifaceted study on the structural dynamics of APbI₃ perovskites. (a) Displays a graph of the Goldschmidt tolerance factor, assessing the structural fit of various A-cations within the perovskite lattice. (b) Illustrates two MAPbI₃ perovskite structures, one with hydrogen bonding causing octahedral tilting and one without hydrogen bonding and no tilting. (c) Depicts hexagonal patterns symbolizing the strain compensation in perovskite materials through compositional engineering. (d) Contains diagrams and a graph showing how external stress like pressure and light affects halide segregation, alongside a schematic of the light-induced lattice expansion mechanism. (e) Provides 3D topography images of perovskite film surfaces, contrasting light-induced corrugation with the smoother surface in absence of light. (f) Shows a graph illustrating how electrical bias and light alter the strain disorder within the perovskite structure.

potentially be controlled by tuning the hydrogen bonding degree and balanced partial substitution of A-site cations with suitable cations of different radii (Figure 4.7c).

Chen et al. demonstrated that the thermal expansion upon illumination in MAPbI₃ perovskites is mainly responsible for the photo-induced strain[75]. Such thermal expansion of unit cells frequently leads to reduced activation energy for halide vacancies, resulting in ion migration and weakened Pb–X bond, hence tensile strain (Figure 4.7d (i)). On the other hand, the optoelectronic properties of the perovskite film can also be positively affected by the photo-induced lattice expansion (Figure 4.7d (ii)), which the authors attributed to the strain relaxation under light (Figure 4.7d (iii)). In addition to the strain disorder induced by the illumination, a positive electrical bias induces the corrugated surface morphology shown in Figures 1.7e, f, while a negative bias leads to the removal of these inhomogeneous features.

Zhu et al. found that the bandgap decreases as the perovskite film experiences compressive strain compared to a strain-free sample, whereas the bandgap increases under exposure to tensile strain, as shown in Figure 4.8a, b[76]. Min et al. found a more stable structure of α-CsPbI$_3$ by introducing methylene ammonium chloride (MDACl$_2$) into the perovskite precursor, owing to the lattice strain relaxation of partial substitution of I site by smaller Cl-ions, as illustrated in Figure 4.8c[77]. Zhang et al. demonstrated an in-situ crosslinking-enabled strain regulating crystallization method by introducing trimethylolpropane Triacrylate (TMTA) into the antisolvent of chlorobenzene (CB) to confine the thermal expansion[78]. The appropriate amount of TMTA in CB can precisely regulate the lattice strain of the perovskite film (shown in Figure 4.8d).

4.1.2.6 Light Influence

Light intensity shows great influence on the phase stability of perovskite. The efficiency of PSCs will degrade under continuous light exposure. Until now, it is not clear which dominating mechanism is responsible for the performance degradation of PSCs due to light influence.

Glowienka et al. utilized he drift-diffusion model to explain the performance loss of PSCs from the charge-carrier recombination and series or shunt resistance[79]. Ji et al. systematically studied CsPbI$_3$-based PSCs with different phases. Figure 4.9a displays the ultraviolet–visible (UV-vis) absorption spectra of perovskite films. The absorption edges of the γ- and β-CsPbI$_3$ are approximately 710 nm and 740 nm, respectively, which is consistent with previous reports in the literature[80]. These measurements additionally confirm that the phases of the CsPbI$_3$ films deposited by solution and evaporation absolutely correspond to the β and γ phases. To investigate the optical properties of the perovskite films, the absorption and photoluminescence (PL) properties were characterized. The steady-state PL measurements are shown in Figure 4.9b. The PL peak positions of γ- and β-CsPbI$_3$ are at approximately 710 nm and 740 nm, respectively, which is consistent with their absorption edges. The PL spectra of the PHJ samples are centered around 740 nm, consistent with light emission only from the low-bandgap β-CsPbI$_3$. Curiously, no emission from the γ-CsPbI$_3$ is observed, even for the PHJ samples with the thickest γ-CsPbI$_3$ layers. Monitoring the intensity of the PL and the corresponding evolution of the photoluminescent quantum yield (PLQY; Figure 4.9c) reveals an intriguing trend. With the introduction of 10 nm of γ-CsPbI$_3$ on top of the β-CsPbI$_3$, the PLQY increases by a factor of two, suggesting a thin layer of γ-CsPbI$_3$ can effectively passivate trap states at the surface of the β-CsPbI$_3$, thus suppressing non-radiative recombination. Figure 4.9d shows the time-resolved luminous decay curves of the common and PHJ-based perovskite films. To further investigate the origin of the improved photovoltaic performance, they focused on the characterization of the β-CsPbI$_3$-based and the PHJ100-based PSCs. Light intensity-dependent open circuit voltage (VOC)

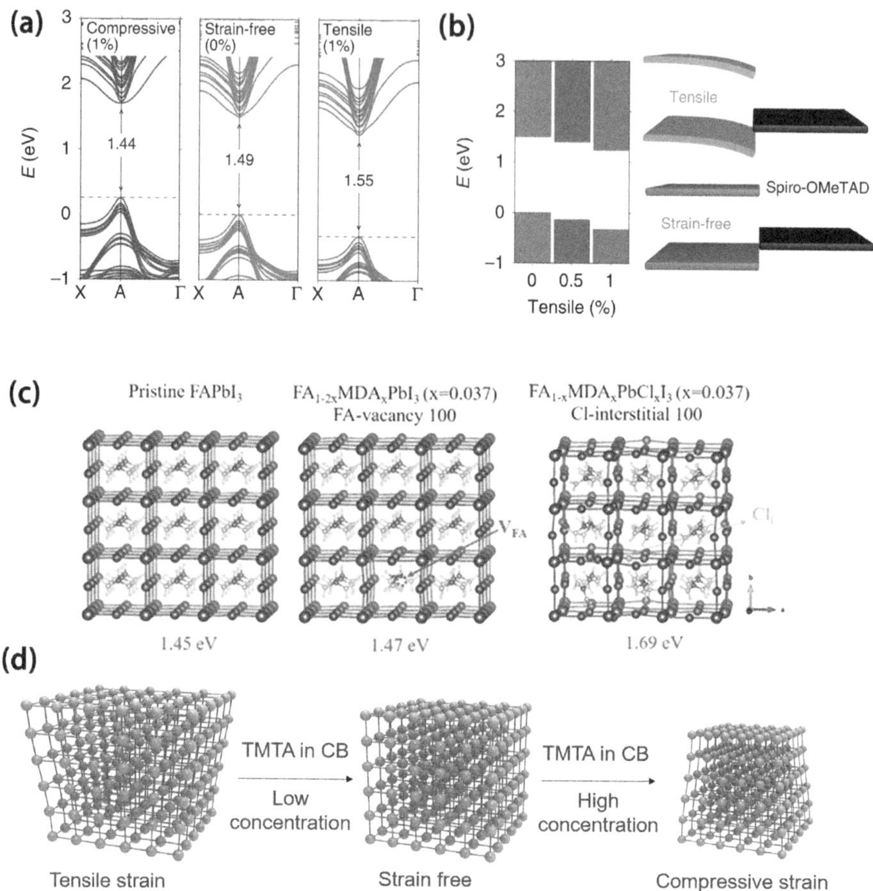

FIGURE 4.8 The image is a comprehensive visualization of the effects of mechanical strain on the electronic properties and structure of perovskite films. (a) Shows energy vs. wavevector graphs for perovskite films under compressive, strain-free, and tensile conditions, with bandgap values indicated, based on the first-principle DFT calculations. (b) Illustrates the change in the bandgap energy of perovskite films with varying strains, highlighting the relationship between strain and electronic structure. (c) Displays atomic structures of pristine FAPbI$_3$ perovskites and two variants with different Cl$^-$ ion compositions, showing lattice adjustments and energy levels. (d) Depicts a schematic of how tensile and compressive strains affect the distribution of TMTA in the CB of the perovskite structure, altering the material's electronic properties. Each section includes a reference to the source of the reproduced content.

measurements (Figure 4.9e) show that the PHJ-based PSCs exhibits an overall lower trap-assisted recombination, as the ideality factor (n) measured for the reference β-CsPbI$_3$ device (n =1.703) is reduced to n =1.163 for the PHJ devices. While it is not trivial to ascribe charge recombination mechanisms from light intensity-dependent VOC and ideality factor measurements in PSCs[79], these measurements are supported by direct characterization of the sub-bandgap trap density in the β-CsPbI$_3$ and PHJ100 films by photothermal deflection spectroscopy (PDS). The PDS spectra

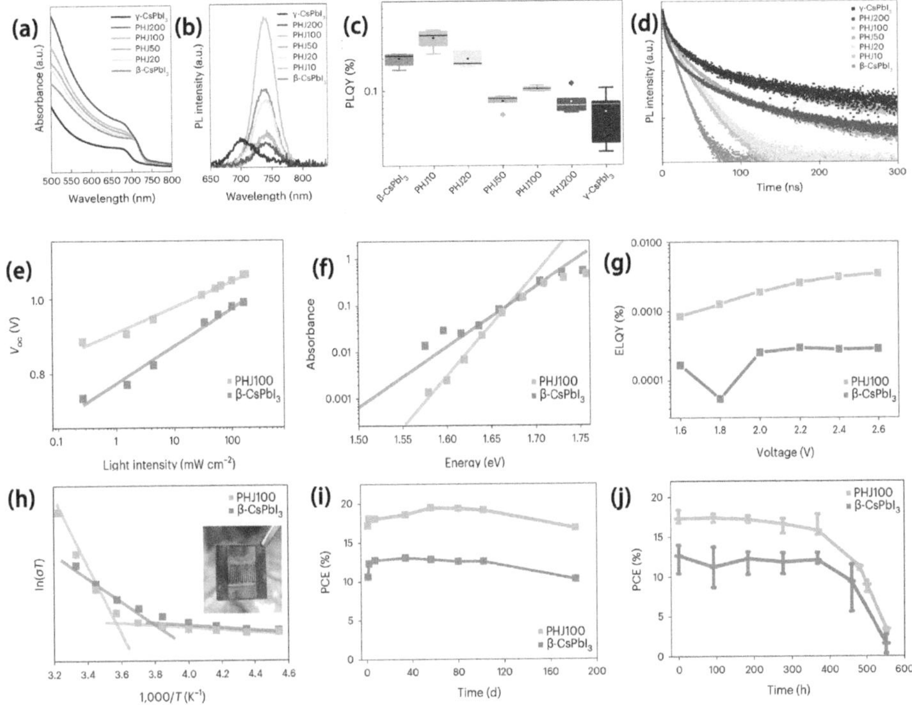

FIGURE 4.9 The figure provides a detailed comparison of the optical properties and photovoltaic performance of different CsPbI$_3$ perovskite films. (a) UV-vis absorption spectra demonstrate the wavelength-dependent light absorption of γ-CsPbI$_3$, β-CsPbI$_3$, and PHJ-based CsPbI$_3$ with varying layer thicknesses. (b) PL spectra show the emission profiles of the same materials. (c) A bar graph compares the PLQY of the films, indicating the efficiency of light emission. (d) PL decay curves illustrate the lifetime of excited states in the materials. Photovoltaic parameters including (e) power conversion efficiency (PCE) vs. light intensity, (f) external quantum efficiency (EQE) vs. energy, (g) current density-voltage (J-V) curves, (h) temperature dependence of PCE, (i) long-term operational stability, and (j) humidity stability for both β-CsPbI$_3$ and PHJ-based PSCs are graphed, with insets showing a device and its stability over time.

(Figure 4.9f) show that the PHJ100 structure results in a notably reduced energetic disorder, corresponding to an Urbach energy of 19.80 meV, in comparison with 33.27 meV for the β-CsPbI$_3$ thin films. The electroluminescence quantum yield (ELQY) of the PHJ-based devices is approximately one order of magnitude higher than that of the common β-CsPbI$_3$-based PSCs (Figure 4.9g). To explore the origin of the reduced hysteresis in the PHJ devices, they characterized the temperature-dependent conductivity of the common β-CsPbI$_3$ and PHJ-based structures integrated in lateral devices, shown in the inset of Figure 4.9h. To evaluate the stability of the PHJ devices, they monitored their performance upon storage under dim light conditions (Figure 4.9i). The PHJ-based device showed improvement efficiency over the first 50 d, reaching a maximum performance of 20.17%. The performance remained stable up to 100 d and decreased only slightly after 180 d. Importantly, throughout the stability study, the PHJ-based devices consistently exhibited superior performance to the common β-CsPbI$_3$-based PSCs. Moreover, the evolution of the PCE of the devices were tracked upon continuous illumination under sun conditions (Figure 4.9j). Both types of devices maintain their performance for approximately 400 h, followed by a significant drop.

4.2 INFLUENCES OF PHASE TRANSITION ON THE PROPERTIES OF HALIDE PEROVSKITES

4.2.1 CRYSTAL STRUCTURE

MAPbI$_3$ has a crystal structure of perovskite with a general chemical formula of ABX$_3$ in a simple cubic lattice. The space group is Pm$\bar{3}$m. In principle, any combination of compatible ions can form a stable perovskite crystal, where the tolerance and octahedral factors are satisfied within an empirically determined range, depending on the ionic radii of the constituents. The anion corner-shared 3D network of the PbI$_6$ octahedra can be viewed in Figure 4.10a[81]. As a simple cubic lattice, the Pb ions are centered in the octahedral, and the MA ions are located between the octahedral building blocks. The highest symmetry phase for MAPbI$_3$ and its related materials is the cubic (Pm$\bar{3}$m) lattice, with sequential transitions lowering the symmetry typically through octahedral tilting. The tilting collectively leads to the phase transition from cubic to tetragonal (see in Figure 4.10b, c)[82]. The collective rotation of the PbI$_6$ octahedra around the c-axis results in the tetragonal space group, I$_4$/mcm, with relatively close packing within the ab plane. Further transition to an orthorhombic phase (space group: Pnma) is accompanied by a tilting of the PbI$_6$ octahedra out of the ab

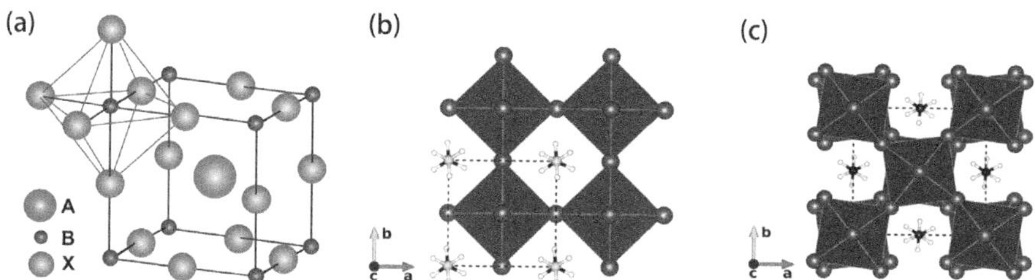

FIGURE 4.10   This image features a series of diagrams illustrating the crystal structures of photovoltaically active perovskites. (a) Depicts the general perovskite structure with a large cation (A), usually the methylammonium ion (MA), a small cation (B), typically lead (Pb), and a halogen anion (X), often iodine (I). The cubic phase of MAPbI$_3$ is stable only above 330 K. Panels (b) and (c) show the structural representation of MAPbI$_3$ in the pseudocubic and tetragonal phases, respectively. The unit cells are outlined by thick dashed lines, with carbon (C) atoms in black, nitrogen (N), hydrogen (H), iodine (I), and lead (Pb). The images convey the transition from cubic to tetragonal phase through octahedral tilting, which reduces the symmetry of the structure.

plane. $MAPbI_3$ transforms from the cubic to the tetragonal phase at temperatures below 315–330 K, and the tetragonal to the orthorhombic phase at ~160 K.

The group of hybrid perovskites exhibits crystallographic phase transitions with varying temperatures and pressures. Heat capacity measurements along with temperature-dependent X-ray diffraction (XRD) studies revealed a crystallographic phase transition from cubic to tetragonal at temperatures of 330.8, 236.1, 177.2, 238, and 403 K for $MAPbI_3$, $MAPbBr_3$, $MAPbCl_3$, $FAPbBr_3$, and $CsPbBr_3$[83], respectively. The cubic structure contains only one formula unit ($Z = 1$); therefore, the noncentrosymmetric MA ions must be randomly oriented to satisfy the Oh symmetry. As the temperature decreases, tetragonal and orthorhombic phases are stabilized with ordering of MA ions[84]. The structural transition from the cubic to tetragonal phase occurs due to the reorientation of MA^+, as observed by NMR studies, where the number of disordered states of MA^+ was lowered from 24 in the cubic phase to 8 in the tetragonal phase. This molecular ordering is completed in the orthorhombic phase such that at 161.8, 149.8, 171.5, 150, and 385 K for $MAPbI_3$, $MAPbBr_3$, $MAPbCl_3$, $FAPbBr_3$, and $CsPbBr_3$, respectively, an orthorhombic Pnma space group is formed. Some studies have reported the crystal structure of $MAPbX_3$ (X = I, Br, Cl) to comprise noncentrosymmetric space groups, such as I4 cm and Pna21 for the tetragonal and orthorhombic phases, respectively. Nandi et al. refined the high-resolution synchrotron XRD data both in centrosymmetric and noncentrosymmetric space groups and reported relatively good reliability parameters in the centrosymmetric structure[85]. Irrespective of the temperature-dependent phase transitions, hybrid perovskites exhibit crystallographic phase transitions with applied pressure. Reportedly, in the literature, the crystal structure of $MAPbI_3$ is tetragonal (I_4/mcm) at ambient temperature and pressure and has a distorted orthorhombic phase with the space group Imm2 when the pressure is above 0.3 GPa[86]. Further, reversible amorphization was realized for this system when a pressure of 10 GPa was applied. However, $MAPbBr_3$ is in the cubic phase with space group $Pm\bar{3}m$ at ambient pressure and has a cubic space group ($Im\bar{3}$) when a pressure of 1.7 GPa is applied[87].

4.2.2 Opto-Electronic Properties

The MA-based perovskite material was first explored for photovoltaic applications. With an ionic radius of 217 pm and a corresponding tolerance factor of 0.91, it crystallized as a tetragonal perovskite at RT (RT), whose octahedra were slightly distorted from the cubic structure[88]. At elevated temperatures (54–57°C), the MA-based perovskite material was found to undergo a phase transition to a cubic structure. The bandgap of the single-crystal $MAPbI_3$ was measured to be 1.51 eV. The effective masses of the charge carriers in $MAPbI_3$ have been obtained either using first-principles calculations of the band curvature or by magneto-absorption measurements, where the low effective masses for both electrons (m_{e*}) and holes (m_{h*}) (around 0.1–0.2 m_0) were correlated with the high charge-carrier mobility (1–70 cm^2/(V s) for polycrystalline films)[89–91]. The organic MA cations in the octahedral cage were found to reorient dynamically within a very short time. At RT, dipolar MA cations reorient between the faces, corners, or edges of the pseudocubic lattice cages of corner-sharing octahedra with a residence time of ~14 ps[92]. The dynamic reorientation of the organic cation is suggested to stabilize the energetic charge carriers by the formation of a large polaron, which, in turn, enhances the lifetime of hot carriers as well as band edge carriers[93]. However, this dynamic disorder of the A cation highlights the relatively weak bond between the A cation and PbX_6 octahedra. Consequently, the degradation of $MAPbI_3$ under illumination or at elevated temperatures is typically triggered by the volatilization of the MA cation.

$FAPbI_3$ crystallizes into a non-perovskite hexagonal phase at RT with a bandgap of 2.43 eV[94]. The hexagonal phase is converted to a pure cubic phase upon annealing at temperatures above 150°C, and the cubic perovskite phase is maintained after cooling to RT[95,96]. The spontaneous crystallization of the non-perovskite hexagonal phase at RT is probably related to the larger ionic radius of the FA cation (253 pm, tolerance factor = 0.99) compared to that of the MA cation[97], resulting in intrinsic lattice strain, which induces the distortion of the corner-sharing octahedra constituting the cubic perovskite lattice to the face-sharing octahedra. Although the cubic perovskite phase was formed upon annealing at relatively high temperatures, the phase was found to be metastable at RT with an activation energy barrier of approximately 0.6 eV for reverse-phase conversion to the hexagonal phase[98]. Owing to the relatively high potential barrier, the as-formed cubic $FAPbI_3$ film was stable under inert atmosphere conditions. However, the reverse-phase conversion was found to be significantly accelerated upon exposure to moisture[99]. Because of the highly polar ionic bonding, water vapor can affect the surficial perovskite phase and induce a significant number of surface defects, which reduce the energy barrier for nucleation of the non-perovskite phase[20]. Consequently, the cubic $FAPbI_3$ film tends to undergo an accelerated phase transition to the non-perovskite phase in humid environments. Nevertheless, the cubic $FAPbI_3$ phase film was found to have a relatively lower bandgap of 1.48 eV compared to that of the $MAPbI_3$ thin film (1.57 eV), while the electronic band structure and, thus, the absorption coefficient, are comparable[100]. Furthermore, the charge-carrier lifetime and diffusion length in the $FAPbI_3$ thin film were found to be much longer than those in the $MAPbI_3$ film. Through solid-state NMR measurements, the reorientation rate of the FA cation in the lattice was found to be much higher (8.7 ± 0.5 ps) than that of the MA cation (108 ± 18 ps), enabling a superior charge-carrier stabilization capability[101]. However, the low phase stability of the $FAPbI_3$ perovskite limits the further enhancement of the PCE in devices based on $FAPbI_3$.

Different from organic cations exhibiting weak secondary hydrogen bonding with the perovskite lattice, Cs forms a strong primary chemical bond with the $CsPbI_3$ perovskite lattice. The strong chemical bond between the Cs cation and perovskite lattice is supported by the considerably high activation energy required for the initiation of the thermal degradation (650 ± 90 kJ/mol) of $CsPbI_3$[102]. However, the relatively small ionic radius of the Cs cation (167 pm) results in a tolerance factor of 0.81, effectuating the low thermodynamic stability of the perovskite phase at RT. In fact, the $CsPbI_3$ film formed by solution processes was found to exist in a non-perovskite orthorhombic phase with a wide bandgap (2.82 eV) at RT; this phase transformed into a cubic perovskite phase at temperatures above 300°C[103]. However, the cubic $CsPbI_3$ film is very unstable under ambient conditions[104]. Consequently, it tends to undergo a phase transition to the orthorhombic phase after a few minutes. The thermodynamic stability of the cubic phase was found to be improved by additive engineering. Eperon et al. found that the addition of hydroiodic acid (HI) into the precursor solution could facilitate the formation of the cubic perovskite phase at 100°C with a bandgap of 1.73 eV. The addition of HI into the precursor solution was found to induce the formation of relatively small grains with a slightly strained lattice, which enables the stabilization of the cubic phase at lower temperatures (approximately 100°C). The enhanced thermodynamic stability with reduced crystal size was also observed for $CsPbI_3$ quantum dots, which enabled the achievement of stable $CsPbI_3$ quantum dot solar cells with efficiencies exceeding 13%[105]. The surface and strain energies were found to be the origin of the enhanced thermodynamic stability with the reduced crystal size. A similar approach was also applied to fabricate stable $FAPbI_3$ quantum dot solar cells. Recently, the development of various additives has enabled noticeable improvement in the PCE. Both zwitterion and 2D perovskite additives induce small grains to stabilize the cubic phase, resulting in efficiencies higher than 11%[106]. Furthermore, the passivation of the crystal surface by organic-halide treatment significantly improves the optoelectronic properties of $CsPbI_3$ films, which results in efficiencies as high as 18%[107]. However, the PCE of $CsPbI_3$ PSCs is still limited due to its relatively high bandgap of ~1.7 eV. It can possibly be applied to the top subcell of tandem solar cells with appropriate bandgap engineering.

4.2.3 THERMAL AND HUMIDITY STABILITY

Depending on the location and architecture of the solar module, the temperature of the solar cell may rise to 95°C[108,109]. Hence, thermal stability represents one of the major concerns in perovskites. The soft matter nature of methylammonium lead halides ($MAPbX_3$) triggers structural transformation under thermal stress, posing an obstacle. Although the $MAPbI_3$ single crystal is stable up to 300°C [8], significant changes already occur during annealing at 85°C for the polycrystalline films due to the partially volatile MA species[110]. In particular, continuous exposure of $MAPbI_3$ at high temperatures meeting the International Standards (85°C) leads to the decomposition of perovskite into volatile gas species like CH_3I and NH_3, as well as reversible decomposition into CH_3NH_2, HI, I_2 and finally non-volatile Pb^0 as shown in Figure 4.11[111]. The material is also prone to degrade to PbI_2 under humidity and light[112].

The stability of the $MAPbI_3$ can be improved by compositional engineering of A cations, introducing Br^- at X sites, grain boundary and surface passivation, and encapsulation[113]. The substitution of MA with larger species can hamper the loss of organic cation. For instance, FA-based perovskites exhibit enhanced thermal stability[114] and decompose at temperatures above 95°C, leading to the formation of sym-triazine[102]. However, pristine $FAPbI_3$ lacks structural (phase) stability at RT[115]. The previous decomposition processes are not only triggered by heat, but also by prolonged UV light exposure, creating a serious problem for perovskite solar cells. Fortunately, these thermal and UV triggered decomposition processes can be reduced by suitable encapsulation[110,111]. Other strategies to increase thermal stability include the use of inorganic cations such as Cs^+[113]. Indeed, fully inorganic perovskites, like cesium lead halides, are thermally stable up to 580°C and show longer ageing stability than $MAPbX_3$[116,117]. With a bandgap of 1.73 eV, 0.22 eV wider than $MAPbI_3$,

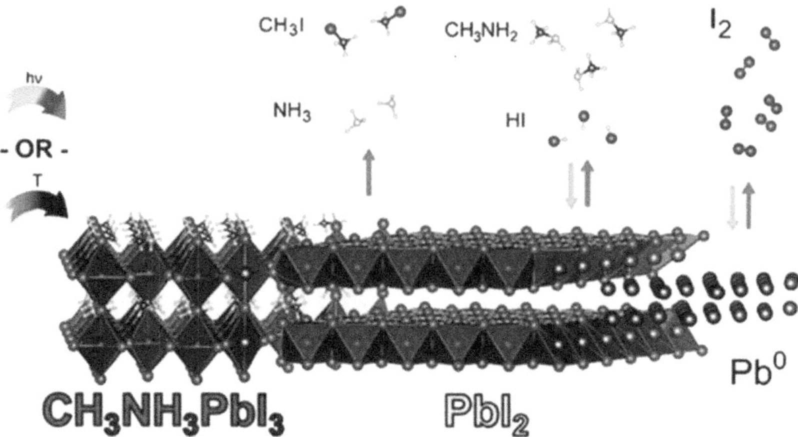

FIGURE 4.11 The diagram illustrates the decomposition pathways of MAPbI$_3$ under the influence of light (hv) or heat (T). MAPbI$_3$ crystals are depicted in black and grey in color, with arrows indicating the release of organic volatile species such as CH$_3$I and NH$_3$, which is an irreversible process. Another decomposition route yields CH$_3$NH$_2$ and HI, shown to be reversible. On the right side, the image shows the evolution of iodine (I$_2$) and the formation of metallic lead Pb^0, both of which can occur reversibly under mild conditions. This visual is a concise representation of photodecomposition and thermal degradation processes in perovskite materials.

CsPbI$_3$ could be an ideal top cell candidate for tandem solar cells to further increase the device performance of silicon or perovskite solar cells. Compared to mixed-halide perovskites, CsPbI$_3$ perovskite can deliberately avoid phase segregation, leading to its outstanding photostability. However, it suffers from structural instability at RT, exhibiting a photoinactive yellow phase. To overcome this problem, structurally and thermally stable perovskites have been found upon integrating mixed FA-Cs cations[99,118]. Double perovskites such as Cs$_2$SnI$_6$, Cs$_2$TiBr$_6$, and Cs$_2$AgBiBr$_6$ are another type of inorganic halide perovskites that could achieve excellent phase and thermal stabilities[119–121]. They are less investigated compared to the ABX$_3$ type perovskites with much worse performance, for example, the record efficiency of Cs$_2$AgBiBr$_6$ solar cell is only 3.11%, and more work should be conducted to improve their optoelectronic performance[122].

$$MAPbI_3\,(s) \overset{hv\,or\,\Delta}{\longleftrightarrow} PbI_2(s) + Pb^0(s) + I_2(g) + CH_3NH_2(g) + CH_3I(g) + HI(g) + NH_3(g)$$

The intrinsic instability of 3D perovskites, which arises from the hygroscopic properties of their organic cations, low formation energy, and phase stabilization, has been the main obstacle to the commercialization of PSCs despite their impressive solar-to-electrical power conversion efficiencies (> 26%). In this sense, it is necessary to understand the possible initiators inducing the degradation process by humidity and their concomitant mechanisms, which is important for the molecular engineering of robust perovskites. Considering the degradation products and rates depending on different humidity parameters (e.g., inert gas, ambient air, vacuum, light, relative humidity (RH), etc.), the adjacent layers (e.g., HTLs, ETLs) have been excluded from current discussion due to some discrepancies in degradation products detected under similar humid conditions[122]. Instead, we have focused on what is mainly responsible for the humidity-induced degradation of perovskites. Although the exact degradation process is still not clear, the plausible intermediates and degradation products depending on the type of initiators can be proposed from previously reported results by combining the experimental and theoretical results. As shown in Figure 4.12, neutral water, hydroxyl radicals ($\cdot$OH), and/or hydroxide (OH$^-$) have been considered as the degradation initiators in perovskites under humid conditions. In general, water from humid air has been suggested to act as a catalyst during the degradation of

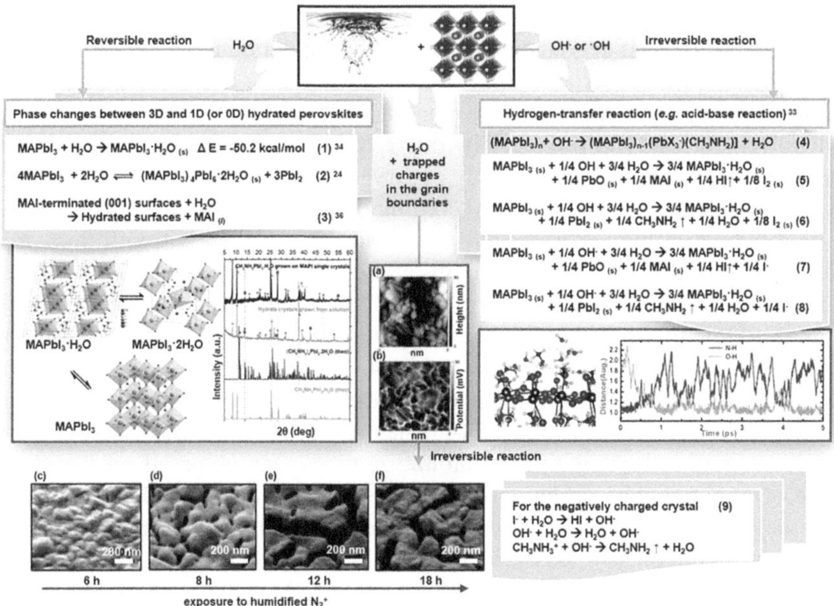

FIGURE 4.12 The composite image outlines potential degradation pathways of perovskite materials under humid conditions. On the left, the hydrated phases of perovskites are shown with a focus on the MAPbI$_3$ structure interacting with water molecules, accompanied by XRD patterns that identify the phase changes between 3D and 1D or 0D hydrated perovskites. In the center, the image details the surface topography and potential profile of an N$_2^+$-treated MA$_{0.6}$FA$_{0.4}$PbI$_{2.9}$Br$_{0.1}$ film, with KPFM measurements indicating changes over time. The right portion of the figure provides a reaction scheme for the interaction between hydroxyl radicals and MA, including ab initio molecular dynamics (AIMD) predictions and bond length trajectories. Below, SEM images illustrate the morphological evolution of the perovskite films at various intervals after exposure to humidified N$_2^+$, showing noticeable changes from solid to liquid states as indicated by (s) for solid and (l) for liquid.

perovskites. Kelly et al. suggested that the formation of a low-dimensional hydrated phase ($MAPbI_3 \cdot nH_2O$) is the first step of the degradation process based on the observation of the isolated PbI_6^{4-} using in-situ GIXRD (cf. eqn. (1) and (2) in Figure 4.12)[123]. They claimed that the acid-base reaction[124] (eqn. (4) in Figure 4.12) was not the first step in the degradation process due to the partial recovery of a black phase from the yellow hydrated phase following flushing dry N_2 gas over the perovskite films degraded under $98 \pm 2\%$ RH. Later, Sit et al. reported ab initio calculations on the adsorption of water on the (110) surface of perovskites, which showed a negative adsorption energy of -40.2 kcal mol^{-1}[125]. However, oxonium ions (H_3O^+), which may be a product from the acid-base reaction between water and methylammonium (MA), were not formed. Their follow-up study on the degradation pathways induced by H_2O showed that the formation of the monohydrate phase (Figure 4.12, $MAPbI_3 \cdot H_2O$(s); s for solids) of perovskites was exothermic, with a reaction energy of -50.2 kcal mol-1 (eqn. (1) in Figure 4.12)[126].

It is interesting to note that the charges trapped along the grain boundaries can trigger the humidity- (or oxygen)-induced degradation of perovskites in the irreversible process, which is similar to the acid–base reaction shown in eqn. (4). To prove the trapped-charge assisted degradation of perovskites under humid air conditions, Choi et al. constructed an experimental set-up employing an ion generator using corona discharge and a chamber connected to the inlet to regulate the humidity[128]. The different polarity of ions was produced by the N_2^+ and H_2^- corona ions. The irreversible decomposition of perovskites was only observed when humidity was present on the charge-deposited perovskite films. According to the evolution of the XRD patterns and morphology of the ion-deposited films under humidified N_2^+ with time, it was found that the degradation process begins at the grain boundaries (Figure 4.12 c, f). Furthermore, the results obtained from KPFM for the negative ion treated perovskite film showed that the surface potential distribution of perovskite surface was consistent with its corresponding topology (Figure 4.12a, b). Eqn. (9) shows the proposed mechanism based on the experimental observations combined with a follow-up modeling study. Briefly, unlike the case of the neutral $MAPbI_3$ surface (without trapped charge), a significant change in the atomic motions was observed on the charged surface of $MAPbI_3$. The evolution of volatile gas (HI and CH_3NH_2) was observed for the negative charged surface, while the positively charged surface induced the solvation of MA^+ (eqn. (3)). Based on the species identified using AIMD, the authors suggest that the degradation of the negatively charged $MAPbI_3$ occurs in a sequential manner involving proton transfer (eqn. (9)), which is similar to the Grotthuss mechanism, rather than direct proton transfer. It is believed that the driving force for the formation of the initially solvated MA or I^- is due to the new electrostatic force field formed by the excessive charge, which helps weaken the existing bond between MA and PbI_6^{4-} or repels the I^- ions away from the $MAPbI_3$ surface.

It is clear that the detection of volatile methylamine was due to hydrogen abstraction from the MA cation. Based on the theoretical results on the degradation of $MAPbI_3$ in the presence of the three possible initiators, hydroxyl and hydroxide (not H_2O) were associated with the hydrogen transfer reaction from MA, which facilitates the desorption of methylamine, resulting in the irreversible decomposition of perovskites (eqn. (6) and (8) in Figure 4.12)[126]. It was predicted that the proton transfers from MA to OH⁻ or ·OH was rapid (< 2 ps from AIMD) and methylamine drifted along the perovskite surface. Furthermore, the infiltration of water was calculated with the aid of CI-NEB, from which the Pb–O bonds formed between the oxygen atoms in ·OH (or OH⁻) and Pb atoms. However, no further exothermic reactions involving methylamine gas were observed during the reaction between water and $MAPbI_3$ perovskites[127,128]. Other studies have shown that the solvation of MAI (eqn. (3) in Figure 4.12) is the first step in the degradation process by using experimental and ab initio molecular dynamics[129,130]. Thus, these findings imply that water from humid air catalyzes the degradation reaction of perovskites by forming hydrated perovskites.

Water can also exist in other forms such as hydroxyl and hydroxide during processing of materials in humid air[131,132]. These species can be formed via dissociation or the reaction between water and oxygen[133]. The presence of hydroxyl or hydroxide on the surface and point defects of perovskites is known to accelerate their decomposition[134,135]. It is apparent that the degradation products are dependent on the different types of initiators (Figure 4.12). In addition to forming perovskite hydrates caused by catalytic water, gaseous species have also been investigated experimentally[136–138]. The formation of the Pb–O bond may be related to degradation products such as PbO and $Pb(OH)_2$ (eqn. (5) and (7) in Figure 4.12), which were detected experimentally[139]. The degradation reactions with the presence of ·OH (or OH–) and gaseous products predicted thermodynamically favorable reaction energies.

Despite the outstanding achievements in the performance of perovskite-based devices, the absence of standard measuring conditions for their stability obscures the main cause of discrepancies in the degradation products and rate. In particular, the degradation rate of perovskites, which undergo a hydrated phase, is difficult to determine without in-situ techniques due to the reversibility between the black and transparent phases. For example, n-butylammonium-treated $MAPbI_3$ shows XRD peaks corresponding to the hydrated phase in the early stage of the degradation process, while some other cation-driven perovskites degrade with a characteristic peak of PbI_2. A precise and full detection for the degradation phenomena (especially, gaseous products) requires additional apparatus (e.g., mass analyzer) and in-situ analytical techniques connected to a humidity chamber. Such an approach helps the prediction of the long-term degradation (lifetime) of perovskites, which will contribute to the construction of a reliable database for the materials properties.

4.3 IMPACTS OF PHASE TRANSITION ON THE PERFORMANCE OF PEROVSKITE PHOTOVOLTAICS

4.3.1 EFFICIENCY

In recent years, the efficiency of PSCs has been rapidly increasing, attracting attention from researchers worldwide[140,141]. Various researches are related to overcoming the phase transition of perovskite to enhance the performance of PSCs. For instance, Yang et al. achieved a PCE of 20.1% by depositing high-quality $FAPbI_3$ films and fabricating PSCs[142]. It has been observed that crystal facets in $FAPbI_3$ significantly affect film stability and device efficiency. Calculations suggest that the (110) facet has a high density of dangling bonds, while the (111) facet is the least stable. By introducing inorganic cesium to triple-cation perovskite compositions, Michael Saliba and his colleagues demonstrated improved stability in PSCs, with fewer phase impurities and less sensitivity to processing conditions, achieving PCEs of 21.1%[143]. State-of-the-art perovskite composition is $FAPbI_3$ with excess methylammonium chloride (MACl) and a low molar ratio of $MAPbBr_3$ for assisting crystallization and stabilizing the perovskite at α-phase[144,145]. The significant prerequisite to obtain high-efficiency PSCs among all these strategies is to control the phase transition for better crystallization of the perovskite. The problem of phase transition of perovskite has been partially solved with precursor additive engineering[146] and interface passivation engineering recently[147].

In Figure 4.13a, a robust phase-structured perovskite was obtained by Zhang et al. by simultaneously excluding alkali halide salts from the precursor solution and surface passivation with CsI[148], resulting in a dramatic improvement in PCE of up to 24.1% (Figure 4.13b). The device stability is also enhanced due to the protection of the passivated perovskite film from the harsh MeOH solvent. Operational stability tests were performed on the control and target devices by subjecting them for 600 h of maximum power point (MPP) tracking under AM 1.5G at RT. The target device retained 90% of its initial efficiency as shown in Figure 4.13c, much better than the control one[9]. To clarify the process of phase transition, several perovskite phases are calculated and three of them are shown to illustrate the process of transition (Figure 4.13d). The bandgap of the cubic phase (1.57 eV) is lower than the orthorhombic one (1.87 eV). The cubic phase with lower bandgap is more efficient for light conversion.

4.3.2 Hysteresis

PSCs are excellent devices for directly converting photon energy into electrical energy. However, some challenges remain to be resolved before the commercialization of this technology[149]. One of the most basic predominant servings as a bottleneck for the commercialization of perovskite photovoltaic technology is ambiguity and inaccurate determination of the PCE for PSCs due to strong J-V scan hysteresis, whose origin (extrinsic and/or intrinsic) is not clear[150–151]. Such hysteresis behaviors seriously influence the facticity of the measured results and the device stability[152]. In PSCs, the presence of hysteresis in (J-V) measurements due to the anomalous dependence on the measuring protocols such as scan rate, scan direction, voltage range, precondition, and device architecture[153]. Measurement protocols widely affect the PV performance of PSCs. Four measuring techniques have been used for PSCs that include conventional J-V curve measurement, MPP tracking, stabilized current at a fixed voltage (SCFV), and dynamic J-V methods. The hysteresis behavior of the PSCs creates complexity to find actual efficiency while measuring it in real-time[154].

Several explanations for hysteresis have been put forward: (i) ion migration during J–V scan, resulting in interfacial band bending[155]; (ii) motion of interfacial defects under the device operating conditions, leading to distinct interfacial electronic dynamics[156]; and (iii) ferroelectricity in the active layer originating from either molecular ferroelectric order (FE) of the organic molecules and/or the ionic FE of the inorganic cages[157]. In the past years, great efforts have been made to eliminate

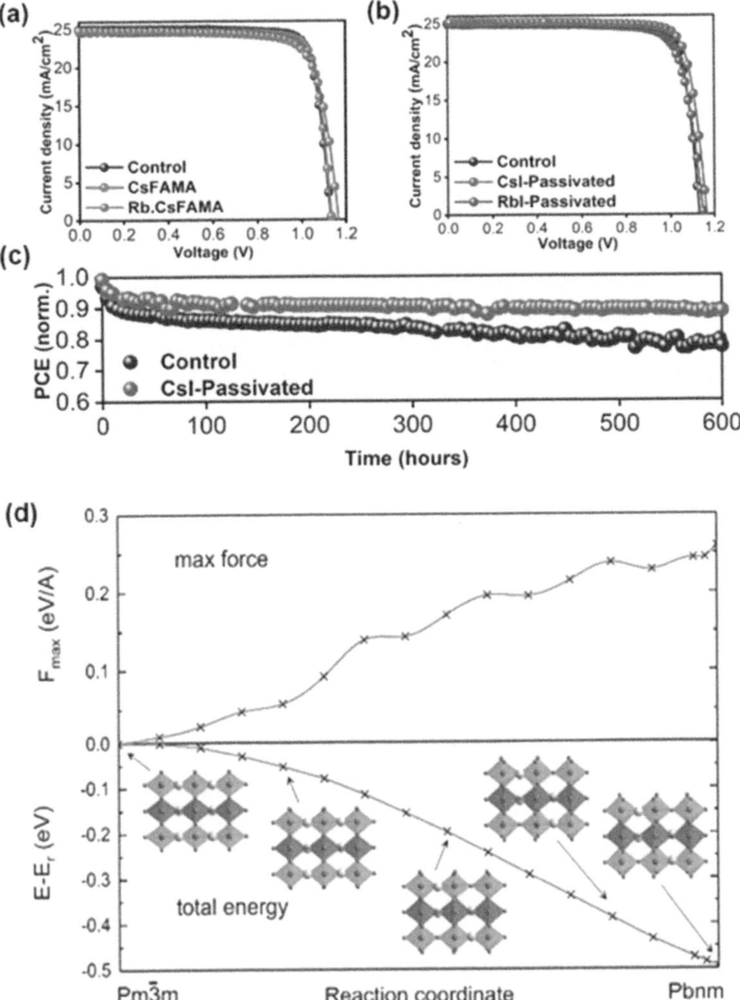

FIGURE 4.13 The image shows graphical data on the performance and stability of perovskite solar cells with various treatments. (a) and (b) display current density-voltage (J-V) curves for control devices, and those treated with CsFAMA and Rb.CsFAMA, as well as CsI-passivated and RbI-passivated devices, indicating the electronic characteristics of each. (c) Illustrates the stability of the normalized PCE over 600 h for control and CsI-passivated devices under MPP tracking in a nitrogen atmosphere with continuous AM 1.5G illumination, demonstrating enhanced stability for the passivated device. (d) Depicts the energy profile along a reaction coordinate, showing the evolution of the maximum force on atoms and total free energy, which provides insight into the mechanical and thermodynamic stability of the material during the operational process.

hysteresis and improve the stability of PSCs from the following aspects: additive modification or treatment of the electron-transporting layer (ETL) to optimize the interfacial electronic dynamics in PSCs[158]; changing the elemental composition of the perovskite intrinsic layer to promote the perovskite quality; and doping of the hole-transporting layer to passive the surface trap states of the perovskite layer[159].

4.3.3 STABILITY OF PSCs

In recent years, with people's unremitting efforts, the efficiency of PSCs has been dramatically enhanced. However, the poor device stability of PSCs remains a big challenge that determines whether exciting achievements can be transferred from the laboratory to industrial applications[160]. Multiple reports have suggested that moisture, oxygen, light, and high temperature are the four key factors affecting the stability of PSCs. Observed (sometimes rapid) degradation occurs when devices are exposed to those environmental factors[161].

Quite a number of researchers have paid attention to the phase stability issue of perovskite. For instance, Wang et al. developed a simple solution-processed CeO_x ($x = 1.87$) as ETL in PSCs, which exhibit superior stability under light soaking compared to TiO_2-based PSCs[162]. Perovskites with mixed-anion have been explored to further improve the performance of $FAPbI_3$-based PSCs[163]. Wang et al. presented PSCs based on $FA_{0.83}Cs_{0.17}Pb(I_{0.6}Br_{0.4})_3$, which shows remarkably enhanced light stability due to enhanced phase stability than perovskite with other composition[164]. Improved photostability and stability upon exposure to moisture were reported for $FA_{0.9}Cs_{0.1}PbI_3$-based PSCs compared with the $FAPbI_3$-based one[165]. Partial replacement of FA with Cs will result in significant improvements in the stability of the dark perovskite phase, as well as significant improvements in device stability[166]. It has also been demonstrated that $FA_{1-x}MA_xPbI_3$-based PSCs exhibited better device stability than those based on $FAPbI_3$, under both light soaking and thermal stress[167]. Li et al. reported that the 1-butyl-3-methylimidazolium-based ILs ([BMIM]X) could react with the excess PbI_2 generating self-assembly supra molecule complex [BMIM] Pb_2X_5, which showed extremely enhanced stability under illumination[168]. Although these ionic liquids ILs can coordinate with the under-coordinated lead by the lone-pair electrons, enhancing the performance and stability together, the composition-purity of perovskite could be decreased[169].

$MA_2Pb_3I_{8.2}DMSO$ is an I-deficient phase that could be converted into perovskite by reacting with organic halides, or annealing to generate the perovskite and PbI_2[170]. As the annealing continues, the solvate phase will react with FAI to form the α-$FAPbI_3$, accompanied by the volatilization of DMSO. However, the $MA_2Pb_3I_8\cdot2DMSO$ intermediate phase exhibits a needle-like morphology[171]. The dissolution of the δ-phase and crystallization process of the α-phase might induce many small crystals and generate more defects at the surface and grain boundaries. With the addition of PACA-ILs (Figure 4.14a), the solvate phase of $MA_2Pb_3I_8\cdot DMSO$ was restrained. Notably, for the methylammonium butyrate (MAB) modified perovskite film, there exist both δ- and α-phase in the wet spin-coated film without annealing, as depicted in Figure 4.14b for MAB-modified film.

$FAPbI_3$-M crystals were found to possess substantially greater α-phase (black) stability in ambient air than neat $FAPbI_3$. The growth of trace regions of a δ-phase (yellow) on the crystal surface was observed via a visible light microscope after only a few hours of storage in the air (Figure 4.14c)[172]. In order to quantitatively track the phase state of the crystal, the Raman spectrum was performed. The absolute Raman intensity of the best-resolved δ-phase peak (at a Raman shift of 108 cm^{-1}, under ambient conditions) was observed (Figure 4.14d). Neat $FAPbI_3$ undergoes detectable

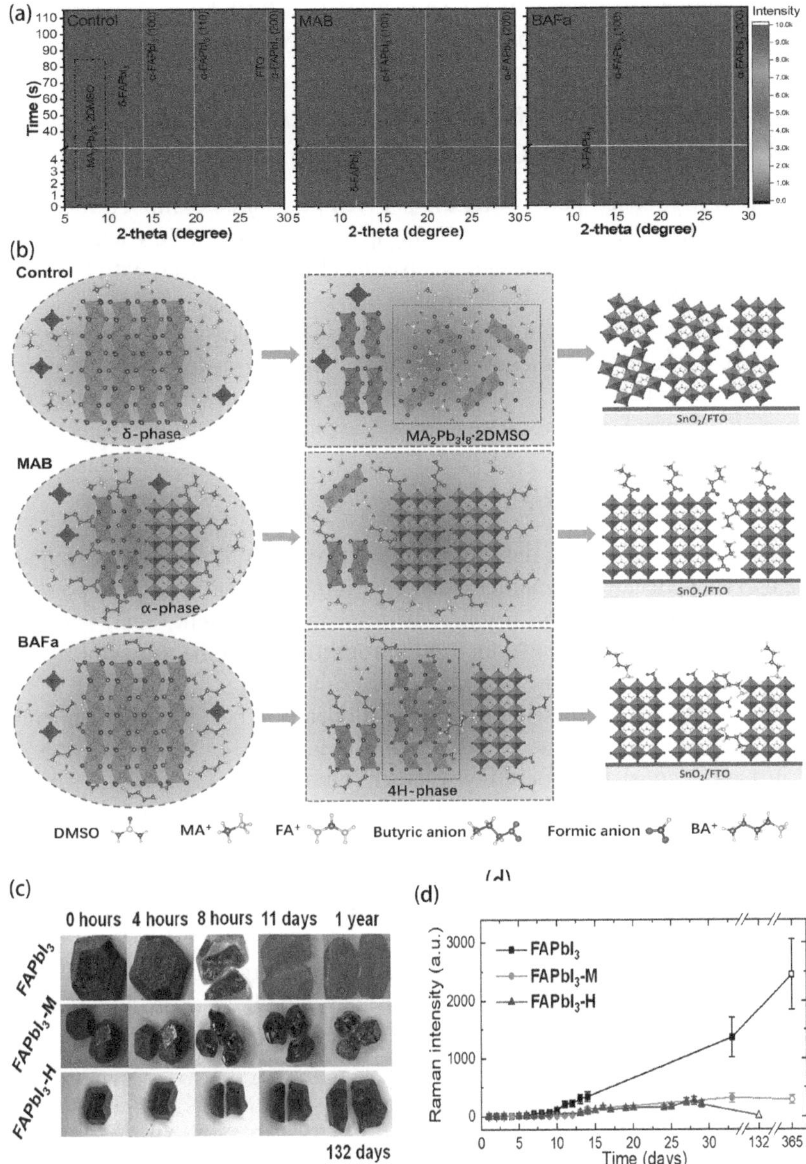

FIGURE 4.14 (a) Shows in-situ XRD patterns for control samples and those modified with MAB and BAFa ILs., displaying characteristic peaks indicative of the perovskite structure. (b) Provides schematic illustrations of the molecular structures during intermediate and perovskite phase transitions, contrasting control samples with those modified by MAB and BAFa ILs. It depicts how additives interact with the perovskite structure. (c) Displays optical microscope images of FAPbI₃, FAPbI₃-M, and FAPbI₃-H crystals over time, showcasing changes from clear to blue, indicating phase degradation. (d) Graphs the temporal evolution of the δ-phase Raman peak intensity for the three types of single crystals, highlighting the difference in stability and degradation over a year.

δ phase formation after 7 d, which occurs after 11 d for $FAPbI_3$-M [173], revealing its enhanced phase stability, compared to the control $FAPbI_3$.

4.4 STRATEGIES TO CONTROL PHASE TRANSITION IN PEROVSKITE PHOTOVOLTAICS

Compared to MA-based perovskite, the FA-rich perovskite exhibited longer charge diffusion length and a more proper bandgap[174]. To achieve PSCs with exceptional efficiency and stability, the photoactive layer has been substituted with FA-rich compositions from the original MA. Unfortunately, the formation of a stable α-phase poses a significant challenge for FA-rich devices. To address this issue, researchers have explored several strategies to regulate the phase transformation in perovskite photoactive layer, including strain engineering, composition engineering, thermal engineering, intramolecular exchange, interface engineering, etc.[175].

4.4.1 STRAIN MANAGEMENT

Strain plays a crucial role in determining the deformation of lattice constants from the ideal crystal structure, whether caused by intrinsic factors such as point defects or orientation inhomogeneity or external influences like light, heat, and pressure. It significantly impacts the optoelectronic properties and inherent stability of poly-crystalline materials. In multilayered and heterojunction structures of PSCs, strain occurs inevitably during their fabrication and operation because of the mismatched thermal expansion coefficients and lattice constants between the perovskites and the substrates and inconsistent operating weather conditions[176]. The strain formation and relaxation in perovskite crystals refers to the extension/narrowing of chemical bonds and expansion/shrinkage of lattice volume with a concomitant increase of defect states, impacting the activation energy for phase transition.

Recent studies have shown that applying appropriate compressive strain to perovskite films can enhance device performance. By regulating the strain, perovskite photovoltaic devices with over 25% efficiency and prolonged stability in different operating modes have been successfully developed. These results demonstrate that strain control plays a crucial role in managing phase transition in perovskite photovoltaics[177].

The manner and magnitude of residual strain in perovskite films depend significantly on the annealing temperature, and strain-free perovskite films can be formed at low temperatures[178]. However, the PCE of the devices based on low-temperature-processed perovskites was relatively low because of the inferior crystal quality compared to those deposited at high temperatures[179]. Therefore, some flash annealing techniques have been developed to facilitate perovskite transformation with suppressed residual strain while preserving superior crystal quality (Figure 4.15)[180]. For instance, a flash thermal shock with instant heating to 453 K followed by an extremely fast cooling was applied to a precursor film, accelerating nucleation and crystal growth. Owing to this flash annealing process, the perovskite exhibited more compressive strain, resulting in a PCE of 22.99% and a T_{80} lifetime exceeding 4,000 h.

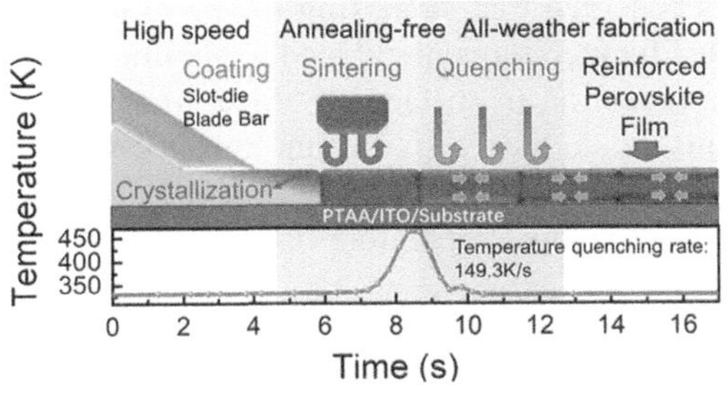

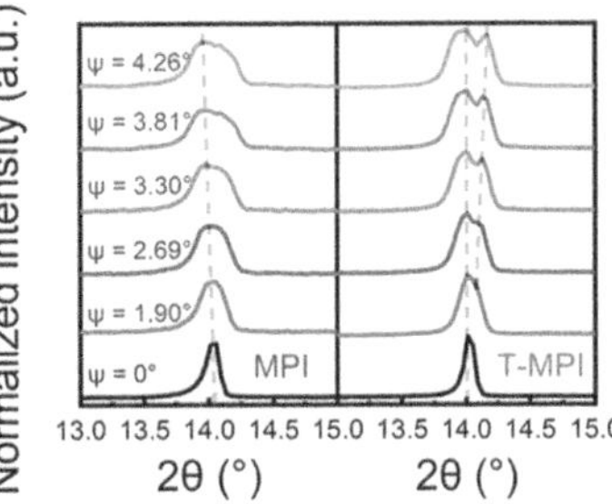

FIGURE 4.15 The image is split into two parts illustrating the process and effects of shock thermal annealing on perovskite films. On the left, a temperature-time graph shows the rapid crystallization process of a perovskite film using a high-speed slot-die coating, followed by annealing-free sintering and quenching, resulting in a reinforced perovskite film. This all-weather fabrication process is visualized with icons representing each step, with the temperature rapidly rising to over 450 K and then cooling down within approximately 16 s. On the right, a series of XRD patterns at different residual strains (u) in the film are displayed, with peak shifts indicating the varying levels of strain from MPI to TMPI, suggesting changes in the crystal lattice due to the shock thermal treatment.

4.4.2 Composition Engineering

Compositional design can also avoid phase transition in perovskite photovoltaics. Due to the limited interaction between the large FA^+ ions and the PbI_6 framework, the phase transition in $FAPbI_3$ could frequently occur in $FAPbI_3$[181]. One of the reliable strategies for control phase transition in perovskites is to adjust the Goldschmidt tolerance factor by alloying small ions, such as MA^+, Rb^+, or Cs^+ in the A-site; Cd^{2+} or Sr^{2+} in the B-site; or BF^{4-}, Br, or Cl- in the X-site[182]–[184]. However, incorporating small ions into $FAPbI_3$ would simultaneously widen its band gap and restrict near-infrared light absorption[185]. Alternatively, large cations such as methylenediammonium (MDA^{2+}), which can strongly associate with the inorganic framework via hydrogen bonding, have been alloyed with FA^+ to stabilize α-$FAPbI_3$ with neglectable changes in band gap[186]. Nevertheless, inserting solely larger or smaller ions can distort the PbI_6 octahedron by tilting the Pb-I-Pb linear bonds, which can be compensated by rationally combining small and large ions together. For example, the phase transition in $FAPbI_3$ was massively reduced by alloying both MDA^{2+} and Cs^+ in the FA-sites, resulting in a device with the PCE up to 25.17% and T_{80} lifetime > 1,300 h at 85°C[187].

4.4.3 Thermal Engineering

Recent research has found that applying suitable compressive strain to perovskite films can improve device performance. Through strain regulation, PSC with a PCE over 25% and long-term stability in diverse operating modes have been successfully obtained. These findings highlight the importance of strain control in managing phase transition in perovskite photovoltaics. Eperon et al. systematically studied the influence of thermal annealing on the morphology control of hybrid perovskite films[188]. Upon their discovery, it was observed that subjecting the perovskite layer to prolonged heat treatment at elevated temperatures could lead to significant damage. This damage occurred due to a reduction in surface coverage. Moreover, the increased thickness of perovskite films would have a significant impact on the film coverage and the device efficiency. Therefore, exploring a novel annealing technique to achieve thick and uniform film coverage at low temperatures is a promising avenue for commercially available PSC.

Bi et al. investigated the relationship between annealing temperature and crystallinity of $((FAI)_{0.85}(MABr)_{0.15}(PbI_2)_{0.85}(PbBr_2)_{0.15})$[189]. They collected SEM images and XRD analysis for hybrid perovskites annealed at temperatures between 80°C and 180°C for 120 min. As shown in Figure 4.16a, the grain size expanded with increased annealing temperature. Except for the film annealed at 80°C, the peak at approximately 12.69° corresponding to the (001) surface reflection of PbI_2 is observed for all samples in the XRD patterns (Figure 4.16b), indicating the formation of PbI_2 phases after releasing the organic species in the annealing process. It is obvious that the full width at half maximum (FWHM) of the (001) surface reflection of PbI_2 decreases with an increase in annealing temperature because of the increased crystallite size[190].

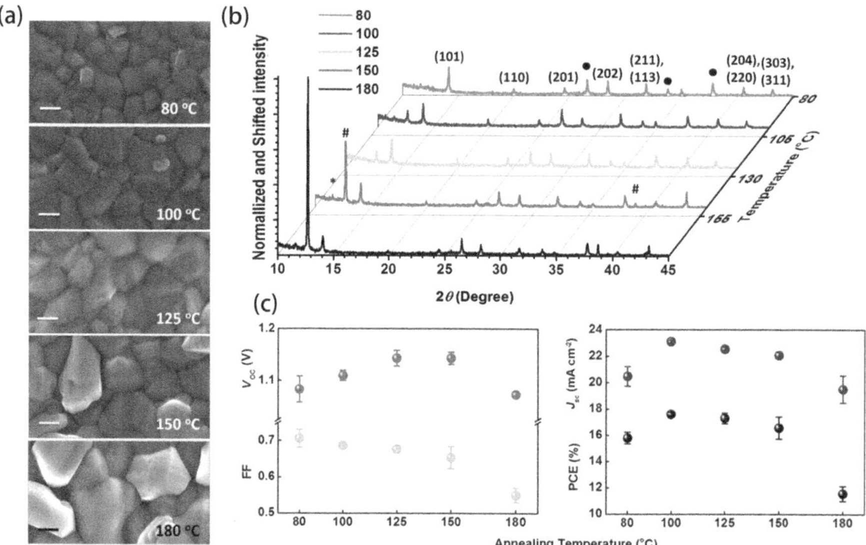

FIGURE 4.16 The image conveys the effect of annealing temperature on the morphology and crystal structure of perovskite films and subsequently on the performance of PSCs. (a) Top-view SEM images show the surface texture of perovskite films annealed at temperatures ranging from 80°C to 180°C, with grain size increasing with temperature. (b) XRD patterns with peaks corresponding to different crystal planes illustrate the crystallinity of the films on meso-TiO_2/c-TiO_2/FTO substrates after annealing. The intensity and position of peaks change with temperature, indicating phase transitions. (c) Plots of photovoltaic parameters such as short-circuit current density (J_{sc}) and open-circuit voltage (V_{oc}) against annealing temperature show how higher temperatures improve J_{sc} but have a varied effect on V_{oc}, reflecting the complex influence of thermal processing on PSC performance.

Furthermore, the FWHM of the (101) perovskite phase reflections showed minimal variation below 125°C. However, increased temperature beyond 125°C would result in a reduction in the perovskite crystallite size. The excess PbI_2 in perovskite films can diminish non-radiative charge carrier recombination[191]. As shown in Figure 4.16c, the champion PSCs fabricated under an annealing temperature of 100°C showed the achievement of the highest average J_{SC} of 23.1 mA cm^{-2}.

The temperature dependence of $FAPbI_3$, $MAPbI_3$, and their mixture has been investigated[192] by A. Dualeh et al. Figure 4.17 shows the SEM images of hybrid perovskite compositions at various annealing temperatures. The best coverage of $MAPbI_3$ films was achieved at 100°C (Figure 4.17a), while the film coverage at 175°C was not as good as that at 100°C (Figure 4.17d), and the $MAPbI_3$ at 230°C decomposes to PbI_2 (Figure 4.17g). The mixture at all annealing temperatures shows good coverage (Figure 4.17b, e, and h). The crystals of $FAPbI_3$ perovskite films are hardly observable at 100°C, while the $FAPbI_3$ crystals at 175°C and 230°C can undoubtedly be recognized (Figure 4.17c, f, and i). As illustrated in Figure 4.17j, the $MAPbI_3$ perovskite showed a dark color up to 200°C, and at 230°C the perovskites started to decompose. Furthermore, the $FAPbI_3$ perovskites had a dark color up to 260°C, then began to decompose. As a result, the optimized annealing temperature for a mixture of $FAPbI_3$ and $MAPbI_3$ is about 175°C. The observation matched the SEM result of hybrid perovskites with side reactions, e.g., the decomposition of $MAPbI_3$ and $FAPbI_3$. Hence, the actual molar ratio between $FAPbI_3$ and $MAPbI_3$ after the annealing process is not 1:1, even though a stoichiometric (1:1) molar ratio of $MAPbI_3$ and $FAPbI_3$ was originally dripped to form the perovskite films. According to the stability test, it is justifiable that the portion of $FAPbI_3$ is a bit larger than that of $MAPbI_3$ due to the smaller cation, MA^+, that is more likely to be diminished in the mixture at a high temperature.

4.4.4 INTRAMOLECULAR EXCHANGE

In the solution-processed perovskite, the chemical compounds of lead halide and organic salts are dissolved in polar aprotic solvents, for example, N, N-dimethylformamide (DMF), N, N-dimethyl sulfoxide (DMSO), γ-butylolactone (GBL), etc. The PbI_2 is known to be a typical Lewis acid while polar aprotic solvents could act as Lewis bases. Hence, the growth of perovskite crystals can be controlled by leveraging the chemical interaction in the Lewis acid-base adduct.

In 2015, Seok et al. first prepared the large-grained, preferred crystallographic orientation $FAPbI_3$ film via the molecular exchange between PbI_2(DMSO) adduct and FAI[193]. Similarly, Shi et al. employed a $MAPbI_3$-FAI-PbI_2-DMSO adduct template to suppress the formation of δ-$FAPbI_3$ and obtained a pure α-$FAPbI_3$ phase perovskite film with improved crystallinity[194]. The resulting device achieves an efficiency of 21.24% under AM 1.5 G illumination with superior thermal and optical stability. Lee et al. replaced the DMSO with N-methyl-2-pyrrolidone (NMP) was able to form a stable intermediate FAI·PbI_2·NMP adduct (Figure 4.18a and b), which showed a stronger coordinative bond between NMP with the FA^+ than DMSO[193]. The stabler intermediate adduct induces uniform nucleation and growth of the film, resulting in the high-quality perovskite active layer.

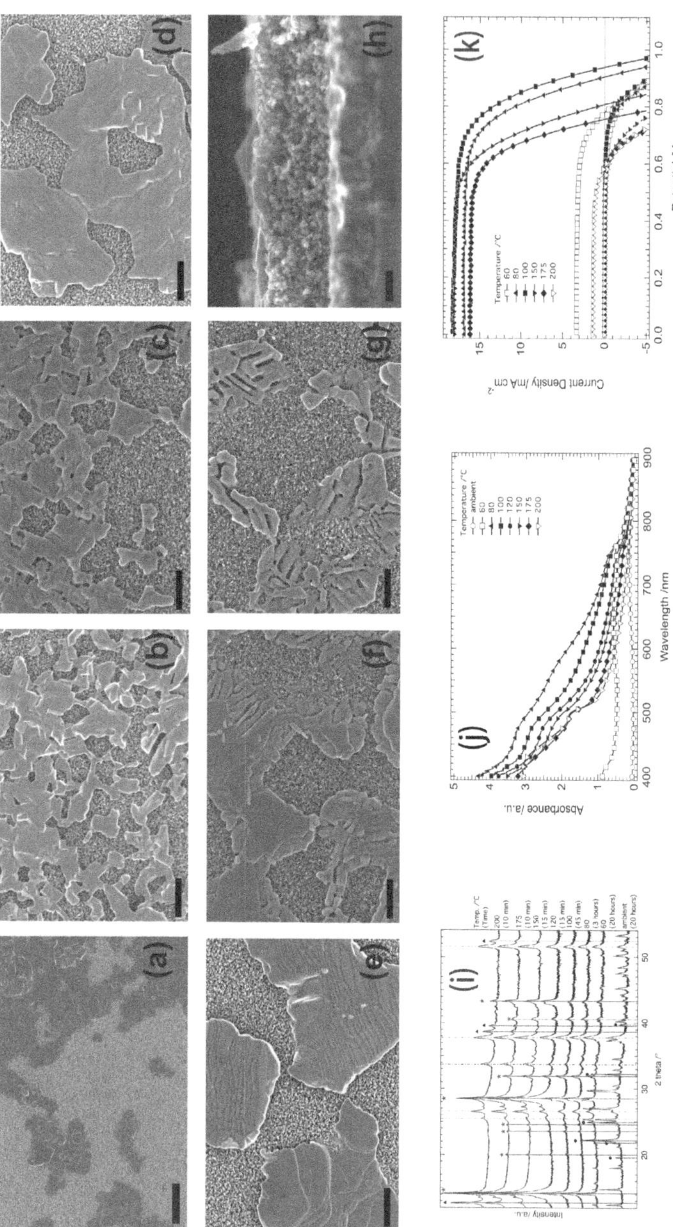

FIGURE 4.17 This composite image showcases SEM images and analytical graphs demonstrating the effects of different annealing temperatures on the morphology of perovskite compositions and their optical properties. (a–h) These panels show SEM images of MAPbI$_3$, and FAPbI$_3$ perovskites, as well as their 1:1 mixture, annealed at 100°C, 175°C, and 230°C. The white scale bar is 5 m, and the orange scale bar (inset) is 1 m. (i) A set of X-ray diffraction patterns for the perovskite films, demonstrating changes in crystallinity and phase composition with varied annealing temperatures. (j) Photos showing the visual appearance of MAPbI$_3$ or FAPbI$_3$ perovskite films on glass substrates post-annealing, highlighting changes in color and opacity that correspond to the processing temperatures. (k) Voltammetry curves for the perovskites, illustrating their electronic properties and the effect of temperature on their performance, potentially in a photovoltaic or electronic application.

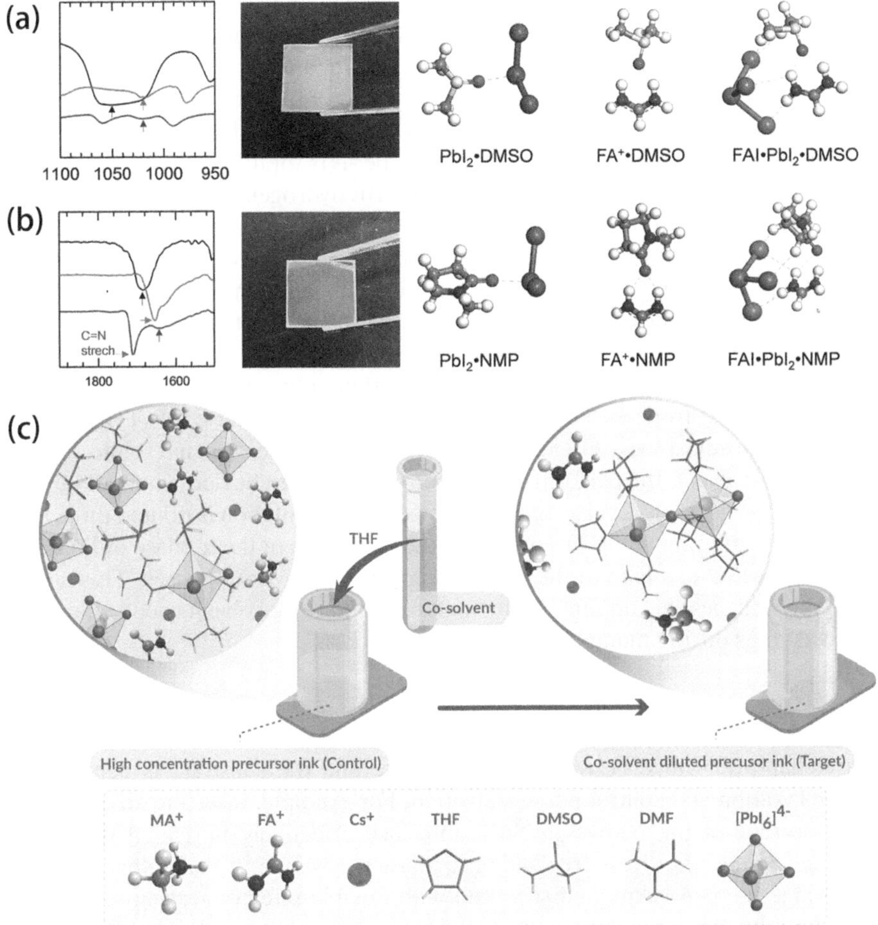

FIGURE 4.18 (a) Shows FTIR spectra highlighting the S=O stretching region, accompanied by a photograph of a shiny, crystalline FAI PbI$_2$ DMSO adduct film, and the corresponding molecular structures from DFT calculations, such as PbI$_2$ DMSO, FA$^+$ DMSO, and FAI PbI$_2$ DMSO. (b) Presents FTIR spectra for the C=O stretching region, a photo of a similar FAI PbI$_2$ NMP adduct film with a golden hue, and DFT-calculated molecular structures including PbI$_2$ NMP, FA$^+$ NMP, and FAI PbI$_2$ NMP. (c) Illustrates the dilution process in a schematic form, using THF as a cosolvent to create a precursor solution for perovskite fabrication, showing a comparison between high concentration precursor control and cosolvent-diluted precursor target, alongside molecular representations of the precursors and solvents involved.

Huang et al. developed a lead halide-templated crystallization strategy for printing FACs-based perovskite films[195]. In this process, the formation of solvent-coordinated perovskite intermediate complex can be suppressed from the stabler PbI$_2$-NMP adduct. Besides, the formation energy of α-FAPbI$_3$ is lower based on this approach, therefore the α-phase rich-FA perovskite can be formed even at RT through the in-situ reaction between PbI$_2$-NMP adduct and embedded FAI/CsI species. Dramatically, the resulting device showed a PCE of 23.35% and enhanced light and thermal stability. In addition, the S-donor Lewis base with a sulfur-bearing lone electron pair was stronger than the O-donor Lewis bases. Lee et al. incorporated thiourea into $FAPbI_3$ precursor solution from a 1:1:1 adduct with FAI and PbI_2, $FAPbI_3$ formed through the adduct displayed larger crystallite size, lower trap states, and enhanced photovoltaic performance[196]. Further studies indicated that the $-NH_2$ of thiourea interacts with I to form hydrogen bonds, and the S-donors can form hydrogen bonds with the $-NH_2$ in FAI or form acid-base adducts with PbI_2. Qin et al. demonstrated the strong interaction between S and Pb in the PbI_6 framework can effectively suppress the formation of δ-phase $FAPbI_3$[197]. Interestingly, Gratzel et al. mentioned a co-solvent dilution approach (Figure 4.18c), which allows for tuning the solvent-solute interaction with an immediate impact on the subsequent conversion and crystallization process of the perovskite phase[198]. Since the energy of interaction between tetrahydrofuran (THF) and Pb^{2+} is not as strong as the interaction between DMSO and DMF, THF is chosen as a dilution to adjust the average interaction between the solvent and solute. By using THF as a co-solvent, the performance of perovskite-based devices can still be maintained at a considerable level even when the perovskite solution is diluted three times. Additionally, diluting with THF promotes the formation of the α-phase of $FAPbI_3$, as excessive DMSO can lead to the formation of non-perovskite phases. This strategy effectively reduces the amount of raw materials and toxic solvents needed, resulting in cost savings during manufacturing.

4.4.5　Interfacial Engineering

A stretchable interface between the perovskite and the substrate is beneficial for stress relaxation and control phase transition. For example, inserting an elastic layer of polystyrene at the perovskite/SnO_2 interface effectively mitigated the residual stress and achieved a stress-free and pure phase perovskite[199-200]. Another alternative is to bridge the perovskite and substrate with flexible interconnecting molecules[200]. Especially, bridging the perovskite and SnO_2 with histamine diiodate (HADI) via $-NH_3-Sn^{4+}$ and $-C=N-Pb^{2+}$ bonds can simultaneously reduce interfacial distortion because of the mismatch between organic and inorganic materials and heal defect states from the buried interface (Figure 4.19a). Given the multifunctional roles of HADI, a flexible PSC with a record PCE of 22.44% was achieved, and more than 90% of the initial PCE remained after 1,000 cycles of perpendicular, parallel, and diagonal bending[201].

Interface integrity by filtrating perovskite into the underlying layer is another way to eliminate interfacial strain. For example, organic halides (*e.g.*, FAI, FACl, MACl) were pre-buried into the ETL precursors (e.g., SnO_2), which can in situ react with upper lead halides to form an interpenetrated perovskite/ETL interface (Figure 4.19b). The residual strain within such an integrated interface was remarkably reduced because of constrained lattice expansion and enhanced interfacial adhesion, leading to high endurance against delamination from external mechanical bending[202]. Similarly, the infiltration of perovskite layers into mesoporous ETL scaffolds or triple porous TiO_2/ZrO_2/carbon structures can also modulate the residual strain by confining perovskite crystallization with imposed compressive stress

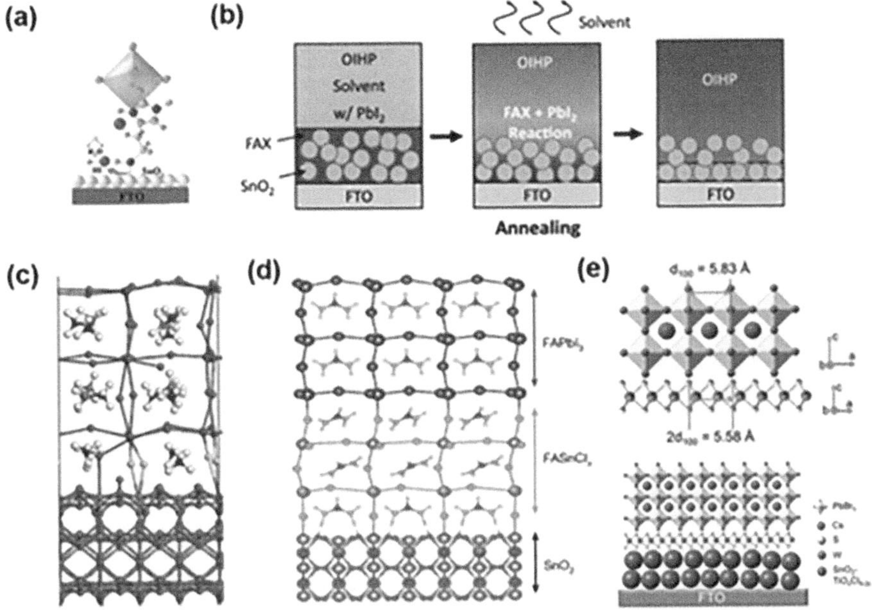

FIGURE 4.19 (a) Depicts a cartoon of a flexible interlayer designed to release strain between layers. (b) Shows a schematic of OIHP film formation, where the integrity of the interface is maintained to constrain lattice expansion during the annealing process. (c) Illustrates a lattice-matched structure where the lattice constants are tuned to match the ETL, depicted with a ball-and-stick model. (d) Presents a side view of a crystal structure where a coherent interface is constructed to release strain, showcasing layers of FAPbI$_3$ and FASnI$_3$ above a SnO$_2$ layer. (e) Demonstrates lattice matching by inserting an interlayer with matching spacing, with graphical representations of the perovskite crystal structure, the interlayer, and the FTO substrate, including distance measurements between atoms in angstroms.

and avoid phase transition, in which the magnitude of residual strain is dependent on the pore size of the micro/nanostructure[203]. The PSCs with interface integrity exhibited operational stability over 9,000 h under international electrotechnical commission standards.

The transition phase of perovskites can be suppressed by lattice matching at their buried interfaces. For instance, the doping of inorganic ETL with small molecules such as NH$_4$Cl, NH$_4$F, NH$_3$, and·H$_2$O can adjust the lattice parameter at the perovskite/ETL interface. Notably, the doping of NH$_4$Cl into a SnO$_2$ layer can generate a Cl-accumulated surface, which was transferred into an NH$_4$Cl-terminated perovskite at the MAPbI$_3$/SnO$_2$ interface, reducing the interfacial lattice mismatch (Figure 4.19c)[204]. Moreover, a coherent interface such as FASnCl$_x$ was constructed when a Cl-containing FAPbI$_3$ perovskite precursor was deposited on chloride-bonded SnO$_2$ (Cl-SnO$_2$)[205], which presented a tetragonal lattice structure with a slightly lower symmetry and smaller lattice parameter than the cubic $FAPbI_3$ (Figure 4.19d)[206]. This atomically interconnected interface can substantially suppress lattice distortion with reduced interface defects, enabling a record PCE of 25.8%. In addition, inserting buffer layers such as WS_2 with lattice constants close to half of $CsPbI_3$ at the perovskite/ETL interface enabled interfacial lattice matching (Figure 4.19e) and a prolonged T_{80} lifetime over 120 d under 80% humidity was achieved[207].

4.4.6 OTHER STRATEGIES

The details of the perovskite active layer markedly influence the optoelectronic properties. Using seed layers to aid in the control of the structure to improve optoelectronic properties is a common strategy in many semiconductor systems[208]. Zhao et al. observed submicron-sized perovskite seed crystals (Figure 4.20a–c) in the pristine PbI_2 film during the two-step deposition process to promote perovskite crystallization during the two-step deposition process[209]. The perovskite film, which grew with the assistance of these seeds, exhibited a lower density of traps, larger grain sizes, preferred crystal orientation, and phase stability. The resulting device achieved a PCE of 21.5% and demonstrated excellent operational stability (Figure 4.20d). Moreover, Zhu et al. developed a self-seeding growth (SSG) strategy to facilitate perovskite film growth that the perovskite film can regrow based on a self-seeding process, resulting in improved phase stability and optoelectronic properties[210]. The SSG devices obtained a PCE of 20.30%. More importantly, even after being stored for 4,680 h in a high-humidity ambient environment, the unencapsulated device maintained over 80% of its initial efficiency. Zhang et al. demonstrated that the PCBM and its derived (α-bis-PCBM) can induce heterogeneous nucleation on the perovskite film, leading to enhanced grain size and a tendency to grow in the preferred direction[211]. The hydrophobic PCBM and α-bis-PCBM passivated the grain boundaries and surface defects to prevent the erosion of moisture from the air, thus greatly improving device stability. This promising approach provides a simple route for fabricating highly efficient and phase-stable PSCs[212].

4.5 CONCLUSIONS AND PERSPECTIVES

Crystal structures, typically dictated by the composition, ionic radius, and valence states of the constituent elements, play an essential role in fundamental properties of perovskites such as optical, electrical, and mechanical capabilities. Manipulation of crystal structure of perovskite via compositional engineering, as well as understanding the link between crystal structure and properties, is critical for developing high-performance PSCs as PSC performance is greatly affected by the provision of diverse phase configurations of perovskite such as cubic, tetragonal, and orthorhombic phases. In this chapter, we reviewed the phase structure of perovskites. Then, the link between phase structure and properties of perovskites such as carrier transportation, electrical structure, stability, and carrier lifetime were discussed. Finally, the relationship of phase transition and the device performance of PSCs is analyzed, and various ways to stabilize the cubic phase to enhance the performance of $FAPbI_3$-based PSCs are systematically summarized.

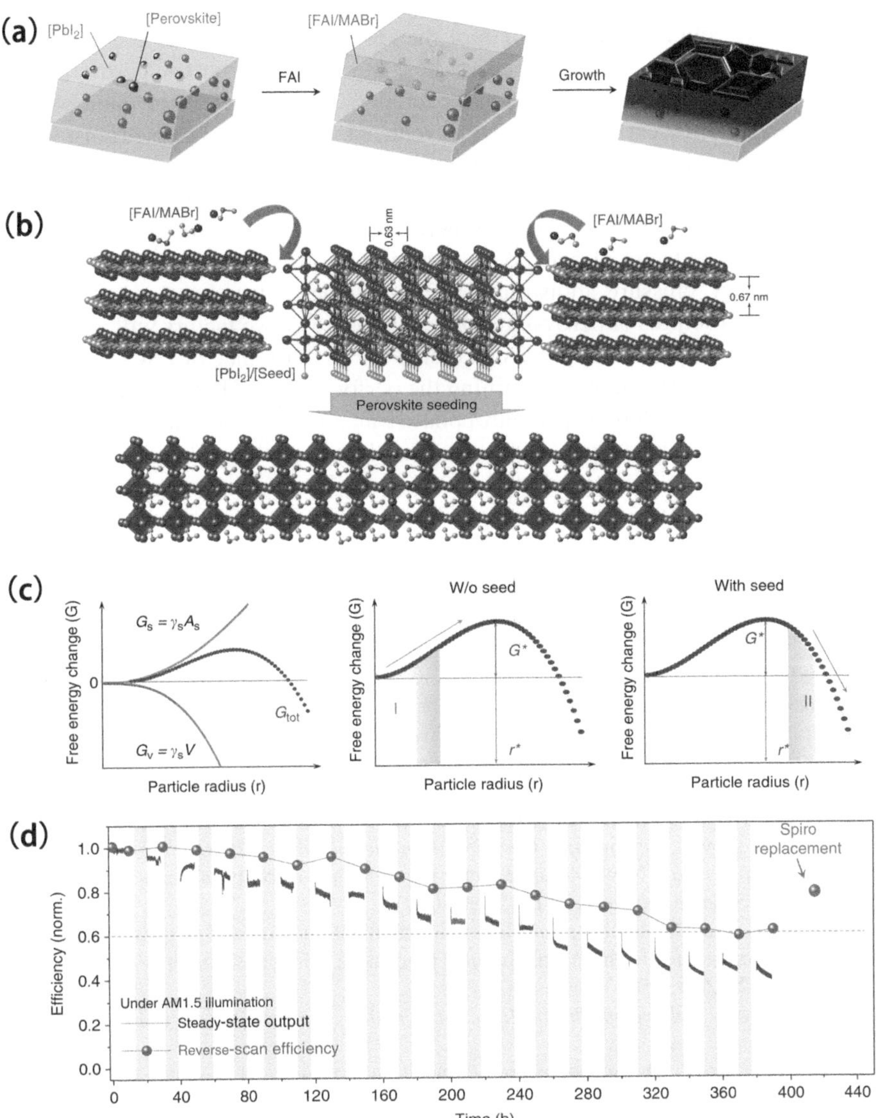

FIGURE 4.20 (a) Shows a three-step schematic: precursor solution deposition, introduction of FAI/MAI for transformation, and the resulting film growth, alongside a photo of the finished perovskite film. (b) Provides a detailed schematic of the perovskite crystallization process, where PbI$_2$ is converted into perovskite structures around initial seed crystals, reducing the energy barrier for crystal growth. (c) Graphs the Gibbs free energy as a function of particle radius, comparing scenarios without and with seed crystals, highlighting how seeds promote spontaneous crystallization by providing a radius larger than the critical threshold. (d) Shows a graph of the long-term operational stability of a perovskite solar cell device under simulated sunlight, measured by steady-state output and reverse-scan efficiency over time.

Perovskites go through several temperature-dependent phase changes, most of which involve rotation and distortion of the $[BX_6]^{4-}$ octahedra. Various factors will influence the phase structures of perovskites. At high temperatures, perovskite has a very symmetric cubic structure; when the temperature is lowered, the cubic structure is deformed and the crystal symmetry is reduced, leading to a phase transition to a tetragonal, orthorhombic, or even hexagonal phase structures. $FAPbI_3$-based PSCs require the black phase (α-phase) as the yellow phase is non-photoactive due to its limited absorption range. Despite temperature, variations in the ionic radius will alter the tolerance factor of perovskite, resulting in different perovskite phase structures. Moreover, pressure, moisture, and light changes also show serious influence on the phase structures of perovskites. Regulating the above factors will stabilize the phase structure of perovskite, thus enhancing the performance of PSCs.

Phase purity is critical in determining the performance of PSCs in operational situations. Significant solutions to control phase transitions are provided in this chapter, including strain management, composition engineering, thermal engineering, intramolecular exchange, interfacial engineering, etc. These perspective methodologies will aid in overcoming the challenges of stability issue and further enhancing the performance of PSCs in the future.

REFERENCES

1. N. Pellet, *et al.*, Mixed-organic-cation perovskite photovoltaics for enhanced solar-light harvesting, *Angewandte Chemie International Edition* **2014,** *53*, 3151.
2. Best cell efficiency chart, www.nrel.gov/pv/assets/pdfs/best-research-cell-efficiencies.pdf
3. S. Pang, *et al.*, $NH_2CH=NH_2PbI_3$: An alternative organolead iodide perovskite sensitizer for mesoscopic solar cells, *Chemistry of Materials* **2014,** *26*, 1485.
4. Y. Dang, *et al.*, Formation of hybrid perovskite tin iodide single crystals by top-seeded solution growth, *Angewandte Chemie International Edition* **2016,** *55*, 3447.
5. C. J. Bartel, *et al.*, New tolerance factor to predict the stability of perovskite oxides and halides, *Science Advances* **2019,** *5*, eaav0693.
6. C. Yi, J. Luo, S. Meloni, *et al.*, Entropic stabilization of mixed A-cation ABX_3 metal halide perovskites for high performance perovskite solar cells, *Energy & Environmental Science* **2016,** *9*, 656.
7. J. Bernstein, *Polymorphism in Molecular Crystals*, Oxford University Press, 2020, 2nd edition.
8. R. Bu, H. Li, C. Zhang, Polymorphic transition in traditional energetic materials: Influencing factors and effects on structure, property, and performance, *Crystal Growth & Design* **2020,** *20*, 3561.
9. E. A. Alharbi, A. Krishna, T. P. Baumeler, M. Dankl, G. C. Fish, F. Eickemeyer, M. Gratzel, Methylammonium triiodide for defect engineering of high-efficiency perovskite solar cells, *ACS Energy Letters* **2021,** *6*, 3650.
10. D. Luo, R. Su, W. Zhang, Q. Gong, R. Zhu, Minimizing non-radiative recombination losses in perovskite solar cells, *Nature Reviews Materials* **2020,** *5*, 44.
11. S. Sood, P. Gouma, Polymorphism in nanocrystalline binary metal oxides, *Nanomaterials and Energy* **2013,** *2*, 82.
12. B. A. Nogueira, C. Castiglioni, R. Fausto, Color polymorphism in organic crystals, *ChemComm* **2020,** *3*, 34.
13. L. Schmidt-Mende, V. Dyakonov, S. Olthof, F. Ünlü, K. M. T. Lê, S. Mathur, C. Draxl, Roadmap on organic-inorganic hybrid perovskite semiconductors and devices, *APL Materials* **2021,** *9*, 109202.

14. R. H. Bube, F. Buch, A. L. Fahrenbruch, Y. Y. Ma, K. W. Mitchell, Photovoltaic energy conversion with n-CdS—p-CdTe heterojunctions and other II-VI junctions, *IEEE Transactions on Electron Devices* **1977**, *24*, 487.

15. R. J. Sutton, M. R. Filip, A. A. Haghighirad, N. Sakai, B. Wenger, F. Giustino, H. J. Snaith, Cubic or orthorhombic? Revealing the crystal structure of metastable black-phase CsPbI3 by theory and experiment, *ACS Energy Letters* **2018**, *3*, 1787.

16. A. Marronnier, G. Roma, S. Boyer-Richard, L. Pedesseau, J. M. Jancu, Y. Bonnassieux, J. Even, Anharmonicity and disorder in the black phases of cesium lead iodide used for stable inorganic perovskite solar cells, *ACS Nano* **2018**, *12*, 3477.

17. T. P. Baumeler, E. A. Alharbi, G. Kakavelakis, G. C. Fish, M. T. Aldosari, M. S. Albishi, M. Grätzel, Surface passivation of FAPbI3-rich perovskite with cesium iodide outperforms bulk incorporation, *ACS Energy Letters* **2023**, *8*, 2456.

18. Y. Gao, H. Raza, Z. Zhang, W. Chen, Z. Liu, Rethinking the role of excess/residual lead iodide in perovskite solar cells, *Advanced Functional Materials* **2023**, *33*, 2215171.

19. J. W. Lee, S. Tan, S. I. Seok, Y. Yang, N. G. Park, Rethinking the A cation in halide perovskites, *Science* **2022**, *375*, eabj1186.

20. S. H. Turren-Cruz, A. Hagfeldt, M. Saliba, Methylammonium-free, high-performance, and stable perovskite solar cells on a planar architecture, *Science* **2018**, *362*, 449.

21. H. Chen, Q. Wei, M. I. Saidaminov, F. Wang, A. Johnston, Y. Hou, Z. Ning, Efficient and stable inverted perovskite solar cells incorporating secondary amines, *Advanced Materials* **2019**, *31*, e1903559.

22. A. Baumann, S. Väth, P. Rieder, M. C. Heiber, K. Tvingstedt, V. Dyakonov, Identification of trap states in perovskite solar cells, *The Journal of Physical Chemistry Letters* **2015**, *6*, 2350.

23. C. Qin, T. Matsushima, T. Fujihara, W. J. Potscavage Jr., C. Adachi, Degradation mechanisms of solution-processed planar perovskite solar cells: Thermally stimulated current measurement for analysis of carrier traps, *Advanced Materials* **2016**, *28*, 466.

24. C. Qin, T. Matsushima, D. Klotz, T. Fujihara, C. Adachi, The relation of phase-transition effects and thermal stability of planar perovskite solar cells, *Advanced Science* **2019**, *6*, 1801079.

25. T. Biakie, Y. Fang, J. M. Kadro, M. Schreyer, F. Wei, S. G. Mhaisalkar, M. Gratzel, T. J. White, Synthesis and crystal chemistry of the hybrid perovskite (CH$_3$NH$_3$)PbI$_3$ for solid-state sensitised solar cell applications, *Journal of Materials Chemistry A* **2013**, *1*, 5628.

26. R. Roesch, T. Faber, E. Hauff, T. M. Brown, M. Lira-Cantu, H. Hoppe, Procedures and practices for evaluating thin-film solar cell stability, *Advanced Energy Materials* **2015**, *5*, 1501407.

27. N. Rolston, K. A. Bush, A. D. Printz, A. Gold-Parker, Y. Ding, M. F. Toney, M. D. McGehee, R. H. Dauskardt, Engineering stress in perovskite solar cells to improve stability, *Advanced Energy Materials* **2018**, *8*, 1802139.

28. R. Beal, *et al.*, Cesium lead halide perovskites with improved stability for tandem solar cells, *The Journal of Physical Chemistry Letters* **2016**, *7*, 746.

29. H. Qiu, J. M. Mativetsky, Nanoscale light-and voltage-induced lattice strain in perovskite thin films, *Nanoscale* **2021**, *13*, 746.

30. J. Wang, *et al.*, Efficient perovskite solar cells by metal ion do, *Energy & Environmental Science* **2016**, *9*, 2892.

31. C. Liang, *et al.*, Two-dimensional Ruddlesden-Popper layered perovskite solar cells based on phase-pure thin films, *Nature Energy* **2021**, *6*, 38.

32. T. Kong, *et al.*, Perovskitoid-templated formation of a 1D@3D perovskite structure toward highly efficient and stable perovskite solar cells, *Advanced Energy Materials* **2021**, *11*, 2101018.

33. Z. Wang, Q. Lin, F. P. Chmiel, N. Sakai, L. M. Herz, H. Snaith, Efficient ambient-air-stable solar cells with 2D–3D heterostructured butylammonium-caesium-formamidinium lead halide perovskites, *Nature Energy* **2017**, *2*, 17135.

34. S. De Wolf, *et al.*, Organometallic halide perovskites: Sharp optical absorption edge and its relation to photovoltaic performance, *The Journal of Physical Chemistry Letters* **2014**, *5*, 1035.

35. L. M. Herz, Charge-carrier mobilities in metal halide perovskites: Fundamental mechanisms and limits, *ACS Energy Letters* **2017**, *2*, 1539.

36. S. Stranks, *et al.*, Electron-hole diffusion lengths exceeding 1 micrometer in an organometal trihalide perovskite absorber, *Science* **2013**, *342*, 341.

37. J. Kim, J.-W. Lee, H. Jung, H. Shin, N. Park, High-Efficiency Perovskite Solar Cells, *Chem. Rev.* **2020**, *120*, 7867.

38. M. Jung, S. Ji, G. Kim, S. Seok, Perovskite precursor solution chemistry: From fundamentals to photovoltaic applications, *Chem. Soc. Rev.* **2019**, *48*, 2011.

39. J.-P. Correa-Baena, M. Saliba, T. Buonassisi, M. Gratzel, A. Abate, W. Tress, A. Hagfeldt, Promises and challenges of perovskite solar cells, *Science* **2017**, *358*, 739–744.

40. J. Zhao, Y. Deng, H. Wei, X. Zheng, Z. Yu, Y. Shao, J. E. Shield, J. Huang, Strained hybrid perovskite thin films and their impact on the intrinsic stability of perovskite solar cells, *Sci. Adv.* **2017**, *3*, eaao5616.

41. C. Stoumpos, C. Malliakas, M. Kanatzidis, Semiconducting tin and lead iodide perovskites with organic cations: Phase transitions, high mobilities, and near-infrared photoluminescent properties, *Inorganic Chemistry* **2013**, *52*, 9019.

42. L. Zhang, Y. Wang, J. Lv, Y. Ma, Materials discovery at high pressures, *Nature Reviews Materials* **2017**, *2*, 17005.

43. B. Rosales, L. Men, S. Cady, M. Hanrahan, A. Rossini, J. Vela, Persistent dopants and phase segregation in organolead mixed-halide perovskites, *Chemistry of Materials* **2016**, *28*, 6848.

44. Z. Ma, *et al.*, Pressure-induced emission of cesium lead halide perovskite nanocrystals, *Nature Communications* **2018**, *9*, 4506.

45. A. Jaffe, Y. Lin, C. Beavers, J. Voss, W. Mao, H. Karunadasa, High-pressure single-crystal structures of 3D lead-halide hybrid perovskites and pressure effects on their electronic and optical properties, *ACS Central Science* **2016**, *2*, 201.

46. M. Szafrański, A. Katrusiak, Mechanism of pressure-induced phase transitions, amorphization, and absorption-edge shift in photovoltaic methylammonium lead iodide, *ACS Central Science.* **2016**, *7*, 3458.

47. C. Chen, A. Herhold, C. Johnson, A. Alivisatos, Size dependence of structural metastability in semiconductor nanocrystals, *Science* **1997**, *276*, 398.

48. R. J. Angel, J. Zhao, N. L. Ross, General rules for predicting phase transitions in perovskites due to octahedral tilting, *Physical Review Letters* **2005**, *95*, 025503.

49. C. J. Howard, B. J. Kennedy, The orthorhombic and rhombohedral phases of $LaGaO_3$- a neutron powder diffraction study, *Journal of Physics: Condensed Matter* **1999**, *11*, 3229.

50. Y. Zhou, J. Kwun, H. F. Garces, S. Pang, N. P. Padture, Observation of phase-retention behavior of the $HC(NH_2)_2PbI_3$ black perovskite polymorph upon mesoporous TiO_2 scaffolds, *Chemical Communications* **2016**, *52*, 7273.

51. S. Hoefler, G. Trimmel, T. Rath, Progress on lead-free metal halide perovskites for photovoltaic applications: A review, *Monatshefte für Chemie - Chemical* **2017**, *148*, 795.

52. T. Jacobsson, J.-P. Correa-Baena, M. Pazoki, M. Saliba, K. Schenk, M. Grätzel, A. Hagfeldt, Exploration of the compositional space for mixed lead halogen perovskites for high efficiency solar cells, *Energy & Environmental Science* **2016**, *9*, 1706.

53. L. Li, N. Liu, Z. Q. Xu, Q. Chen, X. D. Wang, H. P. Zhou, Recent advancement of imidazolate framework (ZIF-8) based nanoformulations for synergistic tumor therapy, *ACS Nano* **2017**, *11*, 8804.

54. Z. Wang, D. Mcmeekin, N. Sakai, S. Reenen, K. Wojciechowski, J. Patel, M. Johnston, H. Snaith, Efficient and air-stable mixed-cation lead mixed-halide perovskite solar cells with n-doped organic electron extraction layers, *Advanced Materials* **2016**, *29*, 1604186.

55. Y. Reyna, M. Salado, S. Kazim, A. Pérez-Tomas, S. Ahmad, M. Lira-Cantu, Performance and stability of mixed FAPbI3(0.85)MAPbBr3(0.15) halide perovskite solar cells under outdoor conditions and the effect of low light irradiation, *Nano Energy* **2016**, *30*, 570.

56. F. Bella, *et al.*, Improving efficiency and stability of perovskite solar cells with photocurable fluoropolymers, *Science* **2016**, *354*, 203.

57. J. Huang, S. Tan, P. Lund, H. Zhou, Impact of H_2O on organic–inorganic hybrid perovskite solar cells, *Energy & Environmental Science* **2017**, *10*, 2284.

58. J. Manser, M. Saidaminov, J. Christians, O. Bakr, P. Kamat, Making and breaking of lead halide perovskites, *Accounts of Chemical Research* **2016**, *49*, 330.

59. E. Duijnstee, *et al.*, Understanding the degradation of methylenediammonium and its role in phase-stabilizing formamidinium lead triiodide, *Journal of the American Chemical Society* **2023**, *145*, 10275.

60. I. Deretzis, E. Smecca, G. Mannino, A. La Magna, T. Miyasaka, A. Alberti, Stability and degradation in hybrid perovskites: Is the glass half-empty or half-full? *The Journal of Physical Chemistry Letters* **2018**, *9*, 3000.

61. M. A. Haque, A., Syed, F. H. Akhtar, R. Shevate, S. Singh, K. V. Peinemann, T. Wu, Giant humidity effect on hybrid halide perovskite microstripes: Reversibility and sensing mechanism, *ACS Appl. Mater. Interfaces* **2019**, *11*, 33, 29821.

62. A. Leguy, Y. Hu, M. Campoy-Quiles, M. Alonso, O. Weber, P. Azarhoosh, P. Barnes, Reversible hydration of CH3NH3PbI3 in films, single crystals, and solar cells, *Chemistry of Materials* **2015**, *27*, 3397.

63. S. Shao, A. Loi, The role of the interfaces in perovskite solar cells, *Advanced Materials Interfaces* **2020**, *7*, 1901469.

64. D. Kubicki, S. Stranks, C. Grey, L. Emsley, NMR spectroscopy probes microstructure, dynamics and doping of metal halide perovskites, *Nature Reviews Chemistry* **2021**, *5*, 624.

65. F. Hao, C. Stoumpos, Z. Liu, R. Chang, G. Kanatzidis, Controllable perovskite crystallization at a gas–solid interface for hole conductor-free solar cells with steady power conversion efficiency over 10%, *Journal of the American Chemical Society* **2014**, *136*, 16411.

66. S. Tolbert, A. Alivisatos, Size dependence of a first order solid-solid phase transition: The wurtzite to rock salt transformation in CdSe nanocrystals, *Science* **1994**, *265*, 373.

67. J. B. Rivest, L. K. Fong, P. K. Jain, M. F. Toney, P. A. Alivisatos, Size dependence of a temperature-induced solid–solid phase transition in Copper(I) sulfide, *Journal of Physical Chemistry Letters* **2011**, *2*, 2402.

68. B. Huang, *et al.*, Ferroic domains regulate photocurrent in single-crystalline $CH_3NH_3PbI_3$ films self-grown on FTO/TiO_2 substrate, *npj Quantum Materials* **2018**, *3*, 30.

69. M. U. Rothmann, W. Li, Y. Zhu, U. Bach, L. Spiccia, J. Etheridge, Y.-B. Cheng, Direct observation of intrinsic twin domains in tetragonal CH3NH3PbI3, *Nature Communications* **2017**, *8*, 14547.

70. W. Rehman, D. McMeekin, J. Patel, R. Milot, M. Johnston, H. Snaith, L. Herz, Photovoltaic mixed-cation lead mixed-halide perovskites: Links between crystallinity, photo-stability and electronic properties, *Energy & Environmental Science* **2017**, *10*, 361.

71. K. Lin, *et al.*, Perovskite light-emitting diodes with external quantum efficiency exceeding 20 per cent, *Nature* **2018**, *562*, 245.

72. J. Feng, *et al.*, Single-crystalline layered metal-halide perovskite nanowires for ultrasensitive photodetectors, *Nature Electronics* **2018**, *1*, 404.

73. G. E. Eperon, S. D. Stranks, C. Menelaou, M. B. Johnston, L. M. Herz, H. J. Snaith, Formamidinium lead trihalide: A broadly tunable perovskite for efficient planar heterojunction solar cells, *Energy & Environmental Science* **2014**, *7*, 982.

74. B. Yang, *et al.*, Strain effects on halide perovskite solar cells, *Chemical Society Reviews* **2022**, *51*, 7509.

75. B. Chen, J. Song, X. Dai, Y. Liu, P. N. Rudd, X. Hong, J. Huang, Synergistic effect of elevated device temperature and excess charge carriers on the rapid light-induced degradation of perovskite solar cells, *Advanced Materials* **2019**, *31*, 1902413.

76. C. Zhu, *et al.*, Strain engineering in perovskite solar cells and its impacts on carrier dynamics, *Nature Communications* **2019**, *10*, 815.

77. T. Liu, *et al.*, Stable formamidinium-based perovskite solar cells via in situ grain encapsulatio, *Advanced Energy Materials* **2018**, *8*, 1800232.

78. H. Zhang, *et al.*, Multifunctional crosslinking-enabled strain-regulating crystallization for stable efficient α-FAPbI$_3$-based perovskite solar cells, *Advanced Materials* **2021**, *33*, 2008487.

79. D. Glowienka, Y. Galagan, Light intensity analysis of photovoltaic parameters for perovskite solar cells, *Advanced Materials* **2022**, *34*, 2105920.

80. R. Ji, Z. Zhang, Y. Hofstetter, R. Buschbeck, C. Hänisch, F. Paulus, Y. Vaynzof, Perovskite phase heterojunction solar cells, *Nature Energy* **2022**, *7*,12, 1170.

81. N.-G. Park, Perovskite solar cells: An emerging photovoltaic technology, *Materials Today* **2015**, *18*, 65.

82. D. A. Egger, L. Kronik, Role of dispersive interactions in determining structural properties of organic–inorganic halide perovskites: Insights from first-principles calculations, *The Journal of Physical Chemistry Letters* **2014**, *5*, 2728.

83. S. Hirotsu, J. Harada, M. Iizumi, K. Gesi, Structural phase transitions in CsPbBr$_3$, *Journal of the Physical Society of Japan* **1974**, *37*, 1393.

84. R. E. Wasylishen, O. Knop, J. B. Macdonald, Cation rotation in methylammonium lead halides, *Solid State Communications* **1985**, *56*, 581.

85. P. Nandi, C. Giri, D. Swain, U. Manju, D. Topwal, Room temperature growth of CH$_3$NH$_3$PbCl$_3$ single crystals by solvent evaporation method, *CrystEngComm* **2019**, *21*, 656.

86. P. Postorino, L. Malavasi, Pressure-induced effects in organic–inorganic hybrid perovskites, *The Journal of Physical Chemistry Letters* **2017**, *8*, 2613.

87. A. Jaffe, Y. Lin, C. M. Beavers, J. Voss, W. L. Mao, H. I. Karunadasa, High-pressure single-crystal structures of 3D lead-halide hybrid perovskites and pressure effects on their electronic and optical properties, *ACS Central Science* **2016**, *2*, 201.

88. T. Baikie, *et al.*, Synthesis and crystal chemistry of the hybrid perovskite (CH$_3$NH$_3$)PbI$_3$ for solid-state sensitised solar cell applications, *Journal of Materials Chemistry A* **2013**, *1*, 5628.

89. A. Miyata, *et al.*, Direct measurement of the exciton binding energy and effective masses for charge carriers in organic–inorganic tri-halide perovskites, *Nature Physics* **2015**, *11*, 582.

90. Z. An, S. Chen, T. Lu, P. Zhao, X. Yang, Y. Li, J. Hou, Interfacial modification via aniline molecules with multiple active sites for performance enhancement of n–i–p perovskite solar cells, *Journal of Materials Chemistry C* **2023**, *11*,12750.

91. G. Giorgi, J.-I. Fujisawa, H. Segawa, K. Yamashita, Small photocarrier effective masses featuring ambipolar transport in methylammonium lead iodide perovskite: A density functional analysis, *The Journal of Physical Chemistry Letters* **2013**, *4*, 4213.

92. A. M. A. Leguy, *et al.*, The dynamics of methylammonium ions in hybrid organic–inorganic perovskite solar cells, *Nature Communications* **2015**, *6*, 7124.

93. T. Chen, *et al.*, Origin of long lifetime of band-edge charge carriers in organic–inorganic lead iodide perovskites, *Proceedings of the National Academy of Sciences USA* **2017**, *114*, 7519.

94. F. Ma, J. Li, W. Li, N. Lin, L. Wang, J. Qiao, Stable α/δ phase junction of formamidinium lead iodide perovskites for enhanced near-infrared emission, *Chemical Science* **2017**, *8*, 800.

95. T. Koh, *et al.*, Formamidinium-containing metal-halide: An alternative material for near-IR absorption perovskite solar cells, *The Journal of Physical Chemistry C* **2014**, *118*, 16458.

96. G. Eperon, S. Stranks, C. Menelaou, M. Johnston, L. M. Herz, H. Snaith, Formamidinium lead trihalide: A broadly tunable perovskite for efficient planar heterojunction solar cells, *Energy & Environmental Science* **2014**, *7*, 982–988.

97. G. Kieslich, S. Sun, A. K. Cheetham, Solid-state principles applied to organic–inorganic perovskites: New tricks for an old dog, *Chemical Science* **2014**, *5*, 4712–4715.

98. T. Chen, *et al.*, Entropy-driven structural transition and kinetic trapping in formamidinium lead iodide perovskite, *Science Advances* **2016**, *2*, e1601650.

99. J.-W. Lee, D.-H. Kim, H.-S. Kim, S.-W. Seo, S. M. Cho, N.-G. Park, Formamidinium and cesium hybridization for photo- and moisture-stable perovskite solar cell, *Advanced Energy Materials* **2015**, *5*, 1501310.

100. J. Lin, *et al.*, Thermochromic halide perovskite solar cells, *Nature Materials* **2018**, *17*, 261.

101. D. J. Kubicki, *et al.*, Formation of stable mixed guanidinium–methylammonium phases with exceptionally long carrier lifetimes for high-efficiency lead Iodide-based perovskite photovoltaics, *Journal of the American Chemical Society* **2018**, *140*, 3345.

102. E. J. Juarez-Perez, L. K. Ono, Y. Qi, Thermal degradation of formamidinium based lead halide perovskites into sym-triazine and hydrogen cyanide observed by coupled thermogravimetry-mass spectrometry analysis, *Journal of Materials Chemistry A* **2019**, *7*, 16912.

103. G. E. Eperon, G. M. Paternò, R. J. Sutton, A. Zampetti, A. A. Haghighirad, F. Cacialli, H. J. Snaith, Inorganic caesium lead iodide perovskite solar cells, *Journal of Materials Chemistry A* **2015**, *3*, 19688.

104. C. C. Stoumpos, C. D. Malliakas, M. G. Kanatzidis, Semiconducting tin and lead iodide perovskites with organic cations: Phase transitions, high mobilities, and near-infrared photoluminescent properties, *Inorganic Chemistry* **2013**, *52*, 9019.

105. A. Swarnkar, *et al.*, Quantum dot–induced phase stabilization of α-CsPbI$_3$ perovskite for high-efficiency photovoltaics, *Science* **2016**, *354*, 92.

106. J. Xue, *et al.*, Surface ligand management for stable FAPbI$_3$ perovskite quantum dot solar cells, *Joule* **2018**, *2*, 1866–1878.

107. Y. Wang, *et al.*, Thermodynamically stabilized β-CsPbI$_3$–based perovskite solar cells with efficiencies >18%, *Science* **2019**, *365*, 591.

108. V. Poulek, A. Khudysh, M. Libra, Innovative low concentration PV systems with bifacial solar panels, *Solar Energy* **2015**, *120*, 113.

109. M. Benghanem, A. A. Al-Mashraqi, K. O. Daffallah, Performance of solar cells using thermoelectric module in hot sites, *Renewable Energy* **2016**, *89*, 51.

110. B. Conings, *et al.*, Intrinsic thermal instability of methylammonium lead trihalide perovskite, *Advanced Energy Materials* **2015**, *5*, 1500477.

111. E. J. Juarez-Perez, L. K. Ono, M. Maeda, Y. Jiang, Z. Hawash, Y. Qi, Photodecomposition and thermal decomposition in methylammonium halide lead perovskites and inferred design principles to increase photovoltaic device stability, *Journal of Materials Chemistry A* **2018**, *6*, 9604.

112. R. K. Misra, S. Aharon, B. Li, D. Mogilyansky, I. Visoly-Fisher, L. Etgar, E. A. Katz, Temperature- and component-dependent degradation of perovskite photovoltaic materials under concentrated sunlight, *The Journal of Physical Chemistry Letters* **2015**, *6*, 326.

113. M. Saliba, *et al.*, Cesium-containing triple cation perovskite solar cells: Improved stability, reproducibility and high efficiency, *Energy & Environmental Science* **2016**, *9*, 1989.

114. A. Binek, F. C. Hanusch, P. Docampo, T. Bein, Stabilization of the trigonal high-temperature phase of formamidinium lead Iodide, *The Journal of Physical Chemistry Letters* **2015**, *6*, 1249.

115. N. J. Jeon, J. H. Noh, W. S. Yang, Y. C. Kim, S. Ryu, J. Seo, S. I. Seok, Compositional engineering of perovskite materials for high-performance solar cells, *Nature* **2015**, *517*, 476.

116. M. Kulbak, S. Gupta, N. Kedem, I. Levine, T. Bendikov, G. Hodes, D. Cahen, Cesium enhances long-term stability of lead bromide perovskite-based solar cells, *The Journal of Physical Chemistry Letters* **2016**, *7*, 167.

117. A. F. Akbulatov, *et al.*, Probing the intrinsic thermal and photochemical stability of hybrid and inorganic lead halide perovskites, *The Journal of Physical Chemistry Letters* **2017**, *8*, 1211.

118. D. P. McMeekin, *et al.*, A mixed-cation lead mixed-halide perovskite absorber for tandem solar cells, *Science* **2016**, *351*, 151.

119. W. Xiang, W. Tress, Review on recent progress of all-inorganic metal halide perovskites and solar cells, *Advanced Materials* **2019**, *31*, 1902851.

120. C. Wu, *et al.*, The dawn of lead-free perovskite solar cell: Highly stable double perovskite $Cs_2AgBiBr_6$ film, *Advanced Science* **2018**, *5*, 1700759.

121. H. Lei, D. Hardy, F. Gao, Lead-free double perovskite $Cs_2AgBiBr_6$: Fundamentals, applications, and perspectives, *Adv. Func. Mater.* **2021**, *31*, 2105898.

122. B. Wang, *et al.*, Chlorophyll derivative-sensitized TiO_2 electron transport layer for record efficiency of $Cs_2AgBiBr_6$ double perovskite solar cells, *Journal of the American Chemical Society* **2021**, *143*, 2207.

123. J. Yang, B. D. Siempelkamp, D. Liu, T. L. Kelly, Investigation of $CH_3NH_3PbI_3$ degradation rates and mechanisms in controlled humidity environments using in situ techniques, *ACS Nano* **2015**, *9*, 1955.

124. J. M. Frost, K. T. Butler, F. Brivio, C. H. Hendon, M. van Schilfgaarde, A. Walsh, Atomistic origins of high-performance in hybrid halide perovskite solar cells, *Nano Letters* **2014**, *14*, 2584

125. L. Zhang, P. H. L. Sit, Ab initio study of interaction of water, hydroxyl radicals, and hydroxide Ions with $CH_3NH_3PbI_3$ and $CH_3NH_3PbBr_3$ surfaces, *The Journal of Physical Chemistry C* **2015**, *119*, 22370.

126. L. Zhang. P. H. L. Sit, Ab initio static and dynamic study of $CH_3NH_3PbI_3$ degradation in the presence of water, hydroxyl radicals, and hydroxide ions, *RSC Advances* **2016**, *6*, 76938.

127. A. M. A. Leguy, *et al.*, Reversible hydration of $CH_3NH_3PbI_3$ in films, single crystals, and solar cells, *Chemistry of Materials* **2015**, *27*, 3397.

128. N. Ahn, *et al.*, Trapped charge-driven degradation of perovskite solar cells, *Nature Communications* **2016**, *7*, 13422.

129. B. Hailegnaw, S. Kirmayer, E. Edri, G. Hodes, D. Cahen, Rain on methylammonium lead Iodide based perovskites: Possible environmental effects of perovskite solar cells, *The Journal of Physical Chemistry Letters* **2015**, *6*, 1543.

130. E. Mosconi, J. M. Azpiroz, F. De Angelis, Ab Initio molecular dynamics simulations of methylammonium lead Iodide perovskite degradation by water, *Chemistry of Materials* **2015**, *27*, 4885.

131. H. K. Adli, T. Harada, W. Septina, S. Hozan, S. Ito, S. Ikeda, Effects of porosity and amount of surface hydroxyl groups of a porous TiO_2 layer on the performance of a $CH_3NH_3PbI_3$ perovskite photovoltaic cell, *The Journal of Physical Chemistry C* **2015**, *119*, 22304.

132. J. A. Aguiar, *et al.*, In situ investigation of the formation and metastability of formamidinium lead tri-iodide perovskite solar cells, *Energy & Environmental Science* **2016**, *9*, 2372.

133. T. Kumagai, *et al.*, H-atom relay reactions in real space, *Nat. Mater.* **2012**, *11*, 167.

134. Y. Cheng, *et al.*, Decomposition of organometal halide perovskite films on Zinc Oxide nanoparticles, *ACS Applied Materials & Interfaces* **2015**, *7*, 19986.

135. J. Yang, B. D. Siempelkamp, E. Mosconi, F. De Angelis, T. L. Kelly, Origin of the thermal instability in $CH_3NH_3PbI_3$ thin films deposited on ZnO, *Chemistry of Materials* **2015**, *27*, 4229.

136. Y. Han, *et al.*, Degradation observations of encapsulated planar $CH_3NH_3PbI_3$ perovskite solar cells at high temperatures and humidity, *Journal of Materials Chemistry A* **2015**, *3*, 8139.

137. G. Niu, W. Li, F. Meng, L. Wang, H. Dong, Y. Qiu, Study on the stability of $CH_3NH_3PbI_3$ films and the effect of post-modification by aluminum oxide in all-solid-state hybrid solar cells, *Journal of Materials Chemistry A*, **2014**, *2*, 705.

138. W.-C. Lin, H.-Y. Chang, K. Abbasi, J.-J. Shyue, C. Burda, 3D in situ ToF-SIMS imaging of perovskite films under controlled humidity environmental conditions, *Advanced Materials Interfaces* **2017**, *4*, 1600673.

139. W. Huang, J. S. Manser, P. V. Kamat, S. Ptasinska, Evolution of chemical composition, morphology, and photovoltaic efficiency of $CH_3NH_3PbI_3$ perovskite under ambient conditions, *Chemistry of Materials* **2016**, *28*, 303.

140. S. H. Turren-Cruz, A. Hagfeldt, M. Saliba, Methylammonium-free, high-performance, and stable perovskite solar cells on a planar architecture, *Science* **2018**, *362*, 449.

141. Z. P. Wang, D. P. McMeekin, N. Sakai, S. van Reenen, K. Wojciechowski, J. B. Patel, M. B. Johnston, H. J. Snaith, Efficient and air-stable mixed-cation lead mixed-halide perovskite solar cells with n-doped organic electron extraction layers, *Advanced Materials* **2017**, *29*, 1604186.

142. W. Meng, K. C. Zhang, A. Osvet, J. Y. Zhang, W. Gruber, K. Forberich, B. Meyer, W. Heiss, T. Unruh, N. Li, C. J. Brabec, Revealing the strain-associated physical mechanisms impacting the performance and stability of perovskite solar cells, *Joule* **2022**, *6*, 458.

143. T. Liu, K. Chen, Q. Hu, R. Zhu, Q. Gong, Inverted perovskite solar cells: Progresses and perspectives, *Advanced Energy Materials* **2016**, *6*, 1600457.

144. J. Zhao, Y. Deng, H. Wei, X. Zheng, Z. Yu, Y. Shao, J. Shield, J. Huang, Strained hybrid perovskite thin films and their impact on the intrinsic stability of perovskite solar cells, *Science Advances* **2017**, *3*, eaao5616.

145. N. Rolston, *et al.*, Engineering stress in perovskite solar cells to improve stability, *Advanced Energy Materials* **2018**, *8*, 1802139.

146. Y. F. Kang, *et al.*, Thermal shock fabrication of ion-stabilized perovskite and solar cells, *Advanced Materials* **2022**, *34*, 2203166.

147. L. N. Wang, *et al.*, Strain modulation for light-stable n–i–p perovskite/silicon tandem solar cells, *Advanced Materials* **2022**, *34*, 2201315.

148. W. X. Zhang, Z. X. Lin, R. Huang, Y. Guo, Effect of phase transition on optical properties and photovoltaic performance in cesium lead bromine perovskite: A theoretical study, *The Journal of Physical Chemistry C* **2019**, *123*, 34, 20764.

149. J. Z. Jiang, *et al.*, Synergistic strain engineering of perovskite single crystals for highly stable and sensitive X-ray detectors with low-bias imaging and monitoring, *Nature Photonics* **2022**, *16*, 575.

150. H. Cheng, C. Liu, J. Zhuang, J. Cao, T. Wang, W. Wong, F. Yan, KBF_4 additive for alleviating microstrain, improving crystallinity, and passivating defects in inverted perovskite solar cells, *Advanced Functional Materials* **2022**, *32*, 2204880.

151. T. Liu, *et al.*, Stable formamidinium-based perovskite solar cells via in situ grain encapsulation, *Advanced Energy Materials* **2018**, *8*, 1800232.

152. T. Liu, *et al.*, Mesoporous PbI_2 scaffold for high-performance planar heterojunction perovskite solar cells, *Advanced Energy Materials* **2016**, *6*, 1501890.

153. Y. Li, *et al.*, Two-step solvent post-treatment on PTAA for highly efficient and stable inverted perovskite solar cells, *Photonics Research* **2020**, *8*, A39.

154. G. C. Xing, *et al.*, Long-range balanced electron- and hole-transport lengths in organic-inorganic $CH_3NH_3PbI_3$. *Science* **2013**, *342*, 344.

155. Y. Li*, L. Zhang, J. Xia, T. Liu, K. Wang, Modifying PTAA/Perovskite Interface via 4-butanediol Ammonium Bromide for Efficient and Stable Inverted Perovskite Solar Cells, *Small*, **2023**, *19*, 2208243.

156. H. Hsu, L. Ji, M. S. Du, J. Zhao, E. T. Yu, A. J. Bard, J. Optimization of PbI_2/MAPbI$_3$ perovskite composites by scanning electrochemical microscopy, *The Journal of Physical Chemistry C* **2016**, *120*, 19890.

157. S. Aharon, A. Dymshits, A. Rotem, L. Etgar, Temperature dependence of hole conductor free formamidinium lead iodide perovskite based solar cells, *Journal of Materials Chemistry A* **2015**, *3*, 9171.

158. Y. Li, G. Xing, D. Cao, Surface passivation of perovskite with organic hole transport materials for highly efficient and stable perovskite solar cells, *Materials Today Advances* **2022**, *16*, 100300.

159. P. J. Shi, *et al.*, Template-assisted formation of high-quality α-phase $HC(NH_2)_2PbI_3$ perovskite solar cells, *Advanced Science* **2019**, *6*, 1901591.

160. Y. Li, *et al.*, Formamidinium-based lead halide perovskites: Structure, properties, and fabrication methodologies, *Small Methods*, **2018**, *2*, 1700387.

161. T. Bu, *et al.*, Lead halide-templated crystallization of methylamine-free perovskite for efficient photovoltaic modules, *Science* **2021**, *372*, 1327.

162. J. Lee, H. Kim, N. G. Park, Lewis acid–base adduct approach for high efficiency perovskite solar cells, *Accounts of Chemical Research* **2016**, *49*, 311.

163. T. Liu, *et al.*, Self-assembled bilayer microstructure improves quasi-2D perovskite light-emitting diodes, *Chemistry of Materials* **2022**, *34*, 10435–10442.

164. Y. Li, *et al.*, A review on morphology engineering for highly efficient and stable hybrid perovskite solar cells, *Journal of Materials Chemistry A*, **2018**, *6*, 12842.

165. J. Wu, *et al.*, A simple way to simultaneously release the interface stress and realize the inner encapsulation for highly efficient and stable perovskite solar cells, *Advanced Functional Materials* **2019**, *29*, 1905336.

166. C. Zhang, S. Yuan, Y. Lou, Q. Liu, M. Li, H. Okada, Z. Wang, Perovskite films with reduced interfacial strains via a molecular-level flexible interlayer for photovoltaic application, *Advanced Materials* **2020**, *32*, 2001479.

167. C. Wang, *et al.*, Observation of electron–phonon coupling and linear dichroism in PL spectra of ultra-small $CsPbBr_3$ nanoparticle solution, *eScience*, **2023**, *1*, 100185, https://doi.org/10.1016/j.esci.2023.100185

168. N. Li, X. Niu, Q. Chen, H. Zhou, Towards commercialization: The operational stability of perovskite solar cells, *Chemical Society Reviews* **2020**, *49*, 8235.

169. S. Ma, *et al.*, Strain-mediated phase stabilization: A new strategy for ultrastable α-CsPbI$_3$ perovskite by nanoconfined growth, *Small* **2019**, *15*, 1900219.

170. Y. Li, J. Liao, H. Pan, G. Xing, Interfacial engineering for high-performance PTAA-based inverted 3D perovskite solar cells, *Solar RRL*, **2022**, *6*, 20200647.

171. J.-L. Xu, K. Thomas, Z. Luo, A. Gowen, FTIR and Raman imaging for microplastics analysis: State of the art, challenges and prospects, *TrAC Trends in Analytical Chemistry* **2019**, *119*, 115629.

172. W. Shen, *et al.*, Protic amine carboxylic acid ionic liquids additives regulate α-FAPbI$_3$ phase transition for high efficiency perovskite solar cells, *Small* **2023**, *19*, 36, 2302194.

173. E. A. Duijnstee, *et al.*, Understanding the degradation of methylenediammonium and its role in phase-stabilizing formamidinium lead triiodide, *Journal of the American Chemical Society* **2023**, *145*, 18, 10275.

174. S. H. Turren-Cruz, A. Hagfeldt, M. Saliba, Methylammonium-free, high-performance, and stable perovskite solar cells on a planar architecture, *Science* **2018**, *362*, 449.

175. Q. Zeng, Y. Li, H. Tang, Y. Fu, C. Liao, L. Wang, G. Xing, Dopant-free dithieno [3′, 2′: 3, 4; 2″, 3″: 5, 6] benzo [1, 2-d] imidazole-based hole-transporting materials for efficient perovskite solar cells, *Dyes and Pigments* **2021**, *188*, 109241.

176. W. Meng, *et al.*, Revealing the strain-associated physical mechanisms impacting the performance and stability of perovskite solar cells, *Joule* **2022**, *6*, 458.

177. G. Kim, H. Min, K. S. Lee, D. Y. Lee, S. M. Yoon, S. I. Seok, Impact of strain relaxation on performance of α-formamidinium lead iodide perovskite solar cells, *Science* **2020**, *370*, 108.

178. J. Zhao, *et al.*, Strained hybrid perovskite thin films and their impact on the intrinsic stability of perovskite solar cells, *Advanced Science* **2017**, *3*, eaao5616.

179. N. Rolston, *et al.*, Engineering stress in perovskite solar cells to improve stability, *Advanced Energy Materials* **2018**, *8*, 1802139.

180. Y. Fu, *et al.*, Advances in hole transport materials for layered casting solar cells, *Sol. Energy* **2021**, *216*, 180.

181. L. Wang, *et al.*, Strain modulation for light-stable n–i–p perovskite/silicon tandem solar cells, *Advanced Materials* **2022**, *34*, 2201315.

182. Z. Li, M. J. Yang, J. S. Park, S. H. Wei, J. J. Berry, K. Zhu, Stabilizing perovskite structures by tuning tolerance factor: Formation of formamidinium and cesium lead oodide solid-state alloys, *Chemistry of Materials* **2016**, *28*, 284.

183. J. Jiang, *et al.*, Synergistic strain engineering of perovskite single crystals for highly stable and sensitive X-ray detectors with low-bias imaging and monitoring, *Nat. Photonics* **2022**, *16*, 575.

184. H. Y. Cheng, C. K. Liu, J. Zhuang, J. P. Cao, T. Y. Wang, W. Y. Wong, F. Yan, KBF_4 additive for alleviating microstrain, improving crystallinity, and passivating defects in inverted perovskite solar cells, *Advanced Functional Materials* **2022**, *32*, 2204880.

185. F. Z. Li, *et al.*, Plasmonic local heating induced strain modulation for enhanced efficiency and stability of perovskite solar cells, *Advanced Energy Materials* **2022**, *12*, 2200186.

186. H. Min, M. Kim, S. U. Lee, H. Kim, G. Kim, K. Choi, J. H. Lee, S. I. Seok, Efficient, stable solar cells by using inherent bandgap of α-phase formamidinium lead iodide, *Science* **2019**, *366*, 749.

187. G. Eperon, V. Burlakov, P. Docampo, A. Goriely, H. Snaith, Morphological control for high performance, solution-processed planar heterojunction perovskite solar cells, *Advanced Functional Materials* **2014**, *24*, 151.

188. G. Xing, *et al.*, Long-Range balanced electron- and hole-transport lengths in organic-inorganic $CH_3NH_3PbI_3$. *Science* **2013**, *342*, 344.

189. D. Bi, J. Luo, F. Zhang, A. Magrez, E. Athanasopoulou, A. Hagfeldt, M. Gratzel, Morphology engineering: A route to highly reproducible and high efficiency perovskite solar cells, *ChemSusChem* **2017**, *10*, 1624.

190. H. Y. Hsu, L. Ji, M. S. Du, J. Zhao, E. T. Yu, A. J. Bard, Optimization of PbI_2/$MAPbI_3$ perovskite composites by scanning electrochemical microscopy, *The Journal of Physical Chemistry C* **2016**, *120*, 19890.

191. Y. Li, B. Wang, T. Liu, Q. Zeng, D. Cao, H. Pan, G. Xing, Interfacial engineering of PTAA/perovskites for improved crystallinity and hole extraction in inverted perovskite solar cells, *ACS Applied Material Interfaces* **2022**, *14*, 3284.

192. A. Dualeh, N. Tetreault, T. Moehl, P. Gao, M. K. Nazeeruddin, M. Gratzel, Effect of annealing temperature on film morphology of organic–inorganic hybrid pervoskite solid-state solar cells, *Advanced Functional Materials* **2014**, *24*, 3250.

193. P. Shi, *et al.*, Template-assisted formation of high-quality α-phase $HC(NH_2)_2PbI_3$ perovskite solar cells, *Advanced Science* **2019**, *6*, 1901591.

194. J. Lee, *et al.*, Tuning molecular interactions for highly reproducible and efficient formamidinium perovskite solar cells via adduct approach, *Journal of the American Chemical Society* **2018**, *140*, 6317.

195. T. Bu, *et al.*, Lead halide–templated crystallization of methylamine-free perovskite for efficient photovoltaic modules, *Science* **2021**, *372*, 1327.

196. J. Lee, H. Kim, N. Park, Lewis acid–base adduct approach for high efficiency perovskite solar cells, *Accounts of Chemical Research* **2016**, *49*, 311.

197. M. Qin, *et al.*, Fused-ring electron acceptor ITIC-Th: A novel stabilizer for halide perovskite precursor solution, *Advanced Energy Materials* **2018**, *8*, 1703399.

198. H. Zhang, *et al.*, A universal co-solvent dilution strategy enables facile and cost-effective fabrication of perovskite photovoltaics, *Nature Communications* **2022**, *13*, 89.

199. J. Wu, *et al.*, A simple way to simultaneously release the interface stress and realize the inner encapsulation for highly efficient and stable perovskite solar cells, *Advanced Functional Materials* **2019**, *29*, 1905336.

200. C. Zhang, S. Yuan, Y. H. Lou, Q. W. Liu, M. Li, H. Okada, Z. K. Wang, Perovskite films with reduced interfacial strains via a molecular-level flexible interlayer for photovoltaic application, *Advanced Materials* **2020**, *32*, 2001479.

201. L. Yang, *et al.*, Record-efficiency flexible perovskite solar cells enabled by multifunctional organic ions interface passivation, *Advanced Materials* **2022**, *34*, 2201681.

202. Q. Dong, *et al.*, Interpenetrating interfaces for efficient perovskite solar cells with high operational stability and mechanical robustness, *Nature Communications* **2021**, *12*, 973.

203. S. Ma, *et al.*, Strain-mediated phase stabilization: A new strategy for ultrastable α-CsPbI$_3$ perovskite by nanoconfined growth, *Small* **2019**, *15*, 1900219.

204. Z. Liu, K. Deng, J. Hu, L. Li, Coagulated SnO$_2$ colloids for high-performance planar perovskite solar cells with negligible hysteresis and improved stability, *Angewandte Chemie International Edition* **2019**, *58*, 11497.

205. E. Jung, *et al.*, Bifunctional surface engineering on SnO$_2$ reduces energy loss in perovskite solar cells, *ACS Energy Lett.* **2020**, *5*, 2796.

206. H. Min, *et al.*, Perovskite solar cells with atomically coherent interlayers on SnO$_2$ electrodes, *Nature* **2021**, *598*, 444.

207. Q. Zhou, J. Duan, X. Yang, Y. Duan, Q. Tang, Interfacial strain release from the WS$_2$/CsPbBr$_3$ van der Waals heterostructure for 1.7v voltage all-inorganic perovskite solar cells, *Angewandte Chemie International Edition* **2020**, *59*, 21997.

208. J. Kim, *et al.*, Nucleation and growth control of HC(NH$_2$)$_2$PbI$_3$ for planar perovskite solar cell, *J. Phys. Chem. C* **2016**, *120*, 11262.

209. Y. Zhao, *et al.*, Perovskite seeding growth of formamidinium-lead-iodide-based perovskites for efficient and stable solar cells, *Nature Communications* **2018**, *9*, 1607.

210. F. Zhang, C. Xiao, X. Chen, B. Larson, S. Harvey, J. Berry, K. Zhu, Self-seeding growth for perovskite solar cells with enhanced stability, *Joule* **2019**, *3*, 1452.

211. F. Zhang, *et al.*, Isomer-pure bis-PCBM-assisted crystal engineering of perovskite solar cells showing excellent efficiency and stability, *Advanced Materials* **2017**, *29*, 1606806.

212. X. Huang, *et al.*, Solvent gaming chemistry to control the quality of halide perovskite thin films for photovoltaics, *ACS Central Science* **2022**, *8*, 1008.

5 Solvent Engineering for Efficient and Stable Perovskite Solar Cells

Ming Luo, Jiangzhao Chen, and Hong Zhang

5.1 INTRODUCTION

Metal halide perovskites are among the most ideal candidates for solar cells because of their high absorption coefficients[1], high ambipolar charge carrier mobility,[2] low exciton binding energy,[3] and long carrier diffusion lengths[4–6]. Great advancements have been made for metal halide perovskite solar cells (PSCs) over the last decade, with the highest certificated power-conversion efficiency (PCE) being 26.1% to date.[7] This PCE is comparable to that of well-established crystalline-silicon solar cells.[8–10] Apart from the impressive PCE of PSCs, another inspiring advantage is the solution processability of perovskite film, which significantly simplifies fabrication processes and lowers the cost of solar cells.[11–13] The solution processability of perovskite also allows high-throughput manufacturing via spray coating, blade coating, slot-die coating, and inkjet printing.[12,14] Up until now, the record PCE of PSCs has been based on solution-processed perovskite film.[15–19] In the solution-processed method, the dissolution of precursors and crystallization of perovskites are carried out in solution that highly depend on the solute–solvent interaction and solvent evaporation, respectively. Therefore, a solvent has great importance because its physical and chemical properties (e.g., boiling point, viscosity, vapor pressure, polarity, coordination ability, etc.) play key roles in the crystallization, film formation, and photovoltaic performance of perovskite films.

Solvents dissolve perovskite precursor materials by coordinating and/or forming hydrogen bonds with perovskite materials, and the solubility varies with the types, functional groups, and stereochemical structures of solvents, affecting the coordination ability.[20–26] For instance, the widely used polar aprotic organic solvents (e.g., dimethylformamide (DMF) and dimethylsulphoxide (DMSO)) can coordinate well with Pb^{2+} to form a Lewis acid-base adduct and show good solubility to PbI_2. Conversely, alcohol-type solvents coordinate weakly with Pb^{2+} and show poor solubility. Compared with conventional organic solvents (e.g., DMSO), ionic liquids (ILs; e.g., methylammonium acetate (MAAc)) show better solubility to perovskite precursors due to the stronger Pb-O coordination and the formation of NH-X hydrogen bond. The coordination ability of solvents also significantly affects the crystallization of perovskite phase, which largely determines the nucleation and crystal growth of perovskite, as well as the morphology of the deposited perovskite film.[12,26]

DOI: 10.1201/9781003400486-5

Therefore, the crystallization dynamics and film quality of solution-processed perovskite films can be effectively controlled by tuning the physical (e.g., volatility and flowability) and chemical (coordination ability) properties of the precursor solvent.

For large-area PSCs, scalable film-deposition techniques (e.g., blade coating, slot-die coating, spray deposition, inkjet printing, and electrodeposition) require the rapid removal of solvent to promote the homogeneity of the perovskite film. Thus, different from laboratory-scale fabrication of perovskite film, decreasing the coordination ability and increasing the volatility of the solvent system are shown to be essential. Moreover, for the future large-scale commercialization of PSCs, the use of highly polar aprotic organic solvents with high toxicity and high boiling point brings non-negligible underlying issues.[14,26–30] Specifically, the high toxicity and high boiling point of organic solvent harm the ecological environment and greatly increase the fabrication cost of PSCs, thereby significantly slowing down the commercialization of PSCs.[30] Therefore, developing less toxic or even nontoxic solvent systems for the solution processing of perovskite film is a growing demand.

Accordingly, great efforts have been made in the solvent engineering of perovskite precursor solutions to regulate film formation with focus on increasing the efficiency and stability of PSCs and reducing the toxicity during perovskite preparation. Currently, advanced solvent systems can be categorized into three major types, namely, organic-solvent system[30–33], IL-based solvent systems[21,23,34,35] and mixed organic/IL solvent systems[24] (Figure 5.1 and Table 5.1.).

In this chapter, we aim to summarize the progress of advanced research on the solvent engineering of perovskite precursor solution towards the industrialization of PSCs with high efficiency, stability, and low toxicity in terms of coordination and solubility regulation. Specifically, we summarize in detail the effects of the physical

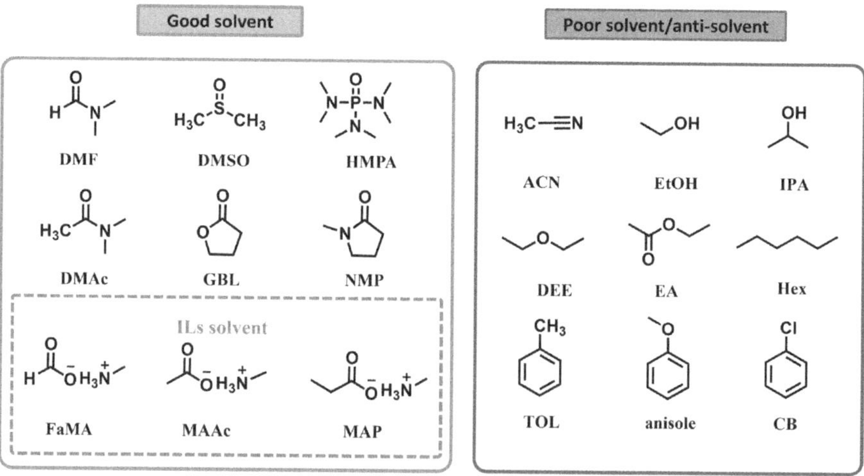

FIGURE 5.1 According to the solubility, representative solvents used for perovskite precursor solutions are divided into two types: good solvents and poor/antisolvents, and their abbreviated name and chemical structures are displayed.

TABLE 5.1

Summary of the Photovoltaic Performance of PSCs Based on Different Solvent Systems

Solvent system		Preparation environment	Film-deposition method	V_{OC} [V]	J_{SC} [mA cm^{-2}]	FF	PCE [%]	Ref.
Organic	DMF	N_2	Two-step spin-coating	1.08	21.92	0.72	17.04	36
	DMF	N_2	One step spin-coating/CB antisolvent	0.99	22	0.71	14.6	108
	H$_2$O/DMF	N_2	Two-step spin-coating	1.03	20.06	85	18	73
	DMSO	N_2	Two-step spin-coating	1.07	22.84	76.11	18.62	80
	DMSO	N_2	Two-step spin-coating	1.02	20.71	64	13.5	77
	DMF/DMSO	N_2	One step spin-coating/antisolvent	1.13	25.92	82.02	24.02	18
	DMF/DMSO	N_2	Two-step spin-coating	1.01	20.82	72	15.14	50
	DMF/DMSO	Ar	One step spin-coating/ANS antisolvent	1.14	23.88	0.76	20.53	114
	DMF/DMSO	75% RH ambient conditions	One step spin-coating/EA antisolvent	-	-	-	14.5	119
	DMF/DMSO	N_2	One step spin-coating	1.11	18.89	49.53	10.41	89
	DMF/DMSO	N_2	One step spin-coating	1.08	19.45	58.22	12.17	90
	H2O/DMF/DMSO	Ambient (16–57 RH%)	One-step spin-coating (vacuum-assist)	1.09	22.1	80.8	20.1	72
	NMP	N_2	One step spin-coating	1.18	24.68	80.46	23.43	44
	NMP/DMF	N_2	One step spin-coating/antisolvent	1.1	23.98	76.0	20.19	75
	GBL/DMSO	N_2	One step spin-coating/TLN antisolvent	1.09	19.5	0.76	16.22	81
	2-ME	N_2	One step spin-coating	0.97	19.32	81.6	15.32	31
	ACN/MA	Air	One-step spin-coating	1.1	22	77	19	33
	ACN/MA	Air	One-step spin-coating	1.05	21.06	72	15.9	103
	ACN/THF/MMA	N_2	One-step spin-coating	1.13	23.83	80.9	21.8	102
	GBL/EtOH/AcOH	N_2	spin-coating	0.88	21.2	71	15.1	30
	GBL/EtOH/AcOH	N_2	Blade-coating on cell	0.89	21.1	73	14.5	30
	GBL/EtOH/AcOH	N_2	Blade-coating on module	3.35	5.1	69	11.9	30
	2-ME/ACN/DMSO	N_2	Blade-coating	1.13	23	81.8	21.3	105
ILs	MMAc	Air	One step spin-coating	1.11	23.16	78.01	20.05	23
mixed organic/ILs	MAP/ACN/DMSO	N_2	One step spin-coating	1.07	23.08	62	15.46	94

and chemical properties of different solvents on various deposition technologies of perovskite film. Some critical insights are provided into the current challenges and opportunities for solvent engineering to achieve highly efficient and stable PSCs.

5.2 COORDINATION CHEMISTRY AND CRYSTALLIZATION OF PEROVSKITE FROM SOLUTION

5.2.1 Perovskite Precursor-Solution Chemistry

Solvents dissolve perovskite precursor primarily by breaking down the van der Waals interaction between (001) interlayers of lead halogen, followed by forming Pb-O coordination to break some chemical bonds inside the (001) interlayer for full dissolution.[36] Perovskite precursor solutions for solar cells are generally colloidal dispersions in a mother solution, with a colloidal size up to the mesoscale[36–38], rather than real solutions, and this is evidenced by typical Tyndall effects. Metal halide perovskite thin-film formation displays the characteristics of a sol-gel process.[38] The colloid is made of a soft coordination complex in the form of a lead polyhalide (such as PbI_3^-, PbI_4^{2-}, PbI_5^{3-}, and PbI_6^{4-}) framework between organic and inorganic components. It can be structurally tuned by the coordination degree, thereby primarily determining the basic film coverage and morphology of deposited thin films. Radicchi et al. obtained the most thermodynamically stable structures of iodoplumbate complexes in various solvent solutions (Figure 5.2) and characterized their

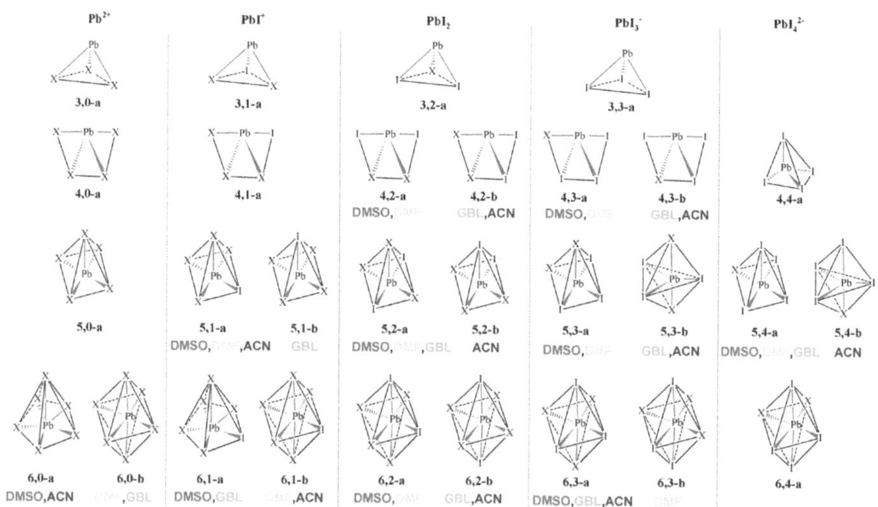

FIGURE 5.2 $[PbI_mX_n]^{2-m}$ ($X =$ solvent) most stable coordination, each row referring to a selected coordination number. Each structure is labelled as **cn, m-a/b**, where **cn** is the coordination number, **m** counts the iodine atoms, and **a/b** refers to different configurations. Adapted with permission.[39] Copyright 2019, American Chemical Society.

optical properties through DFT and TD-DFT calculations.[39] They identified the composition of these complexes involved in the DMSO/DMF solution of $MAPbI_3$ precursors by matching simulated and experimental UV-vis absorption spectra. They found that iodoplumbate-DMSO complexes on average had a higher coordination number and a lower iodide ligand number with respect to the iodoplumbate-DMF ones. The stronger Pb-solvent interaction in DMSO likely slows down the crystallization upon evaporating the solvent. Moreover, the lower tendency to form species with higher number of iodide ligands may also give rise to a less defective structure, which may minimize the incorporation of extra iodine in the lattice in the form of interstitial iodine for example.[40] Therefore, figuring out the coordination abilities of various solvents and their influence on coordination equilibrium in terms of iodoplumbate complex in perovskite sinks has great importance.

The interactions between solute and solvent in perovskite precursor solution have been defined using Mayer bond unsaturation, Hansen's solubility parameters, dielectric constant, and donor number (D_N) (Figure 5.3a).[30,32,40–42] Previous studies have shown that the effects of other parameters are secondary to Lewis acid-base interactions. Based on Lewis acid-base theory, the commonly used strong aprotic solvents (e.g., DMF and DMSO) with lone pair electrons (Lewis base) can coordinate with the electron acceptor Pb^{2+} (Lewis acid) to form Lewis acid-base intermediate adduct, affording a uniform coordinated colloidal solution.[20] Therefore, regulating the coordination interaction between solvent and solute in perovskite precursor solution is recognized as the most effective strategy for controlling the nucleation and growth of perovskite films.[43] Hamill et al. elucidated the influence of Lewis basicity of the processing solvent, quantified by the Gutmann's donor number D_N, on perovskite formation. D_N is a more convincing predictor of a solvent's ability to dissolve perovskite precursors than the dielectric constant (Figure 5.3b).[32] According to their theory, high-D_N solvents ($D_N > 18$ kcal/mol) compete with I^- for coordination sites around Pb^{2+} and consequently suppress the formation of iodoplumbates; such solvents form stable precursor solutions for thin-film processing. By contrast, low-D_N solvents ($D_N < 18$ kcal/mol) coordinate less strongly with Pb^{2+} and thus favor iodoplumbate formation and subsequent single-crystal growth. This understanding allows the judicious selection of processing solvents or solvent additives tailored towards the desired applications. Apart from D_N, the Kamlet–Taft value β is another physicochemical parameter of solvent basicity. Recently, Zheng et al. demonstrated that D_N and β show a competitive selection to determine the structures of solvent-contained intermediates. When $\beta \geq D_N^*$ ($D_N^* = D_N/38.8$), solvents integrate with FA^+ cations to form hydrogen bonds within the (FA⋯solvent)PbI_3 lattice. When $\beta < D_N^*$, solvents coordinate with Pb^{2+} to form the PbI_2^-solvent lattice, but FA^+ cations are excluded (Figures 5.3c and d).[44] Subsequently, the different intermediate structures based on solvent gaming directly affect the thermodynamics and kinetics of α-$FAPbI_3$ formation; meanwhile, the intermediates (FA⋯NMP)PbI_3 and (PbI_2–2DMSO + FAI) can transfer into α-$FAPbI_3$ below the thermodynamic temperature of the traditional δ-to-α phase transition. The hydrogen-bond-favorable intermediates ($\beta \geq D_N^*$) are found to sidestep the breaking of strong coordination bonds and assist the formation of defect-less α-$FAPbI_3$ films.

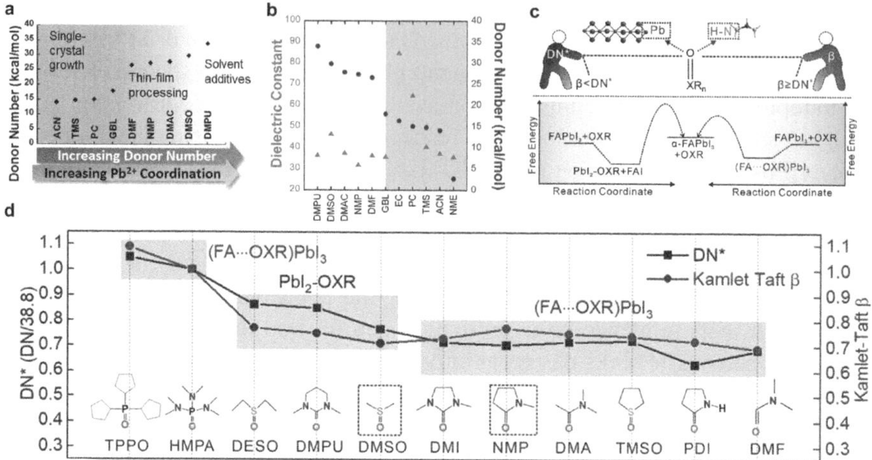

FIGURE 5.3 (a) Effects of donor number of solvent on the perovskite formation. (b) Dielectric constant and Gutmann's donor number (D_N of solvents used for perovskite processing. (a–b) Adapted with permission.[45] Copyright 2017, American Chemical Society. (c) Solvent gaming scheme and two interaction types between Lewis acidic FA^+/Pb^{2+} cations and Lewis basic OXRs, depending on the solvent basicity in terms of D_N^* and Kamlet–Taft β values. OXRs refers to oxygen-containing double bonds (X = C, S, and P; R = alkyl and alkylamine).[44] (d) Summary of solvent gaming results and as-formed intermediate structures. When $\beta < DN^*$, PbI_2–OXR forms; when $\beta \geq DN^*$, $(FA\cdots OXR)PbI_3$ emerges from $FAPbI_3$-based solutions. The mole ratio of Pb and OXR is omitted here, and dashed boxes show the examples studied in details infra. Adapted with permission.[44] Copyright 2022, American Chemical Society.

5.2.2 Crystallization Dynamics of Perovskite Film

To prepare uniform and pinhole-free perovskite film, optimizing the nucleation and growth condition of perovskite crystal is the key, which can be realized by solvent engineering. As mentioned earlier, coordination ability and solubility have great impact on the dissolution process regarding intermediate phase and coordination equilibrium in precursors. The characteristics of solvent (e.g., boiling point, viscosity, vapor pressure, polarity, coordination ability, etc.) play a critical role in determining the crystallization dynamics of perovskite films from precursor solution.

Two classical theories describe crystal nucleation and grain growth. The first one is the LaMer model, as shown in Figure 5.4a, and the mechanism graph presents the concentration change of the perovskite precursor inks as a function of time at a constant and isothermal evaporation rate of the solvent.[46–49] According to the LaMer graph, when the solution concentration reaches the critical level (C_c) and becomes supersaturated, the solute colloid starts to nucleate and grow into crystals. Given that crystal nucleation and growth begin almost simultaneously (t_2), they maintain a competitive relationship in depleting the solute.[12] During this process, the rate of nucleation and growth highly depend on the evaporation rate of the solvent. Under

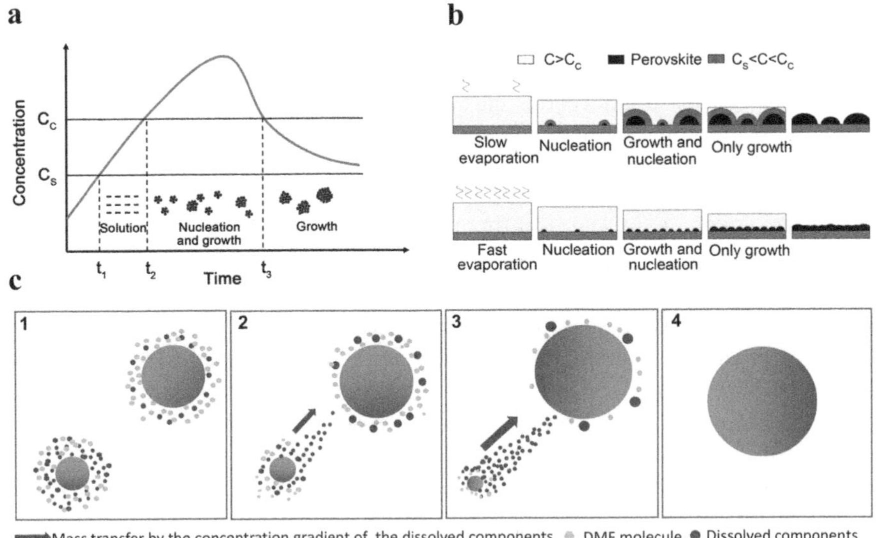

FIGURE 5.4 (a) LaMer diagram for monodispersed particle formation. (b) Model of the nucleation/growth competition of perovskite crystal grown from precursor solution under slow solvent evaporation and fast solvent evaporation. C_C, critical concentration; C_S, saturation concentration. Adapted with permission.[49] Copyright 2013, Royal Society of Chemistry (c) Schematic diagram of the Ostwald ripening model, showing the dynamic coarsening process of the perovskite grains with annealing time: i) initial stage, ii) early stage, ii) middle stage, iv) final stage. (c) Adapted with permission.[53] Copyright 2018, American Chemical Society.

slow solvent evaporation, the concentration of the solution approaches C_c over a long period of time. In this case, the small number of nuclei that initially emerges have enough time and space to grow into large-sized grains. However, the low concentration of nucleus leads to localized solute accumulation on the substrate, resulting in the formed perovskite film having poor surface coverage. By contrast, under the fast evaporation rate, the concentration of the solution rapidly exceeds C_c (Figure 5.4b). In this case, a large number of nuclei form rapidly but have limited time and space to grow, resulting in a fully covered perovskite film with uniform but small grains. Consequently, to obtain perovskite film with high coverage, large crystal size, and uniformity, controlling the evaporation rate of solvent plays a crucial role. The Ostwald ripening model is another fundamental theory that is often used to explain the growth mechanism of crystal. It is a more effective and comprehensible method of describing the growth of small crystals into large ones via a thermodynamically spontaneous process.[50–52] According to this model, small-sized crystal grains with lower surface energy and solubility tend to dissolve and redeposit onto the surface of adjacent large crystal grains with higher surface energy and solubility, leading to the further growth of the larger crystals and the disappearance of smaller particles over time. As a result, crystal grains and solutes in the system decrease, and the average size of the large crystal grains increase (Figure 5.4c).

Specifically, the nucleation mechanism of $MAPbI_3$ perovskite from homogeneous GBL solution is simulated by well-tempered metadynamics (WTMetaD), which uncovers the salient atomistic features.[54] As revealed by Ahlawat et al., it occurs in three stages: (i) dense amorphous clusters primarily consisting of lead and iodide appear from the homogeneous solution; (ii) these clusters evolve into lead iodide (PbI_2)-like structures; and (iii) methylammonium (MA^+) ions diffuse into these PbI_2-like aggregates, triggering a solid–solid transformation into perovskite crystals (Figure 5.5a). They have also identified that the continuous diffusion of MA^+ ions into Pb–I clusters starts the nucleation and growth of perovskite. The formation of a PbI_2-like phase may facilitate the diffusion of MA^+ ions into the Pb–I clusters, and a higher amount of MA^+ ions produce a greater number of perovskite nuclei in a perovskite precursor solution, as demonstrated by the fabricated $MAPbI_3$ film with much smaller grain size.

In fact, perovskite film formation involves a complex interplay of solution flow, solute diffusion, solvent extraction, and the crystal growth. Hu et al. developed a crystallization-depletion mechanism to elucidate the periodic crystallization and the kinetically trapped morphology at a mesoscopic level for slot-die printing.[55] Figure 5.5b shows a series of in-situ optical microscopy images of ripple formation resulting from a periodic precipitation. When the perovskite solution is deposited onto an elevated heating substrate, the evaporation is more rapid. Once the solution reaches supersaturation, nucleation occurs and crystal growth follows. The rapid removal of solvent near the growth front induces liquid convection that carries the solute to the crystal face to allow crystallization, depleting the volume in front of the growth front. This crystallization-depletion mechanism gives rise to a concentration gradient and a periodic crystallization of the perovskite,[56,57] as shown schematically in Figure 5.5c and d.

5.2.3 PEROVSKITE PRECURSOR-SOLUTION STABILITY

Perovskite precursor solutions usually suffer from aging, which impedes the reproducibility of the solution. The reasons for solution aging include colloidal aggregation, decomposition of perovskite component, and solvent deterioration. Understanding the aging mechanism and preventing this phenomenon is needed to obtain high-phase-purity perovskite.

Precursor-solution stability has a great influence on the colloidal size distribution of a solution, which then affects the phase purity of the films and device performance. Boonmongkolras et al. found that a precursor solution, aged in triple-cation lead perovskite for optimum hours, led to the best device efficiency along with the highest reproducibility and is free of large colloidal aggregates. Conversely, others (aged shorter or longer than the optimum hours) contain microsized aggregates as revealed by dynamic light scattering measurements.[58] The absence of the large aggregates in the solutions appears to be correlated with the phase purity of the prepared perovskite films, ultimately leading to the best efficiency, which is close to 17%. Sun et al. demonstrated that adding I_2 into perovskite precursor can sustain the dynamic balance between I^- and I_3^- ions through the reversible reaction $I^- + I_2 \leftrightarrows I_3^-$, which effectively stabilizes the whole I^- ion concentration in perovskite solution and prevents the colloidal particles from aggregating. The result is perovskite films

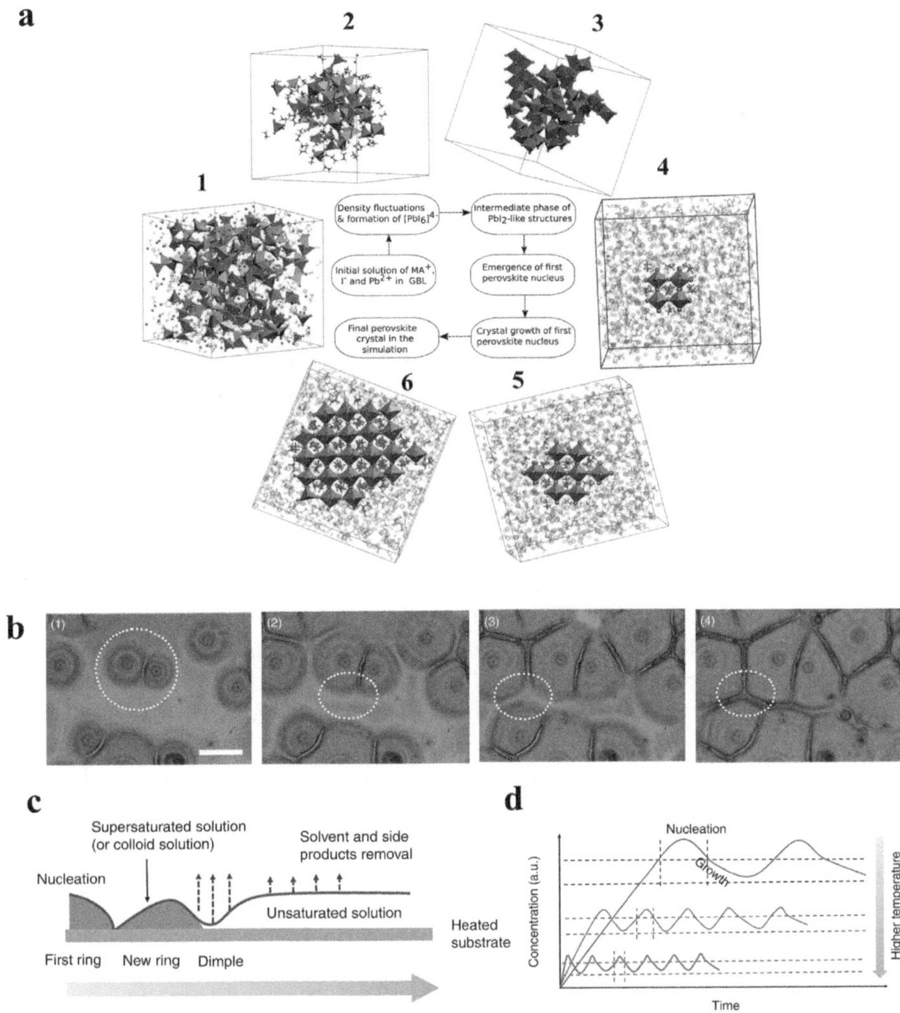

FIGURE 5.5 (a) Full nucleation pathway illustrated with representative snapshots. Pb–I complexes are shown as golden and blue polyhedra with Pb^{2+} in the center and I^- in the corners. Free I^- is shown as pink spheres. Crystalline MA^+ ions are shown in green. (a) Adapted with permission.[54] Copyright 2018, American Chemical Society. (b) The selected in-situ microcopy images of the perovskite films heating at 100°C. The interval time is 0.02 s. The scale bar is 50 mm. (c) The scheme of the periodic and rhythmic crystallisation. (d) The schematic diagram of nucleation and growth with time based on various heating temperatures. (b–d) Adapted with permission.[55] Copyright 2017, Springer Nature.

with better crystallization and less bulk defects.[59] As a result, the devices using the aged solution with the additional tri-iodide ions show a lower efficiency drop from ~19% to ~16% with increased aging time up to 15 days compared with devices using the as-prepared precursor solution, which decreases from ~18% to ~6%. Aging effect is related to temperature because the concentration of iodoplumbate PbI_2S_4 (S = DMF or DMSO) varies with time and temperature.[60] Photocurrent density is commonly observed to be substantially influenced by aging time, whereas voltage is less affected. Fill factor declines with aging time regardless of aging temperature. Therefore, determining an optimal aging time at a certain temperature is conducive to obtaining the best photovoltaic performance.

The conventional wisdom is that MAI and FAI are stable in the solution, but actually they are not.[61,62] Seok et al. demonstrated that the hybrid perovskite precursor solution of $(FAPbI_3)_{0.95}(MAPbBr_3)_{0.05}$ degrades over time through the deprotonation of MA into volatile methylamine (CH_3NH_2).[61] Adding elemental sulfur (S_8) drastically stabilizes the precursor solution owing to amine–sulfur coordination without compromising the PCE of the derived PSCs. Furthermore, sulfur introduction to stabilize the precursor solution results in improved PSC stability. Wang et al. found that CH_3NH_2 triggers subsequent chemical reactions in the aging process of $MA_{0.5}FA_{0.5}PbI_3$ perovskite in DMSO or DMF solution (Figure 5.6a).[62] The methyl group in MA is reportedly an electron-donating group, which enhances the nucleophilicity of the amino group, but the imine group in FA is an electron-withdrawing one with a negative effect on nucleophilicity. The MAI is first deprotonated to form MA, and then MA is reacted with FAI to form two condensation products. This phenomenon explains the low stability of the perovskite precursor solution. They also demonstrated that the impure phase in the perovskite film is a mixed 1D phase with the inclusions of the MFA and DMFA organic cations inside, rather than pure δ-FAPbI$_3$ (Figure 5.6b). To address this issue, triethyl borate can be used as a clean stabilizer and it can be fully removed from the film during the subsequent thermal-annealing treatment. ITIC-Th is found to effectively suppress the formation of the yellow δ-phase in the films made from aged precursor solutions.[63] Consequently, the devices fabricated from the aged precursor solution with ITIC-Th undergoes much less efficiency drop with prolonged precursor aging time, i.e., from 19.20% (fresh) to 16.55% (39 days), compared with the devices made from conventional precursor solutions dropping from 18.07% (fresh) to 1.76% (after 39 days of aging). Characterization studies suggest that ITIC-Th is beneficial for $CH_3NH_3^+$ cations to be incorporated into the crystal structure, thereby facilitating the formation of perovskite phase.

Solvent molecules such as DMF and DMSO are usually highly stable under moderated conditions and in the short term, which explains why they are popular and commonly used in several perovskite system for precursor inks. However, in the long run, solvent molecules are never too stable, let alone in the acid atmosphere caused by MA or FA cation. Thus, the decomposition product from solvent is subsequently incorporated into perovskites as A-site impurities. These impurities affect the stoichiometry, structure, and absorptivity of the thin film that is ultimately produced. Dou et al. found that the hydrolysis of DMF by trace water leads to the degradation of CsFAMA perovskite precursor inks despite the inert storage conditions (Figure 5.6c).[64] The hydrolysis of DMF yields formic acid (HCOOH) and dimethylamine (DMA). It is then converted into HCOO⁻ and DMA⁺ by proton transfer. DMA⁺ in the aged solution can incorporate into the final perovskite film, which along with a reduction in MA⁺ and Cs⁺ cations, resulting in a film that appears yellowish and performs very poorly as a photovoltaic absorber. Fortunately, this adverse effect on stoichiometry can be offset by adding MA⁺, Cs⁺, and I-containing salts. A method of storing ball-milled salts has also been proposed to address the ink aging-induced phase instability in CsFAMA perovskites, which is confirmed by various optoelectronic characterizations

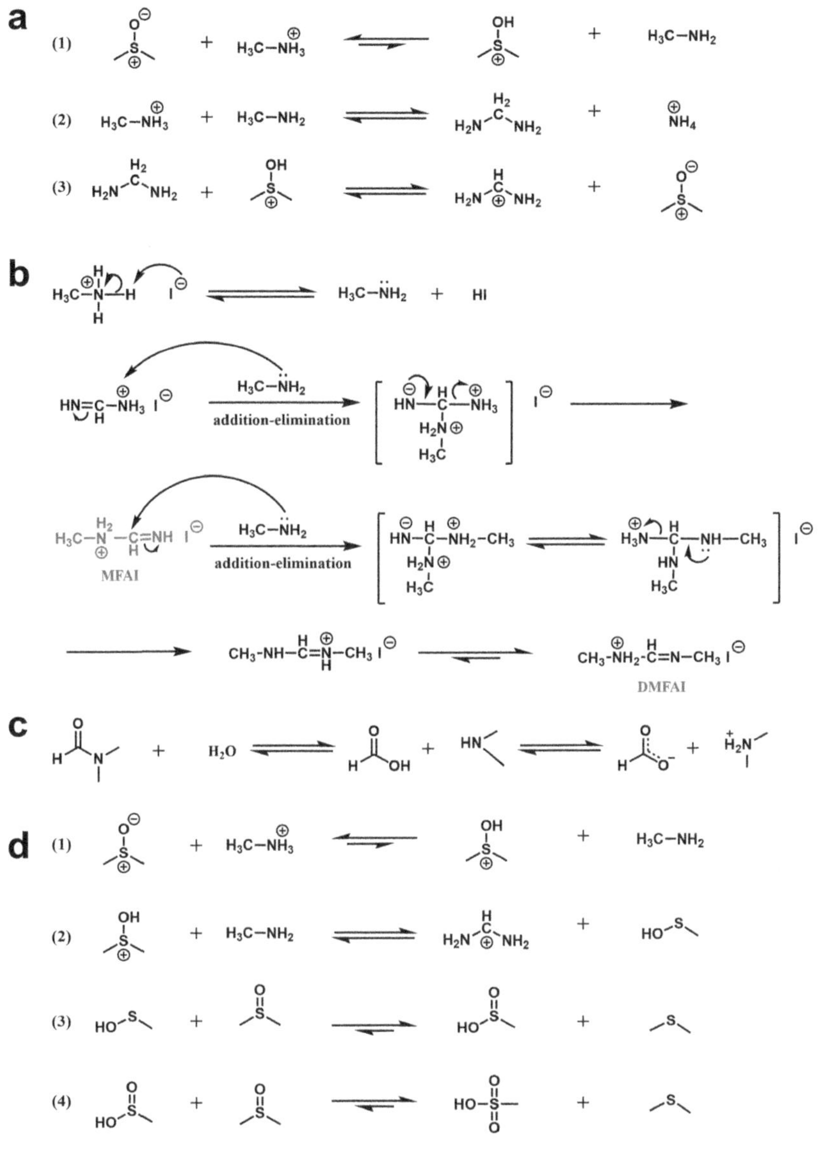

FIGURE 5.6   The aging mechanism of solvent deterioration (a, c, d)[64,65] and decomposition of perovskite component (b).[62] (a, d) Adapted with permission.[64] Copyright 2019, American Chemical Society. (b) Adapted with permission.[62] Copyright 2020, Elsevier. (c) Adapted with permission.[62] Copyright 2018, American Chemical Society.

of perovskite films and devices from ball-milled salts aged for over 30 d. The aging of precursor solution caused by DMSO is also observed (Figure 5.6d).[65] Reactions between $CH_3NH_3^+$ and DMSO lead to the formation of ammonium (NH_4^+) and dimethylammonium ($(CH_3)_2NH_2^+$) reactions cations, which are faster than the hydrolysis that occurs between $CH_3NH_3^+$ and DMF. Proton transfer from $CH_3NH_3^+$ to DMSO is followed either by methyl-group transfer between the resulting CH_3NH_2 and residual $CH_3NH_3^+$ or by transmethylation into CH_3NH_2 from $DMSOH^+$, finally producing NH_4^+ and $(CH_3)_2NH_2^+$. The substitution of $CH_3NH_3^+$ with NH_4^+ and $(CH_3)_2NH_2^+$ in the perovskite crystal results in deviations from the tetragonal structure as expected of phase-pure $CH_3NH_3PbI_3$, with a deleterious effect on the absorptivity of the resulting films. Recently, Zhang et al. reported a cosolvent dilution strategy to tune the precursor colloidal properties and enhance the stability of precursor solution, which extends its processing window and thus minimizes the waste.[66]

5.3 TYPES OF SOLVENT FOR PEROVSKITE PRECURSOR SOLUTION

5.3.1 ORGANIC-SOLVENT SYSTEM

As mentioned earlier, the nature of solvent (e.g., boiling point, viscosity, vapor pressure, polarity, coordination ability, etc.) has great impact on the crystallization and film formation of perovskite. Particularly, the basicity that is measured by Gutmann's donor number D_N and β solvent should be highly valued because it determines the colloid size and intermediate-phase species. According to the coordination ability, organic-solvent systems can be categorized into two groups: the high-D_N one with strong affinity to Pb^{2+} (good solvents) and the low-D_N one with weak affinity to Pb^{2+} (poor solvents or bad solvents, also used as antisolvents). In the following sections, we discuss solvent engineering based on commonly used good solvents and poor solvents/antisolvents.

5.3.1.1 Good Solvents

Inorganic lead halide salts are perovskite precursor materials, but they are barely soluble or even completely insoluble in solvents with low D_N. However, solvents with high D_N exhibit high solubility to perovskites precursor materials, and they are effective for regulating the crystallization of spin-coated perovskite film on small substrate. Notably, good solvents to lead halide salts are mostly strong polar aprotic solvents.

DMF is the solvent most frequently used to prepare precursor inks. The functional group of C=O as Lewis base enables it to coordinate with inorganic lead halide (PbX_2) as Lewis acid and to form PbX_2-DMF intermediate.[67,68] Wakamiya et al. identified the PbI_2-DMF intermediate crystal structure by cooling PbI_2 solution in DMF from 70°C to room temperature and demonstrated its one-dimensional (1D) PbI_2·DMF crystal structure by single-crystal X-ray diffraction (XRD) analysis.[67] The PbI_2·DMF crystals are then converted into three-dimensional (3D) perovskites resulting from the substitution of DMF molecules in PbI_2·DMF after adding CH_3NH_3I (MAI) in isopropanol (IPA) solution. Subsequently, Guo et al. reported the formation of MAI·PbI_2·DMF intermediates (Figure 5.7a).[68] The addition of MAI weakens the interaction between DMF and PbI_2 by intercalation into a closer position to the PbI_2 framework. The interaction of DMF with MAI·PbI_2 is verified by confirming the variation in the stretching vibration of the C=O bond at 1665 cm^{-1} through Fourier transform infrared spectroscopy (FTIR) (Figure 5.7b). The formation of the intermediate

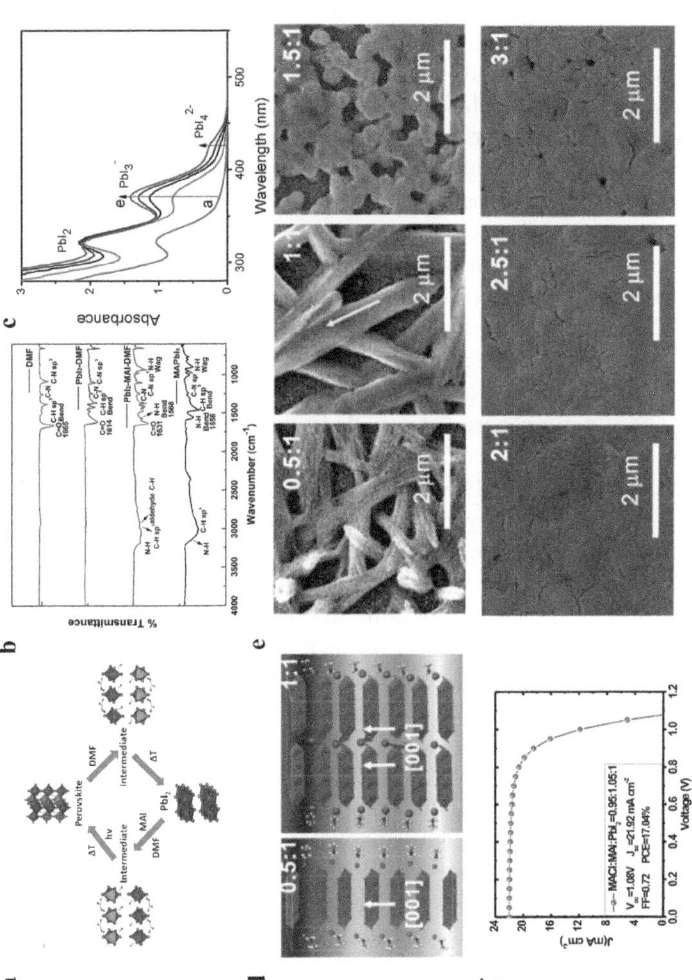

FIGURE 5.7 (a) The formation of MAI·PbI$_2$·DMF intermediates. (b) The FTIR spectrum of pure DMF, PbI$_2$–DMF complex, PbI$_2$–MAI–DMF complex, and MAPbI$_3$. (a–b) Adapted with permission.[68] Copyright 2016, Royal Society of Chemistry. (c) Absorption spectra of 250×10^{-6} M PbI$_2$ solution in DMF with increasing concentration of MAI from 6 to 24×10^{-3} M. Adapted with permission.[69] Copyright 2015, Royal Society of Chemistry. (d) Possible organic–inorganic coordination in perovskite precursors products (DMF as solvent): when the ratios of MAI:PbI$_2$ is 0.5:1, CH$_3$NH$_3$I will selectively coordinate with PbI$_2$ at (001) plane edges at first due to dangling bonds at these sites to form colloidal PbI$_2$ nanorods abiding by trigonal crystallography of PbI$_2$; when the ratios of MAI:PbI$_2$ is 1:1, it further strings the (001) plane stacked nanorods together in the form of nanorod bundles through the atom sharing of halogen between nanorods. (e) SEM micrographs of calcined thin films obtained from perovskite precursors with different ratios of MAI:PbI$_2$. (f) When the ratio of MAI:PbI$_2$ is 0.95:1.05:1), it reached optimal J–V performance. (d-f) Adapted with permission.[36] Copyright 2015, American Chemical Society.

phase is induced even after thermal annealing between 40 and 80°C, indicating its thermodynamically and kinetically metastable nature. Beyond the temperature, annealing the film at 100°C converts the intermediates into perovskite film. Cao et al. suggested that DMF in the intermediates induces the dissolution–recrystallisation process based on the Ostwald ripening model during the thermal annealing of a precursor-deposited substrate, leading to increased grain size.[50] To form a perovskite structure, MAI needs to be introduced into the PbI_2-DMF solution. When the content of MAI gradually increases in the solution, the PbI_2-DMF complex gradually transforms into iodoplumbate coordination complexes (Figure 5.7c).[37,69] Due to the lower coordination ability of DMF with Pb^{2+} than I[-], small $MAPbI_3$ perovskite colloidal clusters form in the precursor solution. Given that MAI coordinates with PbI_2 along the (001) plane edges to form trigonal PbI_2 with nanorod micromorphology (Figure 5.7d), the small perovskite colloidal clusters are needle like, especially when the molar ratio of MAI:PbI_2 is lower than 1:1 (Figure 5.7e). Moreover, the low coordination ability of DMF with Pb^{2+} leads to the rapid evaporation of DMF and the fast nucleation rate of perovskite crystals during film formation. As a result, the deposited perovskite film using DMF as a solvent often shows rough morphology with a large number of pinholes, thereby leading to poor device performance of PSCs.[70,71] To achieve full and even coverage, the preferential orientation, and high purity of planar perovskite thin film, additional MAX (X = Cl, I) over the stoichiometric ratio was added for tuning the coordination degree and mode in the initial colloidal solution, which improved the device performance (Figure 5.7f).[36]

PbI_2 has low solubility in dry DMF. However, when adding a small amount of H_2O into PbI_2/DMF suspension, the polarity, dielectric constant, and solubility parameters of DMF can be changed, producing a homogeneous solution (Figure 5.8a).[72] It can also obviously improve the quality of the perovskite film. In particular, when the amount is 2%, the perovskite film is highly pure and dense even without any pinhole (Figure 5.8b). Thus, inverted PSCs based on such high-quality perovskite film acquires an excellent PCE of 18% with a remarkably high FF of 0.85 (Figure 5.8c). Liu et al. increased the water content to more than 20% (Figure 5.8d), thereby slowing down the drying rate of $CH_3NH_3PbI_{3-x}Cl_x$ film[73] and reducing the nucleation rate of perovskite grains. A dense perovskite film with large grains is obtained (Figures 5.8e and f). The fabricated PSCs show good reproducibility and average PCE of 18.7% with an excellent PCE of 20.1% (Figure 5.8g).

DMAc is one of the homologs of DMF with a suitable boiling point, and it has a lower polarity than DMF and DMSO. Considering these merits, Qiu et al. prepared a highly oriented two-dimensional (2D) Ruddlesden–Popper (RP) perovskite $BA_2MA_3Pb_4I_{13}$ by using DMAc. They found that DMAc can accelerate the crystallization rate of 2D RP perovskite due to its weak coordination with Pb^{2+} and ammonium salts and easy evaporation during film formation.[74] DMAc can also strongly induce the oriented crystallization of 2D RP perovskite. The PCE of their PSC based on 2D RP perovskite can reach 12.15% with a photocurrent density (J_{SC}) of 14.61 mA cm[-2].

Compared with the C=O group in DMF, the S=O group in DMSO possesses a higher electron-cloud density, making DMSO a stronger Lewis base than DMF and leading to shorter Pb-O bond between DMSO and PbI_2 (2.386 Å) than DMF (2.431 Å).[67,75–78] As a result, DMSO can prevent the rapid interaction between MAI and PbI_2 in the precursor

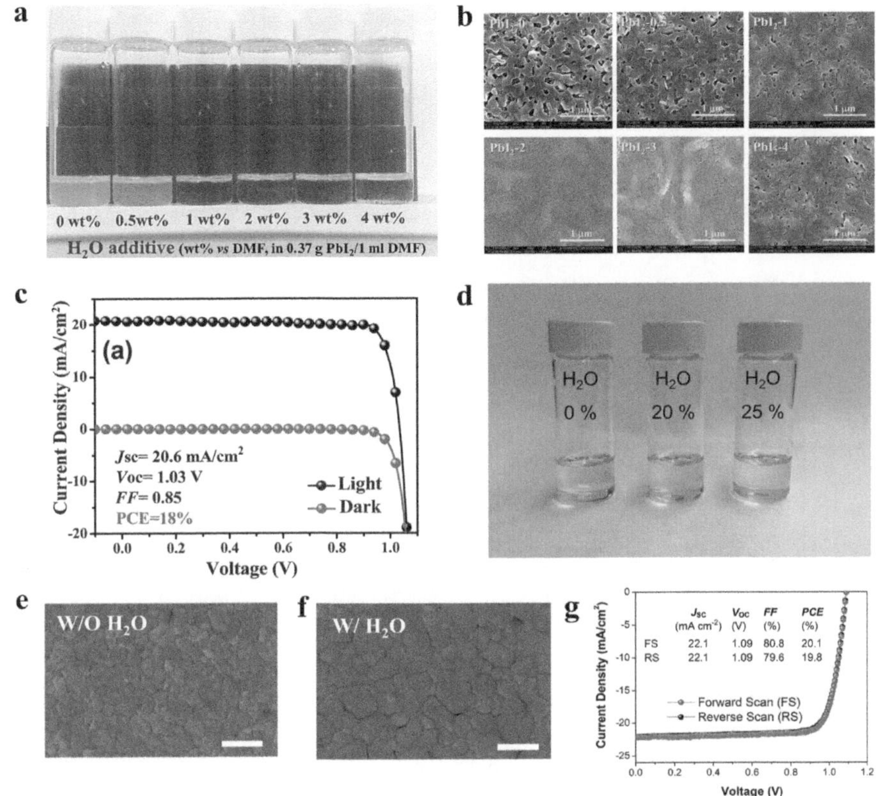

FIGURE 5.8 (a) Photographs of PbI₂/DMF suspensions containing various volume ratio of H₂O to DMF. (b) SEM images of perovskite films prepared from various volume ratio of H₂O to DMF (Prov-0, 0.5, 1, 2, 3, and 4 represent the volume ratio of 0, 0.5, 1, 2, 3, and 4 wt%, respectively). (c) *J–V* curves of the champion PSCs measured under dark and light. Adapted with permission.[72] Copyright 2015, Royal Society of Chemistry. (d) Photograph of perovskite (CH₃NH₃PbI₃₋ₓClₓ) precursor solutions with various H₂O contents. (e, f) SEM images of perovskite film prepared by aqueous-free precursor and aqueous-containing precursor, respectively. Scale bar: 1 μm. (g) *J–V* curves of the champion PSCs device based on H₂O-20% precursor solution measured in reverse and forward scans. (d–g) Adapted with permission.[73] Copyright 2018, Wiley.

solution and form a stable intermediate phase of MAI-PbI₂-DMSO ((MA)₂Pb₃I₈·2DMSO) during film formation (Figure 5.9a and b).[76,78] This stable phase greatly contributes to the retarded nucleation and growth of perovskite crystals.[67,79,80] With increased amount of DMSO in precursor solution, the deposited perovskite film shows a more uniform morphology without pinholes and caves (Figure 5.9c and d). The crystals are also found to be larger in size, making better contact with the substrate, and having better alignment along the vertical direction. Most importantly, no grain boundaries are observed in the horizontal direction of the perovskite film.[79] However, perovskite film shows

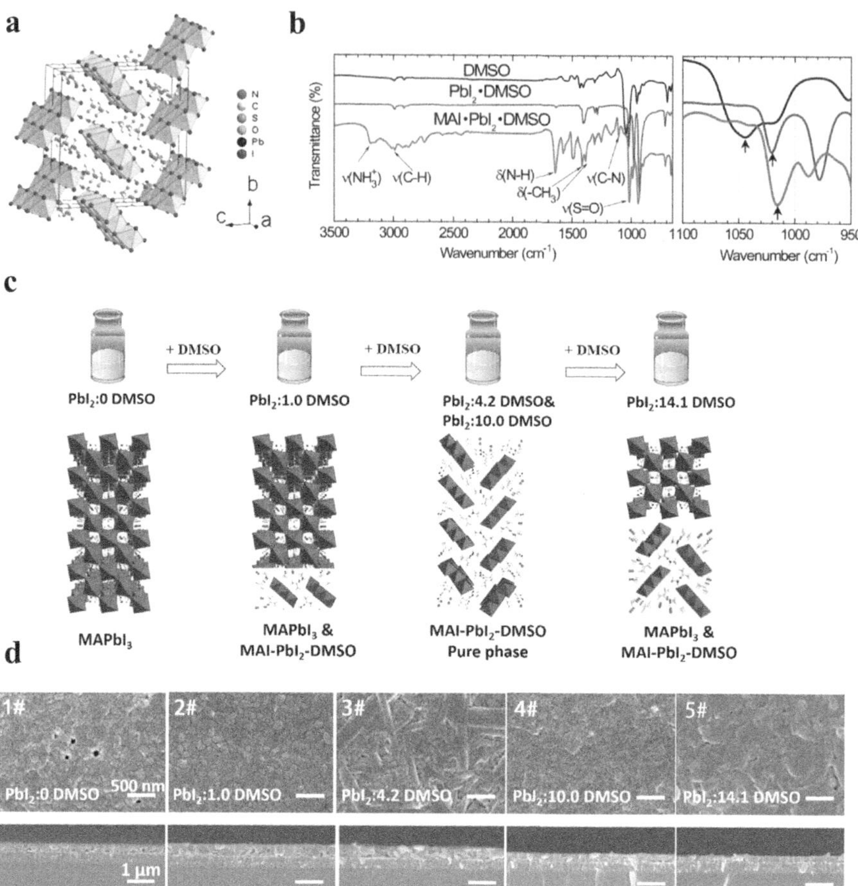

FIGURE 5.9 (a) Crystal structure of $MA_2Pb_3I_8\cdot2DMSO$. Adapted with permission.[78] Copyright 2015, Royal Society of Chemistry. (b) FTIR of DMSO (solution), $PbI_2\cdot DMSO$ (powder), and $MAI\cdot PbI_2\cdot DMSO$ (powder). Adapted with permission.[76] Copyright 2015, American Chemical Society. (c) Schematic illustrating the relation between the amount of DMSO and the component of intermediate film. (d) Scanning electron microscopy (SEM) images and cross-sectional SEM images of $MAPbI_3$ films deposited on NiO_x/FTO substrate with different PbI_2: DMSO ratios in precursor solution (From 1# to 5#: 0:1, 1.0:1, 4.2:1, 10.0:1, and 14.1:1). Adapted with permission.[79] Copyright 2017, Elsevier.

inhomogeneous-sized perovskite crystal grains and high surface roughness if high concentration of DMSO is used as solvent due to the too slow crystallization. Thus, the amount of DMSO should be suitable so as to produce the pure and stable intermediate phase of $MAI\cdot PbI_2\cdot DMSO$ which could facilitate the growth of a perovskite layer with aligned and vertically monolithic crystals on a NiO_x nanocrystal film.

To date, DMF/DMSO mixed solvent is the most commonly used system to prepare high-quality perovskite films towards high-performance PSCs, which may be due to the favorable properties of DMF and DMSO. These properties include polarity, viscosity, and coordination ability with PbI_2, etc.[18,81–88], enabling its use in 2D and 3D perovskites.[89,90] For instance, by adjusting the ratio of DMF to DMSO to control the film formation of 2D RP $PA_2MA_4Pb_5I_{16}$ perovskite film, the PSC yielded a PCE of 10.41%.[89]

NMP is another solvent possessing high coordination capacity, and its D_N is higher than DMF but lower than DMSO, which has significance on the structures of the intermediate phase, crystallization dynamics, and film formation of perovskite. For instance, DMF cannot form a stable adduct with $FAPbI_3$ perovskite, and the unstable $FAI·PbI_2·DMF$ complex with a weak interaction energy induces heterogeneous nucleation. Consequently, the $FAPbI_3$ film prepared with DMF is less uniform and reproducible than its $MAPbI_3$ counterpart. To solve this issue, Huang and Lee et al. used NMP instead of DMSO/DMF to increase the stability of the intermediate adduct phase (NMP-FAI-PbI_2), which is converted into a high-quality $FAPbI_3$ perovskite film with high uniformity and pinhole-free morphology.[75,91] Jo *et al.* claimed that adding NMP into PbI_2/DMF solution induces the formation of a PbI_2-NMP complex.[92] Its presence is confirmed by an XRD peak at 8.11°, compared with its PbI_2-DMSO counterpart from DMF/DMSO solution at 9.81° and PbI_2 from pure DMF solution at 12.71°, respectively. The down-shifted diffraction peak is due to an increase in the interlayer space in between the PbI_2 layered structure by the intercalation of NMP, which has a larger molecular size than DMSO. The $FAPbI_3$ perovskite films converted from PbI_2-NMP intermediate phase shows considerably increased grain size up to approximately 1 mm and prolonged charge-carrier lifetime, resulting in enhanced photovoltaic effciency compared with the PbI_2-DMSO-mediated analog. Zheng et al. further revealed that NMP has a higher β than DMSO and DMF.[44] Thus, in $FAPbI_3$ perovskite precursor solution, NMP molecules tend to interact with FAI via hydrogen bond and form $(FA\cdots NMP)$-PbI_3 consisting of 1D PbI_3^- chains with face-shared $[PbI_6]^{4-}$ octahedra. By contrast, DMSO interaction with PbI_2 comprises 1D single chains made of edge-shared PbI_4O_2 octahedra via coordination. The intermediate crystal had been explored by single crystal X-ray analysis. Meanwhile, both intermediates can evolve into α-$FAPbI_3$ below the δ-to-α thermodynamic temperature, and the hydrogen-bond-favorable kind can form defect-less α-$FAPbI_3$ by sidestepping the break of strong coordination bonds. The disclosed solvent gaming mechanism guides the solvent selection for fabricating high-quality perovskite films and thus high-performance PSCs and modules.[91]

In addition to the coordinating molecules described above, hexamethylphosphoramide (HMPA)[93], γ-butyrolactone (GBL)[94], γ-valerolactone (GVL)[95,96], 2-methoxyethanol[31], and mixed solvents such as DMSO/chlorobenzene (CB)[97], DMA/ethanol (EtOH)[98], and DMF/CB/acetonitrile (ACN)[99] are also used as coordinating molecules to prepare perovskite films with good surface morphology and high crystalline properties.

5.3.1.2 Poor Solvents and Antisolvents

Solvents with high D_N show high solubility to perovskite solution, and the quality of perovskite crystal and thin film can be well tuned by the ratio of different solvents during spin coating. However, these solvents are primarily strong polar aprotic with high boiling point. Too strong coordination and high boiling temperature make it difficult to be fully removed because the perovskite decomposes if the temperature is too high. In particular, large scale-coating technologies (e.g., doctor blade, slot die, bar coating, electrospray, etc.) require the rapid evaporation of solvent, whereas solvents with relatively high boiling point and high coordination ability show slow evaporation rate. The result is the formation of perovskite film with high surface roughness and a large number of pinholes (Figure 5.10d and e). These strong polar aprotic solvents (e.g., DMF, DMSO, and NMP) are also toxic, causing problems in operational safety and environmental protection. Hence, the introduction of poor solvent/antisolvent with low D_N is essential because it facilitates the supersaturation of perovskites precursor solution and the subsequent fast nucleation and crystal growth at a lower temperature, as well as extract high-boiling-point solvent molecules. Consequently, high-quality perovskite films are produced. Furthermore, many poor solvents/solvents are environmentally friendly.

5.3.1.2.1 Poor Solvents

ACN is nontoxic solvent with low D_N, low boiling point, and low viscosity, making it an ideal candidate to replace traditional solvents of perovskite precursor inks. However, due to its low solubility to perovskite precursors, the effective solvents are a combination of ACN with other components that assist dissolution, such as the ACN/MA (gas) and monomethylamine (MMA)/tetrahydrofuran (THF)/ACN cosolvent system.[33,100–103] Noel et al. reported that the black $MAPbI_3$ precipitate in ACN can be dissolved to obtain a stable, clear, colorless precursor solution by passing into MA gas (Figure 5.10a).[33] The novel ACN/MA cosolvent was found to produce dense, uniform, pinhole-free perovskite film with high specularity on a large area of 125 cm^2 through spin coating (Figure 5.10b and c). Notably, PSCs achieve a PCE of over 19% and over 17% for film annealed at 100°C and without thermal annealing, respectively. This universal solvent system is also suitable for the preparing $MAPbBr_3$ perovskite film with smooth, dense, and uniform morphology.[101] Additionally, the resultant perovskite film had a much higher degree of texture intensity of azimuthal angle than the film deposited from DMF, indicating the vertical orientation of the ACN/MA-deposited film, which facilitates carrier transport within the film. In view of these merits of the ACN/MA system, Jeong et al. successfully prepared $MAPbI_3$ perovskite film with an area larger than 100 cm^2 via a simple bar-coating technique.[100] The average grain size of the wire-bar (D-bar) coated $MAPbI_3$ film is about 435 nm (the largest is over 1,000 nm), which was much larger than that of the spin-coated film (ca. 190 nm; Figure 5.10f and g). Notably, an average PCE exceeding 17% and the best PCE approaching 18% are achieved for PSCs with 100 cm^2 area (Figure 5.10h and i). Wang et al. reported the use of MMA/THF/ACN solvent system to obtain nonionic ink of perovskite precursor, $MA(MMA)_nPbI_3$ precursor solution, with less-disordered mobile ions.[102] This ink differs from the perovskite precursor ink prepared in traditional solvents containing highly disordered mobile ions, such as MA^+, Pb^{2+}, and I^-.[104] The nonionic nature of the precursor solution was confirmed by electrochemical analysis based on an electrolytic cell structure: the precipitation of dendritic Pb (reduzate of Pb^{2+}) and I_2 (reduzate of I^-) can be observed in ionic perovskite precursor solution at the cathode and anode, respectively. Conversely, no precipitate was found

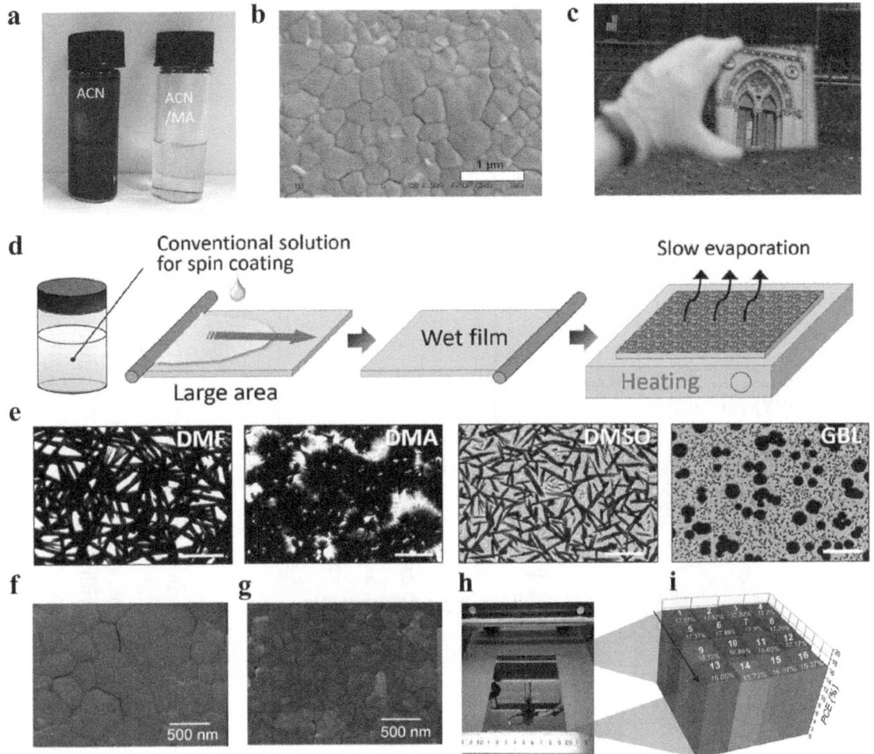

FIGURE 5.10 (a) Photographs of $CH_3NH_3I:PbI_2$ perovskite precursors dissolved in pure ACN and ACN/MA as solvent. (b) SEM image of $MAPbI_3$ film deposited via spin-coating from precursor solution using ACN/MA as solvent. (c) Photograph of an 125 cm², highly specular, and pinhole-free $MAPbI_3$ film. Adapted with permission.[33] Copyright 2017, Royal Society of Chemistry. (d) Schematic illustration of D-bar coating process of large-area $MAPbI_3$ thin film deposited from precursor solution using conventional polar aprotic solvent with high boiling point and high D_N. (e) Optical microscopy images of the dried $MAPbI_3$ films prepared from different precursor solutions using polar aprotic solvents of DMF, DMA, DMSO, and GBL (scale bar is 35 µm). SEM images of $MAPbI_3$ thin films formed (f) by D-bar coating using the ACN/MA solvent system and (g) by spin-coating using the DMSO/DMF-based solution and antisolvent dripping process. (h) Photograph of $MAPbI_3$ film deposited on a SnO_2-coated FTO substrate (area: 10 cm × 12 cm). After annealing at 150°C for 5 min, the substrate was divided into 16 regions with dimension of 2.5 cm × 2.5 cm, where each region contains five individual PSCs. (i) PCEs of best performing PSCs device in every region. (d–i) Adapted with permission.[100] Copyright 2019, American Chemical Society.

in $[MA(MMA)_nPbI_3]$ precursor solution (Figure 5.11a and b). Unlike conventional precursor solution, which requires all disordered ions to rotate and move into the lattice site simultaneously to assemble into the ordered crystal, this nonionic precursor ink can easily from perovskite crystals with fewer defects and preferential orientation along the (110) direction from a relatively ordered state (Figure 5.11c). Most importantly, it enables a one-step deposition of high-quality perovskite thin film without the

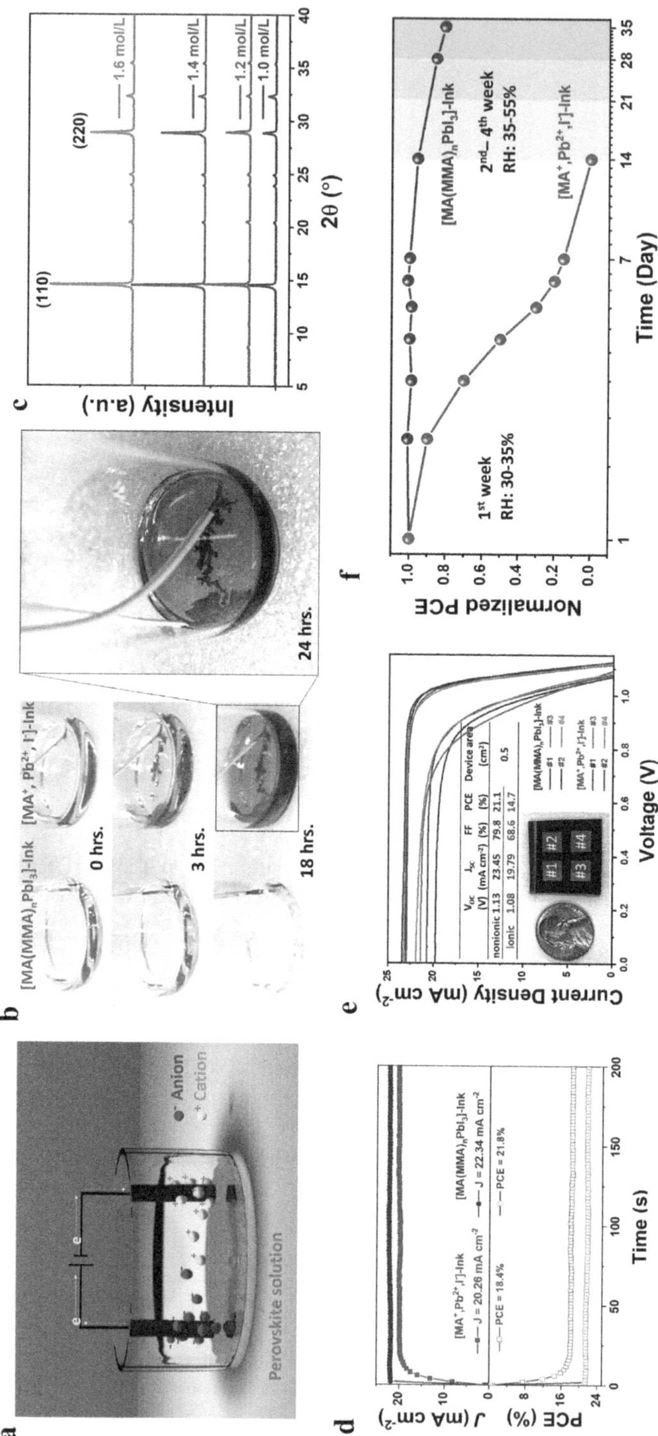

FIGURE 5.11 (a) Schematic electrolytic cell showing the redox reaction appearing in an ionic electrolyte. (b) Difference in behavior in electrolytic cells using different precursor inks as the electrolyte. (c) XRD patterns of perovskite crystals using the nonionic ink of different concentrations from 1.0 to 1.6 M. (d) Static current density and PCE measured as a function of the time for the [MA^+, Pb^{2+}, and I^-]-ink and [$MA(MMA)_nPbI_3$]-ink devices (area of 0.096 cm^2), biased at 0.91 and 0.98 V, respectively. (e) J–V characteristics of PSCs with device area of 0.5 cm^2 using different precursor solutions. (f) Device stability measurement of PSCs using different precursor solutions exposure in ambient atmosphere with increasing relative humidity. Adapted with permission.[102] Copyright 2020, Elsevier.

necessity of antisolvent treatment. Consequently, the resultant $MAPbI_3$ film exhibited an extraordinary long carrier-diffusion length of 4.6 μm at the single-crystal level and achieves a record PCE of 21.8%. Scaling the active area from 0.096 cm² to 0.5 cm² also did not deteriorate cell performance because the efficiency remained over 21% (Figure 5.11d and e). Moreover, cells based on nonionic $[MA(MMA)_nPbI_3]$-ink-based cells displayed much improved stability towards humidity compared to that based on $[MA^+, Pb^{2+}, and\ I^-]$-ink (Figure 5.11f), which could be ascribed to the elimination of DMF/DMSO molecules in the film. Overall, these results demonstrated the great potential of this nonionic ink for upscale film-forming techniques, such as slot die, doctor blading, and electrospray deposition.

Inspiringly, Huang et al. reported that fast blading large-area perovskite films can be produced under ambient conditions by using an ACN/2-ME/DMSO solvent system (Figure 5.12).[105] As mentioned earlier, good solvents such as DMSO and DMF can help obtain large grains but do not favor the fast deposition of a compact and smooth perovskite film at room temperature due to their high coordination ability to Pb^{2+} and low volatility. By contrast, poor solvents ACN/2-ME can be used to obtain perovskite films with highly compact and uniform morphology at high speed and room temperature. However, it also results in low crystallinity and poor contact with substrates. To address this issue, a small amount of DMSO was added into the ACN/2-ME host solvent to tailor the solvent-coordination capability of perovskite solution. In this ACN/2-ME/DMSO solvent, volatile host solvents allow for the quick formation of compact perovskite films with the assistance of N_2-knife, whereas the added coordinating solvents are slowly released from the as-coated solid film, giving sufficient time for the perovskite grains to grow into large sizes (up to 1–2 μm) with high crystallinity and good contact with substrate (Figures 5.12a, 12b). Notably, a coating speed as high as 99 mm/s was realized, yielding perovskite modules with a certified stabilized PCE of 16.4% and an aperture area of 63.7 cm² (Figure 5.12e). The as-fabricated modules also show superior temperature and shading-effect tolerance compared with commercialized PV technologies, including c-Si, CdTe, and CIGS. These features increase the competitiveness of perovskite PV in the future

ACN was also used as weak coordination additive into PbI_2/DMF solution to change the coordination of PbI_2-DMF adduct, thereby affecting the nucleation, grain growth, and film morphology of PbI_2 film, as well as the resultant perovskite film.[106] In addition to cosolvent system, the potential of single solvent for the easy fabrication of perovskite film under ambient condition was also demonstrated, such as 2-ME,[31] in which PSCs showed an excellent PCE of 15.32% with V_{OC} of 0.972 V, J_{SC} of 19.32 mA cm⁻², and FF of 0.816 under one-sun AM1.5G illumination condition.

5.3.1.2.2 *Antisolvents*

Benzene derivatives are extensively used antisolvents, such as chlorobenzene (CB) and toluene. Cheng et al. reported that dropping CB immediately on the wet film from a DMF solution of $MAPbI_3$ during the spinning process produces a highly uniform perovskite film with good reproducibility (Figure 5.13, a-d).[108] The introduction of CB was suggested to readily decrease $MAPbI_3$ solubility in deposited precursor solution, promoting fast nucleation and crystal growth. Chen et al. demonstrated that antisolvent (toluene) surface treatment affects the morphology of the perovskite layer

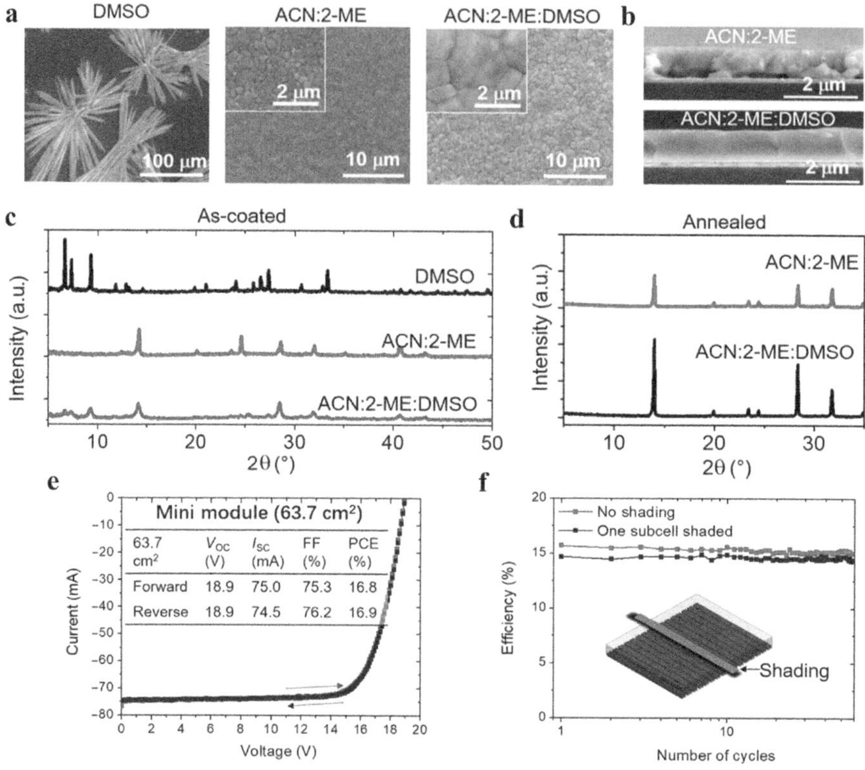

FIGURE 5.12 SEM images (a) and Cross-sectional SEM images (b) of perovskite films prepared with DMSO, ACN:2-ME and ACN:2-ME:DMSO, respectively. XRD spectra of as-coated (c) and annealed (d) perovskite films from DMSO, ACN:2-ME and ACN:2-ME:DMSO, respectively. (e) *J-V* curve of a small-area perovskite solar cell fabricated with the N_2-assisted room-temperature blade-coating method. (f) Efficiencies of a perovskite module with one subcell going through 58 cycles of shading/de-shading. The inset shows schematically how shading is applied over one subcell. Adapted with permission.[107] Copyright 2019, American Association for the Advancement of Science.

and its electronic properties (Figure 5.13, e–k).[109] With the toluene treatment, excess halide and MA ions after the two-step process for the perovskite film deposition were removed from the perovskite surface, leading to a net positive charge on the Pb atoms and a more conductive perovskite surface. This kind of perovskite surface was beneficial for the hole-conductor-free PSCs. Sakai et al. observed that toluene drenching accelerates nucleation at early-stage crystallization in $3MAI:PbCl_2$ precursor, where $MAPbI_3$ and $MAPbCl_3$ nucleation occur and $MAPbCl_3$ is transformed into $MAPbI_3$ phase by anion exchange during crystallization.[110] Toluene treatment strongly affected the ratio of $MAPbI_3$ to $MAPbCl_3$ and nucleation and thus plays a critical role in determining the final film morphology and photovoltaic performance. Wu and co-workers reported that toluene treatment also balanced the competition between

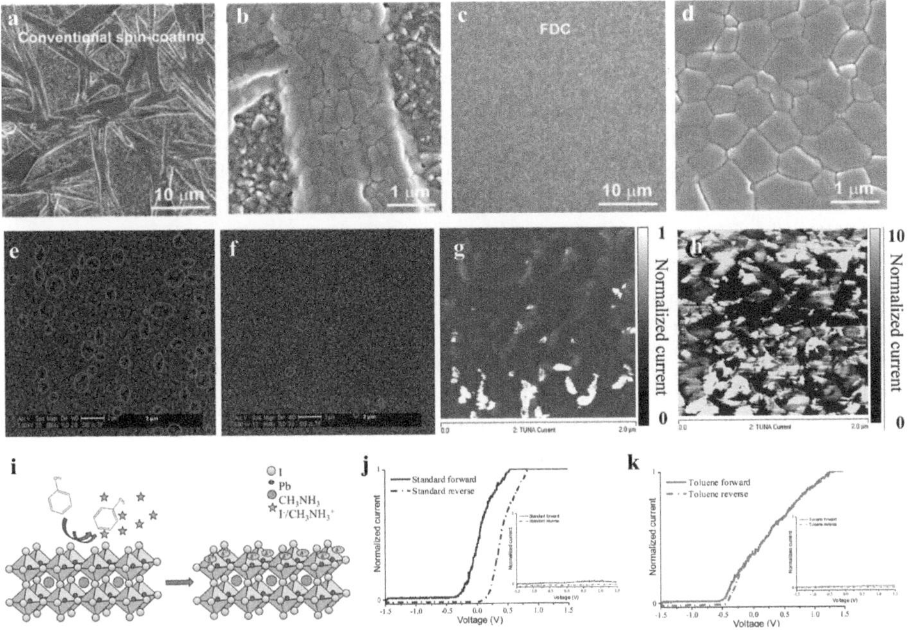

FIGURE 5.13 Low- and high-magnification SEM images of the surface of a $CH_3NH_3PbI_3$ film prepared by (a, b) fast deposition-crystallization (FDC) with the addition of chlorobenzene and (c, d) conventional spin-coating. (a–d) Adapted with permission.[108] Copyright 2014, Wiley. SEM figures of solar cells (e) with and (f) without toluene treatment. Current mapping was measured by cAFM without bias. Conductivity of perovskite films (g) with and (h) without toluene treatment. (i) Schematic illustration of the effect of toluene treatment on the perovskite surface. I–V plots measured on a single perovskite grain by cAFM without (j) and (k) with toluene treatment. Inset: IV plot of single nonconductive grain. (e–k) Adapted with permission.[109] Copyright 2015, American Chemical Society.

nucleation and crystal growth during the formation of a pinhole-free $MAPbI_3$ perovskite film on a planar PEDOT:PSS surface, leading to the efficient dissociation of the excitons at the interface between perovskite and PEDOT:PSS.[111] They also found that iodobenzene (IB) was more efficient as antisolvent alternative to toluene in the washing treatment of $MAPbI_3$ perovskite wet film. By altering the IB volume, the grain size of the resultant $MAPbI_3$ thin film can be regulated to adapt to two different interfacial contacts (perovskite/electron-transport layer (ETL) and perovskite/HTL) for improved exciton dissociation, resulting in higher J_{SC} and PCE.[112,113]

However, halogenated benzenes and toluene are hazardous solvents that can cause systematic health risks for humans, so environment friendly antisolvents need to be developed for scale manufacture technology. Saliba et al. demonstrated a fully non-halogenated solvent, i.e., methoxybenzene (or anisole), which can be used as the antisolvent for perovskite formation and the solvent for the HTM. The state-of-the-art PCEs reach 20.5%, which was the highest value of PSCs fabricated without using toxic CB or toluene at that time.[114] The merits of anisole as antisolvent were also described in other works.[115,116] Other efficient antisolvents based on benzene derivatives such as trifluorotoluene[82] have also been reported.

In addition to benzene derivatives, other organic solvents such as alcohols, ethers, and esters were also widely developed as effective antisolvents.[117] Inspiringly, these organic solvents are much less toxic or even nontoxic compared with those based on benzene derivatives. sec-Butyl alcohol can induce the fast crystallization deposition of $MAPbI_3$, as well as adjust the amount of MAI in $MAPbI_3$ films, which has a positive effect on photovoltaic device performance.[85] Moreover, by adding sec-Butyl alcohol, the rapid crystallization of the Sn-based perovskite layer leading to a poor film morphology and low surface coverage can be improved, thereby affording $MASn_{0.25}Pb_{0.75}I_3$ perovskite films comprising densely packed and uniformly distributed large crystal grains.[118] Different solubilities of PbI_2 and MAI can cause incomplete cover of resultant perovskite film on substrate. The combination of DMSO and antisolvent washing with diethyl ether was found to effectively address this problem by regulating the evaporation rate of solvents.[76] Ethyl acetate and methyl acetate are extensively used during industrial manufacture, including food additives, so are they are real green solvents and used as antisolvents in the film formation of perovskites.[83,84,119,120] Ethyl acetate shows great potential in the scale-up production of perovskite device because it induces the instant crystallization of a $MAPbI_3$ perovskite film without annealing,[120] as well as enables a synergic interface (perovskite/HTL) optimization for large-area PSCs with efficiency and stability over their CB counterparts.[83] Waston et al. also found that ethyl acetate acts as a moisture absorber during spin coating, protecting sensitive perovskite intermediate phases from airborne water during film formation and annealing. They were able to fabricate PSCs under 75% relative humidity (RH) ambient conditions with PCEs over 15%.[119]

Besides single antisolvents, mixed antisolvents have been studied with the aim of obtaining perovskite films with ideal morphology and crystalline properties. For instance, Wang et al. reported that mixing CB with the weakly coordinating IPA could readily remove residual CB, increasing the grain size of perovskite and in turn decreasing defect density.[87] Yu et al. suggested that mixing antisolvents of ethyl ether and n-hexane led to high nucleation density and slows down film formation, resulting in improved grain orientation and a smooth surface.[121]

Considering that various antisolvents are present and applied, and parameters and techniques of antisolvent application including antisolvent volume, temperature, dipping, additives, spin-coating parameters, and atmosphere are diverse, a clear understanding of these variables is urgently needed. Taylor et al. showed that antisolvents generally fall into three categories: those that favor short application times, those that are largely unaffected (Figure 5.14a and b) and those that perform best with longer application times, and they identified two key factors that influence the quality of the perovskite layer: the solubility of the organic precursors in the antisolvent and its miscibility with the host solvent(s) of the perovskite precursor solution.[117] Depending on these two factors, tuning the application time could result in efficient photovoltaic devices from any antisolvent (Figure 5.14c). They also demonstrated that by using

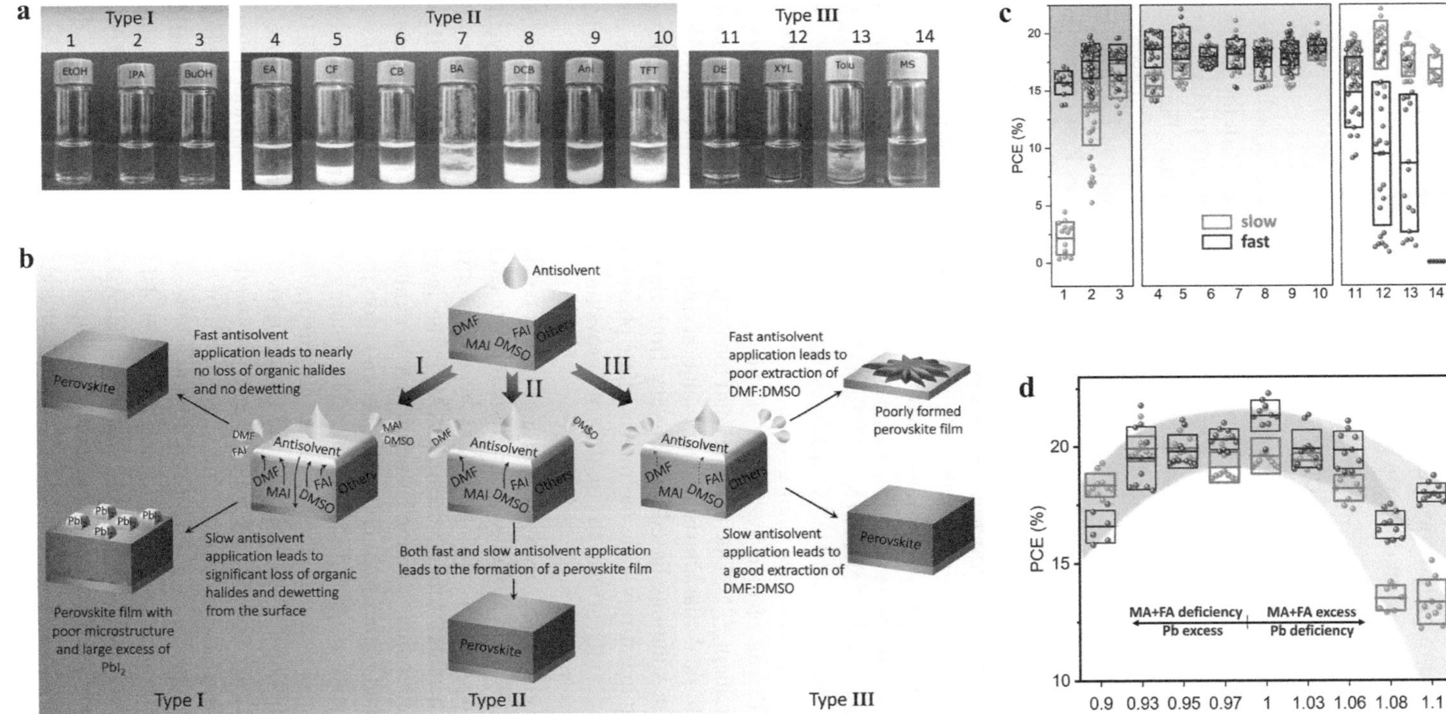

FIGURE 5.14 (a) Solubility of MAI in a solution of DMF:DMSO:antisolvent, meant to simulate the perovskite film intermediate phase during the antisolvent step of fabrication. (b) Summary of the various mechanisms involved in perovskite film formation by the different categories of antisolvents. (c) The PCE of devices resulting from a fast or slow antisolvent application. (d) PCEs of triple-cation (MA-FA-Cs) perovskite solar cells with varying stoichiometric ratios, for fast and slow antisolvent application rates. Adapted with permission.[117] Copyright 2021, Springer Nature.

the optimal application time, it was possible to significantly expand the range of stoichiometries that lead to high device performance, thus eliminating the need for the commonly used PbI_2 excess in the perovskite composition (Figure 5.14d).

5.3.2 Ionic Liquid (IL)-Based Solvent System

To date, high-efficiency PSCs are prepared by perovskite precursor solution with mixed solvents, e.g., DMF and DMSO. However, due to differences in the physical and chemical properties of the solvents, a large number of defects are probably generated during film formation, adversely affecting device performance. Furthermore, polar aprotic solvents with strong coordination ability are highly toxic and even carcinogenic. Alcoholic solvents and ACN may be effective alternatives for toxic solvents, but their use is not completely safe and usually complicates film preparation due to low coordination and poor solubility. The performance of PSCs using these solvents is also not as high as that using organic solvents.

Compared with traditional organic solvents system, ILs as solvents to perovskites have the merits of high solubility, nonhazardous nature, and ability to promote preferred crystal orientation. ILs commonly used in solvent engineering are amine carboxylic acid salt, including special functional groups of -COOH and $-NH_2$ that can form Pb-O coordination and NH-X (X = Cl, Br, I) hydrogen bonds.[21,23,34] The formation of Pb-O coordination and NH-X hydrogen bonds redound to the much higher solubility of ILs to perovskites than conventional organic solvents such as DMF, DMSO, and NMP[21], as verified by ^1H and ^{13}C nuclear magnetic resonance (NMR) and FTIR measurement.[25,34,35] Due to the special interaction between perovskites precursor and ILs, physical characteristics (e.g., solution color and solubility) and chemical interactions in solutions based on ILs vary from those based on conventional solvents, resulting in entirely different intermediate phases and formation process of film, which significantly affects the quality of perovskites and photovoltaic performance of PSCs.

A series of novel amine carboxylic acid salt ILs has been developed for dissolving perovskite materials, such as methylammonium formate (MAFa), MAAc, methylammonium propionate (MAP), ethylammonium acetate, propylammonium acetate, and butylammonium acetate. Among them, MAAc is the most extensively used IL in PSCs. Perovskite precursor solution based on MAAc exhibits more uniform micellar size distribution and higher total colloidal concentration than that based on conventional organic solvents (Figure 5.15 a and b).[23] This finding was attributed to the weak interaction between MAAc and the perovskite precursors PbI_2 and MAX (X = Cl, Br, I) to form agglomerates. Thus, uniform and dense perovskite film using MAAc as a solvent can be easily obtained by a simple one-step method in air without antisolvent treatment or any environmental limitations (humidity, oxygen, and temperature) (Figure 5.15c). MAAc can also effectively dissolve MA-, FA-, and Cs-based halide perovskites by forming Pb–O coordination and N–H···I hydrogen bonds.[21] Results of grazing incidence wide-angle X-ray scattering (GIWAXS) showed that MAAc could make the preferred crystal orientation in multiple directions as additives due to the covalent binding of Pb^{2+} to Ac^- (Figure 5.15d and e).[21] The residual MAAc in the perovskite film after annealing can passivate defects by forming Pb–O, N–H···I,

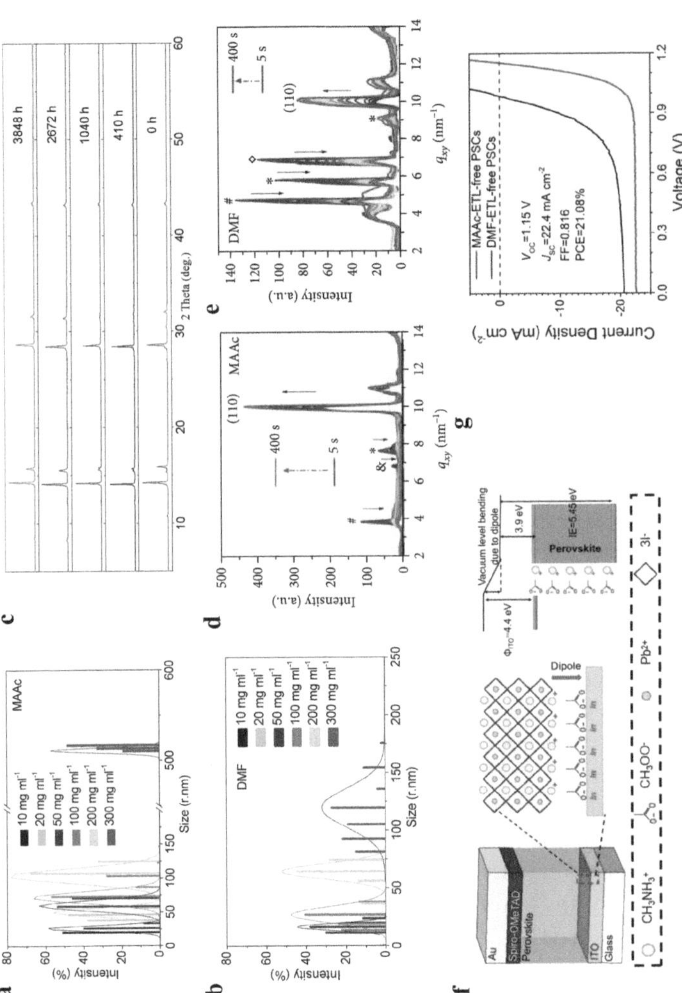

FIGURE 5.15 DLS spectra of (a) MAAc and (b) DMF perovskite precursor solution with different concentrations from 10 to 300 mg mL^{-1}. (c) XRD pattern evolution of perovskite films based on MAAc solution placed in air. (a–c) Adapted with permission.[23] Copyright 2019, Elsevier. (d) 1D GIWAXS diffraction pattern of MAAc thin films in the q_{xy} direction for 40 frames (10 s per frame); (e) 1D GIWAXS diffraction pattern of DMF thin films in the q_{xy} direction for 40 frames (10 s per frame). # and * represent non-perovskite diffraction peaks. (d–e) Adapted with permission.[21] Copyright 2020, Lingfeng Chao et al. (f) Device structure with the MAAc dipole contact and diagram of energy levels. The representations of the dipole are qualitative based on the hypothesized mode of the interaction, and the exact mode needs to be further evaluated. (g) J–V characteristics for the DMF- (black) and MAAc-ETL-free PSCs (red). (f–g) Adapted with permission.[122] Copyright 2020, American Chemical Society.

and N–H···O. As a result, the PCE of the MAAc-based device (21.18%) was slightly higher than that of the DMF-based device (20.49%). Chen et al. reported that MAAc can effectively dissolve a series of non-halide lead sources (e.g., lead acetate ($PbAc_2$), lead sulfate ($PbSO_4$), lead carbonate ($PbCO_3$), lead nitrate ($Pb(NO_3)_2$), lead formate ($Pb(HCOO)_2$), and lead oxalate (PbC_2O_4)). The PCE of PSCs based on $PbSO_4$, $PbAc_2$, and $PbCO_3$ can reach 14.48%, 19.21%, and 20.13%, respectively.[34] However, non-halide lead salts are difficult to dissolve in DMSO–DMF mixed organic solvent except for $PbAc_2$ and $Pb(NO_3)_2$. The strong hydrogen bonding between MAAc and the perovskite precursor facilitates prolonged crystal growth but parasitically restricts mass transport and solvent escape, leading to retarded crystal nucleation and perovskite growth with unintentional defects. Wu et al. found that excess lead halides ($PbCl_2$, $PbBr_2$, PbI_2) can form $MAPbX_{3-x}Ac_x$ (X = Cl, Br, I) nanocrystals with MAAc, thus as the nucleus (in-situ nanocrystal seeding crystallization) to regulate perovskite growth. As a result, high-quality perovskite thin films with simultaneously reduced defect density and shallowed trap depth were obtained. The PCE of the corresponding devices increased from 17.80% to 21.26%.[123] Li et al. demonstrated that MAAc could be physically adsorbed onto indium tin oxide (ITO) electrode and form an interfacial dipole layer, resulting in a well-matched work function and higher electron extraction ability (Figure 5.15f). The PCE of ETL-free devices modified by MAAc reached 21.08% (Figure 5.15g).[122] Wang et al. demonstrated that a stronger Pb–O interaction in MAAc-based solution slowed down crystallization compared with that based on DMF–DMSO mixed solvent and improved the quality of all-inorganic perovskite films.[25] Thus, the PCE of MAAc-based $CsPbI_{2.5}Br_{0.5}$ inorganic PSCs reached 17.10%.

Tuning the chemical properties of IL solvents can also reportedly improve the performance of IL-processed PSCs. For example, Huang et al. demonstrated that adding methylammonium difluoroacetate (MADFA) to MAAc-based perovskite solution could retard perovskite crystallization and improve the stability of the perovskite precursor solution due to the MADFA anchor. As a result, perovskite crystallization slowed down, and the stability of the perovskite precursor solution and Pb–O interaction in precursors improved.[124] With the ILs, the devices had a high efficiency of 21.46 % and excellent stability over 180 days in nitrogen atmosphere at room temperature. Gu et al. used a series of ILs solvents, namely, MAAc, MAP, and methylammonium isobutyrate (MAIB), to study the influence of steric hindrance on the chemical interaction between ILs and PbI_2 in solution.[35] Liquid-state ^{13}C NMR measurements showed that the relative shift of the carbon atom in the carboxylate group followed the trend of MAAc ($\Delta\delta$ = 0.452 ppm) > MAP ($\Delta\delta$ = 0.257 ppm) > MAIB ($\Delta\delta$ = 0.231 ppm). This finding indicated that the interaction between ILs and PbI_2 decreases with increased steric hindrance. The strong interaction between IL and PbI_2 could slow down crystallization and improve the perovskite-film quality, but it was not conducive to solvent removal. Therefore, the MAP-based device achieved the best photovoltaic performance with a PCE of 20.56%.

ILs can also effectively control the phase transition of perovskite, even in harsh environments. Hui et al. found that PbI_2 transformed from a linear Pb–I–Pb structure in the DMF–DMSO system to a corner-sharing one in the MAFa system due to strong interactions between MAFa and PbI_2 in the forms of C=O···Pb chelation and N–H···I hydrogen bonds.[125] The corner-sharing structure notably decreased the formation-energy barriers for FAI penetration into PbI_2, so stable α-$FAPbI_3$ could be prepared under a high RH of 90%. The PCE of devices fabricated in ambient air exceeded 24%. Notably, ILs could produce 3D perovskite film with excellent photovoltaic performance and showed great potential in the field of low-dimensional perovskites, including RP phase and Dion–Jacobson phase.[22,126–129]

5.3.3 Mixed Organic/IL Solvent System

ILs integrate the advantages of high solubility to perovskite precursors and nontoxicity, whereas pure ILs exhibit a high viscosity (e.g., MAP), which is challenging to process a uniform thin film by spin coating at room temperature. To solve this problem, cosolvent systems comprising ILs and volatile organic solvent have been developed (Figure 5.16). For example, after adjusting the perovskite composition,

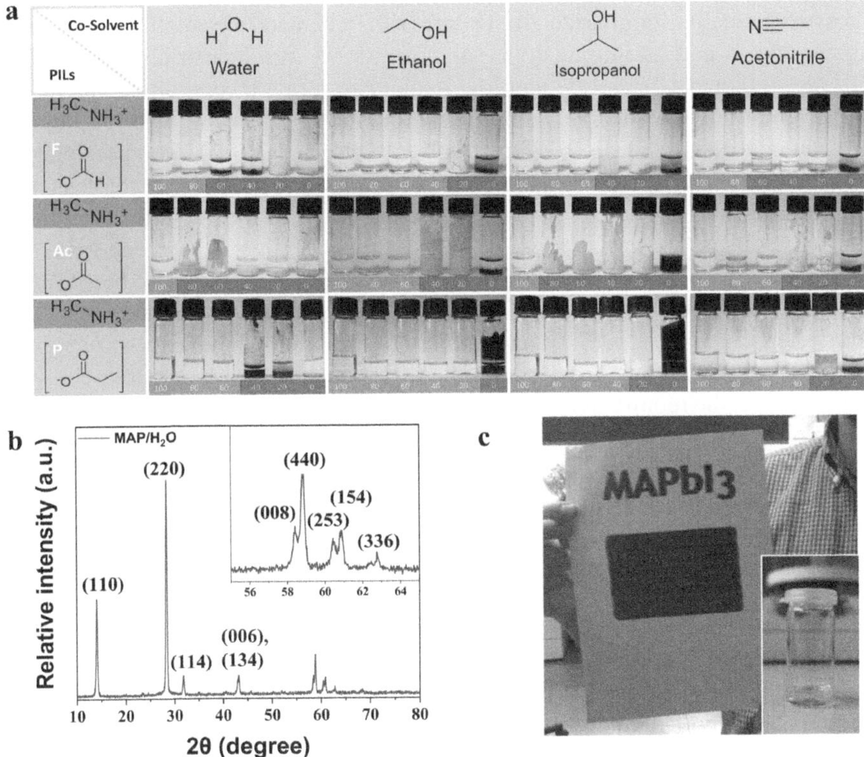

FIGURE 5.16 (a) Photographs of binary perovskite ink-systems (IL/co-solvent) with increasing co-solvent volume. (l-row 1 MAF) methylammonium formate, (l-row 2 MAAc) methylammonium acetate, (l-row 3 MAP) methylammonium propionate. (b) XRD pattern (Mo Kα → Cu Kα) of drop coated $MAPbI_3$ thin film from methylammonium propionate/water ink (3:2, V/V, 0.5 M). (c) Photograph of ink-jet printed $MAPbI_3$ on office paper, inset shows perovskite ink from methylammonium propionate/water (3:2, V/V, 0.5 M). Adapted with permission.[130] Copyright 2018, Elsevier.

a PSC device based on $(MA_{0.15}FA_{0.85})Pb(I_{0.85}Br_{0.15})_3$ mixed-cation perovskite film prepared by sequential deposition method with MAP/ACN/DMSO solvent system achieved an excellent PCE of 15.46%. Additionally, a perovskite precursor solution with MAP/H_2O as solvent was shown to be able to print a graphical perovskite film on regular office paper sheets using a commercial inkjet printer, which is exciting for the fabrication of flexible devices (Figures 5.16b and c).[94] Chen et al. added the weak coordination solvent ACN to an MAAc precursor solution and found that the reduced coordination between the solvent and perovskite solutes $(CH_3COO^-\cdots Pb^{2+})$ facilitated the formation of large micelles by sharing halogens. The strong hydrogen-bond interaction between organic cations and ACN $(CH_3-NH_3^+\cdots CH_3CN)$ also slowed down the growth of perovskite crystals to produce a compact and smooth thin film.[24] As a result, $MAPbI_{2.8}Br_{0.2}$-based PSCs achieved a high PCE of over 21% with long-term stability for over 1,500 h.

5.4 CONCLUSIONS AND OUTLOOK

We systematically summarize the progress of advanced research on the solvent engineering of perovskite precursor solution towards the industrialization of PSCs with high efficiency, stability, and low toxicity in terms of coordination and solubility regulation. Specifically, we discuss in detail the effects of physical properties and chemical interactions of different solvents on various fabrication technologies of perovskite film. Moreover, some critical insights into the current challenges and opportunities in the field of solvent engineering in PSCs for the large-area production of the device are provided. Although significant progress has been made from the perspective of solvent engineering, many challenges still need to be addressed. The fundamental understanding of solvents in the perovskite precursor solution and theirs interactions with perovskites can help figure out the key factors affecting the crystallization dynamics, morphology, and film quality of the prepared perovskite films. To meet the requirements of future large-scale industrialization of PSCs, a few aspects can be considered in terms of solvent engineering.

First, further understanding the effects of solvent–solute interactions in perovskite precursor solution and their impacts on large-area film formation is critical. As the most attractive advantage of PSCs, solution process can significantly reduce the fabrication cost of thin-film photovoltaics. However, current solution-processed PSCs still suffer from obvious efficiency drops upon scaling up because of the non-uniform morphology and numerous point defects in large-area perovskite film. Developing some in-situ characterization techniques such as GIWAXS may help provide researchers more insights into perovskite-film evolution from precursor solution. Theoretical calculations and simulations may also contribute to the fundamental understanding of perovskite precursor properties and accelerate the development of new solvent systems.

Second, developing less-toxic, safe, and low-cost solvent systems is important for scaling up PSCs from the perspective of sustainability. The environmental pollution issue caused by the use of toxic solvents (e.g., DMF) can be a barrier for commercializing PSCs in the future. Some green solvent systems have been developed, such as ILs (e.g., MAAc) and biomass-derived green solvents (e.g., GVL). However, the performance of the fabricated PSCs is still lower than that of DMF-based devices.

Organic/IL mixed-solvent system opens a new avenue to further improve PSC performance through a sustainable manufacturing process. Developing high-throughput experimental techniques with machine-learning algorithms be an effective strategy to identify the optimum composition of green solvent systems.

Third, investigating the stability of perovskite films and the relevant PSCs prepared from different solvent systems is important. Long-term operational stability is the key challenges to PSC commercialization. The degradation mechanism of perovskite films prepared from different types of precursor solution is rarely investigated. Therefore, establishing the relationships between solvent and stability (including phase, ambient, operational, and thermal stabilities) is critical.

This chapter provides a fundamental understanding of the solvent properties of perovskite precursor solution and their impacts on the crystallization and film quality of prepared perovskite films, as well as the relevant PSC performance. Our findings can facilitate the lab-to-fab translation of perovskite technologies through the evolution of solvent engineering.

REFERENCES

1. C. Bi, Q. Wang, Y. Shao, Y. Yuan, Z. Xiao, J. Huang, Non-wetting surface-driven high-aspect-ratio crystalline grain growth for efficient hybrid perovskite solar cells, *Nature Communications* **2015**, *6*, 7747.
2. C. Wehrenfennig, G. E. Eperon, M. B. Johnston, H. J. Snaith, L. M. Herz, High charge carrier mobilities and lifetimes in organolead trihalide perovskites, *Advanced Materials* **2014**, *26*(10), 1584–1589.
3. V. D'Innocenzo, *et al.*, Excitons versus free charges in organo-lead tri-halide perovskites, *Nature Communications* **2014**, *5*(4), 3586.
4. S. D. Stranks, *et al.*, Electron-hole diffusion lengths exceeding 1 micrometer in an orgnometal trihalide perovskite absorber, *Science* **2013**, *342*(6156), 341–344.
5. G. C. Xing, *et al.*, Long-range balanced electron- and hole-transport lengths in organic-inorganic $CH_3NH_3PbI_3$. *Science* **2013**, *342*(6156), 344–347.
6. A. T. Barrows, A. J. Pearson, C. K. Kwak, A. D. F. Dunbar, A. R. Buckley, D. G. Lidzey, Efficient planar heterojunction mixed-halide perovskite solar cells deposited via spray-deposition, *Energy & Environmental Science* **2014**, *7*(9), 2944–2950.
7. Best research-cell efficiency chart, www.nrel.gov/pv/cell-efficiency.html (accessed May 2023).
8. N. H. Tiep, Z. L. Ku, H. J. Fan, Recent advances in improving the stability of perovskite solar cells, *Advanced Energy Materials* **2016**, *6*(3), 1501420.
9. J. Z. Chen, *et al.*, Hole-conductor-free fully printable mesoscopic solar cell with mixed-anion perovskite $CH_3NH_3PbI_{(3-x)}(BF_4)_x$, *Advanced Energy Materials* **2016**, *6*(5), 1502009.
10. T. A. Berhe, *et al.*, Organometal halide perovskite solar cells: Degradation and stability, *Energy & Environmental Science* **2016**, *9*(2), 323–356.
11. B. Li, D. Binks, G. Z. Cao, J. J. Tian, Engineering halide perovskite crystals through precursor chemistry, *Small* **2019**, *15*(47), 1903613.
12. M. Jung, S. G. Ji, G. Kim, S. I. Seok, Perovskite precursor solution chemistry: From fundamentals to photovoltaic applications, *Chemical Society Reviews* **2019**, *48*(7), 2011–2038.
13. Y. H. Cheng, F. So, S. W. Tsang, Progress in air-processed perovskite solar cells: From crystallization to photovoltaic performance, *Materials Horizons* **2019**, *6*(8), 1611–1624.

14. Z. Jiang, *et al.*, Solvent engineering towards scalable fabrication of high-quality perovskite films for efficient solar modules, *Journal of Energy Chemistry* **2023**, *80*, 689–710.

15. Q. Jiang, *et al.*, Surface passivation of perovskite film for efficient solar cells, *Nature Photonics* **2019**, *13*(7), 460–466.

16. Y. Wang, *et al.*, Thermodynamically stabilized beta-CsPbI$_3$-based perovskite solar cells with efficiencies > 18%. *Science* **2019**, *365*(6453), 591–595.

17. J. Qiu, *et al.*, 2D intermediate suppression for efficient Ruddlesden–popper (RP) phase lead-free perovskite solar cells, *ACS Energy Letters* **2019**, *4*(7), 1513–1520.

18. M. Kim, *et al.*, Methylammonium chloride induces intermediate phase stabilization for efficient perovskite solar cells, *Joule* **2019**, *3*(9), 2179–2192.

19. S. Zhang, *et al.*, Minimizing buried interfacial defects for efficient inverted perovskite solar cells, *Science* **2023**, *380*(6643), 404–409.

20. L. Chao, T. Niu, W. Gao, C. Ran, L. Song, Y. Chen, W. Huang, Solvent engineering of the precursor solution toward large-area production of perovskite solar cells, *Advanced Materials* **2021**, *33*(14), e2005410.

21. L. F. Chao, *et al.*, Origin of high efficiency and long-term stability in ionic liquid perovskite photovoltaic, *Research* **2020**, *2020*, 2616345.

22. L. F. Chao, T. T. Niu, Y. D. Xia, X. Q. Ran, Y. H. Chen, W. Huang, Efficient and stable low-dimensional ruddlesden-popper perovskite solar cells enabled by reducing tunnel barrier, *Journal of Physical Chemistry Letters* **2019**, *10*(6), 1173–1179.

23. L. F. Chao, Y. D. Xia, B. X. Li, G. C. Xing, Y. H. Chen, W. Huang, Room-temperature molten salt for facile fabrication of efficient and stable perovskite solar cells in ambient air, *Chem* **2019**, *5*(4), 995–1006.

24. B. Sun, *et al.*, Tuning the interactions of methylammonium acetate with acetonitrile to create efficient perovskite solar cells, *Journal of Physical Chemistry C* **2021**, *125*(12), 6555–6563.

25. X. J. Wang, *et al.*, Tailoring component interaction for air-processed efficient and stable all-inorganic perovskite photovoltaic, *Angewandte Chemie-International Edition* **2020**, *59*(32), 13354–13361.

26. Z. Arain, *et al.*, Elucidating the dynamics of solvent engineering for perovskite solar cells, *Science China-Materials* **2019**, *62*(2), 161–172.

27. T. H. Kim, S. G. Kim, Clinical outcomes of occupational exposure to *N,N*-dimethylformamide: Perspectives from experimental toxicology, *Safety and Health at Work* **2011**, *2*(2), 97–104.

28. J. Galvao, B. Davis, M. Tilley, E. Normando, M. R. Duchen, M. F. Cordeiro, Unexpected low-dose toxicity of the universal solvent DMSO, *FASEB Journal* **2014**, *28*(3), 1317–1330.

29. J. G. Tait, *et al.*, Determination of solvent systems for blade coating thin film photovoltaics, *Advanced Functional Materials* **2015**, *25*(22), 3393–3398.

30. K. L. Gardner, *et al.*, Nonhazardous solvent systems for processing perovskite photovoltaics, *Advanced Energy Materials* **2016**, *6*(14), 1600386.

31. K. H. Hendriks, J. J. van Franeker, B. J. Bruijnaers, J. A. Anta, M. M. Wienk, R. A. J. Janssen, 2-Methoxyethanol as a new solvent for processing methylammonium lead halide perovskite solar cells, *Journal of Materials Chemistry A* **2017**, *5*(5), 2346–2354.

32. J. C. Hamill, J. Schwartz, Y. L. Loo, Influence of solvent coordination on hybrid organic-inorganic perovskite formation, *ACS Energy Letters* **2018**, *3*(1), 92–97.

33. N. K. Noel, *et al.*, A low viscosity, low boiling point, clean solvent system for the rapid crystallisation of highly specular perovskite films, *Energy & Environmental Science* **2017**, *10*(1), 145–152.

34. Y. Chen, *et al.*, A universal ionic liquid solvent for non-halide lead sources in perovskite solar cells, *Journal of Energy Chemistry* **2022**, *71*, 445–451.

35. L. Gu, *et al.*, Designing ionic liquids as the solvent for efficient and stable perovskite solar cells, *ACS Applied Materials & Interfaces* **2022** *14*, 22870–22878.

36. K. Y. Yan, *et al.*, Hybrid halide perovskite solar cell precursors: Colloidal chemistry and coordination engineering behind device processing for high efficiency, *Journal of the American Chemical Society* **2015,** *137*(13), 4460–4468.

37. A. Sharenko, C. Mackeen, L. Jewell, F. Bridges, M. F. Toney, Evolution of Iodoplumbate complexes in methylammonium lead iodide perovskite precursor solutions, *Chemistry of Materials* **2017,** *29*(3), 1315–1320.

38. R. A. Kerner, L. Zhao, Z. Xiao, B. P. Rand, Ultrasmooth metal halide perovskite thin films via sol-gel processing, *Journal of Materials Chemistry A* **2016,** *4*(21), 8308–8315.

39. E. Radicchi, E. Mosconi, F. Elisei, F. Nunzi, F. De Angelis, Understanding the solution chemistry of lead halide perovskites precursors, *ACS Applied Energy Materials* **2019,** *2*(5), 3400–3409.

40. S. Rahimnejad, A. Kovalenko, S. M. Fores, C. Aranda, A. Guerrero, Coordination chemistry dictates the structural defects in lead halide perovskites, *Chemphyschem* **2016,** *17*(18), 2795–2798.

41. M. I. Saidaminov, A. L. Abdelhady, G. Maculan, O. M. Bakr, Retrograde solubility of formamidinium and methylammonium lead halide perovskites enabling rapid single crystal growth, *Chemical Communications* **2015,** *51*(100), 17658–17661.

42. J. Stevenson, *et al.*, Mayer bond order as a metric of complexation effectiveness in lead halide perovskite solutions, *Chemistry of Materials* **2016,** *29*(6), 2435–2444.

43. M. M. Lee, J. Teuscher, T. Miyasaka, T. N. Murakami, H. J. Snaith, Efficient hybrid solar cells based on meso-superstructured organometal halide perovskites, *Science* **2012,** *338*(6107), 643–647.

44. X. Huang, *et al.*, Solvent gaming chemistry to control the quality of halide perovskite thin films for photovoltaics, *ACS Central Science* **2022,** *8*(7), 1008–1016.

45. J. C. Hamill, Jr.; J. Schwartz, Y.-L. Loo, Influence of solvent coordination on hybrid organic–inorganic perovskite formation, *ACS Energy Letters* **2018,** *3*(1), 92–97.

46. V. K. Lamer, R. H. Dinegar, Theory, production and mechanism of formation of monodispersed hydrosols, *Journal of the American Chemical Society* **1950,** *72*(11), 4847–4854.

47. M. Ocana, R. Rodriguez-Clemente, C. J. Serna, Uniform colloidal particles in solution: Formation mechanisms, *Advanced Materials* **1995,** *7*(2), 212–216.

48. J. Ng, S. P. Xu, X. W. Zhang, H. Y. Yang, D. D. Sun, Hybridized nanowires and cubes: A novel architecture of a heterojunctioned TiO_2/$SrTiO_3$ thin film for efficient water splitting, *Advanced Functional Materials* **2010,** *20*(24), 4287–4294.

49. B. Ding, *et al.*, Material nucleation/growth competition tuning towards highly reproducible planar perovskite solar cells with efficiency exceeding 20%. *Journal of Materials Chemistry A* **2017,** *5*(15), 6840–6848.

50. X. B. Cao, *et al.*, Fabrication of perovskite films with large columnar grains via solvent-mediated ostwald ripening for efficient inverted perovskite solar cells, *ACS Applied Energy Materials* **2018,** *1*(2), 868–875.

51. M. J. Yang, *et al.*, Facile fabrication of large-grain $CH_3NH_3PbI_{3-x}Br_x$ films for high-efficiency solar cells via CH_3NH_3Br-selective Ostwald ripening, *Nature Communications* **2016,** *7*, 12305.

52. A. Baldan, Review progress in Ostwald ripening theories and their applications to nickel-base superalloys—part I: Ostwald ripening theories, *Journal of Materials Science* **2002,** *37*(11), 2171–2202.

53. X. Cao, L. Zhi, Y. Li, F. Fang, X. Cui, L. Ci, K. Ding, J. Wei, Fabrication of perovskite films with large columnar grains via solvent-mediated ostwald ripening for efficient inverted perovskite solar cells, *ACS Applied Energy Materials* **2018,** *1*(2), 868–875.

54. P. Ahlawat, M. I. Dar, P. Piaggi, M. Grätzel, M. Parrinello, U. Rothlisberger, Atomistic mechanism of the nucleation of methylammonium lead iodide perovskite from solution, *Chemistry of Materials* **2019,** *32*(1), 529–536.

55. Q. Hu, *et al.*, *In situ* dynamic observations of perovskite crystallisation and microstructure evolution intermediated from $[PbI_6]^{4-}$ cage nanoparticles, *Nature Communications* **2017**, *8*, 15688.

56. S. Huang, *et al.*, Chloroform micro-evaporation induced ordered structures of poly (l-lactide) thin films, *RSC Advances* **2013**, *3*(33), 13705–13711.

57. J. Xu, B.-H. Guo, J.-J. Zhou, L. Li, J. Wu, M. Kowalczuk, Observation of banded spherulites in pure poly(l-lactide) and its miscible blends with amorphous polymers, *Polymer* **2005**, *46*(21), 9176–9185.

58. P. Boonmongkolras, D. Kim, E. M. Alhabshi, I. Gereige, B. Shin, Understanding effects of precursor solution aging in triple cation lead perovskite, *RSC Advances* **2018**, *8*(38), 21551–21557.

59. K. Sun, C. Gao, C. Yang, H. Liu, Z. Hu, J. Zhang, Y. Zhu, Direct formed tri-iodide ions stabilizing colloidal precursor solution and promoting the reproducibility of perovskite solar cells by solution process, *Electrochimica Acta* **2019**, *311*, 132–140.

60. G. S. Shin, S. G. Kim, Y. Zhang, N. G. Park, A Correlation between iodoplumbate and photovoltaic performance of perovskite solar cells observed by precursor solution aging, *Small Methods* **2019**, *4*(5), 1900398.

61. H. Min, *et al.*, Stabilization of precursor solution and perovskite layer by addition of sulfur, *Advanced Energy Materials* **2019**, *9*(17), 1803476.

62. X. Wang, *et al.*, Perovskite solution aging: What happened and how to inhibit? *Chem* **2020**, *6*(6), 1369–1378.

63. M. Qin, *et al.*, Fused-ring electron acceptor ITIC-Th: A novel stabilizer for halide perovskite precursor solution, *Advanced Energy Materials* **2018**, *8*(18), 1703399.

64. B. Dou, *et al.*, Degradation of highly alloyed metal halide perovskite precursor inks: Mechanism and storage solutions, *ACS Energy Letters* **2018**, *3*(4), 979–985.

65. J. C. Hamill, J. C. Sorli, I. Pelczer, J. Schwartz, Y.-L. Loo, Acid-catalyzed reactions activate DMSO as a reagent in perovskite precursor inks, *Chemistry of Materials* **2019**, *31*(6), 2114–2120.

66. H. Zhang, *et al.*, A universal co-solvent dilution strategy enables facile and cost-effective fabrication of perovskite photovoltaics, *Nature Communications* **2022**, *13*(1), 89.

67. A. Wakamiya, M. Endo, T. Sasamori, N. Tokitoh, Y. Ogomi, S. Hayase, Y. Murata, Reproducible fabrication of efficient perovskite-based solar cells: X-ray crystallographic studies on the formation of $CH_3NH_3PbI_3$ layers, *Chemistry Letters* **2014**, *43*(5), 711–713.

68. X. Guo, C. McCleese, C. Kolodziej, A. C. S. Samia, Y. Zhao, C. Burda, Identification and characterization of the intermediate phase in hybrid organic-inorganic $MAPbI_3$ perovskite, *Dalton Transactions* **2016**, *45*(9), 3806–3813.

69. K. G. Stamplecoskie, J. S. Manser, P. V. Kamat, Dual nature of the excited state in organic-inorganic lead halide perovskites, *Energy & Environmental Science* **2015**, *8*(1), 208–215.

70. H. J. Zhang, *et al.*, Synergistic effect of anions and cations in additives for highly efficient and stable perovskite solar cells, *Journal of Materials Chemistry A* **2018**, *6*(19), 9264–9270.

71. Y. J. Jeon, *et al.*, Planar heterojunction perovskite solar cells with superior reproducibility, *Scientific Reports* **2014**, *4*, 6953.

72. C. G. Wu, C. H. Chiang, Z. L. Tseng, M. K. Nazeeruddin, A. Hagfeldt, M. Gratzel, High efficiency stable inverted perovskite solar cells without current hysteresis, *Energy & Environmental Science* **2015**, *8*(9), 2725–2733.

73. D. Y. Liu, *et al.*, Aqueous-containing precursor solutions for efficient perovskite solar cells, *Advanced Science* **2018**, *5*(1), 1700484.

74. J. Qiu, Y. T. Zheng, Y. D. Xia, L. F. Chao, Y. H. Chen, W. Huang, Rapid crystallization for efficient 2D Ruddlesden-popper (2DRP) perovskite solar cells, *Advanced Functional Materials* **2019**, *29*(47), 1806831.

75. J. W. Lee, *et al.*, Tuning Molecular interactions for highly reproducible and efficient formamidinium perovskite solar cells via adduct approach, *Journal of the American Chemical Society* **2018,** *140*(20), 6317–6324.

76. N. Ahn, D. Y. Son, I. H. Jang, S. M. Kang, M. Choi, N. G. Park, Highly reproducible perovskite solar cells with average efficiency of 18.3% and best efficiency of 19.7% fabricated via Lewis base adduct of lead(II) iodide, *Journal of the American Chemical Society* **2015,** *137*(27), 8696–8699.

77. Y. Z. Wu, *et al.*, Retarding the crystallization of PbI_2 for highly reproducible planar-structured perovskite solar cells via sequential deposition, *Energy & Environmental Science* **2014,** *7*(9), 2934–2938.

78. Y. G. Rong, *et al.*, Solvent engineering towards controlled grain growth in perovskite planar heterojunction solar cells, *Nanoscale* **2015,** *7*(24), 10595–10599.

79. Y. Bai, *et al.*, A pure and stable intermediate phase is key to growing aligned and vertically monolithic perovskite crystals for efficient PIN planar perovskite solar cells with high processibility and stability, *Nano Energy* **2017,** *34*, 58–68.

80. J. Cao, *et al.*, Identifying the molecular structures of intermediates for optimizing the fabrication of high-quality perovskite films, *Journal of the American Chemical Society* **2016,** *138*(31), 9919–9926.

81. N. J. Jeon, J. H. Noh, Y. C. Kim, W. S. Yang, S. Ryu, S. I. Seok, Solvent engineering for high-performance inorganic-organic hybrid perovskite solar cells, *Nature Materials* **2014,** *13*(9), 897–903.

82. S. Paek, *et al.*, From nano- to micrometer scale: The role of antisolvent treatment on high performance perovskite solar cells, *Chemistry of Materials* **2017,** *29*(8), 3490–3498.

83. T. L. Bu, *et al.*, Synergic interface optimization with green solvent engineering in mixed perovskite solar cells, *Advanced Energy Materials* **2017,** *7*(20), 1700576.

84. F. Yang, G. Kapil, P. T. Zhang, Z. S. Hu, M. A. Kamarudin, T. L. Ma, S. Hayase, Dependence of acetate-based antisolvents for high humidity fabrication of $CH_3NH_3PbI_3$ perovskite devices in ambient atmosphere, *ACS Applied Materials & Interfaces* **2018,** *10*(19), 16482–16489.

85. F. Zhang, J. R. Lian, J. Song, Y. Y. Hao, P. J. Zeng, H. B. Niu, *sec*-Butyl alcohol assisted pinhole-free perovskite film growth for high-performance solar cells, *Journal of Materials Chemistry A* **2016,** *4*(9), 3438–3445.

86. N. Lin, J. Qiao, H. P. Dong, F. S. Ma, L. D. Wang, Morphology-controlled $CH_3NH_3PbI_3$ films by hexane-assisted one-step solution deposition for hybrid perovskite mesoscopic solar cells with high reproductivity, *Journal of Materials Chemistry A* **2015,** *3*(45), 22839–22845.

87. Y. F. Wang, *et al.*, Stitching triple cation perovskite by a mixed anti-solvent process for high performance perovskite solar cells, *Nano Energy* **2017,** *39*, 616–625.

88. A. Y. Alsalloum, *et al.*, Low-temperature crystallization enables 21.9% efficient single-crystal $MAPbI_3$ inverted perovskite solar cells, *ACS Energy Letters* **2020,** *5*(2), 657–662.

89. P. R. Cheng, *et al.*, Highly efficient Ruddlesden-popper halide perovskite $PA_2MA_4Pb_5I_{16}$ solar cells, *ACS Energy Letters* **2018,** *3*(8), 1975–1982.

90. X. Zhang, *et al.*, Phase transition control for high performance Ruddlesden-popper perovskite solar cells, *Advanced Materials* **2018,** *30*(21), e1707166.

91. T. Bu, *et al.*, Lead halide–templated crystallization of methylamine-free perovskite for efficient photovoltaic modules, *Science* **2021,** *372*(6548), 1327–1332.

92. Y. Jo, K. S. Oh, M. Kim, K.-H. Kim, H. Lee, C.-W. Lee, D. S. Kim, High performance of planar perovskite solar cells produced from PbI_2(DMSO) and PbI_2(NMP) complexes by intramolecular exchange, *Advanced Materials Interfaces* **2016,** *3*(10), 1500768.

93. Y. Zhang, *et al.*, PbI_2-HMPA complex pretreatment for highly reproducible and efficient $CH_3NH_3PbI_3$ perovskite solar cells, *Journal of the American Chemical Society* **2016,** *138*(43), 14380–14387.

94. S. Oz, *et al.*, Protic ionic liquid assisted solution processing of lead halide perovskites with water, alcohols and acetonitrile, *Nano Energy* **2018**, *51*, 632–638.

95. Y. Miao, *et al.*, Green solvent enabled scalable processing of perovskite solar cells with high efficiency, *Nature Sustainability* **2023**, https://doi.org/10.1038/s41893-023-01196-4.

96. C. Worsley, D. Raptis, S. Meroni, A. Doolin, R. Garcia-Rodriguez, M. Davies, T. Watson, γ-Valerolactone: A nontoxic green solvent for highly stable printed mesoporous perovskite solar cells, *Energy Technology* **2021**, *9*(7), 2100312.

97. W. Q. Wu, D. H. Chen, W. A. McMaster, Y. B. Cheng, R. A. Caruso, Solvent-mediated intragranular-coarsening of $CH_3NH_3PbI_3$ thin films toward high-performance perovskite photovoltaics, *ACS Applied Materials & Interfaces* **2017**, *9*(37), 31959–31967.

98. H.-S. Yun, *et al.*, Ethanol-based green-solution processing of α-formamidinium lead triiodide perovskite layers, *Nature Energy* **2022**, *7*(9), 828–834.

99. J. N. Chen, *et al.*, A Ternary solvent method for large-sized two-dimensional perovskites, *Angewandte Chemie-International Edition* **2017**, *56*(9), 2390–2394.

100. D. N. Jeong, *et al.*, Perovskite cluster-containing solution for scalable D-bar coating toward high-throughput perovskite solar cells, *ACS Energy Letters* **2019**, *4*(5), 1189–1195.

101. N. K. Noel, *et al.*, Highly crystalline methylammonium lead tribromide perovskite films for efficient photovoltaic devices, *ACS Energy Letters* **2018**, *3*(6), 1233–1240.

102. K. Wang, *et al.*, A nonionic and low-entropic $MA(MMA)_nPbI_3$-ink for fast crystallization of perovskite thin films, *Joule* **2020**, *4*(3), 615–630.

103. C. C. Wu, *et al.*, Cost-effective sustainable-engineering of $CH_3NH_3PbI_3$ perovskite solar cells through slicing and restacking of 2D layers, *Nano Energy* **2017**, *36*, 295–302.

104. L. G. Wang, *et al.*, A Eu^{3+}-Eu^{2+} ion redox shuttle imparts operational durability to Pb-I perovskite solar cells, *Science* **2019**, *363*(6424), 265–270.

105. Y. Deng, C. H. Van Brackle, X. Dai, J. Zhao, B. Chen, J. Huang, Tailoring solvent coordination for high-speed, room-temperature blading of perovskite photovoltaic films, *Science Advances* **2019**, *5*(12), eaax7537.

106. L. Li, Y. H. Chen, Z. H. Liu, Q. Chen, X. D. Wang, H. P. Zhou, The additive coordination effect on hybrids perovskite crystallization and high-performance solar cell, *Advanced Materials* **2016**, *28*(44), 9862–9868.

107. Y. Deng, C. H. Van Brackle, X. Dai, J. Zhao, B. Chen, J. Huang, Tailoring solvent coordination for high-speed, room-temperature blading of perovskite photovoltaic films, *Science Advances* **2019**, *5*(12), eaax7537.

108. M. D. Xiao, *et al.*, A fast deposition-crystallization procedure for highly efficient lead iodide perovskite thin-film solar cells, *Angewandte Chemie-International Edition* **2014**, *53*(37), 9898–9903.

109. B. E. Cohen, S. Aharon, A. Dymshits, L. Etgar, Impact of antisolvent treatment on carrier density in efficient hole-conductor-free perovskite-based solar cells, *Journal of Physical Chemistry C* **2016**, *120*(1), 142–147.

110. N. Sakai, *et al.*, The mechanism of toluene-assisted crystallization of organic-inorganic perovskites for highly efficient solar cells, *Journal of Materials Chemistry A* **2016**, *4*(12), 4464–4471.

111. K.-F. Lin, *et al.*, Unraveling the high performance of tri-iodide perovskite absorber based photovoltaics with a non-polar solvent washing treatment, *Solar Energy Materials and Solar Cells* **2015**, *141*, 309–314.

112. C. C. Chen, *et al.*, Interplay between nucleation and crystal growth during the formation of $CH_3NH_3PbI_3$ thin films and their application in solar cells, *Solar Energy Materials and Solar Cells* **2017**, *159*, 583–589.

113. S. H. Chang, W. C. Huang, C. C. Chen, S. H. Chen, C. G. Wu, Effects of anti-solvent (iodobenzene) volume on the formation of $CH_3NH_3PbI_3$ thin films and their application in photovoltaic cells, *Applied Surface Science* **2018**, *445*, 24–29.

114. M. Yavari, *et al.*, Nonhalogenated solvent systems for highly efficient perovskite solar cells, *Advanced Energy Materials* **2018,** *8*(21), 1800177.

115. M. Zhang, *et al.*, Green anti-solvent processed planar perovskite solar cells with efficiency beyond 19%. *Solar RRL* **2018,** *2*(2), 1700213.

116. A. Pellaroque, *et al.,* Efficient and stable perovskite solar cells using molybdenum tris(dithiolene)s as p-dopants for spiro-OMeTAD, *ACS Energy Letters* **2017,** *2*(9), 2044–2050.

117. A. D. Taylor, *et al.*, A general approach to high-efficiency perovskite solar cells by any antisolvent, *Nature Communications* **2021,** *12*(1), 1878.

118. L. G. Li, *et al.*, High efficiency planar Sn-Pb binary perovskite solar cells: Controlled growth of large grains via a one-step solution fabrication process, *Journal of Materials Chemistry C* **2017,** *5*(9), 2360–2367.

119. J. Troughton, K. Hooper, T. M. Watson, Humidity resistant fabrication of $CH_3NH_3PbI_3$ perovskite solar cells and modules, *Nano Energy* **2017,** *39*, 60–68.

120. M. S. Yin, *et al.*, Annealing-free perovskite films by instant crystallization for efficient solar cells, *Journal of Materials Chemistry A* **2016,** *4*(22), 8548–8553.

121. Y. Yu, S. W. Yang, L. Lei, Q. P. Cao, J. Shao, S. Zhang, Y. Liu, Ultrasmooth perovskite film via mixed anti-solvent strategy with improved efficiency, *ACS Applied Materials & Interfaces* **2017,** *9*(4), 3667–3676.

122. D. Li, *et al.*, In situ interface engineering for highly efficient electron-transport-layer-free perovskite solar cells, *Nano Letters* **2020,** *20*(8), 5799–5806.

123. W. Wu, *et al.*, *In situ* nanocrystal seeding perovskite crystallization toward high-performance solar cells, *Materials Today Energy* **2021,** *22*, 100855.

124. Y. Wang, *et al.*, Efficient and stable perovskite solar cells by fluorinated ionic liquid–induced component interaction, *Solar RRL* **2020,** *5*(1), 200058.

125. W. Hui, *et al.*, Stabilizing black-phase formamidinium perovskite formation at room temperature and high humidity, *Science* **2021,** *371*(6536), 1359–1364.

126. H. R. Chen, *et al.,* Critical role of chloride in organic ammonium spacer on the performance of low-dimensional Ruddlesden-popper perovskite solar cells, *Nano Energy* **2019,** *56*, 373–381.

127. H. Ren, *et al.*, Efficient and stable Ruddlesden-popper perovskite solar cell with tailored interlayer molecular interaction, *Nature Photonics* **2020,** *14*(3), 154–163.

128. Y. T. Zheng, *et al.*, Oriented and uniform distribution of Dion-Jacobson phase perovskites controlled by quantum well barrier thickness, *Solar RRL* **2019,** *3*(9), 1900090.

129. T. T. Niu, *et al.*, Reduced-dimensional perovskite enabled by organic diamine for efficient photovoltaics, *Journal of Physical Chemistry Letters* **2019,** *10*(10), 2349–2356.

130. S. Öz, *et al.*, Protic ionic liquid assisted solution processing of lead halide perovskites with water, alcohols and acetonitrile, *Nano Energy* **2018,** *51*, 632–638.

6 Additive Engineering in Perovskite Solar Cells

Baomin Xu

6.1 INTRODUCTION

Perovskite solar cells have garnered significant attention in recent years due to their impressive power conversion efficiency (PCE) of 26.1%,[1] making them a key player in the field of photovoltaics. This chapter explores the advancements in additive engineering and its crucial role in optimizing the performance of perovskite solar cells. Despite recent advancements in perovskite solar cells technology, operational stability remains a challenge. The International Electrotechnical Commission (IEC) protocol[2] evaluates the operational stability of solar modules under harsh environmental conditions, and perovskite solar cells still lag behind commercial silicon solar cells in terms of stability.

Defects in perovskite films have a significant impact on charge carrier dynamics, leading to diminished photovoltaic performance. While halide perovskites exhibit a defect-tolerant property, controlling defects is essential for minimizing recombination and improving stability. Several factors affect the stability of perovskite solar cells, including heat, which can deform the crystal structure and dissipate organic substances. The tendency of iodide to oxidize under UV-visible light and sublimate can damage the perovskite crystal structure. Excess photogenerated charge carriers at the surface and grain boundaries can accelerate perovskite film degradation.[3,4]

This chapter emphasizes the role of additive engineering in minimizing defects, enhancing the crystallization kinetics, and stabilizing the phase. Understanding the role of additives in achieving defect-free, high-quality perovskite films is crucial for advancing the field of perovskite solar cells. This chapter provides valuable insights into the design and impact of additives on perovskite film quality and photovoltaic performance.

6.1.1 DEFECT CHEMISTRY IN PEROVSKITE

In the context of solar cell technology, a significant concern is the loss of charge carriers due to trap-mediated non-radiative recombination. This process, illustrated in Figure 6.1a, has a profound impact on the overall energy conversion efficiency.[5] It is particularly problematic when trap states are positioned near the middle of the bandgap, where the energy barrier (ΔE) is much larger than the thermal energy ($k_B T$). Under these conditions, deeply trapped electrons and holes face considerable barriers to thermal excitation back to the conduction band, rendering such transitions highly

DOI: 10.1201/9781003400486-6

improbable. Consequently, the likelihood of non-radiative recombination between these trapped charge carriers significantly increases.

Non-radiative recombination is a highly undesirable phenomenon in solar cells as it imposes limitations on the carrier diffusion length (LD).[5] To ensure the efficient operation of a solar cell device, LD should exceed the thickness of the perovskite film to allow for effective charge extraction. The square of the diffusion length is directly related to the product of charge carrier mobility (μ) and their lifetime (τ). For perovskite material, such as $MAPbI_3$, the diffusion length typically ranges from hundreds of nanometers to microns,[6] which is crucial for enabling the construction of efficient solar cells. The detrimental impact on charge extraction primarily arises from deep traps that facilitate non-radiative recombination. Shallow traps, on the

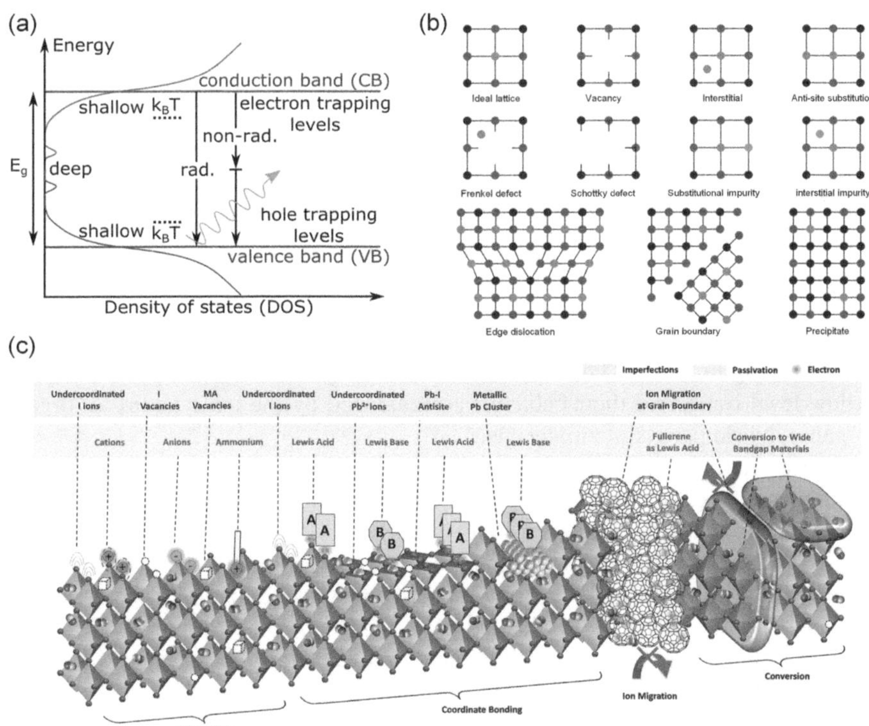

FIGURE 6.1 (a) Schematic representation of state density in a disordered semiconductor, whereby both band-to-band radiative (rad.) and non-band-to-band non-radiative (non-rad.) recombination (vertical arrows) can occur. Reproduced with permission.[5] Copyright 2020, The Royal Society of Chemistry. (b) Defect chemistry in lead halide perovskite. Schematic overview of defect types in lead halide perovskite, relative to an ideal lattice. Reproduced with permission.[7] Copyright 2016, Nature Publishing Group. (c) Imperfections in perovskite film and their passivation by ionic bonding, coordinate bonding, and conversion to wide bandgap materials, and suppression of ion migration at extended defects. Reproduced with permission.[21] Copyright 2019, The Royal Society of Chemistry.

other hand, primarily affect mobility and delay the extraction process, which can suppress the maximum achievable current density in the device.

The $MAPbI_3$ perovskite structure exhibits twelve possible native point defects, including vacancies (V_{MA}, V_{Pb}, V_I), interstitials (MA_i, Pb_i, I_i), and anti-site occupations (MA_{Pb}, MA_I, Pb_{MA}, Pb_I, I_{MA}, I_{Pb}; Figure 6.1b).[7–14] These defects have been extensively studied, and a common understanding is that high-energy defects contribute to deep levels within the bandgap, while low-energy defects create shallow defect states. Deep-level point defects generally do not result in a high density of non-radiative recombination centers, but there is some variation in opinions regarding which defects specifically create deep levels, including I_{Pb}, I_{MA}, Pb_i, Pb_I, V_I, and Pb_{MA}.[9,10,15] Defects like I_{MA}, Pb_I, and V_I, with low formation energies, can significantly contribute to a density of recombination centers.[10,13,14] Under iodine-rich conditions, the PbI anti-site defect has a low formation energy and exhibits stable charge states, making it a potential source of deep-level defects.[10]

Shallow point defects in $MAPbI_3$ include acceptors (V_{Pb}, V_{MA}, MA_{Pb}, I_i) under certain growth conditions with low formation energies, as well as donors (MA_i, V_I, MA_I) of shallow point defects.[9] Donors and acceptors of shallow point defects can influence the material's doping characteristics, potentially converting $MAPbI_3$ from p-type to n-type through intrinsic doping.[9,13] Under stoichiometric growth conditions, cation and anion vacancies may dominate defect formation, and low formation energies can minimize unintentional doping by compensating the vacancies with opposite charge carriers.[16–20]

The provided figure discusses the imperfections and defects present in perovskite crystals, particularly focusing on their impact on the performance of perovskite solar cells (Figure 6.1c).[21] These defects can be categorized into deep-level traps and shallow-level traps, and their behavior is influenced by the ionic nature of organic-inorganic hybrid perovskite materials.

Deep-level traps in perovskite crystals are attributed to various imperfections, including undercoordinated halide ions, undercoordinated Pb^{2+} ions, lead clusters, and intrinsic point defects like Pb-I antisite defects (PbI_3^-). These defects can result from growth or processing conditions. To passivate these defects, unique methods like coordinate bonding and ionic bonding are employed to neutralize and deactivate them.[15,20]

Shallow-level traps, on the other hand, are generally associated with easily formed point defects. While they are not detrimental to the photovoltaic performance of perovskite materials, the migration of these charged point defects under an electric field can lead to unintended doping effects, which can impact device performance.[9,10,22] Their migration, especially through grain boundaries, is a significant concern, necessitating extended defect passivation strategies.

The type and concentration of defects in perovskite solar cells are sensitive to fabrication processes and perovskite compositions.[17,23–28] Modifying precursors, thermal annealing conditions, and incorporating isovalent ions are some of the techniques used to control defect formation. Various passivation methods and agents have been proposed to effectively address these defects.

The techniques for measuring the density of trap states, although pinpointing the type of defects and correlating them with trap depth remains a challenge.[29–31]

Surface-sensitive techniques like scanning tunneling microscopy have been attempted but are complicated in polycrystalline perovskite films. However, the charge of defects can be probed by introducing molecules capable of interacting with these defects – the complex nature of defects in perovskite crystals, their impact on solar cell performance, and the unique passivation methods required due to the ionic nature of these materials. It worth noting that the challenges associated with defect characterization provide insights into strategies for defect control and passivation.[21]

6.1.2 Nucleation and Growth of Perovskite Thin Films

The crystalline quality of perovskite thin films is closely related to the ultimate film defects. Currently, there are several methods commonly employed for the fabrication of perovskite thin films, including vacuum deposition, one-step spin-coating, and two-step spin-coating.[32–34] Among these, the solution-based method is a cost-effective and widely utilized approach. The crystallization of perovskite films during solution processes follows the La Mer mechanism, involving three stages: nucleation, growth, and saturation (Figure 6.2). Nucleation begins when the solution reaches supersaturation, and growth continues as solute is supplied through diffusion. Nucleation and

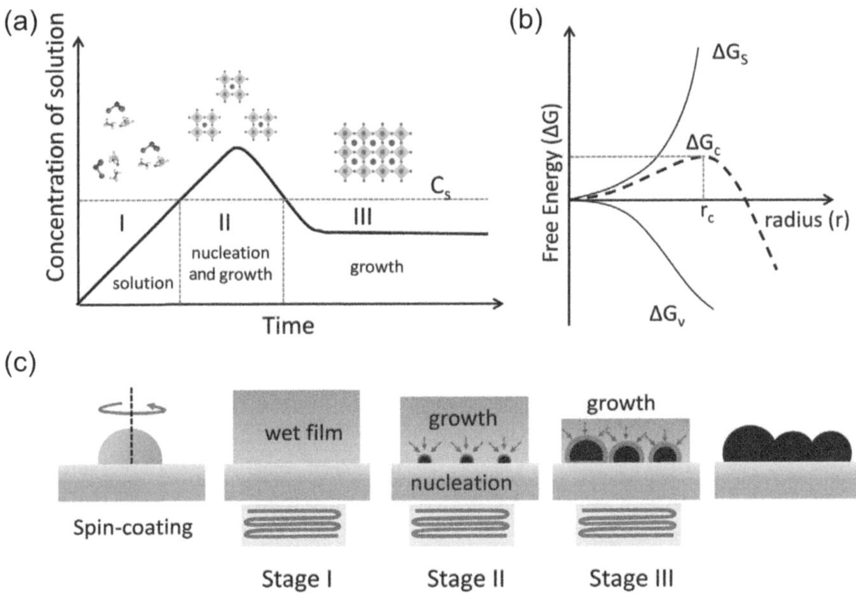

FIGURE 6.2 (a) Schematic representation of the La Mer Model for the nucleation and growth of perovskite thin films. The symbol C_s denotes the supersaturation concentration of a precursor solution. (b) Free energy diagram depicting the nucleation process. ΔG_s represents the surface free energy, ΔG_v corresponds to the bulk free energy, ΔG represents the total free energy, ΔG_c signifies the critical free energy, and r_c denotes the critical radius of the nucleus. (c) Conceptual visualization of the nucleation and growth stages of perovskite films. Reproduced with permission.[38] Copyright 2019, Wiley-VCH.

growth can be controlled to modulate perovskite film crystal growth. Homogeneous nucleation is described by a critical free energy parameter, ΔGc, and an Arrhenius-type equation, where ΔGc depends on surface energy, molar volume, and supersaturation.[35,36] Heterogeneous nucleation can occur on substrates, affected by the contact angle (θ) and other factors. Supersaturation is a key factor influencing nucleation rates and perovskite grain size.

Controlling supersaturation through solution concentration and temperature adjustments significantly impacts nucleation rates and crystal growth.[35,36] Different methods, such as solvent dripping and intermediate phase approaches, can be employed to enhance film uniformity and achieve desirable grain sizes.[37] Understanding nucleation and crystal growth mechanisms is essential to reduce crystallographic and structural defects and improve the crystallinity of halide perovskite materials.

Halide perovskite materials have lower crystallinity and higher defect densities compared to conventional semiconductors. Despite this, they exhibit remarkable charge carrier properties. The nucleation and growth mechanisms in crystallization play a crucial role in controlling the defects and improving crystallinity. Homogeneous and heterogeneous nucleation processes are key factors, influenced by free energy changes, temperature, and preferential sites. These processes can be manipulated to achieve superior optoelectronic properties in perovskite films.[38,39]

6.1.3 Phase Transition Stabilizing the Perovskite Phase

The structural (phase) stability of ABX_3, a generic perovskite composition, is primarily governed by the volumetric ratio between the BX_6 octahedra and the A cation. The formation of the perovskite structure can be ascertained by the Goldschmidt tolerance factor (t), a pivotal parameter described by the equation[40]:

$$t = \frac{R_A + R_X}{\sqrt{2\left(R_B + R_X\right)}}$$

In this equation, R_A, R_B, and R_X represent the ionic radii of A, B, and X site ions, respectively. It is imperative for perovskite stability that the tolerance factor falls within the range of $0.8 < t < 1$. Generally, materials with a tolerance factor in the range of 0.9–1.0 adopt a cubic structure, while a tolerance factor of 0.71–0.9 results in a distorted perovskite structure characterized by tilted octahedra.[41]

Iodide perovskite materials exhibit various optical active polycrystalline phases, including α-cubic, β-tetragonal, and γ-orthorhombic black phases (Figure 6.3).[42] However, at room temperature, the δ-yellow phase exhibits the lowest formation free energy, particularly in the cases of $CsPbI_3$ and $FAPbI_3$.[43,44] With increasing temperature, the non-luminescent δ-phase directly transforms into the α-phase, resulting in an expansion of the unit cell volume. This transition is attributed to enhanced dynamic motion of $[PbI_6]^{4-}$ octahedra, promoting their connectivity.[45] The energy required for the induced angular sharing octahedral geometry is influenced by the Goldschmidt tolerance factor (t)[46] and sphericity factor.

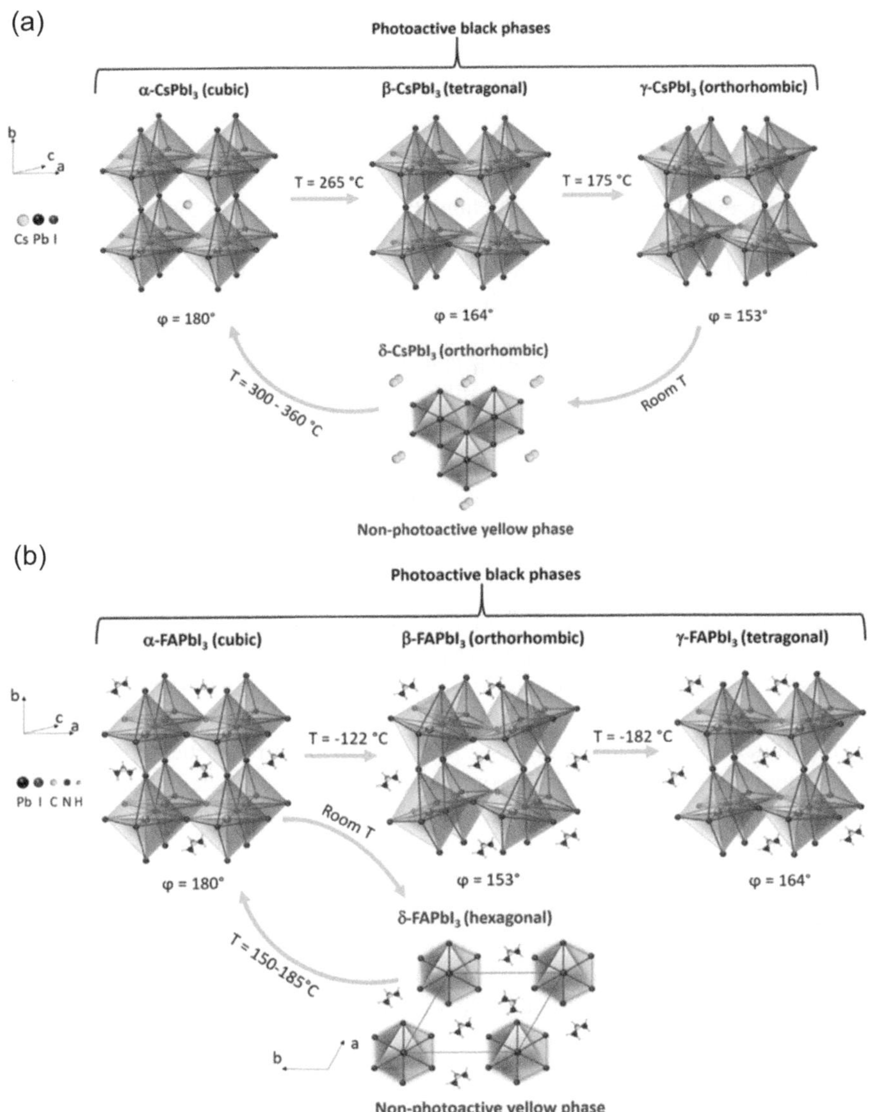

FIGURE 6.3 Crystalline structure and polymorphic phase transitions in (a) $CsPbI_3$ and (b) $FAPbI_3$ perovskites. The nonperovskite δ-yellow phase in both iodide perovskites undergoes a transition to the photoactive black perovskite α-phase at elevated temperatures. Reproduced with permission.[42] Copyright 2020, American Chemical Society.

Specifically, smaller cations like Cs^+ ($t < 0.8$) demand more energy to maintain a cubic structure, while larger cations like FA^+ ($t > 1$) require less energy to preserve the cubic structure. Therefore, α-$CsPbI_3$ forms at higher temperatures ($T = 300$–$360°C$),[47] whereas α-$FAPbI_3$ forms at lower temperatures ($T = 150$–$185°C$).[48] MA^+ cations ($0.8 < t < 1$) also favor the formation of the optical black phase at approximately $100°C$.

It's worth noting that in the case of MAPbI$_3$, even without exposure to air, its thermal stability is relatively low when temperatures exceed 85°C. This is due to the sublimation of MA$^+$ cations, rendering MAPbI$_3$ less ideal for long-term stability.[49]

Despite the ability of calcium titanium perovskites to maintain their black phase post-synthesis, during cooling and storage, the formation of metastable phases is inevitable. Humidity accelerates the transition to the yellow phase,[50] and grain size further influences this process.[51] Mechanistically, this involves the hydrolysis of perovskites and weak interactions between cations and iodide ions within the inorganic octahedra. When exposed to water, the methylammonium cations decompose into ammonia and triazine, whereas the inorganic cesium remains more stable. In the latter case, instability primarily arises from its relatively low tolerance factor (< 0.8). To mitigate this issue, lattice contraction is employed to enhance interactions between organic and inorganic components, reducing dynamic motion at lower temperatures. Consequently, the highly symmetric lattice of black-phase calcium titanium perovskites distorts by decreasing the Pb-I-Pb tilting angle (ϕ), transitioning from the α-phase to the β and γ phases.[44] So stabilizing the photovoltaic phase is ultra important for the high-efficiency perovskite solar cells.

In this chapter, this discourse delves into recent breakthroughs within the realm of additive engineering in the context of solution processing, with the overarching objective of enhancing the quality of perovskite thin films and the characteristics of associated devices. The incorporation of additives into perovskite thin films was through direct addition of the additive to the one-step precursor solution. These methods exhibit several functionality by not only passivating defects but also exerting an influence on the crystallization process and stabilizing the perovskite phase. According to the specific additive composition to the perovskite structure, the additives can be divided into a few common categories: A-site additives, B-site additives, X-site additives, and solvent additives. These additive engineering could be related to the perovskite defects, nucleation and growth as well as the perovskite phase stabilization. After discussing that, we then discuss other specific function additives, such as the surfactant properties, durable properties. This discussion shall culminate in a forward-looking perspective, offering insights into emerging research trends and future prospects in the context of additive engineering for the advancement of perovskite solar cells technology.

6.2 PROGRESS ON ADDITIVE ENGINEERING

6.2.1 A-Site Additive

FAPbI$_3$, MAPbI$_3$, and CsPbI$_3$ are commonly employed as light absorbers. Managing the A-site organic cation represents an effective means of controlling phase and crystal structure, without significantly affecting band properties and band-edge carriers.[52,53] The tolerance factor is determined as 0.99 for FAPbI$_3$, 0.91 for MAPbI$_3$, and 0.84 for CsPbI$_3$, based on cationic radii of 253 pm for FA$^+$ (HC(NH$_2$)$_2^+$), 217 pm for MA$^+$ (CH$_3$NH$_3^+$), and 167 pm for Cs$^+$.[54,55] Consequently, the estimated bandgaps for FAPbI$_3$, MAPbI$_3$, and CsPbI$_3$ are 1.48, 1.55, and 1.73 eV, respectively.[56] This distinct A-site cation effect on phase transition appears to be influenced by thermally active reorientation dynamics of the corresponding A-site cations. The interaction of organic cations with the inorganic lattice is expected to play a significant role in managing crystal structure, phase stability, and purity, thus contributing to defect suppression and improved photovoltaic performance and stability in perovskite solar cells.

In the engineering of perovskite additives, one noteworthy aspect of these emerging developments lies in the strategic utilization of organic additives and inorganic A-site additives. Typically, these additives are characterized by their distinct ion sizes, often referred to as oversized ions ($t > 1$) and undersized ions ($t < 0.8$; Figure 6.4).[57] It is important to note that these ions may not strictly conform to the inherent structural constraints within the bulk crystal lattice. Importantly, these ions may find their accommodating positions at A-sites located on the material's surface or grain boundaries, which are no longer subjected to the geometric constraints of the bulk lattice. In isolated cases, oversize organic cations often play a crucial role in the formation of low-dimensional phases, such as 2D A_2PbI_4 perovskites.[58] In this section, we focus on the practicality of these ions and their functional role in modulating properties within the context of 3D perovskite materials.[57]

6.2.1.1 Organic Additive

Recent years have seen $FAPbI_3$ widely utilized in perovskite solar cells, achieving high efficiency exceeding 20% by incorporating additives into the perovskite precursor solution. For instance, MACl is a common additive for high-efficiency $FAPbI_3$-based perovskite solar cells, and it induces the phase transition from the hexagonal phase to the cubic phase at around 150°C. The addition of MACl stabilizes the cubic phase at a lower temperature when introducing a smaller cation like MA into the FA site. The high efficiency and purity of $FAPbI_3$ perovskites have been demonstrated by including 40 mol.% MACl in the precursor solution. The MA^+ cation and the Cl^- anion stabilize the α-phase of $FAPbI_3$, leading to prolonged charge carrier lifetime and enhanced stability. DFT calculations confirm the role of MACl in stabilizing an intermediate phase during annealing, facilitating the highly pure a-phase $FAPbI_3$

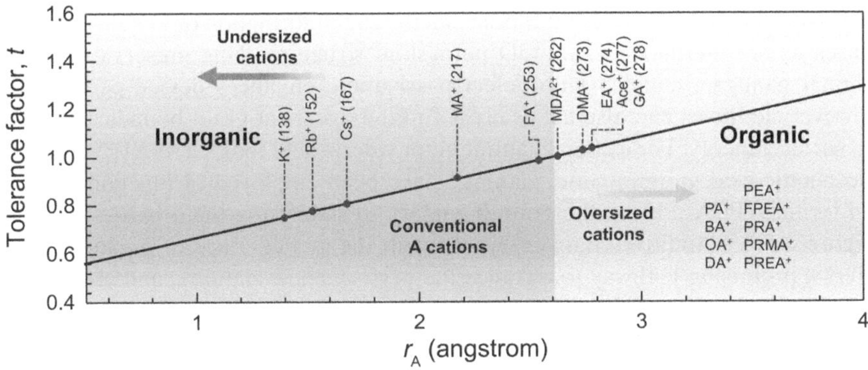

FIGURE 6.4 A-cation selection based on their cationic radius and correlated tolerance factor within the $APbI_3$ lattice framework, with the values in parentheses denoting the cation radius. Reproduced with permission.[57] Copyright 2022, Science Publishing Group.

(Figure 6.5a).[59] The majority of certified perovskite solar cells with over 25% efficiency are based on MACl additive engineering.

Apart from cations smaller than FA, larger cations have also been employed as additives. These large cations can be geometrically spherical or linear in structure. Cations with spherical geometry include guanidinium (GA), imidazolium (IA), and dimethylammonium (DMA), while alkylammonium derivatives are linear in geometry. GA-based additives with an ionic radius of 278 pm have been used to enhance charge carrier lifetime and suppress ion migration in $MAPbI_3$, leading to increased efficiency.[60] These larger cations contribute to the stable perovskite phase by improving the tolerance factor. Large cations have been found to reduce defects and grain boundaries, further enhancing charge carrier lifetime and transport dynamics.

It was reported that the incorporation of n-propylammonium chloride (PACl) as an additive in $FAPbI_3$ exhibited superior performance and enhanced stability in comparison to the traditional MACl,[61] where various cationic chlorides including MA (methylammonium), BA (butylammonium), n-amylammonium (AA), and cyclopropylammonium (CPA) were also assessed. The introduction of 20 mol. % PACl led to a favored orientation of the perovskite, as confirmed by grazing-incidence wide-angle x-ray scattering (GIWAXS) measurements (Figure 6.5b). In contrast to MACl, PACl demonstrated improved charge carrier lifetime and reduced Urbach energy, consequently resulting in enhanced photovoltaic performance.

Additionally, large cations with long alkyl chains can be used for additive engineering to induce low-dimensional perovskite structures. As mono-ammonium salts, phenethylammonium (PEA), 4-fluoro-phenethylammonium (FPEA), butylammonium (BA), and octylammonium (OA) cations have been suggested.[58,62,63] To address stability issues, a careful design for controlling the thickness of the 2D layers is required. Quasi-2D approaches have been employed to achieve the desired control, and different 2D structures can be formed based on the choice of additives. The introduction of these additives helps to enhance charge carrier lifetime, suppress recombination, and trap charge carriers at grain boundaries. Furthermore, by introducing pseudo-2D structures, it is possible to improve crystallinity and stability and enhance resistance to humidity.

These additives, often bulky cations, facilitate the formation of 2D/3D perovskite phases as an alternative to the 3D perovskite structure. This preservation of the original bandgap, coupled with defect passivation, enhances device performance. However, additives can also act as cross-linking agents at grain boundaries, reducing surface energy. For example, ammonium valeric acid iodide (AVAI)[64] and alkyl-phosphonic acid ω-ammonium cations[65] have been employed in this context. PEA can formed $(PEA)_2PbI_4$ in the grain boundary to stabilizing the α perovskite phase (Figure 6.5c).[58] Additive engineering, through the strategic use of organic cations, offers a promising pathway to enhance the performance, stability, and efficiency of perovskite solar cells, ultimately advancing the field of photovoltaics.

For the $CsPbI_3$, DMA cation was found to stabilize β-$CsPbI_3$ at relative lower temperature, where dimethylammonium iodide (DMAI) was suggested as one of the best additives for $CsPbI_3$.[66] The crystallization of $CsPbI_3$ perovskite films can be controlled by forming stable intermediate phase with the DMA cation (Figure 6.5d), where the volatile nature of DMA cation contributes to form stable intermediate phase of $CsPbI_3$·DMAI.

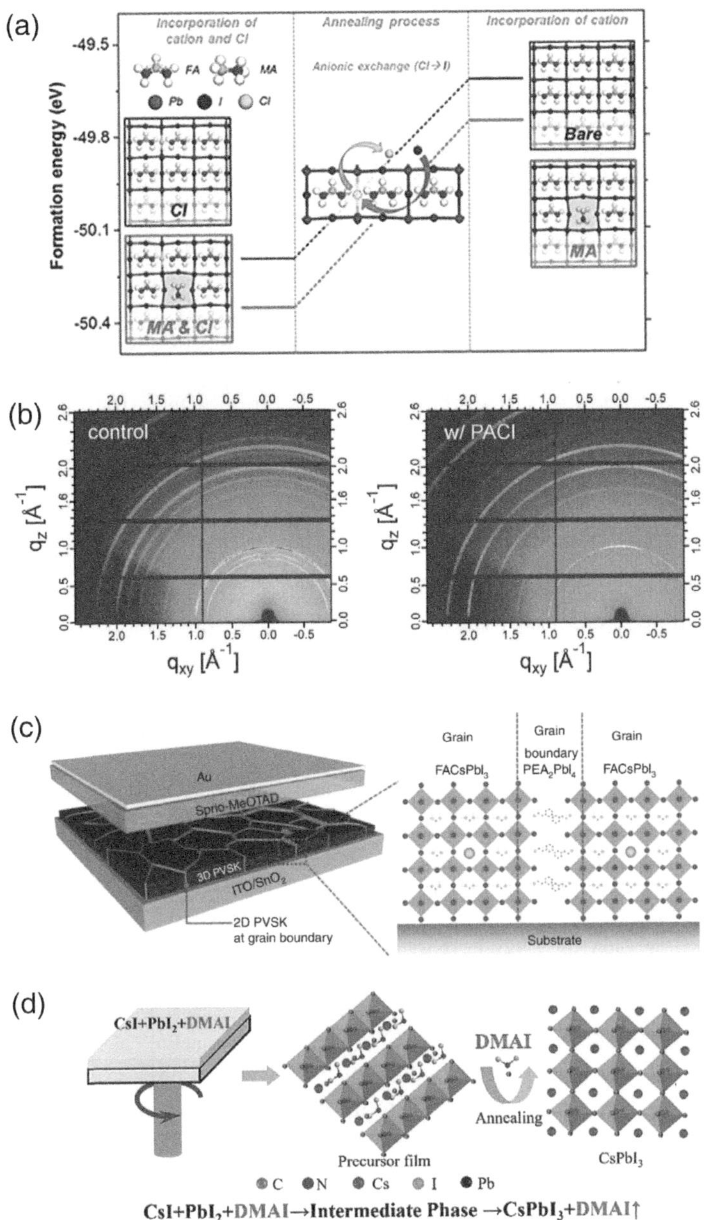

FIGURE 6.5 (a) Theoretical calculation of the α-phase FAPbI$_3$ perovskite structure and that prepared with Cl, MA, or MACl. Formation with bare α-phase FAPbI$_3$ perovskite structure and that with Cl, MA, or MACl. Reproduced with permission.[59] Copyright 2019, Elsevier. (b) GIWAXS pattern of perovskite film with or without 20 mol% PACl. Reproduced with permission.[61] Copyright 2021, Wiley-VCH. (c) The device incorporating polycrystalline 3D perovskite film with 2D perovskite at grain boundaries. Reproduced with permission.[58] Copyright 2018, Nature Publishing Group. (d) Schematic mechanism for DMAI additive induced black phase CsPbI$_3$ formation. Reproduced with permission.[66] Copyright 2019, Wiley-VCH.

6.2.1.2 Inorganic Additive

Except for the organic additive in A site, inorganic cations play a significant role in the properties and stability of perovskite films. They can be involved in direct incorporation into the lattice or doping, which can manipulate the electrical, optical, and structural properties of semiconductor materials. Monovalent alkali metal ions such as cesium (Cs), rubidium (Rb), and potassium (K) cations have been used as additives in perovskite solar cells. Cs cations are known to stabilize the crystal structure and enhance phase purity by incorporating into the perovskite lattice (Figure 6.6a).[67] Rb cations have also been found to improve performance and stability by controlling the tolerance factor and reducing charge trap states. The presence of these inorganic cations in the perovskite film can effectively inhibit ion migration, leading to improved device performance and stability.[68]

The concentration of CsI in the precursor solution affects the performance and stability of perovskite solar cells, with 3 mol. % of Cs^+ leading to a certified PCE of 24.4% and improved thermal stability.[69] The introduction of Cs^+ into perovskite precursor solution helps control the stabilization of the cubic phase of $FAPbI_3$ and improves thermal stability (Figure 6.6c).[70] However, phase segregation towards Cs-rich perovskite under operational conditions needs to be addressed.

These highly positively charged cations (especially K^+, Rb^+) exhibit a pronounced capability to associate with excess and under-coordinated halides, thereby immobilizing them and reducing their undesirable migration (Figure 6.6b).[71] The effectiveness of alkali metal cation doping in enhancing the formation energy of mobile halide interstitial defects is further underscored by theoretical simulations.[72] Additionally, when considering their incorporation at the A-site, this process leads to lattice contraction, resulting in a decreased ion interaction distance, thereby enhancing the structural integrity of the perovskite lattice.[73,74]

However, there remains contention regarding the specific positions of these smaller cations, particularly in the case of Rb^+. Experimental observations provide evidence in support of their incorporation into the lattice A-sites.[73,74] This explanation is somewhat perplexing when considering the ionic radii of these cations, as theoretical assessments suggest that they appear too small to occupy the A-site in the $APbI_3$ lattice. Theoretical calculations provide an alternative perspective, suggesting that occupation of interstitial sites[75–77] is energetically favorable. Nevertheless, results from solid-state nuclear magnetic resonance (NMR) studies indicate the contrary, suggesting that Rb^+ and K^+ are even immiscible in the interstitial sites and instead segregate into minor optically inactive phases (Figure 6.6d).[68,78] This discrepancy may partly arise from differences in solubility limits in different OHP compositions. For instance, the introduction of Br into the chemical stoichiometry of $APbI_{3-x}Br_x$ mixed-halide compounds can facilitate the incorporation of K^+.[79] K additive protects against ion migration and enhances stability and external photoluminescence quantum efficiency (PLQE) in mixed-cation and mixed-halide perovskites. KI additive significantly suppresses phase segregation and enhances both internal and external PLQE in perovskite films. Incorporation of K^+ ion inhibits Frenkel defects and reduces J-V hysteresis in various perovskite compositions (Figure 6.6e; Table 6.1).[75,80]

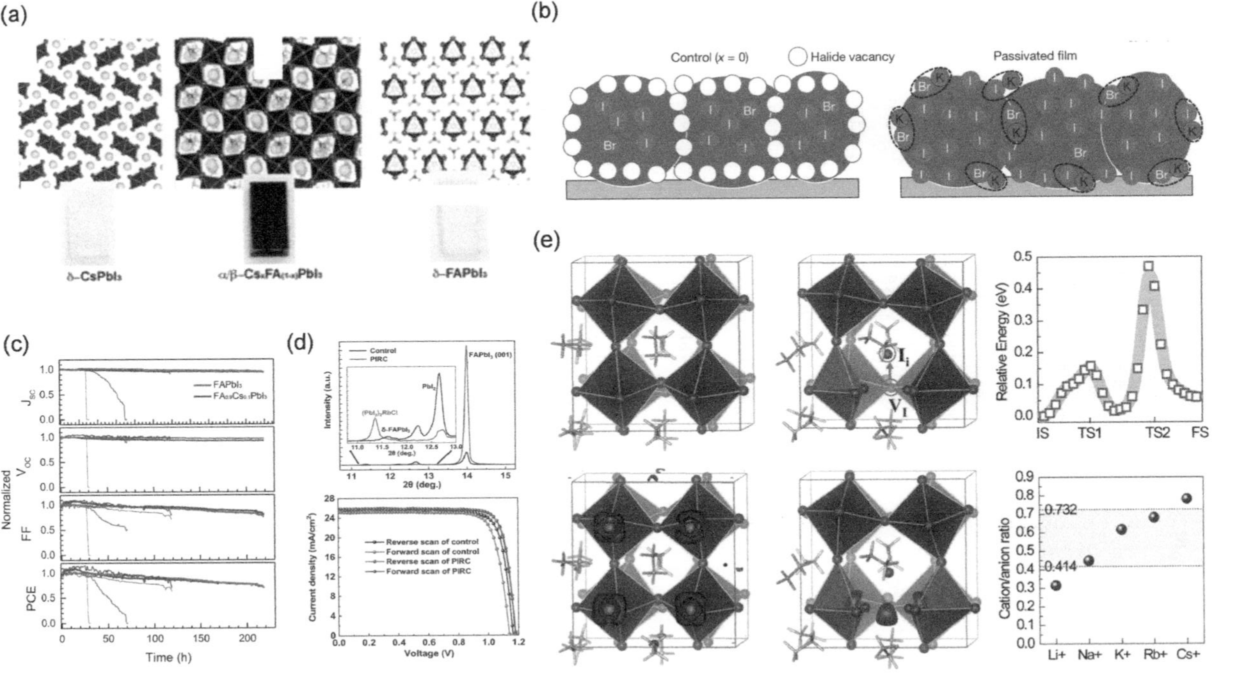

FIGURE 6.6 (a) Lattice structure of δ phase in the FAPbI$_3$ (left top) and CsPbI$_3$ perovskites (right top), the α (center top) FAPbI$_3$. Reproduced with permission.[67] Copyright 2016, The Royal Society of Chemistry. (b) Schematic of a cross-section of a film showing halide-vacancy management in cases of excess halide, in which the surplus halide is immobilized through complexing with potassium into benign compounds at the grain boundaries and surfaces. Reproduced with permission.[71] Copyright 2018, Nature Publishing Group. (c) Normalized photovoltaic parameters stability using 10% Cs doped and pure FAPbI$_3$ perovskite solar cells. Reproduced with permission.[70] Copyright 2015, Wiley-VCH. (d) XRD of perovskite without RbCl and with 5% RbCl (top), I-V curves of the devices with or without RbCl. Reproduced with permission.[68] Copyright 2022, Science Publishing Group. (e) Simulated MAPbI$_3$ perovskite unit cells and (right top) relative energy profile of the iodide Frenkel defect formation (right bottom). Radius ratio vs alkali metal ions. Reproduced with permission.[75] Copyright 2018, American Chemical Society.

TABLE 6.1

Representative A-Site Additives for High-Efficiency ABX_3 Perovskite Solar Cells

Type	Perovskite	Device structure	PCE [%]	Year[Ref.]
KI	$Cs_{0.06}FA_{0.79}MA_{0.15}Pb(I_{0.85}Br_{0.15})_3$	FTO/c-TiO$_2$/m-TiO$_2$/PVSK/ spiro-OMeTAD/Au	21.5	2018[71]
RbI	$(CsFAMA)Pb(IBr)_3$	FTO/c-TiO$_2$/m-TiO$_2$/PVSK/ spiro-OMeTAD/Au	21.8	2016[74]
MACl	$FAPbI_3$	FTO/c-TiO$_2$/mp-TiO$_2$/PVSK/ spiro-OMeTAD/Au	24.02	2019[59]
NH$_4$I	$CsPbI_3$	FTO/compact-TiO$_2$/PCBA/ PVSK/spiro-OMeTAD/Au	18.71	2021[81]
AABr	$MAPbI_3$	FTO/ZnO/MAPbI$_3$/Spiro-OMeTAD/ Au	20.18	2018[82]
NH$_4$Cl	$MAPbI_3$	ITO/PEDOT:PSS/PVSK/PC$_{61}$BM/Al	9.93	2014[83]
NH$_4$Cl+H$_2$O	$MAPbI_3$	FTO/c-TiO$_2$/m-TiO$_2$/ZrO$_2$/PVSK/ carbon	15.6	2017[84]
NH$_4$Cl+NH$_4$SCN	$(PEA)_2(MA)_4Pb_5I_{16}$	ITO/PEDOT:PSS/PVSK/PC$_{61}$BM/ BCP/Ag	14.1	2018[85]
4-ABPACl)	$MAPbI_3$	FTO/c-TiO$_2$/PVSK/spiro-OMeTAD/ Au	16.55	2015[86]
DMAI	$CsPbI_3$	FTO/c-TiO$_2$/PVSK/spiro-OMeTAD/ Ag.	19.03	2019[87]
GABr	$FA_{0.7}MA_{0.3}Pb_{0.7}Sn_{0.3}I_3$	ITO/EMIC-PEDOT:PSS/PVSK/S-acetylthiocholine chloride/C60/ BCP/Ag	20.26	2020[88]
PEAI+HI	$CsPbI_3$	FTO/TiO$_2$/PVSK/PTAA/Au	15.07	2018[89]
PACl	$FAPbI_3$	FTO/SnO$_2$/PVSK/spiro-MeOTAD/Ag	22.2	2021[61]
BAAc	$BA_2MA_{n-1}Pb_nI_{3n+1}$	ITO/SnO$_2$/PVSK/spiro-OMeTAD/ MoO$_3$/Au	16.25	2020[90]
DAP	$MAPbI_3$	ITO/PTAA/PVSK/C60/bathocuproine (BCP)/Cu	21.7	2019[91]
CsI	$CsPbI_3$	ITO/SnO$_2$/ZnO/PVSK/Spiro-OMeTAD/MoO$_3$/Ag	16.39	2019[92]
CsI	$FAPbI_3$	FTO/c-TiO$_2$/SnO$_2$/FAPbI$_3$/Spiro-OMeTAD/Ag	23.3	2021[93]
CsI	$(FAPbI_3)_{0.972}(MDACl_2)_{0.038}$	FTO/bl-TiO$_2$/mp-TiO$_2$/PVSK/Spiro-OMeTAD/Au	25.17	2020[69]
CsCl+MACl	$(FAPbI_3)_{0.96}(MAPbBr)_{0.04}$	FTO/SnO$_2$/PVSK/Spiro-MeOTAD/ Ag	23.22	2021[94]
CsI	$MAPbI_3$	FTO/TiO2/PVSK/Spiro-OMeTAD/ Au	18.43	2021[95]
Na	$CsPbI_3$	FTO/c-TiO$_2$/m-TiO$_2$/CsPbI$_3$/carbon	10.7	2019[96]
Ca	$CsPbI_3$	FTO/c-TiO$_2$/m-TiO$_2$/PVSK/P3HT/Au	13.5	2018[97]
RbCl	$FAPbI_3$	FTO/SnO$_2$/PVSK/Spiro-OMeTAD/ Au	25.6	2022[68]
PEAI	$FA_{0.98}Cs_{0.02}PbI_3$	ITO/SnO$_2$/PVSK/Spiro-MeOTAD/Au	20.64	2018[58]

6.2.2 B-Site Additive

6.2.2.1 Sn Additive

B-site ions play a pivotal role in influencing the bandgap characteristics of ABX_3 perovskite materials. Divalent metal cations, such as Sn, are commonly employed as B-site additives, particularly in the context of lead-based perovskite solar cells, to create lead-tin hybrid perovskite alloys. These group 14 elements exhibit comparable ionic radii when in the +II oxidation state, allowing Sn^{2+} to effectively occupy the B-site without compromising the structural integrity of the perovskite lattice.[98] By modulating the Pb:Sn ratio, it becomes feasible to fine-tune the inherently lower bandgaps exhibited by Pb-Sn mixed perovskites, falling within the range of approximately 1.2 eV to 1.3 eV, a range that pure Pb perovskites cannot attain.[99] This feature renders Pb-Sn perovskites highly appealing candidates for the bottom cells, responsible for absorbing low-energy photons, in all-perovskite tandem solar cells. Such tandem configurations can theoretically achieve a maximum efficiency of 37%.[100],[101] Single-junction Pb-Sn solar cells also align more closely with the ideal Shockley–Queisser limit due to their bandgap proximity to ~1.3 eV. Furthermore, it has been documented that the charge carrier mobility in Sn-based perovskites can surpass that of their Pb-based counterparts.[16]

Intriguingly, the bandgap of the alloy is even lower than that of the individual end compounds (i.e., $MAPbI_3$ and $MASnI_3$, with bandgaps of 1.6 and 1.3 eV, respectively). This reduction is primarily a consequence of the strong nonlinear dependence on chemical composition (as depicted in Figure 6.7a).[102] It arises from the disparity in energy between the s and p atomic orbitals of Pb and Sn, which contribute to the formation of the alloy's band edges. Moreover, the presence of Sn in the alloy leads to an overall deviation of the alloy bandgap from a linear interpolation between the end compounds, a phenomenon known as band gap bowing. The role of Sn in Pb-based perovskite solar cells primarily pertains to its chemical influence, which significantly impacts the electronic properties and bandgap behavior of the material.

However, it is a well-known challenge in Sn-based perovskites that Sn^{2+} is prone to oxidation, leading to its transformation into the more stable Sn^{4+} state. This oxidation process generates a significant number of vacancies and defects, thereby jeopardizing the structural stability of the perovskite material.[103]

6.2.2.2 Other B-Site Additives

Ge is also a group 14 element, however the application of GeI_2 in organic–inorganic perovskite precursors is challenging due to the extremely poor solubility of GeI_2 in hybrid perovskite ink, leading to failure in the formation of uniform films. Ge doping in hybrid perovskite precursors requires the addition of methylammonium chloride (MACl) to improve the solubility of GeI_2 (Figure 6.7b). The resulting Ge-doped mixed cation and mixed halide perovskite films with composition $FA_{0.83}MA_{0.17}Ge_{0.03}Pb_{0.97}(I_{0.9}Br_{0.1})_3$ show superior photoluminescence lifetime, power conversion efficiency above 22%, and greater stability towards illumination and humidity, outperforming photovoltaic properties of perovskite solar cells prepared without the Ge doping (Table 6.2).[104]

The Eu^{3+}-Eu^{2+} ion pair acts as a "redox shuttle" that selectively oxidizes Pb^0 and reduces I^0 defects in the perovskite solar cells, improving their operational durability

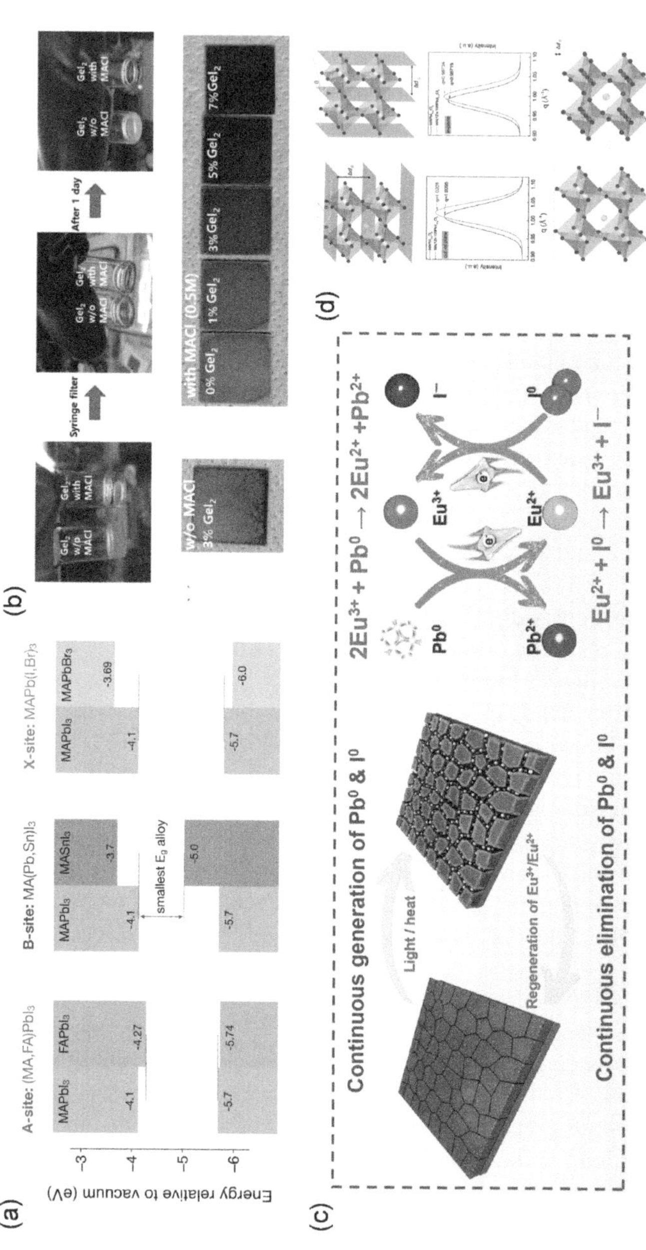

FIGURE 6.7 (a) Band edge positions relative to the vacuum level in pure perovskites constituting the A-, B-, and X-site alloys with MAPbI$_3$ as the reference system. If the band gap in the alloy is derived by the highest valence band edge and the lowest conduction band edge of the constituent pure perovskites, then B-site emerges as the most promising site to achieve smallest band gap via alloying. Reproduced with permission.[102] Copyright 2018, American Chemical Society. (b) GeI$_2$-containing DMF/DMSO organic–inorganic perovskite (3% GeI$_2$, FA$_{0.83}$MA$_{0.17}$Pb$_{0.97}$Ge$_{0.03}$(I$_{0.9}$Br$_{0.1}$)$_3$) inks with and without MACl (0.5M), corresponding photographs of MACl-contained or noncontained perovskite films of FA$_{0.83}$MA$_{0.17}$Pb$_{0.97}$Ge$_{0.03}$(I$_{0.9}$Br$_{0.1}$)$_3$) with various concentration of GeI$_2$. Reproduced with permission.[104] Copyright 2020, Wiley-VCH. (c) Proposed mechanism diagram of cyclically elimination of Pb^0 and I^0 defects and regeneration of Eu^{3+}-Eu^{2+} metal ion pair. Eu^{3+}-Eu^{2+} ion pair promotes the conversion of Pb^0 and I^0 to Pb^{2+} and I$^-$ in solution and perovskite film. Reproduced with permission.[105] Copyright 2019, Science Publishing Group. (d) Schematic architecture of the out-of-plane and in-plane of the MAPbI$_{3-x}$Cl$_x$ and MA(1Zn:100Pb)I$_{3-x}$Cl$_x$ perovskite. Reproduced with permission.[106] Copyright 2018, Elsevier.

TABLE 6.2

Representative B-Site Additives for High-Efficiency ABX$_3$ Perovskite Solar Cells

Type	Perovskite	Device structure	PCE [%]	Year[Ref.]
PbI$_2$	MAPbI$_3$	FTO/TiO$_2$/PVSK/sprio-OMeTAD/Au	12.00	2014[107]
Eu^{3+}-Eu^{2+}	(FA, MA, Cs)Pb(I, Br)$_3$(Cl)	FTO/SnO$_2$/PVSK/spiro-OMeTAD/Au	21.52	2019[105]
MgCl$_2$	MAPbI$_3$	FTO/c-TiO$_2$/m-TiO$_2$/PVSK/spiro-OMeTAD/Au	17.8	2018[108]
NiCl$_2$	MAPbI$_3$	FTO/SnO$_2$/PVSK/spiro-OMeTAD/Au	20.61	2018[109]
Sn	CsPbI$_3$	FTO/c-TiO$_2$/CsPbI$_3$/CuSCN/Au	5.12	2019[110]
Sn+OABF$_3$	Cs$_{0.2}$FA$_{0.8}$Pb$_{0.5}$Sn$_{0.5}$I$_3$	ITO/neutral-PEDOT/Sn-Pb PVSK/C60/BCP/Cu	23.7	2023[111]
Ge	CsPbI$_3$	FTO/c-TiO$_2$/PCBA/PVSK/spiro-OMeTAD/Au	19.52	2022[112]
GeI$_2$	FA$_{0.83}$MA$_{0.17}$Pb(I$_{0.9}$Br$_{0.1}$)$_3$	FTO/c-TiO$_2$/SnO$_2$/PVSK/Spiro-OMeTAD/Au	22.7	2020[104]
ZnCl$_2$	CH$_3$NH$_3$PbI$_3$	ITO/SnO$_2$/PVSK/Poly-TPD/MoO$_3$/Ag	20.06	2018[106]

(Figure 6.7c).[105] The addition of Eu(acac)$_3$ to the CH$_3$NH$_3$I solution eliminates I^0 species absorption peak, indicating the elimination of I^0 defects. The proposed redox shuttle eliminates corresponding defects through two chemical reactions, replenishing each ion in the Eu^{3+}-Eu^{2+} pair during defects elimination.

Zn substitution in perovskite structured materials, such as CH$_3$NH$_3$(Zn:Pb)I$_{3-x}$Cl$_x$, helps to achieve ordered and stable crystal structures, which is important for practical applications of efficient planar perovskite solar cells.[106] Zn substitution releases lattice strain during an appropriate lattice constriction within the BX$_6$ octahedron, resulting in the formation of an ordered and stable perovskite crystal (Figure 6.7d). Zn substitution also leads to a smoother surface roughness of the perovskite film, compared to the blank reference. The in-plane and out-of-plane peaks of the perovskite with Zn substitution are shifted to higher angles, indicating lattice constriction.

6.2.3　X-Site Additive

6.2.3.1　Halide Additive

The introduction of X-site additives exerts a notable influence on the bandgap characteristics of ABX$_3$ perovskite materials. This effect is primarily attributed to variations in the ionic radii of the halides (I, Br, Cl), wherein the bandgap undergoes an increase concomitant with a decrease in ion size (I < Br < Cl). The precise tuning of

X-site constituents assumes a pivotal role in regulating both the bandgap and phase stability of ABX_3 perovskites.

Figure 6.8a illustrates the significance of enhancing electronegativity, which can lead to the formation of robust hydrogen and ionic bonds with the Pb component.[77] The incorporation of a minute quantity of fluoride into the perovskite structure engenders a substantial stabilization of the perovskite surface through the establishment of strong fluoride-organic cation interactions. This interaction effectively mitigates the formation of vacancies associated with organic cations and halide anions. Notably, the exceedingly high electronegativity of fluoride (as depicted in Figure 6.8a) enables the formation of a vigorous ionic bond between fluorine (F) and Pb on the perovskite surface, thereby facilitating the passivation of iodine vacancy defects.[77] The strength of the F-Pb bond, in comparison to the I-Pb bond, serves to immobilize halogen ions and lead ions, effectively curtailing undesired ion migration within the perovskite.

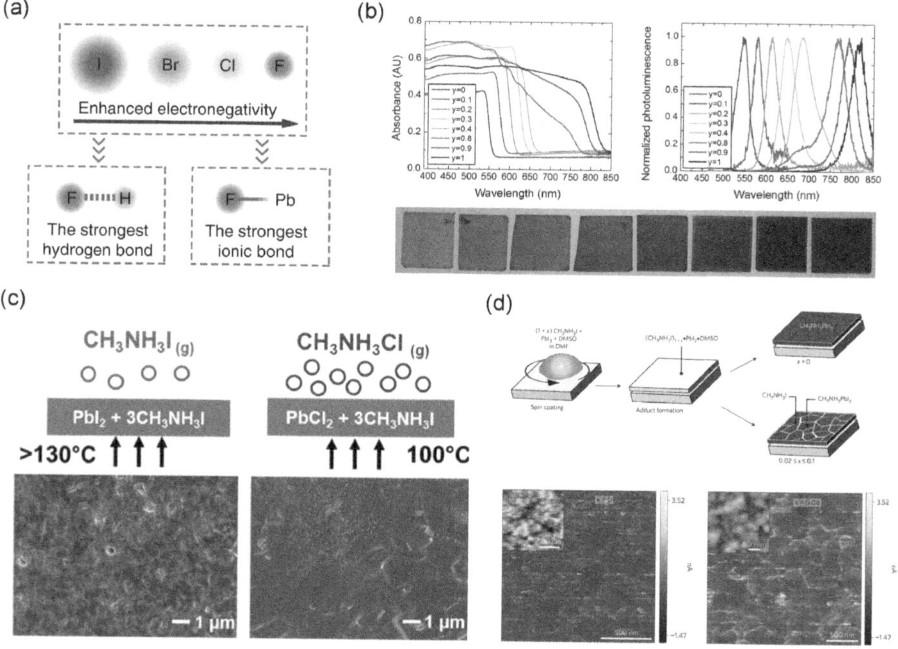

FIGURE 6.8   (a) Schematic illustration of enhancing the hydrogen bond between the halogen and MA/FA ions and strengthening the ionic bond between the halogen and metal ions through increasing the electronegativity of halogen. Reproduced with permission.[77] Copyright 2019, Science Publishing Group. (b) Tunable optical properties of the $FAPbI_yBr_{3-y}$ perovskite system. UV-vis absorbance (left top), photoluminescence (right top), and photographs (bottom) of $FAPbI_yBr_{3-y}$ perovskite system. Reproduced with permission.[56] Copyright 2014, The Royal Society of Chemistry. (c) The mechanism of the Cl in the $MAPbI_3$ perovskite films. Reproduced with permission.[116] Copyright 2015, American Chemical Society. (d) Schematic illustration of the coating process of $MAPbI_3$ with MAI additive (top) and the c-AFM of perovskite with (right bottom) or without (left bottom) MAI. Reproduced with permission.[23] Copyright 2016, Nature Publishing Group.

Furthermore, this incorporation of fluoride fosters the establishment of robust hydrogen bonds (N-H ··F) between fluorine and organic cations, resulting in a bolstering of the formation energy associated with organic cation vacancies, as per density functional theory (DFT) calculations. The overall enhancement of chemical bonding induced by fluoride incorporation is instrumental in achieving exceptional operational stability for perovskite solar cells. In unencapsulated perovskite solar cells, it was observed that these materials retained an impressive 90% of their PCE after 1,000 h of maximum power point (MPP) tracking under 1-sun illumination, using a white LED lamp in a nitrogen atmosphere. In stark contrast, devices lacking fluoride incorporation exhibited a substantial degradation, maintaining only 40% of their original PCE after 600 h under identical conditions. This result underscores the paramount importance of modulating chemical bonds for dual-defect passivation, ultimately leading to the enhanced operational stability of perovskite solar cells.[113]

The partial substitution of I with the comparatively smaller Br ion to create $FAPb(I_{1-x}Br_x)_3$ induces alterations in the average Pb–X bond distance within the perovskite thin films, consequently impacting the bandgap characteristics of the material (refer to Figure 6.8b).[56] This substitution enables a wide-ranging tunability of the bandgap for the mixed halide $FAPbI_yBr_{3-y}$ perovskite system, ranging from 1.48 to 2.23 eV. A study[114] has reported the generation of $MAPbBr_{3-x}Cl_x$ and $MAPbBr_{3-x}I_x$ ($0 \leq x \leq 3$) nanocrystals through halide exchange reactions involving MACl and MAI with $MAPbBr_3$. These nanocrystals were found to exhibit reversible and adjustable bandgaps spanning a broad spectrum from 1.6 to 3.0 eV, which can be achieved by varying the ratios of Br/Cl and Br/I, as also detailed in reference.[114] Additionally, in a separate study, it was observed that the absorption edge of $(MAPbI_{1-x}Br_x)_3$ shifted from 786 to 544 nm as the Br composition increased, as described in reference.[115]

In contrast to $MAPbI_{3-x}Br_x$ perovskites, the incorporation of Cl into iodine-based perovskites has proven to be more challenging, with I replacement for Cl feasible only up to 3–4% of Cl,[116,117] which is elaborated in Figure 6.8c. This limitation can be attributed to the substantial disparity in the ionic radii between the halogens.[118] However, despite this limitation, the introduction of a low concentration of Cl doping has been shown to result in a significant enhancement in carrier mobility.

Regarding iodine as an additive, an excess of MAI, serving as an additive in the precursor solution of MAI and PbI_2, is found to accumulate at the grain boundaries, forming an in-situ passivation layer.[23] Conductive atomic force microscopy (c-AFM) images in Figure 6.8d reveal that this in-situ MAI layer creates a pathway for electron conduction, owing to the ionic character of MAI. This leads to a significant reduction in trap-mediated non-radiative recombination, as demonstrated by time-resolved photoluminescence (TRPL) and transient absorption (TA) measurements, highlighting the role of excess MAI in in-situ grain boundary passivation.

6.2.3.2 Non-Halide Additive

In addition to the halide X-site additive, non-halide ions such as BF_4^-, PF_6^-, $HCOO^-$, and SCN^- have been reported as X-site additives in ABX_3 perovskites. It is noteworthy that the ionic radius of the SCN^- closely approximates that of I^-, measuring 217 pm and 220 pm, respectively. This similarity allows SCN^- to function as a feasible alternative to iodide. Substitution of I^- with SCN^- to form $MAPb(SCN)_2I$ has been shown to significantly enhance the stability of perovskite solar cells. Furthermore, the introduction of a small quantity of $Pb(SCN)_2$ into the precursor solution of $MAPbI_3$ has a remarkable impact on the crystallinity and grain size of perovskite thin films.[119,120] During the annealing process, SCN^- ions are assumed to release gases from the perovskite film, in the form of methylamine (CH_3NH_2) and HSCN. The release of CH_3NH_2 gas during perovskite formation leads to improved crystallinity and larger grain sizes in the films, as illustrated in Figure 6.9a. Notably, a PSC device incorporating 5% $Pb(SCN)_2$ as an additive exhibited exceptional photovoltaic performance, achieving PCE of up to 18.42% with minimal hysteresis.

An innovative anion engineering approach involves the use of the pseudo-halide anion $HCOO^-$ to fill halide vacancy defects located at the grain boundaries and the surface of perovskite films. This approach aims to enhance the crystallinity of $FAPbI_3$-based perovskites. It has been observed that the doping of 2% formate anions leads to an enlargement of grain size to approximately 2 μm, an increase in crystal orientation along the (100) and (200) directions, which are favorable for carrier transport, and the suppression of the formation of the non-photoactive δ-$FAPbI_3$ phase.[121] Theoretical calculations have revealed that formate anions exhibit a greater affinity for iodide vacancy sites compared to other anions such as Cl^-, Br^-, and BF_4^-, owing to the fact that each carboxylate group can form two Pb–O coordination bonds with the lead cations, as depicted in Figure 6.9b. Consequently, perovskite solar cells based on $FAPbI_3$ with pseudo-halide treatment have achieved a record-breaking PCE of 25.6% (certified at 25.2%) and a V_{OC} of 1.19 V.

To enhance anion selectivity and optimize anion sites, molecular anions such as BF_4^- and PF_6^- have been employed to partially or entirely replace halide anions in organic-inorganic hybrid perovskites. For instance, an $FA_{0.88}Cs_{0.12}PbI_{3-x}(PF_6)_x$ interlayer was formed on top of the $FA_{0.88}Cs_{0.12}PbI_3$ perovskite film through an ion exchange reaction,[122] as depicted in Figure 6.9c. This post-treatment with $FAPF_6$ solution resulted in $FA_{0.88}Cs_{0.12}PbI_{3-x}(PF_6)_x$ as a second layer, which induced significant changes in surface morphology and grain boundaries (GB), as observed in scanning electron microscopy (SEM) images. Furthermore, carrier lifetimes increased, and defect density decreased after post-treatment with an optimized concentration of $FAPF_6$, when compared to control perovskite films.

6.2.4 Solvent Additive

The adduct approach represents an effective strategy for the controlled modulation of perovskite film crystallization through the incorporation of Lewis base additives. The formation of a stable intermediate phase, consisting of a Lewis acid-base adduct, plays a pivotal role in inducing uniform crystallization. Conventionally, the polar aprotic solvent DMSO has been widely employed as the Lewis base additive to facilitate the production of high-quality perovskite films, irrespective of the specific perovskite composition (Table 6.3). This is primarily attributed to the ability of the lone

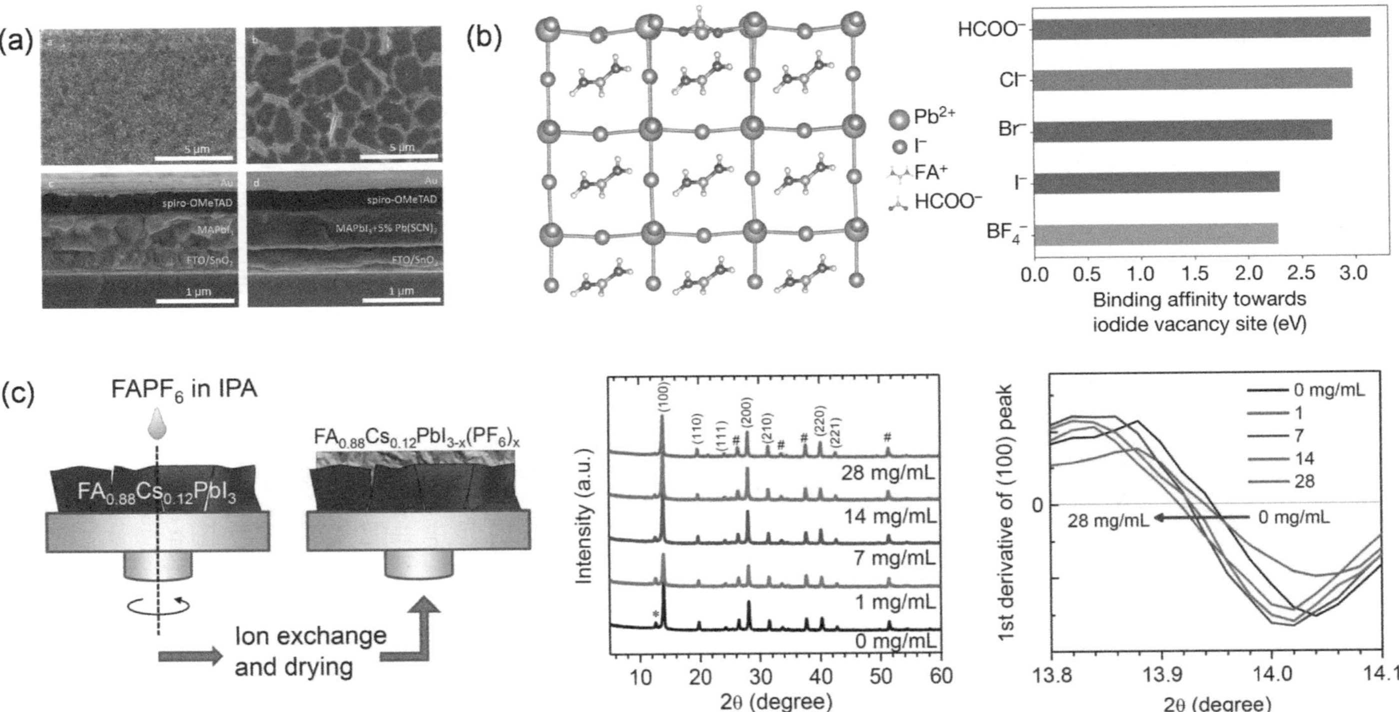

FIGURE 6.9 (a) The cross-section and surface SEM images of MAPbI₃ with (right) and without (left) 5% Pb(SCN)₂. Reproduced with permission.[119] Copyright 2016, Wiley-VCH. (b) Simulated structure (left) of the I-vacancy at the FAPbI₃ by HCOO-anion and the relative interaction strengths (right) of different anions with the I-vacancy at the surface. Reproduced with permission.[121] Copyright 2021, Nature Publishing Group. (c) Schematic illustrate (left) of preparation process of I⁻ and PF₆₋ ion exchange at the upper layer perovskite film FA₀.₈₈Cs₀.₁₂PbI₃₋ₓ(PF₆)ₓ, the (100) XRD peaks shift to low angle with increasing FAPF₆ concentration. Reproduced with permission.[122] Copyright 2018, Wiley-VCH.

TABLE 6.3

Representative X-Site Additives for High-Efficiency ABX_3 Perovskite Solar Cells

Type	Perovskite	Device structure	PCE [%]	Year[Ref.]
MAI	$MAPbI_3$	FTO/c-TiO_2/m-TiO_2/PVSK/spiro-OMeTAD/Au	20.4	2016[23]
PbI_2	$(FAPbI_3)_{0.85}(MAPbBr_3)_{0.15}$	FTO/c-TiO_2/m-TiO_2/PVSK/spiro-OMeTAD/Au	20.1	2016[123]
$Pb(SCN)_2$	$MAPbI_3$	$MAPbI_3$	18.43	2016[119]
NaF	$(Cs_{0.05}FA_{0.54}MA_{0.41})Pb(I_{0.98}Br_{0.02})_3$	FTO/SnO_2/PVSK/spiro-OMeTAD/Au	21.46	2019[77]
Br^-+$PbCl_2$+LiF	$CsPbI_{3-x}Br_x$	ITO/SnO_2/LiF/$CsPbI_3$-xBrx/Spiro-OMeTAD/Au	18.64	2019[124]
Br^-+GA^+	$CsPbI_xBr_{3-x}$	ITO/SnO_2/$CsPbI_{3-x}Br_x$/SIM/Spiro-OMeTAD/Au	18.06	2020[125]
$Pb(SCN)_2$	$CsPbI_3$	FTO/TiO_2/$CsPbI_3$/PTAA/Au	17.04	2019[126]
FAFo	$FAPbI_3$	FTO/m-TiO_2/PVSK/Spiro-OMeTAD/Au	25.59	2021[121]
PF_6^-	$(TMA)_xFA_{1-x}Pb(PF_6)_xI_{3-x}$	FTO/SnO_2/PVSK/PCBM/Ag	21.43	2022[127]
$PbAc_2$	$MAPbI_3$	FTO/SnO_2/$MAPbI_3$/spiro-MeOTAD/Au	15.14	2019[128]
PF_6	$FA_{0.88}Cs_{0.12}PbI_3$	FTO/bl-TIO_2/mp-TiO_2/PVSK/Spiro-MeOTAD/Au	19.3	2018[122]

pair electrons on the oxygen atom of DMSO to interact with the Lewis acid Pb^{2+} of perovskite precursors, resulting in the formation of an MAI PbI_2 DMSO adduct in the as-spun film (Figure 6.10a and 6.10b).[32,129]

Lewis base additives used in the adduct approach encompass small molecules or polymers featuring O-, S-, and N-atoms with strong donor capabilities. Polar solvent additives, such as DMSO, can establish a stable intermediate through the selective removal of the solvent (e.g., DMF) via antisolvent dripping or blowing procedures.

The formation of adducts can be confirmed through Fourier-transform infrared spectroscopy (FT-IR), wherein the stretching frequency of the oxygen-bearing functional group of the Lewis base additive (S=O in the case of DMSO) shifts to a lower wavenumber due to the weakening of the bond strength through adduct formation.[32]

Since the adduct is distinct from the perovskite phase, films composed of adducts exhibit transparency. A set of guidelines for designing adduct formation involves considerations such as minimal steric hindrance around high electron-donating elements of Lewis base additives to enable effective interaction with Lewis acidic species like Pb^{2+}. Additionally, Lewis bases should possess hydrogen bonding capabilities for effective interaction with AX precursors in $APbX_3$ perovskite materials. Furthermore, the selection of Lewis bases should align with the principles of hard and soft (Lewis)

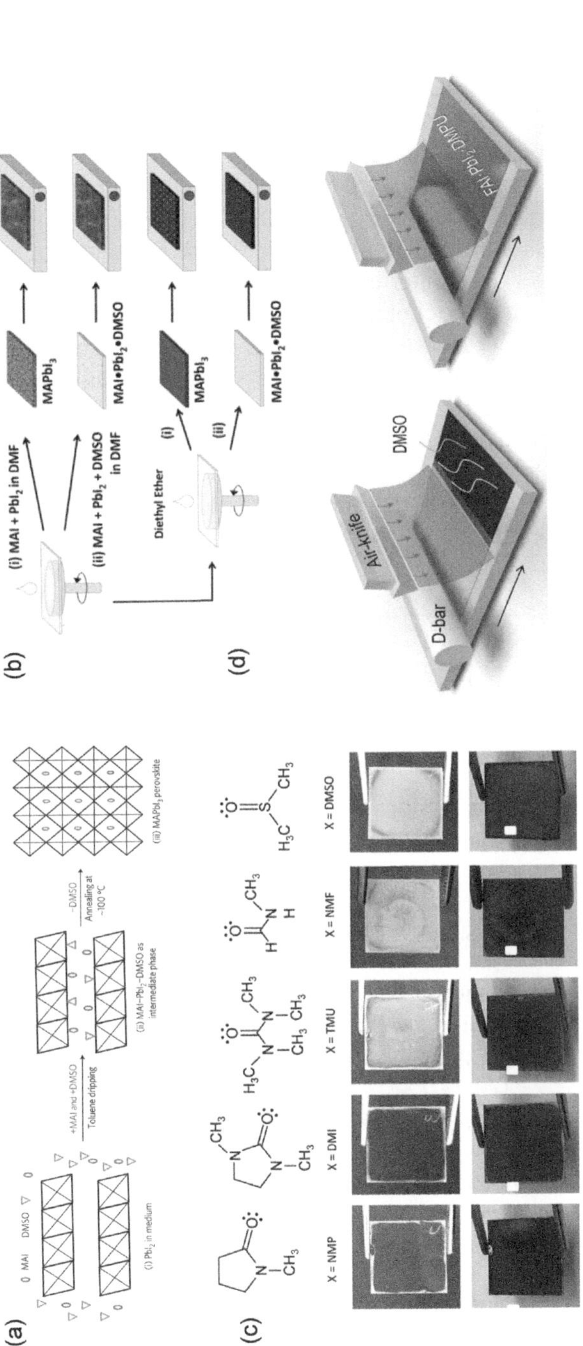

FIGURE 6.10 (a) Scheme for the formation of the MAPbI$_3$ perovskite material with DMSO additive. (i) PbI$_2$, consisting of edge-sharing [PbI$_6$]$^{4-}$ octahedral layers. (ii) MAI and DMSO guest molecules intercalated between the layers, forming the flat MAI–PbI$_2$–DMSO intermediate phase film, when toluene is introduced onto the wet film comprising PbI$_2$, MAI, and DMSO. Here, the positions of the guest molecules are unconfirmed. (iii) Finally, the intermediate phase film is converted to the perovskite phase with corner-sharing octahedra via the extraction of DMSO guest molecules by the annealing process. Reproduced with permission.[133] Copyright 2014, Nature Publishing Group. (b) Schematic representation of fabrication procedure for the MAPbI$_3$ perovskite layers obtained by direct one-step spin-coating of the DMF solution containing DMSO. Reproduced with permission.[32] Copyright 2015, American Chemical Society. (c) Film formation with different Lewis bases. Molecular structures of the Lewis bases (top) and corresponding adduct (middle) and perovskite films (bottom). The perovskite solutions were prepared by mixing 1 mmol of FAI, PbI$_2$ and corresponding Lewis base in 760 mg of DMF. Reproduced with permission.[134] Copyright 2018, American Chemical Society. (d) Schematic of the as-deposited (before annealing) films formed by air-knife-assisted D-bar coating using (left) the DMSO-containing perovskite precursor solution and (right) the DMPU-containing perovskite solution. DMSO was evaporated by Ar blowing, but DMPU stayed in the film due to stronger adduct formation. Reproduced with permission.[132] Copyright 2021, The Royal Society of Chemistry.

acids and bases (HSAB) theory,[130] promoting stable interactions between Lewis acids and bases and facilitating the metathesis reaction for the conversion of adducts into the perovskite phase. While DMSO works effectively for $MAPbI_3$, it may be less suitable for $FAPbI_3$ due to weaker interactions with FAI, resulting in unreacted PbI_2 and poor film morphology. It has been demonstrated that NMP (N-methyl-2-pyrrolidone) is more effective than DMSO for producing high-quality $FAPbI_3$ films due to its stronger interactions with FAI and PbI_2, as confirmed through both DFT calculations and experimental results (Figure 6.10c).[131] Nonetheless, DMSO is still employed for $FAPbI_3$, particularly in conjunction with the MACl additive, as the Cl Lewis base contributes significantly to forming a stable intermediate (Table 6.4).

When considering large-area coating, the choice of solvent and Lewis base should be simultaneous. Vapor pressure becomes a critical parameter in achieving uniform films on extensive substrates. Moreover, an air-blowing process in D-bar coating can replace the antisolvent treatment applied in small-area coating. Lewis bases with low vapor pressure, such as 1,3-dimethyl-3,4,5,6-tetrahydro-2(1H)-pyrimidinone (DMPU; Figure 6.10d),[132] outperform those with high vapor pressure (e.g., DMSO) for large-area coating using an air-knife due to their non-volatility under air-blowing conditions, which aids in selectively removing solvents like DMF. XRD analysis

TABLE 6.4
Representative Solvent Additives for High-Efficiency ABX_3 Perovskite Solar Cells

Type	Perovskite	Device structure	PCE [%]	Year[Ref]
DMSO	$MAPbI_{1-x}Br_x$	FTO/c-TiO_2/m-TiO_2/PVSK/spiro-OMeTAD/Au	16.72	2014[129]
DMSO	$MAPbI_3$	FTO/c-TiO_2/PVSK/spiro-OMeTAD/Au	19.7	2015[32]
NMP	$Cs_{0.02}FA_{0.98}PbI_3$	ITO/SnO_2/PVSK/spiro-OMeTAD/Au	20.19	2018[134]
DIO	$MAPbI_{3-x}Cl_x$	FTO/PEDOT:PSS/PVSK/PC61BM/bis-C_{60}/Ag	12	2014[135]
TA	$MAPbI_3$	ITO/SnO_2/PVSK/spiro-OMeTAD/Ag	15.81	2019[136]
EC	$MAPbI_3$	FTO/c-TiO_2/PVSK/spiro-OMeTAD/MoO_3/Ag	19.41	2019[137]
Urea	$MAPbI_3$	ITO/SnO_2/PVSK/spiro-OMeTAD/Ag	18.25	2017[131]
Pyridine	$Cs_{0.05}(MA_{0.17}FA_{0.83})_{0.95}Pb(I_{0.83}Br_{0.17})_3$	FTO/c-TiO_2/PVSK/spiro-OMeTAD/Au	19.03	2018[138]
TBP	$MAPbI_3$	FTO/c-TiO_2/m-TiO_2/PVSK/spiro-OMeTAD/Au	17.41	2018[139]
DMPU	$(FAPbI_3)_{0.95}(CsPbBr_3)_{0.05}$	FTO/SnO_2/PVSK/spiro-OMeTAD/Au	20.56	2020[132]

reveals the formation of a robust adduct through DMPU, associated with vapor pressure and donor number differences among Lewis bases. This concept results in the production of high-crystallinity large-area perovskite films, consequently enhancing device performance via the stable adduct intermediate formed between DMPU and precursor materials.

6.2.5 OTHER ADDITIVE

6.2.5.1 Small Molecular Additive

Other additives can serve as surfactants in the large-scale coating of perovskite films. Surfactants play a pivotal role in enhancing the adhesion of perovskite solutions to hydrophobic hole transport materials, thereby facilitating the uniform deposition of perovskite films over extensive areas at a high blade-coating rate. These surfactants modify the ink-substrate interface and augment the perovskite ink's affinity for hydrophobic substrates, resulting in comprehensive coverage and improved film quality. The LP surfactant, in conjunction with other surfactants, can be considered a universal additive in perovskite inks for enhancing the quality of perovskite films using diverse scalable fabrication techniques (see Figure 6.11).[140] The incorporation of LP surfactant into the perovskite solution addresses non-wetting issues and enhances the compatibility of the perovskite ink with hydrophobic substrates. This modification effectively alters the ink-substrate interface, leading to the stable positioning of the contact line on the substrate and maintaining nearly 100% coverage of the droplet during the drying process. Consequently, this ensures the consistent coating of perovskite films over extensive areas.

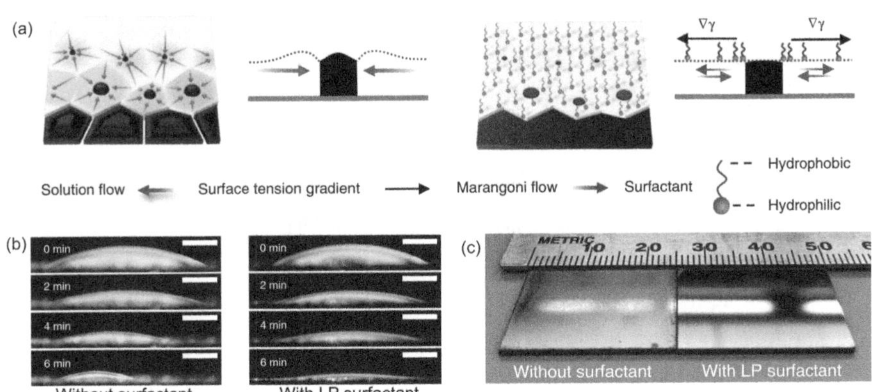

FIGURE 6.11 (a) Schematic illustration of the suppressed solution flow dynamics in the presence of surfactant. (b) The drying processes of perovskite ink droplets on hydrophobic substrates under ambient conditions without and with LP surfactant, respectively. The scale bars are 0.5 mm. (c) A photographic image of blade-coated perovskite films without and with LP surfactant. Reproduced with permission.[140] Copyright 2018, Nature Publishing Group.

LP surfactant contributes to the enhancement of perovskite film quality by mitigating the formation of island structures and reducing thickness variations, resulting in a more uniform and dense perovskite film. Furthermore, LP surfactant can passivate charge traps within bladed perovskite films, resulting in decreased leakage current and improved device performance. As an amphoteric surfactant, LP surfactant not only elevates film quality but also passivates perovskite defects, thereby contributing to the overall enhancement of perovskite solar cell performance.

6.2.5.2 Polymer Additive

Polymer additives represent a promising avenue for enhancing the stretchability of flexible perovskite solar cells (Table 6.5). As reported,[141] a self-healing polyurethane (s-PU) featuring dynamic oxime-carbamate bonds was introduced as a scaffold within the perovskite films, leading to a marked improvement in crystallinity and passivation of grain boundaries in the perovskite films (see Figure 6.12). The inclusion of s-PU as an additive bestowed the PSCs with mechanical self-healing

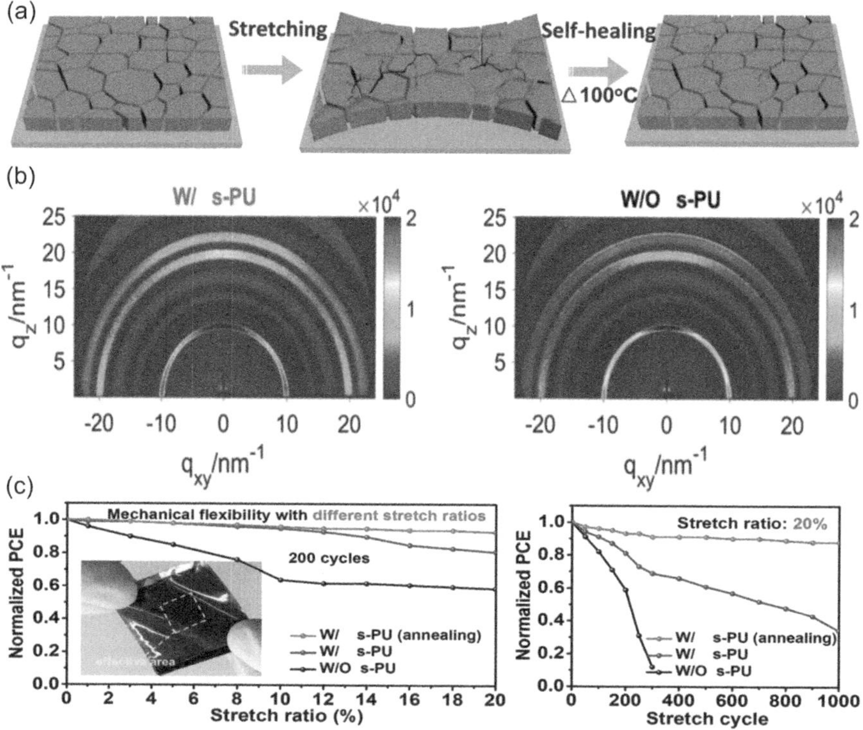

FIGURE 6.12   (a) Schematic illustration of the mechanically self-healing process for perovskite films. (b) GIWAXS data of perovskite films with and without s-PU, respectively. (c) Normalized average PCE of perovskite solar cells under different stretching forces for 200 cycles and normalized average PCE of perovskite solar cells as a function of stretching cycles with 20% stretching. Reproduced with permission.[141] Copyright 2020, Wiley-VCH.

TABLE 6.5

Representative Other Additives for High-Efficiency ABX_3 Perovskite Solar Cells

Type	Perovskite	Device structure	PCE [%]	Year[Ref.]
PCBM	$CsPbI_3$	ITO/SnO$_2$/CsPbI$_3$/Spiro-OMeTAD/Au	16.04	2020[142]
FS	$FAPbI_3$	FTO/SnO$_2$/PVSK/Spiro-OMeTAD/Au	21.72	2023[143]
LP	$MAPbI_3$	ITO/PTAA/MAPbI$_3$/C60/BCP/Cu	20.3	2018[140]
polyurethane	$(FAMA)Pb(IBr)_3$	hc-PEDOT:PSS/PEDOT:PSS AL4083/pvsk/PCBM/PEIE/ hc-PEDOT:PSS	19.15	2020[141]
Thiophene	$MAPbI_3$	FTO/c-TiO$_2$/PVSK/spiro-OMeTAD/Au	15.3	2014[144]
3-alkylthiophene	$Cs_{0.05}FA_{0.81}MA_{0.14}PbBr_{0.45}I_{2.55}$	FTO/c-TiO$_2$/PVSK/spiro-OMeTAD/Au	19.89	2018[145]
PTQ10	$(FAPbI_3)1\text{-}x(MAPbBr_3)x$	FTO/SnO$_2$/PVSK/PTAA/Au	21.21	2018[146]
Thiourea	$MAPbI_3$	FTO/c-TiO$_2$/m-TiO$_2$/PVSK/spiro-OMeTAD/Au	18.46	2017[147]
bis-PCBM+BrPh-ThR	$(FAI)_{0.81}(PbI_2)_{0.85}(MABr)_{0.15}(PbBr_2)_{0.15}$	FTO/c-TiO$_2$/m-TiO$_2$/PVSK/spiro-OMeTAD/Au	21.7	2018[148]
2-pyridylthiourea	$MAPbI_3$	FTO/c-TiO$_2$/m-TiO$_2$/PVSK/spiro-OMeTAD/Au	18.2	2017[149]
SP3	$MAPbI_3$	FTO/c-TiO2/m-TiO2/PVSK/spiro-OMeTAD/Au	20.43	2019[150]
ITIC-Th	$MA_{0.17}FA_{0.83}Pb(I_{0.83}Br_{0.17})_3$	FTO/c-TiO$_2$/PVSK/spiro-OMeTAD/Au	19.20	2018[151]
HMImCl	$MAPbI_3$	FTO/c-TiO$_2$/PVSK/spiro-OMeTAD/Ag	2.44	2015[152]
FIm	$(FAPbI_3)_{0.85}(MAPbBr_3)_{0.15}$	FTO/c-TiO$_2$/m-TiO$_2$/PVSK/spiro-OMeTAD/Au	15.38	2016[153]
CC2	$MAPbI_3$	FTO/c-TiO$_2$/m-TiO$_2$/PVSK/spiro-OMeTAD/Au	19.21	2017[154]
EATZ	$MAPbI_3$	FTO/c-TiO$_2$/PVSK/spiro-OMeTAD/Au	20.03	2019[155]
PMMA	$(FAI)_{0.81}(PbI_2)_{0.85}(MABr)_{0.15}(PbBr_2)_{0.15}$	FTO/c-TiO$_2$/m-TiO$_2$/PVSK/spiro-OMeTAD/Au	21.6	2016[156]
BMIMBF4	$(FA_{0.83}MA_{0.17})0.95Cs_{0.05}Pb(I_{0.9}Br_{0.1})_3$	FTO/NiOx/PVSK/PCBM/BCP/Au	20.0	2019[157]
ETI	$MAPbI_3$	FTO/c-TiO$_2$/m-TiO$_2$/PVSK/spiro-OMeTAD/Au	19.51	2019[158]

(*Continued*)

TABLE 6.5 (Continued)

Representative Other Additives for High-Efficiency ABX_3 Perovskite Solar Cells

Type	Perovskite	Device structure	PCE [%]	Year[Ref.]
PPC polymer	$MAPbI_3$	$ITO/SnO_2/PVSK/Spiro-OMeTAD/Ag$	19.35	2019[159]
Caffeine	$MAPbI_3$	$FTO/c\text{-}TiO_2/PVSK/spiro\text{-}OMeTAD/MoO_3/Ag$	20.25	2019[160]
DR3T	$MAPbI_3$	$FTO/c\text{-}TiO_2/PVSK/spiro\text{-}OMeTAD/Au$	19.28	2018[161]
PFA	$MAPbI_3$	$FTO/c\text{-}TiO_2/PVSK/spiro\text{-}OMeTAD/Au$	16.18	2019[162]
EATZ	$MAPbI_3$	$FTO/c\text{-}TiO_2/PVSK/spiro\text{-}OMeTAD/Au$	20.3	2019[163]
$ILPF_6$	$MAPbI_3$	$FTO/c\text{-}TiO_2/m\text{-}TiO_2/PVSK/carbon$	13.01	2019[164]
$BMIMBF_4$	$(FA_{0.83}MA_{0.17})_{0.95}Cs_{0.05}Pb\,(I_{0.9}Br_{0.1})_3$	$FTO/NiOx/PVSK/PCBM/BCP/Au$	20.0	2019[165]
ETI	$MAPbI_3$	$FTO/c\text{-}TiO_2/m\text{-}TiO_2/PVSK/spiro\text{-}OMeTAD/Au$	19.51	2019[166]
SP3	$MAPbI_3$	$FTO/c\text{-}TiO_2/m\text{-}TiO_2/PVSK/spiro\text{-}OMeTAD/Au$	20.43	2019[167]

capabilities, enabling them to recover an impressive 88% of their original efficiency even after undergoing 1,000 stretching cycles at a 20% strain.

The advantages of perovskite films incorporating s-PU were comprehensively substantiated through a variety of analytical techniques, including the observation of reduced dark current density and enhancements in charge-carrier recombination characteristics. Most notably, perovskite solar cells incorporating s-PU achieved a record-breaking champion PCE of 19.15%, establishing a new milestone for the highest efficiency reported to date for devices constructed on stretchable substrates. Moreover, the introduction of s-PU as an additive not only elevated the PCE but also improved the overall reproducibility of perovskite solar cells, evidenced by a narrower distribution of PCE values and heightened reliability when compared to perovskite solar cells lacking s-PU.

6.3 CONCLUSION AND OUTLOOK

In conclusion, this chapter has provided a comprehensive categorization of additive engineering approaches within the realm of perovskite solar cells, based on the specific additive materials and methods employed. A discernible outcome of these endeavors is the evident enhancement of photovoltaic parameters, primarily attributed to the effective passivation of defects, precise control of crystallization kinetics, grain size enlargement, and phase stabilization. Nonetheless, the field of perovskite solar cells continues to face substantial challenges, with paramount issues centered around ion migration management and the preservation of crystal structures. The pivotal roles played by A-site, B-site, and X-site additives necessitate meticulous scrutiny to discern their impacts and potential risks. The inclusion of solvent additives is pivotal in the growth of perovskite films. In the case of other additives, their development mandates a nuanced approach that encompasses a broad spectrum of defect types and species.

Striving to attain theoretical performance levels places emphasis on minimizing grain boundaries, enlarging grain size, and achieving high crystallinity in perovskite films. Polycrystalline films, despite their merits, confront inherent limitations when juxtaposed with their single-crystal counterparts, primarily stemming from their shorter charge carrier lifetime and diffusion length. While perovskite solar cells have achieved certified power conversion efficiency levels that rival traditional silicon solar cells, their operational stability still remains a challenge. Factors such as moisture, oxygen, heat, light, and thermodynamic phase stability continue to pose formidable obstacles to their durability. Strategies designed to bolster defect-related stability, including the modification of A-site cationic additives, have been advanced. Furthermore, the role of bulk organic cations in ameliorating ion migration and maintaining perovskite phase stability cannot be understated.

This chapter has shed light on the potential of innovative multifunctional additives to facilitate the realization of perovskite films akin to single crystals, with a keen focus on the regulation of defects and crystallization kinetics. The proposal of facet manipulation and epitaxial growth as effective methodologies offers a promising pathway to attain high-quality films characterized by superior charge carrier properties and a reduced defect density.

ACKNOWLEDGMENTS

The authors acknowledge the support of the National Key Research and Development Project funding from the Ministry of Science and Technology of China (Grant No. 2021YFB3800101), the National Natural Science Foundation of China (62104094 and U19A2089), the Guangdong Basic and Applied Basic Research Foundation (2021A1515110606)), and Shenzhen Science and Technology Innovation Committee (JCYJ20200109141014474 and JCYJ20220818100211025)

REFERENCES

1. www.nrel.gov/pv/cell-efficiency.html NREL best research-cell efficiency chart (accessed 22 October 2023).
2. M. V. Khenkin, *et al.*, Consensus statement for stability assessment and reporting for perovskite photovoltaics based on ISOS procedures, *Nature Energy* **2020**, *5*(1), 35–49.
3. R. Brenes, *et al.*, Metal halide perovskite polycrystalline films exhibiting properties of single crystals, *Joule* **2017**, *1*(1), 155–167.
4. Q. Lin, A. Armin, R. C. R. Nagiri, P. L. Burn, P. Meredith, Electro-optics of perovskite solar cells, *Nature Photonics* **2015**, *9*(2), 106–112.
5. H. Jin, *et al.*, It's a trap! On the nature of localised states and charge trapping in lead halide perovskites, *Materials Horizons* **2020**, *7*(2), 397–410.
6. S. D. Stranks, *et al.*, Electron-hole diffusion lengths exceeding 1 micrometer in an organometal trihalide perovskite absorber, *Science* **2013**, *342*(6156), 341–344.
7. J. M. Ball, A. Petrozza, Defects in perovskite-halides and their effects in solar cells, *Nature Energy* **2016**, *1*(11), 16149.
8. J. Kim, S.-H. Lee, J. H. Lee, K.-H. Hong, The role of intrinsic defects in methylammonium lead iodide perovskite, *The Journal of Physical Chemistry Letters* **2014**, *5*(8), 1312–1317.
9. W.-J. Yin, T. Shi, Y. Yan, Unusual defect physics in CH3NH3PbI3 perovskite solar cell absorber, *Applied Physics Letters* **2014**, *104*(6).
10. M. L. Agiorgousis, Y.-Y. Sun, H. Zeng, S. Zhang, Strong covalency-induced recombination centers in perovskite solar cell material CH3NH3PbI3. *Journal of the American Chemical Society* **2014**, *136*(41), 14570–14575.
11. A. Walsh, D. O. Scanlon, S. Chen, X. Gong, S. H. Wei, Self-regulation mechanism for charged point defects in hybrid halide perovskites, *Angewandte Chemie* **2015**, *127*(6), 1811–1814.
12. C. Eames, J. M. Frost, P. R. Barnes, B. C. O'regan, A. Walsh, M. S. Islam, Ionic transport in hybrid lead iodide perovskite solar cells, *Nature Communications* **2015**, *6*(1), 7497.
13. A. Buin, P. Pietsch, J. Xu, O. Voznyy, A. H. Ip, R. Comin, E. H. Sargent, Materials processing routes to trap-free halide perovskites, *Nano Letters* **2014**, *14*(11), 6281–6286.
14. A. Buin, R. Comin, J. Xu, A. H. Ip, E. H. Sargent, Halide-dependent electronic structure of organolead perovskite materials, *Chemistry of Materials* **2015**, *27*(12), 4405–4412.
15. J. Xu, *et al.*, Perovskite–fullerene hybrid materials suppress hysteresis in planar diodes, *Nature Communications* **2015**, *6*(1), 7081.
16. C. C. Stoumpos, C. D. Malliakas, M. G. Kanatzidis, Semiconducting tin and lead iodide perovskites with organic cations: Phase transitions, high mobilities, and near-infrared photoluminescent properties, *Inorganic Chemistry* **2013**, *52*(15), 9019–9038.
17. C. Bi, Y. Shao, Y. Yuan, Z. Xiao, C. Wang, Y. Gao, J. Huang, Understanding the formation and evolution of interdiffusion grown organolead halide perovskite thin films by thermal annealing, *Journal of Materials Chemistry A* **2014**, *2*(43), 18508–18514.

18. Q. Dong, Y. Fang, Y. Shao, P. Mulligan, J. Qiu, L. Cao, J. Huang, Electron-hole diffusion lengths> 175 μm in solution-grown CH3NH3PbI3 single crystals, *Science* **2015**, *347*(6225), 967–970.
19. V. Adinolfi, *et al.,* The in-gap electronic state spectrum of methylammonium lead iodide single-crystal perovskites, *Advanced Materials* **2016**, *28*(17), 3406–3410.
20. Y. Shao, Z. Xiao, C. Bi, Y. Yuan, J. Huang, Origin and elimination of photocurrent hysteresis by fullerene passivation in CH3NH3PbI3 planar heterojunction solar cells, *Nature Communications* **2014**, *5*(1), 5784.
21. B. Chen, P. N. Rudd, S. Yang, Y. Yuan, J. Huang, Imperfections and their passivation in halide perovskite solar cells, *Chemical Society Reviews* **2019**, *48*(14), 3842–3867.
22. K. X. Steirer, P. Schulz, G. Teeter, V. Stevanovic, M. Yang, K. Zhu, J. J. Berry, Defect tolerance in methylammonium lead triiodide perovskite, *ACS Energy Letters* **2016**, *1*(2), 360–366.
23. D.-Y. Son, *et al.*, Self-formed grain boundary healing layer for highly efficient CH3NH3PbI3 perovskite solar cells, *Nature Energy* **2016**, *1*(7), 16081
24. F. Wang, S. Bai, W. Tress, A. Hagfeldt, F. Gao, Defects engineering for high-performance perovskite solar cells, *NPJ Flexible Electronics* **2018**, *2*(1), 22.
25. D. Bi, *et al.*, Efficient luminescent solar cells based on tailored mixed-cation perovskites, *Science Advances* **2016**, *2*(1), e1501170.
26. Q. Chen, *et al.*, Controllable self-induced passivation of hybrid lead iodide perovskites toward high performance solar cells, *Nano Letters* **2014**, *14*(7), 4158–4163.
27. A. Dualeh, N. Tétreault, T. Moehl, P. Gao, M. K. Nazeeruddin, M. Grätzel, Effect of annealing temperature on film morphology of organic–inorganic hybrid pervoskite solid-state solar cells, *Advanced Functional Materials* **2014**, *24*(21), 3250–3258.
28. M. I. Saidaminov, *et al.*, Suppression of atomic vacancies via incorporation of isovalent small ions to increase the stability of halide perovskite solar cells in ambient air, *Nature Energy* **2018**, *3*(8), 648–654.
29. L. K. Ono, Y. Qi, Surface and interface aspects of organometal halide perovskite materials and solar cells, *The Journal of Physical Chemistry Letters* **2016**, *7*(22), 4764–4794.
30. L. She, M. Liu, D. Zhong, Atomic structures of CH3NH3PbI3 (001) surfaces, *ACS Nano* **2016**, *10*(1), 1126–1131.
31. R. Ohmann, *et al.*, Real-space imaging of the atomic structure of organic–inorganic perovskite, *Journal of the American Chemical Society* **2015**, *137*(51), 16049–16054.
32. N. Ahn, D. Y. Son, I. H. Jang, S. M. Kang, M. Choi, N. G. Park, Highly reproducible perovskite solar cells with average efficiency of 18.3% and best efficiency of 19.7% fabricated via Lewis base adduct of lead(II) iodide, *Journal of the American Chemical Society* **2015**, *137*(27), 8696–8699.
33. M. Liu, M. B. Johnston, H. J. Snaith, Efficient planar heterojunction perovskite solar cells by vapour deposition, *Nature* **2013**, *501*(7467), 395–398.
34. J. Burschka, *et al.*, Sequential deposition as a route to high-performance perovskite-sensitized solar cells, *Nature* **2013**, *499*(7458), 316–319.
35. N. T. Thanh, N. Maclean, S. Mahiddine, Mechanisms of nucleation and growth of nanoparticles in solution, *Chemical Reviews* **2014**, *114*(15), 7610–7630.
36. S. G. Kwon, T. Hyeon, Formation mechanisms of uniform nanocrystals via hot-injection and heat-up methods, *Small* **2011**, *7*(19), 2685–2702.
37. N. Ahn, S. M. Kang, J.-W. Lee, M. Choi, N.-G. Park, Thermodynamic regulation of CH 3 NH 3 PbI 3 crystal growth and its effect on photovoltaic performance of perovskite solar cells, *Journal of Materials Chemistry A* **2015**, *3*(39), 19901–19906.
38. J. W. Lee, D. K. Lee, D. N. Jeong, N. G. Park, Control of crystal growth toward scalable fabrication of perovskite solar cells, *Advanced Functional Materials* **2019**, *29*(47), 1807047.
39. D.-K. Lee, N.-G. Park, Additive engineering for highly efficient and stable perovskite solar cells, *Applied Physics Reviews* **2023**, *10*(1).

40. V. M. Goldschmidt, Die gesetze der krystallochemie, *Naturwissenschaften* **1926**, *14*(21), 477–485.

41. Z. Li, M. Yang, J.-S. Park, S.-H. Wei, J. J. Berry, K. Zhu, Stabilizing perovskite structures by tuning tolerance factor: Formation of formamidinium and cesium lead iodide solid-state alloys, *Chemistry of Materials* **2016**, *28*(1), 284–292.

42. S. Masi, A. F. Gualdrón-Reyes, I. Mora-Sero, Stabilization of black perovskite phase in FAPbI3 and CsPbI3. *ACS Energy Letters* **2020**, *5*(6), 1974–1985.

43. J. A. Steele, *et al.*, Thermal unequilibrium of strained black CsPbI3 thin films, *Science* **2019**, *365*(6454), 679–684.

44. D. H. Fabini, *et al.*, Reentrant structural and optical properties and large positive thermal expansion in perovskite formamidinium lead iodide, *Angewandte Chemie International Edition* **2016**, *55*(49), 15392–15396.

45. A. Marronnier, *et al.*, Anharmonicity and disorder in the black phases of cesium lead iodide used for stable inorganic perovskite solar cells, *ACS Nano* **2018**, *12*(4), 3477–3486.

46. C. J. Bartel, *et al.*, New tolerance factor to predict the stability of perovskite oxides and halides, *Science Advances* **2019**, *5*(2), eaav0693.

47. G. E. Eperon, *et al.*, Inorganic caesium lead iodide perovskite solar cells, *Journal of Materials Chemistry A* **2015**, *3*(39), 19688–19695, https://doi.org/10.1039/C5TA06398A.

48. Y. Zhang, S.-G. Kim, D.-K. Lee, N.-G. Park, CH3NH3PbI3 and HC(NH2)2PbI3 powders synthesized from low-grade PbI2: Single precursor for high-efficiency perovskite solar cells, *ChemSusChem* **2018**, *11*(11), 1813–1823.

49. A. Dualeh, P. Gao, S. I. Seok, M. K. Nazeeruddin, M. Grätzel, Thermal behavior of methylammonium lead-trihalide perovskite photovoltaic light harvesters, *Chemistry of Materials* **2014**, *26*(21), 6160–6164.

50. X. Ding, *et al.*, Enhancing the phase stability of inorganic α-CsPbI3 by the bication-conjugated organic molecule for efficient perovskite solar cells, *ACS Applied Materials & Interfaces* **2019**, *11*(41), 37720–37725.

51. S. K. Yadavalli, Y. Zhou, N. P. Padture, Exceptional grain growth in formamidinium lead iodide perovskite thin films induced by the δ-to-α phase transformation, *ACS Energy Letters* **2018**, *3*(1), 63–64.

52. W. Gao, *et al.*, A-site cation engineering of metal halide perovskites: Version 3.0 of efficient tin-based lead-free perovskite solar cells, *Advanced Functional Materials* **2020**, *30*(34), 2000794.

53. F. Ünlü, *et al.*, Understanding the interplay of stability and efficiency in A-site engineered lead halide perovskites, *APL Materials* **2020**, *8*(7).

54. G. Kieslich, S. Sun, A. K. Cheetham, Solid-state principles applied to organic–inorganic perovskites: New tricks for an old dog, *Chemical Science* **2014**, *5*(12), 4712–4715.

55. J. Y. Kim, J.-W. Lee, H. S. Jung, H. Shin, N.-G. Park, High-efficiency perovskite solar cells, *Chemical Reviews* **2020**, *120*(15), 7867–7918.

56. G. E. Eperon, S. D. Stranks, C. Menelaou, M. B. Johnston, L. M. Herz, H. J. Snaith, Formamidinium lead trihalide: A broadly tunable perovskite for efficient planar heterojunction solar cells, *Energy & Environmental Science* **2014**, *7*(3), 982–988.

57. J.-W. Lee, S. Tan, S. I. Seok, Y. Yang, N.-G. Park, Rethinking the A cation in halide perovskites, *Science* **2022**, *375*(6583), eabj1186.

58. J.-W. Lee, *et al.*, 2D perovskite stabilized phase-pure formamidinium perovskite solar cells, *Nature Communications* **2018**, *9*(1), 3021.

59. M. Kim, *et al.*, Methylammonium chloride induces intermediate phase stabilization for efficient perovskite solar cells, *Joule* **2019**, *3*(9), 2179–2192.

60. D. J. Kubicki, *et al.*, Formation of stable mixed guanidinium–methylammonium phases with exceptionally long carrier lifetimes for high-efficiency lead iodide-based perovskite photovoltaics, *Journal of the American Chemical Society* **2018**, *140*(9), 3345–3351.

61. Y. Zhang, Y. Li, L. Zhang, H. Hu, Z. Tang, B. Xu, N. G. Park, Propylammonium chloride additive for efficient and stable FAPbI3 perovskite solar cells, *Advanced Energy Materials* **2021**, *11*(47), 2102538.
62. Z. Wang, Q. Lin, F. P. Chmiel, N. Sakai, L. M. Herz, H. J. Snaith, Efficient ambient-air-stable solar cells with 2D–3D heterostructured butylammonium-caesium-formamidinium lead halide perovskites, *Nature Energy* **2017**, *2*(9), 1–10.
63. M. M. Davy, T. M. Jadel, C. Qin, B. Luyun, G. Mina, Recent progress in low dimensional (quasi-2D) and mixed dimensional (2D/3D) tin-based perovskite solar cells, *Sustainable Energy & Fuels* **2021**, *5*(1), 34–51.
64. A. Q. Alanazi, *et al.*, Atomic-level microstructure of efficient formamidinium-based perovskite solar cells stabilized by 5-ammonium valeric acid iodide revealed by multinuclear and two-dimensional solid-state NMR, *Journal of the American Chemical Society* **2019**, *141*(44), 17659–17669.
65. X. Li, *et al.*, Improved performance and stability of perovskite solar cells by crystal crosslinking with alkylphosphonic acid ω-ammonium chlorides, *Nature Chemistry* **2015**, *7*(9), 703–711.
66. Y. Wang, X. Liu, T. Zhang, X. Wang, M. Kan, J. Shi, Y. Zhao, The role of dimethyl-ammonium iodide in CsPbI3 perovskite fabrication: Additive or dopant? *Angewandte Chemie* **2019**, *131*(46), 16844–16849.
67. C. Yi, *et al.*, Entropic stabilization of mixed A-cation ABX3 metal halide perovskites for high performance perovskite solar cells, *Energy & Environmental Science* **2016**, *9*(2), 656–662.
68. Y. Zhao, *et al.*, Inactive (PbI2) 2RbCl stabilizes perovskite films for efficient solar cells, *Science* **2022**, *377*(6605), 531–534.
69. G. Kim, H. Min, K. S. Lee, D. Y. Lee, S. M. Yoon, S. I. Seok, Impact of strain relaxation on performance of α-formamidinium lead iodide perovskite solar cells, *Science* **2020**, *370*(6512), 108–112.
70. J. W. Lee, D. H. Kim, H. S. Kim, S. W. Seo, S. M. Cho, N. G. Park, Formamidinium and cesium hybridization for photo-and moisture-stable perovskite solar cell, *Advanced Energy Materials* **2015**, *5*(20), 1501310.
71. M. Abdi-Jalebi, *et al.* Maximizing and stabilizing luminescence from halide perovskites with potassium passivation, *Nature* **2018**, *555*(7697), 497–501.
72. L. Qiao, W. H. Fang, R. Long, O. V. Prezhdo, Extending carrier lifetimes in lead halide perovskites with alkali metals by passivating and eliminating halide interstitial defects, *Angewandte Chemie* **2020**, *132*(12), 4714–4720.
73. Y. H. Park, *et al.*, Inorganic rubidium cation as an enhancer for photovoltaic performance and moisture stability of HC (NH2) 2PbI3 perovskite solar cells, *Advanced Functional Materials* **2017**, *27*(16), 1605988.
74. M. Saliba, *et al.*, Incorporation of rubidium cations into perovskite solar cells improves photovoltaic performance, *Science* **2016**, *354*(6309), 206–209.
75. D.-Y. Son, S.-G. Kim, J.-Y. Seo, S.-H. Lee, H. Shin, D. Lee, N.-G. Park, Universal approach toward hysteresis-free perovskite solar cell via defect engineering, *Journal of the American Chemical Society* **2018**, *140*(4), 1358–1364.
76. J. Cao, S. X. Tao, P. A. Bobbert, C. P. Wong, N. Zhao, Interstitial occupancy by extrinsic alkali cations in perovskites and its impact on ion migration, *Advanced Materials* **2018**, *30*(26), 1707350.
77. N. Li, *et al.*, Cation and anion immobilization through chemical bonding enhancement with fluorides for stable halide perovskite solar cells, *Nature Energy* **2019**, *4*(5), 408–415.
78. D. J. Kubicki, D. Prochowicz, A. Hofstetter, S. M. Zakeeruddin, M. Grätzel, L. Emsley, Phase segregation in Cs-, Rb-and K-doped mixed-cation (MA) x (FA) 1–x PbI3 hybrid perovskites from solid-state NMR, *Journal of the American Chemical Society* **2017**, *139*(40), 14173–14180.

79. Y. El Ajjouri, V. S. Chirvony, M. Sessolo, F. Palazon, H. J. Bolink, Incorporation of potassium halides in the mechanosynthesis of inorganic perovskites: Feasibility and limitations of ion-replacement and trap passivation, *RSC Advances* **2018**, *8*(72), 41548–41551.

80. S.-G. Kim, *et al.*, Potassium ions as a kinetic controller in ionic double layers for hysteresis-free perovskite solar cells, *Journal of Materials Chemistry A* **2019**, *7*(32), 18807–18815.

81. S. Tan, *et al.*, Inorganic ammonium halide additive strategy for highly efficient and stable CsPbI3 perovskite solar cells, *Advanced Functional Materials* **2021**, *31*(21).

82. J. Zou, *et al.*, An efficient guanidinium isothiocyanate additive for improving the photovoltaic performances and thermal stability of perovskite solar cells, *Electrochimica Acta* **2018**, *291*, 297–303.

83. C. Zuo, L. Ding, An 80.11% FF record achieved for perovskite solar cells by using the NH4Cl additive, *Nanoscale* **2014**, *6*(17), 9935–9938.

84. Y. Rong, X. Hou, Y. Hu, A. Mei, L. Liu, P. Wang, H. Han, Synergy of ammonium chloride and moisture on perovskite crystallization for efficient printable mesoscopic solar cells, *Nature Communications* **2017**, *8*, 14555.

85. W. Fu, J. Wang, L. Zuo, K. Gao, F. Liu, D. S. Ginger, A. K. Y. Jen, Two-dimensional perovskite solar cells with 14.1% power conversion efficiency and 0.68% external radiative efficiency, *ACS Energy Letters* **2018**, *3*(9), 2086–2093.

86. X. Li, *et al.*, Improved performance and stability of perovskite solar cells by crystal crosslinking with alkylphosphonic acid omega-ammonium chlorides, *Nat Chem* **2015**, *7*(9), 703–711.

87. Y. Wang, X. Liu, T. Zhang, X. Wang, M. Kan, J. Shi, Y. Zhao, The role of dimethylammonium iodide in cspbi3 perovskite fabrication: Additive or dopant? *Angewandte Chemie International Edition* **2019**, *58*(46), 16691–16696.

88. X. Zhou, *et al.*, Highly efficient and stable GABr-modified ideal-bandgap (1.35 eV) Sn/Pb perovskite solar cells achieve 20.63% efficiency with a record small Voc deficit of 0.33 V, *Advanced Materials* **2020**, *32*(14).

89. K. Wang, *et al.*, All-inorganic cesium lead iodide perovskite solar cells with stabilized efficiency beyond 15%, *Nature Communications* **2018**, *9*(1).

90. C. Liang, *et al.*, Two-dimensional Ruddlesden–Popper layered perovskite solar cells based on phase-pure thin films, *Nature Energy* **2020**, *6*(1), 38–45.

91. W.-Q. Wu, *et al.*, Bilateral alkylamine for suppressing charge recombination and improving stability in blade-coated perovskite solar cells, *Science Advances* **2019**, *5*(3), eaav8925.

92. F. Bai, *et al.*, A 0D/3D heterostructured all-inorganic halide perovskite solar cell with high performance and enhanced phase stability, *Advanced Materials* **2019**, *31*(48).

93. H. Xu, *et al.*, CsI enhanced buried interface for efficient and UV-robust perovskite solar cells, *Advanced Energy Materials* **2021**, *12*(2).

94. I. S. Yang, N.-G. Park, Dual additive for simultaneous improvement of photovoltaic performance and stability of perovskite solar cell, *Advanced Functional Materials* **2021**, *31*(20), 2100396.

95. L. Wang, *et al.*, Double-layer CsI intercalation into an MAPbI3 framework for efficient and stable perovskite solar cells, *Nano Energy* **2021**, *86*.

96. S. Xiang, W. Li, Y. Wei, J. Liu, H. Liu, L. Zhu, S. Yang, H. Chen, Natrium doping pushes the efficiency of carbon-based CsPbI3 perovskite solar cells to 10.7%, *iScience* **2019**, *15*, 156–164.

97. C. F. J. Lau, *et al.*, Enhanced performance via partial lead replacement with calcium for a CsPbI3 perovskite solar cell exceeding 13% power conversion efficiency, *Journal of Materials Chemistry A* **2018**, *6*(14), 5580–5586.

98. F. Zuo, S. T. Williams, P. W. Liang, C. C. Chueh, C. Y. Liao, A. K. Y. Jen, Binary-metal perovskites toward high-performance planar-heterojunction hybrid solar cells, *Advanced Materials* **2014**, *26*(37), 6454–6460.

99. R. E. Beal, *et al.*, Cesium lead halide perovskites with improved stability for tandem solar cells, *The Journal of Physical Chemistry Letters* **2016**, *7*(5), 746–751.

100. G. E. Eperon, *et al.*, Perovskite-perovskite tandem photovoltaics with optimized band gaps, *Science* **2016**, *354*(6314), 861–865.

101. B. Zhao, *et al.*, High open-circuit voltages in tin-rich low-bandgap perovskite-based planar heterojunction photovoltaics, *Advanced Materials* **2017**, *29*(2), 1604744.

102. A. Goyal, S. McKechnie, D. Pashov, W. Tumas, M. Van Schilfgaarde, V. Stevanovic, Origin of pronounced nonlinear band gap behavior in lead–tin hybrid perovskite alloys, *Chemistry of Materials* **2018**, *30*(11), 3920–3928.

103. R. M. I. Bandara, S. M. Silva, C. C. Underwood, K. I. Jayawardena, R. A. Sporea, S. R. P. Silva, Progress of Pb-Sn mixed perovskites for photovoltaics: A review, *Energy & Environmental Materials* **2022**, *5*(2), 370–400.

104. G. M. Kim, A. Ishii, S. Öz, T. Miyasaka, MACl-assisted Ge doping of Pb-hybrid perovskite: A universal route to stabilize high performance perovskite solar cells, *Advanced Energy Materials* **2020**, *10*(7), 1903299.

105. L. Wang, *et al.*, A Eu3+-Eu2+ ion redox shuttle imparts operational durability to Pb-I perovskite solar cells, *Science* **2019**, *363*(6424), 265–270.

106. X. Shai, *et al.*, Achieving ordered and stable binary metal perovskite via strain engineering, *Nano Energy* **2018**, *48*, 117–127.

107. Q. Chen, *et al.*, Controllable self-induced passivation of hybrid lead iodide perovskites toward high performance solar cells, *Nano Letters* **2014**, *14*(7), 4158–4163.

108. F. Yang, *et al.*, Magnesium-doped MAPbI(3) perovskite layers for enhanced photovoltaic performance in humid air atmosphere, *ACS Applied Materials & Interfaces* **2018**, *10*(29), 24543–24548.

109. X. Gong, *et al.*, Highly efficient perovskite solar cells via nickel passivation, *Advanced Functional Materials* **2018**, *28*(50).

110. G. Murugadoss, R. Thangamuthu, Metals doped cesium based all inorganic perovskite solar cells: Investigations on Structural, morphological and optical properties, *Solar Energy* **2019**, *179*, 151–163.

111. J. Wang, *et al.*, Enhancing photostability of Sn-Pb perovskite solar cells by an alkylammonium pseudo-halogen additive, *Advanced Energy Materials* **2023**, *13*(15).

112. F. Meng, *et al.*, Ge Incorporation to stabilize efficient inorganic CsPbI3 perovskite solar cells, *Advanced Energy Materials* **2022**, *12*(10).

113. N. Li, X. Niu, Q. Chen, H. Zhou, Towards commercialization: The operational stability of perovskite solar cells, *Chemical Society Reviews* **2020**, *49*(22), 8235–8286.

114. D. M. Jang, *et al.*, Reversible halide exchange reaction of organometal trihalide perovskite colloidal nanocrystals for full-range band gap tuning, *Nano Letters* **2015**, *15*(8), 5191–5199.

115. J. H. Noh, S. H. Im, J. H. Heo, T. N. Mandal, S. I. Seok, Chemical management for colorful, efficient, and stable inorganic–organic hybrid nanostructured solar cells, *Nano Letters* **2013**, *13*(4), 1764–1769.

116. N. Yantara, F. Yanan, C. Shi, H. A. Dewi, P. P. Boix, S. G. Mhaisalkar, N. Mathews, Unravelling the effects of Cl addition in single step CH3NH3PbI3 perovskite solar cells, *Chemistry of Materials* **2015**, *27*(7), 2309–2314.

117. B. Philippe, B.-W. Park, R. Lindblad, J. Oscarsson, S. Ahmadi, E. M. Johansson, H. Rensmo, Chemical and electronic structure characterization of lead halide perovskites and stability behavior under different exposures: A photoelectron spectroscopy investigation, *Chemistry of Materials* **2015**, *27*(5), 1720–1731.

118. S. Colella, *et al.*, MAPbI3-xCl x mixed halide perovskite for hybrid solar cells: The role of chloride as dopant on the transport and structural properties, *Chemistry of Materials* **2013**, *25*(22), 4613–4618.

119. W. Ke, *et al.*, Employing lead thiocyanate additive to reduce the hysteresis and boost the fill factor of planar perovskite solar cells, *Advanced Materials* **2016**, *28*(26), 5214–5221.

120. M. K. Kim, T. Jeon, H. I. Park, J. M. Lee, S. A. Nam, S. O. Kim, Effective control of crystal grain size in CH 3 NH 3 PbI 3 perovskite solar cells with a pseudohalide Pb (SCN) 2 additive, *CrystEngComm* **2016**, *18*(32), 6090–6095.

121. J. Jeong, *et al.*, Pseudo-halide anion engineering for α-FAPbI3 perovskite solar cells, *Nature* **2021**, *592*(7854), 381–385.

122. J. Chen, S. G. Kim, N. G. Park, FA0.88Cs0.12PbI3– x (PF6) x interlayer formed by ion exchange reaction between perovskite and hole transporting layer for improving photovoltaic performance and stability, *Advanced Materials* **2018**, *30*(40), 1801948.

123. Y. C. Kim, *et al.*, Beneficial effects of PbI2 incorporated in organo-lead halide perovskite solar cells, *Advanced Energy Materials* **2015**, *6*(4).

124. Q. Ye, *et al.*, Cesium lead inorganic solar cell with efficiency beyond 18% via reduced charge recombination, *Advanced Materials* **2019**, *31*(49).

125. Y. Zheng, X. Yang, R. Su, P. Wu, Q. Gong, R. Zhu, High-Performance CsPbIxBr3-x All-inorganic perovskite solar cells with efficiency over 18% via spontaneous interfacial manipulation, *Advanced Functional Materials* **2020**, *30*(46).

126. Z. Yao, *et al.*, Pseudohalide (SCN−)-doped CsPbI3 for high-performance solar cells, *Journal of Materials Chemistry C* **2019**, *7*(44), 13736–13742.

127. Z. Zhang, *et al.*, Robust heterojunction to strengthen the performances of FAPbI3 perovskite solar cells, *Chemical Engineering Journal* **2022**, *432*.

128. D.-K. Lee, D.-N. Jeong, T. K. Ahn, N.-G. Park, Precursor engineering for a large-area perovskite solar cell with >19% efficiency, *ACS Energy Letters* **2019**, *4*(10), 2393–2401.

129. N. J. Jeon, J. H. Noh, Y. C. Kim, W. S. Yang, S. Ryu, S. I. Seok, Solvent engineering for high-performance inorganic-organic hybrid perovskite solar cells, *Nature Materials* **2014**, *13*(9), 897–903.

130. R. G. Pearson, Hard and soft acids and bases, *Journal of the American Chemical society* **1963**, *85*(22), 3533–3539.

131. J.-W. Lee, S.-H. Bae, Y.-T. Hsieh, N. De Marco, M. Wang, P. Sun, Y. A Yang, Bifunctional Lewis base additive for microscopic homogeneity in perovskite solar cells, *Chem* **2017**, *3*(2), 290–302.

132. D.-K. Lee, K.-S. Lim, J.-W. Lee, N.-G. Park, Scalable perovskite coating via antisolvent-free Lewis acid–base adduct engineering for efficient perovskite solar modules, *Journal of Materials Chemistry A* **2021**, *9*(5), 3018–3028.

133. N. J. Jeon, J. H. Noh, Y. C. Kim, W. S. Yang, S. Ryu, S. I. Seok, Solvent engineering for high-performance inorganic–organic hybrid perovskite solar cells, *Nature Materials* **2014**, *13*(9), 897–903.

134. J. W. Lee, *et al.*, Tuning molecular interactions for highly reproducible and efficient formamidinium perovskite solar cells via adduct approach, *Journal of the American Chemical Society* **2018**, *140*(20), 6317–6324.

135. P. W. Liang, *et al.*, Additive enhanced crystallization of solution-processed perovskite for highly efficient planar-heterojunction solar cells, *Advanced Materials* **2014**, *26*(22), 3748–3754.

136. L. Guan, N. Jiao, Y. Guo, Trap-state passivation by nonvolatile small molecules with carboxylic acid groups for efficient planar perovskite solar cells, *The Journal of Physical Chemistry C* **2019**, *123*(23), 14223–14228.

137. J. Yang, *et al.*, Extremely low-cost and green cellulose passivating perovskites for stable and high-performance solar cells, *ACS Applied Materials & Interfaces* **2019**, *11*(14), 13491–13498.

138. X. Liu, J. Wu, Y. Yang, T. Wu, Q. Guo, Pyridine solvent engineering for high quality anion-cation-mixed hybrid and high performance of perovskite solar cells, *Journal of Power Sources* **2018**, *399*, 144–150.

139. Y. H. Wu, *et al.*, Incorporating 4-tert-butylpyridine in an antisolvent: A facile approach to obtain highly efficient and stable perovskite solar cells, *ACS Appl Mater Interfaces* **2018**, *10*(4), 3602–3608.

140. Y. Deng, X. Zheng, Y. Bai, Q. Wang, J. Zhao, J. Huang, Surfactant-controlled ink drying enables high-speed deposition of perovskite films for efficient photovoltaic modules, *Nature Energy* **2018**, *3*(7), 560–566.
141. X. Meng, *et al.*, Stretchable perovskite solar cells with recoverable performance, *Angewandte Chemie International Edition* **2020**, *59*(38), 16602–16608.
142. T. Zhang, *et al.*, Mediator–antisolvent strategy to stabilize all-inorganic CsPbI3 for perovskite solar cells with efficiency exceeding 16%, *ACS Energy Letters* **2020**, *5*(5), 1619–1627.
143. Y. Xu, X. Wang, Z. Jin, B. Li, W. An, Q. Zhang, Y. Rui, Fluorocarbon surfactant-assisted large-area uniform and water-resistant perovskite films for efficient photovoltaics, *Solar RRL* **2023**, *7*(15), 2300302.
144. N. K. Noel, *et al.*, Enhanced photoluminescence and solar cell performance via Lewis base passivation of organic–inorganic lead halide perovskites, *ACS Nano* **2014**, *8*(10), 9815–9821.
145. T. Y. Wen, *et al.*, Surface electronic modification of perovskite thin film with water-resistant electron delocalized molecules for stable and efficient photovoltaics, *Advanced Energy Materials* **2018**, *8*(13), 1703143.
146. L. Meng, *et al.*, Tailored phase conversion under conjugated polymer enables thermally stable perovskite solar cells with efficiency exceeding 21%, *Journal of the American Chemical Society* **2018**, *140*(49), 17255–17262.
147. C. Fei, B. Li, R. Zhang, H. Fu, J. Tian, G. Cao, Highly efficient and stable perovskite solar cells based on monolithically grained CH3NH3PbI3 film, *Advanced Energy Materials* **2017**, *7*(9), 1602017.
148. F. Zhang, *et al.*, Suppressing defects through the synergistic effect of a Lewis base and a Lewis acid for highly efficient and stable perovskite solar cells, *Energy & Environmental Science* **2018**, *11*(12), 3480–3490.
149. M. Sun, F. Zhang, H. Liu, X. Li, Y. Xiao, S. Wang, Tuning the crystal growth of perovskite thin-films by adding the 2-pyridylthiourea additive for highly efficient and stable solar cells prepared in ambient air, *Journal of Materials Chemistry A* **2017**, *5*(26), 13448–13456.
150. T. Wu, *et al.*, Efficient defect passivation for perovskite solar cells by controlling the electron density distribution of donor-π-acceptor molecules, *Advanced Energy Materials* **2019**, *9*(17), 1803766.
151. M. Qin, *et al.*, Fused-ring electron acceptor ITIC-Th: A novel stabilizer for halide perovskite precursor solution, *Advanced Energy Materials* **2018**, *8*(18), 1703399.
152. M. Shahiduzzaman, K. Yamamoto, Y. Furumoto, T. Kuwabara, K. Takahashi, T. Taima, Ionic liquid-assisted growth of methylammonium lead iodide spherical nanoparticles by a simple spin-coating method and photovoltaic properties of perovskite solar cells, *RSC Advances* **2015**, *5*(95), 77495–77500.
153. M. Salado, F. J. Ramos, V. M. Manzanares, P. Gao, M. K. Nazeeruddin, P. J. Dyson, S. Ahmad, Extending the lifetime of perovskite solar cells using a perfluorinated dopant, *ChemSusChem* **2016**, *9*(18), 2708–2714.
154. Y. Zhang, *et al.*, A strategy to produce high efficiency, high stability perovskite solar cells using functionalized ionic liquid-dopants, *Advanced Materials* **2017**, *29*(36), 1702157.
155. S. Wang, *et al.*, Water-soluble triazolium ionic-liquid-induced surface self-assembly to enhance the stability and efficiency of perovskite solar cells, *Advanced Functional Materials* **2019**, *29*(15), 1900417.
156. D. Bi, *et al.*, Polymer-templated nucleation and crystal growth of perovskite films for solar cells with efficiency greater than 21%, *Nature Energy* **2016**, *1*(10).
157. S. Bai, *et al.*, Planar perovskite solar cells with long-term stability using ionic liquid additives, *Nature* **2019**, *571*(7764), 245–250.
158. R. Xia, *et al.*, Retarding thermal degradation in hybrid perovskites by ionic liquid additives, *Advanced Functional Materials* **2019**, *29*(22), 1902021.

159. T.-H. Han, *et al.*, Perovskite-polymer composite cross-linker approach for highly-stable and efficient perovskite solar cells, *Nature Communications* **2019**, *10*(1), 520.

160. R. Wang, *et al.*, Caffeine improves the performance and thermal stability of perovskite solar cells, *Joule* **2019**, *3*(6), 1464–1477.

161. T. Niu, *et al.*, Stable high-performance perovskite solar cells via grain boundary passivation, *Advanced Materials* **2018**, *30*(16), e1706576.

162. P. Guo, *et al.*, Surface & grain boundary co-passivation by fluorocarbon based bifunctional molecules for perovskite solar cells with efficiency over 21%, *Journal of Materials Chemistry A* **2019**, *7*(6), 2497–2506.

163. S. Wang, *et al.*, Water-soluble triazolium ionic-liquid-induced surface self-assembly to enhance the stability and efficiency of perovskite solar cells, *Advanced Functional Materials* **2019**, *29*(15).

164. X. Zhou, Y. Wang, C. Li, T. Wu, Doping amino-functionalized ionic liquid in perovskite crystal for enhancing performances of hole-conductor free solar cells with carbon electrode, *Chemical Engineering Journal* **2019**, *372*, 46–52.

165. S. Bai, *et al.*, Planar perovskite solar cells with long-term stability using ionic liquid additives, *Nature* **2019**, *571*(7764), 245–250.

166. R. Xia, *et al.* Retarding thermal degradation in hybrid perovskites by ionic liquid additives, *Advanced Functional Materials* **2019**, *29*(22), 1902021.

167. T. Wu, *et al.*, Efficient defect passivation for perovskite solar cells by controlling the electron density distribution of donor-π-acceptor molecules, *Advanced Energy Materials* **2019**, *9*(17), 1803766.

7 Unraveling the Complex Crystallization Dynamics of Perovskites

Kui Zhao and Yachao Du

7.1 IN-SITU GRAZING-INCIDENCE WIDE-ANGLE X-RAY SCATTERING (GIWAXS), ULTRAVIOLET–VISIBLE (UV-VIS) SPECTROSCOPY, AND PHOTOLUMINESCENCE (PL) CHARACTERIZATION

This section briefly introduces the in-situ grazing-incidence wide-angle X-ray scattering (GIWAXS), ultraviolet–visible (UV-vis) spectroscopy, and photoluminescence (PL) characterization techniques, as they are widely used in studies of perovskite film formation.

7.1.1 IN-SITU GIWAXS

X-rays are electromagnetic waves with wavelengths of 0.1–10 Å, close to atomic dimensions. Therefore, X-rays interact easily with atoms. Following these interactions, X-ray scattering occurs, which can be elastic (e.g., Rayleigh scattering) or inelastic (e.g., Compton scattering). Inelastic scattering occurs in conjunction with energy transfer. An incident X-ray is absorbed by an atom, which emits electrons, thereby generating a lower-energy scattered X-ray.[1] Analysis of such scattered X-rays reveals the electronic states and coordination environments of the target atoms. In contrast, for elastic scattering, there is no energy difference between the incident and scattered X-rays. Elastic scattering is often used to study the atomic arrangements of a sample.

There are two main techniques that exploit X-ray scattering to reveal crystal lattice information: traditional X-ray diffraction (XRD) and grazing-incidence X-ray scattering (GIXS), which is usually GIWAXS. GIWAXS has several advantages over XRD, as detailed later.

1. Compared with XRD, a very small incident angle ($\alpha_i < 1°$) is typically used with GIWAXS. The small α_i ensures that the X-ray travels over a wide sample area. Therefore, when thinner films are targeted for analysis, the detector is provided with more scattering signals. Hence, sufficient and reliable structure information can be obtained.

DOI: 10.1201/9781003400486-7

2. The structure at different depths of a sample crystal can be revealed by changing the GIWAXS α_i, where a larger α_i corresponds to deeper X-ray probing. For example, when α_i is considerably smaller than the critical angle, only the structural information at the sample surface is revealed. For an ideally flat surface, the X-ray penetration depth depends on the X-ray energy (wavelength λ), the critical angle of total reflection, α_c, and the incident angle, α_i, can be estimated using the relation:[2,3]

$$= \frac{\lambda}{4\pi} \sqrt{\frac{2}{\sqrt{\left(\alpha_i^2 - \alpha_c^2\right)^2 + 4\beta^2} - \left(\alpha_i^2 - \alpha_c^2\right)}} \tag{7.1}$$

where β is the imaginary part of the complex refractive index of the compound. Accordingly, for a certain density of the perovskite film, the penetration depth is given as a function of the incidence angle as shown in Figure 7.1.

3. Unlike traditional XRD, for which a point detector is used, a 2D detector is employed for GIWAXS. This detector can reveal the grain orientation information of the sample.
4. Most importantly, for GIWAXS, high-flux and brilliant synchrotron radiation is often used as the X-ray source. Thus, a higher scattering intensity and signal-to-noise ratio can easily be obtained in milliseconds. Therefore, GIWAXS is suitable for in-situ measurements.

Given the aforementioned advantages, GIWAXS is a powerful method for studying the crystallization kinetics that arise during the preparation of perovskite film under various conditions, such as those of slot-die, spin, blade coating, and hot casting.

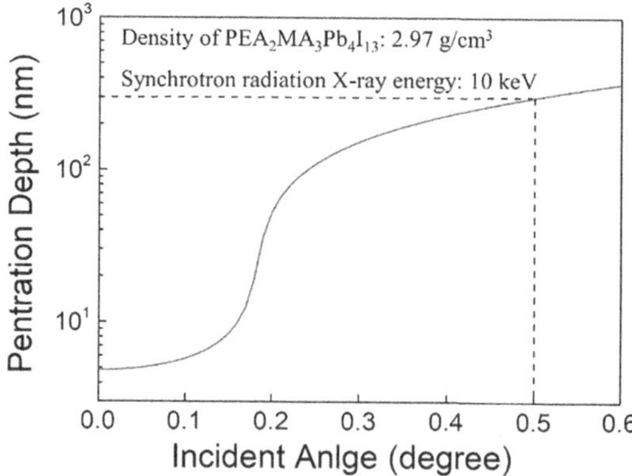

FIGURE 7.1   Correlation between X-ray penetration depth and incident angle. Reprinted in part with permission from Ref.[3]. Copyright 2021, Springer Nature.

7.1.2 IN-SITU UV-VIS ABSORPTION SPECTROSCOPY

Absorption spectroscopy is the most direct method of measuring the band structures of perovskites and other semiconductors. When subjected to electromagnetic radiation, a semiconductor absorbs photons with a certain energy. Electrons from the valence band are then excited to the conduction band or other excited states above the conduction band.[4] The distribution of the electron excitation states can be determined from the absorption spectrum, which indicates not only the bandgap but also certain sub-band or band tail states. Thus, in-situ absorption spectroscopy can be used to record real-time optical evolution during semiconductor fabrication or processing.

7.1.3 IN-SITU PL SPECTROSCOPY

Photoluminescence (PL) refers to light emission following the absorption of external photons with sufficient energy. PL radiation is caused by the transition of electrons from higher occupied electronic states to lower unoccupied states, with the emission of photons.[5] In the case of perovskite film, for a certain radiation intensity, the PL peak intensity is often related to the stacking faults within the film. The PL peak position corresponds to the bandgap and is also affected by the trap states. Generally, spontaneous radiative recombination between the trap states yields an emission peak that is red-shifted compared to that of the band-edge transition. Passivation of these trap states can blue-shift the PL peak.[6] In addition, the Urbach energy, which is related to shallow defects and bandgap fluctuations near the bandgap, can be derived by analyzing the low-energy portion of the PL spectrum. For in-situ PL measurements, high monochromatic-power laser excitation sources are used to ensure good signal-to-noise ratios in a very short time (of millisecond order). In summary, in-situ PL characterization is an effective method of revealing the transformation of stacking faults during perovskite film formation.

Overall, in-situ GIWAXS, UV-vis absorption spectroscopy, and PL spectroscopy measurements are powerful methods for investigating the structural evolutions of perovskite films during their formation.

7.2 COMPLEX CONVERSION DYNAMICS DURING PEROVSKITE CRYSTALLIZATION

This section discusses the complex conversion dynamics of methylammonium and formamidinium lead iodide ($MAPbI_3$ and $FAPbI_3$, respectively), along with all-inorganic lead halide perovskites.

7.2.1 METHYLAMMONIUM LEAD IODIDE ($MAPbI_3$) CONVERSION DYNAMICS

$MAPbI_3$ is a perovskite that was first used as a light-absorption layer for perovskite photovoltaic devices in 2009, owing to its suitable bandgap of ~1.57 eV.[7] Early fabrication of $MAPbI_3$ involved a one-step spin-coating deposition method in which equimolar amounts of lead(II) iodide (PbI_2) and methylammonium iodide (MAI) were dissolved in N, N-dimethylformamide (DMF) as a single solvent. However, the perovskite film fabricated via this method is usually rough and discontinuous, and resulting photovoltaic devices often exhibit low device performance owing to their poor perovskite film quality. Considerable research has been conducted to obtain higher-quality films. In particular, various methods of controlling the conversion dynamics have been considered, including component regulation, additive engineering, and processing procedure regulation.

Amassian et al.[8] investigated the solidification pathways of organometallic halide solutions by varying the halide and MAI contents. As shown in Figure 7.2, they found that the precursor solvates phases that formed during the solvent evaporation process, which were mainly determined by the halide and MAI contents, critically influenced the $MAPbI_3$ conversion dynamics and, thus, the morphology of the final perovskite film. The wet gel subsequently facilitated the formation of ordered (equimolar iodide and chloride solutions) or disordered (equimolar bromide or 3:1 chloride) precursor solvates. The ordered precursor solute phases were stable and retained the solvent for long durations, yielding consistent conversion to the perovskite phase and solar cell performance. Tassone et al.[9] deposited $MAPbI_3$ from lead(II) chloride ($PbCl_2$) + 3MAI in a DMF precursor solution and found that MA_2PbI_3Cl and $MACl$ crystalline precursors formed first. When the $MACl$ release speed was slowed, the precursor

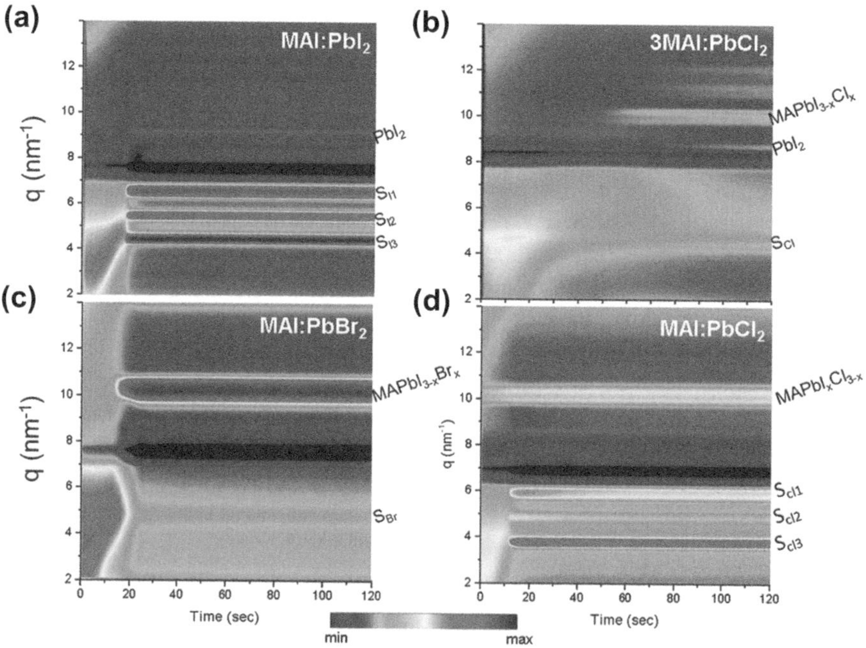

FIGURE 7.2   In-situ GIWAXS measurements of ink-to-solid transformation during one-step spin-coating of perovskite solutions: (a–d) equimolar iodide, iodide:chloride = 3:1, equimolar bromide, and equimolar chloride solutions, respectively. Reprinted in part with permission from Ref.[8]. Copyright 2016, WILEY-VCH.

transformation to perovskite was dramatically delayed, yielding high crystalline quality and long carrier lifetimes.

In addition to adjusting the precursor-solution molecular ratios, additive engineering is an effective method of regulating the crystallization kinetics. For example, by adding an N-cyclohexyl-2-pyrrolidone (CHP) additive to the DMF, the PbI_2:DMF can be remarkably inhibited because of the strong interactions between the CHP molecules and PbI_2. As shown from the in-situ UV-vis absorption results in Figure 7.3, the timing of film formation for w/CHP was 10 s longer than for pristine film, by fitting the results using a biexponential function, as shown in Figure 7.3c and d. The values of t1 and t2 from the bi-exponential fitting were assumed to be respectively strongly related to the evaporation of solvent in the early stage and the formation of solvate in the second stage. The differences in t2 values can be attributed to the sophisticated

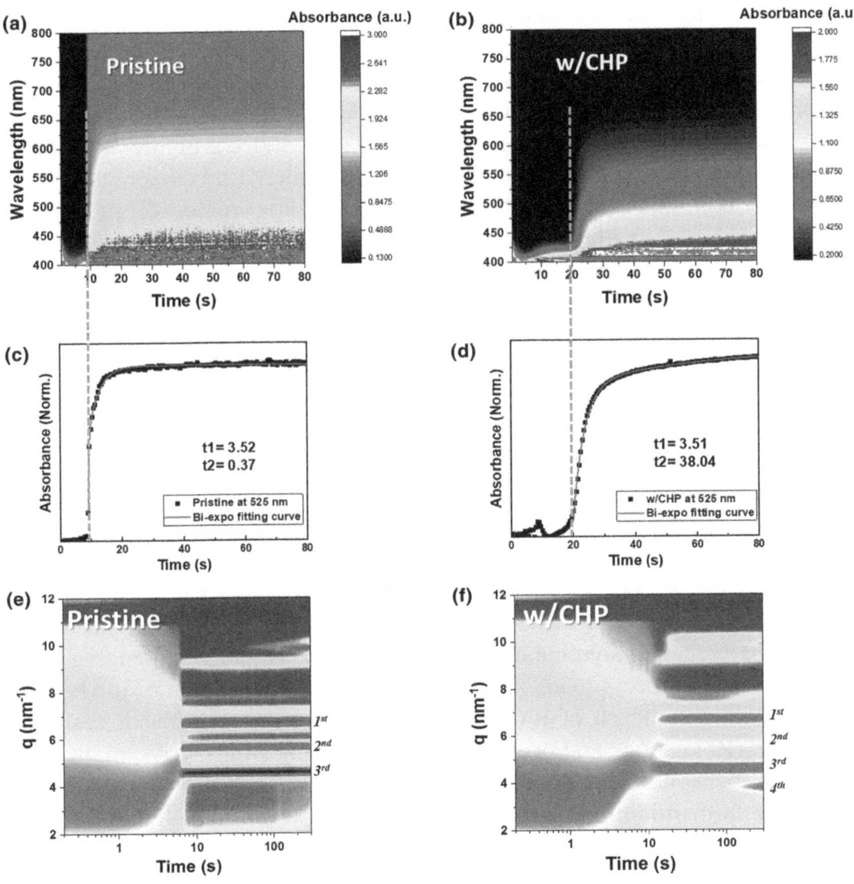

FIGURE 7.3 (a–d) In-situ UV–vis absorption characteristics of (a, c) for pristine and (b, d) for w/CHP films. (e, f) In-situ GIWAXS results of various precursor perovskite films, formation of solvate phase and perovskite crystals of (e) pristine, (f) w/CHP films. Reprinted in part with permission from Ref.[10]. Copyright 2020, Royal Society of Chemistry.

phenomena of solvate phase formation, which was further confirmed by the in-situ GIWAXS results as shown in Figure 7.3e and f; the solvate phase of PbI_2:DMF was remarkably suppressed in the case of w/CHP, which is attributed to the strong Lewis acid–base interaction with PbI_2 molecules. This suppression of the PbI_2:DMF solvate can efficiently balance the crystal growth rate and generate a dense perovskite film morphology.[10]

Our group introduced different semiconducting molecules with Lewis acid or base functional groups to a sol–gel $MAPbI_3$ film.[11] The molecules triggered heterogeneous nucleation and growth of the crystalline solvate ($MAPbI_3$·dimethylsulfoxide (DMSO)) during solution casting, which was explored using in-situ synchrotron-based GIWAXS. These crystalline solvates provided frameworks for the growth of compact films without significant gaps, without detrimentally altering the perovskite-film microstructure and texture.[11]

Furthermore, the coating method and conditions (e.g., the substrate temperature) can influence the crystallization process. In one study, significant differences in the crystallization dynamics were found for samples prepared using spin coating and blade coating at room temperature.[12] The processing temperature affected the structure, composition, and solvent retention of the intermediate solvate. The solvate phase could be avoided at high temperatures (e.g., $> 100°C$). Figure 7.4 shows the real-time perovskite phase transitions from the disordered precursor sol-gel to the solid film for both spin coating and blade coating. The key problem for rational printing design of perovskite films pertains to whether the intermediate phases participate in the phase transition. For both spin coating and blade coating, large-grained dense perovskite films with high crystal quality and good photophysical properties can only be obtained via direct crystallization. This process can be realized by implementing substrate-induced fast removal of solvents during the blade-coating process.

7.2.2   FAPbI$_3$ Conversion Dynamics

Recently, the use of formamidinium (FA$^+$) rather than MA$^+$ as a major organic cation in perovskite formulations has attracted considerable interest because the bandgap for α-FAPbI$_3$ perovskite is lower than that of MA-based perovskite (1.48 vs. 1.58 eV) and closer to the Shockley–Queisser ideal band gap (1.34 eV). This lower bandgap enables the perovskite to harvest and convert more sunlight to electricity.[13,14] However, the precursor solution conversion to FAPbI$_3$ is more challenging than for MAPbI$_3$, and this process tends to be hampered by the formation of the hexagonal non-perovskite polymorph of FAPbI$_3$, which requires the addition of cesium (Cs) or rubidium (Rb) cations and/or extensive thermal annealing.[15,16] As an alternative, FAPbI$_3$ has been mixed with methylammonium lead tribromide (MAPbBr$_3$) to enhance the formamidinium lead halide stability; however, the addition of the bromide compound increases the band gap.

As regards the fabrication of pure FAPbI$_3$ using a one-step spin-coating method, δ-phase FAPbI$_3$ tends to form during the spinning process because of its low formation energy;[17] this has been demonstrated by Zhu et al.[17] using in-situ XRD and by Amassian et al.[18] using in-situ GIWAXS. Amassian et al.[18] further demonstrated that replacing Br with I can facilitate the direct crystallization of the 3C perovskite

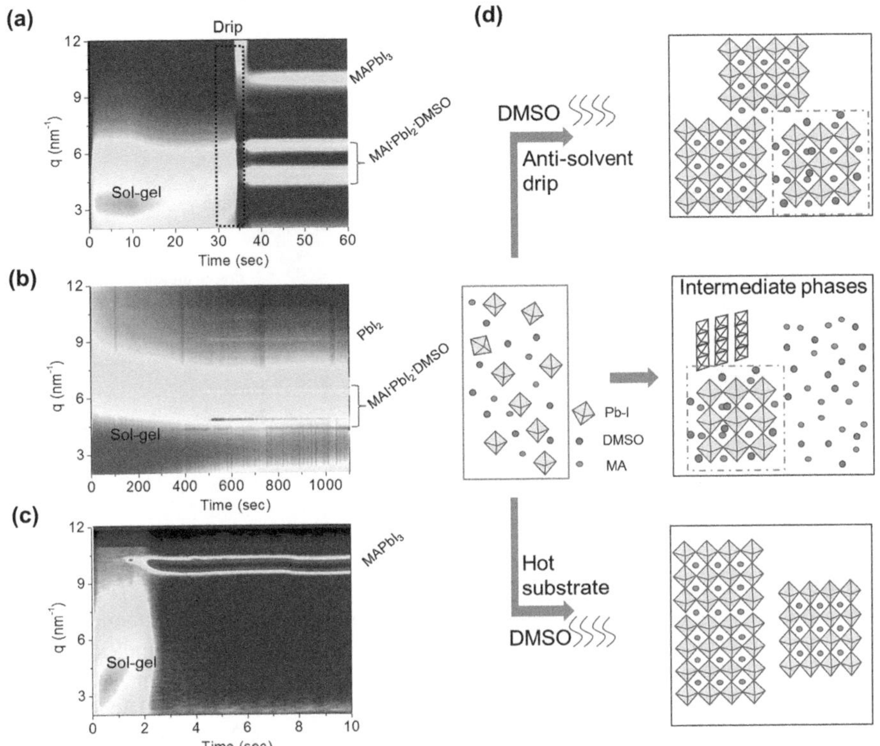

FIGURE 7.4 In-situ GIWAXS measurements of MAPbI₃ for different fabrication conditions: (a) the reference (DMSO:γ-butyrolactone (GBL)–blade–25°C with antisolvent drip); (b) DMSO:GBL–blade–25°C; (c) DMSO:GBL–blade–150°C; (d) schematic models of the MAPbI₃ structural evolution for each of the three cases. Reprinted in part with permission from Ref.[12]] Copyright 2018, Elsevier.

phase. This behavior is ascribed to the presence of Pb–Br bonds, which are stronger than those of Pb-DMSO, as opposed to Pb–I bonds, which are weaker than those of Pb-DMSO.[18]

Cl-based volatile additives are essential for facilitating the crystallization and phase transition of α-FAPbI₃ films. Jen et al.[19] studied the functions of different Cl-based volatile additives in FAPbI₃ films using in-situ PL. As shown in Figure 7.5, ammonium chloride (NH₄Cl) and FACl facilitated crystallization and lowered the phase transition temperatures. MA-based additives (MACl) quickly induced MA-rich nuclei to form pure α-phase FAPbI₃ and dramatically reduced the phase transition temperatures. Furthermore, the volatile MACl had a unique promotion effect on the secondary crystallization growth during annealing. The optimized solar cell exhibited a PCE of 23.1%, which is the highest value reported for FAPbI₃-based inverted PSCs.[19]

Another effective α-FAPbI₃ fabrication method involves regulation of the phase formation path (such as the PbI₂-related intermediate phase) to facilitate the transformation to α-FAPbI₃. Using in-situ GIWAXS measurements (Figure 7.6),[20] Amassian

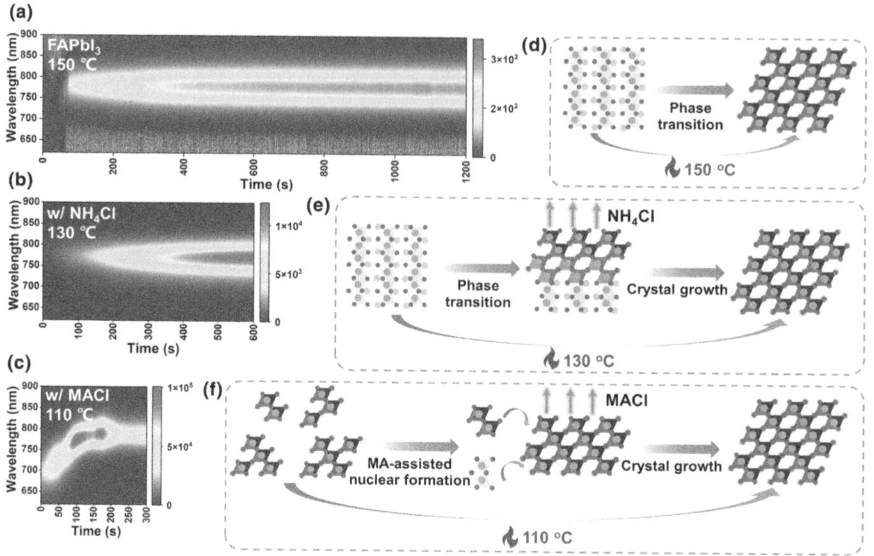

FIGURE 7.5 In-situ PL measurements of FAPbI₃ with different Cl-based volatile additives: heat maps of in-situ PL results for FAPbI₃ films prepared under different annealing temperatures: (a) control, (b) NH₄Cl-containing, and (c) MACl-containing samples; schematic illustrations of crystallization kinetics for transformation to α-FAPbI₃ film under various annealing temperatures for (d) control, (e) NH₄Cl-containing, and (f) MACl-containing samples. Reprinted in part with permission from Ref.[19]. Copyright 2023, American Chemical Society.

et al. compared the conversion behaviors of a PbI_2 precursor from different states, including a disordered colloidal film (P_0) in the presence of DMF, ordered solvated polymorphs with DMF (P_1 and P_2), and crystalline PbI_2. The in-situ results revealed that α-FAPbI₃ formed spontaneously and efficiently from P_2 (a metastable crystalline solvated polymorph of $PbI_2 \cdot DMF$ exhibiting an expanded lattice) at room temperature, with complete conversion after thermal annealing. However, for all other PbI_2 solvate phases, partial conversion of the perovskite phase was obtained together with a high fraction of the hexagonal δ-FAPbI₃ phase; this was despite extensive thermal annealing for conversion to α-FAPbI₃. Perovskite solar cells (PSCs) fabricated from the ordered P_2 exhibited improved semiconductor properties such as a lower recombination rate, higher power conversion efficiency (PCE), and improved reproducibility than devices fabricated using conventionally annealed PbI_2 film. The average PCE of the fabricated solar cells was improved from 16.0 (± 0.32) % using annealed PbI_2 to 17.23 (± 0.28) % for deposition onto P_2, with the PCE of the best-performing device approaching 18%.[20] Therefore, Huang et al.[21] attempted to alter the PbI_2 crystallization process by applying a methylammonium formate (MAFa) ionic liquid as a solvent. The PbI_2@MAFa mixture contained a strong C=O···Pb interaction (where C and O indicate carbon and oxygen, respectively). Moreover, the N–H···I hydrogen bonds (where N and H indicate nitrogen and hydrogen, respectively) between the amino groups of the MAFa and free

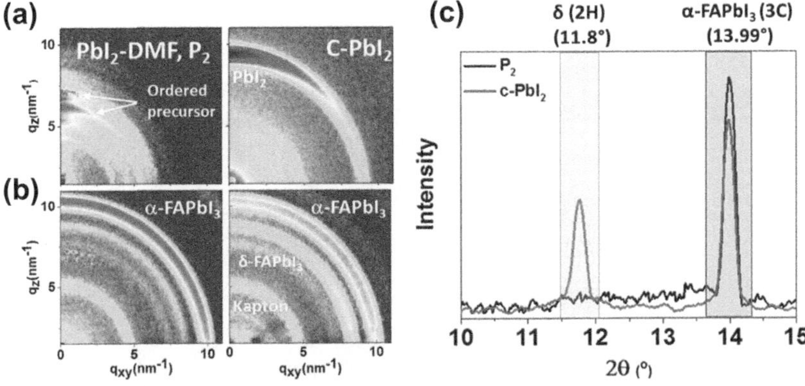

FIGURE 7.6 GIWAXS snapshots obtained during (a) in-situ measurements of PbI_2 from different precursor phases (i.e., metastable P_2 and crystalline PbI_2 (c-PbI_2)) and (b) its conversion to $FAPbI_3$ following loading and spinning of FAI and isopropyl alcohol washing of the samples. (c) XRD patterns of as-cast perovskite films. Reprinted in part with permission from Ref.[20] Copyright 2020, WILEY-VCH.

I^- promoted vertical growth of the PbI_2, which facilitated formamidinium iodide (FAI) insertion into the layered PbI_2 to form α-$FAPbI_3$, rather than interacting with PbI_2 at the edge to form δ-phase $FAPbI_3$.[21]

Furthermore, the addition of MA^+, Cs^+, and Rb^+ has been demonstrated to enhance the FA^+-dominated mixed cations and mixed-halide perovskite phase stability, yielding a more thermally stable, wider antisolvent processing window and efficient multicomponent PSCs. To reveal the relationship between the solidification pathway and the effects of the mixed cations and halides, we compared three perovskite formulations ($FAPbI_3$, $FA_{0.8}MA_{0.15}Cs_{0.05}PbI_3$, and $FA_{0.8}MA_{0.15}Cs_{0.05}PbI_{2.55}Br_{0.45}$) in terms of their representative dynamic microstructural evolution and conversion processes during blade coating using in-situ GIWAXS measurements.[22] As shown in Figure 7.7, for the pure $FAPbI_3$ case, our in-situ results revealed the formation of an intense yellow, non-perovskite 2H (100) phase within 10 s from the disordered sol-gel phase, without perovskite phase formation. The addition of MA^+ and Cs^+ to the $FAPbI_3$ solution, which yielded a $FA_{0.8}MA_{0.15}Cs_{0.05}PbI_3$ composition, promoted the formation of the perovskite 3C (100) phase and the non-perovskite 2H (100) phase from the disordered sol-gel precursor. Through the incorporation of Br− into the previous solution to obtain $FA_{0.8}MA_{0.15}Cs_{0.05}PbI_{2.55}Br_{0.45}$ perovskite ink, the undesired intermediate phases were suppressed and exclusive formation of the desired perovskite 3C phase was enabled.

On the basis of these insights, we achieved maximum and average PCEs of 18.2% and 16.4%, respectively, for $FA_{0.8}MA_{0.15}Cs_{0.05}PbI_{2.55}Br_{0.45}$ based on planar heterojunction n−i−p PSCs made with blade-coated perovskite films in 50% R.H. ambient air. In comparison, we obtained maximum and average PCEs of 16.3% and 14.1%, respectively, for $FA_{0.8}MA_{0.15}Cs_{0.05}PbI_3$, and of 12.4% and 9.0%, respectively, for $FAPbI_3$.[22] Therefore, the phase transition is drastically different in mixed-cation mixed-halide perovskites, with no formation of non-solvated intermediate hexagonal phases that

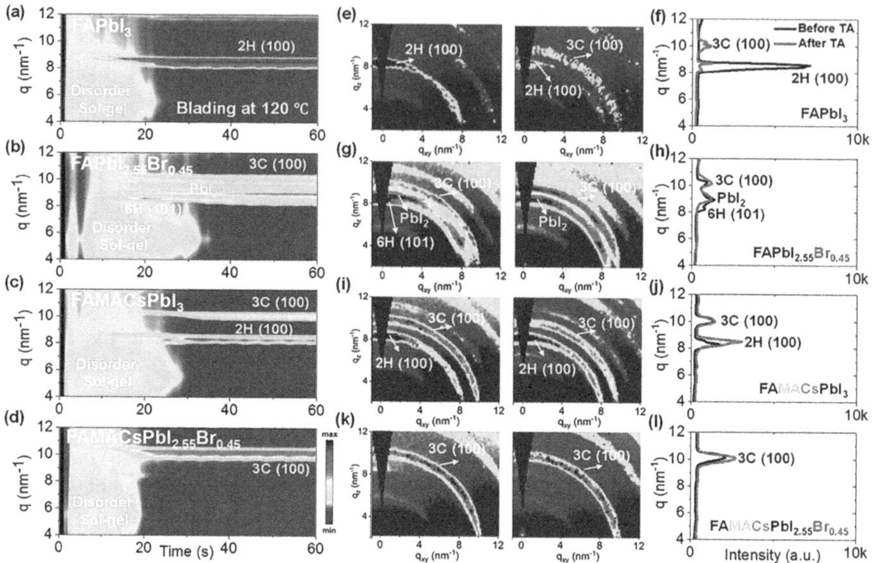

FIGURE 7.7 In-situ 2D GIWAXS scattering evolution vs. time for various perovskite precursor formulations showing ink-to-solid transformation during blade coating for (a) $FAPbI_3$, (b) $FAPbI_{2.55}Br_{0.45}$, (c) $FA_{0.8}MA_{0.15}Cs_{0.05}PbI_3$, and (d) $FA_{0.8}MA_{0.15}Cs_{0.05}PbI_{2.55}Br_{0.45}$. Representative 2D GIWAXS snapshots taken at the final stage of blade coating as well as post-annealing (10 min at 100°C) together with a plot of q vs. intensity are shown for (e and f) $FAPbI_3$, (g and h) $FAPbI_{2.55}Br_{0.45}$, (i and j) $FA_{0.8}MA_{0.15}Cs_{0.05}PbI_3$, and (k and l) $FA_{0.8}MA_{0.15}Cs_{0.05}PbI_{2.55}Br_{0.45}$. Scattering features associated with the disordered sol-gel phase and yellow non-perovskite 2H (100) phase ($q = 8.5$ nm⁻¹) are identified along with the (100) diffraction of PbI_2 ($q = 9.0$ nm⁻¹) and perovskite 3C (110) phase diffraction ($q = 10.2$ nm⁻¹). Reprinted in part with permission from Ref.[22]. Copyright 2020, Royal Society of Chemistry.

promote halide segregation. The ability to control or prevent halide segregation in perovskites, as derived from an in-depth understanding of perovskite ink drying and phase transformation, is crucial for realizing stable and efficient mixed-cation/mixed-halide PSCs.

7.2.3 ALL-INORGANIC LEAD HALIDE PEROVSKITE CONVERSION DYNAMICS

All-inorganic halide perovskites are promising for improving the thermal stability of PSCs.[23,24] PSCs based on $CsPbI_3$ have attained high PCEs exceeding 21.0% when fabricated via the insert spin-coating method and exceeding 20.33% via the ambient printing method.[25–27] Thus, these PSCs exhibit considerable potential for future commercialization. However, the main challenge hindering the practical application of $CsPbI_3$-based PSCs is the poor stability of $CsPbI_3$ itself. In particular, the optically active black $CsPbI_3$ phases (α, β, γ) are easily converted to thermodynamically more stable non-perovskite phases (δ-$CsPbI_3$), with an accompanying loss in optical activity.

Compositional X-site engineering through the introduction of smaller halide ions (namely, Br⁻ or Cl⁻) is a widely used and useful method of adjusting the tolerance factor and, thus, stabilizing the cubic phase of the $CsPbI_3$ film. The roles of I⁻, Br⁻, and Cl⁻ in the phase transition and crystallization kinetics were systematically studied by Lu et al.[28] using synchrotron-based in-situ GIWAXS characterization. They found that Br⁻ plays an important role in regulating the crystallization kinetics, with the ability to promote formation of the photoactive black phase and to suppress the unwanted yellow phase transition. Further introduction of Cl⁻ can also improve the film crystallinity and orientational order.

Another challenge for the practical application of all-inorganic perovskites is the upscaling of uniform and pinhole-free all-inorganic perovskite coatings; this is complicated by the fluid dynamics of the ink, the moisture sensitivity, and its solidification mechanisms. Our group achieved a highly crystalline and uniform pinhole-free all inorganic perovskite $CsPbI_2Br$ film via blade coating under ambient conditions; this was achieved by overcoming the negative influences of moisture and the Bénard–Marangoni instability during drying of the ink at an optimum processing temperature of 80°C.[29] Notably, nucleation-dominated crystallization and sequential crystallization processes were found during film formation using time-resolved GIWAXS (Figure 7.8). The diffraction peak at $q = 10.2$ nm⁻¹ for the 80-°C sample was assigned to the (100) plane of the $CsPbI_2Br$ α-phase, and the interplanar δ spacing was determined to be 0.62 nm. The diffraction peak at $q = 10.8$ nm⁻¹ for the 130°C sample corresponded to the (100) plane of the $CsPbBr_3$ α-phase, with a d spacing of 0.58 nm. The evolution of these two peaks was tracked during crystallization. In the early stage, only Br participated in the crystallization, forming the $CsPbBr_3$ phase. This was followed by an increase in iodide involvement, with the formation of $CsPbI_xBr_{3-x}$ $(0 < x < 2)$ phases. In the final stage, the $CsPbI_2Br$ phase was observed. This sequential crystallization yielded a continuous shift of the perovskite reflection to lower q values in the in-situ GIWAXS patterns. Such sequential crystallization is important for the final film because enhanced mass transport in the drying ink layer may favor high-quality crystallization. In contrast, $CsPbI_2Br$ crystallization that is absent during film formation at 25 and 130°C should occur during high-temperature

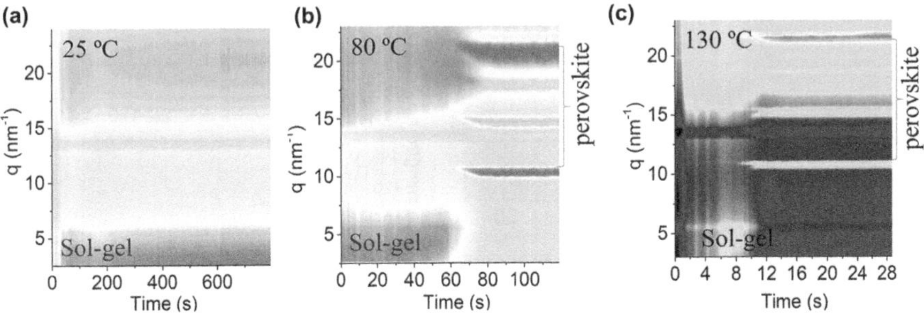

FIGURE 7.8 GIWAXS results showing structural evolution during film formation through blade coating of $CsPbI_2Br$ films at (a) 25, (b) 80, and (c) 130°C. Reprinted in part with permission from Ref.[29]. Copyright 2019, Elsevier.

thermal annealing. In such cases, there may not be sufficient time for a high-quality reaction, as evidenced by the poor morphology and low crystallinity of the resultant film. Ideal sequential crystallization with changing halide composition during film formation was observed for all the inorganic perovskites considered in our study, and the first demonstration of high-efficiency $CsPbI_2Br$ solar cells fabricated via ambient blade coating was achieved.[29] In the case of ambient scalable fabrication of all-inorganic perovskite $CsPbI_3$ solar cells, a slow drying process during low-temperature deposition allows for moisture attack, which suppresses crystallization of the precursors. Such unwanted moisture attack to perovskite films can in principle be suppressed during fast ambient crystallization processes such as those used for scalable manufacturing. However, from a thermodynamics point of view, the fast crystallization generally results in poor film formation and high trap concentrations, which in turn decrease both the efficiency and operating stability of solar cells. Du et al. introduced an ionic liquid to regulated the crystal growth by forming PbI_2–$EMIMHSO_4$ intermediates (Figure 7.9a); as shown in Figure 7.9b, the PbI_2–$EMIMHSO_4$ intermediates can efficiently retard the perovskite crystallization due to the strong interaction between perovskite and $EMIMHSO_4$.[24] The prolonged crystallization could promote the mass transport and diffusion, enlarge the grain size, and reduce the crystal defects, which are essential to obtaining high-quality perovskite film.[24]

7.3 NUCLEATION-GROWTH EQUILIBRIUM DURING CRYSTALLIZATION

A crystal is a solid substance composed of a regular arrangement of constituent molecules, atoms, or ions in a fixed and rigid pattern known as a lattice. Crystal growth involves both nucleation and growth. Nucleation occurs when a sufficient number of atoms or molecules accumulate to form a stable new phase, which is accompanied by a decrease in the free energy of the system. Crystal nucleation proceeds along either a

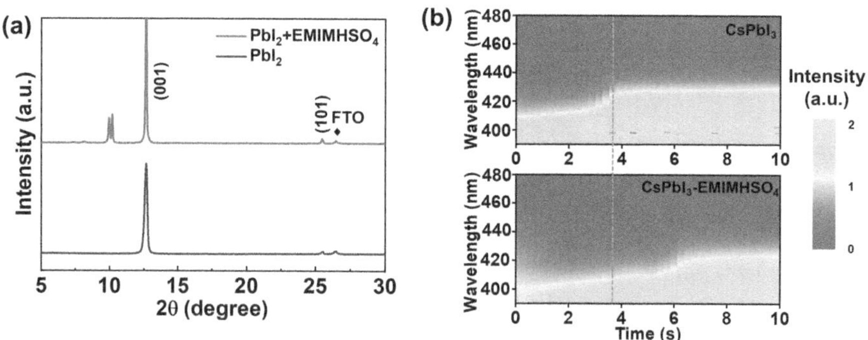

FIGURE 7.9 (a) XRD patterns of neat PbI_2 and PbI_2–$EMIMHSO_4$ mixed films (molar ratio 1/1). (b) In-situ UV-vis absorption spectra during the blade-coating processes for the $CsPbI_3$ and $CsPbI_3$–$EMIMHSO_4$ (0.45 mol%) perovskite films. Reprinted in part with permission from Ref.[24]. Copyright 2022, WILEY-VCH.

homogeneous or heterogeneous pathway. Heterogeneous nucleation occurs at nucleation sites from heterogeneities such as foreign surfaces, impurities, grain boundaries, and dislocations, which preferentially occur in the parent phase.[30] Homogeneous nucleation is the direct formation of nuclei with clusters of atoms in the parent phase and is always spontaneous and random.

The entire nucleation process can be explained using thermodynamic principles by considering the various energy requirements. According to current classical nucleation theory, for the homogeneous nucleation of an aspherical nucleus with radius r, the free energy change of the system ΔG is equal to the sum of the energy change induced by the surface crystal free energy (σ) and the bulk crystal free energy (ΔG_v) in the aspherical nucleus. Thus, the ΔG when a crystal nucleus appears can be expressed as

$$\Delta G = 4\pi r^2 \sigma + \frac{4}{3}\pi r^3 \Delta G_V \tag{7.2}$$

Both variables (σ, ΔG_v) are dependent on the temperature (T).[31] Therefore, at a certain T, ΔG becomes a function of r. Figure 7.10a[32] shows the dependence of ΔG on r. In particular, the ΔG vs. r curve reveals the presence of a critical radius (r_{crit}). When $r < r_{crit}$, the nucleus redissolves in the solution. Its disappearance is energetically more favorable than its growth. When $r \geq r_{crit}$, ΔG_v gradually decreases and the nucleus becomes stable.

However, under experimental conditions, another phase may exist in the parent solution, wherein heterogeneous nucleation can occur because of the presence of impurity-phase particles or interfaces with a vessel, such as a substrate. In the case of heterogeneous nucleation on a flat substrate surface, the nuclei form a spherical crown shape with a certain surface contact angle (θ). The change in the total free energy of the system for a crystal nucleus on a substrate is given by

$$\Delta G_{sub} = \Delta G \frac{2 - 3\cos\theta + \sin^3\theta}{4} \tag{7.3}$$

Figure 7.10b[32] shows the relationship of the ratio of the free energies for homogeneous and heterogeneous nucleation as a function of θ. This plot implies that heterogeneous nucleation can effectively reduce the energy barrier during nucleation by attaching to existing substrate surfaces.

The nucleation and growth of semiconductor nanocrystals follow the main features of the LaMer model,[32] which divides these processes into three basic steps, as shown in Figure 7.10c. First, the solute concentration in the solution gradually increases and approaches the saturation limit. Once saturation is achieved, nucleation occurs in the solution, which rapidly decreases the solute concentration. Finally, under solute diffusion, crystal growth begins at the center of the crystal nuclei.

The Ostwald ripening mechanism is based on the Gibbs–Thompson theory,[33] which implies that tiny particles with high solubility and surface energy rapidly dissolve in solutions. Consequently, a concentration gradient forms in the solution owing to the difference between the solute concentrations in the vicinity of smaller

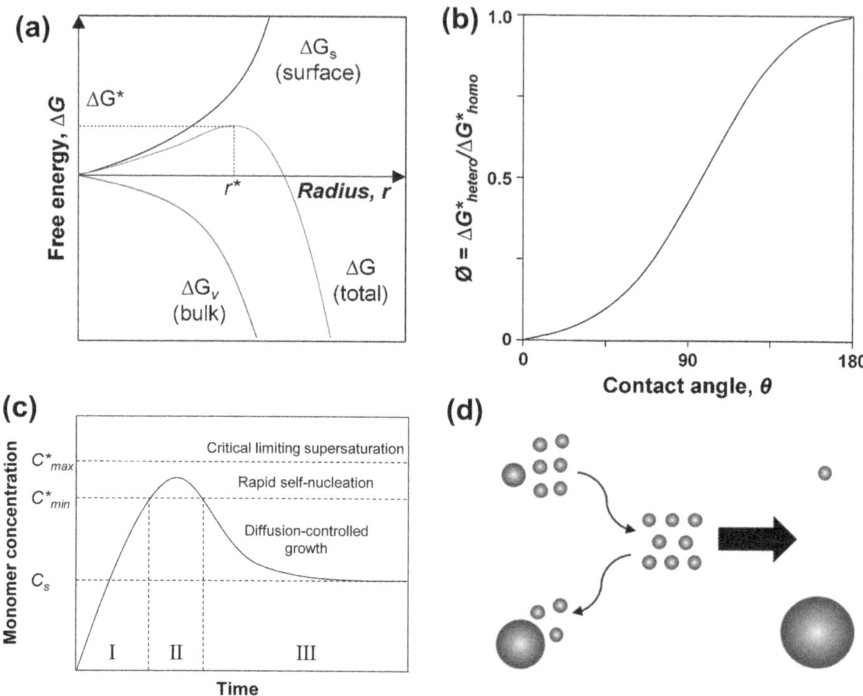

FIGURE 7.10 (a) Schematic of the classical free energy diagram for homogeneous nucleation as a function of particle radius. Reprinted in part with permission from Ref.[32] Copyright 2019 Royal Society of Chemistry. (b) Ratio of the free energies of homogeneous and heterogeneous nucleation as a function of the contact angle (θ). Reprinted in part with permission from Ref.[32]. Copyright 2019 Royal Society of Chemistry. (c) LaMer diagram depicting concentration changes in the perovskite precursor solution as a function of time. Reprinted in part with permission from Ref.[32]. Copyright 2019 Royal Society of Chemistry. (d) Ostwald ripening scheme. Reprinted in part with permission from Ref.[34]. Copyright 1961, Elsevier.

and larger particles; that is, a higher solute concentration is often observed in the vicinity of smaller particles. This concentration gradient promotes movement of the solute towards low-concentration regions. The previous processes cause small particles to precipitate on the surfaces of large particles, thereby increasing the particle sizes, as shown in Figure 7.10d.[34]

In contrast, the theory of digestive ripening, which was proposed by Lee et al.,[35] states that large surface energy-controlled particles redissolve to support the growth of smaller particles. In addition to the LaMer mechanism and Ostwald ripening, other theories, such as the Finke–Watzky two-step mechanism, oriented attachment, and intraparticle growth, have been proposed to explain crystal nucleation and growth. According to the Finke–Watzky two-step mechanism, nucleation and growth occur simultaneously. This phenomenon has been observed in the reduction of transition metal salts by cyclohexene. In this mechanism, the first process involves slow and constant nucleation and the second involves spontaneous growth, which is not controlled by diffusion.[36] The concept of oriented attachment was introduced by Penn and Banfield[37] in their work on the hydrothermal synthesis of titanium dioxide (TiO_2) nanocrystals. In this mechanism, the nanocrystal orientation is constant throughout the growth process. Larger particles can be formed by interconnecting the nanoparticles on a shared crystalline surface.[37] The principles of coalescence and oriented attachment are similar. However, during coalescence, the nanoparticle orientations are inconsistent.[38]

Based on the previous nucleation–growth theory, the key to obtaining pinhole-free and high-quality perovskite films is to control the perovskite nucleation–growth equilibrium during processing. The preparation of perovskite crystals with different morphs can be accomplished through modulating the perovskite nucleation–growth equilibrium. To prepare small perovskite crystals, rapid nucleation process and inhibited growth process are of vital importance. According to LaMer's diagram, the key point is shortening the crystallization time, which can be accomplished by making the concentration of the solution increase and exceed C_{min} quickly. Then the concentration will drop below C_{min} by forming a large number of nuclei and consuming solutes, leading to burst homogeneous nucleation followed by the limited and controlled growth process. On the contrary, to prepare large bulk perovskite crystals, inhibited nucleation process and promoted growth process are significant. However, the insufficient nucleation will leave the substrate uncovered and produce many cracks and pinholes. Thus, during the practical perovskite process, whether they retard or fasten the perovskite crystallization depends on the morphs of the obtained perovskite films; there should be a balance between grain size and film coverage.

For example, promoting the supersaturation of wet films by accelerating the solvent evaporation is an important strategy for improving nucleation to obtain pinhole-free and high-quality perovskite films. The gas-quenching method is a facile and low-cost technique in which a high-speed gas (N_2 or air) is blown onto a wet film through a slot to accelerate solvent evaporation and induce the formation of a large number of nuclei and fine perovskite grains on the dried film.[39] Heating is another method for accelerating solvent evaporation. For example, hot casting, also known as substrate preheating, is a simple solvent evaporation method.[39] For perovskite films, increasing the substrate temperature can suppress the formation of fiber-like intermediates in the films because the hot substrate induces rapid removal of the solvent and intermediate decomposition, thereby converting the ink directly to the perovskite phase.[40] Nie et al.[41] illustrated a solution-based hot-casting technique for growing continuous pinhole-free perovskite films with millimeter-scale crystalline grains. Further, Li et al.[12] investigated the phase transitions of perovskites using real-time measurements and revealed that the key issue in the rational design of printed perovskite films is whether the intermediate phase participates in the phase transition during the drying process.

Solvent engineering is another promising and facile strategy for controlling the nucleation–growth equilibrium for improved perovskite film morphology.[42] The use of different solvents, such as DMF, DMSO, N-methyl-2-pyrrolidone, γ-butyrolactone (GBL), acetonitrile, N,N-dimethylacetamide, hexamethylphosphoramide, methoxyethanol, and their mixtures enables tuning of the physical properties of the solution and the interactions between the solvent and perovskite precursor; hence, the perovskite crystal morphology can be modulated. Combining low boiling-point volatile solvents is an effective approach towards realizing pinhole-free, compact, and uniform perovskite films with high crystallinity, which is key to the reproducible production of large-area perovskite films. Huang et al.[43] reported that volatile noncoordinating solvents for Pb^{2+} and low-volatility coordinating solvents can achieve both fast drying and large perovskite grains at room temperature. Liu et al.[44] designed a low boiling-point precursor solvent consisting of tetrahydrofuran (THF) and methylamine (ME) in ethanol for large-area blade coating of $MAPbI_3$ without post-treatment. The perovskite precursor exhibited relatively high solubility in THF-ME (TME) (> 2 M), which evaporated rapidly and decreased the $MAPbI_3$ nucleation rate. Thus, the formation of an intermediate phase was avoided and large grains were obtained.

7.4 CRYSTALLIZATION CONTROL FOR IMPROVING CHARGE TRANSPORT BEHAVIOR

The detailed balance of the Shockley–Queisser efficiency limit stipulates that all recombinations are radiative,[45] underlining the importance of minimizing the trap-assisted relaxation channels. Poor-quality perovskite films, such as those with small grains or large amounts of grain boundaries, have a relatively high density of structural and energetic disorder, which negatively impacts the charge transport and device performance.

To decrease the number of stacking faults during crystallization, Yang et al.[46] proposed a strategy for anion fixation and synergistic suppression of the formation of anion vacancy defects via a sustainable pinning effect. In this approach, efficient charge transport is retained by using amidino-based organic molecules to construct highly efficient and stable PSCs. Further, 3-amidinopyridine molecules form strong chemical bonds with the Pb–I framework to effectively pin anions and considerably increase the energy barrier for anionic-vacancy formation and migration. In application, this strategy successfully delivered a PCE as high as 25.3% (certified at 24.8%) along with significantly improved operational stability, following the ISOS-L-1 stability protocol.[46]

The addition of 2D perovskite to 3D perovskite also impacts the formation mechanism and crystallization pathways of the latter, thus providing an essential guide for the design and preparation of high-quality perovskite films. Recently, Yang et al.[47] designed a 2D/3D heterostructure based on $FAPbI_3$ 3D perovskites and amidino-based DJ type 2D perovskites ((4AP)PbI_4, 4AP = 4-amidinopyridine). The 2D perovskite-assisted nucleation/growth of 3D perovskites during film formation was revealed via time-resolved optical diagnostics. These processes decrease the void concentrations and bulk defects of perovskite films for efficient charge transport. Planar PSCs based on 2D/3D heterostructures achieved high efficiencies of 24.9% and 22.3% on rigid and flexible substrates, respectively.[47]

The formation of 2D/3D heterostructures is a popular and promising strategy for exploiting the favorable properties of 2D perovskites, such as their excellent structural stability and unique graded quantum effects, which can improve device stability and facilitate interface carrier transport. In addition, manipulation of the formation process of the 2D/3D perovskite heterostructure, including its nucleation/growth dynamics and phase transition pathway, is critical for controlling the charge transport between the 2D and 3D crystals and, consequently, for the scalable fabrication of efficient and stable PSCs. Recently, Du et al.[27] studied the structural evolution and phase-transition pathways of ligand-dependent 2D perovskites on 3D surfaces using time-resolved X-ray scattering. Their results showed that the ligand size and shape critically influenced the final 2D structure. In particular, smaller ligands with more reactive sites tended to form the $n = 1$ phase. Increasing the ligand size and decreasing the number of reactive sites promoted transformation from the 3D structure to the $n = 3$ and < 3 phases. These findings are useful for rational design of the phase distributions in 2D perovskites, to balance the charge transport and stability of the perovskite films. Finally, an improved efficiency of 20.33% was obtained for solar cells based on ambient-printed $CsPbI_3$ with n-butylammonium iodide treatment; this is the highest reported value for printed inorganic PSCs to date.[27]

In addition, the crystal orientation significantly influences the carrier transport characteristics. Seidel and co-workers[48] reported n-type behavior for the (100) facet because of dominant I^- vacancies, whereas the (112) facet exhibited p-type behavior with MA^+ or Pb^{2+} vacancies. Different defects in the facet caused the (100) facet to have higher PL intensity and lower trap density than the (112) facet.[48] Liu et al.[49] further demonstrated (100)- and (111)-facet trap densities of 1.19×10^{15} and 1.49×10^{15} cm^{-3}, respectively, for $CsPbBr_3$ perovskite. The lower trap density of the (100) facet yielded a high carrier mobility of 241 μm^2 V^{-1} $s^{-1}$and a long carrier lifetime of 8.68 ns, sharply contrasting with the low carrier mobility of 49 μm^2 V^{-1} s^{-1} and a short carrier lifetime of 5.88 ns obtained for the (111) facet.[49] To obtain the desired facet orientation and stacking mode, Zhao et al.[50] introduced 2D $(BDA)PbI_4$ perovskite (where BDA = 1,4-butanediamine) as a seed for regulation of the 3D perovskite crystallization kinetics and, thus, facet orientation tailoring.

In addition to 3D-based perovskites, 2D-based perovskites have attracted considerable attention because of their promising stability and excellent optoelectronic properties. The grain quality, phase purity, and quantum-well thickness (n) distribution, which are related to their crystallization kinetics, determine their charge transport behavior and, thus, the final performance of devices containing these materials. Faster solvent removal from the solution via hot casting or antisolvent dripping typically yields a more uniform n distribution. For example, Zhang et al.[51] studied the crystallization behavior of Ruddlesden–Popper reduced-dimensional hybrid perovskite (RDP) using in-situ GIWAXS measurements. They demonstrated the growth of highly crystalline RDPs with high phase purity, preferentially perpendicular crystal orientation, and large grain size through fast and direct transition from a disordered precursor solvate to the perovskite phase.[51] However, at low temperatures, the formation of intermediate phases, including PbI_2 crystals and solvate complexes, slowed the ion intercalation and increased the nucleation barrier, inducing the formation of multiple RDP phases and random orientations, as shown in Figure 7.11.[51] The applicability of the "direct transition strategy" to the Dion–Jacobson (DJ) 2D hybrid perovskite has also been proven. A detailed understanding of the effects of different solution-casting processes

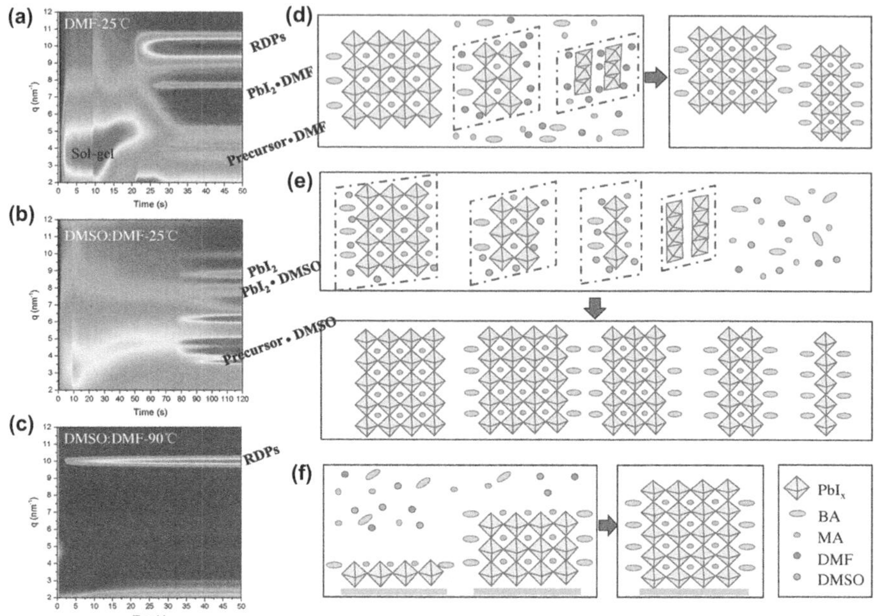

FIGURE 7.11 (a–c) In-situ GIWAXS measurements and (d–f) schematic models showing RDP formation for three distinct fabrication conditions. Reprinted in part with permission from Ref. [51]. Copyright 2018, WILEY-VCH.

on film formation was obtained for the DJ perovskite $(PDMA)(MA)n_{-1}PbnI_3n_{+1}$ ($\langle n \rangle = 4$, PDMA refers to 1,4-phenylenedimethanammonium), for the final film structure and photovoltaic outcomes.[52] Faster removal of the solvent from the solution via hot casting or antisolvent dripping yielded a more uniform n distribution, which eventually enhanced the carrier transport in the perpendicular direction and increased the PCE.[52]

The n distribution can tune the energy landscape of the 2D perovskite components and, thus, influence the charge carrier transport and final device performance. The distribution of the various n phases and the optoelectronic properties can be controlled by tuning the intermediate phases. Zhang et al.[53] applied antisolvent engineering and hot-casting techniques to control the kinetic transformation of the interlayer space (ACI) perovskite $(GA)(MA)nPbnI_3n_{+1}$ ($\langle n \rangle = 3$, denoted as ACI; GA indicates guanidinium) from disordered sol-gel colloidal precursor to the crystalline phase.[53] The intermediate phases, including a crystalline solvate compound and the 2D GA_2PbI_4 perovskite, provided a scaffold for the growth of ACI perovskites during thermal annealing and, finally, yielded an improved n distribution and highly efficient charge transport.[53] Similarly, addition of MACl additive to the ACI 2D perovskite $(GA)(MA)nPbnI_3n_{+1}$ ($\langle n \rangle = 3$) enables improved film quality, higher carrier lifetime, and an improved n distribution. These improvements enable the suppression of nonradiative charge recombination loss and highly efficient charge transport from low-n quantum wells to the final bulk phase.[54]

Thus, controlling the crystallization of perovskite films to obtain high-quality active layers with fewer grain boundaries and improved orientation is an important approach to achieving improved charge transport behavior.

7.5 CRYSTALLIZATION CONTROL FOR LARGE-SCALE SOLAR CELL FABRICATION

Although the fabrication of lab-scale high-performance PSCs is well established, a method of effectively fabricating large-area, pinhole-free, and uniform perovskite thin films remains urgent but challenging. The realization of such thin films will advance the commercialization of PSCs. Several technologies for the preparation of large-area PSCs are available, including blade coating, slot-die coating, spray coating, inkjet printing, screen printing, etc. The photovoltaic parameters of state-of-art printed PSCs are shown in Table 7.1. Effective up-scaling of efficient and stable perovskite layers is one of the main challenges limiting PSC commercialization.

TABLE 7.1

The Photovoltaic Parameters of State-of-the-Art Printed PSCs

Printing methods	Perovskite composition	Area (cm^2)	V_{OC} (V)	J_{SC} (mA cm^{-2})	FF (%)	PCE (%)	Ref.
Spray coating	$Cs_{0.05}FA_{0.81}MA_{0.14}PbI_{2.55}Br_{0.45}$	2.5	1.09	23.1	77	19.4	56
Gravure printing	$(FAPbI_3)_{0.95}(MAPbBr_3)_{0.05}$	0.096	1.17	24.8	81.0	23.5	57
Screen printing	$FA_{0.3}MA_{0.7}PbI_3$	0.05	1.14	23.12	77.9	20.52	58
		16.37	4.71	4.23	59.3	11.8	
Bar-coating	$(FAPbI_3)_{0.95}(MAPbBr_3)_{0.05}$	31	11.79	2.14	80.77	20.40	59
	$FAPbI_3$	1	1.20	24.13	80.43	23.24	60
		36.4	11.79	1.92	79.89	21.23	
Slot-die coating	$FA_{0.91}Cs_{0.09}PbI_3$	0.09	1.121	25.7	78.8	22.7	61
		7.92	4.55	5.657	76.2	19.6	
	$FAPbI_3$	0.09	1.088	24.92	83.1	22.54	62
		12.7	8.36	3.06	66.8	17.1	
Blade coating	$FA_{0.7}MA_{0.3}PbI_3$	0.09	1.13	26.19	81.70	24.31	63
	$FA_{0.3}MA_{0.7}PbI_3$	0.08	1.17	24.39	81.1	23.1	64
		25.03	1.14	22.68	74.6	19.3	
	$Cs_{0.03}(FA_{0.97}MA_{0.03})_{0.97}Pb(I_{0.97}Br_{0.03})_3$	0.04	1.192	24.01	81.04	23.19	65
	$MA_{0.6}FA_{0.4}PbI_3$	0.09	1.17	24.1	84.2	23.8	66
		50.0	16.07	1.527	78.0	19.2	
	$MA_{0.7}FA_{0.3}PbI_3$	0.09	1.17	23.9	83.6	23.4	67
		35.8	11.7	2.045	77.3	18.5	
	$FAPb(I_{0.98}Br_{0.02})_3$	0.09	1.19	25.10	80.20	24.00	68
		25	7.184	3.83	78.50	21.60	
	$CsPbI_3$	0.09	1.23	20.25	81.5	20.33	27

Most state-of-the-art lab-scale PSCs are prepared using spin coating for perovskite film deposition. However, spin coating is not a commercial method for perovskite-layer up-scaling. Large volumes of solution are wasted during spin coating, which increases the cost. Therefore, the most significant challenge for PSC photovoltaic technology is the transition from lab- to fabric-scale manufacturing. The key technology to bridge this gap between PSC coating and printing is to improve the uniformity and quality of printed perovskite films over larger areas. In laboratory-scale devices, the perovskite-film crystallization dynamics can be readily tuned using a precursor ink. For example, the intermediate-phase formation can be controlled by adjusting the solvent characteristics, such as the boiling point, volatility, viscosity, polarity, surface tension, miscibility, and donor number (D_N); this method is of great importance to the phase transition process. Fast evaporation of low boiling-point solvents produces high-density crystal nuclei, reducing the surface tension on the substrate and inhibiting the formation of solvated intermediate phases, which can effectively alter the perovskite-crystal nucleation rate and realize direct crystallization. In addition, low-D_N solvents coordinate less strongly with Pb^{2+}, which facilitates phase transition and subsequent crystalline perovskite growth.[55] Future studies on PSCs should attempt to apply such high-volatility cosolvents to perovskite precursor inks. The results of such investigations can accelerate innovation in printed perovskite films and revolutionize the commercial-scale fabrication of these materials. Further research on printed PSCs should focus on new, green, and controllable solvents with low boiling points to control the intermediate phases under annealing-free conditions.

7.6 CONCLUSIONS AND OUTLOOK

This chapter reviewed various promising processing routes for hybrid perovskites, as analyzed using real-time in-situ techniques. These measurements have revealed important details regarding the complex formation pathways of perovskites. However, there is an ongoing need to combine in-situ GIWAXS, UV-vis absorption spectroscopy, PL spectroscopy, and other in-situ measurement techniques to establish the direct correlation between the crystallization kinetics, band structure transformation, and stacking fault evolution of hybrid perovskites, which remains underexplored. In addition, the development of low-cost and eco-friendly strategies to minimize toxic waste production during the scalable printing of large-area perovskite films is essential. A deeper understanding of perovskite crystallization behavior may signpost valuable research directions towards the optimized preparation of large-area PSCs, ultimately promoting the development of competitive PSCs for the future energy market.

REFERENCES

1. C. Suryanarayana, M. G. Norton, *X-Ray Diffraction A Practical Approach*, Springer, **1998**.
2. S. Shao, *et al.*, Highly Reproducible Sn-Based Hybrid Perovskite Solar Cells with 9% Efficiency. *Advanced Energy Materials* **2018**, *8*, 1702019.

3. Y. Yan, Y. Yang, M. Liang, M. Abdellah, T. Pullerits, K. Zheng, Z. Liang, Implementing an Intermittent Spin-Coating Strategy to Enable Bottom-Up Crystallization in Layered Halide Perovskites. *Nature Communications* **2021**, *12*, 6603.

4. J. I. Pankove, *Opitical Processes in Semiconductor*, Dover Publications, Inc., **2010**.

5. D. Abou-Ras, T. Kirchartz, U. Rau, *Advanced Characterization Techniques for Thin Film Solar Cells*, Wiley-VCH, **2016**.

6. Y. Shao, Z. Xiao, C. Bi, Y. Yuan, J. Huang, Origin and Elimination of Photocurrent Hysteresis by Fullerene Passivation in $CH_3NH_3PbI_3$ Planar Heterojunction Solar Cells. *Nature Communications* **2014**, *5*, 5784.

7. A. Kojima, K. Teshima, Y. Shirai, T. Miyasaka, Organometal Halide Perovskites as Visible-Light Sensitizers for Photovoltaic Cells. *Journal of the American Chemical Society* **2009**, *131*, 2.

8. R. Munir, *et al.*, Hybrid Perovskite Thin-Film Photovoltaics: In Situ Diagnostics and Importance of the Precursor Solvate Phases. *Advanced Materials* **2017**, *29*, 1604113.

9. K. H. Stone, *et al.*, Transformation from Crystalline Precursor to Perovskite in $PbCl_2$-derived $MAPbI_3$. *Nature Communications* **2018**, *9*, 3458.

10. S. Lee, *et al.*, In situ Study of the Film Formation Mechanism of Organic–Inorganic Hybrid Perovskite Solar Cells: Controlling the Solvate Phase Using an Additive System. *Journal of Materials Chemistry A* **2020**, *8*, 7695.

11. T. Niu, *et al.*, Stable High-Performance Perovskite Solar Cells via Grain Boundary Passivation. *Advanced Materials* **2018**, *30*, e1706576.

12. J. Li, *et al.*, Phase Transition Control for HighPerformance Blade-Coated Perovskite Solar Cells. *Joule* **2018**, *2*, 1313.

13. N. Pellet, P. Gao, G. Gregori, T. Y. Yang, M. K. Nazeeruddin, J. Maier, M. Gratzel, Mixed-Organic-Cation Perovskite Photovoltaics for Enhanced Solar Light Harvesting. *Angewandte Chemie International Edition* **2014**, *53*, 3151.

14. C. C. Stoumpos, C. D. Malliakas, M. G. Kanatzidis, Semiconducting Tin and Lead Iodide Perovskites with Organic Cations: Phase Transitions, High Mobilities, and Near-Infrared Photoluminescent Properties. *Inorganic Chemistry* **2013**, *52*, 9019.

15. W. S. Yang, J. H. Noh, N. J. Jeon, Y. C. Kim, S. Ryu, J. Seo, S. Il Seok, High-Performance Photovoltaic Perovskite Layers Fabricated through Intramolecular Exchange. *Science* **2015**, *348*, 1234.

16. W. S. Yang, *et al.*, Iodide Management in Formamidinium-Lead-Halide–based Perovskite Layers for Efficient Solar Cells. *Science* **2017**, *356*, 1376.

17. Z. Li, M. Yang, J.-S. Park, S.-H. Wei, J. J. Berry, K. Zhu, Stabilizing Perovskite Structures by Tuning Tolerance Factor: Formation of Formamidinium and Cesium Lead Iodide Solid-State Alloys. *Chemistry of Materials* **2015**, *28*, 284.

18. K. Wang, *et al.*, Kinetic Stabilization of the Sol–Gel State in Perovskites Enables Facile Processing of High-Efficiency Solar Cells. *Advanced Materials* **2019**, *31*, e1808357.

19. L. Bi, *et al.*, Deciphering the Roles of MA-Based Volatile Additives for α-$FAPbI_3$ to Enable Efficient Inverted Perovskite Solar Cells. *Journal of the American Chemical Society* **2023**, *145*, 5920.

20. D. Barrit, *et al.*, Room-Temperature Partial Conversion of α-$FAPbI_3$ Perovskite Phase via PbI_2 Solvation Enables High-Performance Solar Cells. *Advanced Functional Materials* **2020**, *30*, 1907442.

21. W. Hui, *et al.*, Stabilizing Black-Phase Formamidinium Perovskite Formation at Room Temperature and High Humidity. *Science* **2021**, *371*, 1359.

22. M.-C. Tang, *et al.*, Ambient Blade Coating of Mixed Cation, Mixed Halide Perovskites without Dripping: in situ Investigation and Highly Efficient Solar Cells. *Journal of Materials Chemistry A* **2020**, *8*, 1095.

23. W. Ahmad, J. Khan, G. Niu, J. Tang, Inorganic $CsPbI_3$ Perovskite-Based Solar Cells: A Choice for a Tandem Device. *Solar RRL* **2017**, *1*, 1700048.

24. Y. Du, *et al.*, Ionic Liquid Treatment for Highest-Efficiency Ambient Printed Stable All-Inorganic $CsPbI_3$ Perovskite Solar Cells. *Advanced Materials* **2022**, *34*, e2106750.

25. X. Gu, W. Xiang, Q. Tian, S. F. Liu, Rational Surface-Defect Control via Designed Passivation for HighEfficiency Inorganic Perovskite Solar Cells. *Angewandte Chemie International Edition* **2021**, *60*, 23164.

26. H. Zhang, *et al.*, Fluorine-Containing Passivation Layer via Surface Chelation for Inorganic Perovskite Solar Cells. *Angewandte Chemie International Edition* **2023**, *62*, e202216634.

27. Y. Du, *et al.*, Manipulating the Formation of 2D/3D Heterostructure in Stable High-Performance Printable $CsPbI_3$ Perovskite Solar Cells. *Advanced Materials* **2023**, *35*, e2206451.

28. J. Ma, *et al.*, Unraveling the Impact of Halide Mixing on Crystallization and Phase Evolution in $CsPbX_3$ Perovskite Solar Cells. *Matter* **2021**, *4*, 313.

29. Y. Fan, *et al.*, Scalable Ambient Fabrication of HighPerformance $CsPbI_2Br$ Solar Cells. *Joule* **2019**, *3*, 2485.

30. H. L. Bhat, *Introduction to Crystal Growth: Principles and Practice*, CRC Press Taylor & Francis Group, **2015**.

31. R. Strey, P. E. Wagner, Y. Viisanen, The Problem of Measuring Homogeneous Nucleation Rates and the Molecular Contents of Nuclei: Progress in the Form of Nucleation Pulse Measurements. *The Journal of Physical Chemistry* **1994**, *98*, 7748.

32. M. Jung, S. G. Ji, G. Kim, S. I. Seok, Perovskite Precursor Solution Chemistry: from Fundamentals to Photovoltaic Applications. *Chemical Society Reviews Journal* **2019**, *48*, 2011.

33. T. Sugimoto, F. Shiba, T. Sekiguchi, H. Itoh, Spontaneous Nucleation of Monodisperse Silver Halide Particles from Homogeneous Gelatin Solution I: Silver Chloride. *Colloids and Surfaces A: Physicochemical Engineering Aspects* **2000**, *164*, 183.

34. I. M. Lifshitz, V. V. Slyozov, The Kinetics of Precipitation from Supersaturated Solid Solutions. *Journal of Physics and Chemistry of Solids* **1961**, *19*, 35.

35. W.-r. Lee, M. G. Kim, J.-r. Choi, J.-I. Park, S. J. Ko, S. J. Oh, J. Cheon, Redox– Transmetalation Process as a Generalized Synthetic Strategy for Core– Shell Magnetic Nanoparticles. *Journal of the American Chemical Society* **2005**, *127*, 16090.

36. M. A. Watzky, E. E. Finney, R. G. Finke, Transition-Metal Nanocluster Size vs Formation Time and the Catalytically Effective Nucleus Number: A Mechanism-Based Treatment. *Journal of the American Chemical Society* **2008**, *130*, 11959.

37. R. L. Penn, J. F. Banfield, Morphology Development and Crystal Growth in Nanocrystalline Aggregates under Hydrothermal Conditions: Insights from Titania. *Geochimica et cosmochimica acta* **1999**, *63*, 1549.

38. H. Zheng, R. K. Smith, Y.-w. Jun, C. Kisielowski, U. Dahmen, A. P. Alivisatos, Observation of Single Colloidal Platinum Nanocrystal Growth Trajectories. *Science* **2009**, *324*, 1309.

39. X. Chang, Anthopoulos, S. F. Liu, K. Zhao, *et al.*, Printable $CsPbI_3$ Perovskite Solar Cells with PCE of 19% via an Additive Strategy. *Advanced Materials* **2020**, *32*, e2001243.

40. Y. Deng, E. Peng, Y. Shao, Z. Xiao, Q. Dong, J. Huang, Scalable Fabrication of Efficient Organolead Trihalide Perovskite Solar Cells with Doctor-Bladed Active Layers. *Energy & Environmental Science* **2015**, *8*, 1544.

41. W. Nie, *et al.*, High-Efficiency Solution-Processed Perovskite Solar Cells with Millimeter-Scale Grains. *Science*, **2015**, *347*, 522.

42. L. Chao, T. Niu, W. Gao, C. Ran, L. Song, Y. Chen, W. Huang, Solvent Engineering of the Precursor Solution toward Large-Area Production of Perovskite Solar Cells. *Advanced Materials* **2021**, *33*, e2005410.

  255

43. Y. Deng, C. H. Van Brackle, X. Dai, J. Zhao, B. Chen, J. Huang, Tailoring Solvent Coordination for High-Speed, Room-Temperature Blading of Perovskite Photovoltaic Films. *Science Advances* **2019**, *5*, eaax7537.

44. Q. Liu, *et al.*, A Mixed Solvent for Rapid Fabrication of Large-Area Methylammonium Lead Iodide Layers by One-Step Coating at Room Temperature. *Journal of Materials Chemistry A* **2019**, *7*, 18275.

45. W. Shockley, H. J. Queisser, Detailed Balance Limit of Efficiency of p–n Junction Solar Cells. *Journal of Applied Physics* **1961**, *32*, 510.

46. T. Yang, *et al.*, One-Stone-for-Two-Birds Strategy to Attain Beyond 25% Perovskite Solar Cells. *Nature Communications* **2023**, *14*, 839.

47. T. Yang, *et al.*, Amidino-based Dion-Jacobson 2D Perovskite for Efficient and Stable 2D/3D Heterostructure Perovskite Solar Cells. *Joule* **2023**, *7*, 574.

48. D. Kim, *et al.*, Probing Facet-Dependent Surface Defects in MAPbI$_3$ Perovskite Single Crystals. *The Journal of Physical Chemistry C* **2019**, *123*, 14144.

49. S. Dong, *et al.*, All-Inorganic Perovskite Single-Crystal Photoelectric Anisotropy. *Advanced Materials* **2022**, *34*, e2204342.

50. C. Luo, G. Zheng, F. Gao, X. Wang, Y. Zhao, X. Gao, Q. Zhao, Facet Orientation Tailoring via 2D-Seed-Induced Growth Enables Highly Efficient and Stable Perovskite Solar Cells. *Joule* **2022**, *6*, 240.

51. X. Zhang, et al., Phase Transition Control for High Performance Ruddlesden–Popper Perovskite Solar Cells. *Advanced Materials* **2018**, 30, e1707166.

52. X. Zhang, T. Yang, X. Ren, L. Zhang, K. Zhao, S. Liu, Film Formation Control for High Performance Dion–Jacobson 2D Perovskite Solar Cells. *Advanced Energy Materials* **2021**, *11*.

53. Y. Zhang, *et al.*, Dynamical Transformation of Two-Dimensional Perovskites with Alternating Cations in the Interlayer Space for High-Performance Photovoltaics. *Journal of the American Chemical Society* **2019**, *141*, 2684.

54. T. Luo, *et al.*, Compositional Control in 2D Perovskites with Alternating Cations in the Interlayer Space for Photovoltaics with Efficiency over 18%. *Advanced Materials* **2019**, *31*, e1903848.

55. J. C. Hamill, J. Schwartz, Y.-L. Loo, Influence of Solvent Coordination on Hybrid Organic–Inorganic Perovskite Formation. *ACS Energy Letters* **2017**, *3*, 92.

56. J. E. Bishop, C. D. Read, J. A. Smith, T. J. Routledge, D. G. Lidzey, Fully Spray-Coated Triple-Cation Perovskite Solar Cells. *Scientific Reports* **2020**, *10*, 6610.

57. Y. Y. Kim, T. Y. Yang, R. Suhonen, A. Kemppainen, K. Hwang, N. J. Jeon, J. Seo, Roll-to-Roll Gravure-Printed Flexible Perovskite Solar Cells Using Eco-Friendly Antisolvent Bathing with Wide Processing Window. *Nature Communications* **2020**, *11*, 5146.

58. C. Chen, *et al.*, Perovskite Solar Cells based on Screen-Printed Thin Films. *Nature* **2022**, *612*, 266.

59. J. W. Yoo, *et al.*, Efficient Perovskite Solar Mini-Modules Fabricated via Bar-Coating Using 2-Methoxyethanol-based Formamidinium Lead Tri-Iodide Precursor Solution. *Joule* **2021**, *5*, 2420.

60. J. W. Yoo, *et al.*, R$_4$N$^+$ and Cl$^-$ Stabilized α-Formamidinium Lead Triiodide and Efficient Bar-Coated Mini-Modules. *Joule* **2023**, *7*, 797.

61. M. Du, *et al.*, High-Pressure Nitrogen-Extraction and Effective Passivation to Attain Highest Large-Area Perovskite Solar Module Efficiency. *Advanced Materials* **2020**, *32*, 2004979.

62. J. Li, J. Dagar, *et al.*, Ink Design Enabling Slot-Die Coated Perovskite Solar Cells with >22% Power Conversion Efficiency, Micro-Modules, and 1 Year of Outdoor Performance Evaluation. *Advanced Energy Materials* **2023**, *13*, 2203898.

63. W. Feng, *et al.*, Near-Stoichiometric and Homogenized Perovskite Films for Solar Cells with Minimized Performance Variation. *Angewandte Chemie International Edition* **2023**, *62*, 202300265.

64. M. A. Uddin, *et al.*, Blading of Conformal Electron-Transport Layers in p–i–n Perovskite Solar Cells. *Advanced Materials* **2022**, *34*, 2202954.

65. Q. Liang, *et al.*, Manipulating Crystallization Kinetics in High-Performance Blade-Coated Perovskite Solar Cells via Cosolvent-Assisted Phase Transition. *Advanced Materials* **2022**, *34*, 2200276.

66. S. Chen, X. Dai, S. Xu, H. Jiao, L. Zhao, J. Huang, Stabilizing Perovskite-Substrate Interfaces for High-Performance Perovskite Modules. *Science* **2021**, *373*, 902.

67. S. Chen, X. Xiao, H. Gu, J. Huang, Iodine Reduction for Reproducible and High-Performance Perovskite Solar Cells and Modules. *Science Advances* **2021**, *7*, eabe8130.

68. L. Yin, *et al.*, Crystallization Control for Ambient Printed FA-Based Lead Triiodide Perovskite Solar Cells. *Advanced Materials* **2023**, *35*, 2303384.

8 Interface Engineering for Perovskite Photovoltaics

Pengju Shi, Jiangzhao Chen, Jingjing Xue, and Rui Wang

8.1 ROLES OF INTERFACE ENGINEERING FOR PSCS

8.1.1 MODIFYING BAND ALIGNMENT

In the operational mechanism of perovskite solar cells (PSCs), upon illumination, the perovskite film generates excitons, which readily dissociate into free carriers (electrons and holes) due to the low exciton binding energy of prevalent perovskite absorber materials, such as methylammonium lead iodide ($MAPbI_3$, 16 ± 2 meV). Subsequently, the inherent electric field and elevated carrier mobility of perovskite films facilitate the transportation of electrons and holes to the electron transport layer (ETL)/perovskite interface and perovskite/hole transport layer (HTL) interface, respectively. A well-aligned energy level enables the efficient injection of electrons and holes into the conduction band (CB) and valence band (VB) of the ETL and HTL.[1]

Optimal energy-level alignment can lower the energy barrier for charge collection, enhance the built-in electric field, and effectively suppress charge recombination and accumulation at the perovskite/electrode interface, thereby improving device performance. It is worth noting that specific ionic groups exhibit strong interactions with perovskite materials, enabling interfacial energy level adjustments through the use of appropriate contact materials. This energy-level tuning can also be applied to other functional layers, including charge transport layers and electrodes. Moreover, some bipolar molecules, such as zwitterions and ionic liquids (ILs), exhibit a propensity for organized alignment between electrode and perovskite layers, resulting in the formation of self-assembled dipole layers. These factors collectively contribute to the modification of energy-level alignment.[2]

In PSCs with regular structures, tin dioxide (SnO_2) is the most prevalent ETL due to its advantages over titanium dioxide (TiO_2), including a lower processing temperature, appropriate band level, good conductivity, and cost-effectiveness. Nevertheless, a mismatch remains between the conduction band level of SnO_2 (approximately 4.5 eV, Figure 8.1) and that of perovskite (-3.4 to ~ 3.9 eV), necessitating interfacial modifications. However, the lower CB of SnO_2 may reduce a built-in potential of Schottky barrier between the perovskite and SnO_2 ETLs, accordingly decreasing the voltage of PSCs.

In addition to the ETL interface, energy-level mismatches at the HTL interface, particularly when using spiro-OMeTAD, often lead to considerable performance

DOI: 10.1201/9781003400486-8

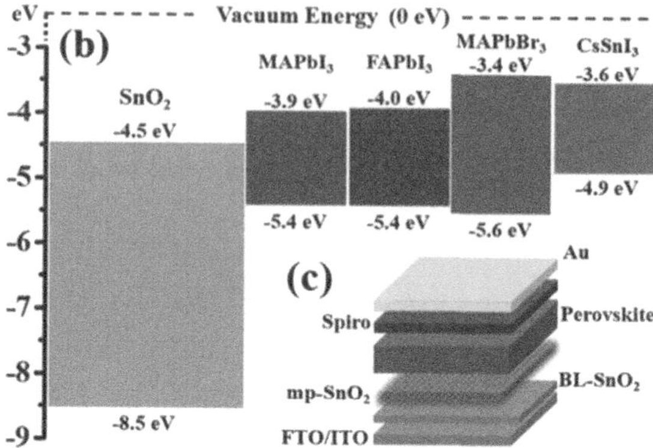

FIGURE 8.1 b) Diagram of energy levels (relative to the vacuum level) of the heterojunction SnO_2/perovskite. c) Schematic representation of the typical full device structure based SnO_2 ETL with an optional mesoporous (mp) SnO_2 layer.[3] Reproduced with permission. Copyright 2018, Wiley-VCH.

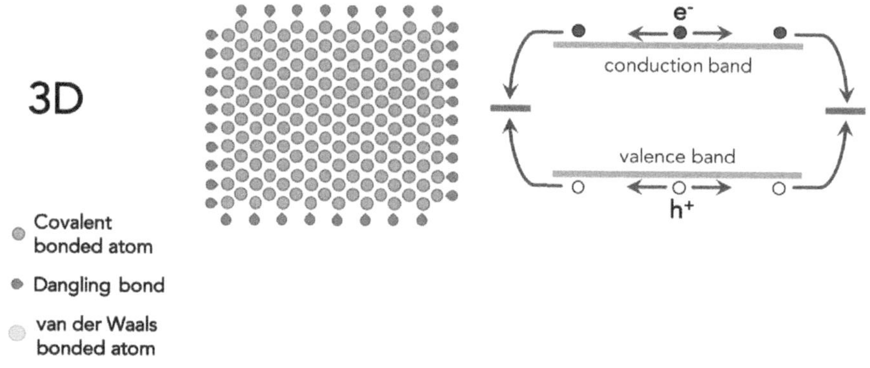

FIGURE 8.2 Crystal structures (left) and energy band diagrams (right) for a poorly passivated 3D crystal that create mid-gap electronic states. Reproduced with permission. Copyright 2021, Elsevier. Dangling bonds are unsaturated Pb- or I- , and vand waals bonds are between N and I.

restrictions. This arises from the presence of dangling bonds on the perovskite surface, which render it energetically unfavorable and more reactive. Consequently, these chemical reactions can produce additional mid-gap states (Figure 8.2), exacerbating recombination processes.[4]

8.1.2 Defects Passivation

At the solar device level, a high absorption coefficient allows for photocurrents near the theoretical maximum without requiring complex light-trapping strategies. Simultaneously, a low recombination velocity is crucial to achieve a minimal open-circuit voltage deficit (W_{OC}), defined as $W_{OC} = E_g/q—V_{OC}$, where E_g represents the bandgap, q denotes the elementary charge, and V_{OC} is the open-circuit voltage.

In a solar cell, the open-circuit voltage (V_{OC}) is directly linked to the separation of electron and hole quasi-Fermi levels in its semiconducting absorber, and consequently, its excess charge carrier densities (Δn and Δp) excited under specific illumination conditions, typically at one sun. Δn (usually, $\Delta n \approx \Delta p$) is directly reliant on the effective free-carrier lifetime (τ_{eff}) via the relation $\tau_{eff} = \Delta n/U_{eff}$, where U_{eff} denotes the effective recombination rate.[5] Notably, τ_{eff} is an experimentally accessible parameter used to evaluate a semiconductor's electronic quality. The marked reliance of τ_{eff} on Δn highlights the importance of measuring τ_{eff} under excitation conditions that resembling actual device operation and consistently reporting the Δn value associated with the measured τ_{eff}. In perovskite absorbers, τ_{eff} depends on both bulk and surface recombination due to the additive nature of recombination rates in semiconductors.

Hybrid perovskites, known for their soft ionic solid nature, possess a considerable number of unintentional defects in solution-processed polycrystalline thin films (Figure 8.3), leading to significant Shockley-Read-Hall (SRH) recombination (ΔE_3).

Surface represent discontinuities in the three-dimensional (3D) semiconductor lattice which can generate large defect densities including strained and unterminated ("dangling") bonds. These physical defects introduce electronic defects and surface states that can mediate recombination of electron-hole pairs and impair device performance.[4,7,8]

To overcome this issue, interfacial engineering offers an exceptional approach to defect passivation, which is crucial in enhancing the optoelectronic quality of the absorber and suppressing recombination losses at the interfaces. Essentially, defect elimination is vital to achieve higher quality perovskite absorbers with long-lived

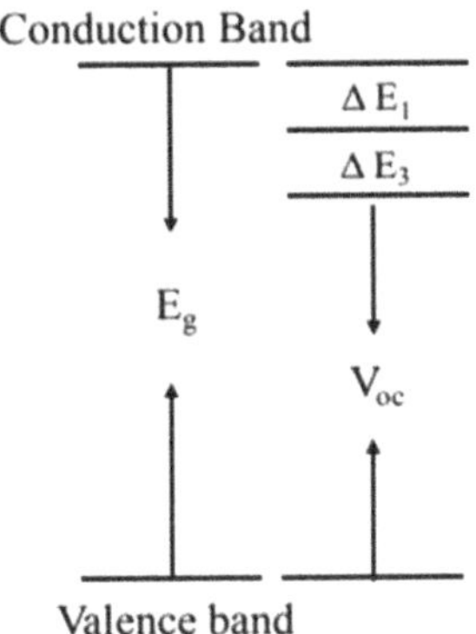

FIGURE 8.3 Energy diagrams of organic solar cells. Energy diagram showing the relationship of optical bandgap (E_{gap}), lowest singlet state (S1), charge transfer (CT) state, V_{oc}, and the three terms of energy losses. $\Delta E1$ is radiative recombination originating from the absorption above the bandgap. The value of ΔE_2 is negligible for PSCs. ΔE_3 represent nonradiative recombination.[6] Reproduced with permission. Copyright 2016, Springer Nature.

photoinduced charge carriers, enabling the V_{oc}, J_{sc}, and PCE of PSCs to approach the thermodynamic limit.[9]

8.1.3 Improved Interfacial Carrier Dynamics

Perovskite semiconductors exhibit semi-intrinsic characteristics, which result from a combination of intrinsic (defect formation chemistry) and extrinsic (film processing environment and interlayer interactions) doping mechanisms. Due to the challenging nature of controlling unintentional intrinsic self-doping in perovskite films, the interlayers (p- and n-type) surrounding the perovskite absorber are vital for enabling photovoltaic action. These interlayers generate the essential built-in electric field (V_{bi}) that drives charge separation, facilitating the movement of electrons and holes towards opposite interfacial contacts. The efficiency of photocurrent extraction relies on a significant disparity between the dynamics of interfacial charge transfer and recombination processes, as they collectively determine the fate of photogenerated carriers.

Charge transfer and transport dynamics can be enhanced to surpass recombination processes by strategically aligning interfacial energy levels with perovskite, ensuring high interlayer conductivity, and minimizing interfacial defects. In PSCs, a high carrier-collection efficiency is favored due to the rapid charge generation, separation, and transfer processes, which occur on timescales shorter than the free-carrier lifetime.[9] Following the injection of electrons and holes into the ETL and HTL, respectively, the charge transport properties of these interfacial materials play a crucial role in determining charge extraction and recombination dynamics. These properties, encompassing conductivity and mobility, significantly impact the overall performance of the system.

Suboptimal conductivity or mobility within the interfacial materials results in elevated series resistance, thereby hindering the charge carriers' ability to reach the electrodes within their respective lifetimes. This, in turn, leads to an accumulation of charge carriers at the perovskite/ETL (HTL) interface, where they undergo recombination with holes (electrons) present in the perovskite active layer. Ultimately, this phenomenon contributes to a marked increase in bimolecular recombination events.[10,11]

8.1.4 Barrier to Moisture Incursion and Ion Migration

At present, the primary hurdle for the commercialization of PSCs lies in their long-term stability. This stability challenge arises not only from the inherent properties of the ETL, HTL, and perovskite materials but also from the degradation of interfaces within the device. Generally, polycrystalline perovskite thin films exhibit a large proportion of uncoordinated ions at the grain boundary and surface. These uncoordinated ions significantly exacerbate ion migration, further complicating the stability issue.[12,13]

In the case of polycrystalline perovskite thin films, for instance, consider the example of MAPbI$_3$ (Figure 8.4). Efficient moisture penetration at the film surface leads to a reaction with the perovskite material, resulting in the formation

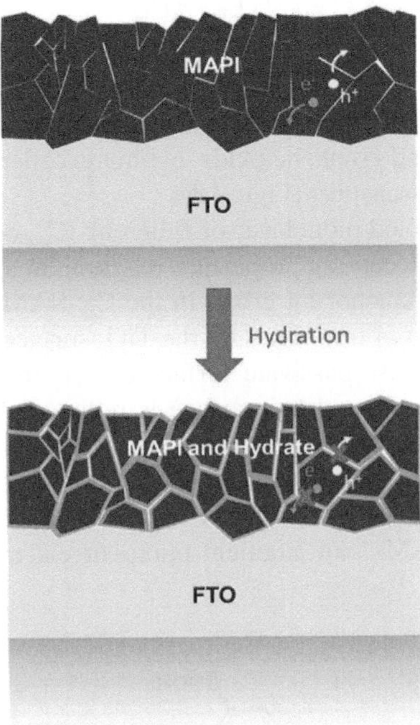

FIGURE 8.4 Presumed microscopic degradation model of $MAPbI_3$ thin films under partial hydration. Reproduced with permission. Copyright 2015, American Chemistry Society.

of monohydrated $CH_3NH_3PbI_3 \cdot H_2O$ (Equation 8.1).[14] Inevitably, upon prolonged exposure to moisture, the reaction progresses further, leading to the formation of needle-shaped dihydrate $(CH_3NH_3)_4PbI_6 \cdot 2H_2O$ (Equation 8.1). This process is concurrent with the generation of lead iodide (PbI_2) and the release of two water molecules.[15,16] The compound $(CH_3NH_3)_4PbI_6 \cdot 2H_2O$ undergoes further decomposition into PbI_2, H_2O, and CH_3NH_3I (Equation 8.2). Simultaneously, the liberated water molecules contribute to the partial self-sustainability of the conversion process, as they can be recycled to transform the remaining perovskites into monohydrate and subsequently dihydrate forms.[17–19] When exposed to dry gases, such as nitrogen, the reaction process can proceed in the reverse direction. However, residual water, as illustrated in Equation (8.1), may persist in dissolving the perovskites and further degrading the structure.[15]

$$4(CH_3NH_3)PbI_3 + 4H_2O \rightleftharpoons 4CH_3NH_3PbI_3 \cdot H_2O \rightleftharpoons$$
$$(CH_3NH_3)_4PbI_6 \cdot 2H_2O + 3PbI_2 + 2H_2O \tag{8.1}$$

$$(CH_3NH_3)_4PbI_6 \cdot 2H_2O(s) \xrightarrow{\text{H}_2\text{O}} 3PbI_2(s) + 2H_2O(l) + 4CH_3NH_3I(aq) \tag{8.2}$$

8.2 THE ETL/PEROVSKITE INTERFACE

8.2.1 Fullerene and Its Derivatives

Carbon-based materials and their derivatives, such as carbon nanotubes, fullerenes, graphdiyne, graphene, and graphene oxide, exhibit excellent conductivity, which can bolster interface charge transport (Figure 8.5).

In 2014, a self-assembled monolayer of fullerene (C_{60}-SAM) was synthesized for its exceptional electron-accepting properties, resulting in a remarkable improvement in electron transfer. The anchoring group in the C_{60}-SAM can effectively passivate or suppress the formation of trap states on the TiO_2 surface, while the fullerene moiety in the C_{60}-SAM can also passivate surface defects in perovskite. Consequently, nonradiative recombination at this interface is reduced. Moreover, the efficacy of C_{60}-SAM was also demonstrated in passivating the SnO_2 surface. In a similar vein, another self-assembled monolayer (SAM) of fullerene derivatives was utilized to diminish hysteresis by lowering capacitive hysteresis. Furthermore, when appropriately functionalized, SAMs can augment photocurrent even in the absence of an

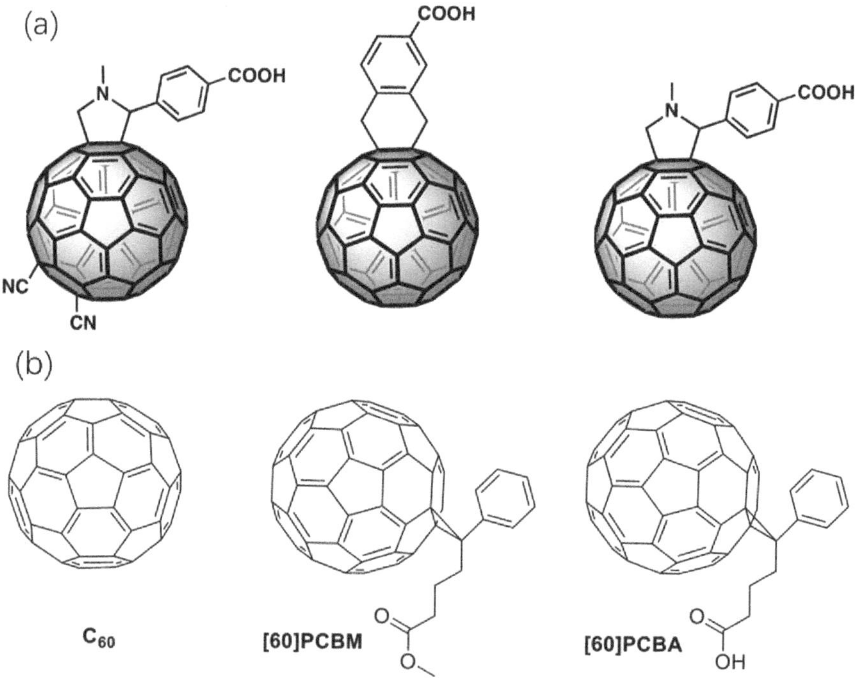

FIGURE 8.5 a) Functionalized fullerenes. Reproduced with permission. Copyright 2016, American Chemical Society. b) Chemical structure of C_{60}, PCBM, and PCBA. Energy diagram of different ETL substrates, assuming a fixed Fermi level. Reproduced with permission. Copyright 2019, Wiley-VCH.

electron selective contact and when a SAM is deposited solely atop the transparent conductive oxide. In Sn-based PSCs, the hydrophilic fullerene derivative C_{60} pyrrolidine tris-acid (CPTA) has been shown to enhance electron transfer at the interface, modulate the conduction band minima to impede hole transport, and suppress charge recombination.[20]

8.2.2 Metal Oxides and Halides

Inorganic salts serve as simple and effective interface passivation materials for modifying the interface and passivating the buried interface, and they have been widely utilized in both regular and inverted PSCs. In particular, the ions of inorganic salts substantially impact the buried interface between inorganic charge transport layers (CTLs) (NiO_x, SnO_2, TiO_2, ZnO) and the perovskite layer. For MAX- and PbX_2-terminated (MA = CH_3NH_3, X = Cl, I) perovskite buried surfaces, the addition of chlorine (Cl) ions could strongly bind with the undercoordinated site at the TiO_2/perovskite interface, reducing the density of interfacial trap states. In addition, chlorine could effectively fill the oxygen vacancy defects in ZnO, suppressing charge recombination and facilitating charge transport at the ZnO/perovskite interface, thereby enhancing power conversion efficiency (PCE) with reduced hysteresis. Simultaneously, chlorination of the ZnO surface could protect the perovskite layer from decomposition, improving device stability. Notably, for the NiO_x HTL interface in the inverted PSCs, alkali chloride has been successfully employed to modify the interface. The introduction of alkali chloride can not only tune the work function of NiO_x but also enhance the ordering of perovskite films. Consequently, the defects/traps density at the buried interface between perovskite and NiO_x is substantially reduced, suppressing interfacial recombination and ion migration and ultimately resulting in improved stability.

Additionally, potassium ions (K^+) have been extensively utilized as a means of passivating interface defects and reducing device hysteresis, akin to their function in bulk materials. On the other hand, potassium ions (K^+) diffuse through the perovskite film, suppressing the recombination activity of grain boundaries and improving the performance of PSCs. Previous research has also shown that the optimal amount of interface potassium plays a significant role in influencing the nucleation and growth of perovskite films, as well as the extraction of interface charges.

Moreover, interface K ions may undergo a substitutional reaction with bromine (Br) ions at the perovskite film surface, passivating halide vacancies at the interface to promote high performance, including hysteresis elimination. Concurrently, an $NdCl_3$ interface layer has been shown to hinder iodine ion migration pathways by filling the initial ion vacancies, resulting in reduced trap densities throughout the entire perovskite region and significantly improved photostability.[21] Another study suggests that potassium thiocyanate (KSCN) offers multiple interface passivation effects, potentially making it a superior choice compared to potassium acetate (KAc) and KCl. Beyond potassium passivation, thiocyanate exhibits dual passivation effects on both perovskite and NiO_x, achieving triple passivation. The robust Ni-N bonding displays strong polar covalent bond properties, causing electrons to deviate from the Ni side. In addition, the powerful electrostatic force between S and Pb in MAPbI$_3$ draws the Pb atomic layer closer to the perovskite, constraining the I atom.[22,23]

In summary, inorganic salt ions have been demonstrated to effectively modify the buried interface.[24] They can be broadly categorized as halogen anions (F$^-$, Cl$^-$, Br$^-$, I$^-$)[25] and other anions (Ac$^-$, SCN$^-$) paired with alkali metal cations (Na$^+$, K$^+$, Rb$^+$, Cs$^+$) and the ammonium ion (NH$_4^+$).[26] These inorganic salt ions can adjust the work function, enhance carrier mobility, and alleviate surface defects of oxide charge transport layers (CTLs). Furthermore, they also passivate bottom traps in perovskite and can even diffuse into the perovskite layer to passivate grain boundary defects. As a result, they not only promote interface carrier extraction and transport but also serve as a buffer layer to prevent interface chemical reactions between oxide CTLs and perovskite.

8.2.3 Zwitterionic Molecules

Structurally, zwitterions comprise two oppositely charged ions covalently linked, one positive (cation) and the other negative (anion), resulting in a net charge of zero for the molecule. Zwitterions exhibit hydrophilic properties due to the presence of ions and hydrophobic characteristics due to their hydrocarbon backbone. Common zwitterions include amino acids, which contain positively charged ammonium ions and negatively charged carboxylate ions, sulfobetaines (SB) featuring ammonium and sulfonate moieties, and carbobetaines (CBB) with carboxylate moieties in place of sulfonate (Figure 8.6).

Zwitterionic materials was integrated into the interface between SnO$_2$ and perovskite to generate interfacial dipoles and facilitate efficient charge extraction, thereby increasing the built-in potential and reducing charge back recombination. Furthermore, the positively charged pyridinium cation passivated Pb–I antisite defects, enhancing stability.[24,27]

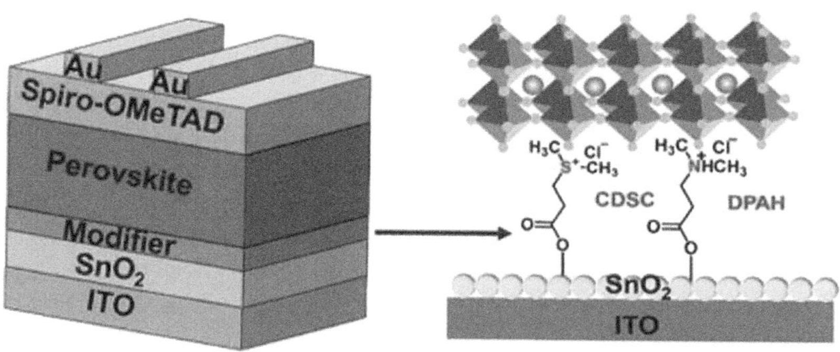

FIGURE 8.6 Schematic illustration of the formation of CDSC or DPAH molecules between SnO$_2$ layer and perovskite layer.[22] Reproduced with permission. Copyright 2022, Elsevier.

8.3   THE PEROVSKITE/HTL INTERFACE

8.3.1   2D PEROVSKITES

Currently, achieving both high efficiency and long-term stability has become an essential requirement for the successful application of PSCs. During the fabrication of perovskite films, a large number of defects form on the surface, which lead to significant non-radiative recombination. Moreover, these defects serve as pathways for ion migration, further contributing to the instability of the devices.[28] 2D perovskites have garnered increasing attention due to their enhanced robustness, as the organic layer acts as a protective barrier to prevent moisture or ion erosion. Inspired by these properties, researchers have begun exploring surface passivation by fabricating 2D perovskites on top of 3D counterparts, which has initiated a new wave of research aimed at achieving both higher efficiency and stability simultaneously.[29,30]

The most widely studied perovskites for solar cells are 3D halide perovskites, generally represented by the chemical formula ABX_3, where A can be Cs^+, $CH_3NH_3^+$, or $HC(NH_2)_2^+$; B can be Pb^{2+} or Sn^{2+}; and X can be Cl^-, Br^-, or I^-. This 3D framework, consisting of $[PbX_6]^{4-}$ octahedra sharing each of their six corners, forms a favorable electronic structure with wide conduction and valence bands, facilitating charge transport. In contrast, 2D perovskites incorporate additional bulky organic ligands, resulting in greater chemical diversity and a wider range of properties. These structures can be considered dimensionally reduced versions of 3D perovskites, with the general chemical formula $A'_m A_{n-1} B_n X_{3n+1}$, where A' can be a monovalent ($m = 2$) or divalent ($m = 1$) cation intercalating between the inorganic layers of $[PbX_6]^{4-}$ octahedra, and n denotes the number of inorganic layers between A' layers. When a monovalent A' cation is used, the 2D perovskite is referred to as the Ruddlesden-Popper (RP) phase, whereas those formed with divalent ($m = 1$) A' cations are called Dion-Jacobson (DJ) 2D perovskites (Figure 8.7).

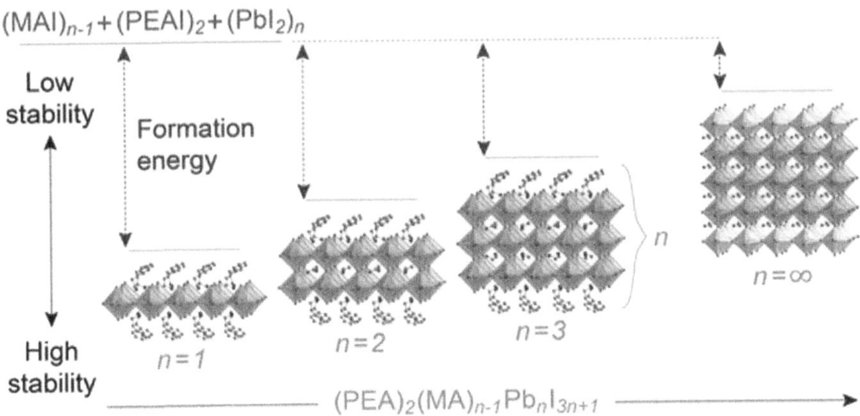

FIGURE 8.7   Crystal and electronic structure of 2D perovskites. Crystal structure evolution from $n = 1$ (2D perovskites) to $n = \infty$ (3D perovskite). Reproduced with permission. Copyright 2016, American Chemistry Society.

Owing to the incorporation of larger, less volatile, and typically more hydrophobic organic cations, 2D perovskites exhibit enhanced thermal, chemical, and environmental stability compared to their 3D counterparts (Figure 8.8). As *n* increases, the properties of 2D perovskites gradually approach those of 3D perovskites. A notable example is the bandgap of these materials, which progressively redshifts with increasing *n*. Concurrently, the stability of 2D perovskites also diminishes monotonically towards that of 3D perovskites as n increases. To clearly distinguish between the two classes of materials, particularly for solar cell applications, 2D perovskites are generally referred to as materials systems with $n \leq 5$.

For the surface treatment or passivation method, a precursor solution containing a spacer cation compound is prepared and spin-coated as a surface layer atop the previously deposited 3D perovskite film (Figure 8.9). This precursor solution comprises passivators, which typically consist of long carbon-chain alkyl or aryl ammonium cations.[31] In the majority of cases, a brief thermal annealing process is employed

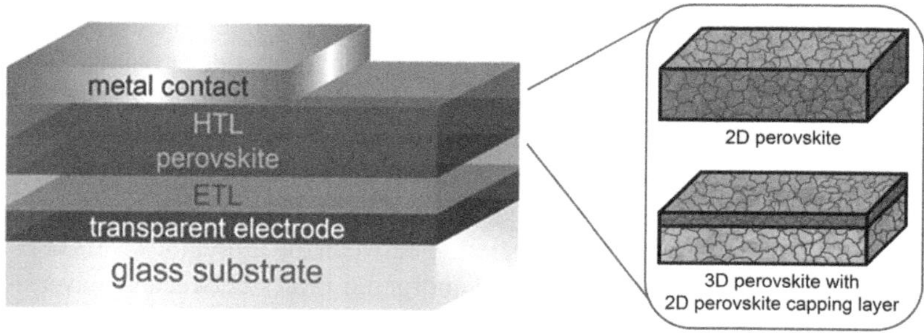

FIGURE 8.8 2D perovskites in solar cells. Scheme illustration of the device stack of a perovskite solar cell (left) and applications of 2D perovskite in solar cells, where the 2D perovskite serves either as the photo-absorber or as a capping layer atop a 3D perovskite absorber (right). Reproduced with permission. Copyright 2019, Wiley-VCH.

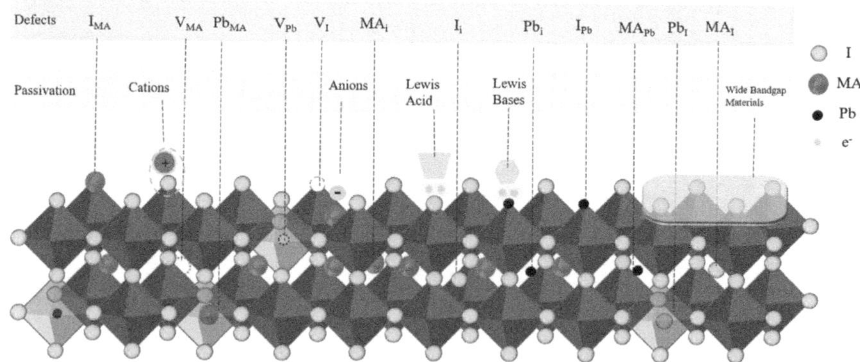

FIGURE 8.9 Examples of common defects and passivation strategies for PSCs.[32] Reproduced with permission. Copyright 2022, Wiley-VCH.

to create the 2D perovskite surface layer. However, there exists some disagreement regarding the optimal solvent choice for the 2D precursor. The most prevalent solvent for interlayer cation compounds is 2-propanol (IPA), but its highly polar nature can dissolve MA⁺ or FA⁺ on the 3D perovskite surface, leading to surface intermixing of 3D and 2D phases at the interface. As an alternative, chloroform-based precursor solutions have been reported to yield smoother films and improved device performance.

It is noteworthy that numerous examples in the literature demonstrate mixed 2D-3D PSCs outperforming reference devices based on 3D perovskite films, with some of the highest reported PSC efficiencies to date achieved using mixed-dimensional perovskite compositions (Figure 8.10). When spacer cations are deposited as a surface layer on a 3D perovskite film, the 2D perovskite mitigates defects such as vacancies

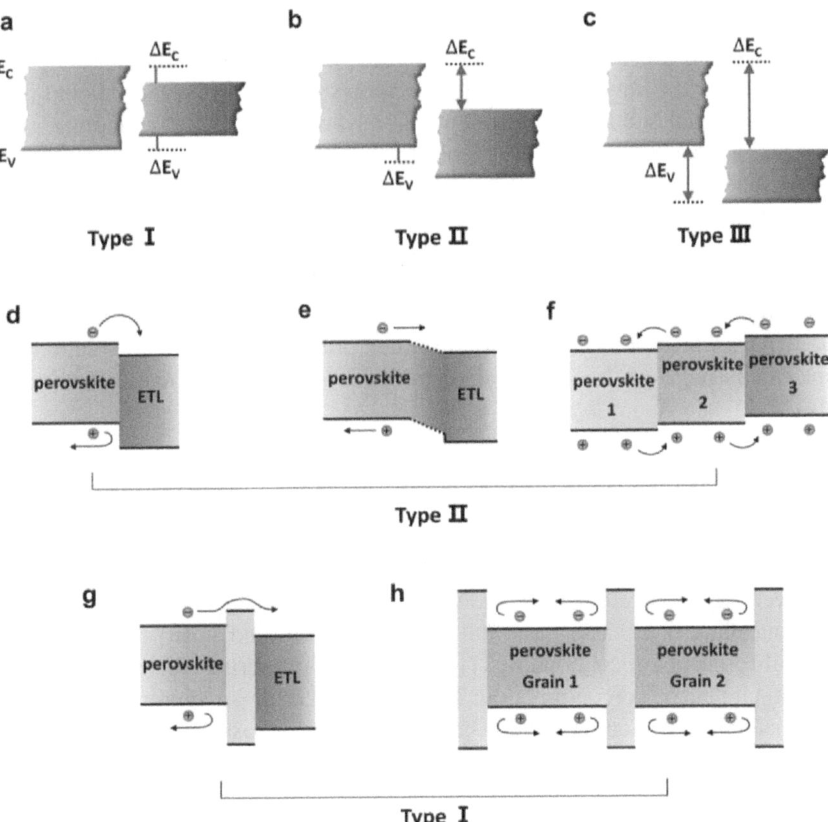

FIGURE 8.10 Three types of heterojunction band alignment: a) type I: straddling gap, b) type II: staggered gap, and c) type III: broken gap. Various cases of type II heterojunctions in PSCs: d) perovskite-ETL planar heterojunction, e) perovskite-ETL gradient heterojunction, and f) perovskite–perovskite multi-graded heterojunction. Various cases of type I heterojunctions in PSCs: g) tunnel junction at the perovskite-ETL interface and h) perovskite-wide bandgap material heterojunction within the active layer. Reproduced with permission. Copyright 2023, Wiley-VCH.

(V_{MA}, V_{Pb}, V_I), interstitials (MA_i, Pb_i, I_i), or anti-site atoms ($MAPbI_3$, e.g., MA_{Pb}, MA_I, Pb_{MA}, Pb_I, I_{MA}, I_{Pb}, where I_{MA}, I_{Pb}, Pb_i, Pb_I, V_I form deep-level defects). These deep-level defects act as primary Shockley–Read–Hall non-radiative recombination centers, potentially causing band bending upwards or downwards and impeding carrier separation.[33] This phenomenon can result in suboptimal energy level alignment between the absorber layer and the carrier transport layer, as well as significant recombination of photogenerated carriers, ultimately inhibiting the extraction and transport of electrons and holes.

In addition to passivation, employing 2D perovskite is an effective strategy for achieving suitable energy-band alignment at the perovskite/HTL interface, which can improve the photovoltaic performance of PSCs. Adjusting the energy level between perovskite and HTL can enhance carrier extraction and reduce energy loss. However, it is crucial to note that a large ΔE_c offset might adversely affect the splitting of the quasi-Fermi level of electrons (E_{Fn}) in the ETL and the quasi-Fermi level of holes (E_{Fp}) in the HTL, which determines the maximum attainable V_{OC} ($V_{OC} = (E_{Fn}-E_{Fp})/q$) in the device. Therefore, the perovskite/HTL interface should be fine-tuned for optimal band alignment. Among various type II heterojunctions, gradient heterojunctions are considered particularly favorable for photocarrier separation.

Nonetheless, it is important to recognize that the properties of the perovskite/HTL interface,[34] which can be significantly influenced by the perovskite surface, also affect carrier management (hole transfer and interfacial recombination) and, ultimately, device performance. Many researchers contend that defects on the perovskite surface induce charge trapping at the perovskite/HTL interface, limiting overall efficiency. As surface passivation of perovskites presents a prominent solution to this issue, extensive efforts have been made in this area, including the use of PEAI, OAI, and BAI.[35,36]

8.3.2 ORGANIC SALTS AND MOLECULES

Instead of forming 2D perovskite interfacial layers, certain insulating organic ammonium salts, such as PEAI, OAI, and 1-naphthylmethylamine iodide (NMAI), can effectively passivate defects and enhance device performance. Ammonium salt-based interface layers can passivate both negatively and positively charged ionic defects through chemical bonding. Their dielectric properties, featuring dipoles, can also mitigate carrier recombination. Therefore, selecting an appropriate ammonium salt is crucial, as it should provide chemical groups for passivation and resist the formation of low-dimensional perovskites. By employing suitable ammonium salts for interface modification, researchers have achieved significant advancements, demonstrated by certified high PCE values.

Furthermore, organic molecules, characterized by their ease of solution-processing and chemical modification, have been effectively utilized to create interface layers. Studies have shown that organic molecules with various functional groups can passivate surface defects through bonding interactions – such as hydrogen bonds – and establish appropriate energy level alignment at the perovskite/HTL interface. Notably, controlling the order, uniformity, and thickness of the molecular layer is essential to avoid compromising device performance.[37,38]

Notably, organic A-site cations usually only exert indirect structural effects because their electronic levels lie far from the band edge. It show that cations with large π-conjugated structures can interact with inorganic frontier molecular orbitals.[39] A surface layer of ethylammonium pyrene, which had an optimal intercalation distance, increased hole mobilities and power conversion efficiencies relative to a reference inorganic perovskite and also improved device stability.[39]

8.3.3 Insulating and Conjugated Polymers

It has been reported that insulating polymers at perovskite/HTL interface not only suppressed defect-induced recombination but also improved device stability. The interface layers exhibit great benefit to increase V_{oc} accompanied by the function of a stable barrier with their long molecular chains. Owing to its insulation property, the interface layer derived with insulating polymer needs to be thin enough to ensure the hole extraction. In this case, PMMA has been used to act as effective interlayer for optimizing the perovskite/HTL interface. Insulative polymer can also be used for modification of the perovskite/HTL interface in inverted p-i-n-based PSCs. The interfacial technique promoted the V_{oc} without seriously destroying the fill factor and achieved an improvement in PCE without hysteresis.

In addition to improving efficiency, the insulative polymer interface layer can enhance the stability of PSCs. A series of insulative polymer such as hygroscopic polymer (PEO) and hydrophobic polymer (PS), was employed to hinder air moisture from surroundings penetrating into perovskite film by absorbing water molecules, thus improving operational stability of the device in high humidity environment. However, in contrast to insulating polymers that induced a trade-off between passivation quality and series resistance, conjugated polymers possess special properties, such as semiconducting feature and facile formation of high-quality film.

8.3.4 Others

8.3.4.1 Metal Salts

Metal salts, such as strontium chloride ($SrCl_2$) and lead sulfate ($PbSO_4$), create robust chemical bonds with perovskite surfaces, effectively impeding ion migration and preventing the loss of unstable components. Additionally, employing these metal salts as interfacial layers offers exceptional protection for perovskite materials against harsh external conditions, including humidity and light exposure.

8.3.4.2 Carbon-Based Materials

Carbon-based materials, characterized by their stable structures and hydrophobic properties, exhibit minimal reactivity with perovskite halides, effectively mitigating ion migration within perovskite layers and serving as water-resistant barriers. Furthermore, these materials demonstrate multifunctionality, as they possess tunable band energy alignments and exceptional electronic properties. Utilizing cost-effective carbon-based materials as interfacial layers in PSCs holds great potential for achieving both high efficiency and long-term stability, paving the way for commercialization.[34]

REFERENCES

1. J. Chen, N.-G. Park, Materials and methods for interface engineering toward stable and efficient perovskite solar cells, *ACS Energy Letters* **2020,** *5*, 2742–2786.
2. X. Wu, *et al.*, Importance and advancement of modification engineering in perovskite solar cells, *Solar RRL* **2022,** *6*, 2200171.
3. L. Xiong, *et al.*, Review on the application of SnO_2 in perovskite solar cells, *Advanced Functional Materials* **2018,** *28*, 1802757.
4. D. L. McGott, *et al.*, 3D/2D passivation as a secret to success for polycrystalline thin-film solar cells, *Joule* **2021,** *5*, 1057–1073.
5. E. Aydin, *et al.*, Defect and contact passivation for perovskite solar cells, *Advanced Materials* **2019,** *31*, 1900428.
6. J. Liu, *et al.*, Fast charge separation in a non-fullerene organic solar cell with a small driving force, *Nature Energy* **2016,** *1*, 16089.
7. J. Xue, R. Wang, Y. Yang, The surface of halide perovskites from nano to bulk, *Nature Reviews Materials* **2020,** *5*, 809–827.
8. R. Wang, *et al.*, Constructive molecular configurations for surface-defect passivation of perovskite photovoltaics, *Science* **2019,** *366*, 1509–1513.
9. A. Rajagopal, *et al.*, Toward perovskite solar cell commercialization: A perspective and research roadmap based on interfacial engineering, *Advanced Materials* **2018,** *30*, 1800455.
10. S. Shao, M. A. Loi, The role of the interfaces in perovskite solar cells, *Advanced Materials Interfaces* **2019,** *7*, 1901469.
11. J. Xue, *et al.*, Crystalline liquid-like behavior: Surface-induced secondary grain growth of photovoltaic perovskite thin film, *Journal of the American Chemical Society* **2019,** *141*, 13948–13953.
12. Y. Zhu, *et al.*, Recent progress on interface engineering for high-performance, stable perovskites solar cells, *Advanced Materials Interfaces* **2020,** *7*, 2000118.
13. Q. Zhuang, *et al.*, Ion diffusion-induced double layer doping toward stable and efficient perovskite solar cells, *Nano Res.* **2022,** *15*, 5114–5122.
14. F. Hao, *et al.*, Controllable perovskite crystallization at a gas-solid interface for hole conductor-free solar cells with steady power conversion efficiency over 10%. *Journal of the American Chemical Society* **2014,** *136*, 16411–16419.
15. A. M. A. Leguy, *et al.*, Reversible hydration of $CH_3NH_3PbI_3$ in films, single crystals, and solar cells, *Chemistry of Materials* **2015,** *27*, 3397–3407.
16. T. M. Koh, *et al.*, Multidimensional perovskites: A mixed cation approach towards ambient stable and tunable perovskite photovoltaics, *ChemSusChem* **2016,** *9*, 2541–2558.
17. D. H. Chun, *et al.*, Grain boundary healing of organic-inorganic halide perovskites for moisture stability, *Nano Letters* **2019,** *19*, 6498–6505.
18. Z. Chu, *et al.*, Impact of grain boundaries on efficiency and stability of organic-inorganic trihalide perovskites, *Nature Communications* **2017,** *8*, 2230.
19. D. Li, *et al.*, Humidity-induced grain boundaries in $MAPbI_3$ perovskite films, *The Journal of Physical Chemistry C* **2016,** *120*, 6363–6368.
20. Z. W. Gao, Y. Wang, W. C. H. Choy, Buried interface modification in perovskite solar cells: A materials perspective, *Advanced Energy Materials* **2022,** *12*, 2104030.
21. H. Bi, *et al.*, Interfacial defect passivation and stress release by multifunctional KPF6 modification for planar perovskite solar cells with enhanced efficiency and stability, *Chemical Engineering Journal* **2021,** *418*, 129375.
22. X. Zuo, *et al.*, Passivating buried interface via self-assembled novel sulfonium salt toward stable and efficient perovskite solar cells, *Chemical Engineering Journal* **2022,** *431*, 133209.

23. C. Zhang, *et al.*, Modulating chemical interaction to realize bottom-up defect passivation for efficient and stable perovskite solar cells, *Solar RRL* **2022**, *6*, 2200512.

24. Bi, H., *et al.*, Multifunctional organic ammonium salt-modified SnO_2 nanoparticles toward efficient and stable planar perovskite solar cells, *Journal of Materials Chemistry A* **2021**, *9*, 3940–3951.

25. D. Gao, *et al.*, Passivating buried interface with multifunctional novel ionic liquid containing simultaneously fluorinated anion and cation yielding stable perovskite solar cells over 23% efficiency, *Journal of Energy Chemistry* **2022**, *69*, 659–666.

26. Q. Zhou, *et al.*, Revealing steric-hindrance-dependent buried interface defect passivation mechanism in efficient and stable perovskite solar cells with mitigated tensile stress, *Advanced Functional Materials* **2022**, *32*, 2205507.

27. J. Chen, *et al.*, Multifunctional chemical linker imidazoleacetic acid hydrochloride for 21% efficient and stable planar perovskite solar cells, *Advanced Materials* **2019**, *31*, 1902902.

28. G. Wu, *et al.*, Surface passivation using 2D perovskites toward efficient and stable perovskite solar cells, *Advanced Materials* **2022**, *34*, 2105635.

29. X. Zhao, T. Liu, Y. L. Loo, Advancing 2D perovskites for efficient and stable solar cells: Challenges and opportunities, *Advanced Materials* **2022**, *34*, 2105849.

30. B. Liu, *et al.*, Simultaneous passivation of bulk and interface defects with gradient 2D/3D heterojunction engineering for efficient and stable perovskite solar cells, *ACS Appl. Mater. Interfaces* **2022**, *14*, 21079–21088.

31. M. A. Mahmud, *et al.*, Origin of efficiency and stability enhancement in high-performing mixed dimensional 2D-3D perovskite solar cells: A review, *Advanced Functional Materials* **2021**, *32*, 2009164.

32. Y. Cao, *et al.*, Defects passivation strategy for efficient and stable perovskite solar cells, *Advanced Materials Interfaces* **2022**, *9*, 2200179.

33. B. Liu, *et al.*, Interfacial defect passivation and stress release via multi-active-site ligand anchoring enables efficient and stable methylammonium-free perovskite solar cells, *ACS Energy Lett.* **2021**, *6*, 2526–2538.

34. J. Chen, S. G. Kim, N. G. Park, $FA_{(0.88)}Cs_{(0.12)}PbI_{(3-x)}(PF(6))_{(x)}$ interlayer formed by ion exchange reaction between perovskite and hole transporting layer for improving photovoltaic performance and stability, *Advanced Materials* **2018**, *30*, 1801948.

35. T. Wang, *et al.*, Recent progress on heterojunction engineering in perovskite solar cells, *Advanced Energy Materials* **2022**, 2201436.

36. J. Chen, J. Y. Seo, N. G. Park, Simultaneous improvement of photovoltaic performance and stability by in situ formation of 2D perovskite at $(FAPbI_3)_{0.88}(CsPbBr_3)_{0.12}$/CuSCN interface, *Advanced Energy Materials* **2018**, *8*, 1702714.

37. G. Qu, *et al.*, Dopant-free phthalocyanine hole conductor with thermal-induced holistic passivation for stable perovskite solar cells with 23% efficiency, *Advanced Functional Materials* **2022**, *32*, 2206585.

38. D. He, *et al.*, Interfacial defect passivation by novel phosphonium salts yields 22% efficiency perovskite solar cells: Experimental and theoretical evidence, *EcoMat* **2021**, *4*, e12158.

39. J. Xue, *et al.*, Reconfiguring the band-edge states of photovoltaicperovskites by conjugated organic cations, *Science* **2021**, *371*, 636–640.

9 Synergistic Modulation of Grain Boundaries and Interfaces in Perovskite Solar Cells

Dong Wei, Ru Li, Guilin Chen, Dongmei He, Xuxia Shai, and Jiangzhao Chen

9.1 INTRODUCTION

Typical perovskite solar cells (PSCs) consist of five distinct polycrystalline layers with four interfaces, more specifically, the perovskite/electron transport layer (ETL; interface i), perovskite/hole transport layer (HTL; interface ii), ETL/electrode (interface iii), and HTL/electrode (interface iv). The four interfaces together with the five functional layers have played a crucial role in the performance of the PSCs. For instance, the morphology and crystallinity of the perovskite layer is largely affected by the wettability of the buried interface and thus greatly affects the short-circuit current density (J_{sc}). The open-circuit voltage (V_{oc}) is mainly determined by the quasi-Fermi-level splitting ($\Delta E_{F, pero}$) splitting between interface i and ii. The fill factor (FF) is mainly dependent on the shunt and series resistance of the complete solar cell. Besides that, the current–voltage (J-V) hysteresis, light soaking degradation, thermal/moisture instability, and electrodes corrosion are heavily affected by the interfaces and functional layer materials. Apart from the interfaces, the grain boundaries (GBs) within the polycrystalline perovskite layer also have important impacts on the V_{oc}, J_{sc}, FF and J-V hysteresis and poor long-term stability.

In order to understand the common issues originated from the interfaces and GBs of PSCs, the stages of photocarrier generation, transport, collection, and recombination within the perovskite and at the interfaces are illustrated in Figure 9.1. As shown in Figure 9.1a and b, with light illumination, free electrons and holes are generated in the conduction band (CB) and valance band (VB), respectively, because of the extremely small exciton binding energy and the high permittivity of perovskites.[1] The population of photocarriers induce the drift of electrons towards the ETL and the holes towards the HTL, respectively. It should be noted that a large number of GBs exist within the bulk of polycrystalline perovskite films. The deep level traps are usually located at GBs, which result in serious trap-assisted nonradiative recombination and accordingly lower V_{oc}. It needs to be stressed that the fundamental role

DOI: 10.1201/9781003400486-9

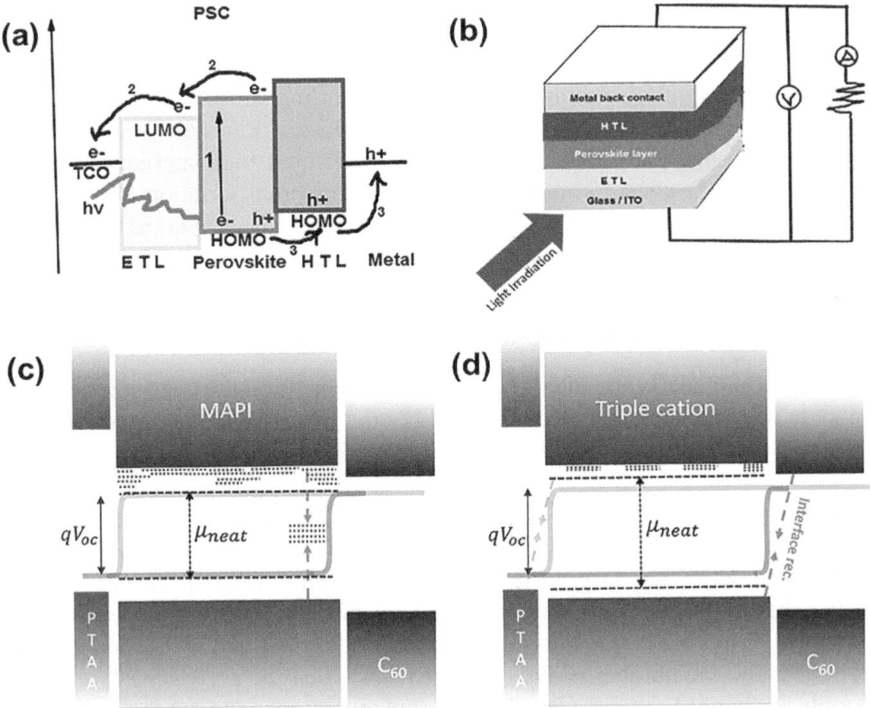

FIGURE 9.1 (a) Schematic working principle of PSC and (b) circuit connection of the perovskite solar cell.[1] Copyright 2022, the authors. Schematic of the nonradiative recombination pathways in (c) MAPI and (d) triple cation perovskite solar cells.[3] Copyright 2020, American Chemical Society.

of the GBs on V_{oc} is still an open question and under debate. Apart from the trap-assisted loss, the second-order band-to-band recombination loss is also able to happen but at a much smaller rate due to the easily formed polarons that arise from strong electron-phonon interaction, which efficiently screens the Coulomb attractions and largely extends the lifetimes of photocarriers.[2] For the commonly used perovskite $FAPbI_3$ (Figure 9.1c and d), the subgap defect states in the bulk and the surfaces limit the quasi-Fermi level in $MAPbI_3$ layer, as shown in Figure 9.1c. The defect-related non-radiative recombination at the interface between the perovskite and the transport layers (HTL and ETL) also restricts the internal quasi-Fermi level splitting and the Voc of perovskite solar cells (Figure 9.1d).[3] Therefore, it is of significant importance to get rid of or suppress the non-radiative recombination loss through interface and GBs engineering in order to obtain high efficiency PSCs. In the rest of this chapter, the common issues related to the GBs and interface are discussed thoroughly, and the corresponding modulation strategies are summarized.

9.2 COMMON ISSUES IN GRAIN BOUNDARIES AND INTERFACES OF PEROVSKITE SOLAR CELLS

The perovskite light absorbing layer is polycrystalline, so the GBs are inevitable. They are the possible reasons for recombination loss, J-V hysteresis and long-term durability. On the other hand, PSCs consists of at least four interfaces, which are responsible for the quasi-Fermi-level splitting, charge transportation, and extraction. Therefore, they are highly related to the recombination loss, ion migration, interface residual strain, hysteresis, and light soaking degradation.

9.2.1 DEFECTS AT GRAIN BOUNDARIES AND INTERFACES

The defects at the GBs and interfaces have huge impacts on the non-radiative recombination, which remarkably lowers the V_{oc}. For the GBs, conflicting results have been reported. In comparison, the role of interface defects is clearer.

9.2.1.1 Recombination at Grain Boundaries

As we know, the GBs consists of dangling bonds and vacancies, which may form carrier traps. However, the point defects, such as the iodine and lead vacancies, only exhibit shallow energy level features, which enables the PSCs a high defect-tolerant advantage. Experimental results show that the V_{oc} of the PSCs is independent of the grain size and the number of GB, around 1.1 V when the perovskite layer is thick enough. It should be noted that the morphology of perovskite is more important than the GBs. The low V_{oc} has been frequently observed for the perovskite layer that is not compact, homogenous, and uniform. Compared to the inorganic solar cells, filled traps usually act as energy barriers for electron transport in PSCs. The majority of the traps at GBs are positively charged iodide vacancies (V_I), which become neutralized when they are filled by electrons and the electron transport is not affected.[4]

9.2.1.2 Recombination at Interfaces

As the electrons and holes transport across the perovskite/ETL and perovskite/HTL interface, respectively, the interfacial traps cause the first-order recombination, while the charge accumulation due to energy barrier and poor charge transport capability leads to second-order recombination. It is found that the interface traps play a more important role than that of the GBs, because the interface trap rate ($kTNT$) competes with the injection rate. For instance, only about 50% electrons are injected into PCBM if the trap density at the perovskite/PCBM interface is more than 5×10^{16} cm^{-3}. Under-coordinated ions are the main defects at the interface, which should be passivated to obtain high-efficiency solar cells.[6] The wettability of the buried interface is found to be important for compact perovskite layer growth. As shown in Figure 9.2, the perovskite layer with small grain size will form on hydrophilic buried interface, while large grain size and compact perovskite layer could be fabricated on hydrophobic buried interface. A large number of bulk traps would form in small grain sized perovskite film, which results in considerable first-order bulk recombination and compete strongly with the interface injection.[5]

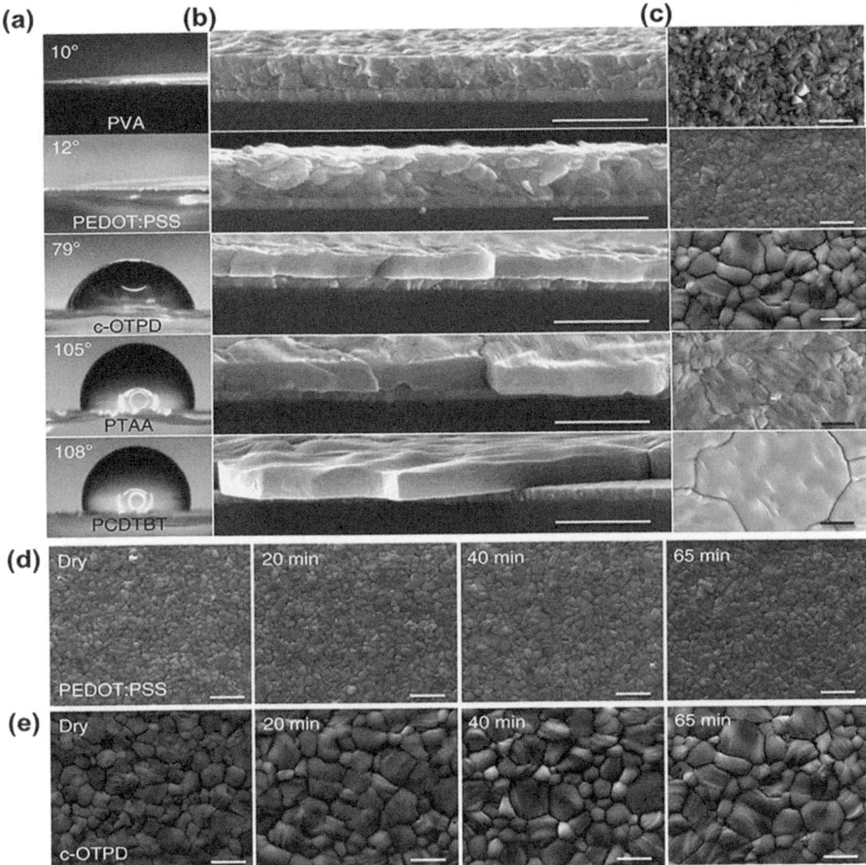

FIGURE 9.2 The contact angle of water on the varied HTLs (a), the cross-section SEM (b), top-view SEM. (c), Scale bars, 1 µm in b, c; (d, e) the top-view SEM images of the MAPbI$_3$ grown on PEDOT:PSS (top row) and c-OTPD (bottom row) right after drying and after 20, 40, and 65 min of thermal annealing at 105 °C. Scale bar, 1 µm.[5] Copyright 2015, Springer Nature.

9.2.2 Ion Migration and Diffusion

The ion migration and diffusion are a phenomenon not observed in traditional silicon or in the second generation CIGS or CdTe thin film solar cells, which is rooted in the soft ionic nature of perovskite crystal structure and has been intensively studied by theoretical calculation and experimental measurements.

9.2.2.1 Ion Migration in Perovskites

Density functional theory (DFT) based calculations show that the I$^-$ ion has the lowest migration energy barrier (E_A) of 0.58 eV, which is less than that of the MA$^+$ ion (0.84 eV), while the Pb^{2+} has the highest E_A of 2.31 eV. Thus, the I$^-$ ion migration becomes the main source for observed hysteresis.[7] The ion transport mechanism is shown in Figure 9.3a. Ion conductivity dependence on temperature shows that the E_A is as low as 0.36 eV, which is not surprising because the DFT calculations are performed for bulk crystal while the experiments are carried out with a polycrystalline sample. The GBs maybe act as the main ion migration path. Consequently, the observed ion migration energy barrier is lowered.

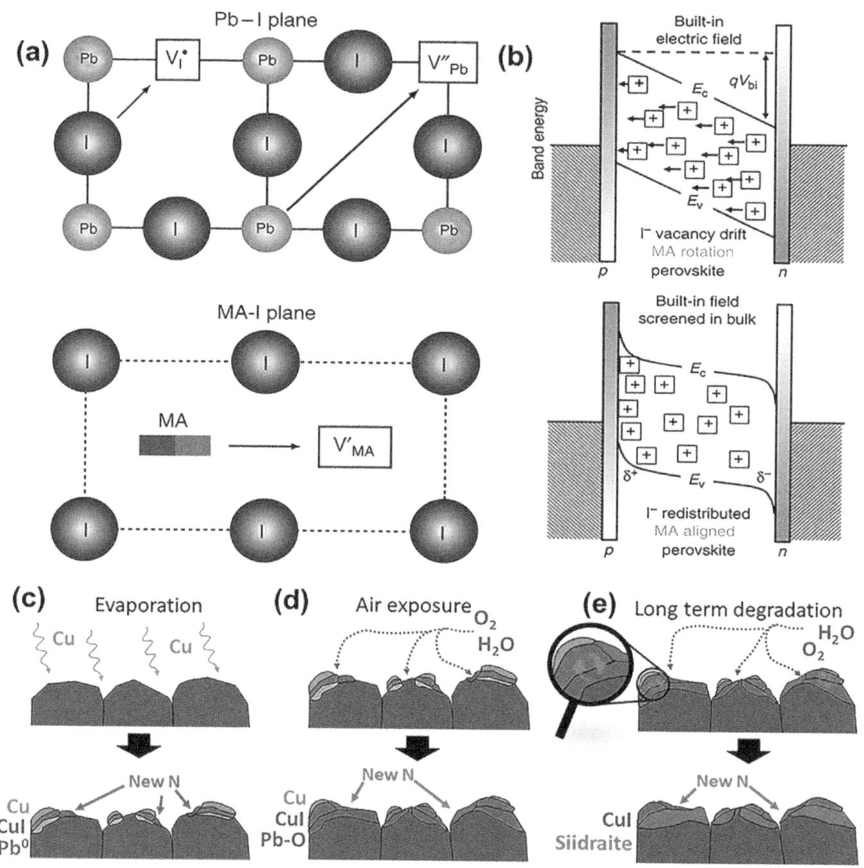

FIGURE 9.3 (a) Ion transportation mechanism in FAPbI$_3$. (b) Schematic diagrams indicating the influence of vacancy drift on the band energies of a p-i-n device at short circuit. EC is the conduction band energy, EV is the valence band energy, and V$_{bi}$ is the built-in potential. Iodide ion vacancies are represented by the squares with "plus" signs. Implicit in the diagram is that the vacancies with effective positive charges are balanced by immobile cation vacancies (not shown) with effective negative charges.[7] Copyright 2015, Springer Nature. Schematic illustration showing the different treatments and their impact on the structure of the Cs$_{0.17}$FA$_{0.83}$PbI$_3$/copper interface. (c) Reactions during and after evaporation of copper. (d) Reactions during initial exposure to air. (e) Long-term reactions leading to the formation of [Pb$_4$(OH)$_4$][Cu$_2$I$_6$] (siidraite).[8] Copyright 2022, the authors.

9.2.2.2 Diffusion between Perovskite and Transport Layers

As the ion migration occurs, interface dipole is established between the perovskite and charge transport layer. Take the I⁻ diffusion as an example; when the cell is short circuited, the built-in electrical field drives the migration of iodide ion vacancies towards the contacts, which partially screens the field and results in less efficient collection of photogenerated charge carriers, as shown Figure 9.3b.[7] However, prolonged poling of the cell under forward or reverses bias reduces the built-in field, which could allow a dissipation of ionic charge from the contacts by diffusion and leads to more efficient collection of photogenerated charges when the device is returned to short circuit, then hysteresis is observed.

9.2.2.3 Diffusion between Perovskite and Electrode

Metal electrodes, such as copper (Cu) and silver (Ag), have the possibility to be corroded by the migrated iodine ions, which has been observed in $PC_{61}BM$-based p-i-n inverted PSCs. The incomplete coverage of rough perovskite surface due to the thin nature of $PC_{61}BM$ enables the easy diffusion of iodine ions from perovskite layer to metal electrode, which should be responsible for the electrode corrosion. Figure 9.3c–e shows the non-uniform chemical corrosion process of copper electrode. It shows that as aging goes on, the corrosion becomes more and more serious.[8]

9.2.3 Charge Transport and Extraction

After the carriers are generated within the perovskite layer, they will transport to the perovskite/ETL interface and perovskite/HTL interface and then are extracted by ETL and HTL, respectively. It needs to be noted that carrier injection and recombination compete with each other at the interface. Therefore, the matched energy level between perovskite layer and charge transport layer is required to guarantee efficient carrier extraction and transfer to suppress carrier recombination.

9.2.3.1 Unsatisfactory Band Arrangement

In principle, an ideal interfacial contact should be a type II heterojunction (i.e., the staggered gap in which electrons and holes are located on the opposite side of a heterojunction) to ensure efficient majority carrier separation yet block minority carriers. For the perovskite/organic interface, the interface band alignments follow a universal alignment rule with three different energy windows: pinning of the lowest unoccupied molecular orbital states (LUMO) to the perovskite Fermi level, band offsets mediated by perovskite work functions, and pinning of the highest occupied molecular orbital states (HOMO) to the perovskite Fermi level, as shown in Figure 9.4.[9] The appropriate ETL and HTL should be carefully designed to match the energy levels of perovskites in order to reduce interfacial energy barrier, facilitate charge extraction and minimize interfacial nonradiative recombination losses. In addition, it should be noted that ion migration, interfacial defects and interfacial modification molecules also would have an influence on interfacial energy band alignment, which thereby affect interfacial carrier dynamics.

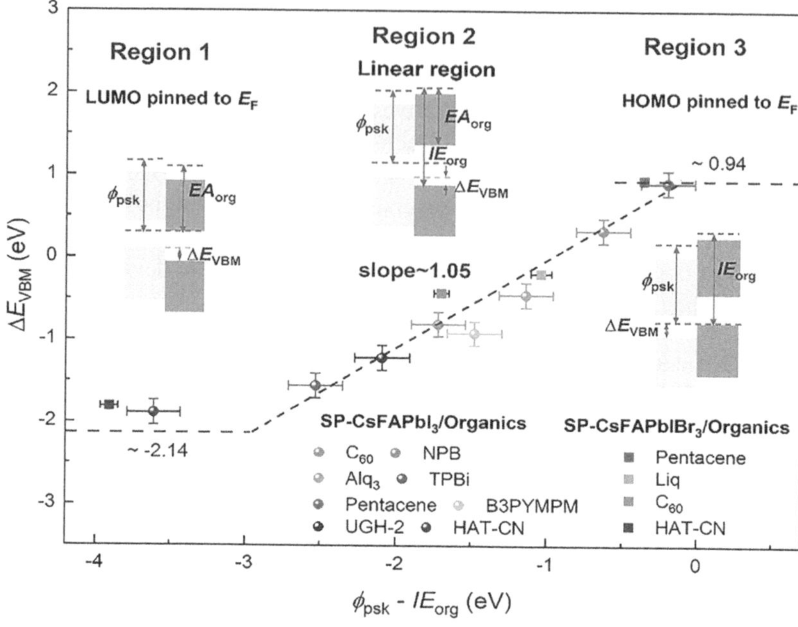

FIGURE 9.4　Perovskite/organic interface energy band alignment. Adapted with permission.[9] Copyright 2023, American Chemical Society.

9.2.3.2　Low Charge Mobility of Perovskite Films and Transport Layers

The reported experimental positive and negative mobilities of typical perovskite measured by methods of THzC, MWC, PLQ, SCLC, Hall, TOF, and FET encompass a considerable range, from 0.2 cm^2 (V s)$^{-1}$ to 2320 cm^2 (V s)$^{-1}$, which might originate from the diverse samples fabricated in different ways. Those samples could be single-crystal with different chemical compositions, polycrystalline with different grain sizes and defect concentrations, or in film and mesoporous morphologies. For the most frequently used FAPbI$_3$ and MAPbI$_3$ perovskites, electron and hole mobility values thus appear to be fundamentally limited to at most ~200 cm^2 (V s)$^{-1}$.[10–13] However, the defects and crystal strain in perovskite films would limit the mobility of carriers.

Low charge mobility of transport layers usually results in large series resistance, which leads to a decreased FF. Therefore, methods of high-temperature annealing, external element doping, and thickness optimization have been proposed to increase the conductivity of the HTL and ETL.

9.2.4　STRAIN ACCUMULATION

Considering the soft nature of perovskite crystal structure, the strain induced by internal and external factors has a dramatic effect on the optoelectronic properties of perovskite. Internal strain is by definition intrinsic in perovskite crystals and is caused by the non-periodicity of crystal lattice in the absence of any external stress, which includes $[BX_6]^{4-}$ octahedra tilting and change in B–X–B bond angle, heterogeneous crystallization. In contrast, external strain is by definition extrinsic in perovskite crystals and is characterized by the distortion of crystal lattice periodicity in the presence of external effects, such as lattice and thermal expansion mismatch and external stress conditions (e.g. light, temperature, external pressure, and applied bias).

9.2.4.1 Lattice Distortion at Grain Boundaries

Heterogeneous crystallization is the main reason for the lattice distortion at the grain boundaries. Slight variation in local environment, for instance, the substrate surface morphology, presence of intermediate complexes, and concentration gradients, would lead to heterogeneous growth of polycrystalline perovskite film. The crystal orientation is highly dependent on their concentrations of the polymeric precursor solution, and strain location occurs. Figure 9.5 a–d shows Pole figures of the crystallographic texture of the different concentrations of mixed-cation perovskite precursors. The 1.2 M sample possesses the most preferred orientation, whereas the 0.4 M shows the most random out-of-plane orientations among the three concentrations measured.[14] The random distribution of grain orientation will increase strain in perovskite films and devices.

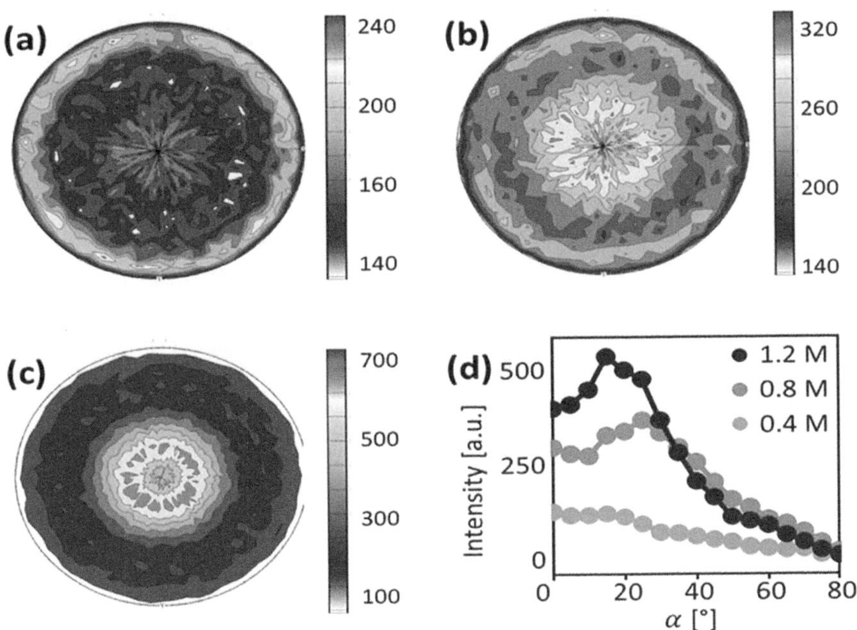

FIGURE 9.5 Pole figures showing the crystallographic texture of the different concentrations of (a) 0.4 M, (b) 0.8 M, and (c) 1.2 M. (d) Line profiles of the diffracted intensity change across the tilt angle α of the different concentrations. The 1.2 M sample is the most preferred-oriented film, whereas the 0.4 M shows the most random out-of-plane orientations among the three concentrations measured. Intensities at α > 80° were not used in the analysis due to the influence of beam defocus at high angles.[14] Copyright 2018, American Chemical Society

9.2.4.2   Tensile and Compressive Strain at Interfaces

As shown in Figure 9.6 a–d, the lattice and thermal expansion coefficient mismatch are the main source of the tensile and compressive strain at the interfaces. For organic and inorganic hybrid perovskite, the annealing temperature is usually in the range of 100–150°C, while it is in the range of 150–240°C for the inorganic counterparts. In general, the inorganic electron transport materials, such as SnO_2, TiO_2 and ZnO, have thermal expansion coefficients, which is smaller than that of the commonly used perovskite, therefore the compressive strain occurs. For the organic hole transport materials, such as P3HT and PDCBT, if they are used in the inverted PSCs, the tensile stress could occur.[15] The strain at the buried interface is usually more serious than that at the top interface, because the top charge transport layers generally are not annealed during the fabrication process.

9.2.5   Instability Induced by Grain Boundaries and Interfaces

Long-term stability is one of the most important factors that hinder the industrial applications of PSCs. The grain boundaries and interfaces are the key sources for the moisture, air, light, and thermal instability issues.

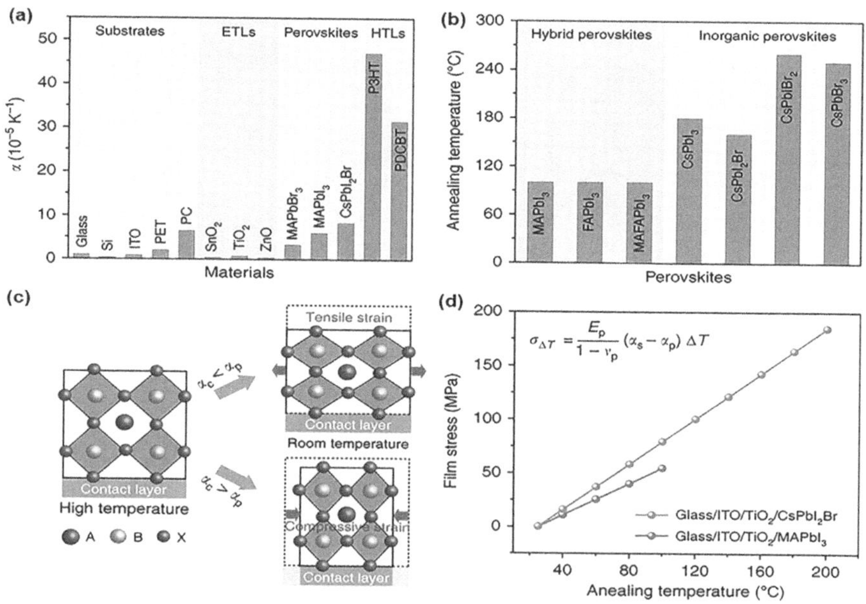

FIGURE 9.6   (a) The common annealing temperature for the organic and inorganic perovskites. (b) thermal elongation coefficient for commonly known substrates, ETLs, HTLs, and perovskites. Adapted with permission. (c) Schematic showing the formation of tensile and compressive strains. (d) The calculated annealing temperature-dependent stress from thermal expansion mismatch between the substrate and perovskites including MAPbI$_3$ and CsPbI$_2$Br.[15] Copyright 2020, Springer Nature.

9.2.5.1 Moisture Instability

Although the different decomposition routes are proposed, there is a consensus that $CH_3NH_3PbI_3$ film decomposes into PbI_2, CH_3NH_3 and I_2 after exposure to moisture. GBs have provided the pathways for moisture penetration (Figure 9.7a).[16] Therefore, the smaller grain size is, the more rapidly the perovskite films degrade. The perovskite films with large grain size and compact surface are favorable for realizing long-term durability. Except for the perovskite layer, the moisture stability and water permeability of interface selective materials are also important when we design high-performance device structure.

9.2.5.2 Air-Related Instability

Experimental measurements have found that the presence of oxygen could lead to degradation of solar cell performance. The atmosphere containing 1 vol% O_2 could decrease the performance to 70% of the initial performance within 24 h, and higher percentage of oxygen induces faster degradation rate.[18] Oxygen enters the sample and diffuses between particles, permeating the inter-particle area within one h. With the reduction of oxygen molecules, the iodide vacancies are occupied and superoxide species immediately begin to form on the particle surface. In a few hours, initial reactions occurred between these superoxide species and the particle surfaces, leading to their passivation. The degradation of the particle surface obstructs the extraction of photo-induced carriers, leading to a rapid decline in device performance such as efficiency (PCE). In a few days, oxygen diffuses into the particles, where it occupies

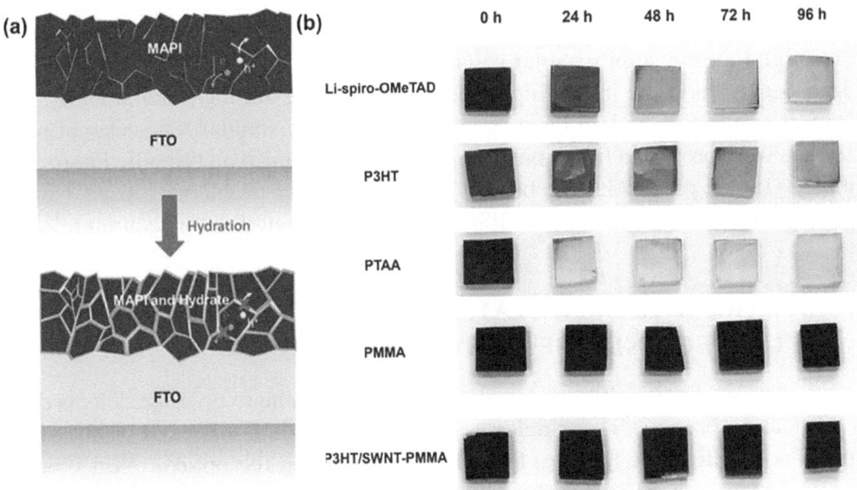

FIGURE 9.7 (a) Illustration of the $MAPbI_3$ hydration when moisture exists. Adapted with permission.[16] Copyright 2015, American Chemical Society. (b) Photographs of the $FAPbI_3$ with different HTL materials, such as Spiro-OMeTAD, P3HT, PTAA, PMMA, and P3HT/SWNT-PMMA. Adapted with permission.[17] Copyright 2014, American Chemical Society.

a large number of iodide vacancies, leading to complete structural degradation of the material.[19] The role of grain boundaries and interface is similar to that of the moisture instability.

9.2.5.3　Light Instability

UV irradiation causes decomposition of the $MAPbI_3$ film in ultrahigh vacuum conditions in the absence of moisture and oxygen. The surface of the perovskite film becomes more *n*-type and metallic Pb appears in the XPS spectrum after 120 min UV illumination, indicating the decomposition of the film. The authors argued that the decomposition is a self-limited process because it stops after 480 min when 33% of the total Pb becomes metallic nature.[20] In general, oxygen or light alone doesn't result in obvious degradation, however, their coexistence could accelerate the degradation process dramatically.

9.2.5.4　Thermal Instability

Thermogravimetric analysis (TGA) of $MAPbI_3$ revealed that the crystal structure is stable up to 300°C but quickly decomposes above this temperature. However, it is frequently observed that the performance of perovskite solar cells drops quickly after 85°C of aging, because the crystal structure has been found to be largely distorted at this temperature. It is observed that the degradation of perovskite thin film begins at GBs, which changes the morphology of perovskite surface from smooth to needle features. As shown in Figure 9.7b, the color of the perovskite films under temperature of 85°C with different exposure time is compared. The perovskite films covered by Spiro-OMeTAD, P3HT, and PTAA decomposed rapidly while the PMMA and P3HT/SWNT-PMMA hole transport material can protect perovskite films from degrading. This indicates that the choice of hole transport material is rather important for high thermal stability solar cells.[17]

In order to study the instability of perovskite devices more systematically, a stability test standard also had been established.[21] In this standard, a series of testing processes were proposed based on the International Summit on Organic Photovoltaic Stability (ISOS) protocols. Accordingly, the instability problems caused from the grain boundaries and interfaces in PSCs could be evaluated more accurately.

9.3　SYNERGISTIC MODULATION STRATEGIES AT GRAIN BOUNDARIES AND INTERFACES

The soft nature of perovskite materials makes it inevitable to produce defects during the fabrication process. These defects mainly exist both at GBs and on the surface of perovskite films. They play a key role in controlling the optical, electronic, and structural properties of perovskite materials and thus affect the performance of devices. Therefore, effective passivation of defects existing at grain boundaries and surfaces is crucial for improving the efficiency and stability of perovskite devices. In the following content, we will discuss the effects of synergistic modulation strategies on these defects at GBs and interfaces of perovskite devices and summarize their modulation mechanisms.

9.3.1 Synergistic Modulation between Grain Boundary and Perovskite/ETL Interface

ETL plays a significant role in PSCs owing to its electron extraction and transport features. High quality of interface between perovskite and ETLs is considered the necessary condition for highly efficient PSCs. Hence, it needs to be modified using suitable approaches to overcome the current challenges, such as low electron mobility, mismatched energy level, and vacancy defects. In this section, we will discuss highly promising modification methods used for synergistic modulation of perovskite grain boundaries and perovskite/ETL interfaces in the state-of-the-art PSCs.

Defects at interfaces and GBs within the perovskite layer can cause the charge recombination and instability in perovskite devices. The evaporation of organic cations and halide anions from GBs or the perovskite surface results in the generation of undercoordinated lead ions (Pb^{2+}). These undercoordinated Pb^{2+} ions exhibit Lewis acid characteristics, enabling them to accept electrons. Consequently, the introduction of Lewis bases to coordinate with these Lewis acids and form Lewis adducts has emerged as an effective passivation technique. These adducts are connected through a coordinate bond, wherein both bonding electrons originate from the same atom in the Lewis base. The improved performance confirms that the passivation of under-coordinated defects through Lewis acid/base strategy is a successful approach for enhancing the efficiency and stability of perovskite photovoltaic devices, as depicted in Figure 9.8.[22,23]

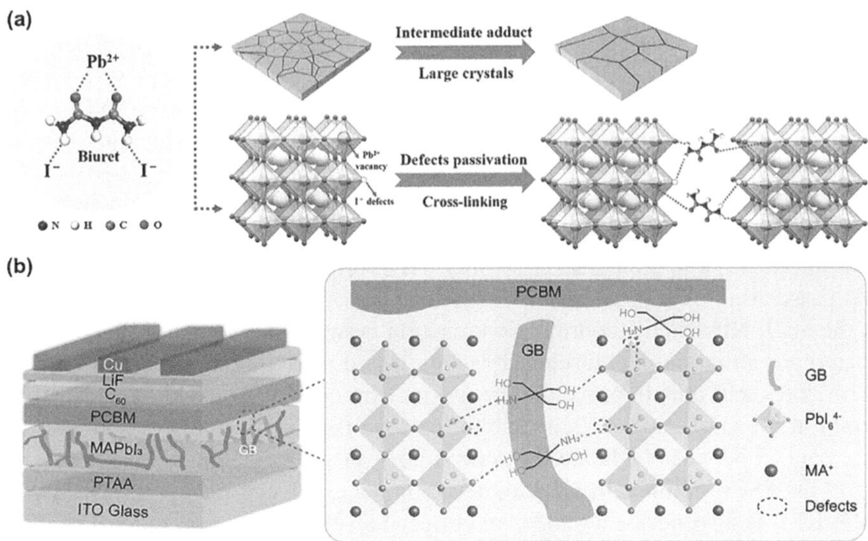

FIGURE 9.8 (a) Schematic illustration of the proposed passivation mechanism and the cross-linked structure of the adjacent crystals with biuret.[22] Copyright 2021, American Chemical Society. (b) Schematic illustration of the PSCs architecture and the TIDP mechanism in the MAPbI3 film.[23] Copyright 2022, American Chemical Society.

Besides passivating the undercoordinated Pb^{2+}, Lewis base molecules also can ameliorate growth kinetics of perovskites crystal, reducing trap-assisted nonradiative recombination of perovskite devices.[24] Many works reported that cyano-based small molecules, such as ITIC or IEICO-4F, can coordinate with Pb atoms and modify the formation kinetics of perovskite films by incorporating into perovskite precursors. As a result, the as-prepared perovskite films possessed dense and uniform[25,26] morphology with fewer defects. In addition, by introducing Lewis base molecules containing different numbers of carbonyl groups (urea, PA, and BA) into SnO_2 colloidal solution, a bottom-up multifunctional modification strategy can be realized.[27] The defect passivation effect can be reasonably modulated by tuning the number of carbonyls and increased as the number of carbonyls is increased. The interfacial energy barrier, charge accumulation, and hysteresis were mitigated after modification, which is principally attributed to improved energy band alignment, reduced interfacial defects, and improved electrical properties of SnO_2 films. Other Lewis bases containing amines (such as $-NH_2$) or nitrogen functionalities have also shown effective passivation of perovskite materials. The trometamol-induced dual passivation (TIDP) strategy can fix both bulk and surface defects of perovskites.[23] The organic small molecule trometamol with an amino group ($-NH_2$) and three hydroxyl groups ($-OH$) served as both chemical additive and surface-modification agent. It was found that trometamol as an additive can effectively reduce ionic defects and enhance the grain size of perovskites through $Pb^{2+}/-NH_2$ coordination bonds and $I^-/-OH$ hydrogen bonds. The formed coordination bonds and hydrogen bonds between trometamol and $CH_3NH_3PbI_3$ ($MAPbI_3$) led to efficient defect passivation of the perovskite layer and suppressed the defect-induced nonradiative recombination.[28]

In addition to the Lewis acid-base interaction aforementioned, other synergistic effects have also been reported to access the synergistic modulation at perovskite surface and perovskite/ETL interfaces. Girard's Reagent T (GRT) with multiple functional groups had been selected to modify SnO_2 nanoparticles (NPs), which significantly reduced the bulk and interfacial nonradiative recombination losses by suppressing nanoparticle agglomeration, improving the electronic property of SnO_2 films, and passivating interfacial defects.[29] The GRT molecules contain several different kinds of functional groups, in which the carbonyl group is expected to be able to block the agglomeration of SnO_2 NPs, promoting their uniform dispersion and passivating oxygen vacancy defects from SnO_2 through bonding to the surface of the SnO_2 NPs via the coordination reaction between $-C=O$ and Sn^{4+}, whereas the quaternary ammonium chloride salt is anticipated to facilitate the crystal growth of perovskites and thus improve the perovskite film quality. The hydrazine functional group in the GRT molecules is able to improve the SnO_2 NP dispersion and interface contact between ETL and the perovskite layer via formation of hydrogen bonds. Thus, the PSCs modified by GRT achieved higher efficiency and longer stability. The TiO_2 ETL is also necessary to be modified. Substitution of undercoordinated titanium atoms by boron species can effectively passivate oxygen vacancy defects in the TiO_2 ETL, leading to increased electron mobility and conductivity, thereby greatly facilitating electron transport.[30]

Ionic bonding involves the complete transfer of one or more valence electrons from one atom to another, resulting in an electrostatic interaction between two ions with complementary charges, namely a positively charged cation and a negatively charged anion. Ionic bonding can be used as another effective passivation technique due to the charged nature of the defects in perovskite. Positively charged defects (halide vacancy, undercoordinated Pb^{2+}) and negatively charged defects (cation vacancy, PbI_{3-} antisite) usually exist in the perovskite films simultaneously. Therefore, the salts containing cations and anions are usually used to simultaneously passivate positively charged and negatively charged defects. Selective introduction of charged complementary ions into a given defect can bind to the defect and effectively eliminate the corresponding trap states. The synergistic modulation of ionic bond has recently attracted much attention owing to its proven ability to improve the stability and efficiency of PSCs (Figure 9.9).

A number of organic ammonium cations have been explored for their ability to enhance device performance and stability, such as butylammonium (BA^+), octylammonium (OA^+), ethylammonium (EA^+), and phenylethylammonium (PEA^+).[33–36] These organic ammonium cations can not only produce perovskite films with significantly improved crystallization quality but also anchor at the GBs and surface to passivate the defects, thus reducing the nonradiative recombination and increasing the carrier lifetime of perovskite films (Figure 9.9a).[31,37] Zwitterion molecules, which possess both positively charged organic cations and negatively charged halide/pseudohalide

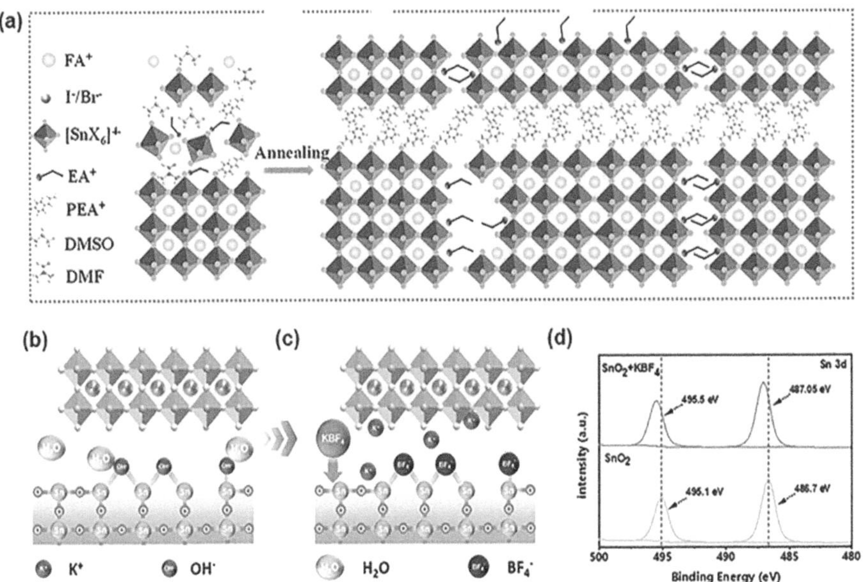

FIGURE 9.9   (a) Schematic of the crystallization process for the control and 5% EABr-containing perovskite films.[31] Copyright 2022, American Chemical Society. (b) Schematic diagram of the mechanism of KBF_4 acting at the SnO_2/perovskite interface. (c) XPS spectra of the reference and experimental SnO_2 films for Sn 3d (d) Chemical structure of 4F-PHCl molecule.[32] Copyright 2022, American Chemical Society.

anions, are one important representative of molecules that can achieve synergistic passivation. The positive organic cations of zwitterion molecules can anchor on the surface of perovskite films and ETLs through hydrogen or ion bonds, which not only passivated the defects in perovskite and ETLs but also reduced the work function (WF) of the ETL surface, yielding barrier-free charge transport.[38–40] The negative-charged halide/pseudohalide anions of zwitterion molecules can regulate the crystallization behavior of perovskites and filling vacancies in perovskites, which is beneficial to obtain pinhole-free perovskite films with less traps and longer carrier lifetimes.[41,42] In addition, Chen et al. reported 4-imidazoleacetic acid hydrochloride (ImAcHCl) to chemically link the perovskite layer and SnO_2 layer.[43] The COOH in ImAcHCl can anchor on the surface of SnO_2 by esterification reaction while the imidazolium cation in ImAcHCl had an electrostatic interaction with iodide anion in perovskite. The chloride anion improved perovskite crystallization. Due to effective defect passivation and crystallization improvement, the nonradiative recombination was much inhibited and the carrier lifetime was much increased, which led to enhanced average PCE from 18.60% ± 0.50% to 20.22% ± 0.34%. Furthermore, the thermal and moisture stability were strengthened after interface modification. Apart from organic ammonium salts, our group also attempted to employ the novel sulfonium salt ((2-carboxyethyl) dimethyl sulfonium chloride, CDSC) to stabilize the buried interface by passivating interfacial defects and optimizing energy level alignment.[44] As a result, the modified device exhibited improved PCE and stability as compared to the control device.

Except for organic salts, inorganic salts are also utilized to simultaneously modulate perovskite layer and ETL. Various alkaline metal cations, such as Li^+, Na^+, K^+, et al., have been proven to be very effective modification materials that can simultaneously achieve surface and interface passivation through the ionic bond.[45–47] For example, Li^+ ions have been proven to diffuse into perovskite films from the perovskite/ETL interface, which contributes to facilitated carrier transport and extraction.[38,39] Besides, the Na_2S interface layer at the perovskite/ETL interface also has been proven to effectively passivate defects in perovskite films and ETLs.[48] The S atom simultaneously bonded with the Sn atoms in SnO_2 and occupied the halogen vacancies in the perovskite and formed a Pb-S-Sn conductive channel, which accelerated carrier extraction at the interface. Moreover, the Na_2S interface layer also affected the growth of the perovskite film and passivated the grain boundaries, resulting in reduced defect density and increased grain size. Except for Li^+ and Na^+, we also reported a bulk and interface defect passivation strategy via the synergistic effect of SO_4^{2-} anions and K^+ cations.[49] The synergistic modification strategy simultaneously realized defect passivation and improved energy band alignment, which resulted in an increase in PCE from 19.45% to 21.18% as well as ameliorated stability. In addition, additives containing F^- ions have been developed to passivate charged defects in perovskite films and improve moisture stability of perovskite devices.[50] Other anions and cations for passivating defects at bulk and interfaces have been proposed as shown in Figure 9.9b–d. The incorporation of anions and cations leads to improved electron mobility, better energy level matching, and reduced nonradiative recombination.[32]

It is highly desirable to achieve simultaneous modulation of GB and interface through enriching chemical bonding modes, such as hydrogen bond, ionic bond, and coordinate bond. To realize this goal, it is imperative to functionalize cation and/or anion with strong coordination functional groups (e.g., F and C=O groups). Based on the previous consideration, our group used KPF_6 to modify SnO_2/perovskite interface. Most K^+ cations diffused into the perovskite layer to passivate the defects at GBs while PF_6^- remained located at interface to passivate the defects from the surface of perovskite and SnO_2 via hydrogen bond, ionic bond, and coordinate bond.[51] KPF_6 modification led to a PCE increase from 19.66% to 21.39%. The unsealed devices demonstrated improved stability, maintaining 80.1% of its initial PCE after 960 h of aging at 60°C and 57.2% after 960 h of aging under one sun illumination, respectively. In addition, we adopted 4-fluoro-phenylammonium tetrafluoroborate (FBABF$_4$) consisting of simultaneously fluorinated anions and cations to modify buried interface.[52] It was revealed that the fluorine atom on BF_4^- can form coordination bond with SnO_2 and the fluorine atom on FBA^+ can form a hydrogen bond with FA^+. BF_4^- can fill halide vacancy defects ionic bond/coordination bond. This work highlights the importance of simultaneous fluorination of anion and cation in the ionic liquid molecule for improving interfacial contact. The uncontrollable crystallization and poor film quality of CsFA-based MA-free perovskites impede the further increase of PCE and stability for inverted MA-free PSCs. Recently, we reported a novel crystallization modulation strategy by incorporating phenformin hydrochloride (PFCl) into the precursor solution to form intermediate phase PFCl·FAI.[53] With forming the intermediate phase, the generation of the δ yellow phase was suppressed and the orientational growth of the α-phase perovskite was promoted. After manipulating crystallization, PFCl can passivate the defects at GBs by hydrogen bond and ionic bond. In order to further improve the quality of perovskite films, multi-active-site S-methylisothiosemicarbazide hydroiodide (SMI) molecules were further employed to post-treat perovskite films. The diverse interfacial defects were effectively healed via hydrogen bond, coordinate bond and ionic bond. In a word, improved crystallization, reduced defects, and released residual stress contributed to much reduced bulk and interfacial trap-assisted nonradiative recombination losses, which led to increased V_{oc} and FF. The device based on GB and interface synergistic modification strategy obtained promising PCEs of 24.67% (0.09 cm^2) and 22.48% (1 cm^2). Meanwhile, the target devices maintained 84% of their initial PCE after 1008 h of continuous light illumination and 90% of their original PCE after 864 h of continuous heating at 85°C. The present work provides a feasible and effective route to minimize trap-mediated nonradiative recombination at GB and interface.

The formation of low-dimensional perovskite at the surface and grain boundary of 3D perovskite is another effective approach to achieve the synergistic modulation in perovskite devices. The low-dimensional perovskites can effectively wrap the grains of 3D phase and form type-I heterojunctions, which can promote the carrier transport and enhance the stability of perovskite devices (Figure 9.10).[54] Hence, the 2D/3D heterostructure is commonly used in PSCs to improve the performance of devices in recent years. The 3D portion of 2D/3D perovskite heterostructures serves as an active material with exceptional optoelectronic characteristics (for example excellent light absorption and an extended carrier lifetime), while its 2D counterpart protects 3D perovskites against environmental conditions (water and oxygen) and increases the stability of the whole structure. The 2D/3D heterostructures with type-I band

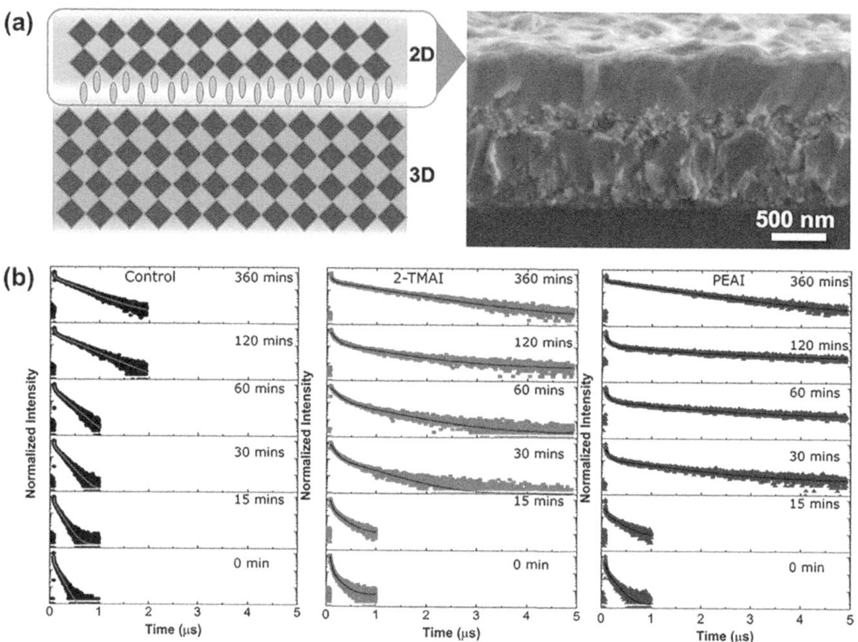

FIGURE 9.10 (a) Cartoon of the 3D/2D interface (left) and the corresponding cross-sectional SEM image of the 3D/2-TMAI 2D interface. (b) TrPL decays (λ_{exc} = 440 nm; F = 9.7 nJ cm^{-2}; λ_{PL} at the maximum of the 3D at 780 nm emission at 50°C in the samples) of the 2D/3D modified perovskites: 3D control, 2-TMAI, and PEAI.[54] Copyright 2022, the authors.

alignment can be formed by incorporation of large-sized alkylamine cation, such as n-butylammonium (BA$^+$) and phenethylammonium (PEA$^+$).[55–57] Compared with 3D perovskites, PSCs based on optimized 2D/3D heterostructures exhibit substantial improvements in photovoltaic efficiency owing to the passivation of cation and halide vacancies on the surface of 3D perovskite, better energy level alignment, longer carrier lifetimes, and fewer defects. The extremely thin film of 2D perovskite on the top of 3D perovskite boosted the stability and reduced the crystal imperfection of 3D perovskites.[58] As the number of carbon atoms in the alkane increases, the water contact angle gradually increases. Moreover, the defect state density is reduced to a certain extent compared to untreated 3D perovskite. The results show that the types of alkyl ammonium salts, including the type of functional groups, the number of carbon atoms in the hydrocarbon chain, as well as the arrangement and distribution of the molecular structure, have a great impact on the performance and stability of films and devices.[59,60] Moreover, the 2D/3D heterostructure is beneficial for the lattice matching and strain releasing in perovskite films, which helps to enhance stability of perovskite devices.[61,62] Our group used amphiphilic organic ammonium salt tert-butylcarbamidine hydrochloride (TBHCl) to post-treat 3D perovskite films.[62] The gradient 2D/3D heterojunction was formed because TBH$^+$ cations diffused to

GBs within perovskite films. By this gradient dimensionality engineering, defect passivation, hole extraction improvement, and moisture stability enhancement were realized. Benefiting from suppressed bulk and interfacial nonradiative recombination, the modified devices showed an enhanced PCE up to 22.54% than that of the control device along with increased stability. 2D perovskites in 2D/3D perovskite heterostructure could be Ruddlesden–Popper (RP), Dion–Jacobson (DJ), or alternating cations in the interlayer space (ACI) perovskites.

In addition, 2D perovskites in 2D/3D perovskite heterostructure could be Ruddlesden–Popper (RP), Dion–Jacobson (DJ), or alternating cations in the interlayer space (ACI) perovskites. Compared with RP and DJ 2D perovskites, ACI 2D perovskites usually possess lower exciton binding energy, higher crystal symmetry, and shorter interlayer distance between neighboring inorganic slabs, which is favorable for carrier transport and transfer. Furthermore, we have demonstrated that multiple-active-site Lewis ligand molecules exhibited strong chemical interaction with perovskites and excellent defect passivation effect.[63] Therefore, synergistic modification of GBs and surface via the combination of ACI 2D perovskites and multiple-active-site Lewis ligands should be capable of further increasing PCE and stability of inverted PSCs. Therefore, our group proposed a synergistic modulation strategy of ACI 2D perovskite and multisite ligand 2-mercapto-1,3,4-thiadiazole (MTD) for preparing high-quality MA-free perovskite films.[64] After the formation of ACI 2D perovskite, the synergistic effect of ACI 2D perovskite and MTD enabled the simultaneous passivation of defects at GBs and surface. The synergistically modified inverted devices achieved an appealing PCE of 24.58%, which is among the highest PCEs reported for inverted PSCs.

9.3.2 Synergistic Modulation between Grain Boundary and Perovskite/HTL Interface

The interface between perovskite and HTL is also essential and very important to obtain high performance PSCs. For the organic HTL, such as Spiro-OMeTAD, PTAA, P3HT, and PEDOT:PSS, the perovskite/HTL interface is suffered by the unstable issues induced by the ion migration between perovskite and these organic materials especially under long-term operation.[65–68] For the inorganic HTL, such as NiO_x, CuO_x, CuI, and MoO_x, the perovskite/HTL interface also faces many issues, such as the high defect density, high transport barrier, and poor film-forming characteristics. The defects at surfaces and GBs of perovskite films make these interface issues more severe. Hence, it is necessary to modulate the perovskite/HTL interface and GBs of perovskite to improve the performance of perovskite devices.

Dipoles refer to a class of molecule with asymmetric chemical bond or asymmetric molecule composed of atoms with unequal electronegativity. When positive and negative charges are distributed asymmetrically in a thin layer, a dipole layer is formed, resulting in a rapid decline of potential from the positive charge side to the negative charge side.[69] This provides an extra driving force to promote charge transport and transfer at the interface of optoelectronic devices. In PSCs, a large number of multifunctional dipoles containing amino groups, halide ions, and pyridine had been reported to form dipole layer to obtain high-performance devices. These dipole

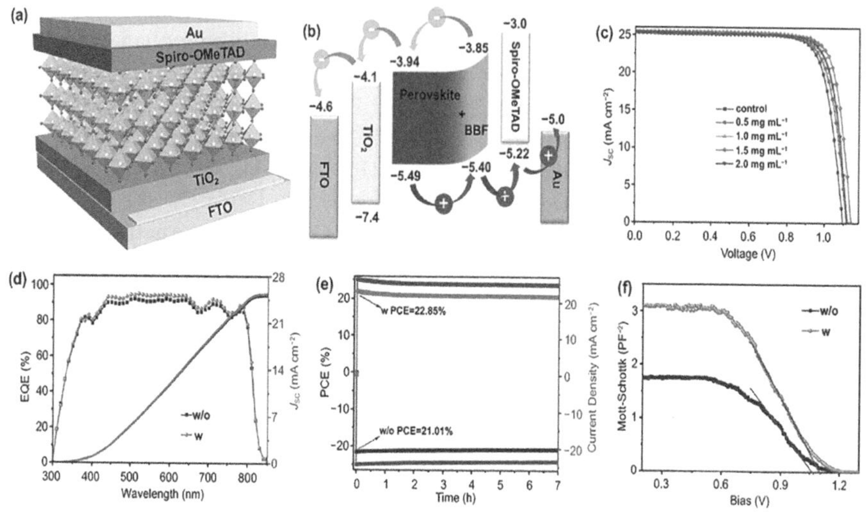

FIGURE 9.11 (a) A schematic image of the hybrid PSC device with the structure FTO/TiO$_2$/perovskite/Spiro-OMeTAD/Au. (b) Schematic energy-level alignment of the hybrid PSCs with BBF additive. (c) J-V curves of the hybrid PSCs with different concentrations of BBFadditive. (d) EQE spectra. (e) Stable output. (f) Mott-Schottky plots.[75] Copyright 2020, the authors.

layers can heal charged defects at interfaces, improve energy level alignment, promote charge extraction, inhibit ion migration, and release surface residual stress, which is beneficial to improve efficiency and stability of PSCs (Figure 9.11).[70–75] Among different kinds of dipole interlayers, self-assembled monolayers (SAMs) have attracted much interest because of their orientation distribution and chemical stability. The interfacial SAMs with a dipole can order the Schottky energy barrier between metal electrodes and organic materials and thus promote the extraction efficiency of electrons or holes.[76] The SAMs with functional groups also can passivate surface uncoordinated Pb^{2+} defects and improve perovskite crystallization.[77] In addition, SAMs with hydrophobic groups, such as alkyl and F groups, can prevent perovskite films from erosion by moisture.[78–84]

Recently, metal-organic frameworks (MOFs) have received significant attention in applications in the charge-transport layers, interlayers, or perovskite layers in PSCs and have proved to be effective to improve performance of PSCs (Figure 9.12).

MOFs possess large number of oxygen sites and anionic porous framework, which can passivate surface defects of perovskite and promote charge transport in perovskite devices.[86] As shown in Figure 9.12 a and b, the hybridization of hygroscopic copper(II) benzene-1,3,5-tricarboxylate metal-organic frameworks (Cu-BTC MOFs) hybrid with a perovskite layer was induced into PSCs, where a moderate level of moisture attracted by Cu-BTC MOFs during the synthesis step leads to enhanced perovskite crystallization. Besides, the perovskite-MOF hybrid facilitates the transfer of photoexcited electrons from the perovskite to TiO$_2$ by providing additional channels for electron extraction.[85] And a (Me$_2$NH$_2$)$^+$-encapsulated indium-based

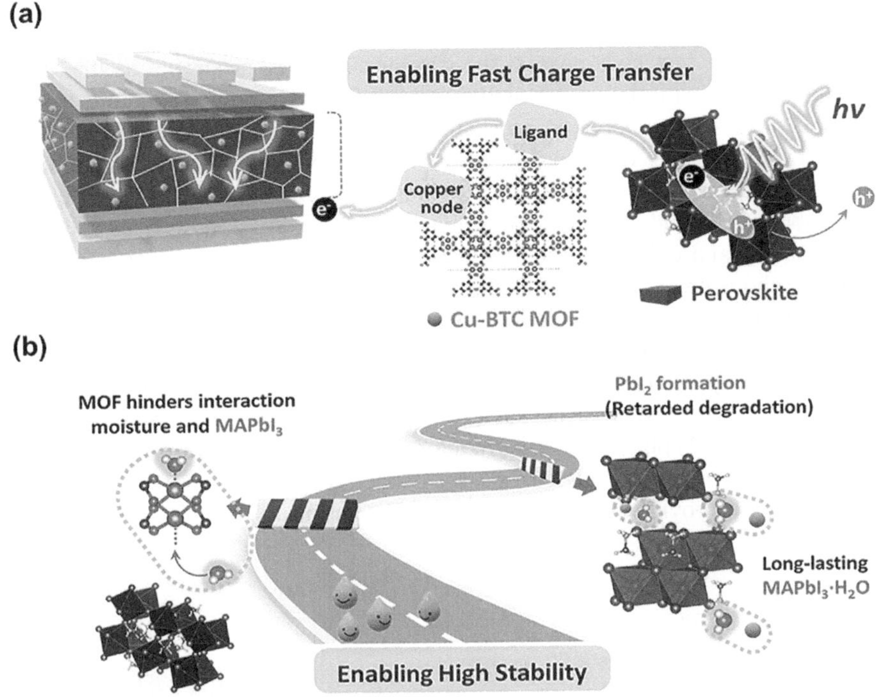

FIGURE 9.12 Schematic design of a light-absorbing perovskite hybrid layer for fast charge transfer and high stability in a PSC. (a) Enhanced charge transfer within the perovskite MOF hybrid layer under illumination, where MOF refers to Cu-BTC MOF. (b) Enhanced moisture stability of the perovskite with Cu-BTC MOFs.[85] Copyright 2022, American Chemical Society.

anionic MOF (FJU-17) [87] had been chosen as a "capsule" into HTM to develop a dual-functional layer (namely, HTM-FJU-17 DFL), which passivated perovskite defects by filling the organic cation vacancies with releasing $(Me_2NH_2)^+$ ions. Furthermore, the HTM-FJU-17 DFL enhanced hole mobility via compensating and stabilizing the oxidized Spiro-OMeTAD$^+$ with the anionic framework of FJU-17 capsule and eliminated lithium ion migration due to the strong binding interaction between Li ion and the anionic framework. Particularly, its ionic characteristics formed by the combination of cations and anions enabled the replacement and/or passivation of the vacancy defects at the surface, GBs, and bulk of perovskite. In addition to MOF materials, ionic liquids (ILs) with large organic cations and various organic or inorganic anions make them have the potential to become the effective passivation materials. As shown in Figure 9.9d-f, 1-butylmethylimidazolium hexafluorophosphate $(BMIMPF_6)$ ionic liquid has been adopted as a surface modifier to passivate perovskite layer through post-treatment.[88] The introduced IL provided a reduced trap density and a suppressed charge recombination via the passivation of the A-site and X-site defects between perovskite and HTL, which was translated into the simultaneous formation of interactions between the cation and anion components. In addition, the nonradiative recombination of the double-passivated devices is significantly suppressed and the migration of charged defects was remarkably hindered.

2D perovskite materials possess the similar chemistry properties as the 3D perovskites and have the layered structure with a hybrid organic-inorganic interface enabling them have synergistic properties, which can alleviate the weaknesses of surfaces and interfaces of perovskite films.[89] In fact, 3D perovskite materials and 2D perovskite materials are compatible in various architectures of perovskite devices.[90] By combining 3D and 2D perovskites, 3D-2D halide perovskite devices can exhibit more stability and efficiency than their constituent materials alone. In addition, by rationally choosing organic cations, the carrier mobility and extraction efficiency can be improved obviously at interfaces in PSCs. For instance, 1,4-butanediammonium diiodide (BDADI), which has a short chain length and diammonium cations, was employed to construct a 3D/2D stacking perovskite solar cells (PSCs). The introduction of BDA^{2+} could passivate surface defects in 3D perovskites by forming a 2D Dion-Jacobson (DJ) phase, resulting in a longer carrier lifetime (Figure 9.13).[91] And, by rationally choosing organic cations, the carrier mobility and extraction efficiency can be improved obviously at interfaces in PSCs. For instance, by introducing CPAH

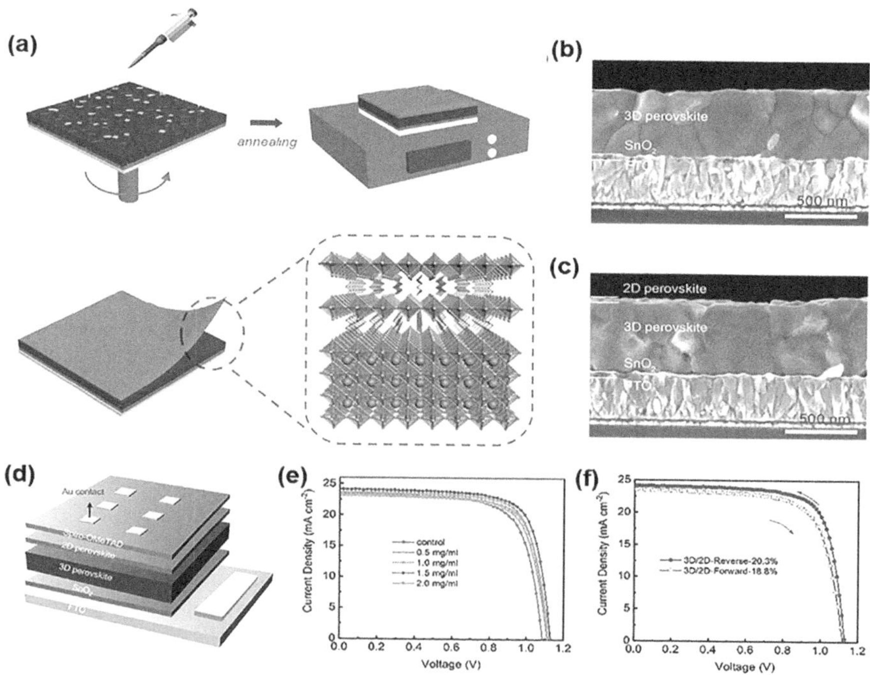

FIGURE 9.13　(a) Schematic diagram of the preparation process of 2D capping layer on top of 3D perovskite films and illustration of 3D/2D structure. Cross-sectional SEM of different perovskite films: (b) 3D and (c) 3D/2D. (d) Schematic diagram of the device structure of 3D/2D PSC. (e) J–V curves of the PSCs based on 3D/2D and 3D perovskites. (f) J–V curves of 3D/2D PSC under reverse and forward scan.[91] Copyright 2021, American Chemical Society

on the surface of phase-pure $FAPbI_3$ perovskite film to form a 2D $(CPA)_2PbI_2Cl_2$ passivation layer, the defects involved with undercoordinated Pb^{2+} can be passivated efficiently (Figure 9.10).[92] CPAH composed of relatively short-chain organic cations offered superior hole transfer, ensuring efficient charge transport. The 2D $(CPA)_2PbI_2Cl_2$ passivation layer contributed to the alleviation of charge recombination and acceleration of charge extraction. Park et al.[93] constructed a versatile ultrathin 2D perovskite $(5\text{-}AVA)_2PbI_4$ (5-AVA = 5-ammoniumvaleric acid) passivation layer that was in situ formed at the interface between $(FAPbI_3)_{0.88}(CsPbBr_3)_{0.12}$ and CuSCN HTL, which effectively passivated the defects on surface of perovskite film and improved the stability of devices. To overcome the issues of inorganic HTM, such as NiO_x-based HTM, 2-aminoindan hydrochloride (AICl) has been introduced into the interface between perovskite and NiO_x layer to form the 2D/3D perovskite heterojunction, which has proved to regulate 3D perovskite crystallization, promote energy level alignment, and enhance the extraction and transport of carriers in p-i-n structural PSCs.[94] The long-term stability and efficiency of PSCs with 2D/3D heterojunction had been improved dramatically.

9.3.3 Synergistic Modulation between Perovskite/ETL Interface and Perovskite/HTL Interface

Defect densities at interfaces of perovskite/ETL and perovskite/HTL are hundreds of times higher than that in perovskite films. Unfortunately, it's not just about defects; the interfacial stress, interfacial ion interdiffusion, and interface energy level mismatch at interfaces of perovskite/ETL and perovskite/HTL also hinder the development of perovskite devices. Thus, the reliable methods to overcome the previous problems at interfaces of perovskite/ETL and perovskite/HTL are very critical to obtain high-performance perovskite devices.

As shown in Figure 9.14a-f, the HTMP molecule, containing imino (−NH) and hydroxyl (−OH) groups, act as a bifunctional interface modifier. HTMP molecules can bond with iodide (I⁻), formamidinium (FA) and methylammonium (MA) ions at the interface and GBs of perovskite by forming bi-directional cross-linking bonds, which can passivate defects effectively. Furthermore, HTMP molecules can delay the crystallization process of perovskite films leading to high-quality films with large grain size.[95] In addition, a depth-dependent manipulation strategy has been developed to modulate both the interface (i.e., the top surface and the buried interface) defects with a binary mixture of guanidine iodine and 4-tert-butyl-phenylmethylammonium iodide, denoted as GuaI and tBPMAI, respectively.[96] Benefiting from the favorable penetrability within the perovskite film, GuaI can penetrate across the bulk directly to the buried interface whereas tBPMAI remained to anchor at the top surface due to the limited penetrability, ensuring the surface defect modulation and being conducive to lowering the defect-assisted nonradiative recombination losses and maximizing the overall device performance. Furthermore, the dual-surface passivation of $CsPbCl_3$ NCs using trivalent metal-chloride salts has been explored, *i.e.*, YCl_3.[97] It demonstrated that Y^{3+} and Cl⁻ ions occupied the surface Pb-Cl ion vacancies and passivated the uncoordinated Pb atoms on the NC surface, reducing the density of states in the CB without creating any mid-gap states. Furthermore, the energy level

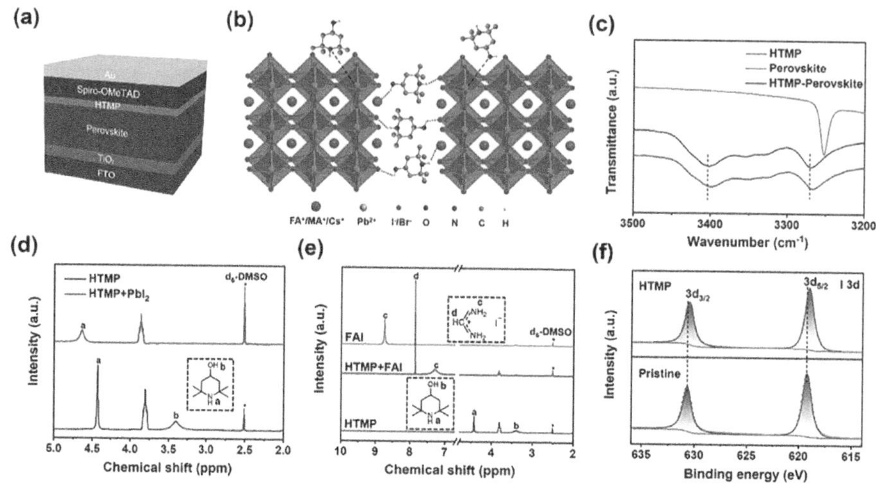

FIGURE 9.14 (a) Structure of the HTMP-modified PSC. (b) Schematic illustration of the possible interaction mechanism between HTMP and perovskite. (c) Fourier transform infrared (FTIR) spectra of HTMP, perovskite, and HTMP-perovskite films. (d) 1H NMR spectra of HTMP and HTMP/PbI$_2$. (e) 1H NMR spectra of HTMP, FAI, and HTMP/FAI dissolve in DMSO-d6. (f) I 3d X-ray photoelectron spectroscopy (XPS) spectra of pristine and HTMP-modified perovskite films.[95] Copyright 2022, American Chemical Society

alignment in PSCs between perovskite films and charge transport layers became more favorable, which is beneficial for the charge extraction.

The synergistic modulation by ionic bond between anion and/or cation and perovskite have been investigated widely and deeply. The guanidinium cations have been proved as one of the most potential cation passivators to realize the synergistic modulation at interfaces of perovskite/HTL and perovskite/ETL (Figure 9.15). The CH$_6$N$_3^+$ cation (guanidinium, abbreviated as Gu), which has an ionic radius (0.278 nm for Gu$^+$) comparable to that (0.253 nm) of formamidinium (HC(NH$_2$)$_2$), abbreviated as FA, had been confirmed to be able to access a cation-diffusion-based double-sided interface passivation scheme, which is compatible with a p-i-n cell architecture.[98] They declared that the partial cation diffusion from the Gu-based passivation layers into the perovskite bulk benefits the suppression of shallow traps in the bulk, and the remaining Gu cations at the perovskite/ETL and perovskite/HTL interfaces are responsible for passivating the relatively deeper surface trap states. The surface passivation of halide perovskites by various surface-passivating materials has proven to be an attractive approach to stabilize perovskites against moisture, heat, and light, keeping intact their structural integrity.[99] By using the cation-diffusion strategy, the passivation molecule on the surface can diffuse to the bottom of the perovskite film, which can realize the synergistic modification at two interfaces.[100, 101] The cyclohexylmethylammonium iodide (CHMAI) molecules can enrich the ratio of the Cl$^-$ ions on the surface of perovskite film and diffuse to the bottom side of the film, which can improve charge transport and suppress the nonradiative recombination in PSCs.

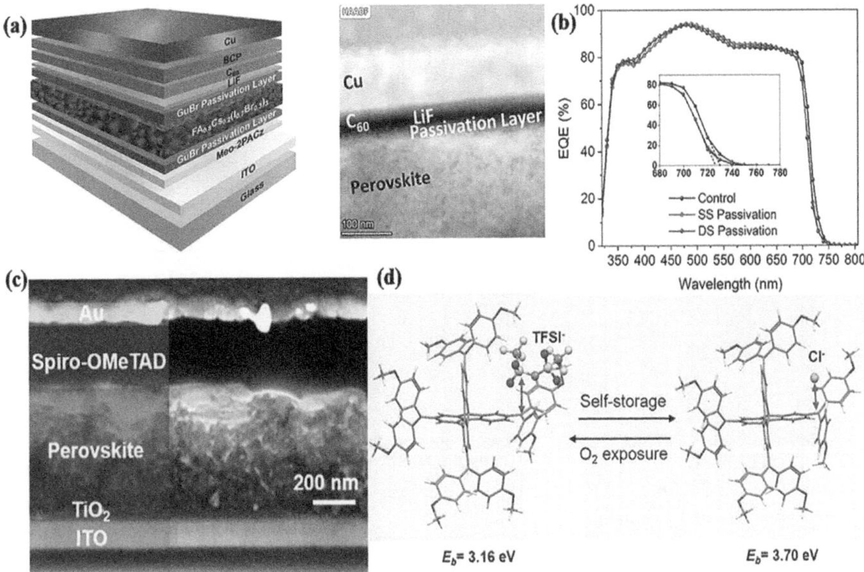

FIGURE 9.15 (a) Schematic diagram showing the cell structure for high bandgap p-i-n perovskite solar cell demonstration using SAM-based HTL and double-sided (DS) passivation layers sandwiching the bulk perovskite film and transmission electron microscopy (TEM) cross-sectional image. (b) External quantum efficiency (EQE) curves.[98] Copyright 2022, the authors. (c) Cross sectional SEM images of the solar cells. (d) 3D molecular structures of the bonded Spiro-OMeTAD-TFSI and Spiro-OMeTAD-Cl during storage and O_2 exposure.[100] Copyright 2019, American Chemical Society.

In order to improve both performance and stability of PSCs, a design was developed by combining the advantages of high-efficiency 3D PSCs and long-term-stability 2D PSCs in the last decade. Engineering 2D/3D perovskite interfaces is a common route to realize efficient and stable PSCs. The role of 2D perovskite has been identified and confirmed by several pioneer works.

The multifunctional (bulk and interface) 2D perovskite passivation approach and graded interface design were developed to reduce the photovoltage loss of the PSCs and enhance the device stability. To form the mixed 2D/3D perovskite heterojunction films, long-chain molecules, such as BABr, were dissolved in isopropyl alcohol at different concentrations and then were spin-coated directly onto the top of the as-prepared 3D perovskite films to form a thin 2D passivation film. GIWAXS indicates that the 2D perovskite layer with a relatively larger bandgap is thin enough for carrier transport and can act as an effective electron-blocking layer to suppress the interfacial charge recombination. In addition, the 2D/3D heterojunction by the diffusion of an amphiphilic spacer cation from the interface to the bulk can increase film and device moisture stability. Additionally, the efficiency was boosted from 21.11% to 22.54% after 2D/3D heterojunction engineering owing to significantly suppressed bulk and interfacial nonradiative recombination losses, as shown in Figure 9.16.[62] By

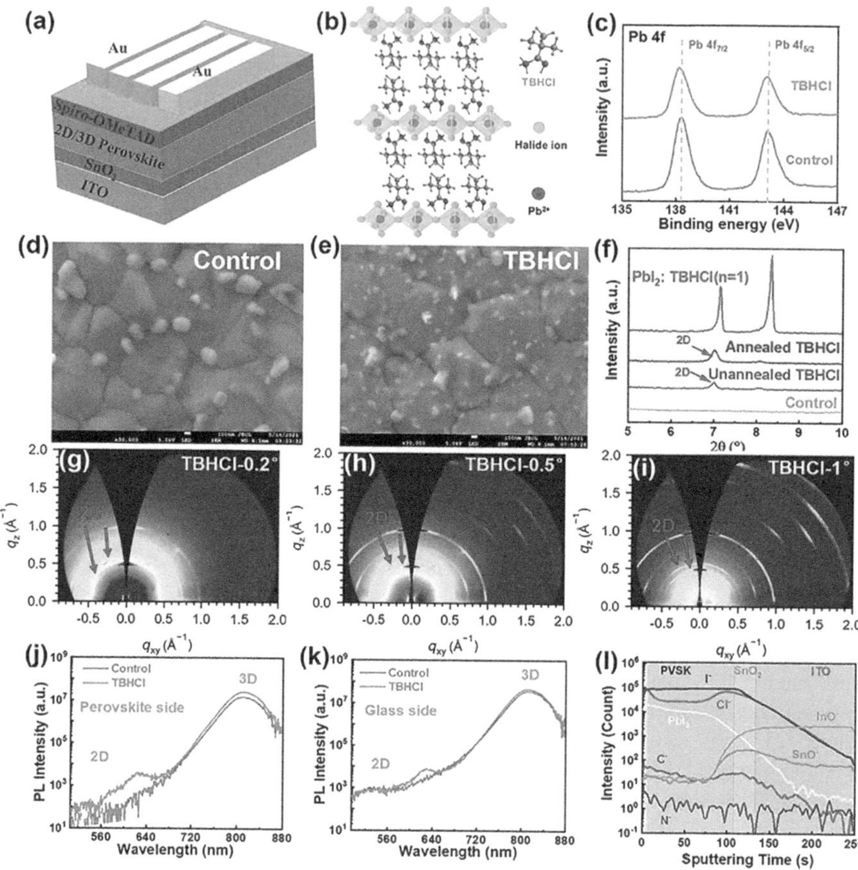

FIGURE 9.16 (a) Schematic illustration of the device architecture with gradient 2D/3D perovskite heterojunction engineering. (b) Schematically illustrated crystal structure diagram of 2D perovskites. (c) Pb 4f XPS spectra of the perovskite films without and with TBHCl. SEM images of the (d) control and (e) TBHCl-treated perovskite films. The scale bar is 100 nm. (f) XRD patterns of the control sample, TBHCl-treated perovskite films without and with annealing as well as pure 2D perovskite film. GIWAXS images of the perovskite film with TBHCl measured at incident angle of (g) 0.2°, (h) 0.5°, and (i) 1°. SSPL spectra of perovskite films without and with TBHCl modification measured from (j) perovskite side and (k) glass side. (l) ToF-SIMS profile of the ITO/SnO₂/PVSK/TBHCl sample. PVSK stands for perovskite film.[62] Copyright 2022, American Chemical Society.

incorporating n-butylammonium bromide (BABr) into the perovskite precursor, the grain-boundary defects also can be passivated efficiently.[102] The hybrid 2D/3D perovskite heterostructure formed through the incorporation of BABr increased crystallinity and mitigated nonradiative recombination, resulting in reduced current-voltage hysteresis, enhanced efficiency, and significantly improved operational stability. Furthermore, by introducing 4-fluoro-phenylethylammonium iodide (FPEAI) as a dual-functional agent to spontaneously form Ruddlesden-Popper 2D perovskite on the grain boundary and surface of 3D crystals, it can passivate defects and protect the perovskite film from moisture erosion as well as suppress ion migration.[103] In addition, the 2D/3D heterostructure induced a matched energy-level alignment, which mitigated the detrimental interfacial charge recombination at the interface of the 3D perovskite and HTL. Two 2D passivation layers also have been confirmed to be a reliable approach to access the synergistic modification. Zhang et al.[104] realized a 2D/3D hybrid perovskite film with a dual (bulk- and surface-) passivation approach and deeply discussed the guanidine halide (GuX) passivation mechanisms. GuCl was used for bulk treatment and GuI was utilized for surface treatment. The former can decrease the nonradiative recombination, and the latter can prevent the halogen migration. The approach can reduce the charge carrier losses in the bulk and prevent decomposition at the surface. Ma et al.[105] introduced oleylammonium (OAm^+) and phenethylammonium (PEA^+) as perovskite bulk and interface passivation agents, respectively. It was found that both OAm^+ and PEA^+ ligands can not only promote crystal growth and vertical orientation but also suppress perovskite bulk and interface defects.

9.3.4 Synergistic Modulation of Grain Boundary, Perovskite/ETL Interface and Perovskite/HTL Interface

To obtain high-performance PSCs, it is very important to decrease the defect density throughout the overall devices. A typical PSC generally consists of a perovskite light absorbing layer, the electron and hole transport layers, and positive and negative electrodes. For the perovskite layer, a large number of defects would be generated at the surface and GBs owing to the soft nature of perovskite materials. Moreover, organic cations and halide anions in perovskite films tend to diffuse and migrate under thermal or illumination conditions producing the unexpected undercoordinated anion and cation vacancy defects. In addition, the interface defects between perovskite and charge transport layers derived from the interfacial stress, ion interdiffusion, and moisture erosion also harms the performance of PSCs. Hence, the synergistic modification in the whole device is an effective way to overcome those issues to further improve the stability and efficiency of PSCs.

A Lewis acid with the capability of accepting a pair of nonbonding electrons has the capability to passivate electron-rich defects, which can also be considered as Lewis bases. The addition of a Lewis acid in perovskite can form a Lewis adduct with the undercoordinated halides and PbI_{3_-} through coordinate bonding or ionic bonding, eliminating the corresponding defects or traps and reducing undesired nonradiative recombination for longer carrier lifetimes (Figure 9.17).

As shown in Figure 9.17 a–c, an organic molecule, tris(5-(((tetrahydro-2H-pyran-2-yl)oxy)pentyl)phosphine oxide (THPPO), is synthesized and introduced as a defect passivation agent in PSCs. The P=O terminal group of THPPO can passivate perovskite surface defects such as undercoordinated Pb^{2+} and improve the photovoltaic performance and device stability of perovskite solar cells.[106] In addition, the non-fullerene n-type organic BTP-4F molecule was introduced in perovskite to modulate crystal growth and passivate defects.[107] The BTP-4F molecules contain a large number of S, O, and N atoms with lone pair electrons, which can passivate

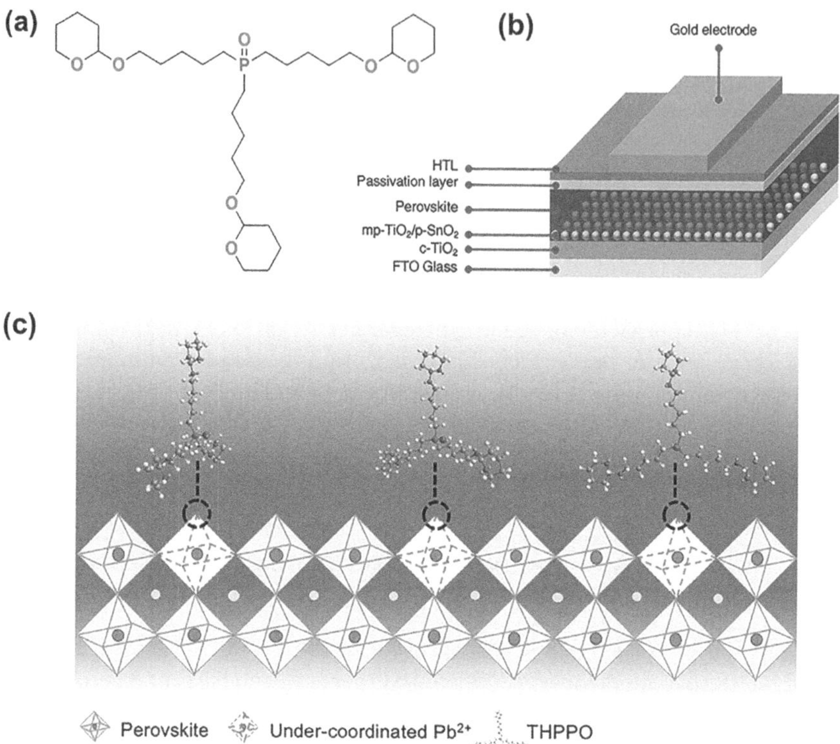

FIGURE 9.17 (a) Molecular structure of THPPO and (b) device structure of n-i-p PSCs employing fluorine-doped tin oxide (FTO), compact TiO$_2$ (c-TiO$_2$), mesoporous TiO$_2$ (mp-TiO$_2$), SnO$_2$, perovskite, THPPO as a passivation layer, HTL, and gold electrode. (c) Cartoon illustration of the proposed passivation mechanism. The concentration of THPPO is 0.030 M.[106] Copyright 2021, American Chemical Society.

under-coordinated Pb^{2+} ions to improve charge transfer and collection, resulting in improved device efficiency. In addition, the hydrophobic properties of BTP-4F molecule can protect perovskite films from moisture, which can greatly enhance the moisture stability of PSCs. Meanwhile, BTP-4F has a series of optimal electrical properties, which are beneficial for electron extraction, thus realizing the simultaneous up-bottom interfacial passivation of perovskite films. Finally, BTP-4F molecule indeed can passivate the Lewis acid traps of perovskite films effectively due to the DAD fused core for BTP-4F molecule, that is, the BTP-4F molecule can diffuse to the interface of SnO$_2$/perovskite to passivate interface defects to decrease charge recombination in the devices. Furthermore, BTP-4F molecules have a strong hydrogen-bonding interaction with perovskite films, which can delay the crystallization process of the perovskite films by immobilizing FA$^+$ ions to further improve the film quality. In addition, Ma et al.[108] developed a multifunctional GB modification strategy where the wide-bandgap hydrophobic PbSO$_4$ modification layer was in situ generated at the GBs of perovskite films through the reaction between methylamine sulfate (MS) and PbI_2. It was demonstrated that the ultrathin $PbSO_4$ GB modification layer contributed to improved crystallization, effective defect passivation, and markedly increased water resistance, which were translated into enhanced efficiency and stability, as shown in Figure 9.14a–c. Zhen et al.[109] proposed an efficient passivation strategy using 4-diaminomethylbenzoic (4-DA) followed by thermal annealing treatment to improve the performance of PSCs. The results show that the contact between crystalline grains was improved, and high quality $MAPbI_3$ films were successfully prepared, which resulted in the elimination of trap states and enhanced performance of the devices. Song et al.[110] employed 4-dimethylaminopyridine (DMAP) as a surface passivator and interlocking linker to selectively heal the defects at the GBs and control topography of the as-prepared perovskite film. The nonradiative recombination at GB was effectively reduced by healing the uncoordinated Pb^{2+} while adhesion strength between the perovskite and the poly(triaryl amine) (PTAA) HTL was significantly enhanced by a mechanical interlock effect, yielding an improved performance and mechanical stability for planar PSCs.

Synergistic modulation of grain boundary, perovskite/ETL interface and perovskite/HTL interface have been widely investigated in many outstanding works. As shown in Figure 9.18, M. Lefler et al.[111] introduced fluorine, owing to its small size and large electronegativity, into perovskite, which can improve the ionic bonding compared to that between the perovskite B-site cation and the other halides, and its electron withdrawing nature can increase dipole moments in the perovskite to improve the intermolecular bonding between the organic cation and BX_6 octahedral framework. Zhao et al.[112] discovered that the proper amount of NiI_2 additive enabled Ni^{2+} doping in the perovskite interstitial lattice, while additional Ni^{2+}

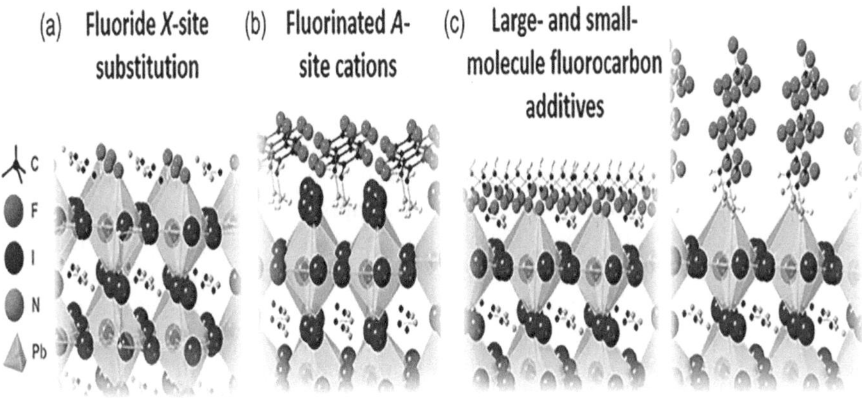

FIGURE 9.18 Approaches to improve perovskite PV performance and stability through passivation with fluorine-containing agents. (a) Terminal X-site substitution with F−. (b) Fluorinated organic cations to occupy terminal A sites (shown: C_6F_5-CH_2-NH^{3+}). (c) Fluorocarbon-based additives such as fluoropolymers (left, shown: polyvinylidene fluoride) and small-molecules (right, shown: 1H, 1H-perfluorohexyl amine) to coordinate perovskite surfaces.[111] Copyright 2020, American Physical Society.

ions were specifically distributed at the upper and lower surfaces of the perovskite film. The special Ni^{2+} distribution resulted in improved perovskite growth, better perovskite film quality, superior charge extraction capability, and effective suppression of interfacial recombination. Wang et al.[113] introduced an organic ionic plastic crystals additive, 1,1 '-Spirobipyrrolidinium hexafluoride ($SBP^+ PF_6^-$), in $CsPbI_2Br$ perovskite, in which PF_6^- molecular anions can fill the vacancy in X-site and passivate the interface defects. Moreover, SBP^+ with high conductivity can be used to regulate A-site defects. This work enhanced the power conversion efficiency and improved the stability by adjusting the crystallization of the $CsPbI_2Br$ perovskite light absorbing layer and passivating the pinhole defects at the grain boundaries. Wu et al.[114] developed an organic ionic compound, 1-ethylpyridinium chloride (EPC), to passivate defects at the surface and grain boundaries of perovskite film, finding that the chlorine anions played an important role in filling iodine vacancies, which significantly reduced the trap density. Ding et al.[115] employed a small quantity of triple cation $NH_3^+C_2H_4NH_2^+C_2H_4NH_3^+$ (denoted as $DETA^{3+}$) to effectively stabilize mutable α-$CsPbI_3$. It was revealed that ionic bonds between NH_3^+ or RNH_2^+ groups in $DETA^{3+}$ and I^- of $[PbI_6]^{4-}$ in $CsPbI_3$ lattices on the surfaces or grain boundaries were expected to cross-link the adjacent $CsPbI_3$ crystal units, avoid the octahedral tilting, and keep the Pb-I-Pb bond at 180°, rendering them less prone to the unwanted phase transition to a δ structure. Xu et al.[116] proclaimed that the combination of APTES and PA ligands can be used as an effective dual passivation system for the synthesis of $CH_3NH_3PbBr_3$ PQDs. Based on a proton exchange reaction between APTES and PA, passivation was achieved mainly through the strong interaction among R-$PO_2(OH)^-$, R-$PO(O)_2^{2-}$, and R-NH_3^+ moieties with MA^+, Pb^{2+}, and Br^- surface defects. This study establishes the combination of PA and APTES ligands as a highly effective dual passivation system for the synergistic passivation of multiple surface defects of PQDs through primarily ionic bonding.

Recently, 2D/3D heterojunction has been considered the most efficient way to realize the synergistic modulation of the whole perovskite devices. Hence, many excellent works have been published to explore its working mechanisms and effects on the performance improvement of perovskite devices (Figure 9.19). Li et al.[117] treated the interface of $MAPbI_3$ perovskite precursor film with a hydrophobic tetra-ammonium zinc phthalocyanine (ZnPc). As expected, a capping layer of 2D $(ZnPc)_{0.5}MA_{n-1}Pb_nI_{3n+1}$ perovskite together with 3D $MAPbI_3$ perovskite was yielded on the top of 3D perovskite layer, thus yielding a 2D/3D graded perovskite interface. Such a modification further achieved the efficient GBs passivation and reduced the defects in GBs, thus resulting in an improved cell performance. More importantly, the significantly increased cell stability was accomplished. Zhou et al.[118] demonstrated a novel strategy to improve the efficiency and stability of 2D/3D perovskites films with modulated diffusion passivation by introducing PEAI and DMF as additive. The DMF solvent additive could promote PEAI to penetrate into the perovskites to form an optimal 2D/3D perovskite heterostructure, which is beneficial for passivating the trap states and enhancing charge transport. As a result, the PSCs based on 2D/3D perovskite films showed significantly improved efficiency and stability. As shown in Figure 9.19a–f, Li et al.[119] introduced 2-(2-pyridyl)ethylamine (2-PyEA) molecules with 2D structure and N atoms with a lone electron pair into perovskite, facilitating

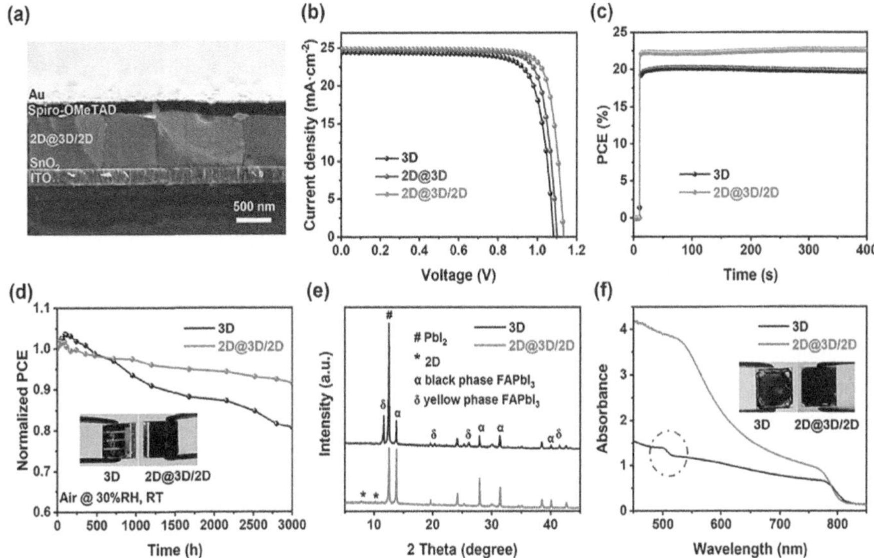

FIGURE 9.19 (a) Cross-view SEM image of the 2D@3D/2D PSC. (b) J–V curves, (c) steady power outputs, and (d) PCE decline of PSCs with different perovskite films (after completely aging, the inset of panel d shows a photograph of PSCs after 3,000 h storing). (e) XRD patterns and (f) UV-vis absorption spectra of 3D and 2D@3D/2D perovskite films stored in a dark air environment at RH = 30 ± 5% for 3,000 h (inset shows a photograph of the perovskite films after 3000 h storing).[119] Copyright 2021, American Chemical Society

the formation of 2D@3D/2D perovskite at the grain boundaries and interfaces. It was verified that defect density was reduced, residual stress was relieved, gradient energy alignment was formed, charge transfer was promoted, and carrier lifetime was extended. Liu et al.[102] fabricated the hybrid 2D/3D perovskite heterostructure through the incorporation of BABr, resulting in increased crystallinity, mitigated nonradiative recombination, reduced current-voltage hysteresis, enhanced efficiency, and significantly improved operational stability. Zhang et al.[61] achieved a highly stable 2D/3D heterostructure by using a PCBM layer as a stress-compensation layer. PCBM counteracted the lattice expansion generated by the migration of MA^+ and I^- ions, while compensating the tensile stress of the 2D perovskite underneath. In addition, a diffusion passivation mechanism of the 2D perovskite induced by thermal annealing was confirmed. The synergistic effect of strain-compensation and diffusion passivation resulted in the suppression of ion migration and of non-radiative recombination. Zhang et al.[120] constructed a gradient-type 2D/3D perovskite structure by regulating the amount of PbI_2 on the surface of the $MAPbI_3$ film, in which the 2D perovskite transition layer formed in the subsurface region of the 3D perovskite film. The gradient-type 2D/3D structure can heal both the surface and subsurface defects, efficiently suppressing the charge recombination and promoting the hole transport/extraction.

9.4 SUMMARY AND OUTLOOK

In summary, we have systematically sorted out the problems existing at perovskite grain boundaries and interfaces in PSCs, such as the recombination centers at surface and interface of perovskite films, the ion migrations, and the charge extraction obstructions. The inner reasons and underlying mechanisms of those problems from the perspectives of perovskite materials and device structures were deeply analyzed. The potential impacts of these problems on device performance, especially on the efficiency and stability of PSCs, also have been discussed comprehensively. Furthermore, the application of synergistic modification strategy for overcoming these problems in PSCs also have been summarized. The synergistic modification can be classified into four species: the modification at grain boundary and interface of perovskite/ETL (such as the trap states reduced and energy level changed), the modification at grain boundary and interface of perovskite/HTL (such as the ion migration inhibited and dipole layer applied), the modification at interfaces of perovskite/ETL and perovskite/HTL (such as the surface defects eliminated and 2D/3D heterojunction structured) and the modification at grain boundary and interfaces of perovskite/ETL and perovskite/HTL (such as the grain crystallization oriented and charge extraction enhanced).

For the synergistic modification at grain boundary and interface of perovskite/ETL, various materials, such as Lewis bases, zwitterion, and fullerene derivative, can be used to reduce the defects at surfaces and grain boundaries of perovskite. The charge extraction at interface between perovskite and ETL also could be improved, obviously. For the synergistic modification at grain boundary and interface of perovskite/HTL, functional materials, such as dipoles, polymers, and alkyl halides, can decrease surface defects of perovskite films and align energy levels between perovskite and HTL, which can improve charge transport properties and inhibit ion migrations in PSCs. For the synergistic modification at interfaces of perovskite/ETL and perovskite/HTL, 2D/3D heterojunction strategy and cation-diffusion strategy have been confirmed to be efficient for decreasing defects at interfaces and enhancing charge extraction. For the synergistic modification at grain boundary and interfaces of perovskite/ETL and perovskite/HTL, chloride salts, long-chain alkyl salts, and 2D/3D heterostructure have been developed, which can modify the whole films and interfaces to promote the performance of PSCs.

By employing the synergistic modification strategy, the efficiency and stability of PSCs have achieved rapid development. The performance of PSCs fabricated in laboratory is close to meeting the standard of industrialization. However, the development of large-area PSCs and flexible and tandem perovskite devices are relatively lagging. Moreover, the reliable approach to eliminate lead leakage from perovskites is still under exploration. The solution of the previous issues still requires synergistic modification strategy to overcome the interface problems, crystallization orientation, ion diffusion/migration, and charge transport properties. Meanwhile, the construction of integrated devices such as photoelectric energy storage, photoelectrochemical energy storage, and flexible photoelectric energy storage will also need the synergistic modification strategy to improve the performance of these devices in the future and meet the commercialization standards.

REFERENCES

1. S. Bello, A. Urwick, F. Bastianini, A. J. Nedoma, A. Dunbar, An introduction to perovskites for solar cells and their characterization, *Energy Reports* **2022**, *8*, 89.
2. F. Zheng, L.-W. Wang, Large polaron formation and its effect on electron transport in hybrid perovskites, *Energy & Environmental Science* **2019**, *12*, 1219.
3. S. Zhang, *et al.*, Defect/interface recombination limited quasi-Fermi level splitting and open-circuit voltage in mono-and triple-cation perovskite solar cells, *ACS Applied Materials & Interfaces* **2020**, *12*, 37647.
4. J.-P. Correa-Baena, *et al.*, Identifying and suppressing interfacial recombination to achieve high open-circuit voltage in perovskite solar cells, *Energy & Environmental Science* **2017**, *10*, 1207.
5. C. Bi, Q. Wang, Y. Shao, Y. Yuan, Z. Xiao, J. Huang, Non-wetting surface-driven high-aspect-ratio crystalline grain growth for efficient hybrid perovskite solar cells, *Nature Communications* **2015**, *6*, 7747.
6. T. S. Sherkar, C. Momblona, L. Gil-Escrig, J. Ávila, M. Sessolo, H. J. Bolink, L. J. A. Koster, Recombination in perovskite solar cells: Significance of grain boundaries, interface traps, and defect ions, *ACS Energy Letters* **2017**, *2*, 1214.
7. C. Eames, J. M. Frost, P. R. F. Barnes, B. C. O'Regan, A. Walsh, M. S. Islam, Ionic transport in hybrid lead iodide perovskite solar cells, *Nature Communications* **2015**, *6*, 7497.
8. S. Svanström, *et al.*, The complex degradation mechanism of copper electrodes on lead halide perovskites, *ACS Materials Au* **2022**, *2*, 301.
9. N. Chen, *et al.*, Universal band alignment rule for perovskite/organic heterojunction interfaces, *ACS Energy Letters* **2023**, 1313.
10. D. Xu, X. Hua, L. Xu, W.-J. Wu, Y.-T. Long, H. Tian, Understanding how ambiance affects the performance of hole-conductor-free perovskite solar cells from a chemical perspective, *ACS Applied Energy Materials* **2019**, *2*, 2387.
11. C. Huo, X. Liu, X. Song, Z. Wang, H. Zeng, Field-effect transistors based on van-der-Waals-grown and dry-transferred all-inorganic perovskite ultrathin platelets, *The Journal of Physical Chemistry Letters* **2017**, *8*, 4785.
12. J. Kim, R. Godin, S. D. Dimitrov, T. Du, D. Bryant, M. A. McLachlan, J. R. Durrant, Excitation density dependent photoluminescence quenching and charge transfer efficiencies in hybrid perovskite/organic semiconductor bilayers, *Advanced Energy Materials* **2018**, *8*, 35.
13. D. Yang, X. Zhou, R. Yang, Z. Yang, W. Yu, X. Wang, C. Li, S. Liu, R. P. H. Chang, Surface optimization to eliminate hysteresis for record efficiency planar perovskite solar cells, *Energy & Environmental Science* **2016**, *9*, 3071.
14. S. Wieghold, *et al.*, Precursor concentration affects grain size, crystal orientation, and local performance in mixed-ion lead perovskite solar cells, *ACS Applied Energy Materials* **2018**, *1*, 6801.
15. D.-J. Xue, *et al.*, Regulating strain in perovskite thin films through charge-transport layers, *Nature Communications* **2020**, *11*, 1514.
16. A. M. A. Leguy, *et al.*, Reversible Hydration of CH3NH3PbI3 in Films, Single Crystals, and Solar Cells, *Chemistry of Materials* **2015**, *27*, 3397.
17. S. N. Habisreutinger, T. Leijtens, G. E. Eperon, S. D. Stranks, R. J. Nicholas, H. J. Snaith, Carbon nanotube/polymer composites as a highly stable hole collection layer in perovskite solar cells, *Nano Letters* **2014**, *14*, 5561.
18. Q. Sun, *et al.*, Role of microstructure in oxygen induced photodegradation of methylammonium lead triiodide perovskite films, *Advanced Energy Materials* **2017**, *7*, 1700977.
19. N. Aristidou, C. Eames, I. Sanchez-Molina, X. Bu, J. Kosco, M. S. Islam, S. A. Haque, Fast oxygen diffusion and iodide defects mediate oxygen-induced degradation of perovskite solar cells, *Nature Communications* **2017**, *8*, 15218.

20. W.-C. Lin, W.-C. Lo, J.-X. Li, Y.-K. Wang, J.-F. Tang, Z.-Y. Fong, In situ XPS investigation of the X-ray-triggered decomposition of perovskites in ultrahigh vacuum condition, *NPJ Materials Degradation* **2021**, *5*, 13.

21. M. V. Khenkin, *et al.*, Consensus statement for stability assessment and reporting for perovskite photovoltaics based on ISOS procedures, *Nature Energy* **2020**, *5*, 35.

22. D. Wang, Z. Zhang, J. Liu, Y. Zhang, K. Chen, B. She, B. Liu, Y. Huang, J. Xiong, J. Zhang, Synergistic effect of defect passivation and crystallization control enabled by bifunctional additives for carbon-based mesoscopic perovskite solar cells, *ACS Applied Materials & Interfaces* **2021**, *13*, 45435.

23. S. Tang, Y. Peng, Z. Zhu, J. Zong, L. Zhao, L. Yu, R. Chen, M. Li, Simultaneous bulk and surface defect passivation for efficient inverted perovskite solar cells, *The Journal of Physical Chemistry Letters* **2022**, *13*, 5116.

24. J. Cheng, *et al.*, Highly efficient planar perovskite solar cells achieved by simultaneous defect engineering and formation kinetic control, *Journal of Materials Chemistry A* **2018**, *6*, 23865.

25. R. Singh, V. K. Shukla, ITIC-based bulk heterojunction perovskite film boosting the power conversion efficiency and stability of the perovskite solar cell, *Solar Energy* **2019**, *178*, 90.

26. H. Aqoma, I. F. Imran, F. T. A. Wibowo, N. V. Krishna, W. Lee, A. K. Sarker, D. Y. Ryu, S. Y. Jang, High-efficiency solution-processed two-terminal hybrid tandem solar cells using spectrally matched inorganic and organic photoactive materials, *Advanced Energy Materials* **2020**, *10*, 2001188.

27. C. Zhang, *et al.*, Modulating chemical interaction to realize bottom-up defect passivation for efficient and stable perovskite solar cells, *Solar RRL* **2022**, *6*, 2200512.

28. J. Ran, *et al.*, Triphenylamine–polystyrene blends for perovskite solar cells with simultaneous energy loss suppression and stability improvement, *Solar RRL* **2020**, *4*, 2000490.

29. H. Bi, *et al.*, Multifunctional organic ammonium salt-modified SnO_2 nanoparticles toward efficient and stable planar perovskite solar cells, *Journal of Materials Chemistry A* **2021**, *9*, 3940.

30. X. Shi, *et al.*, Enhanced interfacial binding and electron extraction using boron-doped TiO_2 for highly efficient hysteresis-free perovskite solar cells, *Advanced Science* **2019**, *6*, 1901213.

31. Y. Chen, K. Wang, H. Qi, Y. Zhang, T. Wang, Y. Tong, H. Wang, Mitigating Voc loss in tin perovskite solar cells via simultaneous suppression of bulk and interface nonradiative recombination, *ACS Applied Materials & Interfaces* **2022**, *14*, 41086.

32. H. Zhang, *et al.*, Dual effect of superhalogen ionic liquids ensures efficient carrier transport for highly efficient and stable perovskite solar cells, *ACS Applied Materials & Interfaces* **2022**, *14*, 28826.

33. Z. Wang, Q. Lin, F. P. Chmiel, N. Sakai, L. M. Herz, H. J. Snaith, Efficient ambient-air-stable solar cells with 2D-3D heterostructured butylammonium-caesium-formamidinium lead halide perovskites, *Nature Energy* **2017**, *2*, 1.

34. M. A. Mahmud, *et al.*, Combined bulk and surface passivation in dimensionally engineered 2D-3D perovskite films via chlorine diffusion, *Advanced Functional Materials* **2021**, *31*, 2104251.

35. E. A. Alharbi, *et al.*, Atomic-level passivation mechanism of ammonium salts enabling highly efficient perovskite solar cells, *Nature Communications* **2019**, *10*, 3008.

36. Q. Ye, *et al.*, Stabilizing γ-CsPbI$_3$ perovskite via phenylethylammonium for efficient solar cells with open-circuit voltage over 1.3 V, *Small* **2020**, *16*, 2005246.

37. B. Chen, P. N. Rudd, S. Yang, Y. Yuan, J. Huang, Imperfections and their passivation in halide perovskite solar cells, *Chemical Society Reviews* **2019**, *48*, 3842.

38. K. Xiao, *et al.*, All-perovskite tandem solar cells with 24.2% certified efficiency and area over 1 cm2 using surface-anchoring zwitterionic antioxidant, *Nature Energy* **2020**, *5*, 870.

39. D. Zheng, *et al.*, Perovskite solar cells: Simultaneous bottom-up interfacial and bulk defect passivation in highly efficient planar perovskite solar cells using nonconjugated small-molecule electrolytes, *Advanced Materials* **2019**, *31*, 1970283.

40. X. Zheng, *et al.*, Dual functions of crystallization control and defect passivation enabled by sulfonic zwitterions for stable and efficient perovskite solar cells, *Advanced Materials* **2018**, *30*, 1803428.

41. Y. Chen, D. Ma, Z. Wang, J. He, X. Gong, W. Wu, Tautomeric dual-site passivation for carbon-based printable mesoscopic perovskite solar cells, *Advanced Materials Interfaces* **2022**, *9*, 2200326.

42. N. F. Jamaludin, *et al.*, Grain size modulation and interfacial engineering of $CH_3NH_3PbBr_3$ emitter films through incorporation of tetraethylammonium bromide, *ChemPhysChem* **2018**, *19*, 1075.

43. J. Chen, X. Zhao, S.-G. Kim, N.-G. Park, Multifunctional chemical linker imidazoleacetic acid hydrochloride for 21% efficient and stable planar perovskite solar cells, *Advanced Materials* **2019**, *31*, 1902902.

44. X. Zuo, *et al.*, Passivating buried interface via self-assembled novel sulfonium salt toward stable and efficient perovskite solar cells, *Chemical Engineering Journal* **2022**, *431*, 133209.

45. Q. Zhuang, H. Wang, C. Zhang, C. Gong, H. Li, J. Chen, Z. Zang, Ion diffusion-induced double layer doping toward stable and efficient perovskite solar cells, *Nano Research* **2022**, *15*, 5114.

46. Z. Li, C. Liu, G. Ren, W. Han, L. Shen, W. Guo, Cations functionalized carbon nano-dots enabling interfacial passivation and crystallization control for inverted perovskite solar cells, *Solar RRL* **2020**, *4*, 1900369.

47. P. Zhu, S. Gu, X. Luo, Y. Gao, S. Li, J. Zhu, H. Tan, Simultaneous contact and grain-boundary passivation in planar perovskite solar cells using SnO_2-KCl composite electron transport layer, *Advanced Energy Materials* **2020**, *10*, 1903083.

48. B. Xiao, X. Li, Y. Qian, Z. Yi, A. Y. Haruna, Q. Jiang, Y. Luo, J. Yang, Simultaneous interface passivation and defect compensation for high-efficiency planar perovskite solar cells, *Applied Surface Science* **2022**, *604*, 154431.

49. C. Zhang, *et al.*, Simultaneous passivation of bulk and interface defects through synergistic effect of anion and cation toward efficient and stable planar perovskite solar cells, *Journal of Energy Chemistry* **2021**, *63*, 452.

50. T. H. Lee, W.-W. Park, S. Y. Park, S. Cho, O.-H. Kwon, J. Y. Kim, Planar organic bilayer heterojunctions fabricated on water with ultrafast donor-to-acceptor charge transfer, *Solar RRL* **2021**, *5*, 2100326.

51. H. Bi, B. Liu, D. He, L. Bai, W. Wang, Z. Zang, J. Chen, Interfacial defect passivation and stress release by multifunctional KPF6 modification for planar perovskite solar cells with enhanced efficiency and stability, *Chemical Engineering Journal* **2021**, *418*, 129375.

52. D. Gao, *et al.*, Passivating buried interface with multifunctional novel ionic liquid containing simultaneously fluorinated anion and cation yielding stable perovskite solar cells over 23% efficiency, *Journal of Energy Chemistry* **2022**, *69*, 659.

53. C. Zhang, H. Li, C. Gong, Q. Zhuang, J. Chen, Z. Zang, Crystallization manipulation and holistic defect passivation toward stable and efficient inverted perovskite solar cells, *Energy & Environmental Science* **2023**, *16*, 3825–3836.

54. A. A. Sutanto, *et al.*, In situ analysis reveals the role of 2D perovskite in preventing thermal-induced degradation in 2D/3D perovskite interfaces, *Nano Letters* **2020**, *20*, 3992.

55. D. S. Lee, *et al.*, Unveiling the importance of precursor preparation for highly efficient and stable phenethylammonium-based perovskite solar cells, *Solar RRL* **2020**, *4*, 1900463.

56. S. H. Kareem, M. H. Elewi, A. M. Naji, D. S. Ahmed, M. K. Mohammed, Efficient and stable pure α-phase FAPbI$_3$ perovskite solar cells with a dual engineering strategy: Additive and dimensional engineering approaches, *Chemical Engineering Journal* **2022**, *443*, 136469.

57. S. Gharibzadeh, *et al.*, Record open-circuit voltage wide-bandgap perovskite solar cells utilizing 2D/3D perovskite heterostructure, *Advanced Energy Materials* **2019**, *9*, 1803699.

58. X. Ye, H. Cai, Q. Sun, T. Xu, J. Ni, J. Li, J. Zhang, Alkyl ammonium salt with different chain length for high-efficiency and good-stability 2D/3D hybrid perovskite solar cells, *Organic Electronics* **2022**, *106*, 106542.

59. X. Jiang, *et al.* Direct surface passivation of perovskite film by 4-fluorophenethylammonium iodide toward stable and efficient perovskite solar cells, *ACS Applied Materials & Interfaces* **2021**, *13*, 2558.

60. J. Suo, *et al.*, Surface reconstruction engineering with synergistic effect of mixed-salt passivation treatment toward efficient and stable perovskite solar cells, *Advanced Functional Materials* **2021**, *31*, 2102902.

61. C. Zhang, *et al.*, Fabrication strategy for efficient 2D/3D perovskite solar cells enabled by diffusion passivation and strain compensation, *Advanced Energy Materials* **2020**, *10*, 2002004.

62. B. Liu, *et al.*, Simultaneous passivation of bulk and interface defects with gradient 2D/3D heterojunction engineering for efficient and stable perovskite solar cells, *ACS Applied Materials & Interfaces* **2022**, *14*, 21079.

63. B. Liu, *et al.*, Interfacial defect passivation and stress release via multi-active-site ligand anchoring enables efficient and stable methylammonium-free perovskite solar cells, *ACS Energy Letters* **2021**, *6*, 2526.

64. Q. Zhuang, H. Li, C. Zhang, C. Gong, H. Yang, J. Chen, Z. Zang, Synergistic modification of 2D perovskite with alternating cations in the interlayer space and multisite ligand toward high-performance inverted solar cells, *Advanced Materials* **2023**, *35*, 2303275.

65. M. Pegu, *et al.*, Dimers of diethynyl-conjugated zinc-phthalocyanine as hole selective layers for perovskite solar cell fabrication, *Journal of Materials Chemistry C* **2022**, *10*, 11975.

66. Y. Du, *et al.*, Polymeric surface modification of NiOx-based inverted planar perovskite solar cells with enhanced performance, *ACS Sustainable Chemistry & Engineering* **2018**, *6*, 16806.

67. J. Jin, *et al.*, Highly efficient and stable carbon-based perovskite solar cells with the polymer hole transport layer, *Solar Energy* **2021**, *220*, 491.

68. B. J. Bruijnaers, E. Schiepers, C. H. Weijtens, S. C. Meskers, M. M. Wienk, R. A. Janssen, The effect of oxygen on the efficiency of planar p-i-n metal halide perovskite solar cells with a PEDOT: PSS hole transport layer, *Journal of Materials Chemistry A* **2018**, *6*, 6882.

69. R. Y. Kezerashvili, V. Y. Kezerashvili, Charge-dipole and dipole-dipole interactions in two-dimensional materials, *Physical Review B* **2022**, *105*, 205416.

70. W. Zhao, *et al.*, A special additive enables all cations and anions passivation for stable perovskite solar cells with efficiency over 23%. *Nano-Micro Letters* **2021**, *13*, 169.

71. J. Chen, S.-G. Kim, X. Ren, H. S. Jung, N.-G. Park, Effect of bidentate and tridentate additives on the photovoltaic performance and stability of perovskite solar cells, *Journal of Materials Chemistry A* **2019**, *7*, 4977.

72. H. Xi, *et al.*, Interfacial dipole poly (2-ethyl-2-oxazoline) modification triggers simultaneous band alignment and passivation for air-stable perovskite solar cells, *Polymers* **2022**, *14*, 2748.

73. F. Li, *et al.*, Regulating surface termination for efficient inverted perovskite solar cells with greater than 23% efficiency, *Journal of the American Chemical Society* **2020**, *142*, 20134.

74. G. Zhang, *et al.*, Improved interfacial property by small molecule ethanediamine for high performance inverted planar perovskite solar cells, *Journal of Energy Chemistry* **2021**, *54*, 467.

75. W. Zhao, *et al.*, A special additive enables all cations and anions passivation for stable perovskite solar cells with efficiency over 23%. *Nano-Micro Letters* **2021**, *13*.

76. J. Han, H. Kwon, E. Kim, D.-W. Kim, H. J. Son, D. H. Kim, Interfacial engineering of a ZnO electron transporting layer using self-assembled monolayers for high performance and stable perovskite solar cells, *Journal of Materials Chemistry A* **2020**, *8*, 2105.

77. G. Yang, *et al.*, Interface engineering in planar perovskite solar cells: Energy level alignment, perovskite morphology control and high performance achievement, *Journal of Materials Chemistry A* **2017**, *5*, 1658.

78. Z. Fan, *et al.*, Simultaneous achievement of defect passivation and carrier transport promotion by using emerald salt for methylammonium-free perovskite solar cells, *Chemical Science* **2022**, *13*, 10512.

79. C. Li, *et al.*, Mitigation of vacancy with ammonium salt-trapped ZIF-8 capsules for stable perovskite solar cells through simultaneous compensation and loss inhibition, *Nanoscale Advances* **2021**, *3*, 3554.

80. L. Zhu, *et al.*, Trap state passivation by rational ligand molecule engineering toward efficient and stable perovskite solar cells exceeding 23% efficiency, *Advanced Energy Materials* **2021**, *11*, 2100529.

81. M. Shibukawa, H. Okutsu, S. Saito, Characterization of the interfacial liquid layer formed on hydrophobic packing material surfaces by liquid chromatographic analysis of the distribution of ions and molecules, *ACS Omega* **2022**, *7*, 15158.

82. X. Jiang, *et al.*, Simultaneous hole transport and defect passivation enabled by a dopant-free single polymer for efficient and stable perovskite solar cells, *Journal of Materials Chemistry A* **2020**, *8*, 21036.

83. M. Lyu, D.-K. Lee, N.-G. Park, Effect of alkaline earth metal chloride additives BCl_2 (B= Mg, Ca, Sr and Ba) on the photovoltaic performance of $FAPbI_3$ based perovskite solar cells, *Nanoscale Horizons* **2020**, *5*, 1332.

84. G. Qu, *et al.*, Dopant-free phthalocyanine hole conductor with thermal-induced holistic passivation for stable perovskite solar cells with 23% efficiency, *Advanced Functional Materials* **2022**, *32*, 2206585.

85. J. Lee, N. Tsvetkov, S. R. Shin, J. K. Kang, Fast charge transfer and high stability via hybridization of hygroscopic Cu-BTC metal–organic framework nanocrystals with a light-absorbing layer for perovskite solar cells, *ACS Applied Materials & Interfaces* **2022**, *14*, 35495.

86. C. Li, *et al.*, Multifunctional anionic metal-organic frameworks enhancing stability of perovskite solar cells, *Chemical Engineering Journal* **2022**, *433*, 133587.

87. J. Zhang, S. Guo, M. Zhu, C. Li, J. Chen, L. Liu, S. Xiang, Z. Zhang, Simultaneous defect passivation and hole mobility enhancement of perovskite solar cells by incorporating anionic metal-organic framework into hole transport materials, *Chemical Engineering Journal* **2021**, *408*, 127328.

88. J. H. Lee, B. Nketia-Yawson, J.-J. Lee, J. W. Jo, Ionic liquid-mediated reconstruction of perovskite surface for highly efficient photovoltaics, *Chemical Engineering Journal* **2022**, *446*, 137351.

89. T. Yang, *et al.*, Amidino-based Dion-Jacobson 2D perovskite for efficient and stable 2D/3D heterostructure perovskite solar cells, *Joule* **2023**, *7*, 574.

90. T. Ye, *et al.*, Efficient and ambient-air-stable solar cell with highly oriented 2D@ 3D perovskites, *Advanced Functional Materials* **2018**, *28*, 1801654.

91. X. Wang, Y. Zhao, B. Li, X. Han, Z. Jin, Y. Wang, Q. Zhang, Y. Rui, Interfacial modification via a 1, 4-butanediamine-based 2D capping layer for perovskite solar cells with enhanced stability and efficiency, *ACS Applied Materials & Interfaces* **2022**, *14*, 22879.

92. J. Wang, *et al.*, Growth of 2D passivation layer in FAPbI$_3$ perovskite solar cells for high open-circuit voltage, *Nano Today* **2022**, *42*, 101357.

93. J. Chen, J. Y. Seo, N. G. Park, Simultaneous improvement of photovoltaic performance and stability by in situ formation of 2D perovskite at (FAPbI$_3$)$_{0.88}$(CsPbBr$_3$)$_{0.12}$/CuSCN interface, *Advanced Energy Materials* **2018**, *8*, 1702714.

94. H. Li, *et al.*, 2D/3D heterojunction engineering at the buried interface towards high-performance inverted methylammonium-free perovskite solar cells, *Nature Energy* **2023**, *8*, 946–955.

95. L. Tan, *et al.*, 4-Hydroxy-2, 2, 6, 6-tetramethylpiperidine as a bifunctional interface modifier for high-efficiency and stable perovskite solar cells, *ACS Applied Energy Materials* **2022**, *5*, 6754.

96. Y. Zhang, *et al.*, Depth-dependent defect manipulation in perovskites for high-performance solar cells, *Energy & Environmental Science* **2021**, *14*, 6526.

97. G. H. Ahmed, *et al.*, Giant photoluminescence enhancement in CsPbCl$_3$ perovskite nanocrystals by simultaneous dual-surface passivation, *ACS Energy Letters* **2018**, *3*, 2301.

98. M. A. Mahmud, *et al.*, Cation-diffusion-based simultaneous bulk and surface passivations for high bandgap inverted perovskite solar cell producing record fill factor and efficiency, *Advanced Energy Materials* **2022**, *12*, 2201672.

99. A. N. Singh, S. Kajal, J. Kim, A. Jana, J. Y. Kim, K. S. Kim, Interface engineering driven stabilization of halide perovskites against moisture, heat, and light for optoelectronic applications, *Advanced Energy Materials* **2020**, *10*, 2000768.

100. H. Zhang, Y. Lv, J. Wang, H. Ma, Z. Sun, W. Huang, Influence of Cl incorporation in perovskite precursor on the crystal growth and storage stability of perovskite solar cells, *ACS Applied Materials & Interfaces* **2019**, *11*, 6022.

101. L. Yuan, *et al.*, Reexamining the post-treatment effects on perovskite solar cells: Passivation and chloride redistribution, *Small Methods* **2023**, *7*, 2201467.

102. X. Liu, C. Qin, X. Liu, H. Ding, X. Du, Y. Cui, Comprehensive insights into defect passivation and charge dynamics for FA$_{0.8}$MA$_{0.15}$Cs$_{0.05}$PbI$_{2.8}$Br$_{0.2}$ perovskite solar cells, *Applied Physics Letters* **2020**, *117*, 013503.

103. Y. Yu, R. Liu, C. Liu, X. L. Shi, H. Yu, Z. G. Chen, Synergetic regulation of oriented crystallization and interfacial passivation enables 19.1% efficient wide-bandgap perovskite solar cells, *Advanced Energy Materials* **2022**, *12*, 2201509.

104. X. Zhang, *et al.*, Dual optimization of bulk and surface via guanidine halide for efficient and stable 2D/3D hybrid perovskite solar cells, *Advanced Energy Materials* **2022**, *12*, 2201105.

105. H. Ju, *et al.*, Roles of long-chain alkylamine ligands in triple-halide perovskites for efficient NiO$_x$-based inverted perovskite solar cells, *Solar RRL* **2022**, *6*, 2101082.

106. A. A. Sutanto, *et al.*, Phosphine oxide derivative as a passivating agent to enhance the performance of perovskite solar cells, *ACS Applied Energy Materials* **2021**, *4*, 1259.

107. Y. Gao, *et al.*, Interface and grain boundary passivation for efficient and stable perovskite solar cells: The effect of terminal groups in hydrophobic fused benzothiadiazole-based organic semiconductors, *Nanoscale Horizons* **2020**, *5*, 1574.

108. X. Ma, *et al.*, Grain boundary defect passivation by in situ formed wide-bandgap lead sulfate for efficient and stable perovskite solar cells, *Chemical Engineering Journal* **2021**, *426*, 130685.

109. Z. He, J. Xiong, Q. Dai, B. Yang, J. Zhang, S. Xiao, High-performance inverted perovskite solar cells using 4-diaminomethylbenzoic as a passivant, *Nanoscale* **2020**, *12*, 6767.

110. S. Song, *et al.*, Selective defect passivation and topographical control of 4-dimethylaminopyridine at grain boundary for efficient and stable planar perovskite solar cells, *Advanced Energy Materials* **2021**, *11*, 2003382.

111. B. M. Lefler, S. J. May, A. T. Fafarman, Role of fluoride and fluorocarbons in enhanced stability and performance of halide perovskites for photovoltaics, *Physical Review Materials* **2020**, *4*, 120301.

112. H. Zhao, *et al.*, Simultaneous dual-interface and bulk defect passivation for high-efficiency and stable $CsPbI_2Br$ perovskite solar cells, *Journal of Power Sources* **2021**, *492*, 229580.

113. C. Wang, B. Zhang, X. Ma, L. Yang, X. Shang, C. Chen, Organic ionic plastic crystals: A promising additive for achieving efficient and stable CsPbI2Br perovskite solar cells, *Journal of Physics and Chemistry of Solids* **2022**, *168*, 110798.

114. X. Wu, *et al.*, Efficient perovskite solar cells via surface passivation by a multifunctional small organic ionic compound, *Journal of Materials Chemistry A* **2020**, *8*, 8313.

115. X. Ding, H. Chen, Y. Wu, S. Ma, S. Dai, S. Yang, J. Zhu, Triple cation additive $NH_3{}^+C_2H_4$ $NH_2{}^+C_2H_4NH_3{}^+$-induced phase-stable inorganic α-$CsPbI_3$ perovskite films for use in solar cells, *Journal of Materials Chemistry A* **2018**, *6*, 18258.

116. K. Xu, *et al.*, Synergistic surface passivation of $CH_3NH_3PbBr_3$ perovskite quantum dots with phosphonic acid and (3-Aminopropyl)triethoxysilane, *Chemistry–A European Journal* **2019**, *25*, 5014.

117. C. Li, X. Lv, J. Cao, Y. Tang, Tetra-ammonium zinc phthalocyanine to construct a graded 2D–3D perovskite interface for efficient and stable solar cells, *Chinese Journal of Chemistry* **2019**, *37*, 30.

118. L. Zhou, *et al.*, Highly efficient and stable planar perovskite solar cells with modulated diffusion passivation toward high power conversion efficiency and ultrahigh fill factor, *Solar RRL* **2019**, *3*, 1900293.

119. G. Li, *et al.*, Efficient and stable 2D@ 3D/2D perovskite solar cells based on dual optimization of grain boundary and interface, *ACS Energy Letters* **2021**, *6*, 3614.

120. Y. Liu, R. Lu, J. Zhang, X. Guo, C. Li, Construction of a gradient-type 2D/3D perovskite structure for subsurface passivation and energy-level alignment of an $MAPbI_3$ film, *Journal of Materials Chemistry A* **2021**, *9*, 26086.

10 Strain Engineering in Perovskite Solar Cells

Yuanyuan Zhao, Lei Gao, Yinping Teng,
Yusheng Cao, Liqiang Bian, and Qunwei Tang

10.1 INTRODUCTION

Perovskites have emerged as one of the most promising photovoltaic materials due to their unique properties, including tunable bandgap[1], excellent charge carrier mobility[2], efficient light absorption[3] and low-cost solution processing[4]. The power conversion efficiency (PCE) of perovskite solar cells (PSCs) has been increased from 3.8%[5] to 26.1%[6], which are already comparable to traditional monocrystalline silicon cells. However, the stability is still a great challenge for the commercialization of PSCs. Due to the intrinsic instability of halide perovskites, it is susceptible to phase transition[7] and degradation[8], especially under external factors such as light, heat, moisture, and oxygen[9]. This is attributed to the soft ionic nature of perovskites, causing a residual strain formation within the perovskite layer and therefore internal defects and phase segregation. Consequently, the stability and photoelectric performances of PSCs are directly affected.[7-9] Therefore, the in-depth understanding on the origins and impacts of strain is a prerequisite to develop effective strain-engineering strategies.

Residual strain in perovskite refers to lattice distortion or deformation caused by internal or external forces, which can be divided into tensile or compressive strain.[10-12] Tensile strain weakens chemical bonds, increases defects, reduces ion migration energy barriers, enhances non-radiative recombination, and accelerates chemical degradation and fracture under light, heat, and moisture. Conversely, slight compressive strain broadens the absorption spectrum, optimizes energy level alignment, extends carrier lifetime, and improves photovoltaic performance. To address the inherent instability caused by residual tensile strain is a major challenge in achieving efficient and stable PSCs. For example, strategies such as ion doping to regulate the perovskite octahedron lattice tilt have been explored to mitigate residual strain generation and enhance stability;[10] external stimuli are also used to control strain within the perovskite film.[13-17] Therefore, a comprehensive understanding of the origins, classical characterization techniques, implications, and strain adjustment strategies is vital to maximize the performance of perovskite.

To address the strain issues in halide perovskite films, scientists have performed systematic studies on strain engineering.[18-22] This review systematically summarizes the origins of strain, the state-of-the-art characterization techniques, the impacts of strain, and different strain-regulation strategies for improving intrinsic photovoltaic

DOI: 10.1201/9781003400486-10

performances. Firstly, strain in perovskite film is induced by internal strain resulting from the tilt of the octahedron in the perovskite lattice as well as external factors.[10–12] The state-of-the-art characterization techniques for strain include X-ray diffraction (XRD), grazing incidence X-ray diffraction (GIXRD), grazing-incidence wide-angle X-ray scattering (GIWAXS), Raman spectroscopy, transmission electron microscopy (TEM), scanning electron diffraction (SED), electron back-scatter diffraction (EBSD), and piezoresponse force microscopy (PFM). These techniques provide in-depth insights into the impacts of strain, including its effects on the defect properties, band structure, and stability (ion migration, phase, and device stability). Finally, this chapter summarizes the recent advances in strain-regulation strategies, such as compositional engineering, heat treatment strategies, interfacial management, and external strain regulation. This perspective aims to inspire further innovations in strain regulation to address the impact of strain on efficiency and stability in PSCs, thereby enhancing the commercial performance of the device.

10.2   STRAIN IN PEROVSKITE FILMS

The strain (ε) in a material is defined as the difference between the lattice constants of a non-strained material (a_0) and a strained material (a), divided by the lattice constant of the non-strained material $\left(\varepsilon = \dfrac{(a_0 - a)}{a_0} \right)$. It represents the relative deformation of a crystal structure caused by external stress and other factors. If the lattice length increases under the influence of stress, it is referred to as tensile strain. Conversely, if the lattice length decreases, it is known as compressive strain. In the case of perovskite, the strain can be categorized into internal strain arising from the crystal lattice tilt and strain induced by external conditions.

10.2.1   INTERNAL STRAIN IN PEROVSKITE FILMS

Internal strain originates within halide perovskite and is the result of non-periodicity within the lattice, without any external stress factors. The primary sources of this strain are $[BX_6]^{4-}$ octahedral tilt and heterogeneous crystallization.

The expression for metal halide perovskites is ABX_3,[23] where A represents a monovalent organic cation or an alkali metal cation, such as methylammonium (MA⁺, $CH_3NH_3^+$), formamidinium (FA⁺, $CH(NH_2)_2^+$), and (Rb⁺, Cs⁺). The B site is typically occupied by bivalent cations, including lead (Pb^{2+}), tin (Sn^{2+}), and germanium (Ge^{2+}) ions.[24] The X site is typically composed of single or multiple halogen anions, such as chloride (Cl⁻), bromide (Br⁻), and iodide (I⁻)[25]] (Figure 10.1a). The internal strain in perovskite lattice is caused by the mismatch of atomic sizes. In order to form a stable symmetric cubic crystal structure, the radius ratio of A, B, and X ions should fall within the range of $0.8 < t = (R_A + R_X)/[\sqrt{2}(R_B + R_X)] < 1$, (Figure 10.1b)[26] where t is the Goldschmidt tolerance factor, and R_A, R_B, and R_X are the respective radii of A, B, and X. A perfect cubic perovskite crystal structure is obtained when $t = 1$, while a square structure occurs when $t < 0.8$, and a hexagonal structure forms when $t > 1$.[27,28] However, when the cation radius of the A site is too large or too small relative

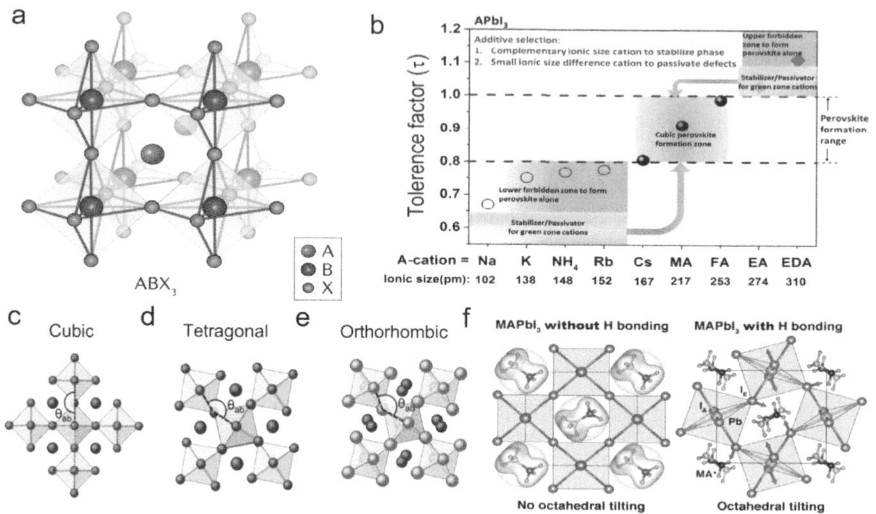

FIGURE 10.1 (a) A general perovskite structure with the stoichiometric formula of ABX₃.[24] Copyright 2019, Springer Nature. (b) Calculated tolerance factors for perovskites with different A-site cations.[26] Copyright 2018, American Chemical Society. Typical crystal phases of ABX₃ Metal Halide Perovskite (MHP) including (c) cubic phase, (d) tetragonal phase, and (e) orthorhombic phase. The symmetry descent is related to small coherent displacements of halides, leading to bending of Pb-X-Pb angles with $\theta_{ab} = \theta_c = 180°$, $\theta_{ab} < 180°$ and $\theta_c = 180°$, and $\theta_{ab} = \theta_c < 180°$.[34] Copyright 2017, American Chemical Society. (f) The effect of hydrogen bonding on the octahedral tilt of MAPbI₃ perovskites.[35] Copyright 2016, American Chemical Society.

to the other parts of the B-X cage, the limited options for A, B, and X ions can lead to cage distortion and octahedral tilt, resulting in local lattice strain.[29] For instance, in lead methylamine iodide (FAPbI₃, where $t > 1$),[30] tensile strain is observed in the (111) plane[31] due to the size difference of the A cation. Conversely, compressive strain is observed in lead methylammonium iodide (MAPbI₃, where $t < 1$)[30] and CsPbI₃ (where $t < 1$).[32] Sargent et al.[33] found that the mixed CsMAFA perovskite showed better stability compared to FAPbI₃. FAPbI₃ is more susceptible to local lattice strain caused by the larger size of FA⁺ cations. This classical ion size mismatch causes the lattice distortion between the FA⁺ cation and the Pb iodine octahedron, as well as the tilt of the PbI₆ octahedron.

It should be noted that the changes in the length and angle of the B-X-B bond can also cause lattice distortion. As the B-X-B bond angle decreases, metal halide perovskites often undergo a transition from a cubic phase to an orthogonal or quadrature phase (Figure 10.1c-e).[34] Additionally, lattice distortion arises from displacement and rotation of A-site cations. While these spatial effects primarily influence the octahedral tilt of inorganic halide perovskites like CsPbX₃, hydrogen bonds between organic A-site cations and [BX₆]⁴⁻ also contribute to the octahedral tilt of hybrid perovskites (Figure 10.1f), such as MAPbI₃ and FAPbI₃.[35] Li et al.[36] compared the mechanical properties of two organic-inorganic hybrid perovskite materials – (GUA)

[Mn(HCOO)$_3$] (GUA = guanidinium) and (AZE) [Mn(HCOO)$_3$] (AZE = azetidinium) – and found significant differences. This variation can be attributed to the distinct hydrogen bonding patterns between their A-site amine cations and the [Mn(HCOO)$_3$] framework.

Local lattice mismatch is a significant cause of internal strain in perovskite materials. Mixed halide perovskites, in particular, experience crystal structure mismatch and local lattice distortion as a result of this local lattice mismatch.[37] The inhomogeneity in mixed perovskites arises from the separation of the Br-rich and I-rich phases when exposed to heat and light, which can be attributed to chemical mismatches between the components and imbalanced growth conditions during film crystallization. Chen et al.[38] found an in-plane residual strain gradient distribution in perovskite films caused by the local lattice mismatch perpendicular to the substrate. Mela et al.[39] investigated Young's modulus (YM) in perovskite films and observed a significant increase at grain boundaries, sometimes exceeding an order of magnitude. This increase in YM is primarily attributed to the preferential accumulation of bromides at morphological grain boundaries but is also observed in specific intra-crystalline regions, leading to material composition heterogeneity. Furthermore, the variations of YM within the same morphological grains suggested the presence of subcrystalline bands and a greater degree of structural inhomogeneity.

In addition to local lattice mismatches, local crystal orientation errors can also cause local strains in perovskite grains. Perovskite films consist of multiple grains with varying orientations and sizes. Diffusion of orientation between grains creates inhomogeneous local strain in halide perovskite films. At the same time, during the nucleation and growth of the film, small changes in the local environment, such as the surface morphology of the substrate, the presence of intermediate compounds, and the concentration gradient, can lead to heterogrowth phenomena.[40,41] During nucleation and growth, small changes in the local environment can induce heterogrowth phenomena. Heterogeneous growth leads to the formation of expansion defects like grains and twin boundaries in the perovskite film. These boundaries serve as stress-concentrated regions with significant mismatches in crystal orientation.[11,42] Jariwala et al.[11], using electron backscattering diffraction (EBSD), demonstrated a strong dependence of this strain on the grain orientation and the grain interface. They also found that higher dislocation angles obtained by EBSD correspond to higher strain degrees.

10.2.2 External Condition-Induced Strain

The crystal lattice can undergo periodic distortion due to external mismatches, resulting in strain induced by external factors. The primary sources of this external strain are lattice and thermal expansion mismatches, as well as external influences such as light, temperature, electrical bias, and external force.

Perovskite films are commonly derived through precursor solution deposition, followed by annealing for crystallization. However, the expansion or contraction of the perovskite film during annealing is limited due to significant differences in thermal expansion coefficients. For instance, the thermal expansion coefficient (α) of MAPbI$_3$ is an order of magnitude larger than the α of the substrate.[43] Xue et al.[20] conducted a comprehensive investigation into residual strain caused by thermal expansion mismatch, revealing various α values for materials used in different functional layers of PSCs (Figure 10.2a). Notably, the widely employed electron transport layer (ETL) exhibits a low α value within the range of 0.37 to 1×10^{-5} K^{-1}. Conversely, perovskites demonstrate significantly higher α values, ranging from 3.3–8.4×10^{-5} K^{-1}.

This substantial thermal expansion mismatch results in significant residual tensile strain. Unfortunately, efficient PSCs require annealing temperatures of at least 100°C for organic-inorganic hybrids (Figure 10.2b). All-inorganic perovskites require even higher annealing temperatures to improve film crystallinity and reduce defects, which inevitably leads to residual strain during the annealing or cooling process.

All-inorganic perovskites necessitate even higher annealing temperatures to enhance film crystallinity and diminish defects, unavoidably leading to residual strain during the annealing and cooling process. The relationship between residual strain (σ) and thermal expansion mismatch can be defined as follows:[46]

$$\sigma_{\Delta T} = \frac{E_p}{1 - v_p}\left(\alpha_s - \alpha_p\right)\Delta T \tag{10.1}$$

where α_s and α_p are the thermal expansion coefficients of the substrate and the perovskite, E_p is the modulus of the perovskite, v_p is Poisson's ratio of the perovskite, and ΔT is the temperature gradient. This indicates that the difference between the perovskite layer and its adjacent layers, as well as the high annealing temperature (ΔT), are the main source of induced strain.

During the cooling process, the difference in α values between the substrate and perovskite material is crucial in determining the remaining strain. When the substrate has a lower α value, it restricts the contraction of the perovskite material during cooling to room temperature. This restriction induces tensile strain along the in-plane direction, as shown in (Figure 10.2c).[20] On the other hand, if a layer with a higher α value is used, it allows for greater contraction of the perovskite film, resulting in compressive strain.

Another important factor contributing to interfacial residual strain is the lattice mismatch between the perovskite material and the substrate, which needs to be considered during device fabrication. Xu et al.[21] conducted epitaxial growth of α-FAPbI$_3$ on various MAPbCl$_x$Br$_{3-x}$ single crystal substrates to investigate the impact of lattice mismatch on strain and the relaxation mechanism. By adjusting the Cl-to-Br ratio, they were able to modify the lattice parameters of the substrate. When the substrate's lattice parameter is smaller than that of α-FAPbI$_3$, it leads to compressive strain at the interface. Conversely, when the substrate's lattice parameter is larger than that of α-FAPbI$_3$, it results in tensile strain at the interface.

Notably, external conditions such as lighting, temperature, electrical bias, and external pressure can generate additional strain in perovskite materials.[22,47] Chen et al.[13] revealed the significant role of thermal expansion induced by illumination in generating photo-induced strain in MAPbI$_3$ perovskites. Under illumination, the hydrogen bond between the MA cation and I weakens, leading to an attenuation of the MA-I interaction. This results in lattice expansion, helping to alleviate the distortion

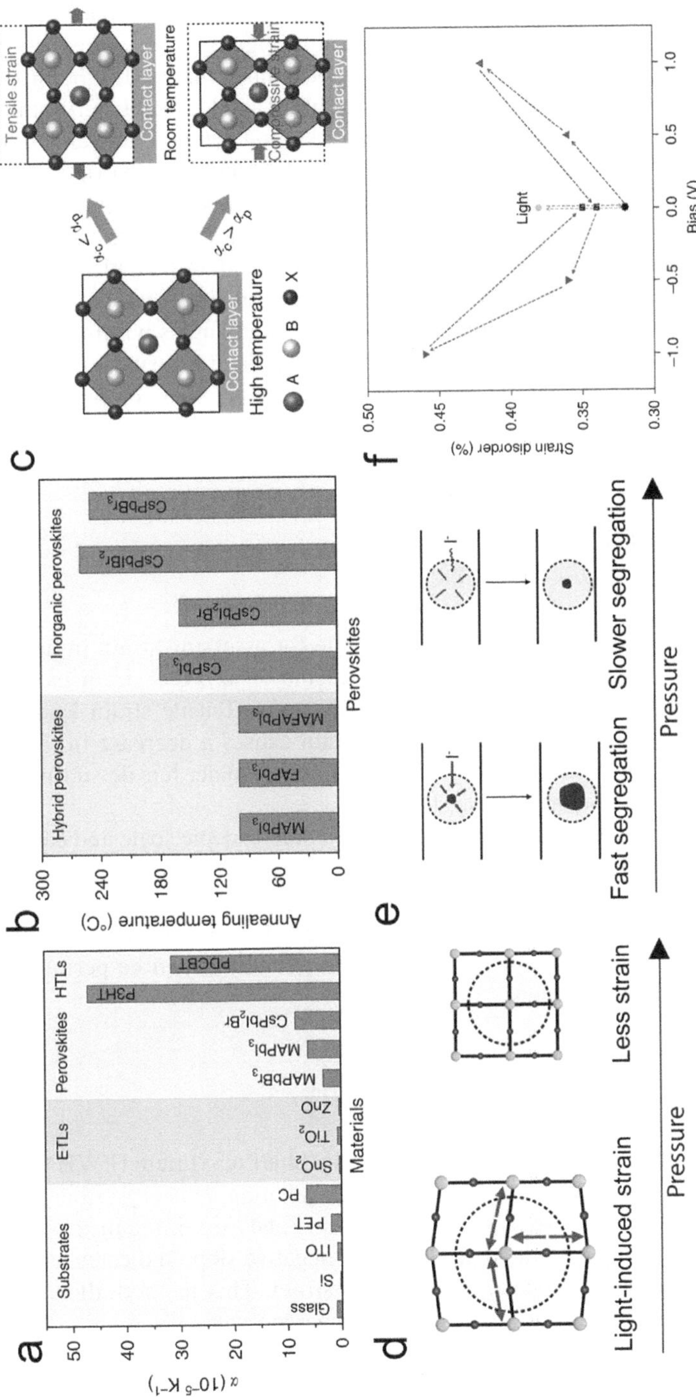

FIGURE 10.2 (a) Thermal expansion coefficients of widely-used functional layers in PSCs including substrates, ETLs, perovskites, and hole transport layers (HTLs). (b) Annealing temperatures of different hybrid and inorganic perovskite films during formation. (c) Schematic showing the formation of tensile and compressive strains.[20] Copyright 2020, Springer Nature. (d) Illustrations of strain pressure and (e) light-induced strain, which affects the rate of halide segregation in mixed-halide Perovskites.[44] Copyright 2020, American Chemical Society. (f) The effect of applied electrical bias and illumination on strain disorder in such perovskite.[45] Copyright 2019, Springer Nature.

of the Pb-I-Pb bond and the iodine octahedron. This unit cell expansion frequently reduces the activation energy for halide vacancies, promoting ion migration, weakening Pb-X bonds, and consequently causing tensile strain (Figure 10.2d, e).[44]

Recently, Liu et al.[48] investigated the influences of temperature and light on lattice expansion. They demonstrated that $MAPbI_3$, $CsPbIBr_2$ and PEA_2PbI_4 perovskites undergo lattice expansion when exposed to both external temperature and light stimuli. However, the lattice of $FAPbI_3$ expands only in response to elevated temperatures. Kim et al.[45] extensively studied the interplay between illumination and electrical bias on perovskite surface strain. Besides the strain disruption caused by illumination, the application of positive electrical bias results in a rippled surface morphology (Figure 10.2f), while negative bias leads to the smoothing of these irregularities. Additionally, the manipulation of bond lengths and angles within the perovskite lattice can be achieved by directly applying hydrostatic pressure to the film or flexing the substrate (in the case of flexible foils), inducing either compressive or tensile strain.[12]

10.3 CHARACTERIZATION METHODS FOR STRAIN IN PEROVSKITE FILM

10.3.1 XRD

XRD is a commonly used characterization technique for assessing strain in halide perovskite materials. By applying Bragg's law ($2d\sin\theta = n\lambda$), the strain can be determined by observing the displacement of XRD peaks. Tensile strain leads to an increase in the d-spacing, while compressive strain causes a decrease in the d-spacing. As a result, XRD peaks shift towards lower angles under tensile strain and towards higher angles under compressive strain.

Standard XRD is suitable for analyzing strain on a macroscopic scale and cannot resolve strain at the micro or nano scale. Additionally, most strains are complex and cannot be simply classified as purely tensile or compressive. In such cases, one indicator of strain analysis is the broadening of XRD peaks. Changes in the d-spacing affect the broadening of Bragg peaks, and micro-strain analysis can be performed using the Williamson-Hall equation:

$$\beta_{hkl} \cos\theta = 4\varepsilon \sin\theta + \frac{K\lambda}{D} \tag{10.2}$$

where β_{hkl} is the width of the XRD peak at full width at half maximum (FWHM), k is the Scherrer constant, λ is wavelength of the X-ray radiation, D is crystallite size, and ε is strain. By plotting $\beta_{hkl}\cos\theta$ as a function of $4\sin\theta$ and examining the slope of the resulting line, the strain ε can be determined. A negative slope indicates tensile strain, while a positive slope indicates compressive strain. This method allows for quantitative analysis of the strain in the material at the nanoscale.

However, there are some limitations to this approach. First, reference standard materials are often used to eliminate instrument broadening. However, these reference standards may differ in form from the actual samples, leading to inaccuracies in the computed results. Second, the distribution of crystal size, shape, and crystal lengths along each crystallographic direction can influence the calculations and affect the measurement of micro-strains. For materials with inherent heterogeneity, errors in calculating grain size can have a cascading effect, further complicating the analysis results. Notably, microstrain values and compressive/tensile strain values are not directly comparable, as they probe different aspects of the material's structure.

XRD analysis includes in-plane and out-of-plane measurements. In-plane measurements assess the inter-plane spacing parallel to the substrate, while out-of-plane measurements characterize the inter-plane spacing perpendicular to the substrate, as depicted in Figure 10.3a. Huang et al.[45] conducted in-plane and out-of-plane XRD analysis on an annealed $MAPbI_3$ film (AF) and out-of-plane XRD analysis on an annealed $MAPbI_3$ powder (SCP; Figure 10.3b). The out-of-plane XRD peak of AF exhibited a larger angular displacement compared to the strain-free SCP, indicating a reduction in the plane spacing perpendicular to the substrate and the presence of compressive strain in this direction. Moreover, the in-plane XRD peak of AF shown a smaller angular displacement compared to the out-of-plane XRD peak, indicating the existence of tensile strain in this direction.

When using XRD for strain measurement, two methods can be employed: (i) Select a single halide perovskite material, as factors like doping and light irradiation-induced boundary effects may influence the accuracy of strain characterization. (ii) Combine out-of-plane and in-plane XRD measurements. Generally, in perovskite materials, if there is tensile strain in one direction, then there is compressive strain in the perpendicular direction, and the shift of in-plane XRD peaks is opposite to the shift of out-of-plane XRD peaks. If the peak shift is solely caused by composition rather than strain-induced displacement, then the XRD results will be the same in both in-plane and out-of-plane measurements.

GIXRD measurements can be employed to monitor both out-of-plane and in-plane strains and study strain distributions at various depths in perovskite films (Figure 10.3c).[38] A higher incidence angle allows for deeper X-ray penetration into the sample. Kong et al.[49] utilized GIXRD to analyze the (012) plane at a diffraction angle of 31.7° and investigate the tensile strain in the perovskite film. Figure 10.3d-e demonstrates that changing the inclination ψ from 0° to 45° induces a gradual leftward shift in the characteristic peaks to indicate lattice expansion in the plane direction. Typically, the relationship between $\sin2\psi$ and 2θ is linear, and a negative slope of the fitting line signifies the presence of tensile stress in the perovskite film.[50] Both the modified and unmodified films show negative slopes (Figure 10.3f), indicating the existence of tensile strain. When the slope is positive, there is compressive strain.

10.3.2 GIWAXS

GIWAXS is an advanced and non-destructive characterization technique widely used in materials science. Recently, there have been reports on the application of GIWAXS in studying metal halide perovskite films. This technique enables precise and high-resolution investigation of the atomic and molecular structures, crystalline ordering, and orientation of films, surfaces, and nanostructured materials.[51] Moreover, in-situ GIWAXS observation of perovskite films allows for tracking

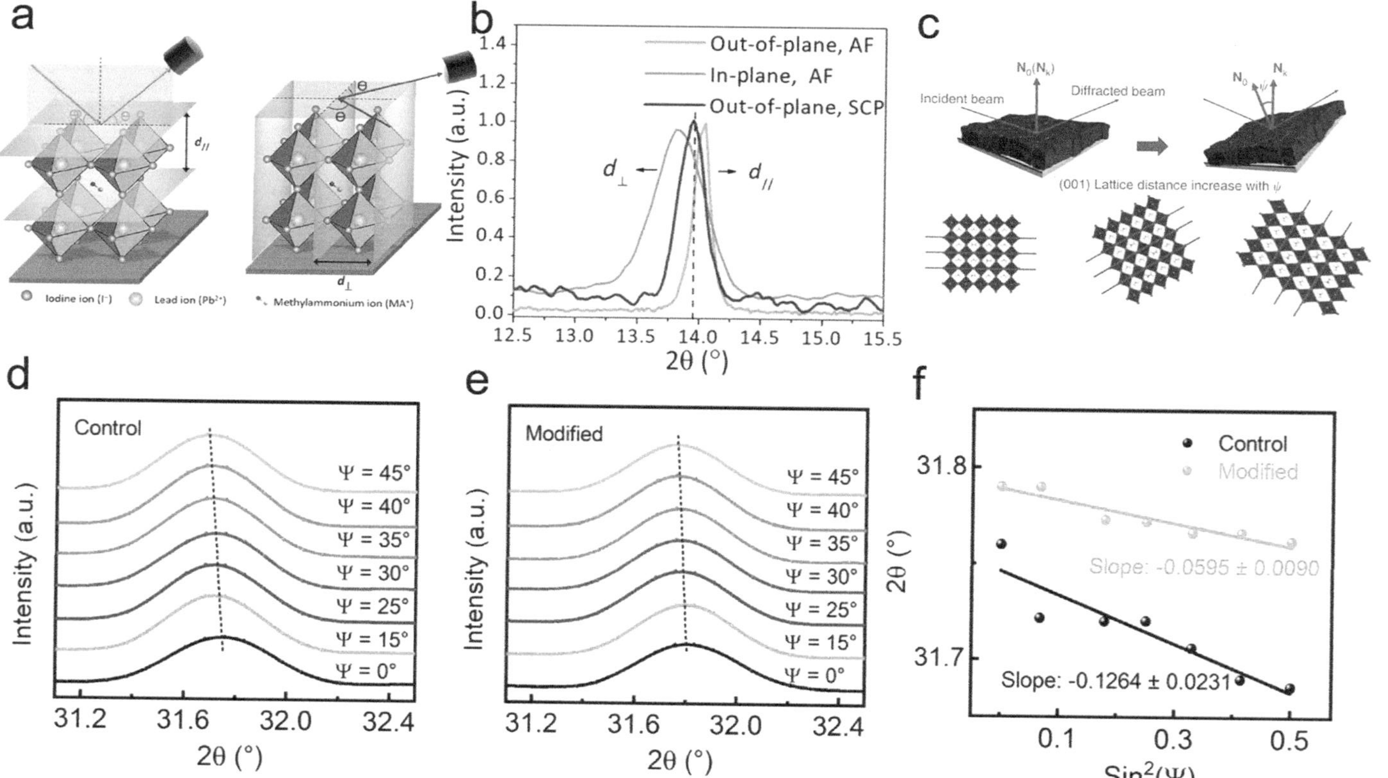

FIGURE 10.3   (a) Differentiation between an out-of- and in-plane diffraction-based characterization. Reprinted with permission.[12] (b) In-plane and out-of-plane XRD of AF and out-of-plane XRD of SCP.[12] Copyright 2017, American Association for the Advancement of Science. (c) Schematic of the strain measurement using GIXRD, where N_0 is the sample normal direction, N_k is the diffraction vector, and ψ is the instrument tilt angle.[38] Copyright 2019, Springer Nature. (d), (e) GIXRD spectra at different tilt angles and (f) the corresponding linear fit of 2θ-$\sin^2(\psi)$.[49] Copyright 2022, American Chemical Society.

microstructure changes such as crystallization and aging processes, providing valuable insights into the kinetics of perovskite materials.[52] The typical measurement configuration of GIWAXS is shown in Figure 10.4a. In this configuration, an X-ray beam with an initial wave vector $\left(\overrightarrow{K_i}\right)$ irradiates the sample surface at a grazing angle (α_i), and the scattered wave $\left(\overrightarrow{K_f}\right)$ is captured by a detector.[53]

By combining high-throughput and high-brightness synchrotron radiation X-ray beams, the bright GIWAXS pattern of perovskite films can be measured under millisecond exposure, enabling real-time observation of rapid crystallization and phase transition process of perovskite. Another advantage of the GIWAXS is the use of a two-dimensional area detector, which allows simultaneous collection of signals in both in-plane and out-of-plane directions. For spin-coated films, which exhibit isotropic behavior in the in-plane direction, the orientation can be described by plotting the $\overrightarrow{q_r}$ and $\overrightarrow{q_z}$ coordinates on a plane (Figure 10.4a). The expression is:

$$\overrightarrow{q_r} = \overrightarrow{q_x} + \overrightarrow{q_y} \tag{10.3}$$

As a result, lattice parameters of perovskite obtain by GIWAXS measurements could offer help to determine the corresponding perovskite phase. For example, Hui et al. investigated the correlation between lattice strain and the occurrence of the δ-phase in $CsPbI_2Br$ perovskite films.[54] Lattice strain often leads to anisotropic crystal growth, resulting in structural instability and the formation of the δ-phase. As depicted in Figure 10.4c, in-situ GIWAXS was utilized to examine the impact of heat on the growth process of all-inorganic $CsPbI_2Br$ perovskite films. The 2D-GIWAXS images demonstrated that increasing the processing temperature to 55°C resulted in excellent crystallinity of the perovskite phase, with the strongest peak at $q = 10$ nm^{-1}, and phase stability without the presence of the δ-phase.

Dang et al. employed time-resolved in-situ GIWAXS to analyze the solution-to-solid transformation of various perovskite precursors for a detailed investigation of the impact of Cs^+ and Rb^+ addition on perovskite film formation.[55] After the addition of a solvent quencher (indicated in Figure 10.4d with a vertical arrow), diffraction peaks corresponding to different phases appeared. By combining these observations with other testing methods, they assigned the distinct diffraction features to the respective precursors and phases.

Wu et al. introduced MXene to regulate SnO_2 dispersion and induce vertical growth of the perovskite. The GIWAXS pattern of SnO_2 MXene/perovskite exhibited a higher intensity at $q \approx 10.0$ nm^{-1}, indicating better crystallinity and out of plane orientation due to the lattice matching between MXene material and perovskite. The introduction of additives enhanced the degree of lattice matching at the SnO_2-perovskite interface, thereby reducing the concentration of interfacial stress (Figure 10.4d).[56] In summary, GIWAXS measurements enable researchers to gain insights into the lattice parameters, crystallographic properties, and grain morphology of perovskite films.

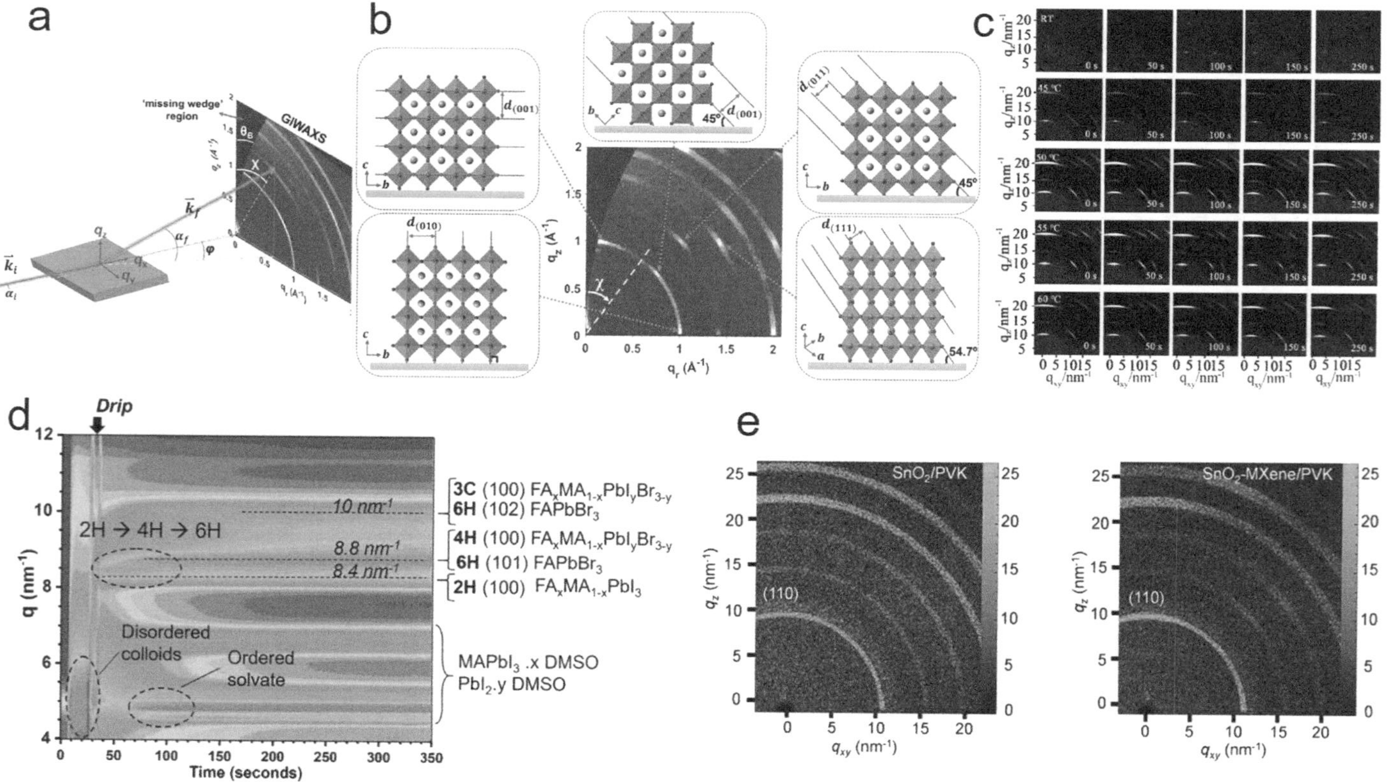

FIGURE 10.4 (a) Schematic diagram of GIWAXS. (b) Schematic illustration of the GIWAXS pattern obtained from a highly oriented 3D perovskite film.[53] Copyright 2021, Wiley-VCH GmbH. (c) Time-varying map of 2D-GIWAXS spectra of perovskite films at different processing temperatures under different annealing time.[54] Copyright 2020, Elsevier. (d) Time evolution of GIWAXS intensity map of a pristine film at different q values.[55] Copyright 2019, Elsevier. (e) GIWAXS patterns of SnO_2/PVK and SnO_2-MXene/PVK film. SnO_2-MXene induces oriented growth of PVK.[56] Copyright 2022, Wiley-VCH GmbH.

10.3.3 Raman Spectroscopy

When an object is illuminated by light, in addition to reflection, absorption, and transmission, a portion of the light scatters in all directions. This scattered light exhibits significant frequency or wave number variations relative to the incident light. If the frequency variation exceeds 3×10^{10} Hz or the wave number variation exceeds 1 cm^{-1}, it is known as Raman scattering. Raman scattering corresponds to transitions between rotational and vibrational energy levels of molecules. Raman spectroscopy is a precise method for measuring the lattice vibration energy of materials. When a material is subjected to stress, its lattice structure undergoes changes, leading to alterations in lattice vibrational energy and, consequently, shifts in Raman frequencies (the change being denoted as $\Delta\omega$, relative Raman shift). By establishing the relationship between stress and the relative Raman shift $\Delta\omega$, one can calculate the internal stress within the crystal.

Raman spectroscopy is a useful tool for characterizing strains in perovskite films. Xu et al.[21] employed Raman spectroscopy to investigate the structural changes in α-FAPbI$_3$ under varying strains from 0% to -2.4%. The peak observed at 136 cm^{-1} initially split into two peaks at 140 cm^{-1} (main peak) and 133 cm^{-1} (shoulder peak; Figure 10.5a–b). As the strain was further increased to -2.4%, these two peaks shifted to 143 cm^{-1} and 130 cm^{-1}, respectively. It is speculated that the compression of in-plane Pb-I bonds leads to the blue shift of the main peak, while the stretching of out-of-plane Pb-I bonds results in the red shift of the shoulder peak.

In addition to providing qualitative characterization of strain in perovskite materials, Raman spectroscopy can also enable quantitative measurements of strain. Badrooj et al.[57] quantified the compressive strain values of perovskite layers by analyzing the displacement of their active Raman modes. In Raman spectroscopy, the variations in vibration frequency can serve as a calibration curve to assess local strain values resulting from residual strain.[58] To explore the influence of strain on the charge transport mechanism, the degree of compressive strain in $CH_3NH_3Sn_xPb_{1-x}I_3$ layers can be estimated using Equation (10.4), utilizing the information extracted from the shifts of vibrational modes observed in the experimental Raman spectra.[59,60]

$$\frac{\Delta\omega^{\pm}}{\Delta\omega_0} = -\gamma\left(1 - \upsilon\right)\varepsilon_z \qquad (10.4)$$

where $\dfrac{\Delta\omega^{\pm}}{\Delta\omega_0}$ and ω_0 are the relative shift of the Raman band and the peak position under zero strain, respectively; γ represents the Gruneisen parameter, which is estimated as 1.6 for polycrystalline perovskite; υ and ε_z are Poisson's ratio and compressive strain. For example, in the diagrams presented in Figure 10.5d–f, the displacements of the three prominent Raman bands (v_1, v_2, v_3 as indicated in Figure 10.5c) were measured against the corresponding compressive strain values for perovskite samples with Sn concentrations of x = 0, 0.5, and 1.

10.3.4 TEM

For characterization of the strain in different layers of materials, a reliable technique with high spatial resolution is required. TEM is an effective technique to provides

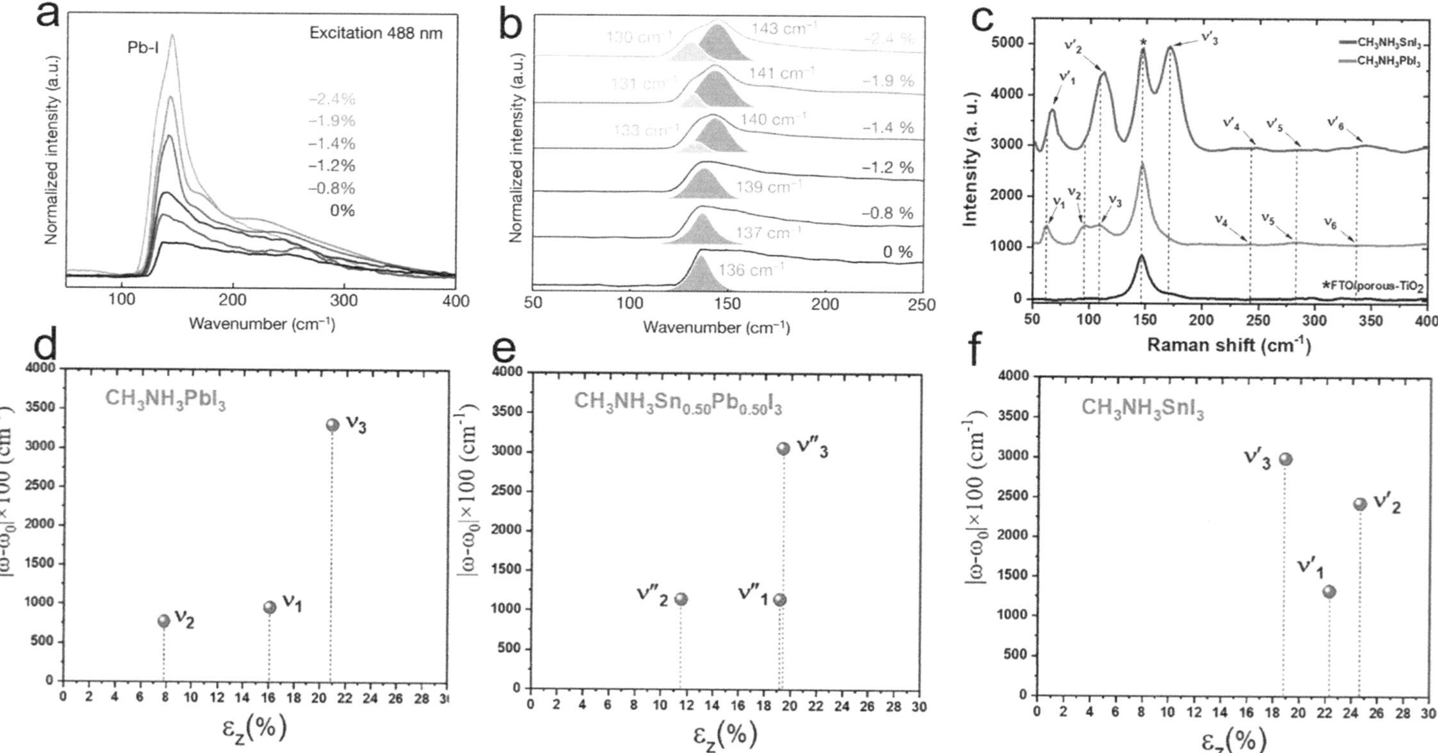

FIGURE 10.5 (a) Confocal Raman spectra of the epitaxial α-FAPbI$_3$ layer at different strains. (b) Fitting analysis of the Raman peaks in (a).[21] Copyright 2020, Springer Nature. (c) Raman spectra obtained from pure MAPbI$_3$, MASnI$_3$, and FTO/porous-TiO$_2$ substrate. Raman band shift of (d) CH$_3$NH$_3$PbI$_3$, (e) CH$_3$NH$_3$Sn$_{0.50}$Pb$_{0.50}$I$_3$, and (f) CH$_3$NH$_3$SnI$_3$ versus the compressive strain.[57] Copyright 2020, American Chemical Society.

insights into the crystallographic, compositional, and defect-related features of perovskite materials. This measurement allows strain measurement at the nanometre scale based on technology of electron diffraction.[61] To measure strain in materials, important TEM techniques for strain analysis include: convergent beam electron diffraction (CBED), high-resolution transmission electron microscopy (HRTEM), nano beam electron diffraction (NBED) and dark-field imaging (DFI).[62]

Within the realm of TEM, CBED is the sole technique capable of yielding quantitative strain information at the nanometer scale. This point-to-point method enables the determination of the strain tensor at each nanoregion region of the sample probed by analyzing the corresponding diffraction pattern.[66] The foundation of the CBED technique lies in the observation of strain-induced shifts in high-order Laue zone (HOLZ) deficiency lines within the central disk of a convergent electron beam pattern. The selection of a zone axis unaffected by dynamical interactions ensures accurate measurement of the strain and lattice parameters through the analysis of these line shifts. Nevertheless, it is crucial to highlight that sufficient HOLZ scattering requires the crystal to be oriented in a relatively high-order zone axis. As a consequence, when structures grow in low-order zone axes, interfaces may suffer from blurring, leading to a reduction in spatial resolution and interpretability of the obtained data. Consequently, NBED technique is gaining increasing popularity over CBED. In the NBED approach, a spot pattern is generated from a small region on the sample. Initially, a reference pattern is acquired from a selected area, typically the substrate, to establish the positions of the spots. Subsequently, patterns are sequentially obtained from distinct areas, enabling the determination of the deformation tensor at each probed position through the analysis of spot position shifts. However, it is noteworthy that the NBED technique exhibits sensitivity to atomic column bending, which may introduce inaccuracies in the obtained results.[67]

HRTEM is an imaging mode of TEM that employs phase-contrast imaging, where both transmitted and scattered electrons are combined to produce the image. In addition to obtaining information from traditional electron microscopy, the most important role of HRTEM is that can obtain high-resolution image information such as lattice stripe images (reflecting crystal plane spacing information) and individual atom images (reflecting the configuration of atomic clusters in the crystal structure). As a result, the use of HRTEM enables the quantification of strain via the measurement of crystal plane spacing. For instance, Zhang et al. utilized HRTEM to confirm the presence of P-Nb_2CT_x and T-Nb_2CT_x MXene nanosheets (Figure 10.6a, b).[63] The lattice parameters of the P-Nb_2CT_x and T-Nb_2CT_x MXene nanosheets are 0.27 nm, corresponding to the (042) plane of Nb_2C. Then, in the study of surface reconstruction on the epitaxial growth of III-Sb on GaAs, Jia et al. used cross-sectional TEM images of III-Sb on GaAs under the Sb absorption temperature of 500°C to prove the interfacial mis fit (IMF) arrays formed successfully. (Figure 10.6c, d).[64] Rothmann et al.[65] used TEM to investigate the character and spatial distribution of misfit dislocations (MDs) that induce non-uniform strain fields (Figure 10.6e, f). In the context of epitaxially grown materials, TEM was utilized to monitor the process of strain relaxation within the epitaxial layer. In cases where epitaxial layers are well-grown, devoid of interfacial imperfections, TEM analysis reveals a lack of defects. Conversely, when uncontrolled strain relaxation occurs, numerous defects emerge and become visually discernible through TEM imaging.[65]

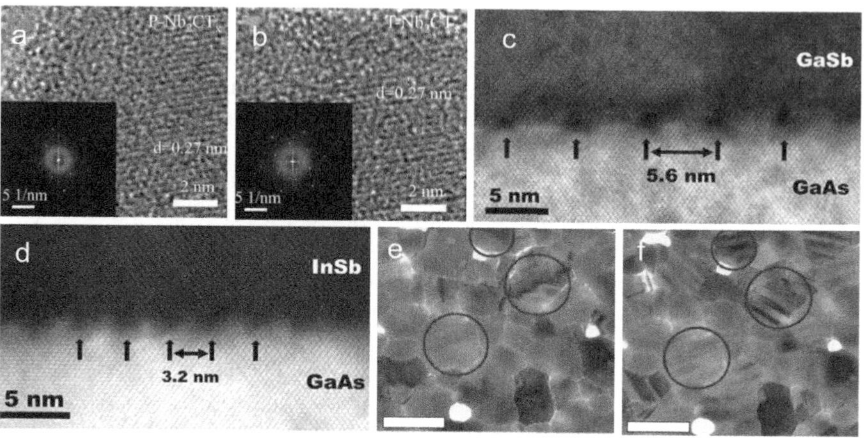

FIGURE 10.6　(a) HRTEM image of P-Nb$_2$CT$_x$ MXene nanosheet (b) HRTEM image of T-Nb$_2$CT$_x$ MXene nanosheet.[63] Copyright 2022, Wiley-VCH GmbH. (c) Cross-sectional TEM image of the interfacial MDs for GaSb.[64] (d) Cross-sectional TEM image of the interfacial MDs for GaSb.[64] Copyright 2017, Elsevier. (e) Bright-field TEM image of pristine CH$_3$NH$_3$PbI$_3$ film at room temperature.[65] (f) Bright-field TEM image of same region after extended electron beam exposure.[65] Copyright 2017, Springer Nature.

10.3.5　SED

SED is a four-dimensional scanning transmission electron microscopy (4D-STEM) technique,[68] which is based on acquiring two-dimensional transmission electron diffraction patterns at each position of a focused electron probe during a two-dimensional scanning. Each two-dimensional electron diffraction pattern contains strong Bragg disks at the scattering angle θ_B (Figure 10.7a–b), and the scattering angle θ_B is related to the atomic plane spacing d_{hkl} according to Bragg's law $\lambda = 2dhkl \sin\theta_B$. The corresponding Bragg disks are related to atomic planes close to the direction parallel to the incident electron beam. Strain alters the spacing of atomic planes, resulting in a shift in the position of Bragg disks.

The technique yields nanoscale spatially resolved diffraction patterns to characterize local changes in d-spacing, grain orientation, and strain (Figure 10.7c–f).[39] For instance, in these SED, scans with a spatial resolution of 5 nm (extracted from individual grains 50 to 200 nm in size), very faint reflections, forbidden from appearing in the $Pm\overline{3}m$ space group (Figure 10.7e), were visible (Figure 10.7c, white arrows). In the realm of 2D strain mapping, SED has emerged as a versatile technique, offering the capability to map strain with resolution down to a few nanometers.[69,70]

10.3.6　EBSD

EBSD obtains crystallographic information from SEM and generates the EBSD map by moving the electron probe point by point at grid position on the surface of bulk sample, which can provide more information (Figure 10.7g). Ginger et al.[11] utilized EBSD to measure local crystal orientations in perovskite films. In the SEM

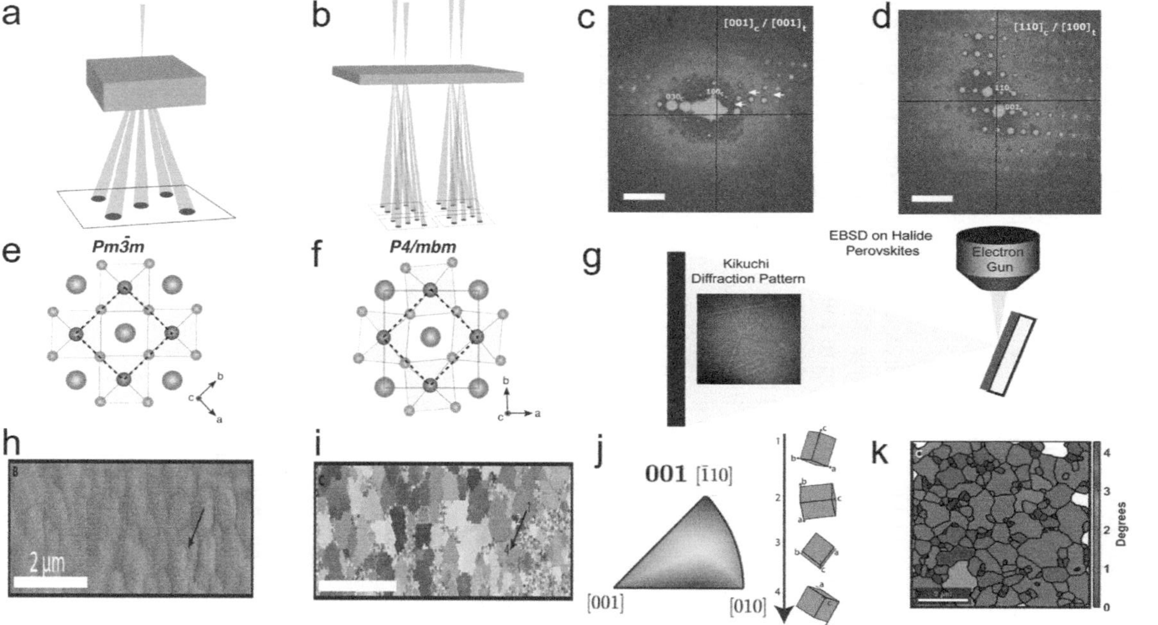

FIGURE 10.7 Schematic illustration of SED. (a) A focused electron probe passes through a sample and a 2D electron diffraction pattern is recorded. If the sample is crystalline then the diffraction pattern comprises intense Bragg disks at positions related to the spacing of atomic planes in the crystal, as shown. (b) The probe is scanned over the sample and a diffraction pattern recorded at each position.[71] Copyright 2020, IOP Publishing. (c) SED patterns of triple/cation perovskite corresponding to [001]c zone axis demonstrating the presence of superlattice reflections (white arrows), which are normally forbidden in a cubic Pm$\bar{3}$m structure, often assumed for halide perovskite materials, as shown in (d). Analyzing SED pattern near [110]c zone axis in (e) allows to conclude that perovskite has a P4/mbm structure shown in (f). Scale bar is 0.5 Å^{-1}.[72] Copyright 2022, American Association for the Advancement of Science. (g) Schematics of EBSD measurement on $CH_3NH_3PbI_3$ films. (h) Top view SEM image and (i) IPF map generated from EBSD of $CH_3NH_3PbI_3$ film with IPF color key. (j) Depiction of changes in local crystal orientation along the black arrow in (h) and (i) as viewed normal to the sample. (k) Plot of grain orientation spread (GOS) showing grain-to-grain heterogeneity in average local misorientation in the same film.[11] Copyright 2019, Elsevier.

morphology image (Figure 10.7h), arrows indicate individual "grain boundaries," but Figure 10.7i (IPF) shows that the film possesses three distinct boundaries. A single arrow in its direction showcases four different local crystal orientation changes. Figure 10.7j exhibits local crystal orientation changes along the arrow directions. The grain orientation spread (GOS) map (Figure 10.7k) is used to illustrate the inter-grain non-uniformity of average local orientation errors in perovskite films generated using EBSD and IPF. High GOS values correspond to high inter-grain strain, while low GOS values correspond to low inter-grain strain. The measured deviations in local crystal orientation confirm the presence of local strains in the perovskite thin film.

10.3.7 PFM

PFM is a specialized mode of AFM that measures the local piezoelectric or ferroelectric response of a sample. It uses a conductive AFM tip to apply an AC voltage to the sample, inducing mechanical deformation in the piezoelectric or ferroelectric domains. The resulting piezoresponse signal is detected as a force variation between the tip and the sample. PFM is mainly applied in the study of ferroelectric and piezoelectric materials, where it provides insights into domain structures, electromechanical coupling phenomena, and local strain variations.

Weber's group reported a typical work using PFM to study crystal strain of $MAPbI_3$ perovskite films.[73] Due to the strain resulting from the cubic to tetragonal phase change, the sub-granular ferroelastic twin domains that make a negative impact on the electronic transport properties could been observed by PFM. As shown in Figure 10.8a, d, they conducted a series of experiments wherein multiple batches of $MAPbI_3$ films were prepared with varying precursor ratios ranging from 9:1 to 6:4. Subsequently, a thorough investigation was carried out to assess the surface topography and lateral piezoresponse of the films using PFM. As a result, it is observed that the width of ferroelastic twin domains increases with the increase of Lead(II) chloride content. Leonhard et al. correlated the local crystal orientation of individual grains with their ferroelectric properties and their electronic surface properties by using PFM.[74] Liu et al. reported that chemical and strain heterogeneity within MHP films can cause differences in the stiffness of the tip-surface contact and result in buckling of the cantilever during scanning over the sample. As a consequence, PFM imaging may exhibit contrast patterns that resemble those observed in ferroelectric domains.[75]

10.4 IMPACTS OF STRAIN

10.4.1 BAND STRUCTURE

The aforementioned strain can have a significant impact on the crystal arrangement and band structure of perovskites. Figure 10.9a shows a schematic diagram of the lattice structure of the strain state, which means the tensile strain (left), non-strain state (middle), and compressive strain (right). In the hybrid perovskite crystal structure, the framework of the corner-sharing PbI_6 octahedron has a great influence on the electronic configuration, especially at the band edge.[76] Under the impaction of

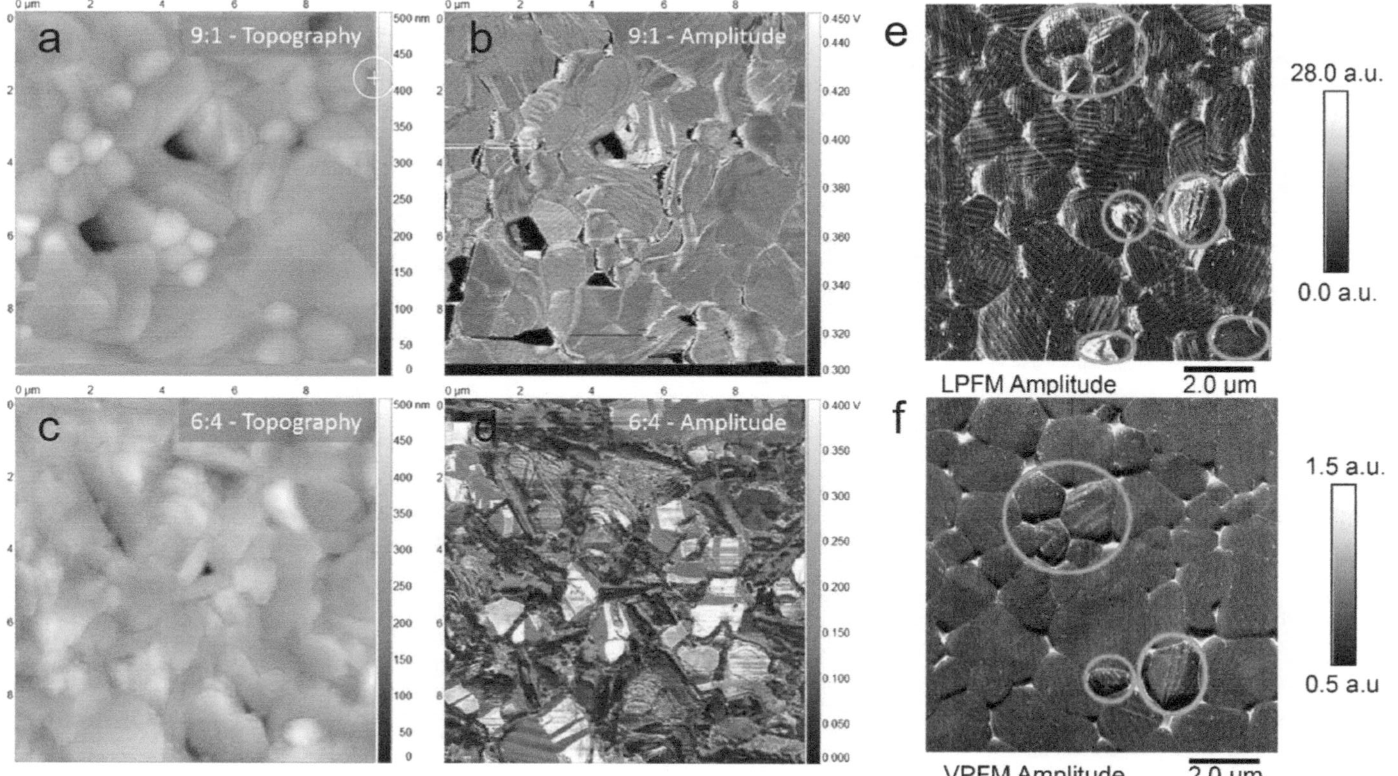

FIGURE 10.8 Topography and PFM amplitude images of MAPbI$_3$ films with Pb(Ac)$_2$/PbCl$_2$ ratios of (a), (c) 9:1, and (b), (d) 6:4.[73] Copyright 2022, Wiley-VCH GmbH. (e) The lateral PFM image (amplitude) of a representative MAPbI$_3$. (f) The corresponding vertical PFM images (amplitude).[74] Copyright 2019, Wiley-VCH GmbH.

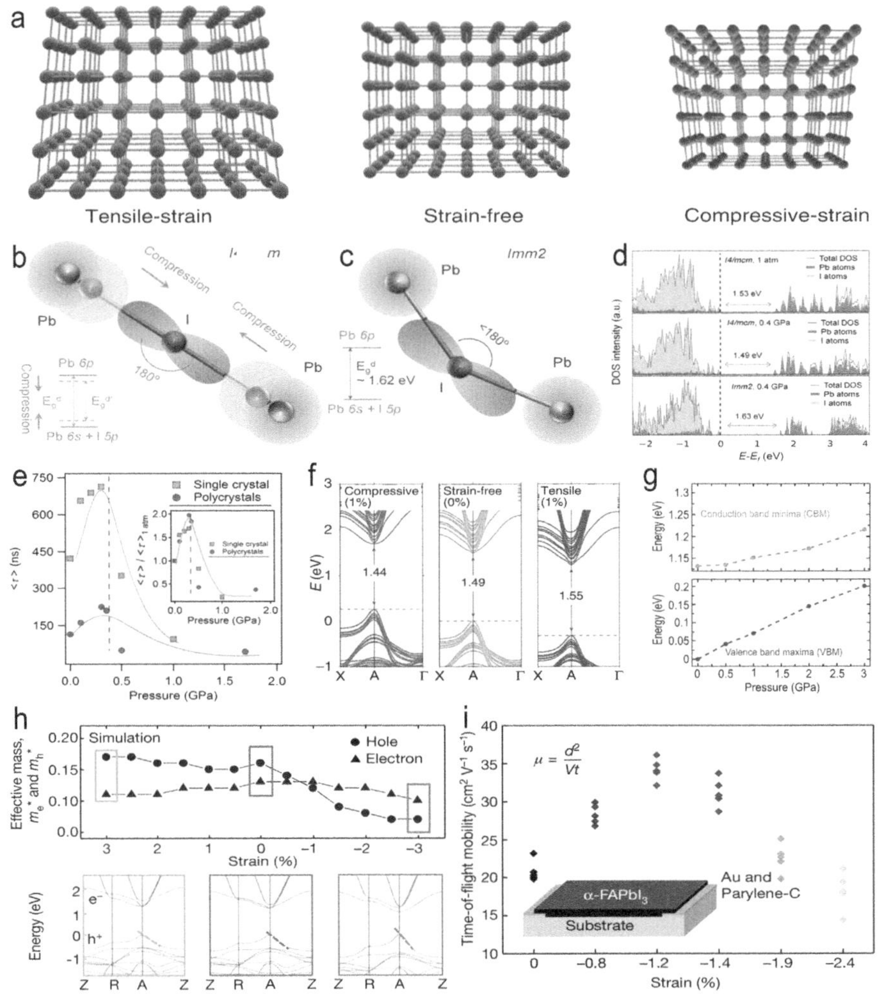

FIGURE 10.9 (a) The schematic representation of the tensile strain state of the film in the top surface.[38] Copyright 2019, Springer Nature. (b, c) Pb-I-Pb bond length and bond angle models for *I4/mcm* and *Imm2* phases. (d) Density of states (DOS) of MAPbI₃ with a structure of *I4/mcm* at 1 atm (top), *I4/mcm* at 0.4 GPa (Middle), and *Imm2* at 0.4 GPa (bottom). (e) Pressure dependence of the mean carrier lifetime for both MAPbI₃ single-crystal and polycrystal samples. Peak values in carrier lifetimes of MAPbI₃ were observed at 0.3 GPa. (Inset) A normalized result.[77] Copyright 2016, National Academy of Sciences. (f) Calculated band structures under biaxial tensile, zero, and compressive strains from first-principal density functional theory (DFT)-based approaches.[38] Copyright 2019, Springer Nature. (g) Modification in energy of CBM and VBM.[78] Copyright 2019, American Chemical Society. (h) Calculated effective masses of the carriers at different strains, and electronic band structures under three strain levels (3%, 0%, and -3%). (i) Plots of calculated carrier mobilities as a function of the strain magnitudes. The inset equation, $\mu = d^2/Vt$. Inset, schematic measurement setup.[21] Copyright 2020, Springer Nature.

stress, the length of Pb-I bond will change, which will affect the bond angle and lead to the tilt of PbI_6 octahedron. Kong et al.[77] explored the impact of varying pressure conditions on the Pb-I-Pb bond length and corresponding valence band structure. In the tetragonal *I4/mcm* symmetric phase at low-pressure ($\leq$ 0.4 GPa), the Pb-I-Pb bond angle remains close to 180°, and the bond length is compressed (Figure 10.9b). This bond length variation enhances the coupling between the Pb s and I p orbitals, resulting in an upward shift of the valence band maximum (VBM). The change in the band gap is chiefly attributed to the variation in the VBM, while the conduction band minimum (CBM) is primarily determined by the non-bonding localized state of the Pb p orbitals, which is insensitive to bond length or pressure. As a result, applying pressure narrows the bandgap, with most of the alteration originating from the VBM. However, when the pressure increases up to 0.4 GPa, the sample undergoes a critical phase change from *I4/mcm* to *Imm2* (the high-pressure orthorhombic phase). This results in a decrease in symmetry from tetragonal to orthorhombic, and the Pb-I-Pb bond angle significantly decreases within the range of 144.0° to 162.0° (Figure 10.9c). Therefore, upon the phase transition, the Pb-I bond is partially fractured and the coupling between Pb s and I p orbitals diminishes, resulting in the widening of the bandgap (Figure 10.9d). The increase of pressure also leads to a significant increase in carrier lifetime. The average carrier life time (Figure 10.9e) is calculated as $<\tau> = [\alpha\tau_1/(\alpha\tau_1+\beta\tau_2)]\tau_1 + [\beta\tau_2/(\alpha\tau_1+\beta\tau_2)]\tau_2$, where τ_1 is the slow-decay component, τ_2 is the fast-decay component, the results show an increase of 70% and 100% in single crystal and polycrystalline samples, respectively.

Chen et al.[38] computationally determined the band structure of $FAPbI_3$ under diverse stress conditions. The results indicated that the bandgap gradually increased from compressive strain to non-strain and ultimately tensile strain (Figure 10.9f). The alteration of the band gap is contingent upon variations in the VBM and CBM. As depicted in Figure 10.9g, under a compressive strain of 3 GPa, the VBM of $FA_{0.75}Cs_{0.25}PbI_3$ was upshifted by 0.2 eV to a higher energy level, while the CBM was likewise upshifted by 0.08 eV to higher energy levels.[78] This also suggests that the compressive strain exerted a greater influence on the VBM relative to the CBM. Xu et al.[21] calculated the electron effective mass $\left(m_e^*\right)$ and hole effective mass $\left(m_h^*\right)$ (upper panel) and three typical electron band structures (lower panel) under different strains (Figure 10.9h). With the change of stress state, the change of m_e^* is very small, while m_h^* decreases significantly with the increase of compressive stress. The functional relationship between the calculated carrier mobility and the applied strain is shown in the Figure 10.9i. The carrier mobility is calculated as follow formula:

$$\mu = \frac{d^2}{Vt} \tag{10.5}$$

where μ is the calculated time-of flight carrier mobility, d is the target region thickness, V is the applied voltage and t is the measured carrier transit time. The film with a strain of -1.2 % has the highest carrier mobility, as the strain increases further, the carrier mobility decreases sharply, which may be because the higher strain level leads to higher dislocation density.

In summary, strain has a significant impact on the band structure of perovskites. Compressive strain upshifts the VBM by changing the Pb-I bond length, angles, and octahedron tilt, thereby narrowing the band gap. Conversely, tensile strain downshifts the valence band, widening the band gap. Electron and hole effective masses and carrier mobility are also affected by strain, exhibiting optimal values under compressive strain. These strain-induced modifications of the electronic structure provide opportunities to engineer the optoelectronic properties of perovskite materials.

10.4.2 Defect Properties

The occurrence of defects in perovskite is a common issue due to the rapid crystal growth process and the high temperatures used during fabrication. These defects can significantly affect the performance of solar cells.[79] The prevalence of intrinsic point defects, including vacancies and interstitials, is tightly correlated with their defect formation energy, which directly reflects the density of point defects in perovskite.[80], [30] Xue et al.[20] employed the DFT approach to calculate the formation energy of halide vacancies across the strain spectrum from compression to tension and the calculated formation energy can reflect the tendency of the formation of halide vacancy defects in perovskite materials. As the stress state alters from compressive strain to tensile strain, the calculated formation energy exhibits a descending tendency (Figure 10.10a), suggesting an enhancing trend in the propensity for halide vacancy formation. They also calculated the activation energy (E_a) for vacancy-assisted migration of halide ions in perovskite under biaxial tensile and compressive strains. Under the strain-free state, the activation energy for halide ion migration is 0.667 eV, which decreases to 0.547 eV at 1.5% tensile strain while increases to 0.794 eV at -1.5% compressive strain.[20] The results demonstrated that compressive strain can not only reduce the formation energy of halide vacancies but also suppress the ion migration in perovskite, thereby enhancing the stability of perovskite.

Islam et al.[78] further calculated the thermodynamic transition levels of the inherent vacancy defects in $FA_{0.75}Cs_{0.25}PbI_3$ and $FAPbI_3$ under diverse strains. Under the influence of applied external pressure (2 GPa), the iodide vacancies (V_I) exhibited a pronounced tendency of experiencing a substantial decrease in their transition energy levels within the bandgap (Figure 10.10b, c). Under the low pressure (≤ 0.5 GPa), the transition state levels of the vacancy defects essentially retained a state of fundamental invariance, which would not exert a notable impact on the lifetime of the charge carriers. Deger et al.[79] systematically investigated the impact of crystal strain on possible intrinsic point defects in $CsPbI_3$, $FAPbI_3$, and $MAPbI_3$ perovskite structures. As the crystal strain changes from -2% (compressive strain) to 2% (tensile strain), the corresponding change trend of point defects of A-site cation antisite $(I_{A\text{-}site})$, Pb antisite (I_{Pb}), A-site cation vacancy $(V_{A\text{-}site})$, I vacancy (V_I), Pb vacancy (V_{Pb}), and I interstitial (I_i) is shown in the Figure 10.10d–f. For $CsPbI_3$, compressive strain can increase the DFE of V_I by 4% (Figure 10.10d), which will reduce the defects. The center of the defect unit cell undergoes a significant contraction in the x-direction due to the contribution of electrons, thereby enhancing the formation energy of the defect through compressive strain. On the contrary, V_{Cs}, I_{Cs}, V_{Pb}, and I_i cause expansion along the x-direction, and their DFE decreases due to compressive strain. $FAPbI_3$

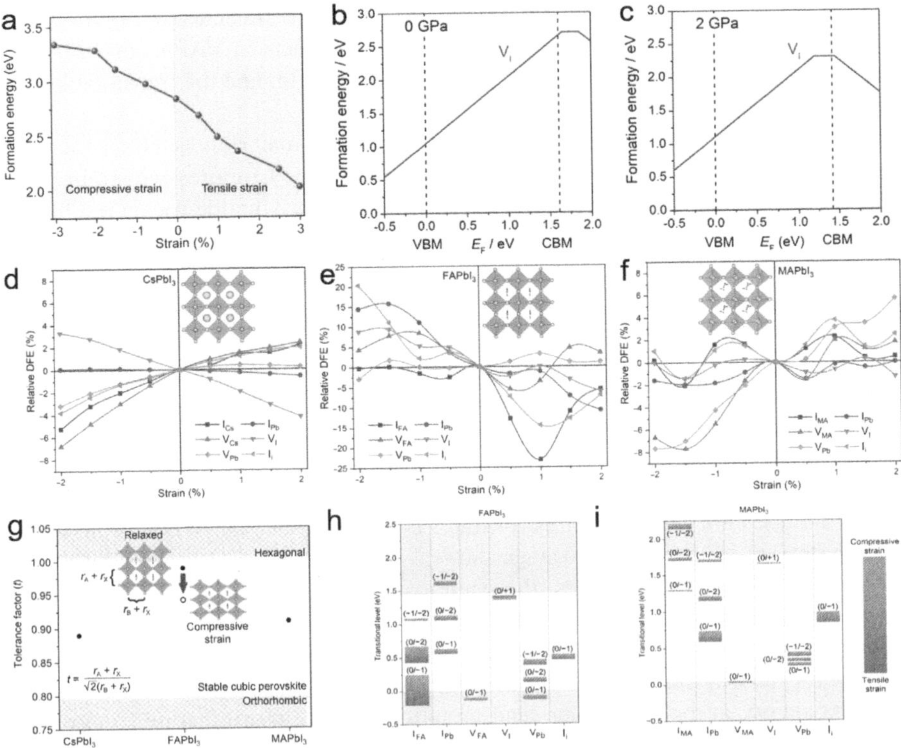

FIGURE 10.10 (a) Calculated strain-dependent formation energies of halide vacancies.[20] Copyright 2020, Springer Nature. Iodide vacancy formation energies in $FA_{0.75}Cs_{0.25}PbI_3$ at the lowest energy equatorial sites. The most stable charge state is shown at (b) 0 and (c) 2 GPa.[78] Copyright 2019, American Chemical Society. The dependence of the defect formation energies (in eV) on the lattice strain and defect formation energy of the perovskites for different growth conditions. The relative DFE with respect to the strain for selected neutral intrinsic defects (I_A, I_{Pb}, V_A, V_I, V_{Pb}, and I_i) in (d) $CsPbI_3$, (e) $FAPbI_3$, and (f) $MAPbI_3$ perovskites under moderate conditions. The negative value of % indicates compressive strain while positive % values correspond to tensile strain along the out-of-plane direction throughout the study. (g) Schematic representation of strain-induced tolerance factor adjustment of $FAPbI_3$. Herein, r_A, r_B, and r_X represent the radii of A-site molecule, B-site metal, and X-site halide in ABX_3 perovskites, respectively. In addition, $r_A + r_X$ shown here is a space-diagonal length. The change in the tolerance factor of $FAPbI_3$ is enlarged to show the influence of the strain. The charge transition energy levels of selected intrinsic defects in (h) $FAPbI_3$ and (i) $MAPbI_3$. The influence of lattice strain on the transition levels is shown by the color gradient. The strain is varied from -2% (compressive) to 2% (tensile).[79] Copyright 2022, Springer Nature.

is most sensitive to strain, and its DFE can be increased to 20% under compressive strain. This behavior may be due to the larger size of FA molecules than Cs and MA, which makes the lattice structure more susceptible to stress. The t of $CsPbI_3$, $FAPbI_3$, and $MAPbI_3$ was shown in Figure 10.10g; the values are 0.89[20], 0.99[81], 0.91[81]. The production of defects in $FAPbI_3$ is more sensitive to crystal distortion. Compressive strain can cause the tolerance factor to move from the hexagonal phase transition region to the stable cubic zone, improving the defect formation energy. Figure 10.10h, i shows the charge transition energy levels of the defects in $FAPbI_3$ and $MAPbI_3$; even deep flaws can become shallow defects due to strain, and the transition level of defects in the band gap can be regulated.

In conclusion, intrinsic point defects in perovskite materials, such as vacancies and interstitials, show a high sensitivity to strain states. Compressive strain generally increases the formation energy of defects, particularly halide vacancies, thereby making their formation more challenging. Furthermore, compressive strain enhances the activation energy for vacancy-assisted halide ion migration. Consequently, the reduced defect concentration resulting from the increased formation energy can effectively suppress ion migration, leading to improved stability. In conclusion, the application of compressive strain emerges as an effective approach for manipulating the defect properties of perovskites, thereby enhancing the performance of optoelectronic devices.

10.4.2.1 Ion Migration

Strain-induced ion migration in perovskites is a common cause of device instability and hysteresis in the current-voltage curve.[30] Due to the soft ionic crystal properties of ABX_3-based perovskites, the chemical bonds among internal A, B, and X ions are relatively weak. The residual strain introduced during device fabrication processes such as annealing and solvent engineering further reduces the activation energy for ion migration.[82–84] External stimuli, such as light, heat, or biased voltage, provide energy to the perovskite, leading to ion migration originating from regions with lower defect formation energy due to stress concentrations.[85,86]

Therefore, various ion management approaches by strain regulation have been developed. Huang et al. first reported a method of suppressing ion migration in perovskites by compressive stress. Under both dark and illuminated conditions, the ion migration energy (E_a) of $MAPbI_3$ perovskite after treatments involving tensile strain, strain free, and compressive strain exhibited a gradually increasing trend (Figure 10.11a).[47] It is shown that in perovskite materials, compressive strain is more advantageous than tensile strain for inhibiting ion migration.

A similar result was reported by Wang et al., where they introduced compressive stress to $Cs_{0.22}FA_{0.78}Pb(I_{0.85}Br_{0.15})_3$ perovskite by adding ATP. The ATP-$Cs_{0.22}FA_{0.78}Pb$ $(I_{0.85}Br_{0.15})_3$ sample shown an improved E_a value from 0.43 to 0.51 eV compared to the control group, indicating that compressive strain can create a stronger barrier against ion migration than strain-free perovskite (Figure 10.11c).[87] The reason why compressive strain can suppress ion migration in perovskites is that it compensates for the tensile strain caused by the mismatch in thermal expansion coefficients between the perovskite and the substrate, thereby increasing E_a. By applying compressive strain, the detrimental effects of this tensile strain can be counteracted, ultimately enhancing the stability of perovskites.

10.4.2.2 Phase Stability

One of the most important advantages of adjusting lattice strain is ensuring the phase stability of metastable perovskite films, because lattice strain can influence the tolerance factor of perovskites by controlling the $[PbX_6]^{4-}$ octahedral structure,

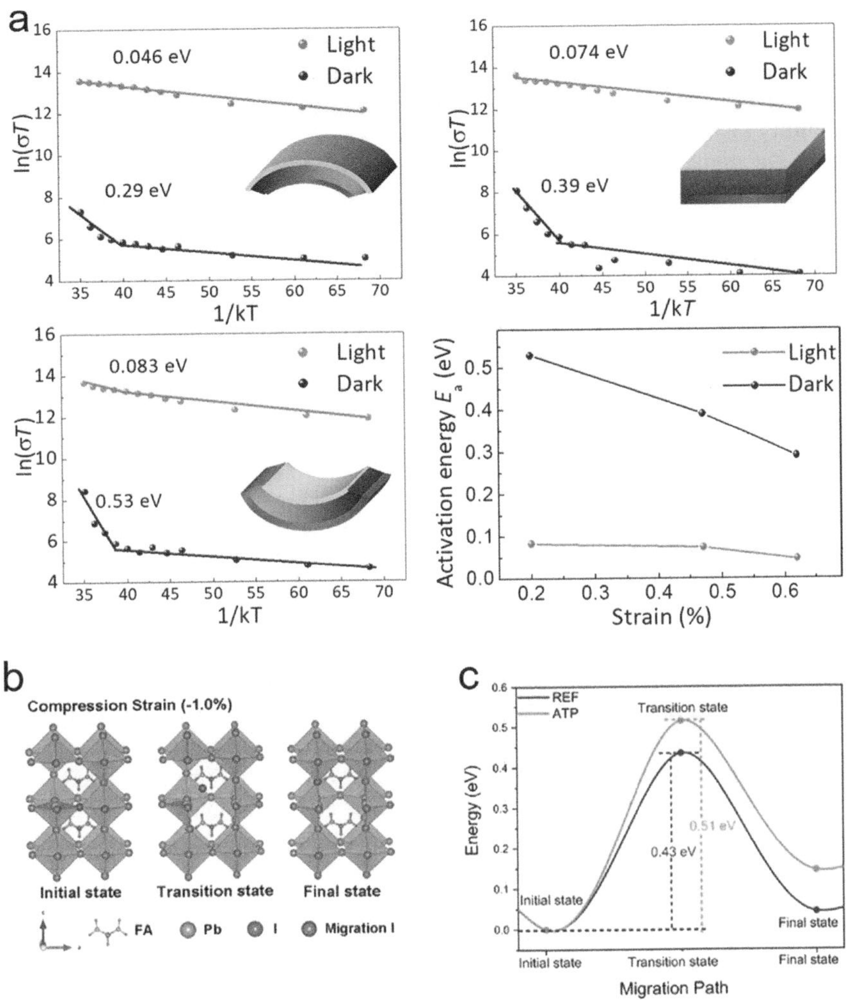

FIGURE 10.11 (a) Ion migration properties of MAPbI$_3$ films with different strains.[12] Copyright 2017, American Association for the Advancement of Science. (b) Scheme of ion migration under compressive strain in the DFT calculation. (c) Strain-dependent activation energies for the vacancy-assisted migration of halide ions for REF and ATP films from DFT calculation.[87] Copyright 2022, Wiley-VCH GmbH.

maintaining t within a stable range of 0.8 to 1.[88] However, the impact of strain on phase stability has been extensively discussed in FAPbI$_3$ perovskite. α-FAPbI$_3$ has a t of 0.987, which is close to 1. The larger FA$^+$ cation induces a metastable tensile-strain lattice, making it prone to spontaneous transformation to the δ non-perovskite phase (Figure 10.12a).[89–90] Notably, the spontaneous transformation from the α phase to the δ phase in FAPbI$_3$ perovskites is often suppressed when various foreign ions (such as MA$^+$ and Cs$^+$) are introduced into the lattice. This is because the smaller ionic radii of MA$^+$ and Cs$^+$ compared to FA$^+$, which lowers the t of FAPbI$_3$ and stabilizes it around 0.912.[88] Therefore, foreign ions doping becomes a method for adjusting the t of perovskite to relieve the internal strain of perovskite in order to address the instability of perovskite at room temperature. Priya et al.[30] introduced a dual-doping strategy at the A and X sites of perovskites using MABr. Due to the smaller ionic radii of MA$^+$ and Br$^+$, their incorporation into the perovskite lattice induces lattice compression to counterbalance tensile strain (Figure 10.12b). XRD spectra clearly reveal a high-angle shift in peaks for MABr-modified perovskite films, indicating the release of residual tensile strain (Figure 10.12c). Finally, MABr mitigates strain-induced phase transitions, significantly enhancing phase stability.

Additionally, even in the absence of foreign ions, residual strain can affect the phase stability of perovskites. For example, Steele et al.[91] demonstrated that the black phase of CsPbI$_3$ can be stabilized at room temperature by interface-induced biaxial strain. Strain interfaces were induced within the perovskite film by annealing with substrate confinement and rapid cooling from 330°C to room temperature, greatly enhancing the thermal stability of black CsPbI$_3$ films (Figure 10.12d–e). Meanwhile, Chen et al.[21] fabricated α-FAPbI$_3$ films with compressive strain (pink lines) and no strain (black lines), and their long-term stability was monitored through XRD (Figure 10.12f), PL (Figure 10.12g), and Raman spectroscopy (Figure 10.12h). In Figure 10.12f, after 360 d under a compressive strain of -2.4%, the <001> and <002> peaks corresponding to the α phase remain unchanged. In contrast, the strain-free α phase transitions into the δ phase within just one day. The enhanced stability of the α-FAPbI$_3$ phase under compressive strain is clearly evident from the PL and Raman results (Figure 10.12g–h). The enhanced stability of α-FAPbI3 can be attributed to the elimination of the phase transition driving force caused by compressive strain, which prevents structural changes associated with the phase transition.

10.4.2.3 Device Stability

Strain has a significant impact on the stability of both perovskite films and devices. The residual tensile strain in perovskite films induces structural defects, weakens chemical bonds, and promotes ion migration, leading to severe non-radiative recombination. Conversely, compressive strain increases the activation energy for ion migration, enhancing the thermal and photostability of PSCs. Xue et al.[20] introduced a strain regulation strategy using a hole transport layer with a high CTE. They compensated for the tensile strain by increasing the annealing temperature of the HTL. After continuous heating at 85°C for 1,000 h, devices with compressive strain and non-strain maintained 96% and 80% of their initial PCE, as shown in Figure 10.13a. In contrast, devices under tensile strain experienced a rapid decline in PCE after only 500 h. They further assessed the photostability of PSCs under continuous one-sun illumination and maximum power point (MPP), which showed a similar trend to the conditions of continuous heating at 85°C (Figure 10.13b). The decrease of PCE in the tensile-strain device can be attributed to phase separation (resulting in Br-rich and I-rich perovskite phases) along with the subsequent transformation from the photoactive α phase to the non-photoactive δ phase.

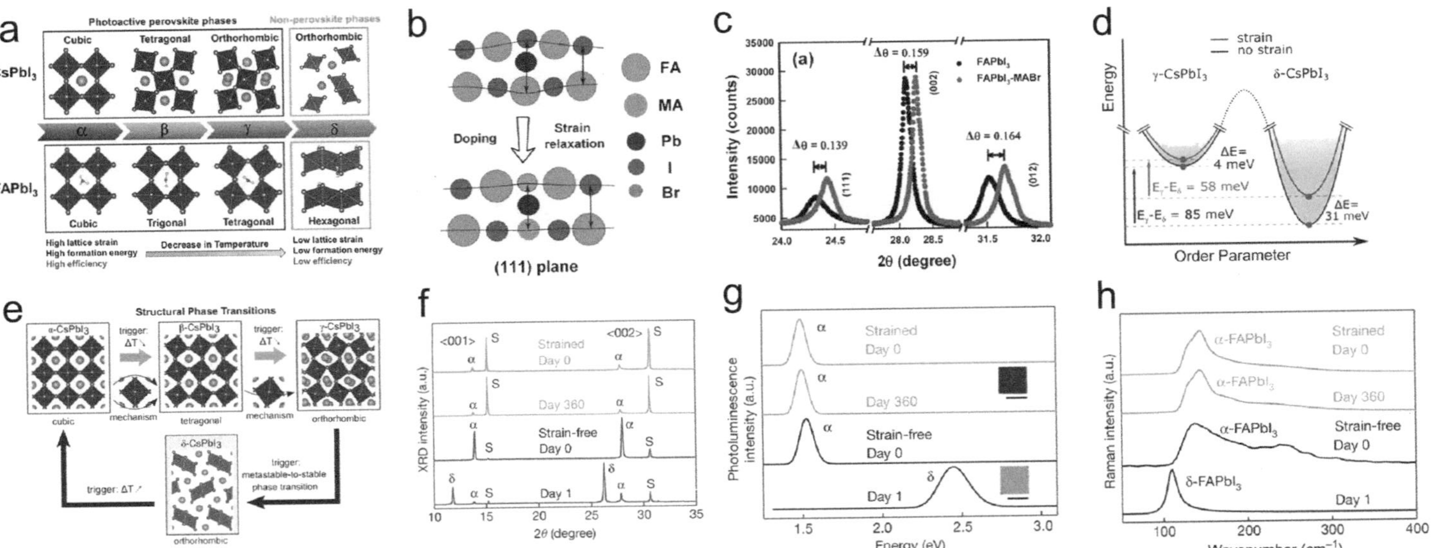

FIGURE 10.12 (a) Structure and transition of $FAPbI_3$ and $CsPbI_3$ in its α-phase, β-phase, γ-phase, δ-phase.[92] Copyright 2021, American Chemical Society. (b) Structure and schematic representation of strain relaxation for $FAPbI_3$ after MABr alloying. (c) Magnification of XRD results of $FAPbI_3$ comparing the peak shift after MABr alloying.[30] Copyright 2016, American Chemical Society. (d) Crystal structure of the different phases and their relative phase transitions. The transitions between the black phases are governed by the local Pb-centered octahedral (black) distortions, depicted here using one lead atom at the center and six iodide atoms at the edges (purple), confining the cesium cations (cyan). (e) Ab initio energy diagram indicating the relative stability (at 0 K) of the black and yellow phases with and without in-plane biaxial strain.[91] Copyright 2019, American Association for the Advancement of Science. (f) Phase stability comparison of thin and thick epitaxial α-$FAPbI_3$ on $MAPbCl_{1.50}Br_{1.50}$ substrates. The thin-strained sample shows better phase stability (red curves) for the thick, strain-free sample. (g) Phase stability study by photoluminescence spectroscopy. Re-measurement of the thin, strained sample after 360 d (lower pink curve) shows no obvious photoluminescence peak shift but does show a slight decrease in peak intensity. (h) Phase stability study by Raman spectroscopy. The Raman characteristics of the thin, strained sample show a peak at 143 cm⁻¹ with no substantial difference after 360 d; the thick, strain-free sample (peak at 136 cm⁻¹) shows signs of a phase transformation to δ-$FAPbI_3$ after 24 h, as revealed by its signature peak at 108 cm⁻¹.[21] Copyright 2020, Springer Nature.

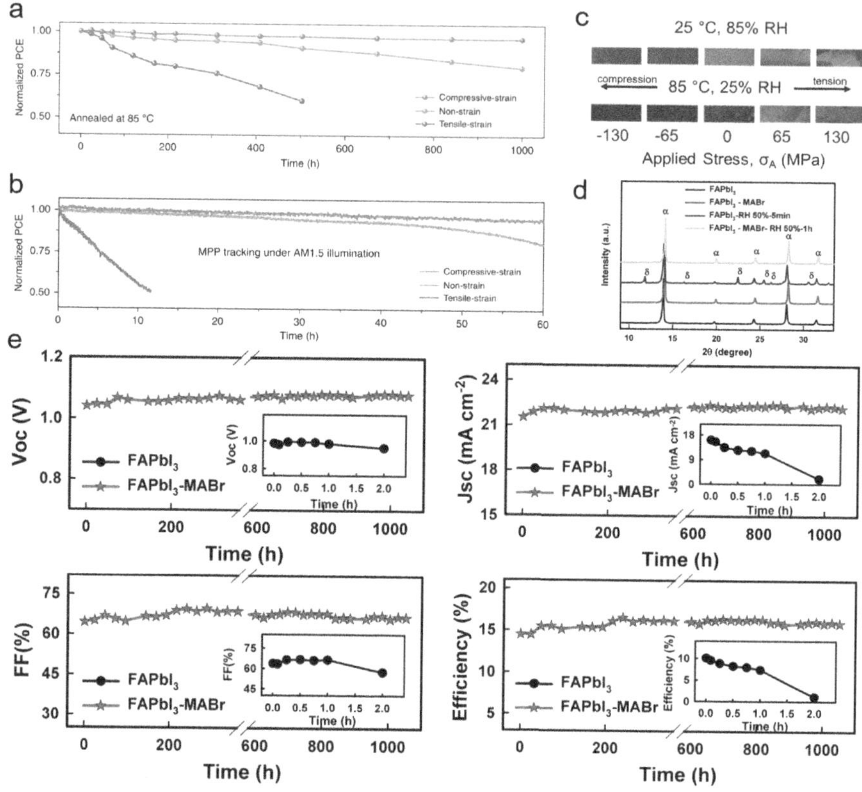

FIGURE 10.13 (a) Evolution of normalized PCEs of PSCs kept at 85°C in a nitrogen atmosphere. (b) Evolution of normalized PCEs under MPP tracking and continuous simulated solar illumination (100 mW cm^{-2}).[20] Copyright 2020, Nature Communications. (c) Photographs of MAPbI$_3$ on PET with externally applied stresses ranging from -130 to 130 MPa after 24 h of damp air aging at 25°C and 85% RH or dry heat aging at 85°C and 25% RH, where applied compressive stress enhanced film stability and applied tensile stress reduced film stability.[46] Copyright 2018, Wiley-VCH GmbH. (d) XRD patterns of FAPbI$_3$ and FAPbI$_3$-MABr films before and after exposure to humid air. (e) Comparative analysis of the performance parameters: V_{oc}, J_{sc}, FF, and PCE of solar cells based on FAPbI$_3$ (insets) and FAPbI$_3$-MABr with time stored in air under a RH of ~50% at 23°C without encapsulation.[30] Copyright 2016, American Chemical Society.

Rolston et al.[46] have altered the strain within MAPbI$_3$ films by applying external tensile and compressive stresses. The substrates were then exposed to either damp air (25°C, 85% RH) or dry heat (85°C, 25% RH) for 24 h and photographed after aging (Figure 10.13c). In both cases, films with applied tensile stress had significant visible degradation with PbI$_2$ formation, whereas films with applied compressive stress resulted in intact perovskite after exposure. Furthermore, Priya et al.[30] achieved remarkable stability in perovskite devices by doping MABr into the perovskite film, introducing smaller A and X site ions and thus achieving a more stable t. Under conditions of 23°C and 50% RH, PSCs with MABr doping maintained nearly the same initial efficiency even after 1,000 h of operation. In contrast, control PSCs exhibited rapid degradation, reaching only 10% of their initial efficiency after 2 h (Figure 10.13d). This underscores the importance of strain modulation for device stability, and obtaining compressive strain in perovskite films through various strategies is one of the most effective means to enhance device stability.

10.5 STRAIN REGULATION STRATEGIES

10.5.1 COMPOSITIONAL ENGINEERING

The intrinsic polycrystalline nature coupled with the flexible lattice characteristics of perovskite films renders the perovskite lattice highly susceptible to deformation. As a result, compositional engineering emerges as the favored approach for enhancing the $[PbX_6]^{4-}$ octahedral structure and mitigating octahedral distortion. This involves skillful ion doping and additive incorporation into the perovskite films, which effectively optimizes the structural integrity and stability of the material.

10.5.1.1 Cation/Anion Doping

The $[PbX_6]^{4-}$ octahedral structure is not only the fundamental framework of perovskite materials but also plays a crucial role in facilitating the efficient transmission of photogenerated carriers, which is closely connected to the material's distinctive properties.[93] The A-cation, acting as a structural stabilizer, critically governs the lattice parameters of perovskite.[88] Intriguingly, any deviation in the A-cation's radius, either too large or too small, induces lattice distortion. An observation from Figure 10.14a demonstrates that the desired cubic phase perovskite can be obtained within the parameter range of $0.8 < t < 1$.[94] Nishimura et al.[95] delved into the impact of A-cation doping, encompassing Na^+, K^+, Cs^+, BA^+, and ethylammonium (EA^+), on the internal strain and carrier mobility of $FA_{0.75}MA_{0.25}SnI_3$ perovskite films. Their findings unveiled a compelling correlation between the proximity of the tolerance factor to unity and diminished lattice strain, with strain size exhibiting a negative association with the magnitude of carrier mobility (Figure 10.14b, c). Remarkably, the introduction of EA doping led to strain reduction and notable enhancement in carrier mobility, culminating in a remarkable champion PCE of 5.41% for $EA_{0.1}(FA_{0.75}MA_{0.25})_{0.9}SnI_3$ films. Furthermore, in SnGe-based devices utilizing EA, the PCE surged from 6.42% to 7.60%. Saidaminov et al.[33] share the perspective that the strain in perovskite arises from the inherent mismatch between the single A-cation and the lead halide cage size. They postulate that this strain induces the formation of internal point defects within the perovskite lattice, facilitating the relaxation of lattice strain (Figure 10.14d). Consequently, such strain relaxation could accelerate the degradation process of perovskite when exposed to oxygen and water molecules.

Abdi-Jalebi et al.[96] demonstrated that the incorporation of undersized alkali cations (e.g., Rb^+, K^+, and Na^+) through doping effectively interacts with and immobilizes excess and uncoordinated halides (Figure 10.14e). As a result, this phenomenon hampers the ion migration within the perovskite lattice and alleviates local strain, leading to a remarkable PCE of 21.5%. Nevertheless, the incorporation of oversized

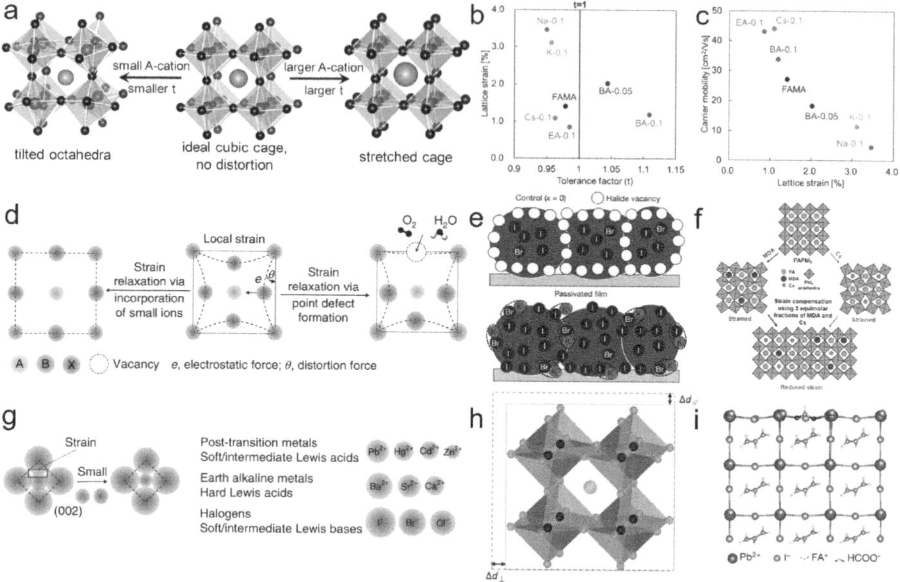

FIGURE 10.14 (a) Crystal structures of ABX_3 perovskites, where smaller or larger A-site cations will make the octahedra slightly tilted.[94] Copyright 2021, American Chemical Society. Relationship between (b) the lattice strain and the tolerance factor, (c) the lattice strain and carrier mobility for various Q-substituted Sn perovskite.[95] Copyright 2019, American Chemical Society. (d) Schematic illustrating the local strain, which is decreased by forming point defects or by doping with small ions.[33] Copyright 2018, Springer Nature. (e) Schematic diagram of halide-vacancy management in film cross-section, using surplus halide to complex with potassium at the grain boundaries and surfaces.[96] Copyright 2018, Springer Nature. (f) Strain compensation strategies via fine-tuning of A-cations.[29] Copyright 2020, American Association for the Advancement of Science. (g) Schematic demonstrating the strain in the (002) plane, which is reduced by incorporation of small B/X-site ions and B/X isovalent candidates for incorporation.[33] Copyright 2018, Springer Nature. (h) The $MA(Zn{:}Pb)I_{3-x}Cl_x$ perovskite contracts in vertical and horizontal directions equivalently.[18] Copyright 2018, Elsevier. (i) Calculated structure illustrating the passivation of an I^- vacancy at the $FAPbI_3$ surface by a $HCOO^-$ anion.[98] Copyright 2021, Springer Nature.

A-cations has also been demonstrated as an effective approach for phase regulation in perovskite materials. For instance, employing the divalent cation methylenediammonium (MDA^{2+}) to forge an increased number of hydrogen bonds with the $[BX_6]^{4-}$ lattice proves beneficial in stabilizing the cubic phase.[97] Remarkably, the presence of oversized A-cations inadvertently induces tensile lattice strain, which can be mitigated by introducing smaller Cs^+ cations (Figure 10.14f).[29]

Lattice strain can also be modulated through B/X site doping. For instance, the regulation of B-cation and X-anion doping effectively relieved residual lattice strain in CsMAFA-based perovskite (Figure 10.14g).[33] Impressively, the resulting unencapsulated PSC devices, utilizing Cd^{2+} and Cl^- doping of the CsMAFA film, retained over 90% of its initial PCE after 30 d of storage in ambient air with a RH of 50%.

Similarly, Shai et al.[18] reported that moderate zinc (Zn) doping in $MAPbI_{3-x}Cl_x$ perovskite fostered a balanced contraction of the $[BX_6]^{4-}$ octahedron in both horizontal and vertical directions, resulting in low-strain perovskite films (Figure 10.14h).

Additionally, the introduction of pseudo-halide anions (e.g., SCN-, HCOO- and BF^{4-}, etc.)[99] with halogen-like characteristics offers an avenue to replace the X site within the $[BX_6]^{4-}$ octahedral, aiming to induce strain relaxation through lattice parameter adjustments. For instance, Jeong et al.[98] employed HCOO- to suppress anion-vacancy defects present at the grain boundary and perovskite surface (Figure 10.14i), achieving a remarkable 25.2% certified PCE. Zhang et al.[100] utilized tetrafluoroborate (BF^{4-}) anion substitutions to enhance the performance of $(FAPbI_3)_{0.83}(MAPbBr_3)_{0.17}$ perovskite films. The successful incorporation of BF^{4-} into the mixed-ion perovskite crystal framework led to lattice relaxation and a prolonged photoluminescence lifetime, resulting in a PCE increase from 17.55% to 20.16%. Furthermore, Cheng et al.[101] introduced a small amount of KBF_4 dopant into triple-cation mixed perovskite films. The longer covalent bonds between BF^{4-} and Pb^{2+} effectively offset the distortion of P-I-Pb bonds caused by replacing larger FA^+ with smaller A-site cations (MA^+ or Cs^+). This interaction between large and small ions indicates the possibility of strain compensation achieved through interaction of large and small ions.

10.5.1.2 Additive Engineering

Recently, additive engineering has emerged as a prevalent approach for modifying the surface and grain boundaries of perovskite films, focusing on external modifications rather than direct incorporation into the perovskite lattice. By carefully selecting specific organic molecules or polymers, this technique effectively regulates the strain in perovskite films. Additive-based strain regulation primarily operates through the following three aspects: (i) influencing the crystallization process of the precursor solution; (ii) tailoring the crystal lattice via crosslinking or hindrance effects; and (iii) adjusting the YM of the perovskite film.

First, the most common additive engineering involves the coordination of Lewis additives with precursors or charged defects to alleviate strain. Zhao et al.[102] demonstrated this approach by incorporating dimethyl sulfoxide (DMSO) with its strong polarity and high boiling point into the $PbBr_2$ precursor. This led to the delayed crystallization of $PbBr_2$ and a reduction in the porosity of the $PbBr_2$ film. During the multistep spin-coating process with CsBr, the formed $PbBr_2$(DMSO) adduct facilitated the intramolecular exchange, promoting the smooth transition of $PbBr_2$ to $CsPbBr_3$. As a result, the resulting perovskite film exhibited an expanded structure with compressive strain, effectively counterbalancing the tensile strain generated during annealing (Figure 10.15a). Consequently, the goal of strain relaxation was achieved, culminating in an impressive 10.11% PCE for the all-inorganic $CsPbBr_3$ PSC with compressive strain. Additionally, Du et al.[103] introduced a dynamic liquid-crystal transition (DLCT) strategy, employing a specifically designed thermotropic liquid crystal molecule (CBO6SS6OCB). The incorporation of liquid-crystal (LC) molecules not only resulted in the delayed crystallization of perovskite precursors but also facilitated interactions with the ETL. During the annealing process, the LC molecules acted as a buffer, effectively overcoming the thermal mismatch and relieving the residual strain existing between the ETL and the perovskite layer (Figure 10.15b).

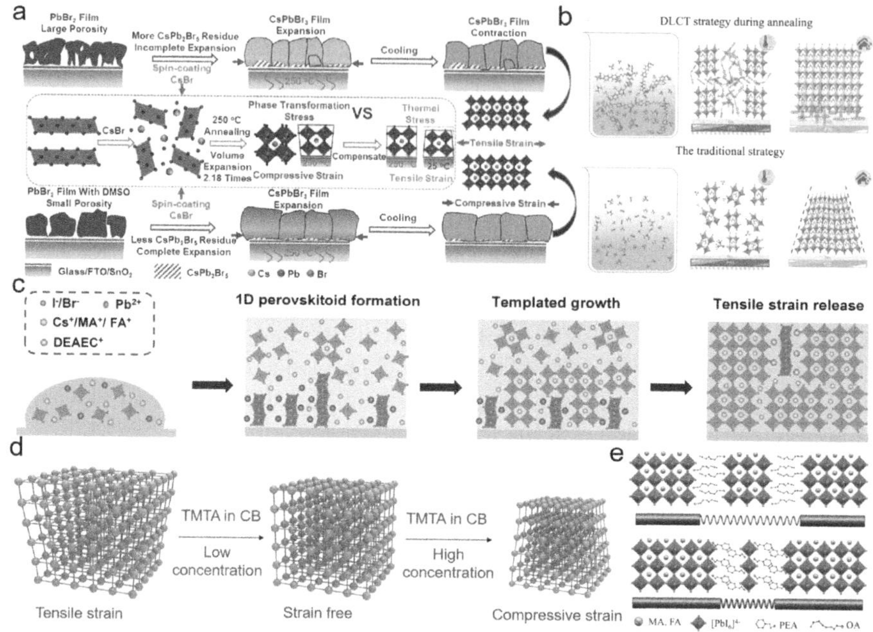

FIGURE 10.15 (a) Schematic diagram of strain regulation mechanism.[102] Copyright 2023, Elsevier. (b) Diagram of the DLCT strategy under different conditions.[103] Copyright 2019, Wiley-VCH GmbH. (c) The schematic illustration of the templated growth process of the 1D@3D perovskite films.[99] Copyright 2021, Wiley-VCH GmbH. (d) The schematic representation of the tensile strain state of the control PSC film regulated by the crosslinking-enabled strain regulating crystallization (CSRC) method with trimethylolpropane triacrylate (TMTA) in different concentrations.[104] Copyright 2021, Wiley-VCH GmbH. (e) Schematic describing the residual stress relaxation with soft and stiff structural subunits.[105] Copyright 2019, Wiley-VCH GmbH.

Furthermore, Kong et al.[99] employed 2-diethylaminoethylchloride hydrochloride (DEAECCl) to fabricate highly crystalline 1D@3D perovskite films, resulting in enhanced stability and a reduction in residual tensile strain (Figure 10.15c). The introduction of 1D perovskite (DEAECPbI$_3$) served as a pre-oriented template, facilitating the directional growth of the 3D perovskite phase. Importantly, this 1D perovskite component also played a critical role in relieving the internal tensile stress within the 1D@3D perovskite film.

Second, the regulation of residual strain in the perovskite film can be achieved through the incorporation of crosslinking molecules. Zhang et al.[104] presented an innovative approach known as in-situ CSRC by introducing TMTA into the antisolvent of chlorobenzene (CB). This strategic addition limited lattice distortion in the top region of the perovskite membrane. Notably, TMTA, with its -OH and C=O functional groups, effectively balanced the overall thermal expansion of the perovskite during annealing by forming cross-links with the perovskite nuclei (Figure 10.15d). As the TMTA concentration increased, the tensile strain in the perovskite film gradually transitioned into compressive strain. This pioneering CSRC approach holds significant potential for precise strain regulation in perovskite films, thereby providing valuable insights into enhancing the performance and stability of perovskite-based devices. In addition, Wang et al.[87] showed the effective use of zwitterions, exemplified by adenosine triphosphate (ATP), at the perovskite boundary to achieve dual interactions with both cations and anions within the perovskite structure. This unique interaction mechanism results in reduced tensile strain through the steric repulsion effect, leading to a transition from tensile stress to compressive stress. Remarkably, this compressive strain hinders device phase separation, leading to a significantly prolonged lifetime of over 2,500 h under 1-sun illumination.

Third, considering the calculation formula of thermal strain, the residual thermal strain is proportionally related to the YM of the perovskite films. Therefore, reducing the YM of perovskite films can effectively alleviate the residual strain within them. Chen et al.[105] reported a stress relaxation strategy for 2D/3D mixed perovskite films by introducing long-chain spacers, such as phenethylammonium (PEA^+) and octylammonium (OA^+), into the 3D perovskite structure. The resulting mixed 2D/3D configuration resembles a "bone-joint" arrangement (Figure 10.15e), enhancing the perovskite films' resistance to external light and thermal stresses through a cushioning effect. The extent of stress release in this setup depends on the length and stiffness of the organic spacer, with OA^+ demonstrating a greater stress-relaxing effect than PEA^+. This behavior is attributed to the fact that the incorporation of larger organic cations leads to the formation of 2D perovskite similar to soft subunits. Specifically, the OA^+ subunits, lacking a rigid phenyl ring, exhibit a "softer" nature compared to PEA^+. As a result, the low YM of 2D perovskite contributes to the overall stress relaxation in the 3D perovskite structure.

10.5.2 Heat Treatment Strategies

The annealing process of perovskite precursor solutions constitutes a pivotal stage in achieving high-performance perovskite films by influencing their supersaturation state. Nonetheless, this process introduces strain due to thermal expansion mismatch between the film and substrate. Therefore, the two most straightforward ways to decrease the tensile strain resulting from the thermal expansion mismatch during the fabrication process are to (i) lower the temperature at which the perovskite film forms or (ii) reduce the thermal coefficient mismatch between the perovskite layer and the substrate.

10.5.2.1 Modified Annealing Process

Huang et al.[45] explored the source of strain within perovskite films, revealing that the $MAPbI_3$ film on an ITO substrate exhibited no strain at 100°C. However, during the cooling process from 100°C to room temperature, XRD peak evolution unveiled gradual emergence of strain in the $MAPbI_3$ films (Figure 10.16a). Since the XRD peak shift corresponds to lattice thermal expansion, the authors measured peak shifts of $MAPbI_3$ SCP across the same temperature range, substantially smaller than those in the film (Figure 10.16b). This discrepancy indicated that the observed peak shift within the perovskite film indeed stemmed from introduced strain. Consequently, the

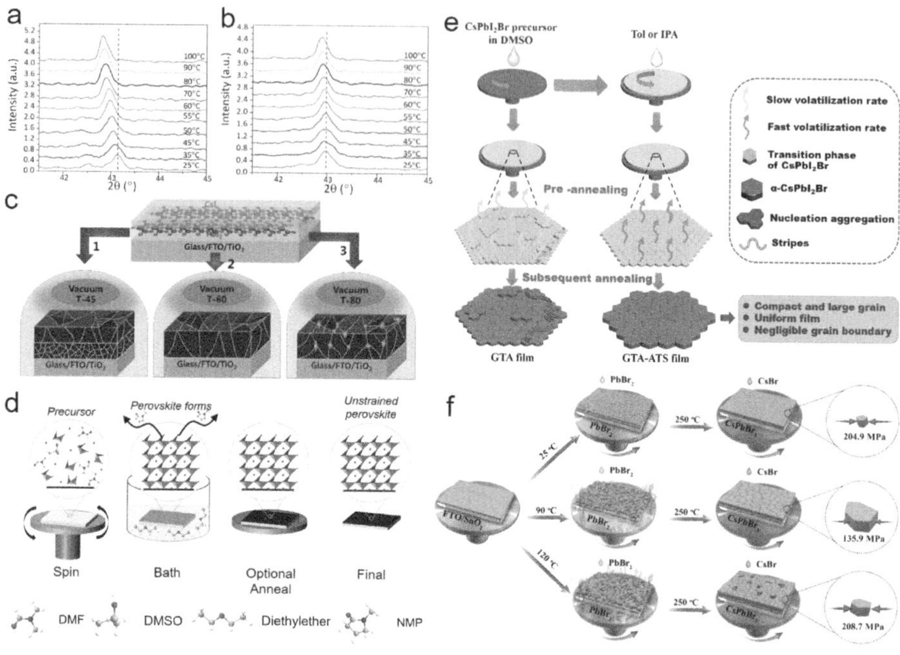

FIGURE 10.16 In-situ out-of-plane XRD of (a) SCP and (b) AF at different temperatures.[12] Copyright 2017, American Association for the Advancement of Science. (c) Schematic diagram of FA-based prelayers by vacuum deposition with annealing temperatures of 45, 60, and 80°C and the corresponding cross-sectional scanning electron microscopy images.[106] Copyright 2021, Royal Society of Chemistry. (d) The formation stages of perovskite films of antisolvent strategy and the bath conversion method.[44] Copyright 2018, Wiley-VCH GmbH. (e) $CsPbI_2Br$ Perovskite Crystallization Process.[107] Copyright 2019, Elsevier. (f) Diagram of converting $PbBr_2$ into perovskites at low, intermediate, and high temperatures as well as compression of perovskites.[108] Copyright 2020, Elsevier.

formation temperature of a perovskite film emerged as a decisive factor influencing its lattice strain. To delve further, the researchers crafted the $MAPbI_3$ perovskite film at room temperature by subjecting it to a 3 d vacuum treatment instead of annealing at 100°C. XRD spectra of the room-temperature-prepared $MAPbI_3$ film aligned with those of unstrained perovskite crystals, solidifying the aforementioned conclusions through direct evidence. Feng et al.[106] unveiled an effective low-temperature annealing deposition technique conducted under vacuum conditions (Figure 10.16c), which yields FA-based perovskite films with fewer defects primarily due to strain alleviation. At temperatures below 45°C, incomplete perovskite reactions lead to the inclusion of small lead iodide particles and a proliferation of grain boundaries in the film. Conversely, temperatures exceeding 80°C induce excess heat, causing perovskite film destruction characterized by prominent cracks and pinholes. Results indicate that annealing at 60°C under a vacuum emerges as the optimal condition for perovskite growth, rendering this technique suitable for producing large-area perovskite films with minimal defects.

Common antisolvent engineering approaches often necessitate annealing to eliminate residual solvents and yield highly crystalline perovskite films. However, such solvent removal usually requires high annealing temperatures, inducing significant tensile strain. Rolston et al.[44] introduced a bath conversion method offering a solution to this challenge by producing less-strained perovskite films at room temperature (Figure 10.16d). In this technique, the spin-coated MAI/PbI_2 film undergoes immersion in diethyl ether (DEE), a low-boiling-point solvent. The darkening of the film indicates that MAI/PbI_2 has transformed into the $MAPbI_3$ perovskite phase. Notably, during subsequent thermal annealing at temperatures of 25°C, 60°C, and 100°C, the residual stress measures 2.6±14.5 MPa, -3.2±19.0 MPa, and 3.0 ± 14.5MPa, respectively. Remarkably, the annealing temperature does not impact the residual strain of the perovskite film prepared through bath conversion. This phenomenon is attributed to a robust chemical interaction between the substrate and the precrystallized perovskite during the bath conversion process. This interaction restricts the volume expansion and contraction of the perovskite crystal that typically occurs during annealing. Furthermore, Chen et al.[107] ascertained that employing isopropanol (IPA) possessing suitable volatility and viscosity facilitates the elimination of residual DMSO from the precursor (Figure 10.16e). This process enables the even deposition and dispersion of the perovskite precursor onto the substrate, thereby enabling meticulous regulation over the consistent crystallization and morphology of $CsPbI_2Br$ films. Consequently, this approach fosters the growth of larger nucleated grains and mitigates the influence of solvent at the film's edges.

Addressing local lattice strain originating from the inhomogeneity in mixed halide perovskites during annealing, Chen et al.[38] devised a novel approach to perovskite annealing and introduced a flipped annealing method to manipulate the gradient in-plane strain of the perovskite film. This pioneering strategy involved the addition of a flipped annealing step to the conventional process, resulting in the introduction of compressive strain into the film. As a result, the lattice structure of the perovskite film exhibited remarkable homogeneity. Ultimately, this strain-alleviated approach yielded a strain-free PSC with a certified PCE of 20.7%. Zhao et al.[108] achieved precise strain regulation within $CsPbBr_3$ perovskite films by meticulously controlling the annealing process of the $PbBr_2$ precursor (Figure 10.16f). This approach effectively diminished strain-induced grain boundaries and mitigated pinhole-induced nonradiative recombination. Central to this strategy was the regulation of the crystallization temperature of $PbBr_2$, ensuring the formation of a suitably porous $PbBr_2$ film. This porous film, attributed to a lattice-volume-expansion effect, served as the foundation for crafting high-quality $CsPbBr_3$ perovskite films. Through adept control over compressive strain induced by volume expansion during the $CsPbBr_3$ phase transition, the authors successfully amplified grain size and reduced grain boundary density. Remarkably, this strain-tailored approach led to the realization of an all-inorganic $CsPbBr_3$ PSC with minimal compressive strain, achieving an impressive champion PCE of 10.71% alongside an ultrahigh voltage of 1.622 V.

10.5.2.2 Heat Treatment Strategies

Post-treatment of perovskite films utilizing HTLs characterized by high thermal expansion coefficients represents a potent avenue for effective thermal treatment.

Xue et al.[20] reported a strain compensation strategy rooted in thermal expansion mismatch. By leveraging an HTL endowed with substantial thermal expansion coefficient and robust binding affinity, they performed post-treatment on $CsPbI_2Br$ films. This process introduced external compressive strain, effectively counteracting the thermally induced tensile strain. Notably, by altering the HTL processing temperature, they adeptly adjusted the internal strain of the $CsPbI_2Br$ film, transitioning from stretching to compression (Figure 10.17a, b). This well-calibrated compressive strain within $CsPbI_2Br$ perovskite films engendered heightened stability. Impressively, the films maintained 95% of their initial power conversion efficiency at the maximum power point for 60 h and retained 96% efficiency after enduring 1,000 h of heating at 85°C.

While the incorporation of a 2D layer atop the 3D perovskite configuration proves effective in minimizing surface defects, the lattice expansion inherent in stretched 2D perovskite layers can compromise the stability of the 2D/3D heterostructure. Addressing this challenge, Zhang et al.[109] demonstrated a strain compensation strategy for 2D PEA_2PbI_4 perovskite layers. They introduced PCBM with a notably lower thermal expansion coefficient than that of PEA_2PbI_4, serving as an external compressive strain layer to counteract lattice expansion. Intriguingly, the $[PbI_6]^{4-}$ octahedra within PEA_2PbI_4, sandwiched between the 3D perovskite and PCBM layer, exhibited minimal movement. Additionally, the PCBM-iodide interaction anchored PEAI, effectively curbing ion migration and stabilizing the 2D/3D heterostructure. The synergistic interplay of diffusion passivation and stress compensation led to an exceptional champion device power conversion efficiency of 21.31%. Impressively, the (002) peak of the 2D perovskite film exhibited only marginal variation after undergoing thermal annealing for 253 h.

10.5.3 INTERFACIAL MANAGEMENT

Although pursuing the fabrication of perovskite films at lower temperatures or employing substrates characterized by elevated thermal expansion coefficients can indeed mitigate thermal strain within perovskite films, achieving both high-quality perovskite films and efficient PSCs using these approaches often presents substantial challenges.[21] Consequently, there exists a compelling need to focus on modifying the interfaces of the ETL/perovskite and perovskite/HTL. Developing suitable interface materials has emerged as a prominent approach for strain regulation in recent years.

10.5.3.1 ETL/Perovskite Interface

To mitigate interfacial strain between the ETL and the perovskite layer, Zhou et al.[110] innovated a $WS_2/CsPbBr_3$ van der Waals heterostructure. This involved integrating a 2D WS_2 nanosheet at the $SnO_2\text{-}TiO_xCl_{4-2x}/CsPbBr_3$ interface, effectively alleviating tensile strain in the $CsPbBr_3$ film. Leveraging matched lattices, a high-quality $CsPbBr_3$ perovskite film grew atop the WS_2 nanoflakes through van der Waals epitaxy (Figure 10.18a). During the cooling process, the feeble $WS_2\text{-}CsPbBr_3$ interaction acted akin to a "lubricant," curbing perovskite lattice contraction and significantly reducing tensile strain (Figure 10.18b). Furthermore, this interface strain regulation concurrently curbed ion migration, enhancing ion migration activation energy.

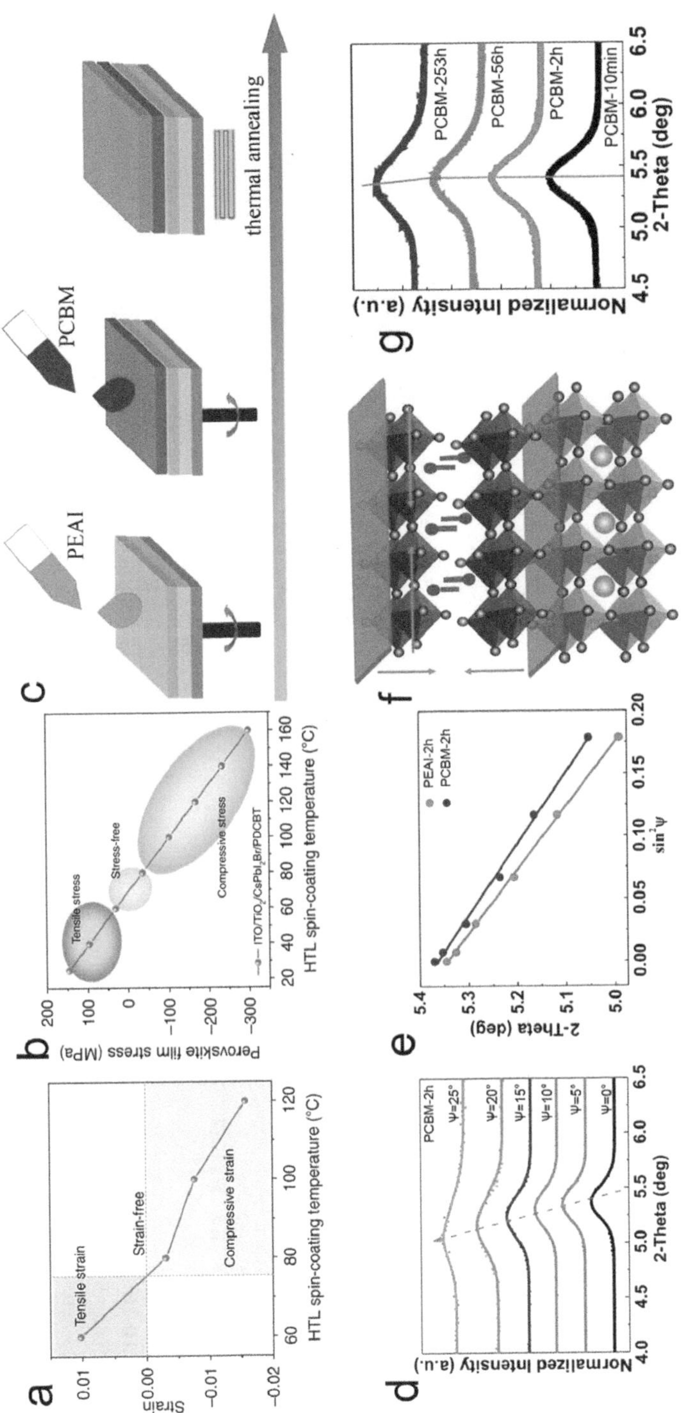

FIGURE 10.17 (a) Measured strain in perovskite films coated with PDCBT HTLs at different spin-coating temperatures. (b) The calculated net average stress in perovskites within structures consisting of ITO/TiO₂/perovskite/PDCBT as a function of PDCBT spin-coating temperature.[20] Copyright 2020, Springer Nature. (c) Optimized 2D/3D perovskite processing procedures. (d) GIXRD spectrums at different tilt angles for the PCBM-2 h film. (e) Residual tensile stresses of PEA₂PbI₄ under different preparation technology. (f) Schematic diagram for stress compensation caused by covering PCBM. (g) Magnified diffraction peaks of (002) plane from 2D perovskite prepared by optimized method under varying annealing time.[109] Copyright 2020, Wiley-VCH GmbH.

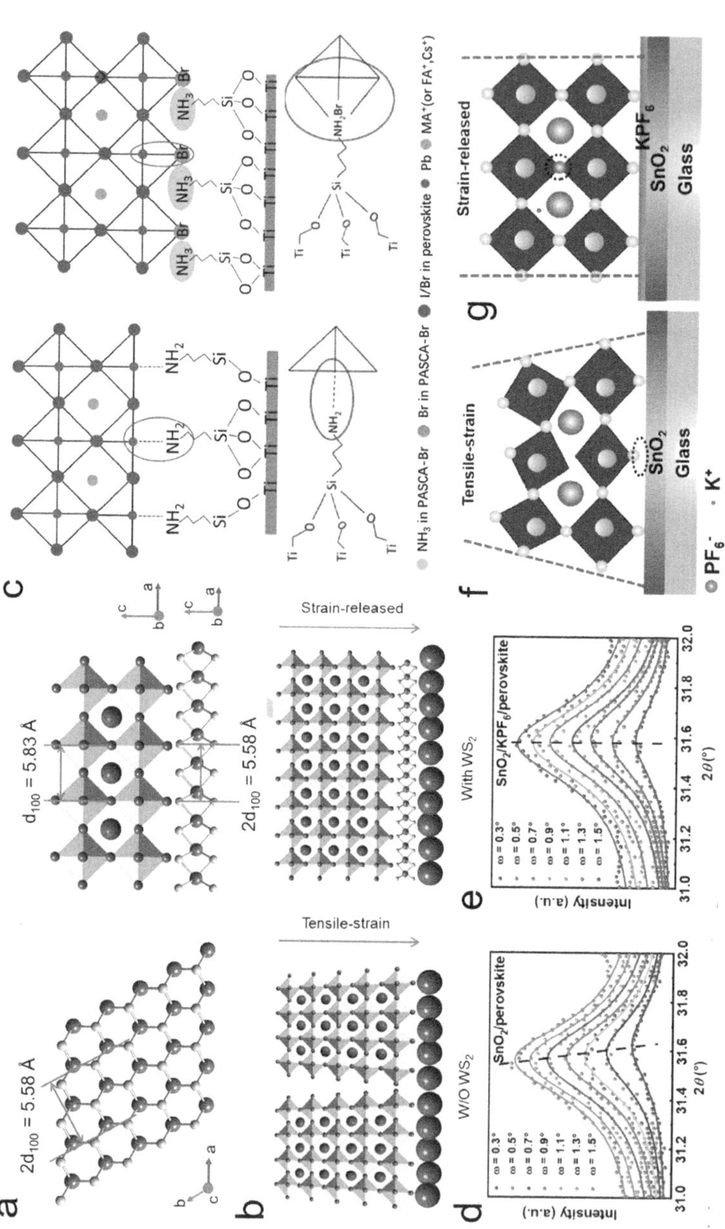

FIGURE 10.18 (a) Atomic crystal structure of the (100) plane of WS_2, and the side view of atomic crystal heterojunction of $CsPbBr_3$ and WS_2. (b) Schematic diagram of residual strain distribution in $CsPbBr_3$ grains with and without WS_2 interlayers.[110] Copyright 2020, Wiley-VCH GmbH. (c) Protonated amino terminals (R–NH_3Br) and lattice structure of APTES and PASCA-Br modified interfaces.[112] Copyright 2020, Wiley-VCH GmbH. Depth-dependent GIXRD patterns of the perovskite films based on (d) SnO_2 and (e) SnO_2/KPF_6. Schematic illustration of residual stress of the perovskite films prepared on (f) SnO_2 and (g) KPF_6-modified SnO_2.[113] Copyright 2021, Elsevier.

Impressively, PSCs with $WS_2/CsPbBr_3$ heterostructures attained a PCE of 10.65%, an ultrahigh V_{oc} of 1.70 V and exceptional stability under both air exposure and continuous light irradiation. Furthermore, Meng et al.[111] explored introducing a compliant buffer layer between SnO_2 and perovskite to mitigate residual tensile strain. By leveraging the flexible attributes of polystyrene (PS) at 150°C, they prevented direct substrate contact, efficiently alleviating thermally induced residual strain within the perovskite film.

The robust anchoring of interface materials via bonding is a notable strain regulation approach. Zhang et al.[112] introduced a protonated amine silane coupling agent (PASCA-Br) as an intermediary layer between TiO_2 and perovskite. PASCA-Br ingeniously engages both layers: Si terminals bonding with TiO_2, while $R-NH_3Br$ terminals mend disrupted perovskite octahedra (Figure 10.18c). Remarkably, the pliable $R-NH_3Br$ not only provides well-matched perovskite growth sites – reducing interfacial strain and subsequent lattice distortion – but also acts as a mechanical buffer during bending, curbing perovskite film cracking. This strategy yielded PSCs with impressive efficiencies of 21.66% and 17.45% on solid and flexible substrates, respectively.

Zhang et al.[49] introduced a molecular tailoring strategy by altering the anionic configuration (CO_3^{2-}, $C_2O_4^{2-}$, and $HCOO^-$) of Li salts and meticulously examined its impact on the $SnO_2/FAPbI_3$ interface. Optimally arranged C-O and C=O groups forge potent bonds with uncoordinated Sn^{4+} and FA^+ species, respectively, liberating residual stress within the $FAPbI_3$ lattice. Notably, CO_3^{2-} outperforms the other anions in efficiently passivating FA^+ defects, owing to the robust interactions stemming from triangular C=O and two C-O configurations with adjacent layers, resulting in the lowest defect density. This defect reduction markedly curtails lattice distortion and effectively mitigates perovskite film strain. Bi et al.[113] reported a versatile interface modification strategy employing KPF_6 molecules for $SnO_2/$perovskite interface enhancement. This strategy involves PF^{6-} positioned at the interface, establishing hydrogen bonding with perovskites and coordinating with SnO_2, while the majority of K^+ ions infiltrate the perovskite layer to neutralize negative defects. GIXRD measurements unveil near-complete release of residual stress in KPF_6-modified perovskite films (Figure 10.18d, e), closely tied to interface defect passivation induced by KPF_6 (Figure 10.18f, g). Ultimately, this KPF_6-modified approach yields an exemplary 21.39% PCE champion device.

10.5.3.2 Perovskite/HTL Interface

From the top surface, the residual strain within perovskite films can be regulated. Liu et al.[114] introduced a strategy for interfacial defect passivation and stress relief by employing (5-Mercapto-1,3,4-thiadiazol-2-ylthio) acetic acid (MTDAA) between the perovskite and HTL. Capitalizing on MTDAA's diverse active sites (N, O, and S electron donors), robust chemical interactions with uncoordinated Pb^{2+} ions on the film surface or at grain boundaries are induced (Figure 10.19a). This interaction substantially reduces perovskite film defect density, leading to full release of interface residual stress. Moreover, device stability against light, heat, and moisture is considerably enhanced. Chen et al.[115] introduced a method utilizing symmetric 1,1'-(Methylenedi-4,1-phenylene) bismaleimide (BMI) molecules interacting with

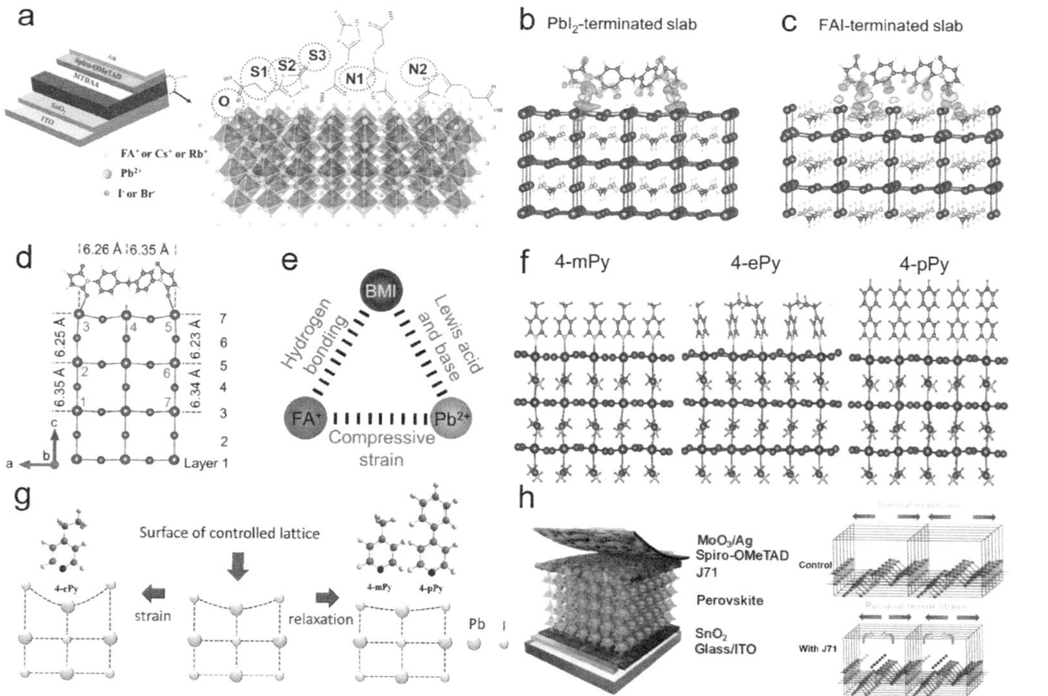

FIGURE 10.19   (a) Device structure and a schematic diagram of defect passivation by MTDAA.[114] Copyright 2021, American Chemical Society. DFT calculations of charge difference between BMI and perovskite with (b) PbI_2-terminated slabs and (c) FAI-terminated slabs. d) Microstructure of Pb-I frame plane perpendicular to the b-axis, where two Pb atoms bind with the BMI. (e) Schematic mechanisms of a stable triangle structure to reinforce the stability of $FAPbI_3$ with BMI.[115] Copyright 2022, Wiley-VCH GmbH. (f) DFT simulation of the steric arrangement of controlled and Pyderivates modified perovskite supercells. (g) Schematic diagram of the surface geometry of $MAPbI_3$ structure treated with different pyridine derivatives.[116] Copyright 2020, Elsevier. (h) Illustration of as-prepared $CsPbI_2Br$ solar cells with the structure of $ITO/SnO_2/perovskite/J71/spiro-OMeTAD/MoO_3/Ag$, molecular formula of J71 Crystal mechanics model of control and $CsPbI_2Br/J71/Spiro-OMeTAD/MoO_3/Ag$.[117] Copyright 2022, Elsevier.

the surface of FA-based perovskite, inducing surface state alterations and converting residual tensile strain to compressive strain. Notably, the length of two maleimides in BMI is twice the spacing of the $FAPbI_3$ lattice, enabling one BMI molecule to bond with two $FAPbI_3$ devices through robust Lewis acid-base interactions and hydrogen bonding with Pb^{2+} and FA^+ (Figure 10.19b, c). The authors computationally assessed structural changes after BMI chemical interaction (Figure 10.19d, e), revealing BMI's impact on the perovskite surface that compresses the distance between Pb atoms, below the pristine unit cell lattice parameter a (6.41 Å), signifying compressed strain along the a-axis. Simultaneously, BMI bears tensile strain from the perovskite, evident by elongation of the distance between BMI's O atoms and Pb to 10.70 Å, contrasting the initial value of 10.63 Å. The BMI-based device achieved a champion PCE of 22.7%, with improved stability against light, moisture, and thermal aging.

By using a series of pyridine derivatives (Py-derivatives), Wu et al.[116] achieved dual control over crystal kinetics and lattice stabilization in $MAPbI_3$. DFT calculations unveiled the geometric layout of various pyridine derivatives (e.g., 4-methylpyridine (4-mPy), 4-ethylpyridine (4-ePy), and 4-phenylpyridine (4-pPy)) on the $MAPbI_3$ surface (Figure 10.19f). Outcomes revealed that planar 4-pPy and 4-mPy demonstrated ordered orientation when anchored on the $MAPbI_3$ surface. The stiffer 4-mPy formed uniform chemical interactions with the $MAPbI_3$ framework, thereby achieving lattice surface strain relaxation. Conversely, the twisted orientation of 4-ePy due to steric effects from alkyl (ethyl) groups between adjacent ligands aggravated lattice strain through asymmetric chemical forces (Figure 10.19g). Consequently, the 4-pPy-based device attained a certified PCE of 20.02% and exhibited enduring stability across diverse harsh environments. Zheng et al.[117] introduced the organic polymer J71 containing N and S atoms to alleviate residual strain in the $CsPbI_2Br$ film (Figure 10.19h). Through J71 modification, the (211) plane d-spacing of $CsPbI_2Br$ decreased from the pristine 4.96 Å to 4.24 Å. This adjustment was due to the N and S elements within J71 bonding with Pb in the perovskite, generating force interactions that effectively counterbalanced the tensile stress within the crystal. Their findings demonstrated that J71 could shift the strain type of CsPbI2Br from tensile to compressive, enabling a strain-free film with an optimal J71 concentration.

10.5.4 EXTERNAL STRAIN REGULATION

Through controlled external strain manipulation, including pressure adjustments through variations in type and magnitude, as well as light management strategies involving intensity and density modifications, it becomes achievable to enhance both the optoelectronic attributes and stability of PSCs. The ensuing discourse offers a systematic exploration, predominantly focused on pressure-assisted approaches and light soaking techniques.

10.5.4.1 External Mechanical Force

Nie et al.[118] introduced mechanical stress into perovskite films to explore its impact on photovoltaic performance and stability. They achieved both tensile and compressive strains in $MAPbI_3$ films through convex and concave bending, respectively, with strain intensifying in tandem with bending curvature (Figure 10.20a). Through

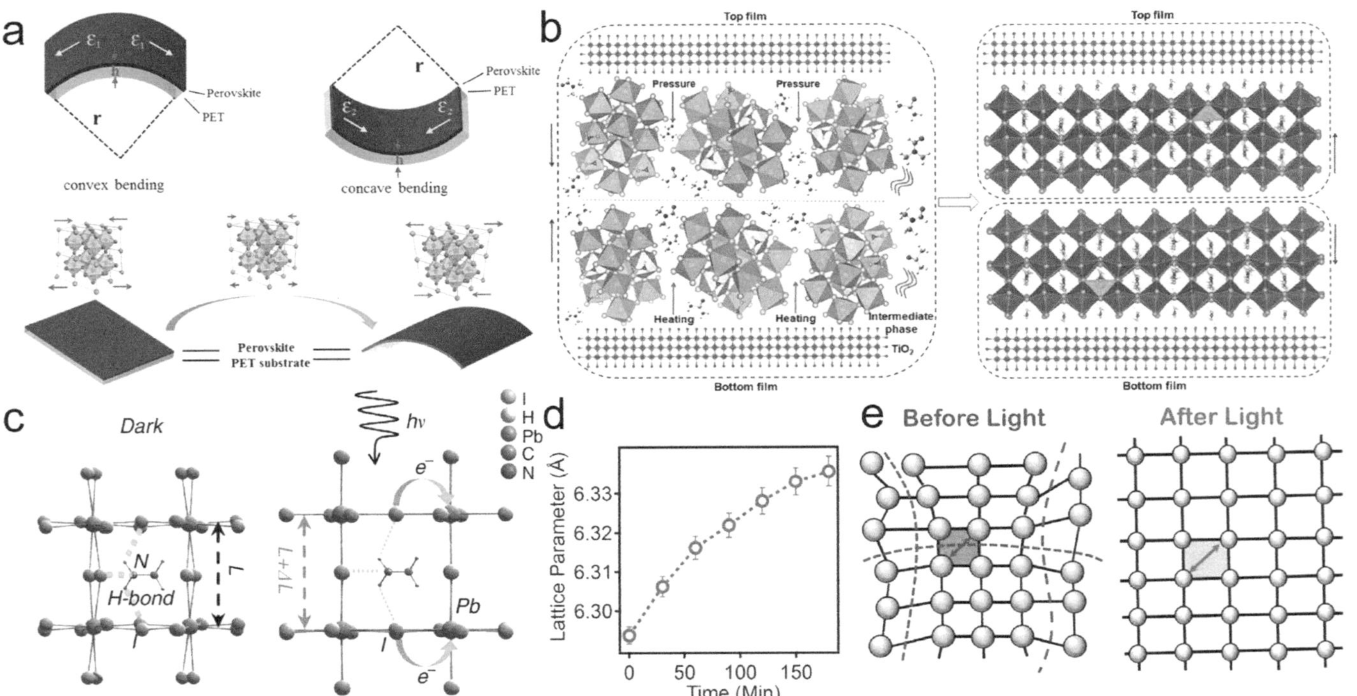

FIGURE 10.20 (a) Schematic diagrams for convex, concave bending, and lattice strain changes during a convex bending process.[118] Copyright 2020, Royal Society of Chemistry. (b) Schematic diagrams for convex, concave bending, and lattice strain changes during a convex bending process.[119] Copyright 2020, Elsevier. (c) Schematic illustrations (not to scale) showing that the weakening of the hydrogen bonding between the amine group and the iodine ion by photo-generated carriers leads to the lattice expansion.[13] Copyright 2016, Springer Nature. (d) Lattice constant as a function of illumination time. (e) Schematic illustration for the crystal structure change before (local distortion) and after illumination (lattice expansion).[120] Copyright 2018, American Association for the Advancement of Science.

time-resolved photoluminescence (TRPL) analysis, they observed prolonged carrier lifetimes under tensile strain and shortened lifetimes under compressive strain. This outcome stemmed from the elongation of lattice constants in the case of tensile strain, enhancing carrier delocalization, while compressive strain led to lattice contraction, curbing carrier delocalization. Thus, the activation energy of ion migration is quantified through the Nernst-Einstein relationship:

$$\sigma_T = \sigma_0 \exp\left(\frac{-E_a}{kT}\right) \tag{10.6}$$

where σ_0 and k are the constants, σ and T are conductivity and temperature, respectively, and E_a is the activation energy for ion migration. E_a is derived from the slope of the $\ln(\sigma_T)$ versus $1/T$ plot. The activation energies for ion migration in MAPbI$_3$ films vary in the dark as a result of strain. The activation energies for tensile-strain, non-strain, and compressive-strain films are 0.29, 0.39, and 0.53 eV, respectively. Upon exposure to white light at an intensity of 25 mW/cm², these values decrease to 0.046, 0.074, and 0.083 eV, respectively. These findings highlight that tensile strain reduces the ion migration activation energy of perovskite films both in dark and illuminated environments, whereas compressive strain adversely impacts device stability.

Luo et al.[121] introduced a pressure-assisted solution treatment (PASP) method to enhance perovskite properties. In this approach, two perovskite precursor films were pressed together under varying pressures, followed by annealing at 110°C for 60 min. The resulting perovskite film was separated from the substrate, leaving the treated bottom layer (Figure 10.20b). By adjusting the applied pressure, nucleation and growth were controlled, leading to larger grain sizes and reduced non-radiative recombination. The PASP-treated PSC achieved a remarkable 20.74% PCE, maintaining over 90% initial PCE even after 60 d of aging and 200 h of continuous 1-sun illumination in an unencapsulated ambient environment.

10.5.4.2 Illumination Force

Light-induced structural changes play a pivotal role in shaping the optoelectronic traits and stability of perovskites. Zhou et al.[13] unveiled a photostrictive phenomenon in MAPbI$_3$, unveiling that intense photons can couple with the lattice, inducing light-induced lattice strain. This occurs as photogenerated carriers weaken the hydrogen bonds, specifically the interaction between the amine group (N) and the iodine ion (I) (Figure 10.20c). Alternatively, the expansion of the MAPbI$_3$ lattice curbs the distortion of the Pb-I-Pb bond and the iodine octahedron, ultimately reducing the rotational barrier of the MA dipole. This reduction in rotational barrier influences the local band structure of MAPbI$_3$, leading to a suppression of electron-hole recombination and enhancing the material's optoelectronic performance. Tsai et al.[122] also documented light-induced lattice strain in FA$_{0.7}$MA$_{0.25}$Cs$_{0.05}$PbI$_3$ perovskite and scrutinized the impact of external light on the perovskite structure. Continuous illumination for 180 min didn't show any phase segregation or degradation. However, the lattice constant d shows a uniform increase from 6.290 ± 0.002 Å to 6.330 ± 0.004 Å with extended illumination, indicating perovskite film lattice expansion and relaxation of local strain (Figure 10.20d). It is suggested that the perovskite film develops local strain due to its distorted lattice accommodating cations of varying sizes. Light exposure prompts lattice expansion, which in turn alleviates the local strain (Figure 10.20e). This reduction in strain weakens non-radiative recombination and enhances V_{oc} and FF, resulting in an elevated PCE of the device, advancing from 18.5% to 20.5%.

10.6　CONCLUSIONS AND PERSPECTIVES

In summary, this comprehensive chapter provides a systematic overview of the definition, origins, characterization methods, implications, and regulation strategies for strain engineering in PSCs. Strain in perovskite arises from both internal and external factors. Internal strain is generated within halide perovskite crystals due to lattice non-periodicity, specifically through $[BX_6]^{4-}$ octahedral tilt and heterogeneous crystallization. External factors such as lattice mismatch, thermal expansion discrepancies, and external stimuli like illumination, temperature, electrical bias, and external pressure can also impose periodic distortion on the crystal lattice. Various characterization methods, including XRD, GIWAXS, Raman spectroscopy, TEM, SED, EBSD, and PFM, can be used to detect and analyze strain in perovskite films. The impacts of strain on perovskite films and device performance are multifaceted, influencing aspects such as band structure, defect properties, ion migration, phase stability, and device stability. This chapter also outlines different approaches and strategies for regulating strain in perovskite films. These strategies can be categorized into compositional engineering, heat treatment strategies, interface management, and external strain regulation methods. Innovating film fabrication methods; introducing additives or alloys; optimizing interface materials; and implementing solvent engineering, antisolvent strategies, component engineering, and additives are some examples of these strategies.

In the pursuit of highly effective and stable PSCs towards ultimate commercialization it becomes imperative to delve deeper into the sources, effects, and regulators of strain in light of the significance of strain for PSCs. Several strain regulation strategies can potentially enhance the stability and performance of PSCs. (i) Innovating film fabrication methods for lattice matching through low-temperature processes or high-α-substrate exploration to mitigate lattice strain during crystallization. (ii) Introducing additives or alloys into perovskite compositions for enhanced defect passivation and stability, along with interface materials innovation for strain adjustment without structural changes. (iii) Integrating process control, component design, interface engineering, and external strain regulation through solvent engineering, antisolvent strategies, component engineering, and additives to alleviate strain. (iv) Employing advanced testing and first-principles analysis to unravel strain release and stability mechanisms across crystal nucleation, growth, and PSCs.

In order to achieve the successful commercialization of PSCs, it is crucial to conduct a comprehensive investigation into the sources, effects, and regulators of strain. This is of particular importance due to the significant impact that strain has on the performance of PSCs. There are several potential strategies for regulating strain that can enhance the stability and efficiency of PSCs. First, innovation in film fabrication methods can be employed to match the lattice structure by utilizing low-temperature processes or exploring high-α-substrates, thereby mitigating lattice strain during crystallization. Second, introducing additives or alloys into the perovskite compositions can enhance defect passivation and stability, while innovations in interface materials can allow for strain adjustment without causing structural changes. Third, integrating process control, component design, interface engineering, and external strain regulation through techniques such as solvent engineering, antisolvent strategies, component engineering, and the use of additives can effectively alleviate strain. Last, employing advanced testing and first-principles analysis can unravel the mechanisms of strain release and stability throughout crystal nucleation, growth, and the functioning of PSCs.

ACKNOWLEDGMENT

This work was supported by the National Natural Science Foundation of China (62104136, 22179051), Project of Shandong Province Higher Educational Young Innovative Team (2022KJ218), China Postdoctoral Science Foundation (2023M732104), Qingdao Postdoctoral Funding Program (QDBSH20220201002), Postdoctoral Innovation Project of Shandong Province (SDCX-ZG-202303032).

REFERENCES

1. S. D. Stranks, H. J. Snaith, Metal-halide perovskites for photovoltaic and light-emitting devices, *Nature Nanotechnology* **2015**, *10*, 391–402.
2. Y. Takahashi, *et al.*, Charge-transport in tin-iodide perovskite $CH_3NH_3SnI_3$: Origin of high conductivity, *Dalton Transactions* **2011**, *40*(20), 5563–5568.
3. J. H. Im, C. R. Lee, J. W. Lee, S. W. Park, N. G. Park, 6.5% efficient perovskite quantum-dot-sensitized solar cell, *Nanoscale* **2011**, *3*(10), 4088–4093.
4. Y. Li, *et al.*, Efficient, stable formamidinium-cesium perovskite solar cells and minimodules enabled by crystallization regulation, *Joule* **2022**, *6*(3), 676–689.
5. A. Kojima, K. Teshima, Y. Shirai, T. Miyasaka, Organometal halide perovskites as visible-light sensitizers for photovoltaic cells, *Journal of the American Chemical Society* **2009**, *131*(17), 6050–6051.
6. L. Chen, *et al.*, Incorporating potassium citrate to improve the performance of tin-lead perovskite solar cells, *Advanced Energy Materials* **2023**, *13*(32), 2301218.
7. W. Mao, *et al.*, Visualizing phase segregation in mixed-halide perovskite single crystals, *Angewandte Chemie International Edition* **2019**, *58*(9), 2893–2898.
8. T. A. S. Doherty, *et al.*, Stabilized tilted-octahedra halide perovskites inhibit local formation of performance-limiting phases, *Science* **2021**, *374*(6575), 1598–1605.
9. C. C. Boyd, R. Cheacharoen, T. Leijtens, M. D. McGehee, Understanding degradation mechanisms and improving stability of perovskite photovoltaics, *Chemical Reviews* **2019**, *119*(5), 3418–3451.
10. J. T. W. Wang, *et al.*, Efficient perovskite solar cells by metal ion doping, *Energy & Environmental Science* **2016**, *9*, 2892–2901.
11. S. Jariwala, *et al.*, Local crystal misorientation influences non-radiative recombination in halide perovskites, *Joule* **2019**, *3*(12), 3048–3060.

12. J. Zhao, *et al.*, Strained hybrid perovskite thin films and their impact on the intrinsic stability of perovskite solar cells, *Science Advances* **2017**, *3*(11), eaao5616.
13. Y. Zhou, *et al.*, Giant photostriction in organic-inorganic lead halide perovskites, *Nature Communications* **2016**, *7*, 11193.
14. B. Conings, *et al.*, Intrinsic thermal instability of methylammonium lead trihalide perovskite, *Advanced Energy Materials* **2015**, *5*(2), 1500477.
15. X. Lü, *et al.*, Enhanced structural stability and photo responsiveness of $CH_3NH_3SnI_3$ perovskite via pressure-induced amorphization and recrystallization, *Advanced Materials* **2016**, *28*, 8663–8668.
16. R. A. Z. Razera, *et al.*, Instability of p-i-n perovskite solar cells under reverse bias, *Journal of Materials Chemistry A* **2020**, *8*(1), 242–250.
17. D. P. Nenon, *et al.*, Structural and chemical evolution of methylammonium lead halide perovskites during thermal processing from solution, *Energy & Environmental Science* **2016**, *9*(6), 2072–2082.
18. X. Shai, *et al.*, Achieving ordered and stable binary metal perovskite via strain engineering, *Nano Energy* **2018**, *48*, 117.
19. T. W. Jones, *et al.*, Lattice strain causes non-radiative losses in halide perovskites, *Energy & Environmental Science* **2019**, *12*(2), 596–606.
20. D. J. Xue, *et al.*, Regulating strain in perovskite thin films through charge-transport layers, *Nature Communication* **2020**, *11*, 1514.
21. Y. Chen, *et al.*, Strain engineering and epitaxial stabilization of halide perovskites, *Nature* **2020**, *577*, 209–251.
22. X. Shai, W. Chen, J. Sun, J. Liu, J. Chen, Origins, impacts, and mitigation strategies of strain in efficient and stable perovskite solar cells, *Small Science* **2023**, *3*, 2300014.
23. J. Sun, Y. Jiang, X. Zhong, J. Hu, L. Wan, Three-dimensional nanostructured electrodes for efficient quantum-dot-sensitized solar cells, *Nano Energy* **2016**, *32*, 130–156.
24. Y. Fu, H. Zhu, J. Chen, M. P. Hautzinger, X. Zhu, S. Jin, Metal halide perovskite nanostructures for optoelectronic applications and the study of physical properties, *Nature Reviews Materials* **2019**, *4*, 169.
25. Y. Jiao, *et al.*, Strain engineering of metal halide perovskites on coupling anisotropic behaviors, *Advanced Functional Materials* **2020**, *31*, 2006243.
26. G. Han, *et al.*, Additive selection strategy for high performance perovskite photovoltaics, *Journal of Physical Chemistry C* **2018**, *122*, 138840–13893.
27. Y. Dang, *et al.*, Bulk crystal growth of hybrid perovskite material $CH_3NH_3PbI_3$. *CrystEngComm* **2015**, *17*, 665.
28. N. K. McKinnon, D. C. Reeves, M. H. Akabas, 5-HT3 receptor ion size selectivity is a property of the transmembrane channel, not the cytoplasmic vestibule portals, *Journal of General Physiology* **2011**, *138*, 453–466.
29. G. Kim, H. Min, K. S. Lee, D. Y. Lee, S. M. Yoon, S. I. Seok, Impact of strain relaxation on performance of α-formamidinium lead iodide perovskite solar cells, *Science* **2020**, *370*, 108–112.
30. W. Travis, E. N. K. Glover, H. Bronstein, D. O. Scanlon, R. G. Palgrave, On the application of the tolerance factor to inorganic and hybrid halide perovskites: A revised system, *Chemical Science* **2016**, *7*, 4548–4556.
31. X. Zheng, C. Wu, S. K. Jha, Z. Li, K. Zhu, S. Priya, Improved phase stability of formamidinium lead triiodide perovskite by strain relaxation, *ACS Energy Letters* **2016**, *1*(5), 1014–1102.
32. D. Liu, W. Zha, Y. Guo, R. Sa, Insight into the improved phase stability of $CsPbI_3$ from first-principles calculations, *ACS Omega* **2019**, *5*(1), 893–896.
33. M. I. Saidaminov, *et al.*, Suppression of atomic vacancies via incorporation of isovalent small ions to increase the stability of halide perovskite solar cells in ambient air, *Nature Energy* **2018**, *3*, 648–654.

34. F. Bertolotti, *et al.*, Coherent nanotwins and dynamic disorder in cesium lead halide perovskite nanocrystals, *ACS Nano* **2017**, *11*, 3819–3831.

35. J. H. Lee, N. C. Bristowe, J. H. Lee, S. H. Lee, P. D. Bristowe, A. K. Cheetham, H. M. Jang, Resolving the physical origin of octahedral tilting in halide perovskites, *Chemistry of Materials* **2016**, *28*, 4259–4266.

36. L. J. Ji, S. J. Sun, Y. Qin, K. Li, W. Li, Mechanical properties of hybrid organic-inorganic perovskites, *Coordination Chemistry Reviews* **2019**, *391*, 15–29.

37. J. Wu, S. C. Liu, Z. Li, S. Wang, D. J. Xue, Y. Lin, J. S. Hu, Strain in perovskite solar cells: Origins, impacts and regulation, *National Science Review* **2021**, *8*(8), nwab047.

38. C. Zhu, *et al.*, Strain engineering in perovskite solar cells and its impacts on carrier dynamics, *Nature Communications* **2019**, *10*(1), 815.

39. I. Mela, *et al.*, Revealing nanomechanical domains and their transient behavior in mixed-halide perovskite films, *Advanced Functional Materials* **2021**, *31*, 2100293.

40. J. P. Correa-Baena, M. Saliba, T. Buonassisi, M. Grätzel, A. Abate, W. Tress, A. Hagfeldt, Promises and challenges of perovskite solar cells, *Science* **2017**, *358*(6364), 739–744.

41. T. W. Jones, *et al.*, Lattice strain causes non-radiative losses in halide perovskites, *Energy & Environmental Science* **2019**, *12*(2), 596–606.

42. D. Liu, *et al.*, Strain analysis and engineering in halide perovskite photovoltaics, *Nature Materials* **2021**, *20*, 1337–1346.

43. V. Craciun, D. Craciun, X. Wang, T. J. Anderson, R. K. Singh, Transparent and conducting indium tin oxide thin films grown by pulsed laser deposition at low temperatures, *Journal of Optoelectronics and Advanced Materials* **2003**, *5*, 401–408.

44. L. A. Muscarella, *et al.*, Lattice compression increases the activation barrier for phase segregation in mixed-halide perovskites, *ACS Energy Letters* **2020**, *5*, 3152–3158.

45. D. Kim, *et al.*, Light- and bias-induced structural variations in metal halide perovskites, *Nature Communications* **2019**, *10*, 444.

46. N. Rolston, *et al.*, Engineering stress in perovskite solar cells to improve stability, *Advanced Energy Material* **2018**, *8*, 1802139.

47. M. K. Khenkin, *et al.*, Consensus statement for stability assessment and reporting for perovskite photovoltaics based on ISOS procedures, *Nature Energy* **2020**, *5*, 35–49.

48. Y. Liu, B. G. Sumpter, J. K. Keum, B. Hu, M. Ahmadi, O. S. Ovchinnikova, Strain in metal halide perovskites: The critical role of A-site cation, *ACS Applied Energy Materials* **2021**, *4*, 2068–2072.

49. Y. Zhang, *et al.*, Molecularly tailored SnO_2/perovskite interface enabling efficient and stable $FAPbI_3$ solar cells, *ACS Energy Letters* **2022**, *7*, 929–938.

50. F. Li, *et al.*, Regulating surface termination for efficient inverted perovskite solar cells with greater than 23% efficiency, *Journal of the American Chemical Society* **2020**, *142*(47), 20134–20142.

51. J. A. Steele, *et al.*, How to GIWAXS: Grazing incidence wide angle X-ray scattering applied to metal halide perovskite thin films, *Advanced Energy Material* **2023**, *13*, 2300760.

52. C. Wang, C. Zuo, Q. Chen, L. Ding, GIWAXS: A powerful tool for perovskite photovoltaics, *Journal of Semiconductor* **2021**, *42*(6), 060201.

53. M. Qin, P. F. Chan, X. Lu, A systematic review of metal halide perovskite crystallization and film formation mechanism unveiled by In situ GIWAXS, *Advanced Materials* **2021**, *33*, 2105290.

54. W. Hui, *et al.*, In situ observation of δ phase suppression by lattice strain in all-inorganic perovskite solar cells, *Nano Energy* **2020**, *73*, 104803.

55. H. Dang, *et al.*, Multi-cation synergy suppresses phase segregation in mixed-halide perovskites, *Joule* **2019**, *3*(7), 1746–1764.

56. C. Wu, *et al.*, MXene-regulated perovskite vertical growth for high-performance solar cells, *Angewandte Chemie International Edition* **2022**, *61*(43), e202210970.

57. M. Badrooj, F. Jamali-Sheini, N. Torabi, Optoelectronic properties of mixed Sn/Pb perovskite solar cells: The study of compressive strain by Raman modes, *The Journal of Physical Chemistry C* **2020**, *124*, 27136.

58. D. A. Strubbe, E. C. Johlin, T. R. Kirkpatrick, T. Buonassisi, J. C. Grossman, Stress effects on the Raman spectrum of an amorphous material: Theory and experiment on a-Si:H. *Physical Review B* **2015**, *92*, 241202.

59. T. M. G. Mohiuddin, *et al.*, Uniaxial strain in graphene by Raman spectroscopy: G peak splitting, Grüneisen parameters, and sample orientation, *Physical Review B* **2009**, *79*, 205433.

60. K. Talit, D. A. Strubbe, Stress effects on vibrational spectra of a cubic hybrid perovskite: A probe of local strain, *The Journal of Physical Chemistry C* **2020**, *124*, 28287.

61. M. Hÿtch, F. Houdellier; F. Hüe, E. Snoeck, Nanoscale holographic interferometry for strain measurements in electronic devices, *Nature* **2008**, *453*, 1086–1089.

62. A. Armigliato, *et al.*, Application of convergent beam electron diffraction to two-dimensional strain mapping in silicon devices, *Applied Physics Letters* **2003**, *82*(13), 2172–2174.

63. J. Zhang, C. Huang, Y. Sun, H. Yu, Amino-functionalized niobium-carbide MXene serving as electron transport layer and perovskite additive for the preparation of high-performance and stable methylammonium-free perovskite solar cells, *Advanced Functional Materials* **2022** *32*(24), 2113367.

64. B. W. Jia, K. H. Tan, W. K. Loke, S. Wicaksono, S. F. Yoon, Effects of surface reconstruction on the epitaxial growth of III-Sb on GaAs using interfacial misfit array, *Applied Surface Science* **2017**, *399*, 220–228.

65. M. U. Rothmann, W. Li, Y. Zhu, U. Bach, L. Spiccia, J. Etheridge, Y. B. Cheng, Direct observation of intrinsic twin domains in tetragonal $CH_3NH_3PbI_3$. *Nature communications* **2017**, *8*(1), 14547.

66. F. Uesugi, A. Hokazono, S. Takeno, Evaluation of two-dimensional strain distribution by STEM/NBD, *Ultramicroscopy* **2011**, *111*(8), 995–998.

67. K. Usuda, T. Numata, T. Irisawa, N. Hirashita, S. Takagi, Strain characterization in SOI and strained-Si on SGOI MOSFET channel using nano-beam electron diffraction (NBD). *Materials Science and Engineering: B* **2005**, *124*, 143–147.

68. C. Ophus, Four-dimensional scanning transmission electron microscopy (4D-STEM): From scanning nanodiffraction to ptychography and beyond, *Microscopy and Microanalysis* **2019**, *25*(3), 563–582.

69. D. Cooper, N. Bernier, J. L. Rouvière, Combining 2 nm spatial resolution and 0.02% precision for deformation mapping of semiconductor specimens in a transmission electron microscope by precession electron diffraction, *Nano Letters* **2015**, *15*(8), 5289–5294.

70. D. Cooper, N. Bernier, J. L. Rouvière, Y. Y. Wang, W. Weng, A. Madan, S. Mochizuki, H. Jagannathan, High-precision deformation mapping in finFET transistors with two nanometre spatial resolution by precession electron diffraction, *Applied Physics Letters* **2017**, *110*(22), 223109.

71. R. Tovey, D. N. Johnstone, S. M. Collins, W. R. Lionheart, P. A. Midgley, M. Benning, C. B. Schönlieb, Scanning electron diffraction tomography of strain, *Inverse Problems* **2020**, *37*(1), 015003.

72. B. Yang, *et al.*, Strain effects on halide perovskite solar cells, *Chemical Society Reviews* **2022**, *51*(17), 7509–7530.

73. Y. Yalcinkaya, I. M. Hermes, T. Seewald, Amann-K. Winkel, L. Veith, L. Schmidt-Mende, S. A. Weber, Chemical strain engineering of $MAPbI_3$ perovskite films, *Advanced Energy Materials* **2022**, *12*(37), 2202442.

74. T. Leonhard, *et al.*, Probing the microstructure of methylammonium lead iodide perovskite solar cells, *Energy Technology* **2019**, *7*(3), 1800989.

75. Y. Liu, P. Trimby, L. Collins, M. Ahmadi, A. Winkelmann, R. Proksch, O. S. Ovchinnikova, Correlating crystallographic orientation and ferroic properties of twin domains in metal halide perovskites, *ACS Nano* **2021**, *15*(4), 7139–7148.

76. Z. Xiao, Y. Yan, Progress in theoretical study of metal halide perovskite solar cell materials, *Advanced Energy Materials* **2017**, *7*(22), 1701136.

77. L. Kong, *et al.*, Simultaneous band-gap narrowing and carrier-lifetime prolongation of organic-inorganic trihalide perovskites, *Proceedings of the National Academy of Sciences* **2016**, *113*(32), 8910–8915.

78. D. Ghosh, A. Aziz, J. A. Dawson, A. B. Walker, M. S. Islam, Putting the squeeze on lead iodide perovskites: Pressure-induced effects to tune their structural and optoelectronic behavior, *Chemistry of Materials* **2019**, *31*(11), 4063–4071.

79. C. Deger, S. Tan, K. N. Houk, Y. Yang, I. Yavuz, Lattice strain suppresses point defect formation in halide perovskites, *Nano Research* **2022**, *15*(6), 5746–5751.

80. L. Xie, *et al.*, Understanding the cubic phase stabilization and crystallization kinetics in mixed cations and halides perovskite single crystals, *Journal of the American Chemical Society* **2017**, *139*(9), 3320–3323.

81. G. Kieslich, S. Sun, A. K. Cheetham, An extended tolerance factor approach for organic–inorganic perovskites, *Chemical Science* **2015**, *6*(6), 3430–3433.

82. G. Divitini, S. Cacovich, F. Matteocci, L. Cinà, A. Di Carlo, C. Ducati, In situ observation of heat-induced degradation of perovskite solar cells, *Nature Energy* **2016**, *1*(2), 1–6.

83. J. Suntivich, K. J. May, H. A. Gasteiger, J. B. Goodenough, Y. Shao-Horn, A perovskite oxide optimized for oxygen evolution catalysis from molecular orbital principles, *Science* **2011**, *334*(6061), 1383–1385.

84. J. Berry, *et al.*, Hybrid organic-inorganic perovskites (HOIPs): Opportunities and challenges, *Advanced Materials* **2015**, *27*(35), 5102–5112.

85. X. Zhu, J. Lee, W. D. Lu, Iodine vacancy redistribution in organic-inorganic halide perovskite films and resistive switching effects, *Advanced Materials* **2017**, *29*(29), 1700527.

86. D. W. DeQuilettes, *et al.*, Photo-induced halide redistribution in organic-inorganic perovskite films, *Nature Communications* **2016**, *7*(1), 11683.

87. L. Wang, *et al.*, Strain modulation for light-stable n-i-p perovskite/silicon tandem solar cells, *Advanced Materials* **2022**, *34*(26), 2201315.

88. J. Y. Kim, J. W. Lee, H. S. Jung, H. Shin, N. G. Park, High-efficiency perovskite solar cells, *Chemical Reviews* **2020**, *120*(15), 7867–7918.

89. G. Kieslich, S. Sun, A. K. Cheetham, An extended tolerance factor approach for organic-inorganic perovskites, *Chemical Science* **2015**, *6*(6), 3430–3433.

90. Q. Han, *et al.*, Single crystal formamidinium lead iodide (FAPbI$_3$): Insight into the structural, optical, and electrical properties, *Advanced Materials* **2016**, *28*(11), 2253–2258.

91. J. A. Steele, *et al.*, Thermal unequilibrium of strained black CsPbI$_3$ thin films, *Science* **2019**, *365*(6454), 679–684.

92. Y. An, *et al.*, Structural stability of formamidinium-and cesium-based halide perovskites, *ACS Energy Letters* **2021**, *6*(5), 1942–1969.

93. R. Prasanna, *et al.*, Band gap tuning via lattice contraction and octahedral tilting in perovskite materials for photovoltaics, *Journal of the American Chemical Society* **2017**, *139*(32), 11117–11124.

94. S. Jin, Can we find the perfect a-cations for halide perovskites? *ACS Energy Letters* **2021**, *6*(9), 3386–3389.

95. K. Nishimura, *et al.*, Relationship between lattice strain and efficiency for Sn-perovskite solar cells, *ACS Applied Materials Interfaces* **2019**, *11*(34), 31105–31110.

96. M. Abdi-Jalebi, *et al.*, Maximizing and stabilizing luminescence from halide perovskites with potassium passivation, *Nature* **2018**, *555*(7697), 497–501.

97. H. Min, *et al.*, Efficient, stable solar cells by using inherent bandgap of α-phase formamidinium lead iodide, *Science* **2019**, *366*(6466), 749–753.

98. J. Jeong, *et al.*, Pseudo-halide anion engineering for α-FAPbI$_3$ perovskite solar cells, *Nature* **2021**, *592*(7854), 381–385.

99. T. Kong, *et al.*, Perovskitoid-templated formation of a 1D@ 3D perovskite structure toward highly efficient and stable perovskite solar cells, *Advanced Energy Materials* **2021**, *11*(34), 2101018.

100. J. Zhang, S. Wu, T. Liu, Z. Zhu, A. K. Y. Jen, Boosting photovoltaic performance for lead halide perovskites solar cells with BF4⁻ anion substitutions, *Advanced Functional Materials* **2019**, *29*(47), 1808833.

101. H. Cheng, C. Liu, J. Zhuang, J. Cao, T. Wang, W. Y. Wong, F. Yan, KBF_4 additive for alleviating microstrain, improving crystallinity, and passivating defects in inverted perovskite solar cells, *Advanced Functional Materials* **2022**, *32*(36), 2204880.

102. L. Gao, Y. Zhao, Y. Chang, K. Gao, Q. Wang, Q. Zhang, Q. Tang, In-situ compensation for thermal strain by phase transformation strain in all-inorganic perovskite solar cells, *Chemical Engineering Journal* **2023**, *469*, 143881.

103. X. Du, *et al.,* Spontaneous interface healing by a dynamic liquid-crystal transition for high-performance perovskite solar cells, *Advanced Materials* **2022**, *34*(49), 2207362.

104. H. Zhang, *et al.*, Multifunctional crosslinking-enabled strain-regulating crystallization for stable, efficient α-$FAPbI_3$-based perovskite solar cells, *Advanced Materials* **2021**, *33*(29), 2008487.

105. H. Wang, *et al.*, Interfacial residual stress relaxation in perovskite solar cells with improved stability, *Advanced Materials* **2019**, *31*(48), 1904408.

106. J. Feng, *et al.*, High-throughput large-area vacuum deposition for high-performance formamidine-based perovskite solar cells, *Energy Environmental Science* **2021**, *14*(5), 3035–3043.

107. W. Chen, H. Chen, G. Xu, R. Xue, S. Wang, Y. Li, Y. Li, Precise control of crystal growth for highly efficient $CsPbI_2Br$ perovskite solar cells, *Joule* **2019**, *3*(1), 191–204.

108. Y. Zhao, J. Duan, Y. Wang, X. Yang, Q. Tang, Precise stress control of inorganic perovskite films for carbon-based solar cells with an ultrahigh voltage of 1.622 V, *Nano Energy* **2020**, *67*, 104286.

109. C. Zhang, *et al.*, Fabrication strategy for efficient 2D/3D perovskite solar cells enabled by diffusion passivation and strain compensation, *Advanced Energy Materials* **2020**, *10*(43), 2002004.

110. Q. Zhou, J. Duan, X. Yang, Y. Duan, Q. Tang, Interfacial strain release from the WS_2/ $CsPbBr_3$ van der Waals heterostructure for 1.7 V voltage all-inorganic perovskite solar cells, *Angewandte Chemie* **2020**, *132*(49), 22181–22185.

111. J. Wu, *et al.*, A simple way to simultaneously release the interface stress and realize the inner encapsulation for highly efficient and stable perovskite solar cells, *Advanced Functional Materials* **2019**, *29*(49), 1905336.

112. C. C. Zhang, S. Yuan, Y. H. Lou, Q. W. Liu, M. Li, H. Okada, Z. K. Wang, Perovskite films with reduced interfacial strains via a molecular-level flexible interlayer for photovoltaic application, *Advanced Materials* **2020**, *32*(38), 2001479.

113. H. Bi, B. Liu, D. He, L. Bai, W. Wang, Z. Zang, J. Chen, Interfacial defect passivation and stress release by multifunctional KPF_6 modification for planar perovskite solar cells with enhanced efficiency and stability, *Chemical Engineering Journal* **2021**, *418*, 129375.

114. B. Liu, *et al.*, Interfacial defect passivation and stress release via multi-active-site ligand anchoring enables efficient and stable methylammonium-free perovskite solar cells, *ACS Energy Letters* **2021**, *6*(7), 2526–2538.

115. C. Shi, *et al.*, Molecular hinges stabilize formamidinium-based perovskite solar cells with compressive strain, *Advanced Functional Materials* **2022**, *32*(28), 2201193.

116. J. Xu, *et al.*, Local nearly non-strained perovskite lattice approaching a broad environmental stability window of efficient solar cells, *Nano Energy* **2020**, *75*, 104940.

117. K. Zheng, *et al.*, Interfacial engineering strategy based on polymer modification to regulate the residual stress in $CsPbI_2Br$ based perovskite solar cells, *Chemical Engineering Journal* **2022**, *446*, 137307.

118. C. Wang, *et al.*, Balanced strain-dependent carrier dynamics in flexible organic-inorganic hybrid perovskites, *Journal of Materials Chemistry C* **2020**, *8*(10), 3374–3379.

119. J. Luo, *et al.,* A pressure process for efficient and stable perovskite solar cells, *Nano Energy* **2020**, *77*, 105063.

120. H. Tsai, *et al.*, Light-induced lattice expansion leads to high-efficiency perovskite solar cells, *Science,* **2018**, *360*(6384), 67–70.

121 C. J. Bartel, C. Sutton, B. R. Goldsmith, R. Ouyang, C. B. Musgrave, L. M. Ghiringhelli, M. Scheffler, New tolerance factor to predict the stability of perovskite oxides and halides, *Science Advances* **2019**, *5*(2), eaav0693.

122. M. R. Filip, G. E. Eperon, H. J. Snaith, F. Giustino, Steric engineering of metal-halide perovskites with tunable optical band gaps, *Nature Communications* **2014**, *5*(1), 5757.

11 Ion Migration and Mitigation in Halide Perovskite Solar Cells

Huan Li and Longbin Qiu

11.1 INTRODUCTION

Due to its appealing optoelectronic properties such as adjustable band gap, long electron-hole diffusion length, ultrahigh light absorption coefficient, and low exciton binding energy, the organic-inorganic hybrid metal halide perovskite material (OIHP) has been widely investigated in the field of perovskite solar cells (PSCs) over the past decades.[1-5] Perovskite materials used in optoelectronics usually have a formula of ABX_3, where A refers to a monovalent organic cation such as methylammonium $CH_3NH_3^+$ (MA$^+$), formamidinium NH_2-CH=NH_2^+ (FA$^+$), cesium (Cs$^+$), etc.; B stands for divalent metal cation such as lead (Pb) and tin (Sn); X is a series of halide anions such as chlorine (Cl$^-$), bromide (Br$^-$), or iodide (I$^-$).[1-5] The tolerance factor t is used to depict the structural stability of the perovskite materials, which is evaluated from the formula of $\dfrac{(R_A + R_X)}{\sqrt{2}(R_B + R_X)}$, where the R_A, R_B, and R_X are ionic radii of A, B, and X. In general, the halide perovskite with stable 3D-crystal structure

could be formed when the t value is in the range of $0.81 < t < 1$. The PSCs based on OIHP have raised a lot of attention, owing to a significant increase in the power conversion efficiency (PCE) reached over 26%,[6] which makes the PSCs greatly promising for commercialization.

There is, however, a primary challenge needed to be addressed that the stability of PSCs is still inferior to that of commercial silicon solar cells, which can last for more than 25 years, restricting their commercialization.[7] For silicon solar cells, the atoms are bonded together covalently, while component elements of halide perovskite are linked by the relatively weak interaction, containing ionic bonds, hydrogen bonds, and van der Waals forces.[8] Moreover, halide perovskites have a soft lattice because the compositions, such as MA$^+$, I$^-$ and Pb^{2+}, are large and easily polarized. Attributed to these features, the perovskite lattice is inclined to deform or break under the conditions of moisture, oxygen, heat, light, and electric field, which leads to the perovskite decomposition, lattice destroyed, phase segregation, etc., thereby deteriorating the photoelectric performance of PSCs. The issue of degradation of PSCs induced by moisture and oxygen could be solved by employing encapsulation techniques. However, the instability owing to the heat, light, and electric field is still

DOI: 10.1201/9781003400486-11

challenging because ions will migrate under these conditions. Furthermore, the soft lattice and ionic characteristic of perovskite result in the lower formation energy of imperfections, which further serve as ion migration channels. Thus, ion migration is considered as one of the intrinsic causes of instability of PSCs.

Ion migration implies that the ions such as halide anions could move throughout the perovskite film under an applied electric field or light illumination, and the migrated ions will generate vacancies or other imperfections, which serve as the trap-assisted recombination centers.[9,10] And a rise in recombination centers would adversely impact the performance of PSCs. The intrinsic factors of ion migration seem to be difficult to be resolved when the PSCs are working for a long time. During operation, the ions in PSCs will migrate and accumulate at the interfaces or grain boundaries, which results in the local distortion of crystal structure, leading to further deteriorate the perovskite absorber, as well as the hole/electron transport layers (HTL/ETL) and the electrodes. Eventually, the operational stability of PSCs will be severely degraded.[11–14] Suppression of ion migration is an effective strategy to prevent the deformation of crystal structure and stoichiometry change of halide perovskite under harsh operational conditions, thereby impeding the degradation of photoelectric performance for PSCs. Thus, a deep investigation of the origin and ion migration is significant for developing the PSCs with high efficiency and stability.

In this chapter, a comprehensive overview of recent studies about ion migration in PSCs is described. First, the types of defects, origin, and channels of ion migration are introduced. Most importantly, the deep mechanisms of ion migration are discussed, which is expected to provide more insight into alleviating ion migration and improving the performance of PSCs. Next, the extensively used characterization methods of ion migration are summarized. Then, the effects of ion migration on the stability of PSCs are discussed in detail. Specifically, the ion migration severely impacts the photoelectric performance, such as hysteresis, phase segregation, conductivity, etc., thereby deteriorating the operational stability. Furthermore, the effective strategies for mitigating ion migration are systematically investigated through different aspects such as composition modulation, additive regulation, and interface passivation. Finally, together with such detailed studies of ion migration, a research direction and an outlook on further developing high efficiency and outstanding long-term stability PSCs are outlined in both characterization and regulation aspects of ion migration.

11.2 DEFECT TYPES AND EXPERIMENTAL RESULTS

Although the PSCs have many overwhelming advantages, the instability is still a main obstacle to restrict the commercialization of perovskite devices. And the instability is heavily influenced by ion migration, which is strongly related to the defects or imperfections in the perovskite materials owing to the intrinsically soft structure. Classification of imperfections is usually made according to the location or dimensionality of the defect. On the one hand, according to the energy level location of defects, the defects could be divided into the deep and shallow level defects that are formed during the crystal growth and/or post-treatment process. The deep level defects are located in the middle of the forbidden band and the captured carriers mainly induce the non-radiative recombination in the perovskite film. The shallow level defects near the conduction band minimum (CBM) or valence band maximum (VBM) may not the major non-radiative recombination sites, but the migration of the charged shallow level defects in perovskite films plays a crucial role in the carrier extraction efficiency (Figure 11.1a).[15] On the other hand, according to the dimensionality of defects, they could be divided into zero-dimensional (0D), one-dimensional (1D), two-dimensional (2D), and three-dimensional (3D) defects. 0D defects refer to the point defects related to one or two atomic positions. 1D defects mean dislocations, where several atoms are misaligned. 1D defects are linear defects centering on the line along the edges where the crystal structure is broken. The localized lattice distortion exists around the linear defects. The 2D defects contain the undercoordinated ions (halides ions, Pb^{2+}), defects arrays, and the combination of point defects and dislocations on the crystal surface and interface. And the 2D defects are usually observed at the crystal surface, such as the external surface and the grain boundaries. The 3D defects involve pinholes, cracks, and impurities on the perovskite film that are normally introduced during the preparation or post-treatment steps. Any imperfections in perovskite materials results in localized structure distortion, which significantly influence the photovoltaic performance of PSCs. This section mainly focuses on three-point defects (0D) and two typical extrinsic defects involving grain boundary and surface (1D, 2D, 3D).

11.2.1 Point Defects

The point defects formed in the crystalline structure include vacancies, interstitials, and anti-site substitutions (Figure 11.1b). These point defects are intrinsically generated or via introducing the impurities and dopants into the crystal structure. Moreover, all point defects could create a transition energy level in the forbidden band region, which further determines the type of defects. Take methylammonium lead iodide perovskite ($MAPbI_3$) as an example, which is widely used perovskite. Twelve possible intrinsic point defects are observed in the $MAPbI_3$ crystal structure, like three vacancies (V_{MA}, V_{Pb}, V_I), three interstitials (i_{MA}, i_{Pb}, i_I), and six substitutions (MA_{Pb}, Pb_{MA}, MA_I, Pb_I, I_{MA}, I_{Pb}).[16] It is well known that the theoretical calculation is a useful method to investigate the photovoltaic properties of PSCs affected by the imperfections, as well as guiding the defects modulation. The formation energy of defects determined by the calculation is an important parameter to explore the origin of defects and ion migration in perovskite materials. Based on the density functional theory (DFT), the formation energy of point defects in $MAPbI_3$ was calculated, and these point defects have the lowest formation energy.

Owing to different models of theoretical calculation, there are different and even contradictory results about the point defects and whether they can cause deep transition level in the forbidden region of perovskite. Yan et al. supposed that the point defects belong to the shallow-level defects, and the deep-level defects are hardly generated in the $MAPbI_3$.[16] However, Agiorgousis and Wang et al. believed that V_I could form deep-level defects by inducing the generation of lead dimers.[17–19] It is worth mentioning that the formation energy of point defects is closely related to the growth conditions of perovskite film including the precursor concentrations, precursor composition, solvents, and temperature. It is reported that the formation energy of Pb_I

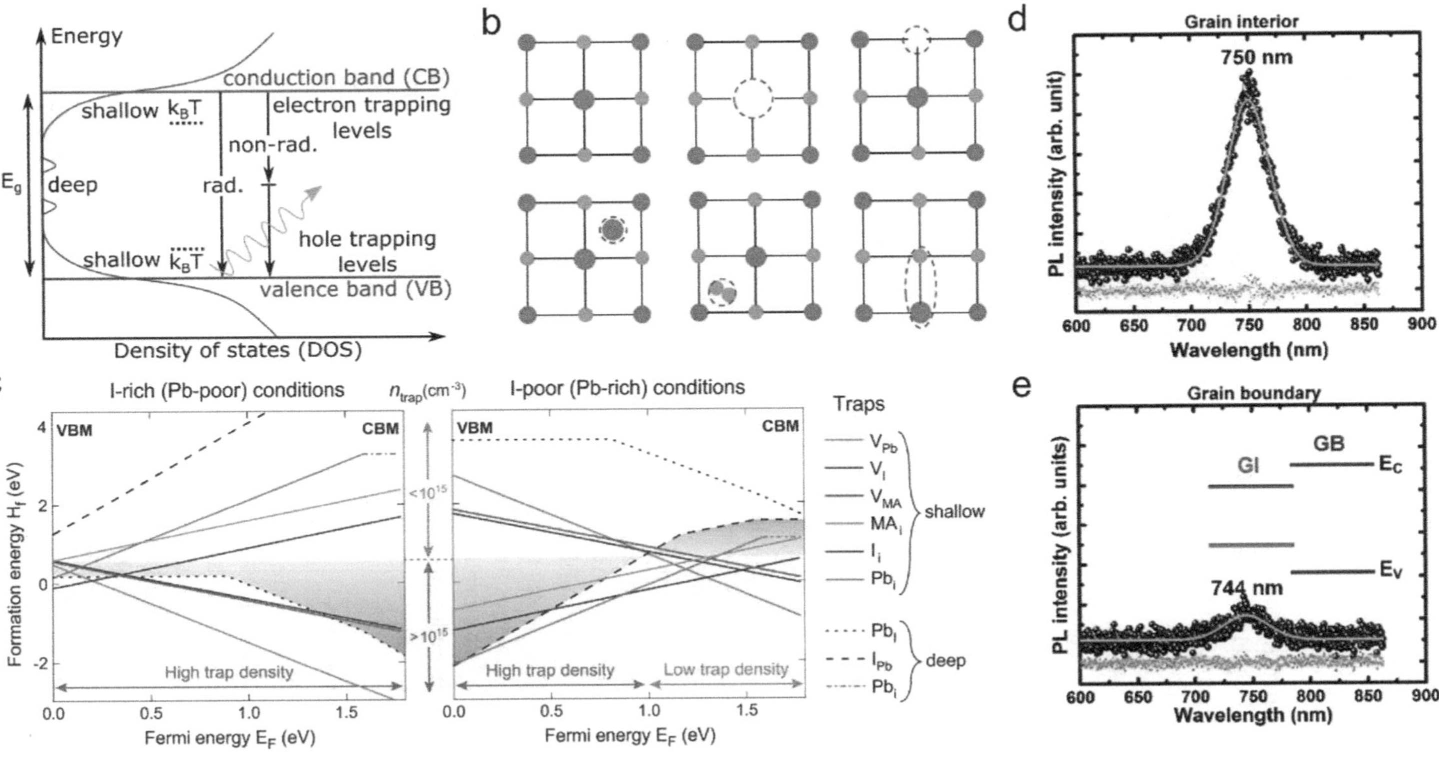

FIGURE 11.1 (a) Schematic diagram of defect energy levels[15]. Copyright 2020, The Royal Society of Chemistry. (b) Illustration of a perfect lattice, vacancy, interstitial, anti-site substitution. (c) Defect formation energies for iodide-poor and iodide-rich growth conditions. Solid (dashed) lines denote shallow (deep) traps[20]. Copyright 2014, American Chemical Society. PL spectra of (d) the grain interior and (e) the grain boundary fitted by a single Gaussian function[22]. Copyright 2017, Royal Society of Chemistry.

anti-site deep defects in MAPbI$_3$ under I-rich condition is much lower than that under the I-poor condition (Figure 11.1c).[20]

11.2.2 Grain Boundary

It is well known that the perovskite film is polycrystalline and contains abundant grain boundaries, which play a significant role in the long-term stability and carrier dynamics of perovskite. Walsh et al. reported that the iodine interstitials could segregate to grain boundaries in the CsPbI$_3$ crystal. Meanwhile, they confirmed that the grain boundaries serve as reservoirs for point defects, which can affect the ion transport and current-voltage hysteresis of PSCs. The iodine interstitial defects act as the recombination centers in bulk crystal and are prone to accumulate at grain boundaries, indicating that the defect density at the grain boundary is higher than that in the bulk. Iodine interstitial defects could form split interstitial, I-I-I trimers, and Pb-I bonds with an undercoordinated Pb atom where split interstitial and I-I-I trimers can promote carrier recombination at the grain boundary.[21] In addition, the first-principles calculations suggest that increased structural relaxation of the defects at grain boundaries induces an enhancement of defect stability (higher concentration) and deeper trap states (faster recombination).[21]

Besides the theoretical calculation, the results of microscale photoluminescence (PL) measurements manifest that there are more defects and faster non-radiative recombination at the grain boundary (Figures 11.1d and e).[22] Moreover, the grain boundaries probably have a significant effect on the ion migration and photovoltaic properties owing to the relatively open structure. Huang et al. uncovered that a larger hysteresis of local dark current was detected at the grain boundaries than grain interiors, which could be explained by faster ion migration at the grain boundaries.[23,24] On the other hand, some researchers proposed that the grain boundaries are not detrimental to the performance of PSCs. Koster et al. suggested that the photogenerated carriers could fill the defects at the interfaces and grain boundaries. Thus, the defects could not capture the charges for non-radiative recombination, leading to highly efficient PSCs with compact morphologies.[25] For MAPbI$_3$ perovskite film, PbI$_2$ could be formed at the grain boundaries during the annealing process, suppressing the carrier recombination and increasing the device performance.[26]

11.2.3 Surface Defects

PSCs has a sandwich structure with layer stacking, and the photo-induced carriers need transfer from perovskite film to the electrode through ETL and HTL. Thus, the interface configuration is important for the carrier transport dynamic. However, the perovskite film is polycrystalline, leading to the presence of plentiful defects at the surface, such as uncoordinated atoms, dangling bonds, surface dislocation, and chemical impurities (lead cluster). The surface defects could change the properties of the surface, eventually impeding the transport of carriers. Beard et al. studied the surface carrier dynamics in MAPbI$_3$ perovskite polycrystalline film by employing transient reflection spectroscopy. It is found that the recombination at the surface is more serious than that inside grain or grain boundaries.[27] Moreover, it is proved that the surface defect density is 1-2 orders of magnitude larger than that of bulk.[28,29] They demonstrated that the intrinsic defect densities in perovskite film are associated with the fabrication process and film composition. In particular, the defect densities at the interface of perovskite/PTAA are higher than that at the perovskite/C_{60} interface, due to the generation of small crystal close to the perovskite/PTAA interface.[29] As mentioned earlier, the uncoordinated atoms and dangling bonds are detected on the surface, which increases the point defect densities and changes their formation energy. Particularly, Banerjee et al. analyzed the electronic structure of the (100) perovskite surface under the circumstance of presence of I_{Pb} anti-site defects based on the DFT calculations. They indicated that the I_{Pb} anti-site defects on the MAI terminated surface could form shallow-level defects, while both the deep and shallow-level defects are generated on the PbI_2 terminated surface. These defects originate from the generation of new bonds between defect atoms and other chemicals, which is unfavorable to both the stability and efficiency of PSCs.[30] Therefore, much effort has been made to reduce the defects on the surface.

In 2014, Huang et al. proposed that the charge defects on the surface could be effectively passivated by depositing the fullerene layer (PCBM) on the perovskite film, eliminating the photocurrent hysteresis.[31] Sargent et al. reported that the I_{Pb} anti-site defects cold be dramatically decreased by the PCBM passivation.[32] Very recently, Wang et al. employed the strategy of surface reconstruction to transform the extra PbI_2 into lead sulfate-silica bi-layer. This method eliminates the under-coordinated Pb^{2+} and modified Pb_I anti-site defects on the surface.[33] Besides, the interstitial defects on the surface are also more active than that in bulk, since the steric hindrance in the crystal lattice is serious, while it can be mitigated on the surface. On the basis of DFT calculations, the interstitial halogen atoms are prone to bind to the undercoordinated Pb on the surface.[34] Because of the presence of point defects on the surface, the surface also plays an important role in the behavior of ion migration. In conclusion, the defects on the surface are kind of driving force to form the deep level defects and the path of ion migration. Surprisingly, the passivation strategy for surface defects have been widely employed to reduce the defect densities and achieve breakthroughs. Consequently, the photocurrent hysteresis and photovoltage parameters have been significantly alleviated and improved, respectively.

11.3 MECHANISM AND CHANNELS OF ION MIGRATION

11.3.1 ION MIGRATION MECHANISM

Hybrid perovskite materials usually are composed of a complex network of cations, anions, and metal atoms, displaying ionic properties. Considerable works based on both theoretical calculations and experimental results have proved that the ions in perovskite have a low activation energy (E_a) and moderate ion diffusion coefficients. Here, the E_a refers to the lowest free energy barrier that the ions could exceed to transfer from its initial position to adjacent vacancy sites. This motion process could be easily understood by the hopping mechanism mediated by vacancy or interstitial diffusion. Moreover, the velocity of ion migration (k) could be estimated by the Nernst-Einstein relation and Arrhenius equation, $\sigma(T) = \dfrac{\sigma_0}{T}\exp\left(-\dfrac{E_a}{k_B T}\right)$ and $k = \dfrac{k_B T}{\hbar}\exp\left(-\dfrac{E_a}{RT}\right)$, where σ, σ_0, k需要斜体, B需要下标, and T refer to conductivity, a constant, Boltzmann constant and temperature, respectively.[35] And $\hbar$ and R refer to the reduced Planck constant and the ideal gas constant, respectively.[36] Clearly, the migration rate is related to the E_a and temperature. It is well established that vacancy-mediated diffusion is dominated in hybrid perovskite materials. The DFT calculations show that the Schottky disorder is easily generated when the vacancies concentration of I^-, Pb^{2+}, and MA^+ exceed 0.4% at room temperature.[37]

Furthermore, the migrated pathways of I^-, Pb^{2+} and MA^+ were investigated elaborately (Figure 11.2a). The I^- migrates along an octahedron edge, the Pb^{2+} moves along the diagonal (<110> directions) of the cubic unit cell, and the MA^+ migrates into an adjacent vacant A-site cage. The I^- migration has the lowest Ea among three types of ions, implying inclined vacancy-mediated diffusion of iodide ions, which is deeply influenced by the intrinsic vacancies of iodide ions. More interestingly, the paths of I^- migration is not linear but is a curved path between iodine sites. The Pb^{2+} vacancy migration with high energy barrier shows constant Pb sublattice. And the MA^+ ions hardly diffuse in perovskite structure because of a small diffusion coefficient.[38] There is an unambiguous agreement that the A and X ions are easier to leave their original sites than B ions due to the lower E_a. On the other hand, the ion migration in perovskite materials is highly dependent on the nature of the cations and anions present in the crystal lattice. For example, materials containing small cations, such as MA^+, exhibit faster ion migration compared to materials containing large cations, such as lead. That means the ion migration is not absolute but relative. In conclusion, the ion migration mechanisms in perovskite materials are complex and highly dependent on composition of perovskite materials. The presence of point defects, such as vacancies and interstitials, can significantly impact the ion migration kinetics. Further research is needed to fully understand the underlying mechanisms and to develop strategies for controlling and optimizing the ion migration in these materials. Given the exceptional optoelectronic properties of perovskites, such efforts could lead to the development of new materials with improved performance in a variety of applications.

11.3.2 Ion Migration Channels

Ion migration in perovskite films could occur under electric field, illumination, and heat conditions. It is of great importance to understand how the ions move under different conditions, because it could provide hints to suppress the ion migration to develop applicable PSCs that are highly effective and stable. As mentioned in the last section, ion migration process is dominated by way of defect-mediated hopping mechanism. Thus, the defects in the perovskite play a crucial role in migration process, because of the difficulty of ion migration in the perfect crystal. Normally, the defects in the crystal involve Schottky defects and Frenkel defects. The former refers to the fact that the anions and cations with equal number leave their lattice sites together, leading to an electrical neutrality in the crystal and the formation of vacancies that are usually generated during perovskite fabrication. The Frenkel defect is an ion vacancy where the ion moves out of its initial position, leaving a vacancy and

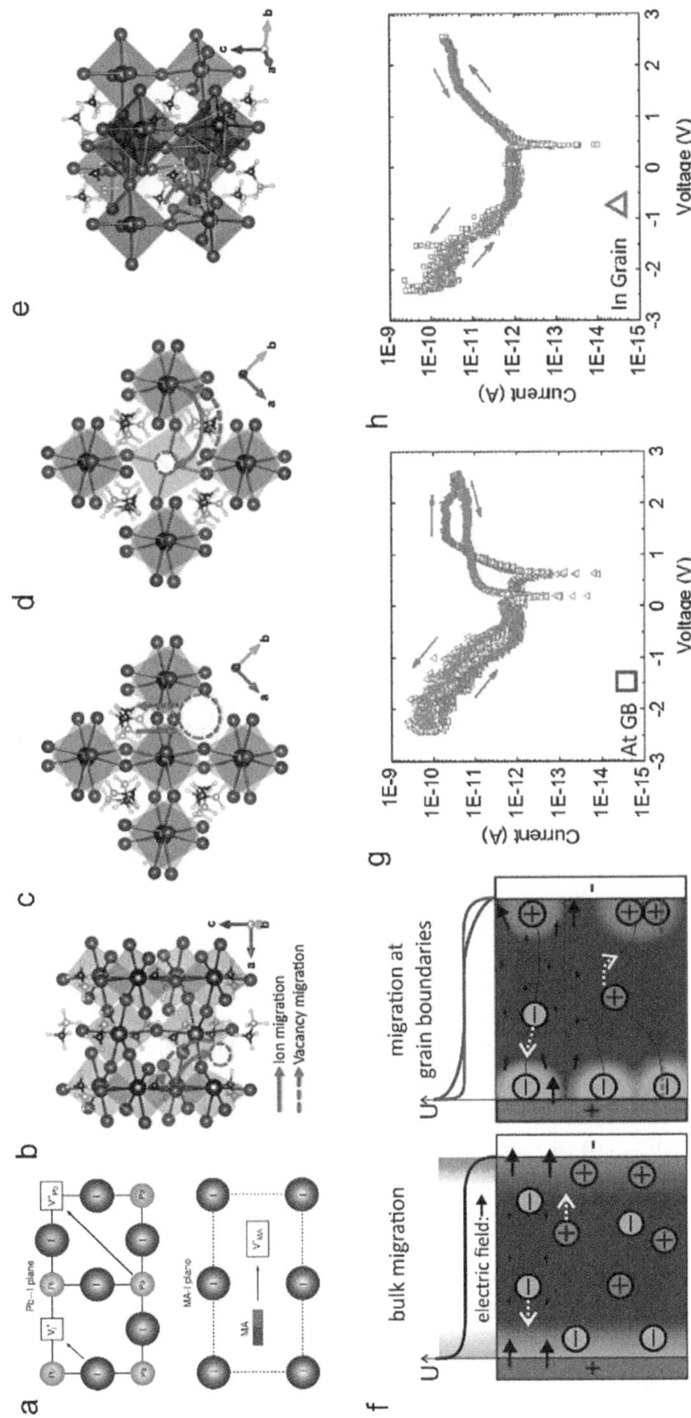

FIGURE 11.2 (a) The migrated mechanisms of I⁻, Pb^{2+} and MA^+[38]. Copyright 2015, Springer Nature Limited. Migration channels of I vacancy (b), MA vacancy (c), Pb vacancy (d), and I interstitial (e)[36]. Copyright 2015, Royal Society of Chemistry. (f) Schematic diagram of uniform electrostatic double layers[39]. Copyright 2018, The Royal Society of Chemistry. Local dark current measured at the grain boundaries (g) and on the grain surface (h)[24]. Copyright 2016, The Royal Society of Chemistry.

instead occupying another intermediary vacant site. These interstitials and vacancies can be generated in pairs by thermal excitation or recombination, which makes the concentration of interstitials and vacancies reach dynamic equilibrium. The ion vacancies and interstitials primarily provide channels for ion migration because the anti-site substitutions are considered as unstable defects energetically and kinetically, and they split into the vacancies and interstitials. Angelis et al. analyzed the possible migration pathways of iodine, MA, Pb vacancies, and iodine interstitials in the $MAPbI_3$ perovskite, as shown in Figures 11.2b–e. For iodine vacancy, they create the vacancy in an equatorial position and the vacancy moves towards an axial site. Considering the MA vacancy, the vacancy hops between neighboring cavities of the inorganic scaffold lying in the *ab* plane and moves across a Pb-I framework. The Pb vacancy migrates along the side of square constituted by four Pb and four I atoms. Similar to iodine vacancy, the iodine interstitials move along the *c*-axis and embed between a pair of axial/equatorial iodine atoms with nearly same distance. Based on the migration process, they also calculated the E_a of mobile ions in the $MAPbI_3$ and $MAPbBr_3$ with tetragonal structure by using the first principle computational analyses. They demonstrated that the iodine vacancies and interstitials have the lowest E_a of ~0.1 eV compared to the MA (~0.5 eV) and Pb (~0.8 eV) vacancies. For $MAPbBr_3$, it's total E_a of a full hop for bromide vacancy is estimated to be lower than that of iodine vacancy, whereas a higher E_a of MA vacancy migration is suggested, indicating that it is more difficult for MA vacancy to pass two-unit cells of $MAPbBr_3$. Besides, they proposed that the charged defects could move and pack at the interface between perovskite and the selective contacts under working conditions, forming an electrostatic potential throughout the perovskite film. In that case, the electronic structure of perovskite was modified, hampering charge extraction and negatively affecting the performance of PSCs.[36]

In addition to point defects mainly containing vacancies and interstitials, the grain boundaries and surface also act as ion migration pathways. In the normally used model, the ions are uniformly distributed and move freely within the entire perovskite film, forming uniform electrostatic double layers (EDLs) at either electrode (Figure 11.2f). However, if the ions move preferentially at the grain boundaries, the EDLs and the distribution of electric field will be heterogeneous. Berger et al. measured the evolution of the local contact potential difference (CPD) using the time-resolved Kelvin probe force microscopy (KPFM) to detect the charge redistribution upon applying an electric field or a light pulse. The CPD results reveal that it took hundreds of milliseconds (ms) to accomplish an equilibrium potential distribution throughout the perovskite film, which was ascribed to the slow ion migration inside the perovskite film. They also proposed that the free iodide/iodide-vacancy pairs are formed close to the interface under an external electric field, and the iodide could migrate through interstitials/Frenkel defects.[39] Moreover, Huang et al. measured the local current hysteresis at the grain boundary and surface to study ion migration rate using conductive atomic force microscopy (C-AFM), seen from Figures 11.2g and h. Note that a larger current hysteresis would imply faster ion migration during the current measurement. It is clear that the dark-current measured at grain boundary shows obvious hysteresis, while that measured within grain interiors displays a negligible hysteresis, thereby clarifying the quicker ion migration at the grain boundaries. When the device was illuminated, the short circuit current at the grain boundary increased gradually, but that measured atop of grain remains almost constant, indicating that the photoinduced ion migration could improve the efficiency of memristors. They also found that the ions move faster in perovskite with smaller grains, and the single crystal shows hysteresis-free current and negligible ion migration.[24]

Furthermore, there are other probabilities for the generation of ion migration paths, such as the local lattice distortion caused by mesoporous scaffold confinement,[40] accumulated charges,[41] uneven strain,[42] dissolved impurities,[43] other introduced atoms[44] and a softened lattice induced by the light illumination.[23] For the $MAPbI_3$ perovskite within a mesoporous structure of TiO_2, approximately 70% $MAPbI_3$ was significantly disordered, which might influence the device performance.[40] Mathews et al. investigated the origin of hysteresis in PSCs by an electro-optical analysis, suggesting that the lattice distortion occurs in the $MAPbI_3$ perovskite upon applying an electric field due to charge accumulation, which plays a dominating role in the hysteresis.[41] Huang et al. found that the electric field-induced strain was not uniform in the single crystal of $MAPbI_3$, resulting in the formation of a crystal-plane displacement. During the poling process, the weakly bonded planes (100) could slip, which caused the generation of parallel strip lines on the crystal surface. These slipping-formed interfaces/grain boundaries and dislocation lines could serve as the ion migration channels, which is the main restriction for device properties.[42] Additionally, the DFT calculations performed by Zaban et al. suggested that the MA^+ ions rotate freely and arrange along the electric field under illumination, changing the structure of the inorganic scaffold. The ions alignment is relatively fast, while the inorganic scaffold could not adjust within a short time, leading to the slow photoconductivity response. The soften lattice induced by illumination also could increase the defect generation and thus ion migration.[45]

11.4 CHARACTERIZATION OF ION MIGRATION

Characterizing ion migration in perovskite materials is crucial to understand the underlying mechanisms that govern their optoelectronic properties. Over the years, various techniques have been developed to study ion migration in these materials, each with its strengths and limitations. Here we will review some of the most popular techniques for characterizing ion migration in perovskite materials.

From the perspective of experiment, there are two most widely used methods for characterizing ion migration. The first method constitutes the imaging and elemental characterization techniques, which can visually observe the process of ion migration. One of extensively employed means is KPFM, which can provide the work function of perovskite top-surface associated with the p- or n-doping induced by ion drift. In 2014, Huang et al. discovered a switchable photocurrent in the vertical device with a structure of $ITO/PEDOT:PSS/MAPbI_3/Au$ and proposed that the vacancies or positively charged ions migrate to the Au electrode side under positive bias and aggregate there, resulting in n-doping in perovskite. And the residual negative space charge layer could lead to the p-doping in perovskite near to PEDOT:PSS layer, creating a p-i-n homojunction. The p-i-n structure can change to n-i-p structure under a reverse bias. The KPFM measurement verified the aforementioned scenario because of the discrepancy in work function after applying the electric field. Meanwhile, optical microscopy was also used to monitor the ion migration (Figures 11.3a and b), apparently manifesting the ion drift from the anode side.[46] Later on, they designed a device with the lateral structure where the perovskite film was directly deposited on the glass substrates. Through a combination of energy dispersive X-ray (EDX), KPFM and C-AFM techniques, they observed the macroscopic migration of iodide ions within the $MAPbI_3$ perovskite film under external bias. In particular, a thread is generated close to the anode region after applying an electric field, and the thread gradually migrates to the cathode side along the direction of applied bias, which confirms the migration of I^-. The EDX results uncover massive I^- transport, while the Pb^{2+} migration is negligible. The composition of the thread is also confirmed to be PbI_2 using EDX and XRD. To clarify the properties of different areas of A, B, C in the device seen from Figure 11.3c, the KPFM and C-AFM measurements were conducted. The A, B, and C areas corresponded to fresh perovskite, thread, and reformed perovskite. The schematic illustration of KPFM measurement is shown in Figure 11.3d. For KPFM measurement, a uniform current would flow across the measured region, if a constant electric-field was applied between two electrodes. The surface potential gradually decreases, as seen in Figures 11.3e and f. The decrease of potential across the thread is much faster than that across the fresh perovskite film, showing that the applied potential also completely drops at the thread. Furthermore, they employed the C-AFM to identify the conductivity of the materials determined from the potential-position plot. The device structure is shown in Figure 11.3g, where the Au electrode was used as a ground electrode, and the electric field was applied between C-AFM tip and Au electrode. In this way, the current of the perovskite device at different areas was detected (Figure 11.3h). The current of refresh and reformed perovskite is higher than that of PbI_2 thread. The results of KPFM and C-AFM demonstrated that the conductivity of PbI_2 thread is lower than the $MAPbI_3$ perovskite film.[47]

Another visualization characterization is photothermal induced resonance (PTIR) microscopy, which integrates an AFM with infrared absorption spectroscopy. Herein, the local infrared absorption spectrum could be measured by a pulsed infrared laser with tunable wavelength, and the AFM tip detects the induced photothermal expansion because of the light absorption into cantilever oscillations. Finally, a diode laser could reflect into the AFM 4-quadrant detector in the far-field (Figure 11.4a).[48,49] PTIR could provide the information of the distribution for given chemical groups because the chemical composition is specific in infrared spectroscopy. In 2015, Huang et al. employed this technique to detect the spatial distribution of MA^+ in the lateral $MAPbI_3$ PSCs under an applied electric field. When applying the electric field, the MA^+ ion gradually migrates towards the cathode, indicating the redistribution of MA^+ ion (Figures 11.4b and c). This result is direct and effective evidence to characterize the migration of MA^+ ion. However, the redistribution of Pb^{2+} and I^- was not observed in the lateral structure due to the low sensitivity of inorganic species in IR microscopy.[49,50] Thus, the nanoprobe X-ray fluorescence (nano-XRF) technique was developed, and the heavy elements could be detected with a sensitivity lower than parts per million. The core-shell electrons of the atoms are excited using a focused

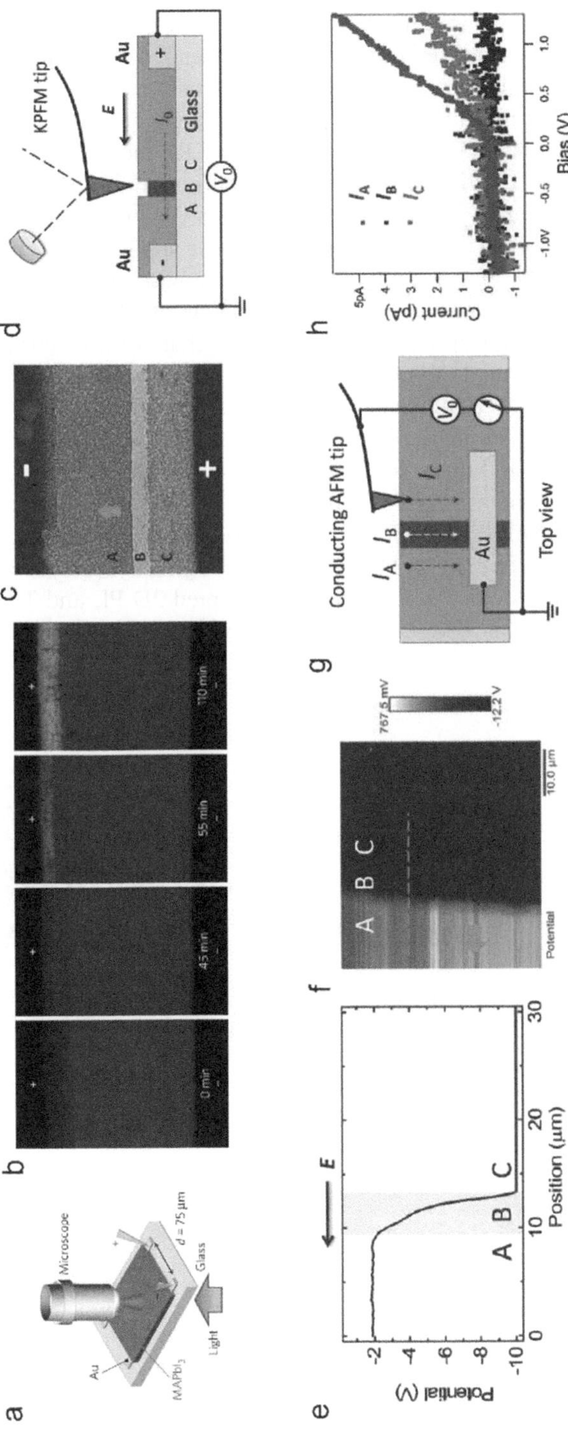

FIGURE 11.3　(a) Illustration of set-up for in-situ monitoring of the poling process using a lateral structure device. (b) Snapshots of in-situ recorded video[46]. Copyright 2015, Springer Nature Limited. (c) The optical images of MAPbI₃ PSC. (d) Schematic illustration of the KPFM measurement around the PbI₂ thread. (e) Surface potential curve around the PbI₂ thread for the position marked in 3f. (g) Schematic illustration of the C-AFM measurement around the PbI₂ domain. (h) Local current-voltage curve measured with the C-AFM tip in contacting with fresh MAPbI₃[47]. Copyright 2016, Wiley-VCH GmbH.

X-ray beam. When the relaxation of an electron at outer shell fills a core-hole, a fluorescent photon will be emitted and its energy can recognize the element. To build the direct relationship between chemical composition and optoelectronic performance, Fenning et al. combined the nano-XRF and spatially resolved PL spectroscopy to detect the bromide migration in the $MAPbBr_3$ single crystal under applied bias (Figure 11.4d). As shown in Figure 11.4e, the Nano-XRF measurement suggests that the Br:Pb distribution throughout the single crystal is initially uniform, but its distribution changes when applying the electric field. On the other hand, they also measured the spatially resolved relative PL intensity under applied bias with respect to the unbiased PL map. The PL intensity rises in the area with higher potential (Br rich region), while it decreases in the area with low potential (Br deficient region). Overall, the direction of Br ions migration is opposite to that of the electric field.[51]

In addition to aforementioned techniques, the time of flight secondary ion mass spectrometry (TOF-SIMS) has also been employed to explore the distribution of elements and composition of perovskite film. As shown in Figures 11.5a-d, the I^- can permeate through the HTL and move into the Ag side in both $MAPbI_3$ and $FAPbI_3$; also Ag^+ could migrate slightly in the $MAPbI_3$, indicating that the degradation rate in $FAPbI_3$ is slower compared to $MAPbI_3$.[52] Similarly, the Au could migrate into perovskite layer through HTL, Spiro-MeOTAD, when aging the device at 70 °C, shown in Figure 11.5e and f.[53] In 2022, Chen et al. adopted 2D PL and TOF-SIMS techniques to investigate the individual cation migration in $FA_{0.9}Cs_{0.1}PbI_3$ perovskite

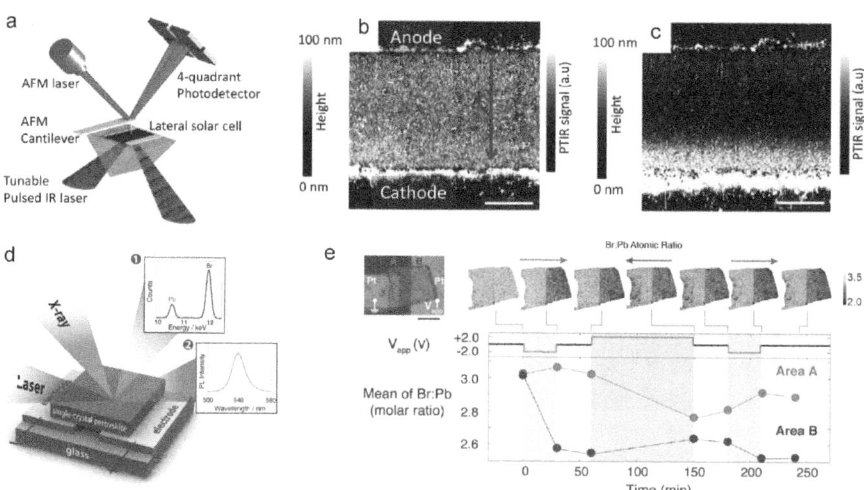

FIGURE 11.4 (a) Schematic illustration of the PTIR measurement. PTIR images of $MAPbI_3$ film before (b) and after electrical poling for 200 s (c)[49]. Copyright 2015, Wiley-VCH GmbH. (d) Schematic of nanoprobe X-ray fluorescence and spatially resolved PL measurement. (e) Nano-XRF measurement of the changes in elemental distribution in a $MAPbBr_3$ single crystal under bias[51]. Copyright 2017, Wiley-VCH GmbH.

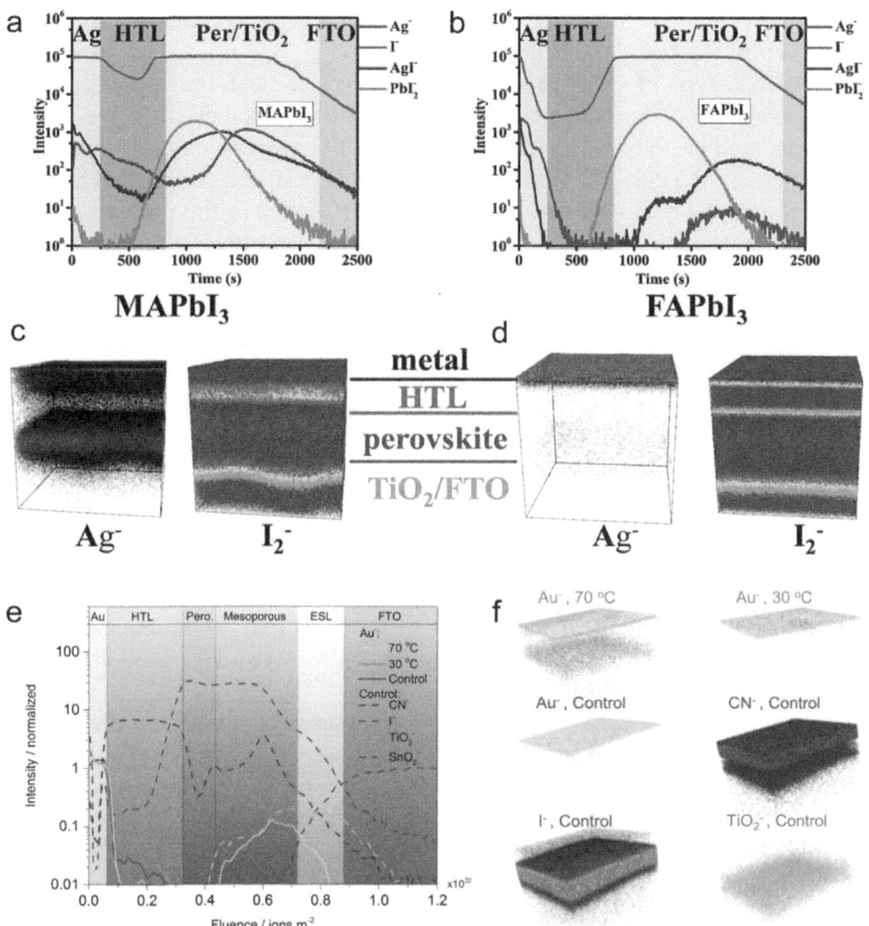

FIGURE 11.5 The TOF-SIMS depth profiles of degraded MAPbI₃ (a, c) and FAPbI₃ (b, d)[52]. Copyright 2017, The Royal Society of Chemistry. (e) TOF-SIMS depth profiles of the aged perovskite devices. (f) Reconstructed elemental 3D maps of Au ions traced in the depth profile[53]. Copyright 2016. American Chemical Society.

under continuous light illumination at 80 °C. It is found that the Cs⁺ and FA⁺ could migrate to different regions, leading to severe phase segregation.[54] Very recently, Zhao et al. discovered that the iodide ions in a device could diffuse through the ETL and accumulate on the ITO after operating for a period of time, evidenced by TOF-SIMS.[55] In conclusion, imaging and elemental characterization techniques play a significant role in visualizing the ion migration behavior. And it is also useful to understand the underlying mechanisms that govern their optoelectronic properties, like hysteresis, phase segregation, and degradation of PSCs. However, these methods can only detect the ion migration qualitatively, a quantitative characterization of the migration process still remains necessary.

The second type of experimental methods are electrical, optical, and structure measurements. For structure measurement, the X-ray photoelectron spectroscopy (XPS) was used to study the distribution of I^- and Pb^{2+} under long-term electric field, where a lateral $MAPbI_{3-x}Cl_x$ perovskite device was designed (Figures 11.6a and b). The XPS analysis demonstrated that I/Pb ratio increased at the positive electrode, implying the accumulation of I^- ions. On the other hand, the I/Pb ratio decreased at the negative electrode, an indication of the deficiency of I^- ions. Because the E_a of Pb^{2+} transportation is higher, they presumed that the change of I/Pb ratio may be due to the I^- migration.[56] To verify that the iodide ions could migrate to the ITO side, the PSCs without the ETL were fabricated by Zhao et al. and operated with illumination for 10 h. Then, they removed the perovskite and other layers to characterize the ITO surface by XPS, as shown in Figure 11.6c. The iodide ions could be obviously observed on the ITO surface, while it is not detected in the original ITO, indicative of the migration of iodide ions from ETL. They further proposed that the downward migration of ion could change the work function of ITO surface, increasing the energy gap between ITO and SnO_2, leading to decreased V_{OC}.[55] For electrical and optical measurement, the physical properties related to ion migration should be measured at different temperature, such as ionic conductivity and capacitance. Then, the E_a of ion migration could be estimated using the Arrhenius equation, as shown in Table 11.1.[8]

Generally, the ion conductivity is easily measured to determine the E_a of ion migration. In 2015, Huang et al. evaluated the changes in conductivity of $MAPbI_3$ perovskite under 0.35 V μm^{-1} bias at various temperatures. And the E_a could be estimated using Nernst-Einstein relation mentioned earlier, which could be simplified as $\ln(\sigma T) = C - E_a/(k_B T)$. Specifically, the E_a can be extracted from the slope of the $\ln(\sigma T)$-$1/k_B T$. The plot of lateral conductivity for 250-nm-thick $MAPbI_3$ under the darkness (blue) and illumination (red) is shown in Figure 11.6d. In the dark, there are two linear regions: the first region (100–270 K) displays an E_a of 0.035 eV corresponding to charge carrier transportation; the second region (290–350 K) shows

---

TABLE 11.1

Methods for Experimental Measurement of E_a[56]

Method	Equation	Parameter
Ionic conductivity	$\ln(\sigma) = C - E_a/(k_B T)$	Conductivity (σ), temperature (T)
Electrochemical impedance spectroscopy	$\ln[1/(T T_W)] = C - E_a/(k_B T)$	Warburg time constant (T_W), temperature (T)
Chronophotoamperometry	$\ln(k) = C - E_a/(k_B T)$	Relaxed rate (k), temperature (T)
Thermal admittance spectrosocopy	$\ln[\omega_{peak}/(T^2)] = A - E_a/(k_B T)$	Frequency of inflection (ω_{peak}), temperature (T)
Photoluminescene spectral evolution	$\ln(r_g) = C - E_a/(k_B T)$	Initial growth rate (r_g), temperature (T)

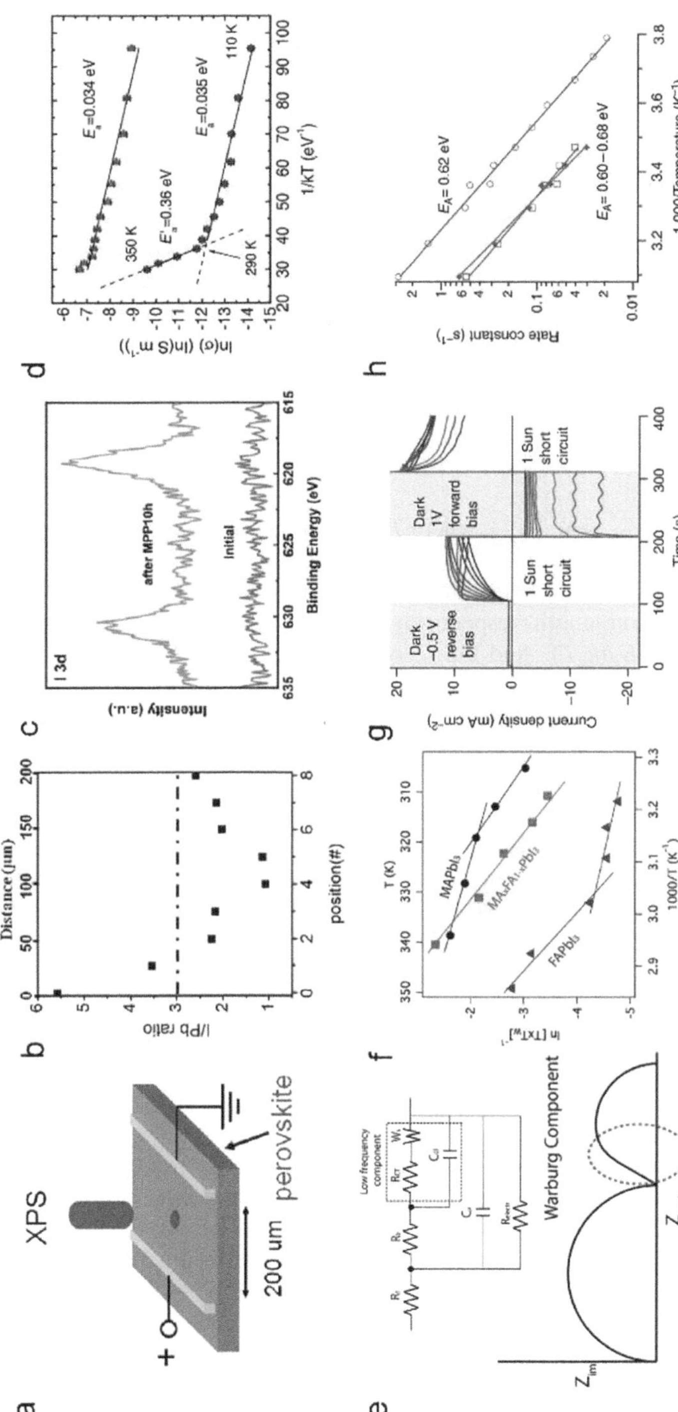

FIGURE 11.6 (a) Illustration of the set-up for characterization of elements distribution using a lateral structure. (b) Distribution of the ratio of I/Pb between the electrode pairs after applying bias between two lateral electrodes[56]. Copyright 2016, Wiley-VCH GmbH. (c) I 3d XPS spectra, of initial ITO and ITO from PSC after maximum power point tracking (MPPT) for 10 h[55]. Copyright 2023. Wiley-VCH GmbH. (d) Temperature-dependent conductivity of the MAPbI$_3$ film under dark and illumination[49]. Copyright 2015, Wiley-VCH GmbH. (e) The Nyquist plot for mixed conductor system. Inset: equivalent circuit diagram of PSCs. (f) Arrhenius-like plot of $1/(TT_W)$ versus T^{-1} for MAPbI$_3$, FAPbI$_3$, and MA$_x$FA$_{1-x}$PbI$_3$[58]. Copyright 2015, American Chemical Society. (g) Temperature-dependent chronophotoamperometry measurements of a MAPbI$_3$-based cell. (h) Arrhenius plot of the photocurrent relaxation rates[38]. Copyright 2015, Springer Nature Limited.

a larger E_a of 0.36 eV assigned to the electromigration of MA$^+$.[49] In the following year, they studied the effect of light on ion migration by calculating the E_a of ion conduction in both single crystal and polycrystalline films of MAPbI$_3$ under darkness and illumination. The E_a decreases dramatically under illumination, quantitatively implying the easier migration of ions under light. The largest energy barrier of ion migration is suggested in the single crystal, indicating that the grain boundaries could assist the ion migration.[35] Besides, Zhao et al. also demonstrated a reduction in E_a of ion migration under illumination on the basis of the ion conductivity.[57]

Electrochemical impedance spectroscopy (EIS) is another common method to characterize the ion migration in perovskite devices. Venkataraman et al. investigated the kinetics of ion transport of perovskite devices using EIS technique under dark and AM 1.5G illumination conditions. For MAPbI$_3$ devices, the Nyquist plot unequivocally shows a linear component in low-frequency regime, which can be attributed to Warburg ion diffusion simulated as Warburg element (Figure 11.6e). According to the equivalent circuit model, the E_a of ion diffusion could be determined based on the

equation of $\dfrac{1}{T_w} = \dfrac{k_B T}{h} \dfrac{a^2 \alpha}{L_D} \exp\left(\dfrac{-E_a}{k_B T}\right)$, where T_w, h, a, α, and L_D refer to the Warburg

time constant, Planck's constant, lattice parameter of perovskite, coordination factor and effective ion diffusion length, respectively. This equation also could be simplified as $\ln(1/TT_w) = C - E_a/(k_B T)$. And the E_a of ion diffusion for MAPbI$_3$, FAPbI$_3$, and MA$_x$FA$_{1-x}$PbI$_3$ could be easily calculated from the plot of $\ln[1/(T \cdot T_w)]$ vs T^{-1} via measuring the temperature-dependent impedance spectra (Figure 11.6f). The E_a of ion diffusion for MA$_x$FA$_{1-x}$PbI$_3$ and FAPbI$_3$ is higher than that for MAPbI$_3$, indicating higher stability in mixed ion perovskite and FAPbI$_3$. They also assumed that the discontinued Arrhenius-like behavior of ion diffusion in MAPbI$_3$ is related to the discontinuity in volume change caused by phase transition.[58] In fact, the photocurrent could be generated under illumination, and it can relax to equilibrium after removing the light. By measuring the photocurrent at different temperatures and times, the E_a of the relaxation of the device could be evaluated (Figure 11.6g and h). Islam et al. conducted the temperature-dependent chronophotoamperometry measurement of a perovskite device based on the structure of FTO/TiO$_2$/MAPbI$_3$/Spiro-OMeTAD/Au. The E_a of short-circuit photocurrent relaxation from both reverse (-0.5 V) and forward (1 V) bias was estimated to be 0.60–0.62 eV and 0.68 eV, respectively. This value of E_a is very close to that of the migration of I$^-$ vacancies, 0.58 eV, calculated by themselves.[38]

In addition to ionic conductivity, impendence, and photocurrent, capacitance measurement is also an ordinary method to quantify the ion migration by calculating the value of E_a. In 2019, Ehrler et al. performed the transient ion drift measurements by probing the change of capacitance with a voltage pulse at different temperatures based on the inverted planar PSCs architecture (ITO/NiO$_x$/MAPbI$_3$/C$_{60}$/BCP/Ag or Cu). It is clear that the behavior of capacitance transient is different at various temperatures, as shown in Figure 11.7a. And the capacitance could be estimated using the

formula of $C(t) = C(\infty) + \Delta C\left(1 - e^{\frac{-t}{\tau}}\right)$, where ΔC is the capacitance change because

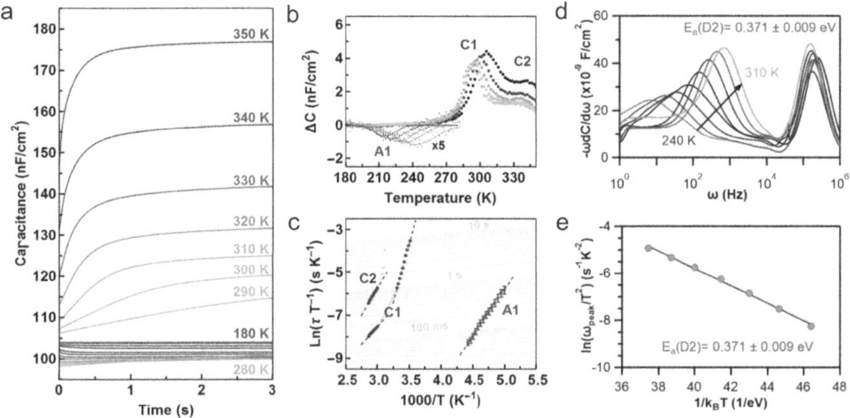

FIGURE 11.7 (a) Capacitance transient measurements at the temperature range of 180–350 K. (b) Rate-window plot of measured capacitance transients. (c) Arrhenius plot of the observed thermal emission rates versus temperature[59]. Copyright 2019, Royal Society of Chemistry. (d) -$\omega dC/d\omega$ versus ω spectra and (e) corresponding Arrhenius plot of PSC based on PTAA HTL[60]. Copyright 2020, Elsevier Inc.

of the ions drift towards the interfaces, $C(\infty)$ refers the steady-state capacitance, and τ is a time constant determined by $\tau = \dfrac{k_B T \varepsilon_0 \varepsilon}{q^2 DN}$. Herein, k_B, T, q, D, and N refer to Boltzmann constant, temperature, elementary charge, ion-diffusion coefficient, and doping density, respectively. The D could be calculated based on the equation of $D = \dfrac{\vartheta_0 d^2}{6} \exp\left(-\dfrac{E_a}{k_B T}\right)$, where ϑ_0, d refers to the attempt-to-escape frequency of an ionic jump and jump distance, respectively. The analysis of the rate window plot suggested three distinct processes labeled as A1, C1, and C2 (Figure 11.7b), where A1 can be assigned to the I⁻ ion migration and C1 and C2 are associated with the MA⁺ ion migration. Then the E_a of mobile ions were extracted based on the aforementioned formulas, as shown in Figure 11.7c. The E_a of A1, C1 and C2 were calculated to be 0.29 eV, 0.90 eV, and 0.39 eV, respectively, indicating that the I⁻ ions migration is the easiest in the MAPbI$_3$ perovskite device.[59]

Another capacitance-based technique is thermal admittance spectroscopy (TAS), which could provide the low-frequency capacitance feature. For TAS measurement, a small AC modulation voltage is imposed to trap and de-trap the charge carriers at fixed frequencies and temperatures, and their E_a could be extracted from the capacitance-frequency-temperature spectra (C-f-T). Specifically, the E_a can be estimated by fitting the formula of $\ln(\omega_{peak}/T^2) = A - E_a/(k_B T)$, where A, T, ω_{peak} refer to a constant, temperature, and frequency of inflection in the capacitance-angular frequency-temperature (C-ω-T) spectra that is manifested as peaks in the -$\omega dC/d\omega$ versus ω spectra (Figure 11.7d and e). Yan et al. performed the TAS measurement of inverted perovskite devices with different HTL (PEDOT:PSS and PTAA) or without

HTL. The E_a of 0.363 eV, 0.371 eV and 0.365 eV for ion migration can be assigned to the devices using PEDOT:PSS, PTAA, and without HTL, respectively. Through analysis, they demonstrated that the electrical transport characteristic cannot be evaluated using the low-frequency capacitance signatures owing to the coupling with ETL and HTL in n-i-p PSCs. Nevertheless, the ETL and HTL do not have an effect on the low-frequency capacitance in p-i-n PSCs, indicating that it is reasonable to characterize the ion migration by the calculated E_a.[60] On the other hand, the evolution of PL spectra under illumination at variant temperatures could also be used to investigate the photo-induced ion migration.[61]

In conclusion, the imaging and elemental characterizations and the temperature-dependent transient measurements are undoubtedly effective techniques to qualitatively and quantificationally study the electrical transport properties, i.e., ions migration of PSCs. By using a combination of these techniques, researchers can obtain a more complete understanding of ion migration in perovskite materials and develop strategies for optimizing the photoelectric performance of PSCs. However, it is very difficult to exactly distinguish which ion migrates and where it migrates in mixed perovskite materials. In addition to the key parameter of E_a, other parameters may play a significant role in understanding the ions migration, such as the concentration of mobile ions and their diffusion coefficient. Although the efficiency of PSCs has achieved unprecedented development, the perovskite absorber also has become complicated compared to the $MAPbI_3$. So far, extensive works of ion migration are based on the simple perovskite; it is still useful in mixed perovskite. But the migration of various ions is substantially influenced by the chemical composition and actual experimental environment. Thus, the investigation of ion migration in the mixed perovskite is still limited.

11.5   EFFECT OF ION MIGRATION ON STABILITY

It is well known that degradation caused by moisture and oxygen has been a major issue for long-term stability of PSCs. Nevertheless, these factors may not be important if the device is encapsulated in the nitrogen-protected glove box. The ion migration becomes more serious under the operated conditions. For example, the illumination, heat, and built-in electric field provide extra force for ion migration, which will produce adverse effects on the operational stability. On the other hand, the ion will migrate and accumulate at the interfaces of perovskite film and charge transport layer or penetrate into other layers, which is also detrimental to the long-term stability of PSCs. As previously mentioned, ion migration within the perovskite has significant effects on the PSCs, such as switchable photocurrent under applied bias,[46] photoinduced phase segregation,[54] and notorious current-voltage hysteresis.[62] In particular, the ion migration could generate more defects in the perovskite, which may be the centers of non-radiative recombination, increasing the energy loss and impairing the stability of PSCs. Meanwhile, mobile ions could accumulate at the interface of perovskite/HTL or perovskite/ETL under continued illumination and external electric field, producing a reverse polarization electric field to adversely influence the carrier dynamic in PSCs. It should be noted that the mobile ions have very low threshold electric field to move, i.e., 0.3 V μm^{-1}, which is far less than the built-in electrical field in PSCs, indicating faster ion migration under external bias.[49] Furthermore, the ion migration may change the morphology of perovskite film and accelerate the degradation of perovskite, which is harmful to the stability of PSCs. Thus, the operational stability of PSCs is significantly impacted by ion migration under illumination and electric field or other harsh conditions.

Although a plethora of works about ion migration has been reported, more deeply understanding is anticipated to provide a universal strategy to not only make a breakthrough for efficiency, but also ameliorate the stability which could meet the standards of commercialization. The impact of ion migration on the stability of PSCs will be introduced elaborately and systematically in the sections that follow, which involve current-voltage hysteresis, the changed electronic structure including the band bending, the electron-hole recombination rate, and the conductivity, phase segregation, formation of nanoscale clusters, and morphology changes of perovskite film (the formation of pinholes).

First of all, the current-density hysteresis has been a severe issue for practical application, leading to the imprecise estimation of efficiency and inferior photoelectrical properties. The current density-voltage (J-V) hysteresis refers to the phenomenon that the PCE in the reverse scan direction is usually higher than the forward one. Three mechanisms have been conjectured for this hysteresis: (i) charge recombination induced by the defects of perovskite, (ii) the ion migration during the measurement process, and (iii) the intrinsic ferroelectric effect of perovskite.[63,64] Barnes et al. applied one-dimensional time-dependent drift diffusion model to confirm the speculation that both interfacial recombination and ion migration contribute to hysteresis.[62] Moreover, Yang et al. suggested that hysteresis is induced by ion migration under external electric field and light illumination. Experimentally, they verified the reduced hysteresis with introduction of Br in the $MAPbI_3$ lattice. The DFT simulation further revealed an increase of E_a for iodine migration after partly substitution of Br, indicating that the ion migration is a plausible reason for the reduced hysteresis in $MAPbBr_xI_{3-x}$ device.[65] Overall, ion migration definitely plays a vital role in carrier dynamics within perovskite, thus affecting the J-V hysteresis. As for PSCs, serious hysteresis could cause the instability of devices.[66] Therefore, it is essential to design hysteresis-free perovskite devices to achieve high efficiency and stability of PSCs.

Moreover, the performance of PSCs gradually increases and eventually saturates after exposure to light illumination. Particularly, the short-circuit current density (J_{SC}), open-circuit voltage (V_{OC}), fill factor (FF), and the PCE exhibit an apparent enhancement. This increase behavior is called the light-soaking effect, which is related to ion migration. The ions will migrate and accumulate to the interfaces under light illumination (ions redistribution), altering the landscape of the built-in electric field (E_{bi}), thereby influencing the carrier collection. Actually, the E_{bi} drives the ions to drift and the ions accumulate close to the interface in the dark,[38] even without performing any measurements. The perovskite layer could be divided into three different regions (I, II, III). For I and III regions, ion migration results in n-doping and p-doping near the ETL and HTL, respectively, increasing the interfacial field and bending the energy bands.[46,47] Besides, the halide ions could migrate towards the HTL layer and undermine their electronic properties, thus affecting the device's stability. For example, Kim et al. investigated the conductivity of pure Spiro-OMeTAD film and Spiro-OMeTAD on $MAPbI_3$ at 85 °C, as shown in Figure 11.8a and b. The conductivity of pure Spiro-OMeTAD film increased with increasing the treatment time due to the oxidation of Spiro-OMeTAD, while the conductivity of Spiro-OMeTAD on $MAPbI_3$ decreased, implying that the oxidation process was impeded owing to the interfacial reaction. They suggested that the iodide ions could diffuse into the Spiro-OMeTAD side and hinder the oxidation of Spiro-OMeTAD under thermal condition, which deteriorates the conductivity of Spiro-OMeTAD. On the other hand, the energy barrier formed at the interface of $MAPbI_3$ and Spiro-OMeTAD due to the energy band bending, which hinders the carrier dynamics, thereby resulting in the degradation of Spiro-OMeTAD and the stability of devices.[68] Moreover, the iodide ions can migrate to the HTL irreversibly and react with Ag electrode, then generate the stable AgI compound, resulting in the degradation of PSCs.[11] Zhao et al. found that MA^+ ions could come cross the Spiro-OMeTAD during the operation and induce the degradation of Spiro-OMeTAD.[69] The halide ions could also move into the PTAA layer, which significantly reduces its p-doping property and adversely affects the extraction and transport of the hole.[70] All these results imply that the severe ion migration could change the electronic properties of the perovskite layer, charge transport layer, or interface between them, deteriorating the hole transport ability and performance of PSCs. When considering the stability of PSCs, it is necessary to pay more attention to the effect of ion migration on the absorber and transport layers.

Ion migration could lead to the change of stoichiometry and the redistribution of element in perovskite, further causing severe phase segregation, thereby impacting the device stability. Generally, the bandgap of $MAPb(Br_xI_{1-x})_3$ perovskite will expanded with increasing the content of Br, leading to an enhancement of V_{OC}.[71] Nevertheless, McGehee et al. observed that the PCE of mixed $MAPb(Br_xI_{1-x})_3$ perovskite device was reduced due to the light-induced halide segregation. In other words, the ion migration led to the formation of the bromide-enriched majority and iodide-rich minority domains. The latter domains may act as carrier traps, causing non-radiative charge recombination.[72] This phase segregation may be ascribed to the localized strain due to the interaction of a single photoexcited charge with soft and ionic perovskite lattice.[73] It should be noted that the light-induced halide segregation might be observed when the VBM is deeply tuned enough by changing the bromide content. Because the deep VBM could activate the pathways of iodide oxidation, there is a threshold bromide content for photo-induced segregation (< 20% for $MAPb(Br_xI_{1-x})_3$).[74] Besides the localized strain, Rand et al. pointed out that the iodide oxidation and its mobilization also play a crucial role in phase segregation of mixed-halide perovskite. And the rate of iodide oxidation is related to the oxidation potential of iodide ligands, which is set by the non-bonding orbitals.[74] This light-induced instability of perovskite film is expected to impact the operational stability of PSCs based on this kind of material. Besides the hybrid halide system (Br and I alloying) discussed earlier, Zhou et al. studied the performance of PSCs based on $FA_{0.9}Cs_{0.1}PbI_3$ perovskite under different conditions, including dark, heat, illumination, and stabilized power output (SPO).

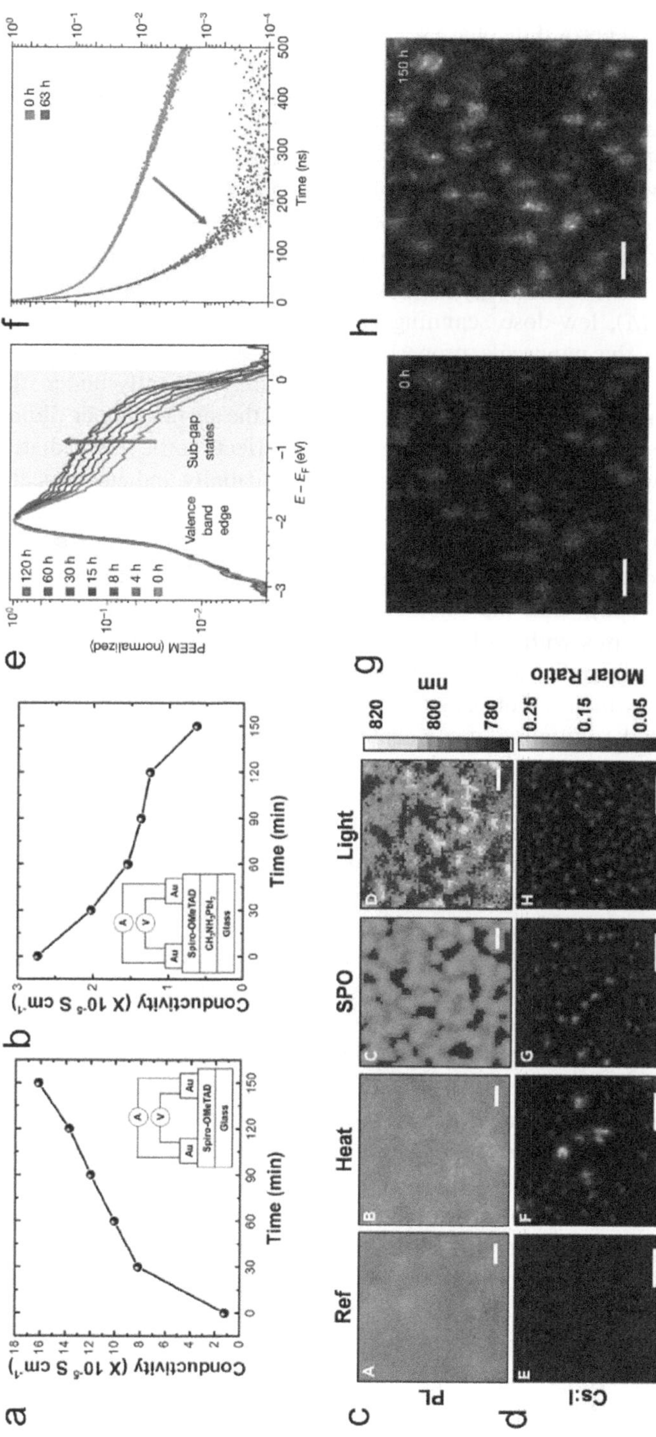

FIGURE 11.8 Dark conductivity of (a) spiro-OMeTAD and (b) MAPbI$_3$/spiro-OMeTAD at 85 °C with increasing the treatment time. Inset shows structures for the measurement of conductivity of glass/spiro-OMeTAD/Au and glass/MAPbI$_3$/spiro-OMeTAD/Au[68]. Copyright 2017, Springer Nature Limited. (c) The PL mapping images of perovskite films originated from PSCs and (d) X-ray fluorescence (XRF) mapping of perovskite devices that are subjected to ref, heat, light, and SPO stressing conditions. The scale bar is 5 mm[75]. Copyright 2020, Elsevier Inc. (e) Spatially averaged photoemission spectra at different time snapshots of in-situ solar-equivalent-illumination dose. (f) TRPL decay before and after 63 h of solar-equivalent illumination. Spatially resolved PEEM intensity at sub-bandgap energy $E - E_F = -0.83 \pm 0.15$ eV after 0 h (g) and 150 h (h) of in-situ solar-equivalent illumination[76]. Copyright 2020, Springer Nature Limited.

It was found that the performance of PSCs degraded under illumination and SPO conditions, because the initial perovskite phase segregated into FA-rich ($FA_{>0.9}Cs_{<0.1}PbI_3$, yellow regions of PL maps) and Cs-rich ($FA_{<0.9}Cs_{>0.1}PbI_3$, brown regions of PL maps) regions shown in Figure 11.8c. And the nano-XRF measurement (Figure 11.8d) further verifies the formation of Cs-rich clusters, which is photoinactive and current blocking, mainly contributing to the performance degradation.

Moreover, the DFT calculations demonstrated that the phase separation was induced by a thermodynamic driving force, which is caused by the migration of photo-induced carriers into FA-rich regions under continuous illumination.[75] Very recently, Stranks et al. used microscopy techniques, including photoemission electron microscopy (PEEM), low-dose scanning electron diffraction (SED), and PL mapping to investigate the nanoscale properties of perovskite film composed of $Cs_{0.05}FA_{0.78}MA_{0.17}Pb(I_{0.83}Br_{0.17})_3$ and their electronic structure transiently under illumination. It is found that the nanoscale clusters formed on the surface under illumination and acted as surface traps of hole charge carriers, affecting the non-radiative recombination of perovskite film. The PEEM was used to spatially and energetically detect valence band and occupied sub-bandgap states on the surface of perovskite film. The SED and PL mapping were used to spatially characterize the structure properties and non-radiative performance losses, respectively. Spatially averaged photoemission spectra show that the photoemission intensity at energies in the sub-bandgap range increases with prolonging the illumination time (Figure 11.8e). The increase of sub-bandgap distribution is related to a reduction in the PL lifetime shown in Figure 11.8f, which is consistent with an enhancement of non-radiative recombination sites for the initially defective sites after exposure to light condition (Figure 11.8g and h).

Moreover, the morphology of perovskite was also changed after illumination, and the pinholes were preferentially formed at the grain boundaries. They also suggested that the pinholes appear to generate at the trap clusters. Such morphological imperfections will accelerate the degradation of perovskite film, thereby deteriorating the stability of PSCs. The low-dose SED measurements further show obvious structure changes in local areas after exposure to light for 1 h, indicating a slight reorientation of grain in the pristine perovskite film. By contrast, for the phase impurities region, its crystal structure substantially changes. From the diffraction pattern of phase impurity, a large amount of streaking in the (002) direction was observed, attributing to the multiple degradation products and/or high density of new structural defects. Based on these results, they concluded that the accumulated trap clusters and their related trapping of photoexcited carriers play a crucial role in local perovskite layer degradation.[76] Such phase segregation and phase impurities under illumination are considered to be a more serious issue for stability of PSCs because of the ionic nature of perovskite and the interaction between perovskites and light, such as light-induced ion migration and defect generation. In conclusion, ion migration is a key factor that affects the stability of perovskite materials and PSCs and can cause significant impacts on the electronic structure, morphological structure, performance of films, and devices. Understanding the mechanisms of ion migration and developing strategies for mitigating its effects are crucial for the optimization and improvement of the stability of PSCs.

11.6 MITIGATION AND SUPPRESSION OF ION MIGRATION

As mentioned earlier, the ion migration and accumulation could significantly impact the photoelectric properties of PSCs. Numerous studies have demonstrated that the ion migration could lead to J-V hysteresis, band bending, phase segregation, etc., eventually causing the degradation of PSCs. In particular, the long-term instability of PSCs is mainly attributed to the ion migration because the ions easily move under operational conditions, like continuous light illumination and electric field. Therefore, suppressing the ion migration is of great importance to improve the long-term stability of PSCs. Moreover, the ion migration is dependent on the defect in halide perovskite, as well as the grain boundaries and surface. Thus, it will be helpful to increase the formation energy of vacancies, passivating the defects at the grain boundaries and surface to mitigate the ion migration. This section will provide several strategies to effectively mitigate the intrinsic and extrinsic ion migration from the aspects of composition modulation, additive regulation, and interface passivation.

11.6.1 COMPOSITION MODULATION

The Goldschmidt tolerance factor is an important parameter to evaluate the structural stability of perovskite materials, which is strongly dependent on the ionic radius of component in perovskite. Moreover, the formation energies of defect and E_a of ion movement for halide perovskite are highly related to the Coulomb interaction between crystal lattice and ions. As a result, it is imperative to precisely modulate the perovskite composition to increase the binding energy of each component and inhibit the ion migration, thereby improving the structural stability of PSCs. In 2016, Karunadasa et al. observed the light-induced PL redshift and appearance of new peak at 0.6–1.6 GPa in $MAPb(Br_xI_{1-x})_3$ powders, which is attributed to the halide migration under illumination. It is also found that the PL intensity decreases with increasing pressure. At 0.9 GPa, only the new peak could be observed, and the PL intensity remains unchanged with increasing light-exposure time, indicating that compressed perovskite materials could hamper the halide segregation due to the mitigation of halide movement.[77] However, the commercial PSCs could not operate under high pressure since the cost will be increased. Therefore, an alternative method of chemical pressure is suggested, which means the ions in perovskites are substituted by other ions with larger ionic radii than the original ions in perovskite.

Many endeavors of the A-site cation substitution have been made to improve the efficiency and stability of PSCs. For instance, the MA^+ ions in $MAPbI_3$ perovskite could be partially substituted by the larger organic cations of guanidinium (GA^+), leading to more stable and efficient PSCs.[78–80] As reported, the ionic radius of GA^+ is ~278 pm, which sightly exceeds the upper limit of the tolerance factor ($t = $ ~1.03), destructing the 3D perovskite phase and forming low-dimensional perovskite.[78,81] Thus, it is strict to manage the concentration of GA^+. In 2017, Nazeeruddin et al. reported that the 3D perovskite of $MA_{1-x}GA_xPbI_3$ ($0 < x < 0.25$) with improved thermal and environmental stability could form when the GA^+ was inserted into the crystal lattice of $MAPbI_3$. XRD measurement was conducted to characterize the change of the crystal lattice, as shown in Figure 11.9a. The diffraction intensity of $MAPbI_3$

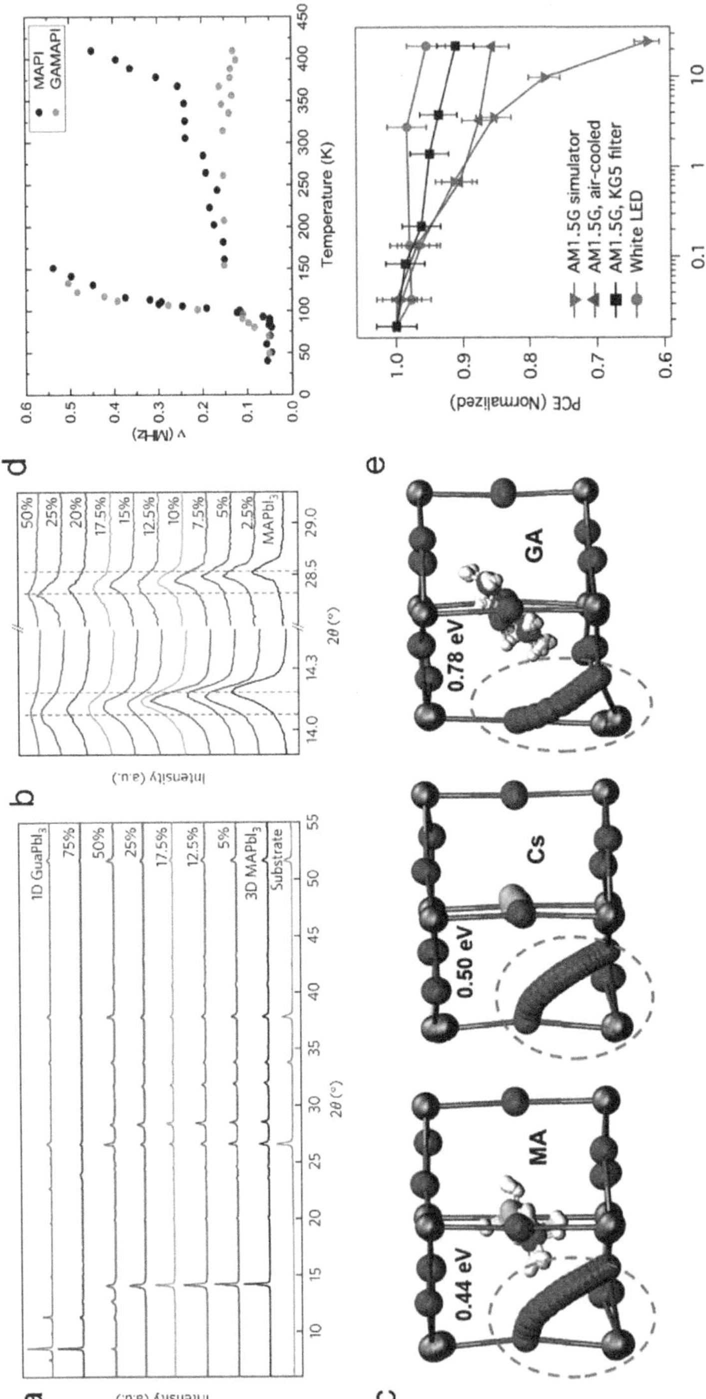

FIGURE 11.9 (a) Normalized XRD data for the mixed MA/GA perovskite films containing different percentages of GA. (b) Magnification of the XRD peaks at (220) (left) and (440) (right) on variation of the GA content (indicated in %)[79]. Copyright 2017, Springer Nature Limited. (c) Ab initio simulations of the activation energies of GA substitution. (d) Temperature dependence of muon spins fluctuation rate for $MAPbI_3$ and $MA_{0.95}GA_{0.5}PbI_3$[80]. Copyright 2019, Royal Society of Chemistry. (e) $MAPbI_3$ perovskite device stability (normalized PCE) under different consitions[58]. Copyright 2015, American Chemical Society.

phase shows a gradual decrease and new diffraction peaks appear with increasing the GA^+ percentage, indicating the formation of 1D $GAPbI_3$ phase. It is worth noting that the $MAPbI_3$ tetragonal phase could be maintained when the percentage of GA^+ is less than or equal to 25%. From the magnification of XRD peaks corresponding to (110) and (220) lattice planes (Figure 11.9b), the peaks gradually shift to lower angles, which means the expansion of crystal lattice, implying that a mixed perovskite $MA_{1-x}GA_xPbI_3$ could be formed by substituting the MA^+ by larger GA^+ cations. The lattice parameters of a and b could be determined from the XRD patterns. The a value gradually increases from 8.838 Å to 8.902 Å and remains constant when the content of GA^+ is larger than 25%. However, the c value does not change obviously for all percentages of GA^+. These results suggest that the crystal distortion mainly impacting a lattice parameter occurs at $x_{GA+} < 0.25$, indicating that further increment of GA^+ would not insert into the 3D perovskite network.[79] Overall, this distortion of crystal lattice caused by the introduction of GA^+ substantially improved the stability of PSCs and suppressed the iodine ions migration in the halide perovskite.[79,80]

The theoretical simulations indicated that the cations substitution in $MAPbI_3$ could improve the E_a of iodide ion movement. Particularly, just 5% GA^+ substitution could steeply increase the E_a from 0.44 eV to 0.78 eV, significantly inhibiting the iodide ion transport compared to Cs^+ substitution. Moreover, the simulations also show a curved migration channel for iodide migration in $MAPbI_3$ and Cs-substituted perovskite, and the displacement of surrounding Pb ion is small. Nevertheless, for GA-substituted perovskite, the local distortion of the inorganic framework causes a larger displacement of surrounding Pb ions, leading to a curved path bowed out of the Pb/I plane (Figure 11.9c), thereby inducing a higher E_a. The increased E_a in GA-substituted perovskite is mainly ascribed to the local deformation of Pb/I cage induced by the GA^+ cation size mismatch with MA^+. Experimentally, they further studied the iodide ion diffusion in pure $MAPbI_3$ and GA-substituted perovskite by using the muon spin relaxation measurement (μSR). In a μSR measurement, muons are implanted within the sample. On the decay of muon, a positron is most likely emitted along the direction of the muon spin. The local magnetic field on the sample could impact the muon spins, which is detected by asymmetry signal in the positron counts. By fitting the asymmetry data at different temperatures, the fluctuation rate (v) could be calculated. Figure 11.9d shows the temperature dependence of v for $MAPbI_3$ and GA-substituted perovskite. For $MAPbI_3$, there are two clear increases in v. They suggested that the first increase of v at 100 K is due to the improved movement of molecular cation within the Pb/I cages. When the temperature is higher than 150 K, the v decreases because of phase transition of perovskite phase where the cation migration is too fast to be observed by μSR. They proposed that the low temperature process is caused by molecular cation reorientation within the Pb/I cage. For GA-substituted perovskite, the v remains constant above room temperature, indicating that the iodide ions migration could be suppressed at higher temperature and no iodide ion transport is found above 420 K.[80] In a word, the GA^+-doped perovskite displays better long-term stability of devices due to the reduction of iodide ion migration.

Furthermore, the incorporation of FA^+ cation into $MAPbI_3$ perovskite could effectively suppress the ion migration and enhance operational stability. As reported, the degradation of $MAPbI_3$ device under AM 1.5G illumination is associated with the thermal activation of the visible light-driven ion migration and lattice expansion, both of which are induced by the infrared radiation (IR) of the solar spectrum. According to this result, it is found that the operational stability of $MAPbI_3$ PSCs is substantially promoted under while-LED light, AM 1.5G solar simulator with an IR cutoff filter or a fan cooler, as shown in Figure 11.9e. Through the XRD and EIS results, they concluded that the crystal lattice of FA^+-doped perovskite shows quite low thermal expansion under light illumination compared to $MAPbI_3$. And a higher E_a barrier and lower diffusion coefficient of ions in FA^+-doped perovskite were suggested, contributing to the improved stability of the device.[58]

According to the theory of the Goldschmidt tolerance factor, the halide perovskite could form photoactive cubic phase when the tolerance factor is in the range of 0.8–1.0. If the tolerance factor is too high or too low, the non-perovskite phase with the structure of either hexagonal or orthorhombic will be generated. The tolerance factor of $APbI_3$ perovskite (A refers to alkali metal ions or MA^+ or FA^+) is shown in Figure 11.10a. The pure $FAPbI_3$ with a tolerance factor of 1.0 could not display a stable cubic perovskite phase at room temperature. Though the tetragonal $MAPbI_3$ phase has a more suitable tolerance factor, its structure could be destroyed by environmental conditions. As discussed previously, the combination of MA^+ and FA^+ could effectively improve the structural stability. However, phase segregation still occurs in the MAFA-mixed perovskite because of the similar cation radius of MA^+ (270 pm) and FA^+ (279 pm).[82] In addition to the larger-size cation of GA^+ and FA^+, the alkali metal ions, such as cesium (Cs^+), rubidium (Rb^+), potassium (K^+), sodium (Na^+), and Lithium (Li^+), have been widely inserted into perovskite lattice to prevent the ion transport and improve the stability of the device.[82–87] Based on the calculated tolerance factor (Figure 11.10a), the $CsPbI_3$ perovskite shows a tolerance factor of 0.81. Thus, the Cs^+ with an ionic radius of 167 pm could be regarded to substitute the A-site cation. And the combination of Cs^+ and other organic cations has been considered to be a useful method to obtain appropriate tolerance factor and strengthen structural stability. Yuan et al. reported that the incorporation of 5% Cs^+ in $MAPbI_3$ perovskite lattice could suppress the ion migration and the formation of reactive superoxide ions which react with the organic cation in perovskite under oxygen and light, resulting in the high efficiency of PSCs with improved stability. As seen from Figure 11.10b and c, the activity energy in Cs^+-doped $MAPbI_3$ film was increased to 0.50 eV from 0.37 eV in $MAPbI_3$ film, indicating enhanced ions migration barrier which may be associated with the rigidized inorganic PbI_6 octahedra networks. Moreover, it can be observed that the activity energy of Cs^+-doped $MAPbI_3$ remains almost constant after treatment with light and oxygen, implying superior photo-oxygen stability. Based on the theoretical calculation, the increased activity energy is attributed to more narrow channels of ion migration because of the repelling effects induced by the incorporation of Cs^+.[83] When the MA^+ cation is fully substituted by Cs^+, all-inorganic perovskite ($CsPbI_2Br$), the light-induced ion migration can be significantly eliminated, causing ultra-stable PSCs. As shown in Figure 11.10d–f, the ion migration barrier for $MAPbI_3$ reduces from 0.67 eV to 0.07 eV with increase in light intensity from 0.1 to 25 mW/cm^2, implying more severe ion migration. In contrast, the ion migration barrier for $CsPbI_2Br$ remains invariable

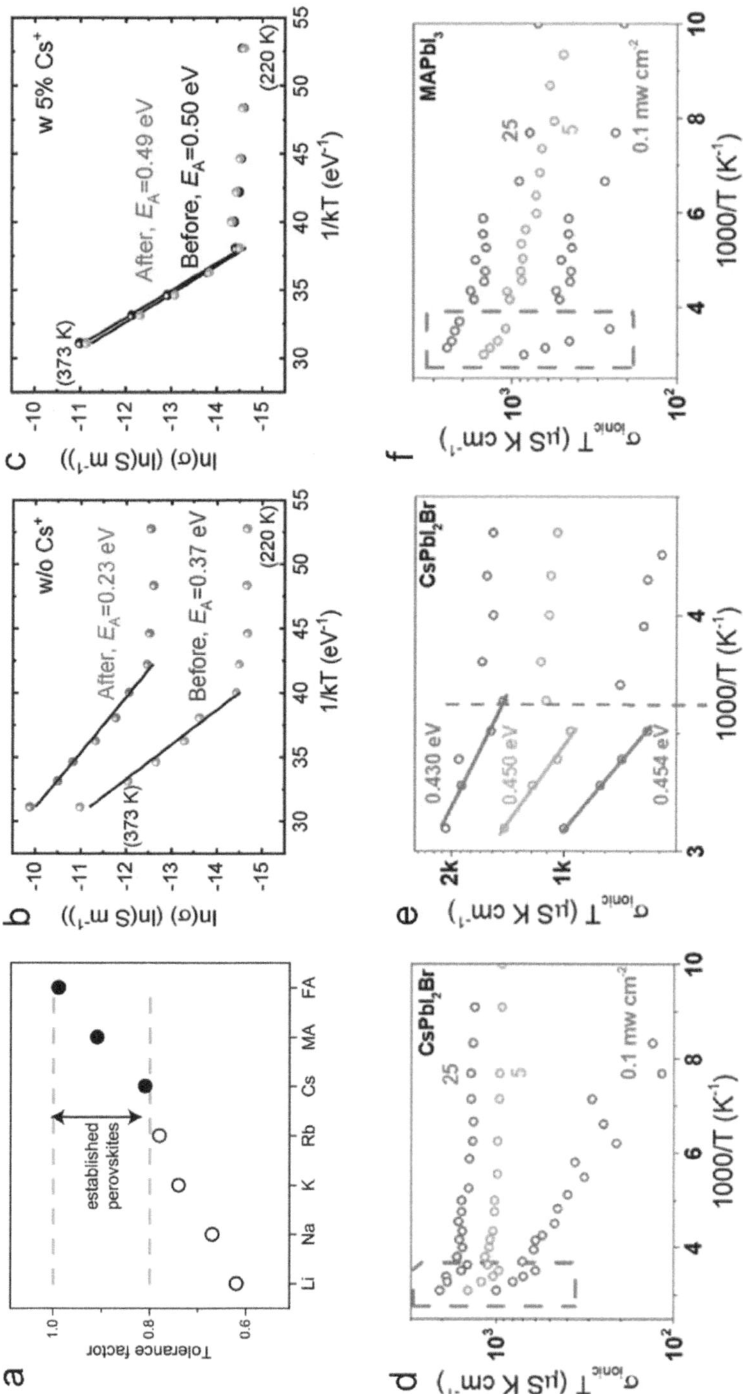

FIGURE 11.10 (a) Tolerance factor of APbI₃ perovskite with the oxidation-stable A (Li, Na, K, Rb, or Cs) and MA or FA[82]. Copyright 2017, American Association for the Advancement of Science. Temperature-dependent conductivity of MAPbI₃ (b) and MA₀.₉₅Cs₀.₀₅PbI₃ films (c) before and after 1 h of oxygen and photo-induced decay, respectively[83]. Copyright 2021, Wiley-VCH GmbH. Ionic conductivity of CsPbI₂Br (d) and MAPbI₃ (f) films σ_ion T with different light intensities. (e) Magnified data of the dashed box in (d)[84]. Copyright 2017, American Chemical Society.

(0.45–0.43 eV) under various light intensities, indicating the elimination of light-induced ion migration. Experimentally, the J-V hysteresis and the operational stability could be effectively ameliorated.[84]

As for the alkali metal cation of Rb^{2+}, the tolerance factor is calculated to be 0.78 based on a smaller ionic radius of 152 nm, leading to the different nucleation and growth process because of a mismatch with the stable perovskite phase. When annealing the $RbPbI_3$ film at different temperatures, the film is yellow at 25 °C. Upon heating to 460 °C, the film still keeps yellow, rather than black phase, which could explain why $RbPbI_3$ has not been applied for PSCs. But it could be embedded into multiple A-cation perovskite phase to stabilize the FA-based perovskite, despite not being adequate as a pure $RbPbI_3$ perovskite.[82] Besides, whether the Rb^+ cation occupies the perovskite lattice is still up for debate because of the different ionic radius. Therefore, it is important and meaningful to reveal the occupied positions of alkali ions with different ionic radii, which play an unexpected role in the crystal structure and operational stability. Zhao et al. investigated the size-dependent interstitial occupancy of Rb^+, K^+, Na^+, and Li^+ in a triple cation perovskite of $Cs_{0.11}MA_{0.15}FA_{0.74}Pb(Br_{0.17}I_{0.83})_3$. Experimental results displayed that the incorporation of Rb^+, K^+, Na^+ and Li^+ cations led to lattice expansion, which is typically related to the interstitial occupancy of ions (Figure 11.11a). Figure 11.11b shows the formation energy of $FAPbI_3$ and alkali cation doped $FAPbI_3$ at zero temperature. It can be found that the larger the alkali cations, the smaller the formation energy, as for A-site substitution. For interstitial occupancy, the formation energy decreases with decreasing the ionic radius and becomes larger upon increasing the concentration of 0.5% to 6%. For alkali cations with the smallest ionic radius, Li^+ and Na^+, these cations prefer to occupancy the interstitial sites when the doping concentration is low, implying more difficulty in the incorporation of Li^+ and Na^+ into the perovskite lattice. To further investigate how the interstitial ions of K^+ impact the ion migration, the iodide ion migration barriers of different transport paths were calculated using DFT calculation (Figure 11.11c). For all transport paths, the incorporation of K^+ cation increases the diffusion barrier, meaning that the ion migration was suppressed, which is also evidenced by the increased E_a determined from temperature-dependent conductivity measurement (Figure 11.11d). Moreover, extrinsic alkali ions doped perovskite also shows highly suppressed phase segregation, indicating the improved photostability which plays a key role in achieving high-performance PSCs.[85] Therefore, the incorporation of alkali cation into perovskite lattice could be a general method for mitigating the ion migration and improving the stability in PSCs. It is noted that the doping concentration and the type of alkali cation should be carefully considered to eliminate the negative effects on the 3D perovskite lattice and additional migration of ions.

Besides, the substitution of B-site ions could change the Coulomb force, increasing the E_a of ion migration. Jen et al. reported that the wide bandgap perovskite of $MAPb(I_{0.6}Br_{0.4})_3$ could be stabilized by partially substituting the Pb^{2+} with Sn^{2+} to form $MAPb_{0.75}Sn_{0.25}(I_{0.6}Br_{0.4})_3$. They hypothesized that the change of internal bonding environment (increased crystallite size and macrostrain) raises the energy barrier for ion diffusing and inhibits phase segregation under illumination.[88] Recently, Sirringhaus and colleagues visualized the suppression of ionic migration by the substitution of Pb with Sn. From the PL mapping results of $CsFAPbI_3$ (Figure 11.11e–g),

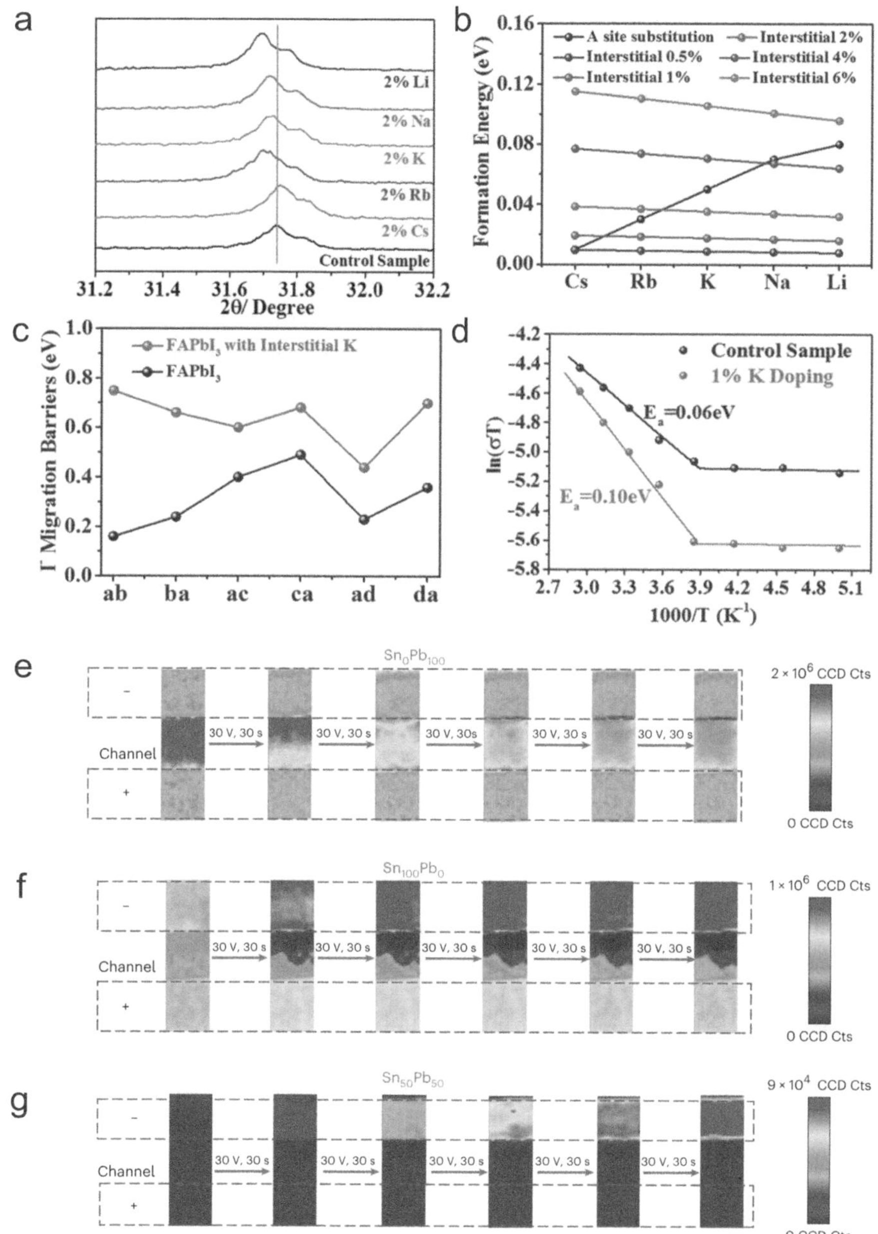

FIGURE 11.11 (a) The (310) peak of the XRD patterns of 2% Cs^+, Rb^+, K^+, Na^+, or Li^+ embedded Cs/MA/FA perovskite. (b) Formation energy of alkali cation incorporation at an A site (being constant for concentration from 0.5 to 6%) and at an interstitial site. (c) Comparison of the I^- diffusion barriers in $FAPbI_3$ with and without interstitial K^+. (d) Extraction of the activation energy from the temperature-dependent conductivity measurement of Cs/MA/FA perovskite and 1% K^+-doped Cs/MA/FA perovskite[85]. Copyright 2018, Wiley-VCH GmbH. PL mapping performed on lateral channel devices (L = 20 μm) of $Cs_{0.15}FA_{0.85}PbI_3$ (e), $Cs_{0.15}FA_{0.85}SnI_3$ (f) and $Cs_{0.15}FA_{0.85}Pb_{0.5}Sn_{0.5}I_3$ (g) perovskites upon multiple biasing (30 V, 30 s)[89]. Copyright 2023, Springer Nature Limited.

the devices without bias displays a homogeneous PL intensity, while the PL intensity under bias is weakened near the positively biased electrode, which can be assigned to the halide migration upon applying the electronic filed. For $CsFASnI_3$ devices, the PL intensity shows a change after the first bias and keeps constant. The PL intensity of $CsFAPb_{0.5}Sn_{0.5}I_3$ device does not show appreciable variation upon bias, indicating the suppressed migration of ions under external bias.[89] For single crystal $CsPbBr_3$ perovskite, Ag^+ and Bi^{3+} replace the Pb^{2+}, forming $Cs_2AgBiBr_6$. This could effectively increase the E_a of ion migration, indicating that ion migration in $Cs_2AgBiBr_6$ is more difficult than that in $MAPbBr_3$. From theoretical calculations, they further verify that the diffusion barrier for Br vacancies in $Cs_2AgBiBr_6$ is calculated to be 0.33 eV, which is higher than in $MAPbBr_3$ with diffusion barrier of 0.2 eV, expecting higher stability.[90] Based on these investigations, the substitution of B-site ions could also be a valid method to inhibit ion diffusion under illumination or an electric field.

As discussed earlier, the iodide migration is easily observed in halide perovskite. Thus, it is reasonable to replace the X-site to suppress the ion migration. For instance, the incorporation of Br content in $MAPbI_3$ perovskite is verified to reduce hysteresis significantly. Based on the DFT calculations (Figure 11.12a–c), the E_a of MA^+ migration is relatively similar for $MAPbI_3$ and $MAPbBr_xI_{3-x}$. The E_a of Br^- for $MAPbBr_xI_{3-x}$ is calculated to be 0.326 eV, similar to that of I^- for $MAPbI_3$. However, the E_a of I^- for $MAPbBr_xI_{3-x}$ is increased to 0.46 eV, shown in Figure 11.12c. The less mobile I^- is likely ascribed to stronger confinement of the surrounding atoms in $MAPbBr_xI_{3-x}$ lattice. Besides, the formation energy of iodide vacancies for $MAPbBr_xI_{3-x}$ (3.6 eV) is much higher than that for $MAPbI_3$ (2 eV), corresponding to reducing the trap density of iodide vacancies. These results demonstrated suppressed ion migration and improved structural stability for $MAPbBr_xI_{3-x}$ perovskite.[65] Nevertheless, phase segregation is prone to occur for the mixed-halide perovskites with the incorporation of Br atom, which is still an obstacle limiting the stability of PSCs. Thus, replacing the X-site may not be an appropriate method to mitigate ion diffusion and improve stability.

On the other hand, tremendous studies have proposed that low-dimensional perovskites have better stability than 3D perovskite both under illumination and dark conditions, due to higher formation energy of imperfections in low-dimensional perovskites.[91,92] Most importantly, large ammonium cations incorporated into the low-dimensional perovskites were regarded as spacers, which not only hinders the penetration of moisture but also mitigates the migration of interstitial ions. Huang et al. first reported the suppressed ion migration in low-dimensional perovskites. They calculated the E_a of ion migration for 3D $MAPbI_3$ and 2D $(BA)_2(MA)_3Pb_4I_{13}$ (BA: n-butylammonium) by measuring the temperature-dependent electrical conductivity under 0.25-sun illumination and in the dark (Figure 11.12d–f). The conductivity at low temperature is mainly attributed to the electronic conduction, because of the mitigation of ion migration. For 3D perovskite, there is a clear transition process over 240 K, meaning that the ionic conductivity is dominated for total conductivity. However, such a transition process from electronic to ionic conductivity was not found in 2D $(BA)_2(MA)_3Pb_4I_{13}$ perovskite, indicating the absence of obvious ion migration, which will alleviate the degradation process related to ion migration.[93] Later, they further studied the ion migration properties for quasi-2D $(BA)_2(MA)_3Pb_3I_{10}$ ($n = 3$) single crystal, showing similar behavior for temperature-dependent conductivity

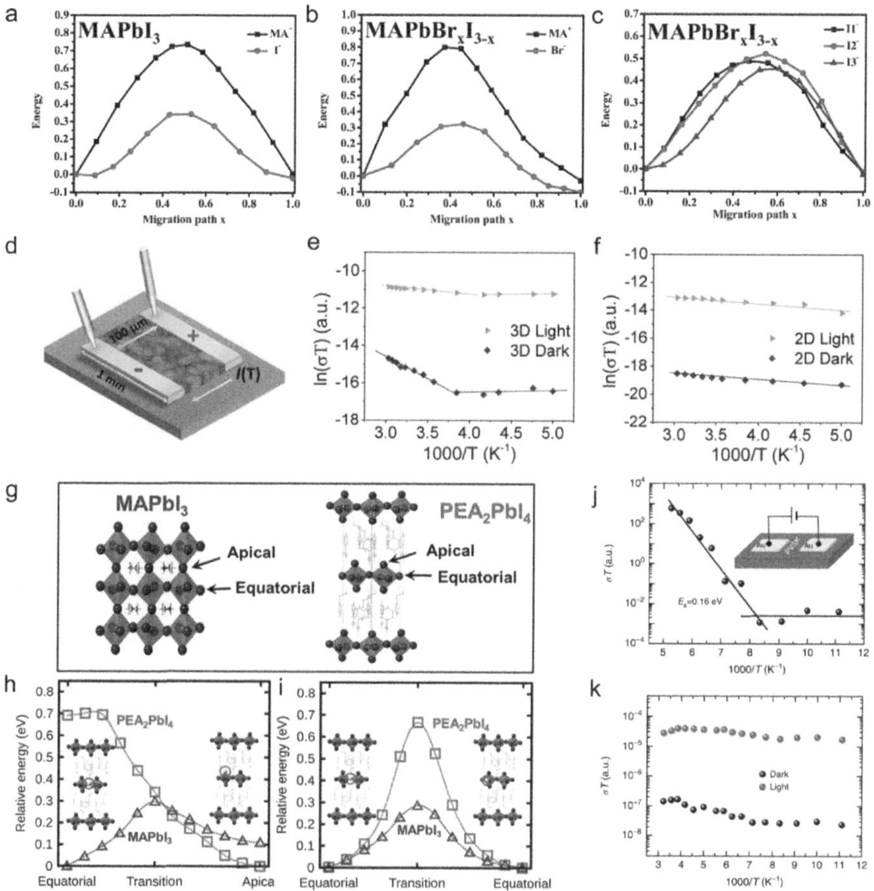

FIGURE 11.12 Energy profiles for I^-, Br^- and MA^+ migration in $MAPbI_3$ (a) and $MAPbBr_xI_{3-x}$ (b, c)[65]. Copyright 2016, Elsevier Ltd. Temperature-dependent conductivity measurement geometry (d) and the results of $MAPbI_3$ (e) and $(BA)_2(MA)_3Pb_4I_{13}$ (f) polycrystalline films[93]. Copyright 2017, American Chemical Society. (g–i) Energetics of iodide vacancies in $MAPbI_3$ and PEA_2PbI_4 with different migration channels using DFT calculations[95]. Copyright 2017, American Chemical Society. Conductivity of a bare $FAPbI_3$ film (j) and with 1.67 mol% 2D PEA_2PbI_4 perovskite (k) under different temperature[96]. Copyright 2018, Springer Nature Limited.

measurement. Similarly, the transition from electronic to ionic conductivity was not observed, implying the insignificance of ion migration induced current. The ions migrate through the vacancies in the quasi-2D layer, since the grain boundaries are negligible in quasi-2D single crystal. And the ion migration is mainly mediated by the vacancies. Therefore, they calculated the formation energy of I and MA vacancies to evaluate the ion migration. It is found that the formation energies for both vacancies are lower in $MAPbI_3$ than that in quasi-2D $(BA)_2(MA)_3Pb_3I_{10}$, meaning much lower vacancy states in low-dimensional perovskite, thereby generating a larger barrier for ion diffusion to improve the intrinsic stability.[94]

Because of 2D perovskite with higher stability, considerable efforts have been made to incorporate the 2D perovskite into 3D perovskite to improve the performance of PSCs. For example, a small amount of 2D perovskite of PEA_2PbI_4 (PEA = $C_6H_5(CH_2)_2NH_3$) was introduced into the $MAPbI_3$ perovskite precursor to prepare $(PEA_2PbI_4)_{0.017}(MAPbI_3)$ perovskite film, and the 2D perovskite is likely located at the grain boundaries suppressing the iodide migration, which results in reduced J-V hysteresis, improved efficiency and stability of PSCs. Specifically, the DFT results (Figure 11.12g) display the energy barrier of I vacancy transporting from an equatorial site to an apical site. It can be observed that the I vacancies in $MAPbI_3$ prefer to occupy the equatorial position rather than apical position due to the higher energy of I vacancies on apical position (0.11 eV). However, the I vacancies in PEA_2PbI_4 preferentially sit on apical position, because the energy of I vacancies on equatorial position is 0.69 eV which is much higher than on the apical position. And the energy barrier is only 0.01 eV migrating from an equatorial site to the apical site, meaning spontaneous migration along this path. Figures 11.12h and i show the energy barrier of I vacancy moving from an equatorial position to another equatorial position. It is clear that movement between two equatorial positions shows a higher energy barrier of 0.67 eV in PEA_2PbI_4, implying the significantly limited migration between equatorial sites. Based on DFT analysis, they proposed that I vacancy in $(PEA_2PbI_4)_{0.017}(MAPbI_3)$ will preferably migrate from the bulk to an apical site, while the migration between two equatorial positions is considerably prevented. These results explain the suppression of iodide migration into Ag electrode and improved stability.[95] Similarly, the 2D perovskite PEA_2PbI_4 with a concentration of 1.67 mol% was incorporated into the $FAPbI_3$ perovskite precursor, where 2D perovskite is spontaneously formed at the grain boundaries to hinder moisture penetration and improve carrier collection. From the temperature-dependent conductivity measurement (Figures 11.12j and k), the conductivity of $FAPbI_3$ perovskite shows an exponential enhancement at around 130 K, corresponding to ionic conductivity. The E_a of migration in $FAPbI_3$ is estimated to be 0.16 eV, meaning that the active ions are dominated at room temperature, which induces the degradation of perovskite film and device. For $FAPbI_3$ perovskite incorporated 2D PEA_2PbI_4, the transition from electronic to ionic conductivity is not observed with increasing temperature.[96] Therefore, incorporating 2D perovskite passivated the grain boundaries of 3D perovskite, likely suppressing the ion migration.

11.6.2 Additive Regulation

Given that the ions are prone to migrate along the grain boundaries, thereby the reduced grain boundaries are conducive to suppressing ion diffusion. Therefore, another method that has been substantially investigated is the elimination of grain boundaries, namely enlarging the grain size. The grain boundaries are negligible for single perovskite crystals, which could dramatically impede ion migration. For instance, Huang et al. investigated the temperature-dependent conductivity of large-grain film (~1 μm), small-grain film (~0.3 μm), and single crystal under light illumination and in the dark, as shown in Figures 11.13a–c. The ion migration is easier for all samples under light illumination because of the decreased E_a. It is apparent that the E_a for ion migration in the single crystal is larger than that in polycrystalline films, demonstrating that the increased grain size is conducive to suppressing ion diffusion.[35] And the grain size is deeply impacted by the additive in the perovskite precursor, which could regulate the crystal growth. MACl is a widely used additive to increase the grain size of perovskite film by forming the intermediate phase.[97–99] As reported, the grain size and crystallinity of $FAPbI_3$ gradually increase with increasing the concentration of MACl to 40% mol, leading to high efficiency and stability of PSCs.[97]

Apart from the MACl additive, the $Pb(SCN)_2$ and trimethylammonium chloride (TACl) were also reported to increase the grain size of perovskite film.[100–102] The $CH_3NH_3^+$ cations react with SCN^- to form CH_3NH_2 and HSCN gases because of the low stability of $CH_3NH_3 \cdot SCN$ adduct. And the CH_3NH_2 gas is known to increase the grain size of perovskite film.[100] Zhao and colleagues developed a seed-assisted method to obtain extremely large gain size. They incorporated the cesium served as nuclei into the PbI_2 precursor solution to facilitate the crystallization. It is observed that the grain size is over 3 μm by seed-induced strategy, exhibiting a lower defect density due to reduced grain boundaries.[103] In conclusion, a number of additives could enlarge the grain size and increase the crystalline quality of perovskite film. And decreasing the grain boundaries is an effective method to reduce imperfections and J-V hysteresis, thereby obtaining more stable PSCs. Enlarging the grain could limit the channels of ion migration instead of suppressing the ion migration intrinsically.

Incorporating the additive involving inorganic structures and organic/polymer molecules into the perovskite is considered to be an effective mean to passivate the grain boundaries, stabilize the perovskite lattice, and impede ion migration. The majority of molecules function as electron donors or acceptors, interacting with positively or negatively charged defects to block the pathway of ion migration. As general knowledge, the inorganic cesium lead halide perovskites are more thermally stable than organic-inorganic halide perovskites. They display outstanding operational stability under illumination for hundreds of hours because of light-independent ionic diffusion in cesium perovskite.[104] Zhao and colleagues introduced $CsPbBr_3$ into the $Cs_{0.05}FA_{0.80}MA_{0.15}PbI_{2.55}Br_{0.45}$ (CsFAMA) precursor solution to suppress ion migration and phase segregation (Figure 11.13d). The $CsPbBr_3$ clusters are accumulated at the interface/surface of bulk CsFAMA grains and are detected around CsFAMA grains, forming the quasi-core-shell structure. It is found that the CsFAMA perovskite film with 3% mol of $CsPbBr_3$ exhibits increased PL lifetime and decreased defect densities. What is more, they conducted the cryogenic Galvanostatic with voltage-current measurements at different temperatures to determine the ion migration barrier. The E_a of CsFAMA device is about 0.17 eV, which is smaller than that of the sample with $CsPbBr_3$ (0.23 eV), indicating that the ion migration is easier in the sample without $CsPbBr_3$ (Figure 11.13e). Consequently, the modified devices show improved efficiency and stability, mainly attributed to the inhibited ion migration.[105]

As we know, there are two main mobile ions of organic cations and halide ions in organic-inorganic halide perovskite and the organic cations are more sensitive to external environmental factors. Particularly, the hydrogen bonding between Pb-I framework and organic cations could be weakened under light illumination, leading to the migration of the organic cations, which further accelerates the degradation of PSCs. Therefore, it is imperative to impede the movement

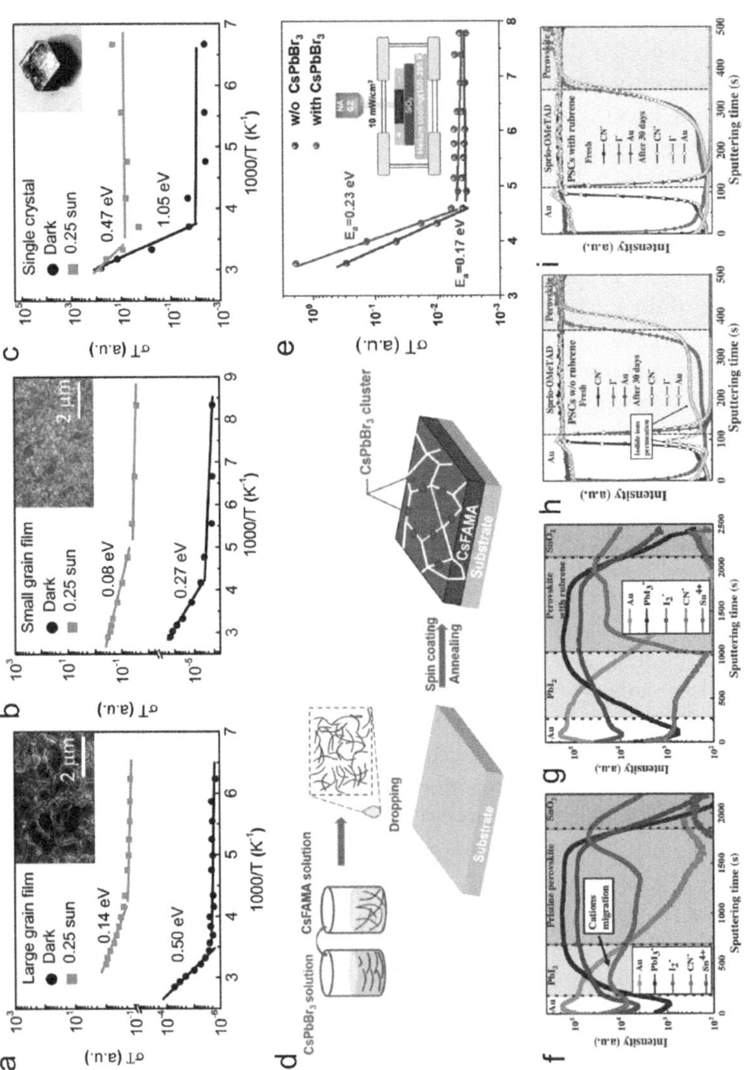

FIGURE 11.13 Conductivity of a small-grain film (a), large-grain film (b), and a single crystal (c) under different temperature[35]. Copyright 2016, Royal Society of Chemistry. (d) Scheme of the device fabrication process of CsPbBr$_3$ passivated perovskite films. (e) Temperature-dependent conductivity of samples w/o CsPbBr$_3$ (dark grey) and with CsPbBr$_3$ (3% mol; light grey) under light illumination. Inset: scheme of experimental set-up[105]. Copyright 2019, Wiley-VCH GmbH. TOF-SIMS elemental depth profiles of aged devices without (f) and with (g) rubrene. TOF-SIMS elemental depth profiles of CN$^-$, I$^-$, and Au of devices without rubrene (h) and with rubrene (i) as freshly prepared and after 30 d operation[106]. Copyright 2018, Wiley-VCH GmbH.

of organic cations to improve the optoelectronic properties and operational stability of PSCs. Li et al. added rubrene additive into the perovskite precursor to achieve the cation-π interaction which is a noncovalent bond between the face of an electron-rich π structure and a surrounding cation. This interaction could be detected between aromatic π group and organic cation in organic-inorganic perovskite system. Based on the DFT calculation, the interaction energy of MA^+ cation and rubrene were calculated to be -1.53 eV, which is much higher than that of common cation-π interaction. The strong interaction is expected to anchor the MA^+ cation firmly. The interaction between FA^+ cations and rubrene was also verified. The TOF-SIMS was conducted to trace the migration of organic cation. After continues operation for 72 h with 1 V DC bias, the CN^- ions of control film remarkably migrated into the adjacent PbI_2 layer, indicative of severe diffusion of organic cations (Figure 11.13f). In contrast, the migration of CN^- in modified film could be negligible, as seen from Figure 11.13h, demonstrating that the cation could be efficiently immobilized by the incorporation cation-π interaction. Furthermore, they also investigated the stability of PSCs after operation for 30 d. From the TOF-SIMS results (Figures 11.13h and i), the I^- ions in PSC without rubrene apparently moved to the Spiro-OMeTAD and Au layers, while the migration of I^- ions in modified PSC is negligible, indicative of suppressed ion migration because of decreased defects at the surface and grain boundaries serving as dominant channels for ion migration. In summary, the incorporation of rubrene, cation-π interaction, could remarkably inhibit the migration of organic cation and iodide ions, enabling high-efficiency PSCs with improved stability.[106]

The organic small molecule is also considered to be an appropriate additive to improve the PSCs' performance. In 2018, Fang et al. incorporated a cross-linkable monomer trimethylolpropane triacrylate (TMTA) into $MAPbI_3$ precursor, as shown in Figure 11.14a. The liquid TMTA can be automatically expelled to grain boundaries and anchor to the grain boundaries due to the weak interaction between TMTA and PbI_2, thus leading to passivation of defects. Then the in-situ cross-linking polymerization would be occurred by three alkenyl groups in TMTA during thermal treatment to form robust and continuous network polymer, establishing a blocking channel for ion migration. To verify the inhibition of ion migration, the measurement of E_a is conducted, as shown in Figures 11.14b and c. The E_a of lateral device with TMTA is increased to 0.48 eV from 0.21 eV. Additionally, the temperature of transition from electronic to ionic conductivity is also raised to 263 K from 246 K in a cross-linked sample. The increased E_a and threshold temperature demonstrate that it is more difficult for ions movement in modified devices than in $MAPbI_3$, leading to an improvement in operational stability.[107]

It is commonly believed that the carbonyl and amino groups could interact with organic and inorganic ions of perovskite, resulting in the passivation of defects and suppression of ion migration. In 2020, Liu et al. selected a multifunctional molecule, 2,2-difluoropropanediamide (DFPDA), to control the crystallization rate of perovskite and obtain high-quality perovskite film with reduced defects. The carbonyl groups in DFPDA could form chemical bonds with undercoordinated Pb^{2+} and passivate the defects in bulk. Meanwhile, the amino groups not only interact with the iodide to impede its diffusion but also increase the coordination ability of carbonyl group and Pb^{2+} to further improve the passivation effect. The DFT calculation manifests the increased formation energies of iodide vacancies, lead vacancies, and Pb-I anti-sites, which is conducive to suppressing the generation of defects. Thus, the channels of ion migration are blocked, leading to improved long-term stability.[108] Zhu and colleagues incorporated a trace amount of phenylmethylammonium iodide (PMAI) into $FA_{1-x}MA_xPMA_yPbI_3$ precursor to achieve ultralong carrier lifetimes (6,214 nm), prolonged electron (5,712 nm) and hole (6,078 nm) diffusion lengths, indicative of less trap density in modified perovskite. The PMAI molecule could electrostatically interact with abundant defects at the grain boundaries and eliminate the ionic vacancies, thereby weakening the ion migration.[109] Besides, the bifunctional urea is incorporated into the precursor solution to slow down the crystal growth and enlarge the grain size. Thus, the addition of urea significantly reduces the defect density at the grain boundaries and in the grain interior, which may block the ion migration.[110]

The divalent anion Se^{2-} originating from PbSe could also form chemical bond with Pb^{2+} with high covalency to improve the hydrogen-bonding-like interaction between the inorganic framework and organic cations. Consequently, the migration of iodide is mitigated, impeding the degradation of Ag electrode because of formation of AgI.[111] Unfortunately, the perovskite crystal structure with the incorporation of Se^{2-} is not stable due to the smaller size of Se^{2-} (198 pm) compared to the I^- (220 pm).[112] Very recently, Li et al. developed an ionic-liquid polymer (poly[Se-MI][BF_4]), including a phenyl vinyl selenide-alt-N-phenylmaleimide (PVSe-alt-NPMI) copolymer backbone, a phenyl vinyl selenide (PVSe) ionized backbone, and coordinated tetrafluoroborate (BF_4^-), to stabilize perovskite films and devices. The carbonyl group and selenium ion could coordinate with Pb^{2+} and iodide to improve the stability of the lead polyiodide colloids and suppress the migration of iodide. As shown in Figure 11.14d and e, the KPFM measurements under bias were conducted to evaluate the stability of film. For pristine perovskite, the halide ions are redistributed and accumulated at the grain boundaries because the positive bias attracts negative ions. These accumulated defects would change the CPD of grain boundaries. In stark contrast, the accumulation of defects is not clear in modified film and the CPD does not show obvious variation, indicating that few charged defects move to grain boundaries. The E_a of ion migration in modified film increases to 0.304 eV from 0.216 eV in pristine film, meaning the suppressed migration of charged defects. Additionally, the I^- ions in the control devices migrate to the Spiro-OMeTAD layer and accumulate at the Au electrode after aging for two weeks, as shown in the Figure 11.14f. However, the intensity of I^- ions in the modified devices decreases, demonstrating the impeded ion migration. As a result, the modified PSCs exhibits an outstanding operational stability with 90.2% of its initial PCE after illumination for 2,200 h.[113] Apart from the inorganic and organic molecules mentioned earlier, other molecules such as PCBM, pyridinic nitrogen-doped graphdiyne with 2D structure, self-polymerized methyl methacrylate (sMMA), fluorinated perylenediimide (F-PDI) 1-butyl-3-methylimidazolium tetrafluoroborate ($BMIMBF_4$), and 1-butyl-1-methylpiperidinium tetrafluoroborate ($BMPBF_4$) are also studied to reduce the trap state, inhibit ion migration and improve the stability under operation.[32,114–118]

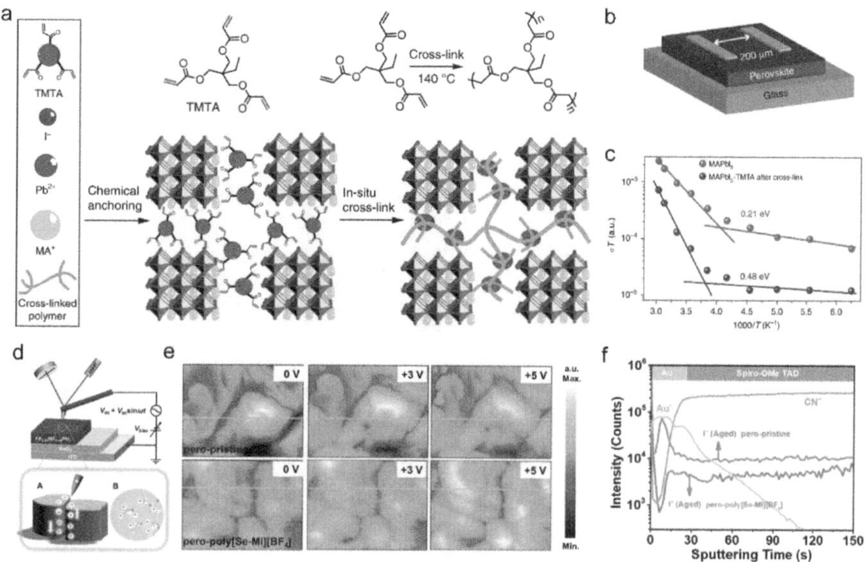

FIGURE 11.14 (a) Chemical structure of TMTA, cross-linking polymerization of TMTA under thermal conditions and working mechanism of TMTA in PSCs: TMTA chemically anchors to the grain boundaries of MAPbI$_3$ and then in situ cross-links to a continuous network polymer. (b) Device structure used in activation energy measurement. (c) Temperature-dependent conductivity of perovskite films without and with cross-linked TMTA[107]. Copyright 2018, Springer Nature Limited. (d) Schematic illustration of the KPFM measurements performed under bias voltages. A: directional migration of the charged defects in the perovskite film bulk; B: surface ion migration at the GBs. (e) KPFM measurements of the pero-pristine and pero-poly[Se-MI][BF$_4$] films under dark conditions after applying different bias voltages. (f) TOF-SIMS results of the devices fabricated using the fresh perovskite precursor inks after operating at the MPPT under a 100 mW cm^{-2} white LED for two weeks (at 50 ± 5 °C)[113]. Copyright 2023, Wiley-VCH GmbH.

11.6.3 Interface Passivation

It is widely accepted that the interfacial defects play a more important role in impacting the performance of PSCs compared to the bulk defects. And the E_a of ion migration at the interface or grain boundaries are relatively low because the interface or boundaries has much higher defect densities. The ions accumulated at the interface of perovskite/HTL or perovskite/ETL account for the deterioration of PSCs. The migrated ions could be trapped by interface modification, leading to improved operational stability. Thus, interface passivation is another extensively used method to inhibit ion migration by reducing the charged defects and blocking the pathway of ion migration. As we know, PCBM has been widely used as ETL to assist the electron transport. In 2014, Huang et al. adopted double fullerene layer of PCBM and C$_{60}$ to synergistically reduce the trap density of states in perovskite film, which is evidenced by the results of TAS measurement (Figure 11.15a).[119] Then, they further investigated the mechanism of the interaction between PCBM and perovskite. They

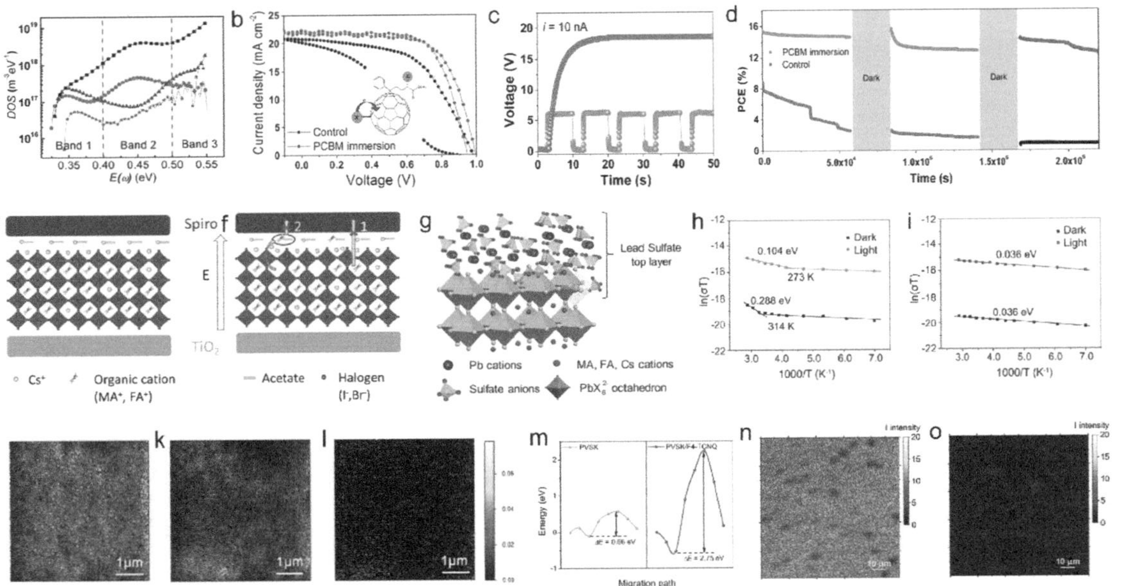

FIGURE 11.15   (a) Trap density of states (tDOS) for devices passivated by single PCBM layer (blue triangles), single C_{60} layer (red dots), PCBM/C_{60} double fullerene layers (grey stars); black squares represent the devices without fullerene passivation[119]. Copyright 2014, The Royal Society of Chemistry. (b) *J-V* curves of PSCs without and with PCBM. (c) Voltage-time (V-t) characteristics for the perovskite film with and without PCBM treatment, under Galvanstatic measuring mode. (d) Steady-state efficiency at MPPT as a function of time for devices with and without PCBM immersion treatment[9]. Copyright 2016, American Chemical Society. Schematic diagram of cesium acetate modified PSC before (e) and after (f) stability test[120]. Copyright 2018, American Chemical Society. (g) Illustration of in-situ formation of a lead sulfate layer on the perovskite surface. Temperature-dependent conductivity of PTAA/CsFAMA perovskite/PCBM films without (h) and with (i) methylammonium sulfate treatment[121]. Copyright 2019, American Association for the Advancement of Science. Mapping of TOF-SIMS signals of I⁻ in the HTLs for (j) perovskite/Spiro-MeOTAD, (k) perovskite/GO/Spiro-MeOTAD, and (l) perovskite/Cl-GO/Spiro-MeOTAD.[70] Copyright 2019, American Association for the Advancement of Science. (m) Migration energy curves of iodine ion of the perovskite surface without and with F4-TCNQ passivation, where the migration energy is marked by the arrows. Mapping of TOF-SIMS signals of I in the HTLs for perovskite/Spiro-OMeTAD (n) and perovskite/F4-TCNQ/spiro-OMeTAD (o)[122]. Copyright 2021, Elsevier Inc.

proposed that the PCBM could be incorporated into the interstitial space among grains and passivate the Pb-I anti-site defects, resulting in the inhibition of iodide migration.[32] Zhao et al. found that the PCBM could penetrate into the perovskite film through grain boundaries and chemically anchor on the perovskite surface. In this scenario, the ion diffusion near the surface of perovskite/TiO$_2$ is impeded, which can be attributed to the formation of donor-acceptor complex between PCBM and perovskite. Experimentally (Figures 11.15b and c), the sample with PCBM treatment shows negligible *J-V* hysteresis and faster voltage response time compared to control sample, suggesting a significant suppression of ion movement by PCBM. As a result, the devices with PCBM treatment show higher steady-state efficiency and better long-term stability than control devices, as shown in Figure 11.15d.[9] In short, the fullerene layer of PCBM or C$_{60}$ could reduce the channels of ion migration, thereby impeding the ion migration and alleviating the current hysteresis.

Apart from the fullerene layer served as ETL, an additional layer can be also incorporated between perovskite and HTL/ETL to mitigate ion diffusion. Zhao et al. introduced cesium acetate as a passivation layer into the interface of perovskite and Spiro-OMeTAD. The cesium acetate (CsCH$_3$COO) could interact with perovskite to form the Cs-rich perovskite at the surface, reducing the defects density. After stability test (Figures 11.15e and f), the Cs could suppress ion migration at the first place. On the other hand, the residual CsCH$_3$COO prones to immobilized organic ions at the interface through the interaction of FA$^+$/MA$^+$ + CH$_3$COO$^-$ = (FA/MA)CH$_3$COO. The FACH$_3$COO and MACH$_3$COO are considered as immobile molecules, leading to suppressed ion migration. Consequently, the PSCs with CsCH$_3$COO treatment show improved operational stability remaining 80% of its initial PCE after MPPT tracking for 4,500 min.[120] In 2019, Huang et al. proposed a general passivation method to stabilize the perovskite surface and reduce the defect density at the perovskite surface. As shown in Figure 11.15g, the water-insoluble lead sulfate with a layer thickness of < 5 nm could be formed on the surface of lead halide perovskite by the reaction of (C$_8$H$_{17}$NH$_3$)$_2$SO$_4$ and the perovskite surface, indicating a strong interaction between Pb in MAPbI$_3$ and SO$_4^-$. Additionally, the temperature-dependent conductivity of the PTAA/CsFAMA perovskite/PCBM films was measured to calculate the E_a of ion migration, as shown in Figures 11.15h and i. In the case of pristine film, the ionic conductivity starts to dominate the total conductivity at the temperature of 314 K in the dark condition, and E_a is determined to be 0.288 eV. Under 0.1-sun illumination, both the transition temperature and E_a are reduced, demonstrating that the light would accelerate the ion migration. By contrast, such transition temperature of the perovskite film with methylammonium sulfate is not detected upon increasing the temperature to 330 K under illumination and in the dark, indicating that the ion migration is significantly impeded, which may be ascribed to immobilized surface defects by the lead sulfate layer. And the E_a of 0.036 eV is obtained, which may be ascribed to the electronic conduction.[121] Similarly, Han and colleagues constructed strong Pb-Cl and Pb-O bonds between perovskite and chlorinated graphene oxide (Cl-GO) layer to prevent the loss of the perovskite component and improve the stability of organic HTL. The measurement of TOF-SIMS was carried out to analyze the distribution of perovskite component within Spiro-OMeTAD. As shown in Figures 11.15j–l, the signal of I$^-$ for Cl-GO-treated sample is extensively reduced after 200 h aging at the MPPT at 60 °C, suggesting that the ion migration to Spiro-OMeTAD is greatly blocked.[70] In 2021, Chen and colleagues proposed a strategy of the sandwiched electrode buffer (F4-TCNQ/HTL/F4-TCNQ)) to bridge the perovskite to rear electrode through multiple bonding. DFT calculations suggest that the migration energy of iodide ions greatly increases to 2.75 eV from 0.66 eV when employing F4-TCNQ, as shown in Figure 11.15m. Meanwhile, they confirmed alleviated ion migration in aged films using TOF-SIMS, showing a low signal intensity of iodide in the aged sample with F4-TCNQ treatment (Figure 11.15n and o).[122]

Constructing 2D/3D interface is universal method to passivate the interfacial defects and decrease the non-radiative recombination.[123,124] Actually, such interface could also block ion migration to boost the stability of device. Grancini and colleagues employed a series of thiophene-terminated cations such as 2-thiophenemethylammonium iodide (2-TMAI), 3-thiophenemethylammonium iodide (3-TMAI), and 2-thiopheneethylammonium iodide (2-TEAI) to engineer the 2D/3D interface (Figure 11.16a). Based on the results of grazing incidence X-ray diffraction, they discovered that different 2D perovskite could be generated depending on the organic molecules. The majority of 2D perovskites with $n = 2$, only $n = 2$, and mostly $n = 1$ is observed for 2-TMAI, 3-TMAI, and 2-TEAI, respectively. Then, they monitored the evolution of photovoltaic performances for the devices when aging in the dark and dry air atmosphere. A monotonic enhancement in PCE is found for the devices with 2D/3D structure. They proposed that the 2D layer obtained from 2-TMAI and 3-TMAI could incorporate and immobilize the MA cation to form a quasi-2D phase. In the case of 2-TEAI, the 2D layer only physically impedes ion migration and preserves the initial 2D phase with $n = 1$. Such 2D phase is a structurally robust 2D layer, contributing to improved stability with 90% of the initial PCE with illumination over 1,000 h (Figure 11.16b).[125] In 2020, Yan et al. introduced arylammonium cation of phenethylammonium iodide (PEAI) into the interface of perovskite/ETL to reduce the V_{OC} deficit and suppress ion migration (Figure 11.16c). They measured the low-frequency capacitance under different temperatures to estimate the E_a of ion movement. As shown in Figure 11.16d, the device with PEAI treatment shows a higher E_a of 0.905 eV compared to the control device with E_a of 0.680 eV, indicative of inhibited ion migration after the treatment with PEAI. In addition, they further performed the DFT calculations to investigate the behaviors of ions defect. The energy barrier of the movement of I interstitial in control devices is lower than that in PEAI-modified devices, suggesting that the PEAI may play a key role in mitigating halide ion migration at the surface.[126]

The same year, Park et al. adopted hydrophilic 2-aminoethanol hydroiodide (2AEI) or 4-amino-1-butanol hydroiodide (4ABI) to improve the stability of PSCs. Over 90% of initial PCE could be preserved after storage with relative humidity of 20% for 1,000 h (Figure 11.16e). According to the results of temperature-dependent electrical conductivity (Figure 11.16f), the E_a of control film is estimated to be 83 meV, whilst the perovskite films with 2AEI and 4ABI show higher E_a of 265 and 295 meV, respectively. The increased E_a indicates that the ions migration is significantly suppressed, which is associated with the reduced defects and J-V hysteresis.[127] Very recently, Li et al. introduced a cross-linked polymer (CLP) on the top of a 3D perovskite to avoid the degradation of 3D/2D interface. The CLP could impede the migration of FA$^+$ and 4F-PEA$^+$ between 3D and 2D perovskite to achieve stable

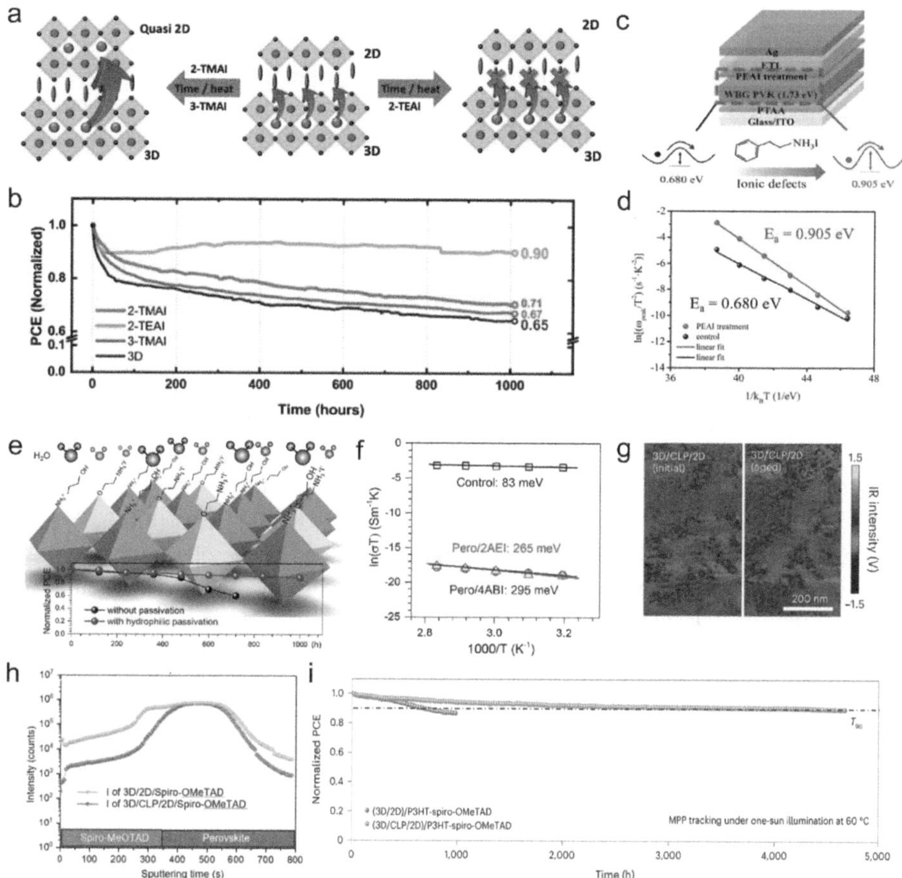

FIGURE 11.16 (a) Cartoon illustrating the proposed interfacial mechanism using thiophene-terminated cations. (b) Stability test for freshly prepared devices under continuous 1-sun illumination for 1,000 h in an Ar atmosphere without any encapsulation[125]. Copyright 2020, The Royal Society of Chemistry. (c) Device structure of PEAI-treated perovskite. (d) Arrhenius plots of the characteristic transition frequencies derived from the admittance spectra[126]. Copyright 2020, American Chemical Society. (e) Schematic of 4ABI and 2AEI passivating perovskites. (f) Temperature-dependent conductivity of perovskite devices before and after post-treatment with 4ABI and 2AEI[127]. Copyright 2020, American Chemical Society. (g) AFM-IR (recorded at 1,712 cm^{-1} corresponding to the symmetric C = NH$_2^+$ bending of FA$^+$) images of 3D/CLP/2D perovskite heterostructures before and after aging at 100 °C for 120 min. (h) TOF-SIMS of I ions in Spiro-OMeTAD and the perovskite layer. (i) The operational stability of PSCs using a mixed P3HT-Spiro-OMeTAD HTL under continuous illumination in N$_2$ at 60 °C [128]. Copyright 2023, Springer Nature Limited.

3D/2D heterostructure. After the 3D/2D heterostructures are heated at 100 °C for 120 min, the 3D/CLP/2D structure displays an almost invariant signal of FA$^+$, confirming the suppressed ion diffusion, as shown in Figure 11.16g. For a 1,000 h aged sample with CPL, the migration of iodide from the perovskite to the Spiro-OMeTAD layer is effectively mitigated (Figure 11.16h), which can be attributed to the synergetic effect of CPL and 2D layer. It is exciting that the device with CPL and 2D layer displays an extraordinary stability with 90% of initial PCE after operating at MPPT for 4,390 h (Figure 11.16i).[128]

For small-area PSCs, ions usually migrate perpendicularly to the substrate, which is called vertical migration. Tremendous works have been concentrated on impeding the vertical diffusion in small-area PSCs, as discussed earlier. However, the ions in module devices could also migrate parallelly to the substrate at the interconnecting sites between subcells, called lateral migration, generating severe degradation issues. To realize high-performance PSC modules, Han and colleagues introduced the low-dimensional diffusion barriers (DBLs) including stable metal oxide nanoparticles (0D), inert silicon-based organic polymers (1D), and nano-structured inorganic material (2D) between subcells (Figure 11.17a). From results of TOF-SIMS (Figures 11.17b-d), the diffusion of iodide and Ag in film with 2D-DBL is negligible compared to the films with 0D and 1D treatment. Finally, the module with 2D-DBL shows 95% of its initial PCE after heating aging for 1,000 h at 85 °C with relative humidity of about 85% and 91% after illumination at MPPT for 1,000 h.[129] These works indicate that the 2D layer has an important effect on the operational stability of device, and the improved stability is deeply related to the organic molecular and the structure of 2D/3D layer. It is still challenging to employ an ultrathin 2D layer to completely inhibit ion migration. Therefore, it is imperative to carefully choose appropriate organic molecules to balance the composition of 2D and 3D perovskite.

Apart from passivating the top surface of perovskite, the buried interface has also been studied. For instance, Hao et al. demonstrated an effective strategy using KCl to impede the recombination at the interface of perovskite/SnO_2. The Cl in KCl layer could bond strongly with Pb, leading to reduced density of interfacial defects. And the K ions could occupy the lattice sites and fill the cation vacancies. Therefore, the KCl layer can efficiently passivate the charged defects at the buried interface, mitigating the J-V hysteresis of devices (Figure 11.17e).[86] Besides, $NdCl_5$ is incorporated into the interface between perovskite and SnO_2 to form ultrathin Nd_2O_5 layer and reduce oxygen vacancies at the buried side. The halide-related defects could also be passivated by forming Pb-Cl bond (Figure 11.17f). The decreased trap states are conducive to blocking the pathway of ion migration.[130] Recently, Zhao and colleagues incorporated porous organic cage (POC) composed of discrete organic molecules with cavities between perovskite and SnO_2 to reduce the defect density at the buried interface. The POC has abundant chelation sites to interact with different ions or molecules by hydrogen bonding, coordination, and host-guest interaction, etc. By host-guest interaction, the POC strongly anchors iodide ions and impedes their migration to the buried interface of an ITO. TOF-SIMS results show that the intensity of iodide signal decreased in POC treatment film (Figures 11.17g and h). From the rectification effect test in the dark, the reduced breakdown voltage further implies the inhibition of mobile ions in POC-device (Figure 11.17i). To visualizing the ion migration, an in-situ confocal laser scanning microscope (CLSM) was used in lateral structure devices under light illumination and electric field, as shown in Figures 11.17j and k. With increasing the time, the degradation gradually extends throughout control device, indicative of severe ion migration. In stark contrast, the

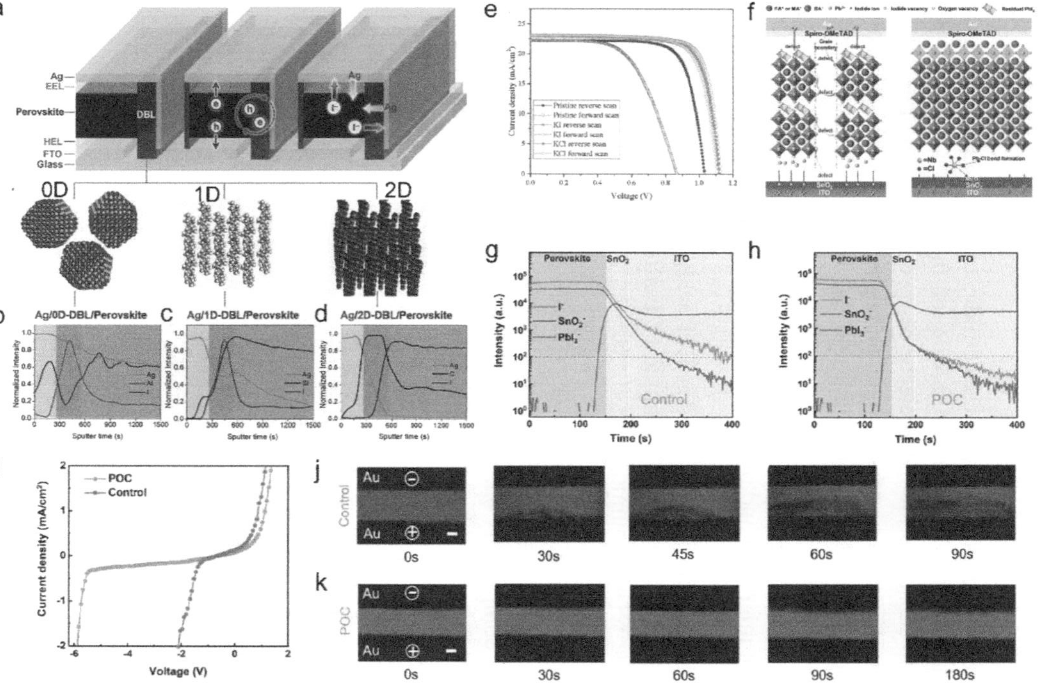

FIGURE 11.17 (a) Illustration of the diffusion process and interfacial charge transfer (solid lines) and recombination (dotted circle) in PSC (Al_2O_3 nanoparticles for 0D-DBL; Polydimethylsiloxane (PDMS) for 1D-DBL; graphitic carbon nitride (g-C_3N_4) for 2D-DBL). The depth profile of diffused iodide and Ag in (b) Ag/0D-DBL/perovskite film with aluminum signal, (c) Ag/1D-DBL/perovskite film with silicon signal, and (d) Ag/2D-DBL/perovskite film with carbon signal, respectively[129]. Copyright 2019, Elsevier Inc. (e) *J-V* characteristics of Cs-FAMA cells with underlying passivation.[86] Copyright 2018, Wiley-VCH GmbH. (f) Schematic illustration of channels for iodide ion diffusion provided by the vacancy defects inside the device and the effect of the internal encapsulation[130]. Copyright 2022, Wiley-VCH GmbH. TOF-SIMS depth profiles with the structure ITO/SnO_2/perovskite (g) and ITO/SnO_2/POC/perovskite (h). (i) Current-voltage curves of the devices in the dark state. (j) Time-dependent PL images of a control sample (Au/perovskite/Au) using CLSM. (k) Time-dependent PL images of a POC-sample (Au/perovskite with POC/Au) using CLSM. Scale bars, 20 μm[55]. Copyright 2023, Wiley-VCH GmbH.

PL images of POC-device remain almost invariant from 0 s to 180 s. These results undoubtedly verify the suppression of ion migration in POC-device owing to the strong interaction between POC and iodide ions. As a result, the POC-device displays a best PCE of 24.13% with excellent stability remaining 93% of initial PCE after 5,000 h under ambient conditions.[55] There are also a large number of defects at the buried interface of perovskite that could be effectively reduced by interfacial passivation. Therefore, the channels of ion migration could be blocked and the long-term stability and efficiency could be improved.

It is well known that ions could migrate into electrodes to deteriorate the PSCs. And anchoring the ions is a useful method to impede ion migration towards electrodes. Another way is to substitute the metal electrode by corrosion-resistant electrodes. The carbon electron is used to substitute the metal electrode because carbon materials are cheap, stable, insensitive to ion migration, and intrinsically water resistant.[131,132] Nazeeruddin and colleagues developed HTL-free PSCs and modules by replacing the HTL with carbon electrode. The hydrophobic and thick carbon layer is a blocker to inhibit ion migration and prevent the ingress of moisture, resulting in improved long-term and humidity stability.[133] In the following year, Yang et al. developed all-carbon based PSCs by using C_{60} served as ETL with the device structure of $FTO/C_{60}/MAPbI_3/carbon$, showing 95% of its initial PCE after continuous working for 180 h in air with the relative humidity of 40–60%.[134] Apart from the carbon electrode, metal oxides are also used as alternative electrodes because of their chemical stability. McGehee et al. employed sputtered ITO with good moisture-resistant properties as top electrodes of PSCs. The dense ITO layer could not only prevent the volatilization of organic components but also could impede the diffusion of halide ions.[135] Although the stability could be improved using the electrodes of carbon or metal oxides, the efficiencies of these devices are still inferior to those used metal electrodes. There are several factors such as insufficient conductivity, complicated processing, and difficult monolithic integration, limiting the performance of PSCs with carbon or metal oxide electrodes.[10]

In summary, the ion migration in organic-inorganic is extensively suppressed by incorporating passivation layer between perovskite and ETL or HTL. The interfacial defects and grain boundaries are significantly reduced by the interaction between passivated materials and perovskite, blocking the pathway of ion migration. Although 2D perovskite could passivate the interfacial defects, it is necessary to carefully control the dimensionality of 2D, which is deeply related to the extraction of photoinduced carriers.

11.7 CONCLUSION AND OUTLOOK

In this chapter, we elaborately overviewed the current developments in understanding the ion migration in organic-inorganic perovskite materials or PSCs from the perspective of experiment and DFT simulation, which covers the defects, origin of migrating ions, mechanism, migrated channels, and characterization techniques. Which ions easily migrate within perovskite film or PSCs is also systematically investigated. And various characterization techniques used to detect the ion migration are evaluated. In addition, we thoroughly elaborate on a plethora of works associated with ion migration to demonstrate its impacts on stability and corresponding coping strategies for suppressing ion migration, such as composition modulation, additive regulation, and interface passivation. These strategies are effective but are sufficient for the commercialization of PSCs. Although the migrated ions, origin, and channels of ion migration have been discussed to a certain extent, a more thorough understanding of migration mechanisms in complicated perovskite components under the actual environment is still lacking. The structure and quality of perovskite are sensitive to the experimental and measured conditions. Thus, there are a lot of opportunities to develop collaborative characterizations for in-situ study of the intrinsic ion migration. Moreover, the negative impacts of ion migration are merely suppressed, and more efforts should be made to completely remove the ion migration in organic-inorganic perovskite systems. In fact, many issues of ion migration still need to be resolved, such as the following:

1. The vertical migration of ions has been extensively investigated, while scarce works studied the lateral migration of ions which not only exists in module cells but also in small-area PSCs. What is the mechanism and impact of lateral migration of ions in organic-inorganic PSCs? Do the lateral migrated ions improve or damage the efficiency and stability of PSCs?
2. The exact pathway of each ion in perovskite is still ambiguous. Which ion could penetrate into the ETL or HTL, and which ion only migrates within perovskite? Do all ions in perovskite have an adverse effect on the device performance?
3. Is there any chemical reaction between migrated ions and individual layers including perovskite, ETL, and HTL? And how could the reaction alter the physical and chemical properties of individual layers, such as work function, carrier transport mobility, and stability?
4. Last but not least, is the behavior of ion migration reversible or not? If the migration is reversible, will the device performance be recovered? What are the conditions for such recovery?

To further understand ion migration and achieve high-efficiency PSCs with improved intrinsic stability, we briefly demonstrate some research directions from the aspect of suppressed ion migration.

1. Component engineering and device structure innovation. In general, the triple or binary cation and anion perovskite materials are used as the light absorber to achieve the PSCs with high efficiency and stability. The cation and anions are considered to diffuse under the working conditions, resulting in phase segregation and deterioration of the perovskite film as well as devices. Thus, simplifying the perovskite composition may be appropriate to reduce the mobile ions in perovskite. Recently, $FAPbI_3$ perovskite has exhibited high efficiency with PCE of over 25%, serving as a proper candidate to make PSCs access to commercialization.[2] However, there are rare investigations about ion migration in $FAPbI_3$. Therefore, it is meaningful to provide

an in-depth study of the migrated mechanism and migrated impacts, and suppression of ion migration in $FAPbI_3$. On this basis, researchers should pay more attention to identifying the evolution of local composition and structure as well as the behavior of charge carriers, thereby analyzing the degree of phase segregation. The relationship between electron and ion as well as the underlying kinetic process of ion migration in $FAPbI_3$ should be clarified. The quantitative correlation among ion migration quantity, local ion bias, undesired recombination, and device performance should also be studied to illustrate the synergistic effect of ion migration and charge carrier transport on performance of PSCs. Developing new structures of the device is another way to inhibit ion migration. Designing novel materials or physical blockers probably confines the behavior of ion migration, avoiding the damage of HTL, ETL, and electrodes.

2. The mechanism of reducing the defects and interfacial passivation. As discussed in this chapter, there are a large number of defects in perovskite materials, which may be generated during the preparation process. The defects and grain boundaries provide the pathway of ion migration. Thus, it is significant to reduce defects and grain boundaries to suppress ion migration. However, the strategies to reduce defects and grain boundaries are not universal. The method would not be appropriate if the perovskite composition and structures change. Thus, researchers should develop a universal method or passivation materials to block the pathway of ion migration.

3. Quantify the E_a of ion migration under actual environment. The temperature-dependent conductivity measurement and first-principles calculations have been widely employed to evaluate the E_a of ion migration in the lateral device, which is an intuitive parameter to illustrate the ability of ion migration. However, the lateral devices are different from the PSCs under working conditions. And the actual environment is more complicated than that in the lab. Thus, it is urgent to quantity the E_a of ion migration under extreme conditions. Meanwhile, in-situ characterization techniques should be developed to further study the complex migration of ions and their E_a in perovskite film and PSCs under actual operating conditions. Additionally, understanding the impacts of the external environment on E_a of ion migration is also crucial. In this regard, the intrinsically changed behavior of migration E_a could be uncovered to elucidate the underlying mechanism of interaction and coupling among migrated behaviors of various ions.

4. The scalable PSCs and modules are the future of photovoltaic field. Most of the research has been focused on studying ion migration in small-area solar cells, but the behavior of ion migration in modules remains unclear. It is necessary to investigate the complicated ion migration in modules by employing in-situ characterization techniques to elaboratively understand the vertical and lateral migration of ions. Meanwhile, the real-time monitoring technique should be developed to characterize the real-time evolution of ion migration in scalable PSCs and modules.

REFERENCES

1. G. E. Eperon, S. D. Stranks, C. Menelaou, M. B. Johnston, L. M. Herz, H. J. Snaith, Formamidinium lead trihalide: A broadly tunable perovskite for efficient planar hetero-junction solar cells, *Energy & Environmental Science* **2014**, *7*, 982.
2. J. J. Yoo, *et al.*, Efficient perovskite solar cells via improved carrier management, *Nature* **2021**, *590*, 587–593.
3. R. J. Sutton, *et al.*, Bandgap-tunable cesium lead halide perovskites with high thermal stability for efficient solar cells, *Advanced Energy Materials* **2016**, *6*, 1502458.
4. S. D. Stranks, *et al.*, Electron-hole diffusion lengths exceeding 1 micrometer in an orga-nometal trihalide perovskite absorber, *Science* **2013**, *342*, 341–344.
5. T. M. Brenner, D. A. Egger, L. Kronik, G. Hodes, D. Cahen, Low-cost semiconductors with intriguing charge-transport properties, *Nature Reviews Materials* **2016**, *1*, 15007.
6. NREL best research-cell efficiency chart, www.nrel.gov/pv/cell-efficiency.html, Accessed: (2023).
7. T. A. Berhe, *et al.*, Organometal halide perovskite solar cells: degradation and stability, *Energy & Environmental Science* **2016**, *9*, 323–356.
8. H. Zai, Y. Ma, Q. Chen, H. Zhou, Ion migration in halide perovskite solar cells: Mecha-nism, characterization, impact and suppression, *Journal of Energy Chemistry* **2021**, *63*, 528–549.
9. Y. Zhao, *et al.*, Correlations between immobilizing ions and suppressing hysteresis in perovskite solar cells, *ACS Energy Letters* **2016**, *1*, 266–272.
10. E. Bi, Z. Song, C. Li, Z. Wu, Y. Yan, Mitigating ion migration in perovskite solar cells, *Trends in Chemistry* **2021**, *3*, 575–588.
11. C. Besleaga, *et al.*, Iodine migration and degradation of perovskite solar cells enhanced by metallic electrodes, *The Journal of Physical Chemistry Letters* **2016**, *7*, 5168–5175.
12. D. Di Girolamo, *et al.*, Ion migration-induced amorphization and phase segregation as a degradation mechanism in planar perovskite solar cells, *Advanced Energy Materials* **2020**, *10*, 2000310.
13. Y. Zhou, *et al.*, Enhancing chemical stability and suppressing ion migration in $CH_3NH_3PbI_3$ perovskite solar cells via direct backbone attachment of polyesters on grain boundaries, *Chemistry of Materials* **2020**, *32*, 5104–5117.
14. Y. Cheng, *et al.*, Revealing the degradation and self-healing mechanisms in perovskite solar cells by sub-bandgap external quantum efficiency spectroscopy, *Advanced Materials* **2021**, *33*, e2006170.
15. H. Jin, E. Debroye, M. Keshavarz, I. G. Scheblykin, M. B. J. Roeffaers, J. Hofkens, J. A. Steele, It's a trap! On the nature of localised states and charge trapping in lead halide perovskites, *Material Horizons* **2020**, *7*, 397–410.
16. W.-J. Yin, T. Shi, Y. Yan, Unusual defect physics in $CH_3NH_3PbI_3$ perovskite solar cell absorber, *Applied Physics Letters* **2014**, *104*, 063903.
17. M. L. Agiorgousis, Y. Y. Sun, H. Zeng, S. Zhang, Strong covalency-induced recombi-nation centers in perovskite solar cell material $CH_3NH_3PbI_3$, *Journal of the American Chemical Society* **2014**, *136*, 14570–14575.
18. W. Li, Y. Y. Sun, L. Li, Z. Zhou, J. Tang, O. V. Prezhdo, Control of charge recombination in perovskites by oxidation state of halide vacancy, *Journal of the American Chemical Society* **2018**, *140*, 15753–15763.
19. J. Wang, W. Li, W. J. Yin, Passivating detrimental DX centers in $CH_3NH_3PbI_3$ for reduc-ing nonradiative recombination and elongating carrier lifetime, *Advanced Materials* **2020**, *32*, e1906115.
20. A. Buin, P. Pietsch, J. Xu, O. Voznyy, A. H. Ip, R. Comin, E. H. Sargent, Materials pro-cessing routes to trap-free halide perovskites, *Nano Letters* **2014**, *14*, 6281–6286.

21. J.-S. Park, J. Calbo, Y.-K. Jung, L. D. Whalley, A. Walsh, Accumulation of deep traps at grain boundaries in halide perovskites, *ACS Energy Letters* **2019**, *4*, 1321–1327.
22. A. A. Mamun, T. T. Ava, H. J. Jeong, M. S. Jeong, G. Namkoong, A deconvoluted PL approach to probe the charge carrier dynamics of the grain interior and grain boundary of a perovskite film for perovskite solar cell applications, *Physical Chemistry Chemical Physics* **2017**, *19*, 9143–9148.
23. Y. Yuan, J. Huang, Ion migration in organometal trihalide perovskite and its impact on photovoltaic efficiency and stability, *Accounts of Chemical Research* **2016**, *49*, 286–293.
24. Y. Shao, *et al.*, Grain boundary dominated ion migration in polycrystalline organic-inorganic halide perovskite films, *Energy & Environmental Science* **2016**, *9*, 1752–1759.
25. T. S. Sherkar, *et al.*, Recombination in perovskite solar cells: significance of grain boundaries, interface traps, and defect ions, *ACS Energy Lett.* **2017**, *2*, 1214–1222.
26. Q. Chen, *et al.*, Controllable self-induced passivation of hybrid lead iodide perovskites toward high performance solar cells, *Nano Letters* **2014**, *14*, 4158–4163.
27. Y. Yang, M. Yang, D. T. Moore, Y. Yan, E. M. Miller, K. Zhu, M. C. Beard, Top and bottom surfaces limit carrier lifetime in lead iodide perovskite films, *Nature Energy* **2017**, *2*, 16207.
28. B. Wu, H. T. Nguyen, Z. Ku, G. Han, D. Giovanni, N. Mathews, H. J. Fan, T. C. Sum, Discerning the surface and bulk recombination kinetics of organic-inorganic halide perovskite single crystals, *Advanced Energy Materials* **2016**, *6*, 1600551.
29. Z. Ni, *et al.*, Resolving spatial and energetic distributions of trap states in metal halide perovskite solar cells, *Science* **2020**, *367*, 1352–1358.
30. M. F. N. Taufique, R. Khanal, S. Choudhury, S. Banerjee, Impact of iodine antisite (I_{Pb}) defects on the electronic properties of the (110) $CH_3NH_3PbI_3$ surface, *Journal of Chemical Physics* **2018**, *149*, 164704.
31. Y. Shao, Z. Xiao, C. Bi, Y. Yuan, J. Huang, Origin and elimination of photocurrent hysteresis by fullerene passivation in $CH_3NH_3PbI_3$ planar heterojunction solar cells, *Nature Communications* **2014**, *5*, 5784.
32. J. Xu, *et al.*, Perovskite-fullerene hybrid materials suppress hysteresis in planar diodes, *Nature Communications* **2015**, *6*, 7081.
33. L. Ye, *et al.*, Managing secondary phase lead iodide in hybrid perovskites via surface reconstruction for high-performance perovskite solar cells with robust environmental stability, *Angewandte Chemie International Edition* **2023**, e202300678.
34. D. Meggiolaro, F. De Angelis, First-principles modeling of defects in lead halide perovskites: best practices and open issues, *ACS Energy Letters* **2018**, *3*, 2206–2222.
35. J. Xing, Q. Wang, Q. Dong, Y. Yuan, Y. Fang, J. Huang, Ultrafast ion migration in hybrid perovskite polycrystalline thin films under light and suppression in single crystals, *Physical Chemistry Chemical Physics* **2016**, *18*, 30484–30490.
36. J. M. Azpiroz, E. Mosconi, J. Bisquert, F. De Angelis, Defect migration in methylammonium lead iodide and its role in perovskite solar cell operation, *Energy & Environmental Science* **2015**, *8*, 2118–2127.
37. A. Walsh, D. O. Scanlon, S. Chen, X. G. Gong, S. H. Wei, Self-regulation mechanism for charged point defects in hybrid halide perovskites, *Angewandte Chemie International Edition* **2015**, *54*, 1791–1794.
38. C. Eames, J. M. Frost, P. R. Barnes, B. C. O'Regan, A. Walsh, M. S. Islam, Ionic transport in hybrid lead iodide perovskite solar cells, *Nature Communications* **2015**, *6*, 7497.
39. S. A. L. Weber, *et al.*, How the formation of interfacial charge causes hysteresis in perovskite solar cells, *Energy & Environmental Science* **2018**, *11*, 2404–2413.
40. J. J. Choi, X. Yang, Z. M. Norman, S. J. Billinge, J. S. Owen, Structure of methylammonium lead iodide within mesoporous titanium dioxide: active material in high-performance perovskite solar cells, *Nano Letters* **2014**, *14*, 127–133.

41. B. Wu, K. Fu, N. Yantara, G. Xing, S. Sun, T. C. Sum, N. Mathews, Charge accumulation and hysteresis in perovskite-based solar cells: an electro-optical analysis, *Advanced Energy Materials* **2015**, *5*, 1500829.

42. Q. Dong, J. Song, Y. Fang, Y. Shao, S. Ducharme, J. Huang, Lateral-structure single-crystal hybrid perovskite solar cells via piezoelectric poling, *Advanced Materials* **2016**, *28*, 2816–2821.

43. N. J. Jeon, J. H. Noh, Y. C. Kim, W. S. Yang, S. Ryu, S. I. Seok, Solvent engineering for high-performance inorganic-organic hybrid perovskite solar cells, *Nature Materials* **2014**, *13*, 897–903.

44. K. Wang, *et al.*, Short-range migration of a-site cations inhibit photoinduced phase segregation in $FA_xMA_yCs_{1-x-y}PbI_{3-z}Br_z$ single crystals, *The Journal of Physical Chemistry C* **2021**, *125*, 23050–23057.

45. R. Gottesman, *et al.*, Extremely slow photoconductivity response of $CH_3NH_3PbI_3$ perovskites suggesting structural changes under working conditions, *The Journal of Physical Chemistry Letters* **2014**, *5*, 2662–2669.

46. Z. Xiao, Y. Yuan, Y. Shao, Q. Wang, Q. Dong, C. Bi, P. Sharma, A. Gruverman, J. Huang, Giant switchable photovoltaic effect in organometal trihalide perovskite devices, *Nature Materials* **2015**, *14*, 193–198.

47. Y. Yuan, Q. Wang, Y. Shao, H. Lu, T. Li, A. Gruverman, J. Huang, Electric-field-driven reversible conversion between methylammonium lead triiodide perovskites and lead iodide at elevated temperatures, *Advanced Energy Materials* **2016**, *6*, 1501803.

48. A. Dazzi, F. Glotin, R. Carminati, Theory of infrared nanospectroscopy by photothermal induced resonance, *Journal of Applied Physics* **2010**, *107*, 124519.

49. Y. Yuan, J. Chae, Y. Shao, Q. Wang, Z. Xiao, A. Centrone, J. Huang, Photovoltaic switching mechanism in lateral structure hybrid perovskite solar cells, *Advanced Energy Materials* **2015**, *5*, 1500615.

50. T. Glaser, *et al.*, Infrared spectroscopic study of vibrational modes in methylammonium lead halide perovskites, *The Journal of Physical Chemistry Letters* **2015**, *6*, 2913–2918.

51. Y. Luo, *et al.*, Direct observation of halide migration and its effect on the photoluminescence of methylammonium lead bromide perovskite single crystals, *Advanced Materials* **2017**, *29*, 1703451.

52. T. Zhang, *et al.*, Profiling the organic cation-dependent degradation of organolead halide perovskite solar cells, *Journal of Materials Chemistry A* **2017**, *5*, 1103–1111.

53. K. Domanski, J *et al.*, Not all that glitters is gold: metal-migration-induced degradation in perovskite solar cells, *ACS Nano* **2016**, *10*, 6306–6314.

54. Y. Bai, *et al.*, Initializing film homogeneity to retard phase segregation for stable perovskite solar cells, *Science* **2022**, *378*, 747–754.

55. F. Gao, *et al.*, Porous organic cage induced spontaneous restructuring of buried interface toward high-performance perovskite photovoltaic, *Advanced Functional Materials* **2023**, 2211900.

56. C. Li, *et al.*, Iodine migration and its effect on hysteresis in perovskite solar cells, *Advanced Materials* **2016**, *28*, 2446–2454.

57. Y. C. Zhao, W. K. Zhou, X. Zhou, K. H. Liu, D. P. Yu, Q. Zhao, Quantification of light-enhanced ionic transport in lead iodide perovskite thin films and its solar cell applications, *Light: Science & Applications* **2017**, *6*, e16243.

58. M. Bag, *et al.*, Kinetics of ion transport in perovskite active layers and its implications for active layer stability, *Journal of the American Chemical Society* **2015**, *137*, 13130–13137.

59. M. H. Futscher, *et al.*, Quantification of ion migration in $CH_3NH_3PbI_3$ perovskite solar cells by transient capacitance measurements, *Material Horizons* **2019**, *6*, 1497–1503.

60. R. A. Awni, *et al.*, Influence of charge transport layers on capacitance measured in halide perovskite solar cells, *Joule* **2020**, *4*, 644–657.

61. D. W. deQuilettes, *et al.*, Photo-induced halide redistribution in organic-inorganic perovskite films, *Nature Communications* **2016,** *7,* 11683.

62. P. Calado, A. M. Telford, D. Bryant, X. Li, J. Nelson, B. C. O'Regan, P. R. Barnes, Evidence for ion migration in hybrid perovskite solar cells with minimal hysteresis, *Nature Communications* **2016,** *7,* 13831.

63. M. De Bastiani, *et al.*, Ion migration and the role of preconditioning cycles in the stabilization of the *J-V* characteristics of inverted hybrid perovskite solar cells, *Advanced Energy Materials* **2016,** *6,* 1501453.

64. D. H. Kang, N. G. Park, On the current-voltage hysteresis in perovskite solar cells: dependence on perovskite composition and methods to remove hysteresis, *Advanced Materials* **2019,** *31,* e1805214.

65. T. Zhang, *et al.*, Understanding the relationship between ion migration and the anomalous hysteresis in high-efficiency perovskite solar cells: a fresh perspective from halide substitution, *Nano Energy* **2016,** *26,* 620–630.

66. N.-G. Park, M. Grätzel, T. Miyasaka, K. Zhu, K. Emery, Towards stable and commercially available perovskite solar cells, *Nature Energy* **2016,** *1,* 16152.

67. J. Liu, M. Hu, Z. Dai, W. Que, N. P. Padture, Y. Zhou, Correlations between electrochemical ion migration and anomalous device behaviors in perovskite solar cells, *ACS Energy Letters* **2021,** *6,* 1003–1014.

68. S. Kim, *et al.*, Relationship between ion migration and interfacial degradation of $CH_3NH_3PbI_3$ perovskite solar cells under thermal conditions, *Scientific Reports* **2017,** *7,* 1200.

69. Y. Zhao, *et al.*, Mobile-ion-induced degradation of organic hole-selective layers in perovskite solar cells, *Journal of Physical Chemistry C* **2017,** *121,* 14517–14523.

70. Y. Wang, *et al.*, Stabilizing heterostructures of soft perovskite semiconductors, *Science* **2019,** *365,* 687–691.

71. J. H. Noh, S. H. Im, J. H. Heo, T. N. Mandal, S. I. Seok, Chemical management for colorful, efficient, and stable inorganic-organic hybrid nanostructured solar cells, *Nano Letters* **2013,** *13,* 1764–1769.

72. E. T. Hoke, D. J. Slotcavage, E. R. Dohner, A. R. Bowring, H. I. Karunadasa, M. D. McGehee, Reversible photo-induced trap formation in mixed-halide hybrid perovskites for photovoltaics, *Chemical Science* **2015,** *6,* 613–617.

73. C. G. Bischak, C. L. Hetherington, H. Wu, S. Aloni, D. F. Ogletree, D. T. Limmer, N. S. Ginsberg, Origin of reversible photoinduced phase separation in hybrid perovskites, *Nano Letters* **2017,** *17,* 1028–1033.

74. R. A. Kerner, Z. Xu, B. W. Larson, B. P. Rand, The role of halide oxidation in perovskite halide phase separation, *Joule* **2021,** *5,* 2273–2295.

75. N. Li, *et al.*, Microscopic degradation in formamidinium-cesium lead iodide perovskite solar cells under operational stressors, *Joule* **2020,** *4,* 1743–1758.

76. S. Macpherson, *et al.*, Local nanoscale phase impurities are degradation sites in halide perovskites, *Nature* **2022,** *607,* 294–300.

77. A. Jaffe, Y. Lin, C. M. Beavers, J. Voss, W. L. Mao, H. I. Karunadasa, High-pressure single-crystal structures of 3D lead-halide hybrid perovskites and pressure effects on their electronic and optical properties, *ACS Cent. Sci.* **2016,** *2,* 201–209.

78. C. M. M. Soe, *et al.*, New type of 2D perovskites with alternating cations in the interlayer space, $(C(NH_2)_3)(CH_3NH_3)_nPb_nI_{3n+1}$: structure, properties, and photovoltaic performance, *Journal of the American Chemical Society* **2017,** *139,* 16297–16309.

79. A. D. Jodlowski, *et al.*, Large guanidinium cation mixed with methylammonium in lead iodide perovskites for 19% efficient solar cells, *Nature Energy* **2017,** *2,* 972–979.

80. D. W. Ferdani, *et al.*, Partial cation substitution reduces iodide ion transport in lead iodide perovskite solar cells, *Energy & Environmental Science* **2019,** *12,* 2264–2272.

81. G. Kieslich, S. Sun, A. K. Cheetham, Solid-state principles applied to organic-inorganic perovskites: new tricks for an old dog, *Chemical Science* **2014,** *5,* 4712–4715.

82. M. Saliba, *et al.*, Incorporation of rubidium cations into perovskite solar cells improves photovoltaic performance, *Science* **2016**, *354*, 206–209.

83. D. Lin, *et al.*, Ion migration accelerated reaction between oxygen and metal halide perovskites in light and its suppression by cesium incorporation, *Advanced Energy Materials* **2021**, *11*, 2002552.

84. W. Zhou, Y. Zhao, X. Zhou, R. Fu, Q. Li, Y. Zhao, K. Liu, D. Yu, Q. Zhao, Light-independent ionic transport in inorganic perovskite and ultrastable Cs-based perovskite solar cells, *The Journal of Physical Chemistry Letters* **2017**, *8*, 4122–4128.

85. J. Cao, S. X. Tao, P. A. Bobbert, C. P. Wong, N. Zhao, Interstitial occupancy by extrinsic alkali cations in perovskites and its impact on ion migration, *Advanced Materials* **2018**, *30*, e1707350.

86. X. Liu, *et al.*, Exploring inorganic binary alkaline halide to passivate defects in low-temperature-processed planar-structure hybrid perovskite solar cells, *Advanced Energy Materials* **2018**, *8*, 1800138.

87. J. K. Nam, *et al.*, Potassium incorporation for enhanced performance and stability of fully inorganic cesium lead halide perovskite solar cells, *Nano Letters* **2017**, *17*, 2028–2033.

88. Z. Yang, A. Rajagopal, S. B. Jo, C. C. Chueh, S. Williams, C. C. Huang, J. K. Katahara, H. W. Hillhouse, A. K. Jen, Stabilized wide bandgap perovskite solar cells by tin substitution, *Nano Letters* **2016**, *16*, 7739–7747.

89. S. P. Senanayak, *et al.*, Charge transport in mixed metal halide perovskite semiconductors, *Nature Materials* **2023**, *22*, 216–224.

90. W. Pan, *et al.*, $Cs_2AgBiBr_6$ single-crystal X-ray detectors with a low detection limit, *Nature Photonics* **2017**, *11*, 726–732.

91. I. C. Smith, E. T. Hoke, D. Solis-Ibarra, M. D. McGehee, H. I. Karunadasa, A layered hybrid perovskite solar-cell absorber with enhanced moisture stability, *Angewandte Chemie International Edition* **2014**, *126*, 11414–11417.

92. D. H. Cao, C. C. Stoumpos, O. K. Farha, J. T. Hupp, M. G. Kanatzidis, 2D homologous perovskites as light-absorbing materials for solar cell applications, *Journal of the American Chemical Society* **2015**, *137*, 7843–7850.

93. Y. Lin, Y. Bai, Y. Fang, Q. Wang, Y. Deng, J. Huang, Suppressed ion migration in low-dimensional perovskites, *ACS Energy Letters* **2017**, *2*, 1571–1572.

94. X. Xiao, *et al.*, Suppressed ion migration along the in-plane direction in layered perovskites, *ACS Energy Letters* **2018**, *3*, 684–688.

95. J. Chen, D. Lee, N. G. Park, Stabilizing the ag electrode and reducing *J-V* hysteresis through suppression of iodide migration in perovskite solar cells, *ACS Applied Materials & Interfaces* **2017**, *9*, 36338–36349.

96. J. W. Lee, *et al.*, 2D perovskite stabilized phase-pure formamidinium perovskite solar cells, *Nature Communications* **2018**, *9*, 3021.

97. M. Kim, *et al.*, Methylammonium chloride induces intermediate phase stabilization for efficient perovskite solar cells, *Joule* **2019**, *3*, 2179–2192.

98. M. Li, B. Li, G. Cao, J. Tian, Monolithic $MAPbI_3$ films for high-efficiency solar cells via coordination and a heat assisted process, *Journal of Materials Chemistry A* **2017**, *5*, 21313–21319.

99. M. Yang, *et al.*, Perovskite ink with wide processing window for scalable high-efficiency solar cells, *Nature Energy* **2017**, *2*, 17038.

100. W. Ke, *et al.*, Employing lead thiocyanate additive to reduce the hysteresis and boost the fill factor of planar perovskite solar cells, *Advanced Materials* **2016**, *28*, 5214–5221.

101. F. Cai, Y. Yan, J. Yao, P. Wang, H. Wang, R. S. Gurney, D. Liu, T. Wang, Ionic additive engineering toward high-efficiency perovskite solar cells with reduced grain boundaries and trap density, *Advanced Functional Materials* **2018**, *28*, 1801985.

102. Q. Jiang, D. Rebollar, J. Gong, E. L. Piacentino, C. Zheng, T. Xu, Pseudohalide-induced moisture tolerance in perovskite $CH_3NH_3Pb(SCN)_2I$ thin films, *Angewandte Chemie International Edition* **2015**, *54*, 7617–7620.

103. Y. Zhao, *et al.*, Perovskite seeding growth of formamidinium-lead-iodide-based perovskites for efficient and stable solar cells, *Nature Communications* **2018**, *9*, 1607.

104. Q. Zeng, *et al.*, Inorganic CsPbI$_2$Br perovskite solar cells: the progress and perspective, *Solar RRL* **2019**, *3*, 1800239.

105. W. Zhou, *et al.*, Constructing CsPbBr$_3$ cluster passivated-triple cation perovskite for highly efficient and operationally stable solar cells, *Advanced Functional Materials* **2019**, *29*, 1809180.

106. D. Wei, *et al.*, Ion-migration inhibition by the cation-π interaction in perovskite materials for efficient and stable perovskite solar cells, *Advanced Materials* **2018**, *30*, e1707583.

107. X. Li, W. Zhang, Y. C. Wang, W. Zhang, H. Q. Wang, J. Fang, In-situ cross-linking strategy for efficient and operationally stable methylammoniun lead iodide solar cells, *Nature Communications* **2018**, *9*, 3806.

108. Y. Cai, *et al.*, Multifunctional enhancement for highly stable and efficient perovskite solar cells, *Advanced Functional Materials* **2021**, *31*, 2005776.

109. X. Yang, *et al.*, Superior carrier lifetimes exceeding 6 micros in polycrystalline halide perovskites, *Advanced Materials* **2020**, *32*, e2002585.

110. J.-W. Lee, S.-H. Bae, Y.-T. Hsieh, N. De Marco, M. Wang, P. Sun, Y. Yang, A bifunctional lewis base additive for microscopic homogeneity in perovskite solar cells, *Chem* **2017**, *3*, 290–302.

111. J. Gong, M. Yang, D. Rebollar, J. Rucinski, Z. Liveris, K. Zhu, T. Xu, Divalent anionic doping in perovskite solar cells for enhanced chemical stability, *Advanced Materials* **2018**, e1800973.

112. J. Gebhardt, A. M. Rappe, Mix and match: organic and inorganic ions in the perovskite lattice, *Advanced Materials* **2019**, *31*, e1802697.

113. Y. Shen, *et al.*, Functional ionic liquid polymer stabilizer for high-performance perovskite photovoltaics, *Angewandte Chemie International Edition* **2023**, *62*, e202300690.

114. W. Fan, *et al.*, Grain boundary perfection enabled by pyridinic nitrogen doped graphdiyne in hybrid perovskite, *Advanced Functional Materials* **2021**, *31*, 2104633.

115. X. Duan, X. Li, L. Tan, Z. Huang, J. Yang, G. Liu, Z. Lin, Y. Chen, Controlling crystal growth via an autonomously longitudinal scaffold for planar perovskite solar cells, *Advanced Materials* **2020**, *32*, e2000617.

116. J. Yang, *et al.*, High-performance perovskite solar cells with excellent humidity and thermo-stability via fluorinated perylenediimide, *Advanced Energy Materials* **2019**, *9*, 1900198.

117. S. Bai, *et al.*, Planar perovskite solar cells with long-term stability using ionic liquid additives, *Nature* **2019**, *571*, 245–250.

118. Y. H. Lin, *et al.*, A piperidinium salt stabilizes efficient metal-halide perovskite solar cells, *Science* **2020**, *369*, 96–102.

119. Q. Wang, Y. Shao, Q. Dong, Z. Xiao, Y. Yuan, J. Huang, Large fill-factor bilayer iodine perovskite solar cells fabricated by a low-temperature solution-process, *Energy & Environmental Science* **2014**, *7*, 2359–2365.

120. Y. Zhao, Y. Zhao, W. Zhou, Q. Li, R. Fu, D. Yu, Q. Zhao, In situ cesium modification at interface enhances the stability of perovskite solar cells, *ACS Applied Materials & Interfaces* **2018**, *10*, 33205–33213.

121. S. Yang, *et al.*, Stabilizing halide perovskite surfaces for solar cell operation with wide-bandgap lead oxysalts, *Science* **2019**, *365*, 473–478.

122. H. Zai, *et al.*, Sandwiched electrode buffer for efficient and stable perovskite solar cells with dual back surface fields, *Joule* **2021**, *5*, 2148–2163.

123. H. Kim, S. U. Lee, D. Y. Lee, M. J. Paik, H. Na, J. Lee, S. I. Seok, Optimal interfacial engineering with different length of alkylammonium halide for efficient and stable perovskite solar cells, *Advanced Energy Materials* **2019**, *9*, 1902740.

124. Q. Jiang, *et al.*, Surface passivation of perovskite film for efficient solar cells, *Nature Photonics* **2019**, *13*, 460–466.

125. A. A. Sutanto, *et al.*, Dynamical evolution of the 2D/3D interface: a hidden driver behind perovskite solar cell instability, *Journal of Materials Chemistry A* **2020**, *8*, 2343–2348.

126. C. Chen, *et al.*, Arylammonium-assisted reduction of the open-circuit voltage deficit in wide-bandgap perovskite solar cells: the role of suppressed ion migration, *ACS Energy Letters* **2020**, *5*, 2560–2568.

127. C. Ma, N.-G. Park, Paradoxical approach with a hydrophilic passivation layer for moisture-stable, 23% efficient perovskite solar cells, *ACS Energy Lett.* **2020**, *5*, 3268–3275.

128. L. Luo, *et al.*, Stabilization of 3D/2D perovskite heterostructures via inhibition of ion diffusion by cross-linked polymers for solar cells with improved performance, *Nat. Energy* **2023**, *8*, 294–303.

129. E. Bi, *et al.*, Efficient perovskite solar cell modules with high stability enabled by iodide diffusion barriers, *Joule* **2019**, *3*, 2748–2760.

130. Y. Ge, *et al.*, Internal encapsulation for lead halide perovskite films for efficient and very stable solar cells, *Advanced Energy Materials* **2022**, *12*, 2200361.

131. H. Wang, H. Liu, W. Li, L. Zhu, H. Chen, Inorganic perovskite solar cells based on carbon electrodes, *Nano Energy* **2020**, *77*, 105160.

132. H. Chen, S. Yang, Methods and strategies for achieving high-performance carbon-based perovskite solar cells without hole transport materials, *Journal of Materials Chemistry A* **2019**, *7*, 15476–15490.

133. G. Grancini, *et al.*, One-year stable perovskite solar cells by 2D/3D interface engineering, *Nature Communications* **2017**, *8*, 15684.

134. X. Meng, J. Zhou, J. Hou, X. Tao, S. H. Cheung, S. K. So, S. Yang, Versatility of carbon enables all carbon based perovskite solar cells to achieve high efficiency and high stability, *Advanced Materials* **2018**, *30*, e1706975.

135. K. A. Bush, *et al.*, Thermal and environmental stability of semi-transparent perovskite solar cells for tandems enabled by a solution-processed nanoparticle buffer layer and sputtered ito electrode, *Advanced Materials* **2016**, *28*, 3937–3943.

12 Origins and Elimination of Hysteresis in Perovskite Solar Cells

Rui Lin, Yibing Wu, and Xinhua Ouyang

12.1 INTRODUCTION

Perovskite solar cells (PSCs) have been demonstrated to be one of the most promising photovoltaic technologies, attributed to their ease of processing and high performance. In the past decade, it has experienced dramatic development with a certified power conversion efficiency (PCE) over 25%,[1,2] which is comparable with their silicon counterpart (26.7%).[3] Multiple efforts have been devoted to overcoming the challenges of PSCs, such as suppressing recombination, optimizing interlayers, and improving device stability.[4–6] The current density-voltage (J-V) hysteretic behavior is one of the critical issues for the research of PSCs. Eliminating hysteresis is an effective strategy to improve the performance of PSCs. In the highly efficient PSCs, negligible hysteresis can be observed.[7–9] Therefore, a full understanding of J-V hysteresis is essential to minimize its negative effects.

The J-V measurement is a diagnostic characteristic for solar cells, which is normally applied to determine the PCE. However, the J-V curves of PSCs show strong dependence on the scanning conditions of the J-V measurements, such as scan directions and scan rates (Figure 12.1). This phenomenon is well known as J-V hysteresis. It creates a great challenge in determining the accurate PCEs of PSCs.[10] Snaith et al.[11] first issued this anomalous J-V hysteresis in 2014. Since then, many studies have been focused on revealing the nature of hysteresis, such as ferroelectricity[12,13], interfacial charge accumulation,[14,15] ion migration,[16–18] etc.

12.1.1 TYPES OF HYSTERESIS

The J-V curves are generally obtained by measuring the current density while changing the applied voltage. The forward scan of the J-V measurements from reverse voltage bias to forward voltage bias go through short circuit condition (0 V) followed by open circuit voltage (V_{OC}), while the reverse scan is from forward voltage bias to reverse voltage bias. Due to the presence of hysteresis in PSCs, there exists a mismatch between the J-V curves measured under the forward scan and the reverse scan. It categorizes the hysteresis into three types: normal hysteresis, free hysteresis, and inverted hysteresis.[20] In the normal hysteresis, the photovoltaic performance, such as

DOI: 10.1201/9781003400486-12

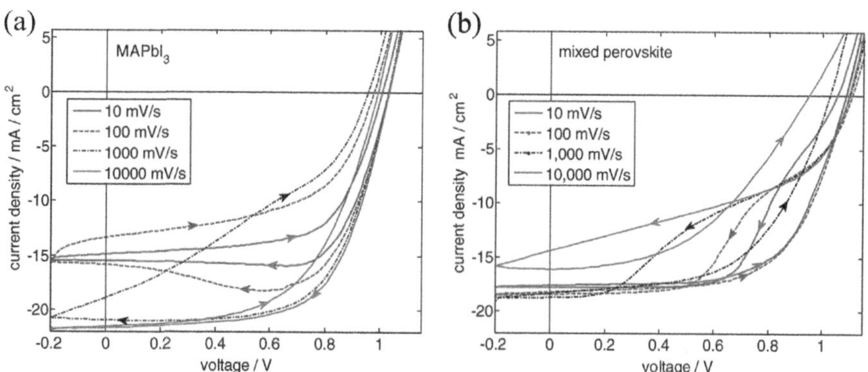

FIGURE 12.1 (a) Normal hysteresis of a CH$_3$NH$_3$PbI$_3$ (MAPbI$_3$) perovskite solar cell and (b) inverted hysteresis of a mixed perovskite solar cell with a precursor solution of FAI, PbI$_2$, MABr and PbBr$_2$ with a ratio of 1.0: 1.1: 0.2: 0.22.[19] Copyright 2016, WILEY-VCH.

PCE and photocurrent, under reverse scan is better than that under the forward scan (Figure 12.1a). Most of the reported PSCs present a photovoltaic performance with a normal hysteresis.[21–23] The free hysteresis refers to the negligible mismatch between the J-V curves under the forward and reverse scan, which can be achieved by precise optimization.[24–26] As displayed in Figure 12.1b, the inverted hysteresis is opposite to the normal hysteresis, in which PSCs under the forward scan performs better.[27–29]

12.1.2 Hysteresis Index

The hysteresis can be qualified by hysteresis index (HI), which indicates the difference between J-V curves under the forward and reverse scan. The HI was first defined by Sanchez et al.[30] using Equation (12.1) with a α value of 0.5. It was applied as an indicator of the photocurrent hysteresis. A HI with a value of 0 refers to no observable hysteresis. As shown in Figure 12.1, the hysteresis is more prominent near the voltage of maximum power point, which is around 0.8 V$_{OC}$. Therefore, the α value of Equation (12.1) can be adjusted to 0.8.[31] In the inverted hysteresis, the J-V curve under forward scan possesses a higher photocurrent. Rong et al.[32] modified this HI to Equation (12.2). Wei et al.[33] defined another HI by comparing the PCE under the forward and reverse scan (Equation 12.3). As the PCE is the key factor for the solar cell performance, it is the most widely used HI in current research.[34–36] No matter how the HI is defined, it only describes the discrepancy between the photovoltaic performance under different scan directions without considering the scan rate (Figure 12.1). Therefore, it must be noticed that the degree of hysteresis should not be solely determined by the HI.[37]

$$\mathrm{HI} = \frac{J_{RS}\left(\alpha V_{OC}\right) - J_{FS}\left(\alpha V_{OC}\right)}{J_{RS}\left(\alpha V_{OC}\right)} \tag{12.1}$$

$$HI = \frac{J_{FS}\left(0.8V_{OC}\right) - J_{RS}\left(0.8V_{OC}\right)}{\left(J_{FS}\left(0.8V_{OC}\right) + J_{RS}\left(0.8V_{OC}\right)\right)/2} \tag{12.2}$$

where $J_{RS}(\alpha V_{OC})$ is the current density achieved with a voltage bias of αV_{OC} under the reverse scan and $J_{FS}(\alpha V_{OC})$ is the current density achieved with a voltage bias of αV_{OC} under the forward scan. The α is a constant $(0 < \alpha < 1)$.

$$HI = \frac{PCE_{RS} - PCE_{FS}}{PCE_{RS}} \tag{12.3}$$

where PCE_{RS} is the PCE obtained under the reverse scan and PCE_{FS} is the PCE obtained under the forward scan.

12.2 ORIGINS OF HYSTERESIS

When the hysteresis was first issued, three possible origins were proposed by Snaith et al.[11] It hypothesized that ferroelectric property, the defeats within the bulk or near the surface and ion migrations, may attribute to the anomalous J-V hysteresis.[38] In order to get a clear picture of its origins, these hypotheses need to be proved.

12.2.1 Ferroelectric Property

Many materials with the perovskite structure (ABX_3) have been demonstrated to have ferroelectricity.[39] According to the model developed by Frost et al.,[40] the $CH_3NH_3^+$ (MA^+) ions inside the $MAPbI_3$ are able to rotationally mobile at room temperature (Figure 12.2), which will be polarized by the applied electric field. The interactions between these MA^+ ions help the formation of polar domains, which results in a ferroelectric response. Wei et al.[33] detected this weak ferroelectric effect in the $MAPbI_{3-x}Cl_x$ perovskite film through the polarization-electric field (P-E) loop measurements. Although different ferroelectric properties are observed,[41-44] the $MAPbI_3$ perovskite film has been demonstrated to be a ferroelectric semiconductor.[45] A clear polarization behavior of the $MAPbI_3$ film can be detected using piezoelectric force microscopy

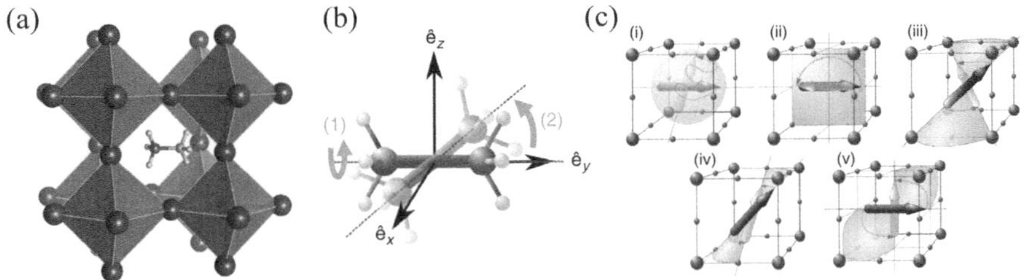

FIGURE 12.2 (a) $MAPbI_3$ crystal structure and (b) the rotation modes of MA^+ ion. (c) Possible reorientation pathways for MA^+ ions: (i) free rotation, (ii) π-flips, (iii) corner-to-corner hops, (iv) edge-to-edge hops, and (v) face-to-face hops.[51] Copyright 2015, Springer.

(PFM), which changes with the applied bias voltage (Figure 12.3a–d).[46] A hysteretic response of this polarization with 180° phase switching is observed (Figure 12.3e–h). This ferroelectric polarization may be one of the mechanisms that causes the J-V hysteresis.[47] However, the J-V hysteresis can be also detected in a non-ferroelectric PSC.[48] This statement should be carefully interrogated. The hysteretic behavior is strongly dependent on the frequency, which is an uncharacteristic of the ferroelectric response.[49] The polarization time for the rotation of a single MA^+ ion is on the pico-second timescale.[50] As a result, the estimated timescale for the polarization of all the ferroelectric domains across a 200-nm-thick PSC is around 0.1~1 ms.[51] Compared with the characteristic time taking for the J-V measurement (e.g. ~0.1 ~ 100 s in Figure 12.1), it is too short. These results completely rule out the ferroelectricity as the cause for the J-V hysteresis.

12.2.2 TRAPPING AND DETRAPPING OF DEFECTS

The polycrystalline perovskite films possess large density of charge defects, such as interstitials, substitutional defects, and vacancies, present within the bulk of the grains and at the brain boundaries.[52,53] For instance, the $MAPbI_3$ film possesses a high concentration of charge vacancies and defects,[54,55] which are mostly located at the crystallite surfaces and interfaces.[56] Under the short circuit conditions (or reverse bias), the electrons and holes will be trapped by the defects. When applied to a forward bias, they will escape from the trap states of these defects.[57] This charge trapping and detrapping (releasing) process of the defects strongly affects the charge collection depending on the voltage,[58] which partially contributes to the notorious hysteretic effect of J-V curves.[59] Shao et al.[60] reported that the trap filling process of the trap states leads to a delay of photocurrent response, which results in a long charge carrier transit mobility and a clear hysteretic behavior (Figure 12.4). They demonstrated that the large trap densities inside the PSCs play an important role in the photocurrent hysteretic behavior. Multiple strategies which reduce the defects of PSCs, such as increasing crystal size,[31] defect passivation,[61,62] and interfacial engineering,[63,64] have been successfully applied to eliminate the J-V hysteresis. These results provide direct evidence that the charge defects inside PSCs will lead to the J-V hysteresis.

12.2.3 ION MIGRATIONS

Due to the low defect formation energies,[65–67] there exist intrinsic ions inside the perovskites, which are halogen anions (e.g., I^-, Br^-), Pb^{2+}, MA^+ (or formamidinium cation, FA^+) and their related vacancies (e.g., V_I^+ or V_{Br}^+, V_{Pb}^{2-} and V_{MA^-} or V_{FA}^-). These ions possess low activation energies of migration, which facilitates them to diffuse inside the perovskite film (Figure 12.5a and b).[68–70] And the extrinsic ions, such as Li^+ cations from spiro-OMeTAD, are also capable to migrate into the perovskites.[71] These ion migrations have been demonstrated to influence the photovoltaic performance and device stability.[72,73] Unger et al.[74] reported that the J-V characteristics of PSCs are sensitive to the pre-conditioning bias before measurements. They proposed that this phenomenon is attributed to the ion migrations. When a positive bias is applied,

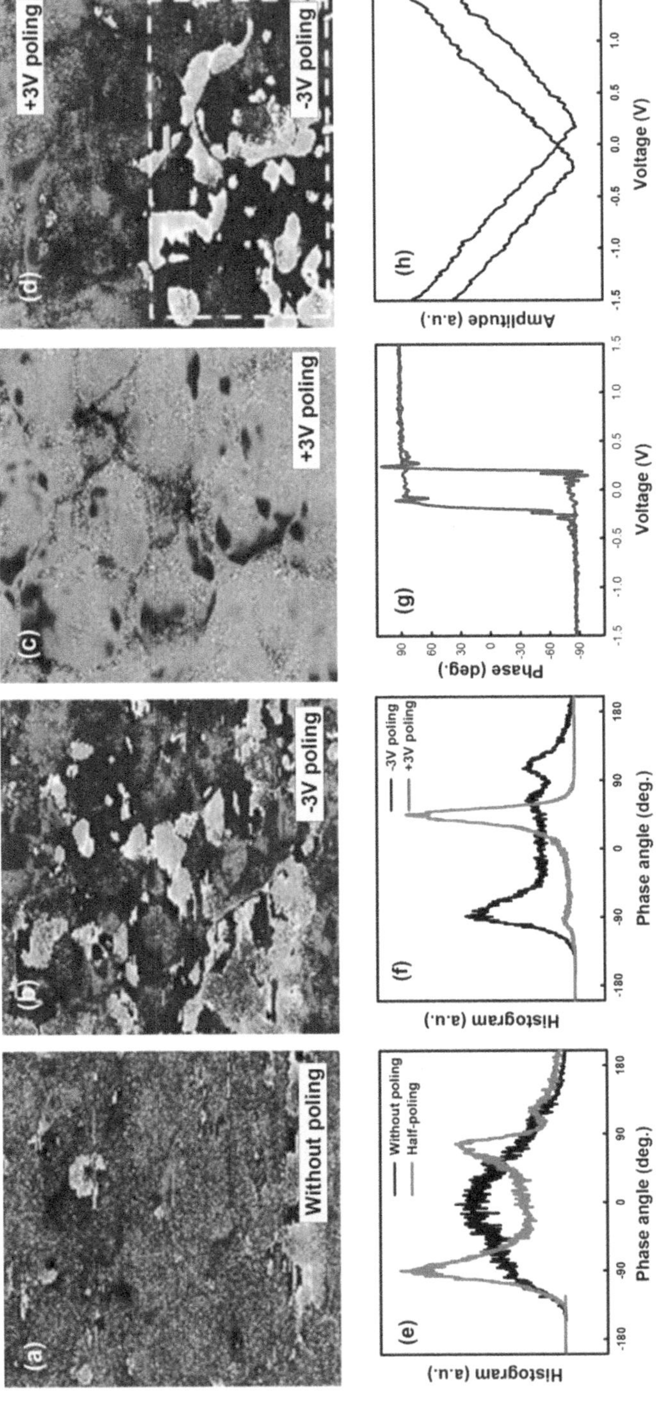

FIGURE 12.3 PFM phase image of MAPbI$_3$ film under different poling conditions: (a) without poling, (b) -3 V poling, (c) +3 V poling, and (d) half-poling with +3 V poling on the top and -3 V poling on the bottom. (e) Domain phase angel distribution for the MAPbI$_3$ film without poling and under half-poling; (f) Domain phase angel distribution for the MAPbI$_3$ film under -3 V poling and +3 V poling. (g) Piezoresponse phase and (h) piezoresponse amplitude at different bias voltages.[46] Copyright 2015, Elsevier.

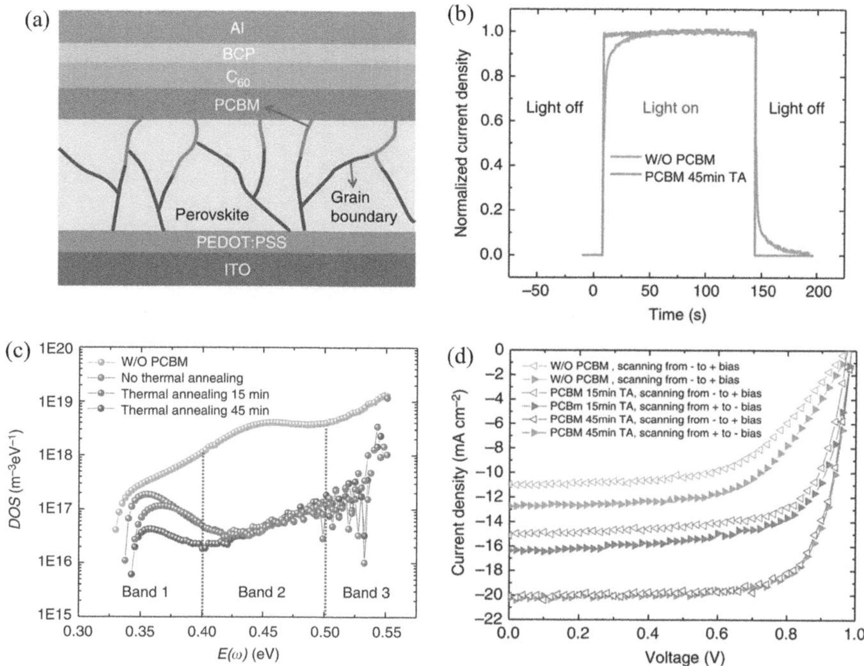

FIGURE 12.4 (a) Schematic of the perovskite photovoltaic device structure with a PCBM layer. (b) Photocurrent rising process of the PSCs with/without the PCBM layer through turning on and turning off the incident light. c) Trap density of states (tDOS) and (d) *J*-V curves of the PSCs with/without the PCBM layer and the PSCs with PCBM layer under different thermal annealing conditions. Both the forward and reverse scan are recorded for the *J*-V measurement.[60] Copyright 2014, Springer.

the anions (negative charges) are driven towards the hole transport layer (HTL) and the cations (positive charges) accumulate in the perovskite near the interface of the electron transport layer (ETL). This unbalanced distribution of ions leads to the formation of interfacial charge, which affects the band bending of the perovskite near the interfaces (Figure 12.5c).[75,76] It enhances the build-in potential (V_{bi}) of PSCs. In contrast, the V_{bi} is reduced under a negative bias. In the reverse scan, an initial positive bias is generally applied before the measurement, while a zero or negative bias is applied before the forward scan. It creates a different interfacial charge status before the voltage sweeping.[77] As a result, the photovoltaic performance under the reverse scan should be better than that under the forward scan, which is the typical feature of normal hysteresis. The rate dependence is another feature of the hysteresis, which becomes more obvious at a slower scan rate. A slow field-introduced process should be responsible for this rate-dependent hysteric performance.[78] The ions start to migrate and accumulate to form interfacial charges within 10 ms after applying the bias.[79] However, it needs a much longer time to become stabilized (> 10^3 s).[80] Thus, the ion migration is a slow field-introduced dynamic process. Almora et al.[81,82] reported that the *J*-V hysteresis is related to the slow capacitive mechanisms,

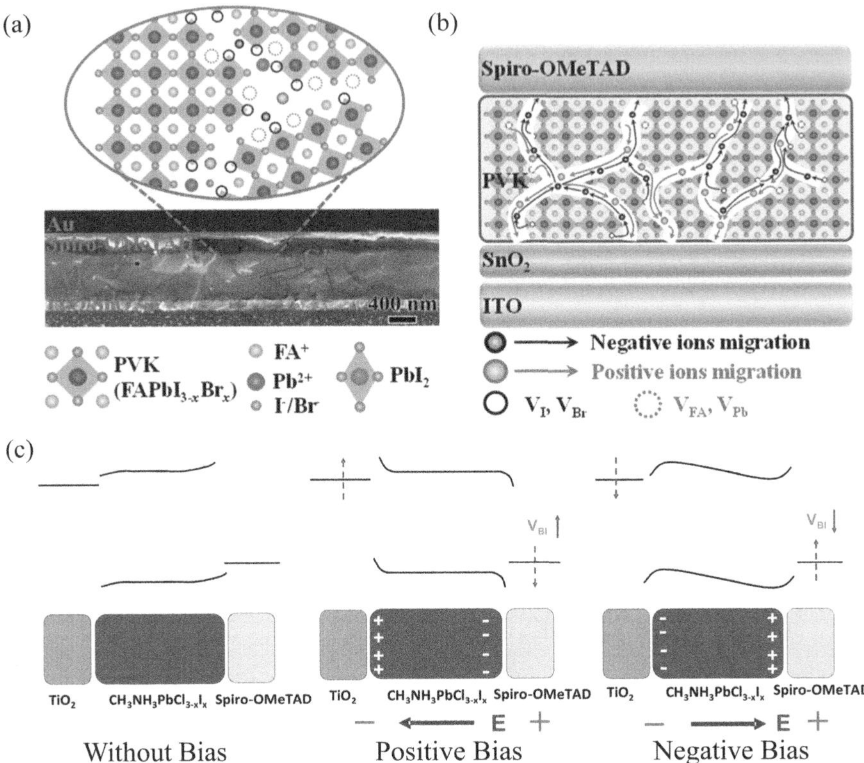

FIGURE 12.5 (a) Cross section scanning electron microscopy (SEM) image and the schematic of grain boundaries of PSCs and (b) schematic illustration of ion migrations.[76] Copyright 2022, WILEY-VCH. (c) Schematic diagrams of band bending in the PSCs under different applied bias. The E presents the direction of electrical field caused by the applied bias.[91] Copyright 2016, WILEY-VCH.

which are mostly caused by the ion accumulation. These results suggest that the ion migration can be the origin of hysteresis.[83] Shao et al.[84] used conductive atomic force microscopy (c-AFM) to detect the difference between the hysteretic behavior at the grain boundaries and on the grains. The ions move much faster at the grain boundaries than through the grain interiors.[85] As a consequence, the clear hysteresis observed at the grain boundaries demonstrates the key role of ion migrations in the J-V hysteresis.

The inverted hysteresis is also attributed to this slow field-introduced ion redistribution. Under a positive bias, the piling up of cations and anions can be observed in the side of perovskite near the ETL and HTL, respectively. It will create a reversed electric field close to the contacts, which hampers the charge extraction and increases surface recombination.[19, 86] Therefore, the charge transfer is suppressed in the reverse scan, which leads to poor performance. It creates an inverse hysteresis detected in few cases of PSCs. The ion accumulation status strongly relies on the biasing conditions. Both the normal hysteresis and the inverted hysteresis can be obtained in the same samples under different pre-poling bias (V_{pol}),[87] as shown in Figure 12.6. Stabilizing the devices at a large positive bias will eliminate the detrimental effects of the ion accumulation, which contributes to an inverted hysteresis at fast scan rates.[88,89] All these studies demonstrate that the J-V hysteresis is mainly limited by the ion migration.[90]

12.2.4 Charge Carrier Transport and Extraction

The organic-inorganic metal halide perovskite materials possess unprecedented ambipolar charge transport property, which enables them to transport both holes and

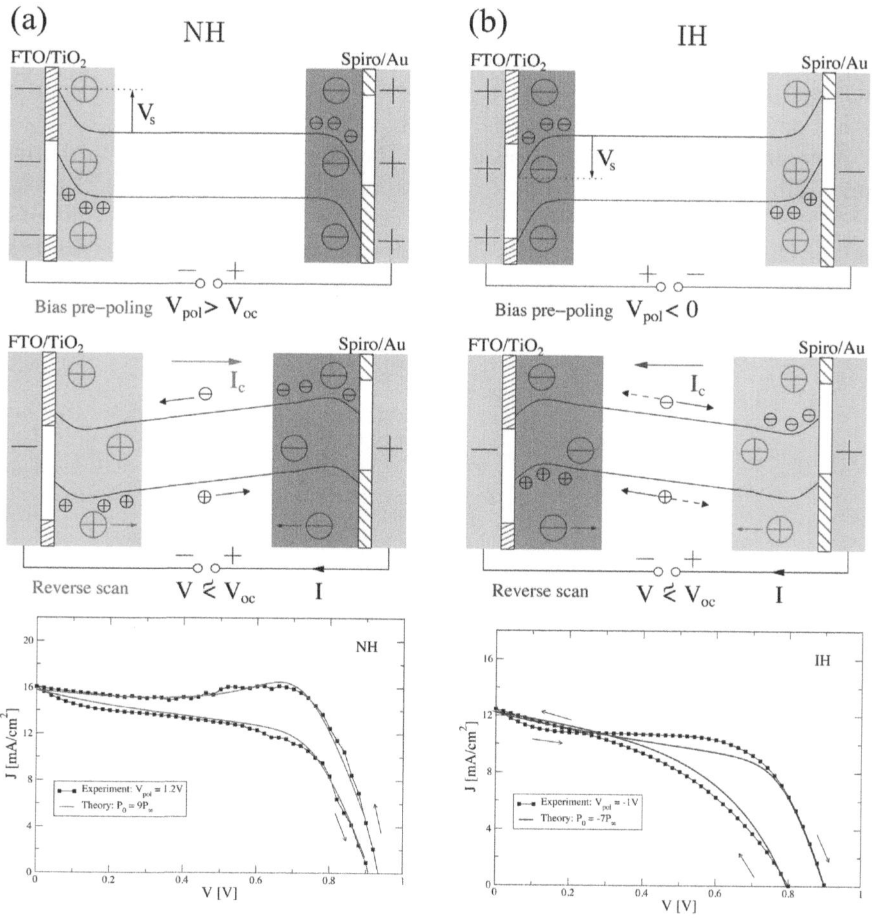

FIGURE 12.6 Schematics of band diagrams under different pre-poling bias (top), followed by the reversed scan starting from the V_{OC} (middle) and the J-V curves (bottom) of the samples exhibit (a) normal hysteresis and (b) inverted hysteresis.[87] Copyright 2017, American Chemical Society.

electrons.[92] Due to the lower electron mobility,[93] the holes transport more efficiently inside the perovskites. And the electrons may accumulate at the perovskite interface near the cathode. This interfacial charge accumulation gives rise to the capacitive current, which should be responsible for the hysteresis response.[94] Therefore, the ETL is normally required to promote the electron transport and reduce charge accumulation.[95,96] However, the electron mobility of the conventional metal oxide ETL (e.g., $TiO_2 \approx 10^{-4}$ cm$^2 \cdot$V$^{-1} \cdot$s^{-1}) is lower than the hole mobility of HTL (e.g., spior-OMeTAD $\approx 10^{-3}$ cm$^2 \cdot$V$^{-1} \cdot$s^{-1}).[97–99] Wu et al.[100] demonstrated that the inefficient electron transport and extraction will cause pronounced hysteresis. It causes an unbalanced charge carrier transport, which leads to charge accumulation near the perovskite/ETL interfaces. Increasing the conductivity of ETL helps to achieve a more balanced charge transport and reduce the hysteresis.[101] Ravishankar et al.[102] investigated the charge accumulation in the PSCs with different ETLs (ETL free, compact TiO_2, and mesoporous TiO_2 on compact TiO_2). The mesoporous TiO_2-based device presents the minimum hysteresis response (Figure 12.7), due to its reduced charge carrier accumulation. Thus, it can be concluded that the unbalanced charge carrier transport and extraction contribute to the hysteresis.

According to the experiment results and the modeling analysis, both the ionic and electronic process are involved into the formation of hysteresis.[103–105] In summary, the J-V hysteresis is a combined effect of the ionic process (ion migration) and the electronic process (trapping of carriers and charge transport), as illustrated in Figure 12.8.[106] These processes are all strongly correlated with each other. For

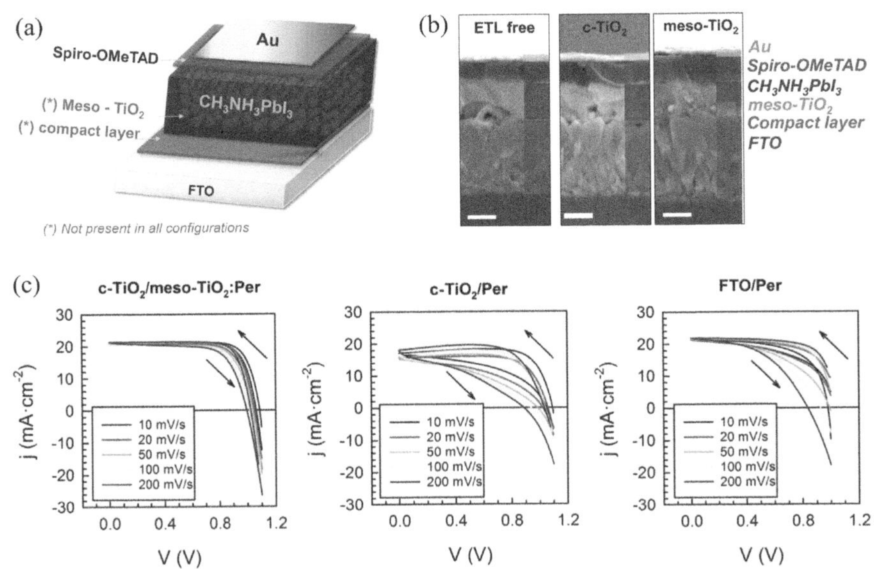

FIGURE 12.7 (a) Schematic of the PSC architecture; (b) cross-section SEM images (1 μm scale bar) of the PSC without ETL and the PSCs with compact TiO_2 (c-TiO_2) or mesoporous TiO_2 on c-TiO_2 (c-TiO_2/m-TiO_2) ETL. (c) The J-V curves of the PSCs without ETL and the PSCs with c-TiO_2 or c-TiO_2/m-TiO_2 ETL.[102] Copyright 2018, Elsevier.

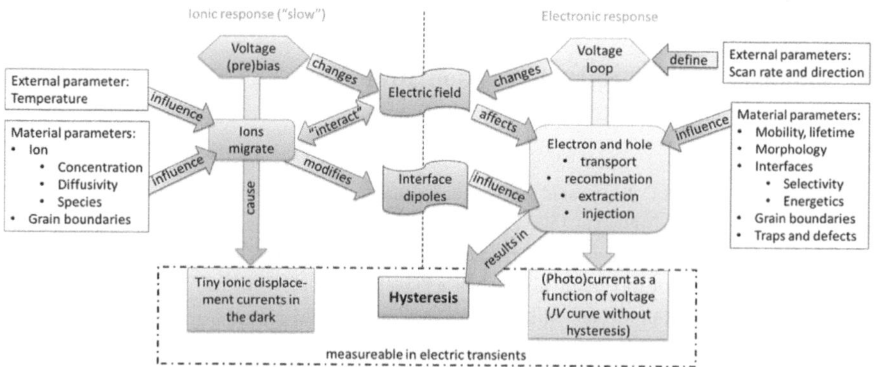

FIGURE 12.8 Schematic illustrations of the origins of the hysteresis.[106] Copyright 2017, American Chemical Society.

example, the ion migration will leave defects for carrier trapping.[107] The defects for carrier trapping, especially the surface defects, will significantly influence the charge accumulation at the interface, which facilitates the ion migration.[108] These defects are also one of the main cause for the stability issue of PSCs.[109,110] And the degradation of perovskite will, in turn, produce defects, which makes ion migration more serious. As a result, a stronger hysteresis can be observed in an aging device.[78] Therefore, the elimination of hysteresis, mostly achieved by reducing defects and suppressing ion migration, is crucial for the device performance and stability.

12.3 ELIMINATION OF HYSTERESIS

The hysteretic performance mostly originates from ion migration, charge trapping, and unbalanced charge transport. Therefore, the strategies for hysteresis elimination, displayed in Table 12.1, are all centered around the optimization of these processes, such as enhancing film quality, defect passivation, and interfacial engineering.

12.3.1 Enhancing Film Quality

The perovskite is an ionic material with fragile chemical bonds,[65] which promotes the formation of mobile ions and charge defects. The ion migrations are strongly correlated with the formation of defects, which offers the transport channels for the mobile ions.[111, 112] Therefore, the suppressions of ion migration and charge defects share almost the same strategies. The primary attempt for reducing ion migrations and charge defects is to improve the crystallinity of the perovskites. Increasing the crystallinity can enlarge the grain size of the perovskite, which reduces the grain boundaries. Thus, the defects on the grain boundaries and the migration routes for the ions will be reduced.[113] McGovern et al.[114] suggested that the suppression of ion migration in the PSCs with large grains is due to their high activation energy for mobile ions. Both the defect density and ion migration are reduced with the increasing of grain

TABLE 12.1

The Summary of Hysteresis Elimination Strategies

Strategies		$HI_{control}$[a] (%)	$HI_{optimized}$[b] (%)
Perovskite Film Modification	$FAPbI_{3-x}Br_x$ treated with different annealing time from 5 to 30 min at 150°C to enhance crystallinity[76]	22.43	1.04
	$MAPbI_3$ treated with different reaction temperature from 150 to 200°C to enhance crystallinity[117]	10.48	3.10
	$MAPbI_3$ treated with solvent annealing by DSMO[120]	4.96	2.16
	$MAPbI_{3-x}Cl_x$ processed with dimethylformamide solvent annealing to enhance crystallinity[121]	10.50	5.99
	$FA_{1-x}MA_xPbI_3$ with dimethyl itaconate to enhance crystallinity[122]	7.66	1.74
	$(FAPbI_3)_{1-x}(MAPbBr_3)_x$ varied with different perovskite components ($x = 0–1$)[127]	39.32	6.53
	$FA_{1-x}MA_xPbI_3$ varied with different perovskite components ($x = 0–1$)[130]	36.47	8.63
	$FA_{1-x}Cs_xPbI_3$ varied with different perovskite components ($x = 0–0.2$)[131]	28.74	26.84
	$FA_{0.85}MA_{0.1}Cs_{0.05}PbI_{2.7}Br_{0.3}$ with KI as the additive to enhance crystallinity [132]	16.18	1.26
	$MA_{0.17}FA_{0.83}PbI_{2.5}Br_{0.5}$ with KI as the additive to enhance crystallinity[133]	11.88	0.78
	$Cs_{0.05}(FA_{0.85}MA_{0.15})_{0.95}Pb(I_{0.85}Br_{0.15})_3$ with KI as the additive to enhance crystallinity[135]	19.21	2.14
	$FA_{0.85}MA_{0.15}Pb(I_{0.85}Br_{0.15})_3$ with KI as the additive to enhance crystallinity[136]	17.10	5.60
	$MAPbI_3$ with $Pb(SCN)_2$ as the additive to enhance crystallinity[137]	19.50	4.85
	$MAPbI_3$ with aluminium acetylacetonate as the additive to enhance crystallinity[138]	0.96	0.46
	$MAPbI_3$ with KSCN as the additive to enhance crystallinity[141]	23.73	2.55
	$FA_{0.83}Cs_{0.17}PbI_3$ with CuI as the additive to enhance electronic properties[142]	4.68	−0.47
	$MAPbI_3$ with poly(ethylene oxide) as the additive to passivate the defects[143]	4.79	0.05
	$FA_{0.85}MA_{0.15}PbI_3$ with a small organic molecule with the core architecture of indaceno[1,2-b:5,6-b'] dithiophene and bilateral arms of oxindole as the additive to passivate the defects[34]	10.68	3.67
	$MAPbI_3$ with 1,4,8,11,15,18,22,25-octakis(hexylthio) phthalocyaninato oxotitanium(IV) as the additive to passivate the defects[61]	16.86	5.47

TABLE 12.1 (*Continued*)
The Summary of Hysteresis Elimination Strategies

Strategies		$HI_{control}$[a] (%)	$HI_{optimized}$[b] (%)
	$FA_{0.5}MA_{0.5}Pb_{0.5}Sn_{0.5}I_3$ with ionic imidazolium tetrafluoroborate as the additive to passivate the defects[62]	2.31	0.52
	$FA_{0.92}MA_{0.08}PbI_3$ with poly(vinylidene fluoride):dabcoHReO$_4$ additive to enhance the build-in electric field[21]	4.83	3.81
ETL Optimization	Mesoporous TiO_2 doped with lithium in MAPbI$_3$ to enhance electron transport[147]	14.04	1.55
	Mesoporous TiO_2 treated with sodium in MAPbI$_3$ to enhance electron transport[148]	21.32	4.61
	Mesoporous TiO_2 doped with boron in MAPbI$_3$ to enhance electron transport[149]	13.06	0.83
	Mesoporous TiO_2 doped with Li_2CO_3 in FAPbI$_3$ to enhance electron transport[150]	4.78	4.07
	Apply C60 pyrrolidine tris-acid as the ETL in MAPbI$_3$[153]	39.93	2.07
	TiO_2/Trimethylamine oxide in $FA_{0.85}MA_{0.15}PbI_{2.55}Br_{0.45}$ to introduce interfacial dipolar[15]	8.09	1.29
	C_{60}/phenethylammonium iodide and ethylenediamine diiodide in $FA_{0.8}MA_{0.2}Pb_{0.8}Sn_{0.2}I_3$ to passivate defects[36]	9.14	3.51
	TiO_2/PCBM in MAPbI$_3$ to enhance the contact[157]	61.64	6.10
	Compact TiO_2/amorphous SnO_2 in $(FAPbI_3)_{0.85}(MAPbBr_3)_{0.15}$ to enhance electron extraction[159]	3.52	1.36
	SnO_2/$NdCl_3$ in $FA_{1-x}MA_xPbI_3$ to passivate the defect[160]	9.66	1.22
HTL Optimization	PTAA doped with a fluorine-containing hydrophobic Lewis acid dopant in MAPbI$_3$ to enhance hole mobility[163]	14.01	12.84
	Spiro-MeOTAD doped with Zn-TFSI$_2$ to enhance hole mobility (CsI: FAPbI$_3$: MAPbBr$_3$ = 0.3: 5: 1)[164]	3.34	1.91
	Spiro-MeOTAD doped with Zn-TFSI$_2$ to enhance hole mobility (CsI: FAPbI$_3$: MAPbBr$_3$ = 0.3: 5: 1)[165]	−1.19	0.21
	Spiro-MeOTAD doped with H_3PO_4 in $FA_{0.85}Cs_{0.15}PbI_3$ to enhance the conductivity[168]	14.00	4.00

[a] HI of the control device extracted using **Equation (12.3)**.

[b] HI of the optimized device extracted using **Equation (12.3)**.

size.[115,116] Therefore, a better crystallinity of perovskite normally indicates a smaller hysteresis.[117] Controlling the annealing process is the most easy way to enhance the crystallinity and enlarge the grain size.[118–120] Zuo et al.[121] reported that the crystal size of $MAPbI_{3-x}Cl_x$ becomes significantly larger after a solvent annealing treatment with dimethylformamide. Zhao et al.[122] reported a polymerization-assisted grain growth strategy to promote the crystallization process, which increases the average grain size of $FA_{1-x}MA_xPbI_3$ from 0.75 µm to 1.25 µm. The HIs of these reported PSCs are all significantly reduced. Thus, improving crystallinity and increasing grain size will help to reduce the J-V hysteresis.

Optimizing the perovskite composition is another effective approach to suppress the hysteresis. It has been demonstrated that there exists a direct correlation between the ion migration and the perovskite composition.[123] Kim et al.[124] increased the mole ratio between PbI_2 and MAI (or FAI) in the recipe of $MAPbI_3$ (or $(FAPbI_3)_{0.85}(MAPbBr_3)_{0.15}$) to suppress ion migration and hysteresis (Figure 12.9). It suggests that the additional amount of PbI_2 can passivate the defects and enhance the film quality.[125,126] Zhang et al.[127] varied the composition of mixed perovskite $(FAPbI_3)_{1-x}(MAPbBr_3)_x$ through a dual ion exchange method to enhance crystallinity and suppress ion migration. This mixed-cation strategy produces a high-quality perovskite film and makes the hysteresis negligible.[128,129] By changing the perovskite recipe from MA-rich to FA-rich stoichiometries, Cui et al.[130] illustrated the strong relation between the organic cation composition and J-V hysteresis. They attributed the reduced hysteresis to the combined effect of PbI_2 passivation and preferred crystal orientation, which is formed during the thermodynamically driven perovskite formation process. Lee et al.[131] suggested that the incorporation of Cs cation into $FAPbI_3$ will suppress the charge recombination. Thus, the $FA_{0.9}Cs_{0.1}PbI_3$-based device turns out to have a smaller hysteresis than the $FAPbI_3$-based device. The exploring of perovskite recipe enhances the film quality with fewer defects, which leads to a less hysteretic performance.

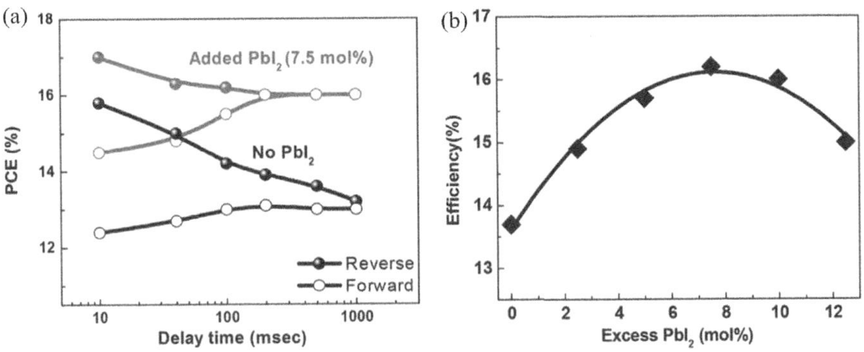

FIGURE 12.9 (a) PCE trend for the PSCs fabricated with and without adding PbI_2 into $MAPbI_3$, measured by forward and reverse scans under AM 1.5G illumination. (b) The variation in PCE of the PSCs when adding excessed PbI_2 in the range from 0 to 12.5 mol%.[124] Copyright 2016, WILEY-VCH.

12.3.2 DEFECT PASSIVATION

The defect passivation has been demonstrated to be an effective approach to decrease the *J*-V hysteresis. It can reduce the trap states for charge trapping and hamper the ion migration. As discussed before, both enhancing the crystallinity and optimizing the composition of perovskite show strong ability to eliminate the defects. The defect passivation, mostly obtained by the incorporation of additive, is another effective approach. Son et al.[132] have conducted a series investigation of hysteresis by adding alkali metal iodides (LiI, NaI, KI, RbI and CsI) into the perovskite films of $MAPbI_3$, $FAPbI_3$, $FA_{0.85}MA_{0.15}PbI_{2.55}Br_{0.45}$ and $FA_{0.85}MA_{0.1}Cs_{0.05}PbI_{2.7}Br_{0.3}$ (Figure 12.10). The hysteresis disappears after the incorporation of KI, due to the reduced bulk and interfacial trap densities. The K^+ cations either enter the interstitial sites or substitute the A-site cations of the perovskite crystal,[133,134] which is able to enhance the crystallization. Incorporating potassium into $Cs_{0.05}(FA_{0.85}MA_{0.15})_{0.95}Pb(I_{0.85}Br_{0.15})_3$ will increase the grain size of perovskites up to around 1 µm.[135] Tang et al.[136] reported that the K^+ doping helps to reduce the electron extraction barrier between the perovskite and TiO_2, which minimizes the hysteresis with an enhanced electron transportation. The additives, such as KSCN, $Pb(SCN)_2$, and CaI_2, have also been demonstrated to be effective in enhancing the crystalline quality of perovskite, which remarkably

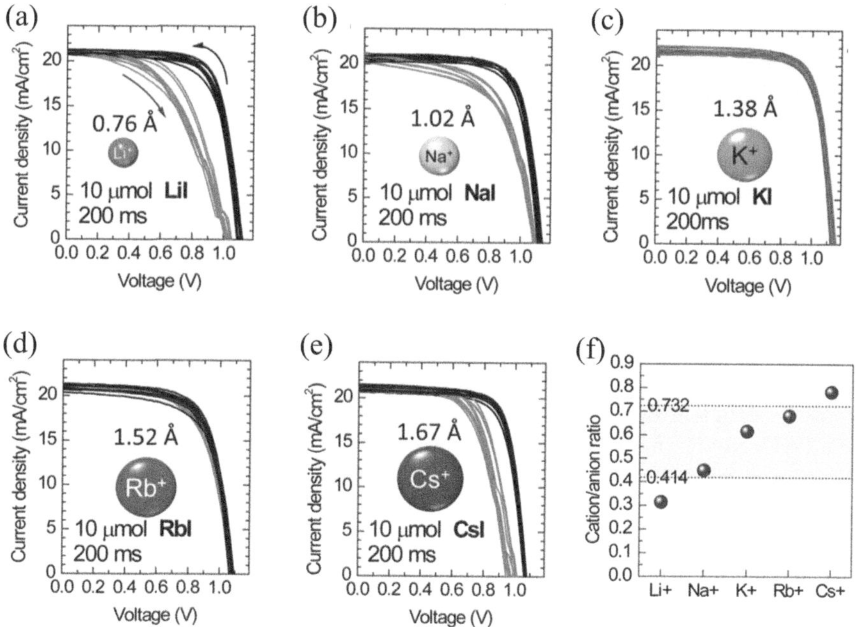

FIGURE 12.10 (a–e) *J*-V curves of the $(FAPbI_3)_{0.875}(CsPbBr_3)_{0.125}$ PSCs doped with 10 µmol of LiI, NI, KI, RbI, and CsI, respectively. The *J*-V characteristic is conducted under a forward and reverse scan with a scan rate of 130 mV/s and AM1.5G illumination. (f) Cation/anion radius ratio between the alkali metal cation of the dopant and I⁻ anion of the perovskite.[132] Copyright 2018, American Chemical Society.

reduces the defect density.[137–142] Other additive strategies, such as anchoring the ions at their related crystal sites,[143] filling the grain boundaries to block the routes for ion migrations,[144] and adjusting the tolerance factor to stable crystal structure,[145] have been reported to eliminate the hysteresis.

12.3.3 Interfacial Engineering

The interfacial engineering has been applied to promote the charge carrier transport and alleviate the charge accumulation near the interface. Due to the inefficient electron transport in perovskites, promoting the electron extraction ability of ETL is critical for the suppression of hysteresis.[146] The addition of dopant has been widely used to increase the conductivity of the interlayer.[147] Doping TiO_2 with sodium, boron or Li-salt will significantly enhances the electrical properties of TiO_2.[148–150]. Akin et al.[151] doped the SnO_2 ETL with ruthenium to reduce the contact barriers and increase electron mobility. These results demonstrate that the doping of ETL can significantly relieve the J-V hysteresis. The fullerene-based materials have been widely used to optimize the ETL due to its appropriate energy level and high electron mobility.[152] Replacing the traditional metal oxide-based ETL with the fullerene derivative (C60 pyrrolidine tris-acid or PCBM) can effectively suppress the hysteresis.[153,154] These fullerene-based materials will fill into the grain boundaries, which passivates the surface defects and accelerates the electron transfer.[155,156] Kegelamnn et al.[157] reported a double-layer ETL strategy using TiO_2-PCBM and SnO_2-PCBM, as shown in Figure 12.11. The additional fullerene layer improves the interfacial contact, which endows a faster electron extraction.[158] Tavakoli et al.[159] inserted an amorphous SnO_2 layer between the c-TiO_2 ETL and the perovskite, which helps to adjust the energy level alignment and decrease the surface recombination. Recently, Xiong et al.[160] reported a double-layer ETL of $SnO_2/NdCl_3$, which is able to prevent the iodide ion migration by reducing the iodide vacancies and increasing activation energy. Therefore, a reduced hysteretic effect can be obtained through a double-layer ETL. It should be noted that the modification of HTL also has impacts on the hysteretic performance of PSCs. The HTL should have proper energy level alignment with both the perovskite and the electrode to avoid the interfacial charge accumulation.[161,162] And it should also possess a high hole mobility to ensure efficient hole extraction.[163] The conductivity of the classic HTL, spiro-OMeTAD, can be improved by doping with additives, such as bis(trifluoromethane) sulfonimide lithium salt (Li-TFSI), Zn-TFSI$_2$, and 4-tert-butylpyridine (t-BP).[164–167] Li et al.[168] enhanced the conductivity of HTL by adding an acid additive (H_3PO_4) into spiro-OMeTAD, which remarkably reduced the HI from 0.14 to 0.04. Similar to the ETL, the double-layer strategy is also applicable for the HTL in the elimination of hysteresis. Wang et al.[169] deposited an ultrathin poly(triarylamine) (PTAA) layer sandwiched between the perovskite layer and PEDOT:PSS HTL to passivate the interfacial defects and achieve a device with negligible hysteresis. Therefore, optimizing the interfacial quality of the PSCs, such as achieving energy level alignment, passivating the defects, increasing conductivity, and improving interfacial contact, will help to remove the hysteresis.

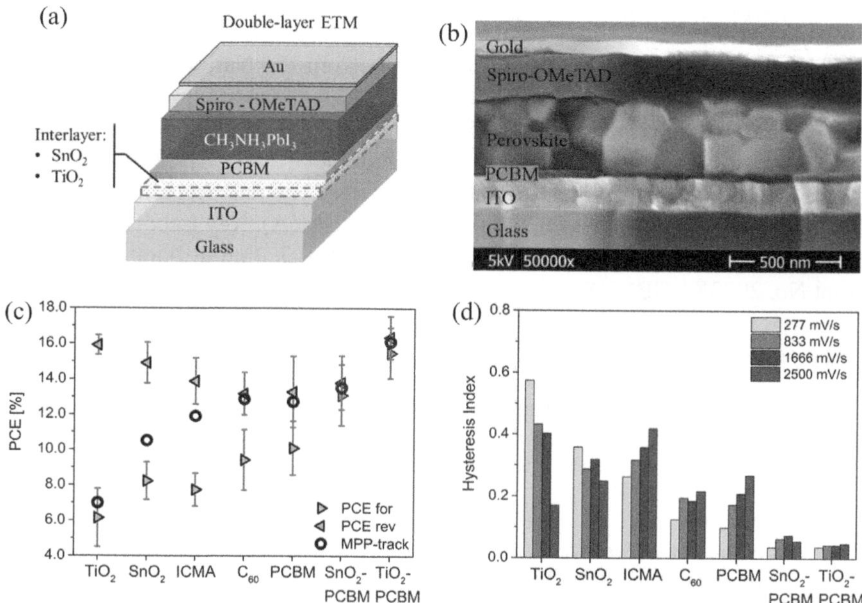

FIGURE 12.11 (a) Schematic illustration of the device architecture with a double-layer electron contract structure; (b) cross section SEM image of a PSC with PCBM as the single layer ETL; (c) PCEs of the PSCs with a single-layer ETL of TiO_2, SnO_2, ICMA, C_{60}, and PCBM and the PSCs with a double layer ETM of SnO_2-PCBM and TiO_2-PCBM; (d) calculated HI at different scan rates of the PSCs with a single-layer ETL of TiO_2, SnO_2, ICMA, C_{60}, and PCBM and the PSCs with a double-layer ETM of SnO_2-PCBM and TiO_2-PCBM.[157] Copyright 2017, American Chemical Society.

12.4 SUMMARY AND OUTLOOK

Due to the presence of hysteresis, the J-V curves are strongly relied on the pre-measurement condition and the measuring procedure of J-V characterization. It causes a challenge in determining the accurate PCE. Since first issued in 2014,[11] the mechanism of hysteresis has been widely studied. A combined effect of ion migration, charge trapping, and charge carrier transport leads to the hysteretic effect. Thus, the elimination of hysteresis can be achieved by film quality improvement, defect passivation, and interfacial engineering. It shares almost the same strategies in improving PCE and device stability, which leads to an overall enhancement in the performance of PSCs.[170] As a result, negligible hysteresis can be detected in the highly efficient PSCs.[1, 2] With a good understanding of hysteresis, the hysteretic behavior is no longer an obstacle for the research of PSCs since 2019. And now, it has become an indicator (e.g., HI) for the photovoltaic performance to evaluate the suppression of the notorious ionic and electronic process, such as ion migration and charge recombination.171 This chapter offers a full insight into the origins and elimination of hysteresis in PSCs, which may help to achieve a completely hysteresis-free PSCs with a high PCE.

ACKNOWLEDGMENT

The authors acknowledge the financial support from the National Natural Science Foundation of China (Grant No. 22375043, 21674123), Guangdong Basic and Applied Basic Research Foundation (2021A1515110720), Fujian Provincial Department of Finance for the research of organic photovoltaic solar cell (KLe20001A), Special Fund for Innovative Science and Technology of Fujian Agriculture, Forestry University (KFb22084XA, KFb22086XA, KFb22118XA), Key Laboratory of Green Perovskite Application of Fujian Provincial Universities, Fujian Jiangxia University (Grant No. 2023KLGPA01).

REFERENCE

1. M. Kim, *et al.*, Conformal Quantum Dot-SnO$_2$ Layers as Electron Transporters for Efficient Perovskite Solar Cells, *Science* **2022**, *375*, 302.
2. Y. Zhao, *et al.*, Inactive (PbI$_2$)$_2$RbCl Stabilizes Perovskite Films for Efficient Solar Cells, *Science* **2022**, *377*, 531.
3. K. Yoshikawa, *et al.*, Silicon Heterojunction Solar Cell with Interdigitated Back Contacts for A Photoconversion Efficiency over 26%, *Nature Energy* **2017**, *2*, 17032.
4. Z. Guo, *et al.*, The High Open-Circuit Voltage of Perovskite Solar Cells: A Review, *Energy & Environmental Science* **2022**, *15*, 3171.
5. G. Ren, *et al.*, Strategies of Modifying Spiro-OMeTAD Materials for Perovskite Solar Cells: A Review, *Journal of Materials Chemistry A* **2021**, *9*, 4589.
6. M. A. Mahmud, *et al.*, Origin of Efficiency and Stability Enhancement in High-Performing Mixed Dimensional 2D-3D Perovskite Solar Cells: A Review, *Advanced Functional Materials* **2021**, *32*, 2009164.
7. S. Wang, *et al.*, Over 24% Efficient MA-Free Cs$_x$FA$_{1-x}$PbX$_3$ Perovskite Solar Cells, *Joule* **2022**, *6*, 1344.
8. M. Wang, *et al.*, Rational Selection of the Polymeric Structure for Interface Engineering of Perovskite Solar Cells, *Joule* **2022**, *6*, 1032.
9. M. T. Neukom, *et al.*, Why Perovskite Solar Cells with High Efficiency Show Small IV-Curve Hysteresis, *Solar Energy Materials and Solar Cells* **2017**, *169*, 159.
10. N. J. Jeon, *et al.*, Solvent Engineering for High-Performance Inorganic–Organic Hybrid Perovskite Solar Cells, *Nature Materials* **2014**, *13*, 897.
11. H. J. Snaith, *et al.*, Anomalous Hysteresis in Perovskite Solar Cells, *The Journal of Physical Chemistry Letters* **2014**, *5*, 1511.
12. W. Ma, *et al.*, Reducing Anomalous Hysteresis in Perovskite Solar Cells by Suppressing the Interfacial Ferroelectric Order, *ACS Applied Materials & Interfaces* **2020**, *12*, 12275.
13. P. Nandi, *et al.*, Organic-Inorganic Hybrid Lead Halides as Absorbers in Perovskite Solar Cells: A Debate on Ferroelectricity, *Journal of Physics D: Applied Physics* **2020**, *53*, 493002.
14. T. Chen, *et al.*, Correlating Hysteresis Phenomena with Interfacial Charge Accumulation in Perovskite Solar Cells, *Physical Chemistry Chemical Physics* **2020**, *22*, 245.
15. Y. Yang, *et al.*, Eliminating Charge Accumulation via Interfacial Dipole for Efficient and Stable Perovskite Solar Cells, *ACS Applied Materials & Interfaces* **2019**, *11*, 34964.
16. J. Haruyama, *et al.*, First-Principles Study of Ion Diffusion in Perovskite Solar Cell Sensitizers, *Journal of the American Chemical Society* **2015**, *137*, 10048.
17. Y. Zhang, *et al.*, Charge Selective Contacts, Mobile Ions and Anomalous Hysteresis in Organic–Inorganic Perovskite Solar Cells, *Materials Horizons* **2015**, *2*, 315.
18. I. Levine, *et al.*, Interface-Dependent Ion Migration/Accumulation Controls Hysteresis in MAPbI$_3$ Solar Cells, *Journal of Physical Chemistry C* **2016**, *120*, 16399.

19. W. Tress, *et al.*, Inverted Current–Voltage Hysteresis in Mixed Perovskite Solar Cells: Polarization, Energy Barriers, and Defect Recombination, *Advanced Energy Materials* **2016**, *6*, 1600396.

20. P. Liu, *et al.*, Fundamental Understanding of Photocurrent Hysteresis in Perovskite Solar Cells, *Advanced Energy Materials* **2019**, *9*, 1803017.

21. W. Chen, *et al.*, High-Polarizability Organic Ferroelectric Materials Doping for Enhancing the Built-In Electric Field of Perovskite Solar Cells Realizing Efficiency over 24, *Advanced Materials* **2022**, *34*, 2110482.

22. Y. Zhang, *et al.*, Mechanochemistry Advances High-Performance Perovskite Solar Cells, *Advanced Materials* **2022**, *34*, 2107420.

23. Q. Liang, *et al.*, Manipulating Crystallization Kinetics in High-Performance Blade-Coated Perovskite Solar Cells via Cosolvent-Assisted Phase Transition, *Advanced Materials* **2022**, *34*, 2200276.

24. T. Bu, *et al.*, Universal Passivation Strategy to Slot-Die Printed SnO_2 for Hysteresis-Free Efficient Flexible Perovskite Solar Module, *Nature Communications* **2018**, *9*, 4609.

25. L. Wang, *et al.*, Interface Regulation Enables Hysteresis Free Wide-Bandgap Perovskite Solar Cells with Low V_{OC} Deficit and High Stability, *Nano Energy* **2021**, *90*, 106537.

26. P. Wang, *et al.*, Seed-Assisted Growth of Methylammonium-Free Perovskite for Efficient Inverted Perovskite Solar Cells, *Small Methods* **2022**, *6*, 2200048.

27. G. A. Sepalage, *et al.*, Copper(I) Iodide as Hole-Conductor in Planar Perovskite Solar Cells: Probing the Origin of J–V Hysteresis, *Advanced Functional Materials* **2015**, *25*, 5650.

28. E. Erdenebileg, *et al.*, Very Small Inverted Hysteresis in Vacuum-Deposited Mixed Organic–Inorganic Hybrid Perovskite Solar Cells, *Energy Technology* **2017**, *5*, 1606.

29. F. Wu, *et al.*, Inverted Current–Voltage Hysteresis in Perovskite Solar Cells, *ACS Energy Letters* **2018**, *3*, 2457.

30. R. S. Sanchez, *et al.*, Slow Dynamic Processes in Lead Halide Perovskite Solar Cells. Characteristic Times and Hysteresis, *Journal of Physical Chemistry Letters* **2014**, *5*, 2357.

31. H.-S. Kim and N.-G. Park, Parameters Affecting I–V Hysteresis of $CH_3NH_3PbI_3$ Perovskite Solar Cells: Effects of Perovskite Crystal Size and Mesoporous TiO_2 Layer, *Journal of Physical Chemistry Letters* **2014**, *5*, 2927.

32. Y. Rong, *et al.*, Tunable Hysteresis Effect for Perovskite Solar Cells, *Energy & Environmental Science* **2017**, *10*, 2383.

33. J. Wei, *et al.*, Hysteresis Analysis Based on the Ferroelectric Effect in Hybrid Perovskite Solar Cells, *Journal of Physical Chemistry Letters* **2014**, *5*, 3937.

34. W. Zhao, *et al.*, Symmetrical Acceptor–Donor–Acceptor Molecule as a Versatile Defect Passivation Agent toward Efficient $FA_{0.85}MA_{0.15}PbI_3$ Perovskite Solar Cells, *Advanced Functional Materials* **2022**, *32*, 2112032.

35. D. Zheng, *et al.*, Controlling the Formation Process of Methylammonium-Free Halide Perovskite Films for a Homogeneous Incorporation of Alkali Metal Cations Beneficial to Solar Cell Performance, *Advanced Energy Materials* **2022**, *12*, 2103618.

36. Z. Liang, *et al.*, A Selective Targeting Anchor Strategy Affords Efficient and Stable Ideal-Bandgap Perovskite Solar Cells, *Advanced Materials* **2022**, *34*, 2110241.

37. S. N. Habisreutinger, *et al.*, Hysteresis Index: A Figure without Merit for Quantifying Hysteresis in Perovskite Solar Cells, *ACS Energy Letters* **2018**, *3*, 2472.

38. D.-H. Kang and N.-G. Park, On the Current–Voltage Hysteresis in Perovskite Solar Cells: Dependence on Perovskite Composition and Methods to Remove Hysteresis, *Advanced Materials* **2019**, *31*, 1805214.

39. A. M. Glazer, The Classification of Tilted Octahedra in Perovskites, *Acta Crystallographica Section B* **1972**, *28*, 3384.

40. J. M. Frost, *et al.*, Molecular Ferroelectric Contributions to Anomalous Hysteresis in Hybrid Perovskite Solar Cells, *APL Materials* **2014**, *2*, 081506.

41. Z. Fan, *et al.*, Ferroelectricity of $CH_3NH_3PbI_3$ Perovskite, *Journal of Physical Chemistry Letters* **2015**, *6*, 1155.

42. A. Pecchia, *et al.*, Role of Ferroelectric Nanodomains in the Transport Properties of Perovskite Solar Cells, *Nano Letters* **2016**, *16*, 988.

43. D. Rossi, *et al.*, On the Importance of Ferroelectric Domains for the Performance of Perovskite Solar Cells, *Nano Energy* **2018**, *48*, 20.

44. X.-L. Xu, *et al.*, Molecular Ferroelectrics-Driven High-Performance Perovskite Solar Cells, *Angewandte Chemie International Edition* **2020**, *59*, 19974.

45. H. Röhm, *et al.*, Ferroelectric Properties of Perovskite Thin Films and Their Implications for Solar Energy Conversion, *Advanced Materials* **2019**, *31*, 1806661.

46. B. Chen, *et al.*, Interface Band Structure Engineering by Ferroelectric Polarization in Perovskite Solar Cells, *Nano Energy* **2015**, *13*, 582.

47. H.-W. Chen, *et al.*, Emergence of Hysteresis and Transient Ferroelectric Response in Organo-Lead Halide Perovskite Solar Cells, *Journal of Physical Chemistry Letters* **2015**, *6*, 164.

48. A. K. Jena, *et al.*, The Interface between FTO and the TiO_2 Compact Layer Can Be One of the Origins to Hysteresis in Planar Heterojunction Perovskite Solar Cells, *ACS Applied Materials & Interfaces* **2015**, *7*, 9817.

49. J. Beilsten-Edmands, *et al.*, Non-Ferroelectric Nature of the Conductance Hysteresis in $CH_3NH_3PbI_3$ Perovskite-Based Photovoltaic Devices, *Applied Physics Letters* **2015**, *106*, 173502.

50. S. Meloni, *et al.*, Ionic Polarization-Induced Current–Voltage Hysteresis in $CH_3NH_3PbX_3$ Perovskite Solar Cells, *Nature Communications* **2016**, *7*, 10334.

51. A. M. Leguy, *et al.*, The Dynamics of Methylammonium Ions in Hybrid Organic-Inorganic Perovskite Solar Cells, *Nature Communications* **2015**, *6*, 7124.

52. E. Aydin, *et al.*, Defect and Contact Passivation for Perovskite Solar Cells, *Advanced Materials* **2019**, *31*, 1900428.

53. F. Gao, *et al.*, Recent Progresses on Defect Passivation toward Efficient Perovskite Solar Cells, *Advanced Energy Materials* **2020**, *10*, 1902650

54. A. Walsh, *et al.*, Self-Regulation Mechanism for Charged Point Defects in Hybrid Halide Perovskites, *Angewandte Chemie International Edition* **2015**, *54*, 1791.

55. T. Leijtens, *et al.*, Electronic Properties of Meso-Superstructured and Planar Organometal Halide Perovskite Films: Charge Trapping, Photodoping, and Carrier Mobility, *ACS Nano* **2014**, *8*, 7147.

56. X. Wu, *et al.*, Trap States in Lead Iodide Perovskites, *Journal of the American Chemical Society* **2015**, *137*, 2089.

57. B. Ray, *et al.*, Defect Characterization in Organic Semiconductors by Forward Bias Capacitance–Voltage (FB-CV) Analysis, *Journal of Physical Chemistry C* **2014**, *118*, 17461.

58. W. Xiang, *et al.*, Interfaces and Interfacial Layers in Inorganic Perovskite Solar Cells, *Angewandte Chemie International Edition* **2021**, *60*, 26440.

59. D.W. Miller, *et al.*, Defect States in Perovskite Solar Cells Associated with Hysteresis and Performance, *Applied Physics Letters* **2016**, *109*, 153902.

60. Y. Shao, *et al.*, Origin and Elimination of Photocurrent Hysteresis by Fullerene Passivation in $CH_3NH_3PbI_3$ Planar Heterojunction Solar Cells, *Nature Communications* **2014**, *5*, 5784.

61. Q. Hu, *et al.*, Dual Defect-Passivation Using Phthalocyanine for Enhanced Efficiency and Stability of Perovskite Solar Cells, *Small* **2021**, *17*, 2005216.

62. H. Kim, *et al.*, Synergistic Effects of Cation and Anion in an Ionic Imidazolium Tetrafluoroborate Additive for Improving the Efficiency and Stability of Half-Mixed Pb-Sn Perovskite Solar Cells, *Advanced Functional Materials* **2021**, *31*, 2008801.

63. J. Q. Zhang, *et al.*, Elimination of Interfacial Lattice Mismatch and Detrimental Reaction by Self-Assembled Layer Dual-Passivation for Efficient and Stable Inverted Perovskite Solar Cells, *Advanced Energy Materials* **2022**, *12*, 2103674.

64. Z. Li, *et al.*, Organometallic-Functionalized Interfaces for Highly Efficient Inverted Perovskite Solar Cells, *Science* **2022**, *376*, 416.
65. W.-J. Yin, *et al.*, Unusual Defect Physics in $CH_3NH_3PbI_3$ Perovskite Solar Cell Absorber, *Applied Physics Letters* **2014**, *104*, 063903.
66. J. Kim, *et al.*, The Role of Intrinsic Defects in Methylammonium Lead Iodide Perovskite, *Journal of Physical Chemistry Letters* **2014**, *5*, 1312.
67. A. Oranskaia, *et al.*, Halogen Migration in Hybrid Perovskites: The Organic Cation Matters, *Journal of Physical Chemistry Letters* **2018**, *9*, 5474.
68. J. M. Azpiroz, *et al.*, Defect Migration in Methylammonium Lead Iodide and Its Role in Perovskite Solar Cell Operation, *Energy & Environmental Science* **2015**, *8*, 2118.
69. C. Eames, *et al.*, Ionic Transport in Hybrid Lead Iodide Perovskite Solar Cells, *Nature Communications* **2015**, *6*, 7497.
70. C.-J. Tong, *et al.*, Role of Methylammonium Orientation in Ion Diffusion and Current–Voltage Hysteresis in the $CH_3NH_3PbI_3$ Perovskite, *ACS Energy Letters* **2017**, *2*, 1997.
71. Z. Li, *et al.*, Extrinsic Ion Migration in Perovskite Solar Cells, *Energy & Environmental Science* **2017**, *10*, 1234.
72. J. Thiesbrummel, *et al.*, Universal Current Losses in Perovskite Solar Cells Due to Mobile Ions, *Advanced Energy Materials* **2021**, *11*, 2101447.
73. Y. Yuan and J. Huang, Ion Migration in Organometal Trihalide Perovskite and Its Impact on Photovoltaic Efficiency and Stability, *Accounts of Chemical Research* **2016**, *49*, 286.
74. E. L. Unger, *et al.*, Hysteresis and Transient Behavior in Current–Voltage Measurements of Hybrid-Perovskite Absorber Solar Cells, *Energy & Environmental Science* **2014**, *7*, 3690.
75. H. Zhang, *et al.*, Dynamic Interface Charge Governing the Current–Voltage Hysteresis in Perovskite Solar Cells, *Physical Chemistry Chemical Physics* **2015**, *17*, 9613.
76. Y. Chen, *et al.*, In Situ Management of Ions Migration to Control Hysteresis Effect for Planar Heterojunction Perovskite Solar Cells, *Advanced Functional Materials* **2022**, *32*, 2108417.
77. S. Ravishankar, *et al.*, Surface Polarization Model for the Dynamic Hysteresis of Perovskite Solar Cells, *Journal of Physical Chemistry Letters* **2017**, *8*, 915.
78. W. Tress, *et al.*, Understanding the Rate-Dependent J–V Hysteresis, Slow Time Component, and Aging in $CH_3NH_3PbI_3$ Perovskite Solar Cells: the Role of a Compensated Electric Field, *Energy & Environmental Science* **2015**, *8*, 995.
79. S. A. L. Weber, *et al.*, How the Formation of Interfacial Charge Causes Hysteresis in Perovskite Solar Cells, *Energy & Environmental Science* **2018**, *11*, 2404.
80. K. Domanski, *et al.*, Migration of Cations Induces Reversible Performance Losses over Day/Night Cycling in Perovskite Solar Cells, *Energy & Environmental Science* **2017**, *10*, 604.
81. O. Almora, *et al.*, Capacitive Dark Currents, Hysteresis, and Electrode Polarization in Lead Halide Perovskite Solar Cells, *Journal of Physical Chemistry Letters* **2015**, *6*, 1645.
82. O. Almora, *et al.*, Noncapacitive Hysteresis in Perovskite Solar Cells at Room Temperature, *ACS Energy Letters* **2016**, *1*, 209.
83. G. Richardson, *et al.*, Can Slow-Moving Ions Explain Hysteresis in the Current–Voltage Curves of Perovskite Solar Cells?, *Energy & Environmental Science* **2016**, *9*, 1476.
84. Y. Shao, *et al.*, Grain Boundary Dominated Ion Migration in Polycrystalline Organic–Inorganic Halide Perovskite Films, *Energy & Environmental Science* **2016**, *9*, 1752.
85. J. S. Yun, *et al.*, Critical Role of Grain Boundaries for Ion Migration in Formamidinium and Methylammonium Lead Halide Perovskite Solar Cells, *Advanced Energy Materials* **2016**, *6*, 1600330.
86. D. A. Jacobs, *et al.*, Hysteresis Phenomena in Perovskite Solar Cells: the Many and Varied Effects of Ionic Accumulation, *Physical Chemistry Chemical Physics* **2017**, *19*, 3094.

87. G. A. Nemnes, *et al.*, Normal and Inverted Hysteresis in Perovskite Solar Cells, *Journal of Physical Chemistry C* **2017**, *121*, 11207.

88. H. Shen, *et al.*, Inverted Hysteresis in $CH_3NH_3PbI_3$ Solar Cells: Role of Stoichiometry and Band Alignment, *Journal of Physical Chemistry Letters* **2017**, *8*, 2672.

89. F. Wu, *et al.*, Bias-Dependent Normal and Inverted J–V Hysteresis in Perovskite Solar Cells, *ACS Applied Materials & Interfaces* **2018**, *10*, 25604.

90. C. Li, *et al.*, Origins and Mechanisms of Hysteresis in Organometal Halide Perovskites, *Journal of Physics: Condensed Matter* **2017**, *29*, 193001.

91. C. Li, *et al.*, Iodine Migration and its Effect on Hysteresis in Perovskite Solar Cells, *Advanced Materials* **2016**, *28*, 2446.

92. J. M. Ball, *et al.*, Low-Temperature Processed Meso-Superstructured to Thin-Film Perovskite Solar Cells, *Energy & Environmental Science* **2013**, *6*, 1739.

93. Y. Chen, *et al.*, Efficient and Balanced Charge Transport Revealed in Planar Perovskite Solar Cells, *ACS Applied Materials & Interfaces* **2015**, *7*, 4471.

94. G. Garcia-Belmonte and J. Bisquert, Distinction between Capacitive and Noncapacitive Hysteretic Currents in Operation and Degradation of Perovskite Solar Cells, *ACS Energy Letters* **2016**, *1*, 683.

95. E. Edri, *et al.*, Why Lead Methylammonium Tri-Iodide Perovskite-Based Solar Cells Require a Mesoporous Electron Transporting Scaffold (but Not Necessarily a Hole Conductor), *Nano Letters* **2014**, *14*, 1000.

96. H. Oga, *et al.*, Improved Understanding of the Electronic and Energetic Landscapes of Perovskite Solar Cells: High Local Charge Carrier Mobility, Reduced Recombination, and Extremely Shallow Traps, *Journal of the American Chemical Society* **2014**, *136*, 13818.

97. I. Abayev, *et al.*, Electronic Conductivity in Nanostructured TiO_2 Films Permeated with Electrolyte, *Phys Status Solidi A* **2003**, *196*, R4.

98. D. Yang, *et al.*, Hysteresis-Suppressed High-Efficiency Flexible Perovskite Solar Cells Using Solid-State Ionic-Liquids for Effective Electron Transport, *Advanced Materials* **2016**, *28*, 5206.

99. C. Huang, *et al.*, Dopant-Free Hole-Transporting Material with a C_{3h} Symmetrical Truxene Core for Highly Efficient Perovskite Solar Cells, *Journal of the American Chemical Society* **2016**, *138*, 2528.

100. B. Wu, *et al.*, Charge Accumulation and Hysteresis in Perovskite-Based Solar Cells: An Electro-Optical Analysis, *Advanced Energy Materials* **2015**, *5*, 1500829.

101. J. H. Heo, *et al.*, Hysteresis-Less Inverted $CH_3NH_3PbI_3$ Planar Perovskite Hybrid Solar Cells with 18.1% Power Conversion Efficiency, *Energy & Environmental Science* **2015**, *8*, 1602.

102. S. Ravishankar, *et al.*, Influence of Charge Transport Layers on Open-Circuit Voltage and Hysteresis in Perovskite Solar Cells, *Joule* **2018**, *2*, 788.

103. P. Calado, *et al.*, Evidence for Ion Migration in Hybrid Perovskite Solar Cells with Minimal Hysteresis, *Nature Communications* **2016**, *7*, 13831.

104. S. van Reenen, *et al.*, Modeling Anomalous Hysteresis in Perovskite Solar Cells, *Journal of Physical Chemistry Letters* **2015**, *6*, 3808.

105. R. Gottesman, *et al.*, Dynamic Phenomena at Perovskite/Electron-Selective Contact Interface as Interpreted from Photovoltage Decays, *Chem* **2016**, *1*, 776.

106. W. Tress, Metal Halide Perovskites as Mixed Electronic–Ionic Conductors: Challenges and Opportunities—From Hysteresis to Memristivity, *Journal of Physical Chemistry Letters* **2017**, *8*, 3106.

107. C.-J. Tong, *et al.*, Synergy between Ion Migration and Charge Carrier Recombination in Metal-Halide Perovskites, *Journal of the American Chemical Society* **2020**, *142*, 3060.

108. N. E. Courtier, *et al.*, How Transport Layer Properties Affect Perovskite Solar Cell Performance: Insights from a Coupled Charge Transport/Ion Migration Model, *Energy & Environmental Science* **2019**, *12*, 396.

109. T. Xu, *et al.*, Strategic Improvement of the Long-Term Stability of Perovskite Materials and Perovskite Solar Cells, *Physical Chemistry Chemical Physics* **2016**, *18*, 27026.
110. Z. Fan, *et al.*, Layer-by-Layer Degradation of Methylammonium Lead Tri-iodide Perovskite Microplates, *Joule* **2017**, *1*, 548.
111. A. J. Barker, *et al.*, Defect-Assisted Photoinduced Halide Segregation in Mixed-Halide Perovskite Thin Films, *ACS Energy Letters* **2017**, *2*, 1416.
112. Z. Xiao, *et al.*, Giant Switchable Photovoltaic Effect in Organometal Trihalide Perovskite Devices, *Nature Materials* **2015**, *14*, 193.
113. J.-W. Lee, *et al.*, Verification and Mitigation of Ion Migration in Perovskite Solar Cells, *APL Materials* **2019**, *7*, 041111.
114. L. McGovern, *et al.*, Grain Size Influences Activation Energy and Migration Pathways in $MAPbBr_3$ Perovskite Solar Cells, *Journal of Physical Chemistry Letters* **2021**, *12*, 2423.
115. H. D. Kim, *et al.*, Photovoltaic Performance of Perovskite Solar Cells with Different Grain Sizes, *Advanced Materials* **2016**, *28*, 917.
116. D. Meggiolaro, *et al.*, Formation of Surface Defects Dominates Ion Migration in Lead-Halide Perovskites, *ACS Energy Letters* **2019**, *4*, 779.
117. X. Ren, *et al.*, Modulating Crystal Grain Size and Optoelectronic Properties of Perovskite Films for Solar Cells by Reaction Temperature, *Nanoscale* **2016**, *8*, 3816.
118. D. Khatiwada, *et al.*, Efficient Perovskite Solar Cells by Temperature Control in Single and Mixed Halide Precursor Solutions and Films, *Journal of Physical Chemistry C* **2015**, *119*, 25747.
119. Z. Xiao, *et al.*, Solvent Annealing of Perovskite-Induced Crystal Growth for Photovoltaic-Device Efficiency Enhancement, *Advanced Materials* **2014**, *26*, 6503.
120. Y. Wang, *et al.*, Solvent Annealing of PbI_2 for the High-Quality Crystallization of Perovskite Films for Solar Cells with Efficiencies Exceeding 18%, *Nanoscale* **2016**, *8*, 19654.
121. L. Zuo, *et al.*, Morphology Evolution of High Efficiency Perovskite Solar Cells via Vapor Induced Intermediate Phases, *Journal of the American Chemical Society* **2016**, *138*, 15710.
122. Y. Zhao, *et al.*, A Polymerization-Assisted Grain Growth Strategy for Efficient and Stable Perovskite Solar Cells, *Advanced Materials* **2020**, *32*, 1907769.
123. N. F. Montcada, *et al.*, Photo-Induced Dynamic Processes in Perovskite Solar Cells: the Influence of Perovskite Composition in the Charge Extraction and the Carrier Recombination, *Nanoscale* **2018**, *10*, 6155.
124. Y. C. Kim, *et al.*, Beneficial Effects of PbI_2 Incorporated in Organo-Lead Halide Perovskite Solar Cells, *Advanced Energy Materials* **2016**, *6*, 1502104.
125. T. Zhang, *et al.*, A Controllable Fabrication of Grain Boundary PbI_2 Nanoplates Passivated Lead Halide Perovskites for High Performance Solar Cells, *Nano Energy* **2016**, *26*, 50.
126. T. J. Jacobsson, *et al.*, Unreacted PbI_2 as a Double-Edged Sword for Enhancing the Performance of Perovskite Solar Cells, *Journal of the American Chemical Society* **2016**, *138*, 10331.
127. T. Zhang, *et al.*, Crystallinity Preservation and Ion Migration Suppression through Dual Ion Exchange Strategy for Stable Mixed Perovskite Solar Cells, *Advanced Energy Materials* **2017**, *7*, 1700118.
128. D. Bi, *et al.*, Efficient Luminescent Solar Cells Based on Tailored Mixed-Cation Perovskites, *Science Advances* **2016**, *2*, 1501170.
129. M. Saliba, *et al.*, Cesium-Containing Triple Cation Perovskite Solar Cells: Improved Stability, Reproducibility and High Efficiency, *Energy & Environmental Science* **2016**, *9*, 1989.
130. Y. Cui, *et al.*, Correlating Hysteresis and Stability with Organic Cation Composition in the Two-Step Solution-Processed Perovskite Solar Cells, *ACS Applied Materials & Interfaces* **2020**, *12*, 10588.

131. J.-W. Lee, *et al.*, Formamidinium and Cesium Hybridization for Photo- and Moisture-Stable Perovskite Solar Cell, *Advanced Energy Materials* **2015**, *5*, 1501310.

132. D.-Y. Son, *et al.*, Universal Approach toward Hysteresis-Free Perovskite Solar Cell via Defect Engineering, *Journal of the American Chemical Society* **2018**, *140*, 1358.

133. D. Yao, *et al.*, Hindered Formation of Photoinactive δ-FAPbI$_3$ Phase and Hysteresis-Free Mixed-Cation Planar Heterojunction Perovskite Solar Cells with Enhanced Efficiency via Potassium Incorporation, *Journal of Physical Chemistry Letters* **2018**, *9*, 2113.

134. J. K. Nam, *et al.*, Potassium Incorporation for Enhanced Performance and Stability of Fully Inorganic Cesium Lead Halide Perovskite Solar Cells, *Nano Letters* **2017**, *17*, 2028.

135. T. Bu, *et al.*, A Novel Quadruple-Cation Absorber for Universal Hysteresis Elimination for High Efficiency and Stable Perovskite Solar Cells, *Energy & Environmental Science* **2017**, *10*, 2509.

136. Z. Tang, *et al.*, Hysteresis-Free Perovskite Solar Cells Made of Potassium-Doped Organometal Halide Perovskite, *Scientific Reports* **2017**, *7*, 12183.

137. W. Ke, *et al.*, Employing Lead Thiocyanate Additive to Reduce the Hysteresis and Boost the Fill Factor of Planar Perovskite Solar Cells, *Advanced Materials* **2016**, *28*, 5214.

138. J. T.- W. Wang, *et al.*, Efficient Perovskite Solar Cells by Metal Ion Doping, *Energy & Environmental Science* **2016**, *9*, 2892.

139. Y. Wu, *et al.*, Thermally Stable MAPbI3 Perovskite Solar Cells with Efficiency of 19.19% and Area over 1 cm^2 achieved by Additive Engineering, *Advanced Materials* **2017**, *29*, 1701073.

140. C. Chen, *et al.*, CaI$_2$: a More Effective Passivator of Perovskite Films Than PbI$_2$ for High Efficiency and Long-Term Stability of Perovskite Solar Cells, *Journal of Materials Chemistry A* **2018**, *6*, 7903.

141. R. Zhang, *et al.*, A Potassium Thiocyanate Additive for Hysteresis Elimination in Highly Efficient Perovskite Solar Cells, *Inorganic Chemistry Frontiers* **2019**, *6*, 434.

142. N. D. Pham, *et al.*, Tailoring Crystal Structure of FA$_{0.83}$Cs$_{0.17}$PbI$_3$ Perovskite Through Guanidinium Doping for Enhanced Performance and Tunable Hysteresis of Planar Perovskite Solar Cells, *Advanced Functional Materials* **2019**, *29*, 1806479.

143. K. Wang, *et al.*, High Performance Perovskites Solar Cells by Hybrid Perovskites Co-Crystallized with Poly(ethylene oxide), *Nano Energy* **2020**, *67*, 104229.

144. Y. Ma, *et al.*, Suppressing Ion Migration across Perovskite Grain Boundaries by Polymer Additives, *Advanced Functional Materials* **2021**, *31*, 2006802.

145. C. Xu, *et al.*, Interpretation of Rubidium-Based Perovskite Recipes toward Electronic Passivation and Ion-Diffusion Mitigation, *Advanced Materials* **2022**, *34*, 2109998.

146. G. A. Elbaz, *et al.*, Unbalanced Hole and Electron Diffusion in Lead Bromide Perovskites, *Nano Letters* **2017**, *17*, 1727.

147. F. Giordano, *et al.*, Enhanced Electronic Properties in Mesoporous TiO$_2$ via Lithium Doping for High-Efficiency Perovskite Solar Cells, *Nature Communications* **2016**, *7*, 10379.

148. X. Li, *et al.*, Synergistic Effect to High-Performance Perovskite Solar Cells with Reduced Hysteresis and Improved Stability by the Introduction of Na-Treated TiO$_2$ and Spraying-Deposited CuI as Transport Layers, *ACS Applied Materials & Interfaces* **2017**, *9*, 41354.

149. X. Shi, *et al.*, Enhanced Interfacial Binding and Electron Extraction Using Boron-Doped TiO$_2$ for Highly Efficient Hysteresis-Free Perovskite Solar Cells, *Advanced Science* **2019**, *6*, 1901213.

150. M. Kim, *et al.*, Enhanced Electrical Properties of Li-Salts Doped Mesoporous TiO$_2$ in Perovskite Solar Cells, *Joule* **2021**, *5*, 659.

151. S. Akin, Hysteresis-Free Planar Perovskite Solar Cells with a Breakthrough Efficiency of 22% and Superior Operational Stability over 2000 h, *ACS Applied Materials & Interfaces* **2019**, *11*, 39998.

152. L.-L. Deng, *et al.*, Fullerene-Based Materials for Photovoltaic Applications: Toward Efficient, Hysteresis-Free, and Stable Perovskite Solar Cells, *Advanced Electronic Materials* **2018**, *4*, 1700435.

153. Y.-C. Wang, *et al.*, Efficient and Hysteresis-Free Perovskite Solar Cells Based on a Solution Processable Polar Fullerene Electron Transport Layer, *Advanced Energy Materials* **2017**, *7*, 1701144.

154. D. Bryant, *et al.*, Observable Hysteresis at Low Temperature in "Hysteresis Free" Organic–Inorganic Lead Halide Perovskite Solar Cells, *Journal of Physical Chemistry Letters* **2015**, *6*, 3190.

155. J. Xu, *et al.*, Perovskite–Fullerene Hybrid Materials Suppress Hysteresis in Planar Diodes, *Nature Communications* **2015**, *6*, 7081.

156. C.-H. Chiang and C.-G. Wu, Bulk Heterojunction Perovskite–PCBM Solar Cells with High Fill Factor, *Nature Photonics* **2016**, *10*, 196.

157. L. Kegelmann, *et al.*, It Takes Two to Tango—Double-Layer Selective Contacts in Perovskite Solar Cells for Improved Device Performance and Reduced Hysteresis, *ACS Applied Materials & Interfaces* **2017**, *9*, 17245.

158. H. Jiang, *et al.*, High Current Density and Low Hysteresis Effect of Planar Perovskite Solar Cells via PCBM-doping and Interfacial Improvement, *ACS Applied Materials & Interfaces* **2018**, *10*, 29954.

159. M. M. Tavakoli, *et al.*, Surface Engineering of TiO_2 ETL for Highly Efficient and Hysteresis-Less Planar Perovskite Solar Cell (21.4%) with Enhanced Open-Circuit Voltage and Stability, *Advanced Energy Materials* **2018**, *8*, 1800794.

160. Q. Xiong, *et al.*, Rear Interface Engineering to Suppress Migration of Iodide Ions for Efficient Perovskite Solar Cells with Minimized Hysteresis, *Advanced Functional Materials* **2022**, *32*, 2107823.

161. A. Guerrero, *et al.*, Switching Off Hysteresis in Perovskite Solar Cells by Fine-Tuning Energy Levels of Extraction Layers, *Advanced Energy Materials* **2018**, *8*, 1703376.

162. Z. Liu, *et al.*, Rationally Induced Interfacial Dipole in Planar Heterojunction Perovskite Solar Cells for Reduced J–V Hysteresis, *Advanced Energy Materials* **2018**, *8*, 1800568.

163. J. Luo, *et al.*, Toward High-Efficiency, Hysteresis-Less, Stable Perovskite Solar Cells: Unusual Doping of a Hole-Transporting Material Using a Fluorine-Containing Hydrophobic Lewis Acid, *Energy & Environmental Science* **2018**, *11*, 2035.

164. J.-Y. Seo, *et al.*, Novel p-Dopant toward Highly Efficient and Stable Perovskite Solar Cells, *Energy & Environmental Science* **2018**, *11*, 2985.

165. H.-S. Kim, *et al.*, Power Output Stabilizing Feature in Perovskite Solar Cells at Operating Condition: Selective Contact-Dependent Charge Recombination Dynamics, *Nano Energy* **2019**, *61*, 126.

166. Q. Lou, *et al.*, Multifunctional CNT:TiO_2 Additives in Spiro-OMeTAD Layer for Highly Efficient and Stable Perovskite Solar Cells, *EcoMat* **2021**, *3*, e12099.

167. T. Zhang, *et al.*, Ion-Modulated Radical Doping of Spiro-OMeTAD for More Efficient and Stable Perovskite Solar Cells, *Science* **2022**, *377*, 495.

168. Z. Li, *et al.*, Acid Additives Enhancing the Conductivity of Spiro-OMeTAD toward High-Efficiency and Hysteresis-Less Planar Perovskite Solar Cells, *Advanced Energy Materials* **2017**, *7*, 1601451.

169. M. Wang, *et al.*, Defect Passivation Using Ultrathin PTAA Layers for Efficient and Stable Perovskite Solar Cells with a High Fill Factor and Eliminated Hysteresis, *Journal of Materials Chemistry A* **2019**, 7, 26421.

170. B. Li, *et al.*, Defect Engineering toward Highly Efficient and Stable Perovskite Solar Cells, *Advanced Materials Interfaces* **2018**, *5*, 1800326.

171. J. Wu, *et al.*, Using Hysteresis to Predict the Charge Recombination Properties of Perovskite Solar Cells, *Journal of Materials Chemistry A* **2021**, *9*, 6382.

Index

Note: First use only recorded in the index.

0–9

2D perovskites, 60–64, 66, 69–71, 74–78, 80, 81, 83, 85, 88, 89, 91, 93, 95, 97, 102–106, 265

A

ABX$_3$ crystal structure, 5
additive engineering, 196, 339
A-site additive, 202

B

band alignment, 257
band structure, 326
B-site additive, 209

C

carbon-based materials, 269
carrier dynamics, 260
cation/anion doping, 337
channels of ion migration, 367–369
characterization methods for strain, 316
characterizations of ion migration, 370–378
common issues, 274
component engineering, 33
components and structure, 30
compositional engineering, 337
crystallization dynamics, 23, 32, 38, 252
crystal structure, 126

D

decomposition, 120
defect chemistry, 196
defect properties, 330
defects passivation, 258
defect types, 362–365
device stability, 334
donor number, 163

E

EBSD, 324
electronic dimensionality, 42
elimination of hysteresis, 423
evaporation deposition, 15
external condition-induced strain, 313
external strain regulation, 349

F

fullerene, 262

G

GIWAXS, 317

H

heat treatment strategies, 341
hysteresis, 134, 414
hysteresis index (HI), 415

I

impacts of strain, 326
in-situ measurements, 234, 241
interfacial engineering, 144
interfacial management, 344
internal strain, 311
intramolecular exchange, 141
ionic liquid, 184
ion migration, 260, 332

L

LaMer model, 164
LD-3D perovskite solar cells, 43
lead polyhalide, 162
Lewis acid/base additive, 297

M

machine learning, 13
mechanism of ion migration, 365–366
mesoscopic-structured devices, 18
metal oxides and halides, 263
metal salts, 269
multi-dimensional perovsites, 36

O

optoelectronic physical properties, 10
organic amines, 62–66, 69, 71, 73, 75, 80, 81, 83, 84, 86, 88, 91, 104–106
organic salts and molecules, 268

origin of hysteresis, 416
Ostwald ripening model, 165

P

performance, 59, 61, 63, 65, 66, 69–71, 75, 76,
    80, 84, 88, 94, 97, 101, 105, 106
perovskite, 234–252
perovskite/ETL interface, 283
perovskite/HTL interface, 289
perovskite photovoltaics, 133
perovskite solar cells, 3
PFM, 326
phase stability, 332
phase transition, 111
photostability, 130
planar-structured devices, 19
polymers, 269
polymorphism, 113

R

Raman spectroscopy, 321

S

SED, 324
solar cell, 236, 239, 240, 244, 249, 251
solar energy, 1

solution aging, 166
solution-processed deposition, 16
solvent additive, 214
stability, 59, 60, 62–66, 70, 74, 76, 78, 81, 83–85,
    87–89, 94, 95, 101, 105, 106, 136
stability induced by ion migration, 378–382
strain engineering, 310
strain regulation strategies, 337
strategies of mitigating ion migration,
    383–404
structure, 59–65, 67–74, 76–78, 80, 81, 83–86,
    88–95, 97–102, 105, 106
synergistic modulation, 272

T

tandem devices, 21
TEM, 321
temperature change, 115
types of hysteresis, 414

X

XRD, 316
X-site additive, 211

Z

zwitterionic molecules, 264